Grundlehren der mathematischen Wissenschaften 326

A Series of Comprehensive Studies in Mathematics

Springer
Berlin
Heidelberg
New York
Barcelona
Hong Kong
London
Milan
Paris
Singapore
Tokyo

Michel Waldschmidt

Diophantine Approximation on Linear Algebraic Groups

Transcendence Properties of the Exponential Function in Several Variables

Springer

Michel Waldschmidt
Institut de Mathématiques de Jussieu
Université Pierre et Marie Curie (Paris VI)
Case 247
75252 Paris Cedex 05, France
e-mail: miw@math.jussieu.fr

Cataloging-in-Publication Data applied for

Die Deutsche Bibliothek - CIP-Einheitsaufnahme

Waldschmidt, Michel:
Diophantine approximation on linear algebraic groups : transcendence properties of the exponential function in several variables / Michel Waldschmidt. - Berlin ; Heidelberg ; New York ; Barcelona ; Hong Kong ; London ; Milan ; Paris ; Singapore ; Tokyo : Springer, 2000
(Grundlehren der mathematischen Wissenschaften ; 326)
ISBN 3-540-66785-7

Mathematics Subject Classification (1991):
11-02, 11Jxx, 14Lxx, 20Gxx, 33B10

ISSN 0072-7830

ISBN 3-540-66785-7 Springer-Verlag Berlin Heidelberg New York

Springer-Verlag is a company in the BertelsmannSpringer publishing group.

Printed in Germany

Cover design: MetaDesign plus GmbH, Berlin
Typeset by the author using the MathTime fonts.
Printed on acid-free paper SPIN: 10691667 41/3143LK-5 4 3 2 1 0

Preface

A transcendental number is a complex number which is not a root of a polynomial $f \in \mathbb{Z}[X] \setminus \{0\}$. Liouville constructed the first examples of transcendental numbers in 1844, Hermite proved the transcendence of e in 1873, Lindemann that of π in 1882. Siegel, and then Schneider, worked with elliptic curves and abelian varieties. After a suggestion of Cartier, Lang worked with commutative algebraic groups; this led to a strong development of the subject in connection with diophantine geometry, including Wüstholz's Analytic Subgroup Theorem and the proof by Masser and Wüstholz of Faltings' Isogeny Theorem.

In the meantime, Gel'fond developed his method: after his solution of Hilbert's seventh problem on the transcendence of α^β, he established a number of estimates from below for $|\alpha_1^\beta - \alpha_2|$ and $|\beta_1 \log \alpha_1 - \log \alpha_2|$, where α_1, α_2 and β are algebraic numbers. He deduced many consequences of such estimates for diophantine equations. This was the starting point of Baker's work on measures of linear independence of logarithms of algebraic numbers. One of the most important features of transcendental methods is that they yield quantitative estimates related to algebraic numbers. This is one of the main reasons for which "there are more mathematicians who deal with the transcendency of the special values of analytic functions than those who prove the algebraicity"[1]. A first example is Baker's method which provides lower bounds for nonvanishing numbers of the form

$$|\alpha_1^{b_1} \cdots \alpha_m^{b_m} - 1|,$$

when $\alpha_1, \ldots, \alpha_m$ are algebraic numbers and $b_1, \ldots, b_m$ rational integers. Such estimates, which are of central interest, have a wide range of applications. A second important example is Schmidt's Subspace Theorem, which extends the Thue-Siegel-Roth Theorem to simultaneous diophantine approximation; its range of application is wider than Baker's Theorem, but, in contrast with Baker, Schmidt's result is so far not effective.

This subject is growing so fast that it is hard to give a report on the state of the art which covers all aspects. Our concern here is with *commutative linear algebraic groups*. A connected and commutative algebraic subgroup of GL_n splits over a finite extension; over an algebraically closed field it is a product of additive and multiplicative groups. Hence the algebraic groups we consider are $\mathbb{G}_{\mathrm{a}}^{d_0} \times \mathbb{G}_{\mathrm{m}}^{d_1}$, with

[1] G. Shimura, Duke Math. J. **44**, No 2 (1977), p. 365.

$d_0 \geq 0$ and $d_1 \geq 0$. In terms of analytic functions, our main object of study is the usual exponential function. We discuss the qualitative as well as the quantitative aspects of the subject. The latter is not restricted to measures of linear independence of logarithms of algebraic numbers, but includes also simultaneous diophantine approximation results leading to statements of algebraic independence for values of the exponential function, in either one or several variables.

We do not consider elliptic curves, abelian varieties, and more generally nonlinear algebraic groups; we do not consider either elliptic functions, Weierstraß zeta functions, theta functions nor abelian functions. A lot of results in this book have already been extended to the more general set-up of commutative algebraic groups, but a few items are specific to the linear ones. An example of a feature particular to linear algebraic groups is the Fourier-Borel duality, which relates Gel'fond's method to Schneider's. Moreover, restricting ourselves to the linear case enables us to compute more easily all constants.

Among the recent developments of the subject is the introduction, by M. Laurent, of interpolation determinants. They replace the constructions of auxiliary functions. Instead of solving some system of equations, we only consider the determinant of a matrix corresponding to this linear system. There is no need any more to appeal to Dirichlet's box principle (or pigeonhole principle, alias Thue-Siegel's Lemma). Here, we use this approach in most proofs.

The above-mentioned matrix is associated to the linear system with respect to given bases. A further step has been performed by J-B. Bost, using Arakelov theory, where he considers directly the related linear map without selecting bases. This approach will certainly be more efficient for nonlinear algebraic groups, and we mention it in passing, but we do not follow it here.

A central result in this book is the *Linear Subgroup Theorem,* which occurs in two forms. The qualitative one (Chapter 11) is a lower bound for the dimension n of the $\mathbb{C}$-vector subspace of $\mathbb{C}^d$ spanned by points η whose coordinates are either algebraic numbers, or else logarithms of algebraic numbers. The images of such points η under the exponential map of some commutative linear algebraic group are algebraic over the field of algebraic numbers. Hence the Linear Subgroup Theorem deals with n-parameter subgroups of linear algebraic groups, and involves functions of n complex variables.

The quantitative version of the Linear Subgroup Theorem concerns the simultaneous approximation of such points η. Linear combinations of logarithms of algebraic numbers arise in several ways as special cases of this general setup.

The main conjecture is that linearly independent logarithms of algebraic numbers should be algebraically independent. As a matter of fact, so far all known partial results on this topic are consequences of the Linear Subgroup Theorem.

There is a strong contrast between the simplicity of the conjectures, both for qualitative and quantitative statements, and currently known results. A comparison between the conjecture on algebraic independence of logarithms (Conjecture 1.15) on one hand, the Linear Subgroup Theorem of Chapter 11 (Theorem 11.5) on the other, illustrates this point for the qualitative aspect. For the quantitative one, an example of this contrast is illustrated by comparing the known measures of linear independence of logarithms (Theorem 9.1) with the conjectural ones (Conjectures 1.11 and 14.25).

We very much expect that, once the theory is more highly developed, the results will be simpler to state, but we have far from reached this stage at present and the statements of the results of the last chapters are not as simple as we would wish. The quantitative version of the Linear Subgroup Theorem in Chap. 13 (Theorem 13.1) is by no means a simple statement; on the other hand it includes a lot of diophantine estimates, as shown in Chap. 14. The large amount of corollaries it contains may be an excuse for its lack of simplicity, but it remains a challenge to get simpler statements which are as powerful.

The first chapters may serve as an introduction to the subject of transcendental numbers. For instance the first three chapters do not require much preliminary knowledge and include already complete proofs of a number of classical transcendence results.

Three proofs of Baker's transcendence theorem on linear independence of logarithms of algebraic numbers are given: in Chap. 4 we follow an argument of Bertrand and Masser who derived Baker's Theorem from the Schneider-Lang criterion concerning algebraic values of meromorphic functions on Cartesian products. In Chap. 6 (and Chap. 9 for the nonhomogeneous case) we extend Schneider's method, and in Chap. 10 we explain Baker's argument which extends Gel'fond's solution of Hilbert's seventh problem. We give also several measures of linear independence of logarithms of algebraic numbers: a comparatively simple proof is given in Chap. 7, and refined estimates are proved in Chap. 9 and 10.

We do not consider applications of such estimates to diophantine equations, but we give further examples of diophantine approximation results (in Chap. 14) together with consequences (in Chap. 15). This last chapter deals with algebraic independence; it does not cover the subject in an exhaustive way; a more complete introduction to this topic is [NeP 2000], which includes transcendence criteria with proofs.

Several results presented here are new, and the full details have not appeared in print before.

Our emphasis is not only on the results, but also on the methods; this is why we give several proofs of the same results. In the same spirit, sometimes we also propose several choices of the parameters which occur in the transcendence arguments. It turns out that the freedom in this choice is closely related to the quality of the quantitative refinements: if the proof of the qualitative transcendence result can be achieved with a broad range of choice for the auxiliary parameters, then one should expect a sharp diophantine estimate.

Another goal is to describe some of the main tools which are available. We make no attempt to be complete. In [FNe 1998] the reader will find some items which are not discussed here. An important example of a missing item is Nesterenko's proof [Ne 1996] of the algebraic independence of π and e^{π}.

Writing this book took more than 10 years. The first written parts were notes of lectures given at the Institut Henri Poincaré in the 80's for several courses of the DEA (Diplôme d'Études Approfondies) of Mathématiques at the Université P. et M. Curie (Paris VI). In 1992, I was invited by R. Balasubramanian to give a series of lectures at the MathScience Institute of Madras, and I took this opportunity to write down a preliminary version of some of the chapters below (more or less the seven first chapters). These notes were published in [W 1992]. A chapter on zero estimates by D. Roy was included, as well as an appendix by M. Laurent [Lau 1994]. The present book grew out of these Lecture Notes; the material of the last eight chapters includes a multiplicity estimate (again written by D. Roy), the Linear Subgroup Theorem (both in qualitative and quantitative form), as well as results of simultaneous approximation and algebraic independence. Some of these results are due to D. Roy (like the Strong Six Exponentials Theorem of § 11.6), others (mainly in the last two chapters) have been obtained in joint papers with D. Roy.

The influence of Damien Roy on this book is important; not only did he write two chapters, but he also contributed to the proof of many results quoted in this book, and furthermore his many comments have been very influential.

Many other colleagues and friends also sent me comments, remarks and suggestions along the many years which have been needed to complete this book. Even though I do not mention them all, I am deeply thankful to them.

Special thanks are due to Guy Diaz who sent me a long list of comments on a preliminary version of this book. I wish also to express my gratitude to Francesco Amoroso, Yann Bugeaud, François Gramain, and Paul Voutier.

The help of Sinnou David during the last stage of the TEXnical preparation of this volume is also gratefully acknowledged.

We consider mainly the Archimedean situation; the same topic has been investigated in the ultrametric domain also, and this would have deserved consideration also. In fact my main motivation to study this subject arose from Leopoldt's Conjecture on the p-adic rank of the units of algebraic number fields (solved by Ax-Baker-Brumer for abelian extension). I wish to take this opportunity to thank Jean Fresnel, who suggested this topic to me thirty years ago, and helped me take my first steps in mathematical research.

Paris, January 2000 *Michel Waldschmidt*

Table of Contents

Part III. Multiplicities in Higher Dimension

Part IV. The Linear Subgroup Theorem

Part V. Simultaneous Approximation of Values of the Exponential Function in Several Variables

Prerequisites

In this book, an *algebraic number* is a complex number which is algebraic over the field of rational numbers. Given a (commutative) ring A and a subring k which is a field, an element θ in A is *algebraic* over k if there exists a nonzero polynomial $P \in k[X]$ such that $P(\theta) = 0$. An element of A is *transcendental* over k if it is not algebraic over k. Hence a *transcendental number* is a complex number which is not algebraic.

We denote by $\mathbb{N} = \{0, 1, 2, \ldots\}$ the set of nonnegative integers, by $\mathbb{Z}$ the ring of rational integers and by $\mathbb{Q}$, $\mathbb{R}$, $\mathbb{C}$ the fields of rational numbers, real numbers and complex numbers respectively.

The set of algebraic numbers is a subfield of $\mathbb{C}$: it is the algebraic closure of $\mathbb{Q}$ into $\mathbb{C}$ (see [L 1993], Chap. V § 2). This field will be denoted by $\overline{\mathbb{Q}}$. We shall need a few properties of algebraic numbers and number fields which will be recalled in Chap. 3.

Given elements $\theta_1, \ldots, \theta_n$ in our ring A, we say that they are *algebraically dependent over* k if there exists a nonzero polynomial $P \in k[X_1, \ldots, X_n]$ such that $P(\theta_1, \ldots, \theta_n) = 0$. Otherwise they are *algebraically independent over* k. The *transcendence degree* of A over k is the maximal integer n such that there exist n elements in A which are algebraically independent over k. We denote it by $\mathrm{trdeg}_k(A)$. For $k_1 \subset k_2 \subset k_3$, we have (see for instance [L 1993], Chap. VIII)

$$\mathrm{trdeg}_{k_1}(k_3) = \mathrm{trdeg}_{k_1}(k_2) + \mathrm{trdeg}_{k_2}(k_3).$$

Any element of k_2 is algebraic over k_1 if and only if

$$\mathrm{trdeg}_{k_1}(k_2) = 0;$$

in this case we say that k_2 is an *algebraic extension* of k_1. As a consequence, for complex numbers, the concept of algebraic independence over $\mathbb{Q}$ or over $\overline{\mathbb{Q}}$ is the same: we shall just speak of *algebraically dependent* or *independent* numbers.

We shall use the basic notions of linear algebra. The dimension of a k-vector space V will be denoted by $\dim_k(V)$, the rank of a $\mathbb{Z}$-module M by $\mathrm{rank}_{\mathbb{Z}}(M)$ or simply $\mathrm{rank}(M)$. An abelian group is nothing else than a $\mathbb{Z}$-module; when it is written multiplicatively, one speaks of *multiplicatively dependent* or *independent* elements (which means $\mathbb{Z}$-linearly dependent or independent elements in the abelian group). For instance if k is a field and $\gamma_1, \ldots, \gamma_m$ elements in $k^\times = k \setminus \{0\}$, then $\gamma_1, \ldots, \gamma_m$

are multiplicatively dependent if and only if there exists $\underline{b} = (b_1, \ldots, b_m) \in \mathbb{Z}^m \setminus \{0\}$ such that the number

$$\underline{\gamma}^{\underline{b}} = \gamma_1^{b_1} \cdots \gamma_m^{b_m}$$

is 1.

The rank of a matrix M will be denoted $\operatorname{rank}(\mathsf{M})$: this is the largest integer r for which there exists a regular $r \times r$ submatrix of M.

For a ring B, a subring A and a subset E of B, we denote by $A[E]$ the subring of B generated by $A \cup E$, namely the intersection of all subrings of B containing A and E. For a field K, a subfield k and a subset E of K, we denote by $k(E)$ the subfield of K generated by k and E. When $E = \{\gamma_1, \ldots, \gamma_m\}$ is finite, we write simply $A[\gamma_1, \ldots, \gamma_m]$ and $k(\gamma_1, \ldots, \gamma_m)$. In particular $\mathbb{Q}(\gamma)$ (resp. $\mathbb{Q}(\underline{\gamma})$) denotes the field generated by an element $\gamma \in \mathbb{C}$ (resp. by a tuple $\underline{\gamma} = (\gamma_1, \ldots, \gamma_m) \in \mathbb{C}^m$).

For U and V vector spaces over a field k, $\operatorname{Hom}_k(U, V)$ will denote the space of k-linear mappings $U \to V$.

The basic facts from algebraic geometry and commutative algebra which are needed are recalled in §§ 5.2 and 8.2 respectively.

A useful tool is *Dirichlet's box principle*, also called *Dirichlet's pigeonhole principle* (Schubfachprinzip). One of the many equivalent statements is:

- *A mapping $E \to F$ between two finite sets E and F with*

$$\operatorname{Card}(E) > \operatorname{Card}(F)$$

is not injective.

An important application of it is Thue-Siegel's Lemma (see § 4.5). We shall not need the more sophisticated version of Thue-Siegel's Lemma in [BoVa 1983], based on an idea of Mahler using geometry of numbers, but Minkowski's Theorem (see for instance [Sc 1991], Chap. I) will be used in § 7.8 for the proof of Lemma 7.19.

The notion of algebraic independence will be needed not only for numbers, but also for functions. In a single variable we take for k the field $\mathbb{C}(z)$ of rational functions and for A either the ring of analytic (i.e. holomorphic) functions over a domain (= connected open subset) D of $\mathbb{C}$, or the field of meromorphic functions over D. A function $f \in A$ is called *transcendental* if it is transcendental over $\mathbb{C}(z)$, *algebraic* otherwise. An *entire function* is a function which is analytic in the whole of $\mathbb{C}$. It is easy to check that an entire function is algebraic if and only if it is a polynomial, and that a meromorphic function in $\mathbb{C}$ is algebraic if and only if it is a rational function, i.e. an element of $\mathbb{C}(z)$.

According to the general definition, analytic functions $f_1, \ldots, f_d$ of n variables are algebraically independent over $\mathbb{C}$ if and only if, for any nonzero polynomial $P \in \mathbb{C}[X_1, \ldots, X_d]$, the function $P(f_1, \ldots, f_d)$ is not the zero function. Also $f_1, \ldots, f_d$ are algebraically independent over $\mathbb{C}(z_1, \ldots, z_n)$ if and only if, for any nonzero polynomial P in the ring of polynomials $\mathbb{C}[X_1, \ldots, X_n, Y_1, \ldots, Y_d]$ in $n+d$ variables, the function

$$P\bigl(z_1, \ldots, z_n, f_1(\underline{z}), \ldots, f_d(\underline{z})\bigr)$$

is not the zero function.

A function f is called *transcendental* if the $n+1$ functions $z_1, \dots, z_n, f(\underline{z})$ are algebraically independent: this means that f is transcendental over the field $\mathbb{C}(z_1, \dots, z_n)$.

The exponential function

$$1 + z + \frac{z^2}{2} + \frac{z^3}{6} + \cdots = \sum_{n=0}^{\infty} \frac{z^n}{n!}$$

is denoted either by e^z or by $\exp(z)$, and

$$e = \exp(1) = 2.71828182\dots$$

is the natural basis of Napierian logarithms. For $\alpha \in \mathbb{C}^\times$, a determination of the logarithm of α is any complex number λ such that $\exp(\lambda) = \alpha$. For a given $\alpha \in \overline{\mathbb{Q}}^\times$, the set of λ in $\mathbb{C}$ with $\alpha = e^\lambda$ is a whole class of the additive group $\mathbb{C}$ modulo $2i\pi\mathbb{Z}$. In order to avoid confusion, we shall not use too often the notation $\log \alpha$ which depends on the choice of the branch of the logarithmic function. Nevertheless we remark that the $\mathbb{Q}$-vector space of logarithms of nonzero complex algebraic numbers

$$\mathcal{L} = \exp^{-1}(\overline{\mathbb{Q}}^\times) = \left\{\lambda \in \mathbb{C}\,;\ e^\lambda \in \overline{\mathbb{Q}}^\times\right\}$$

is the set of all numbers of the form $\log \alpha$ where α runs over the set of nonzero complex algebraic numbers and where we take all possible values for log:

$$\mathcal{L} = \{\log \alpha; \alpha \in \overline{\mathbb{Q}}^\times\}.$$

When a determination λ of the logarithm of α is chosen, for $\beta \in \mathbb{C}$ we write α^β in place of $\exp(\beta\lambda)$.

We shall say that a complex function f of one variable is *analytic in a closed disc* $\{z \in \mathbb{C}\,;\ |z| \le R\}$ *of* $\mathbb{C}$ if f is continuous on this disc and analytic in the open disc $|z| < R$. In this case we denote by $|f|_R$ the number $\sup\{|f(z)|\,;\ |z| \le R\}$. By *maximum modulus principle* we also have

$$|f|_R = \sup\{|f(z)|\,;\ |z| = R\}.$$

We shall also work with functions of several variables. For $\underline{z} = (z_1, \dots, z_n) \in \mathbb{C}^n$ (and therefore also for $\underline{z}$ in $\mathbb{N}^n$ or in $\mathbb{Z}^n$), we set

$$|\underline{z}| = \max_{1 \le i \le n} |z_i| \quad \text{and} \quad \|\underline{z}\| = |z_1| + \cdots + |z_n|.$$

If, further, $\underline{\sigma} = (\sigma_1, \dots, \sigma_n) \in \mathbb{N}^n$, then we define

$$\underline{z}^{\underline{\sigma}} = z_1^{\sigma_1} \cdots z_n^{\sigma_n}, \qquad \underline{\sigma}! = \sigma_1! \cdots \sigma_n!$$

(with $0! = 1$) and

$$\mathcal{D}^{\underline{\sigma}} = \left(\frac{\partial}{\partial z_1}\right)^{\sigma_1} \cdots \left(\frac{\partial}{\partial z_n}\right)^{\sigma_n}.$$

For $\underline{z}$ and $\underline{z}'$ in $\mathbb{C}^n$, let

$$\underline{z}\underline{z}' = z_1 z_1' + \cdots + z_n z_n'$$

denote the standard scalar product.

To each $\underline{w} = (w_1, \ldots, w_n) \in \mathbb{C}^n$ we attach a *derivative operator of order* 1:

$$\mathcal{D}_{\underline{w}} = w_1 \frac{\partial}{\partial z_1} + \cdots + w_n \frac{\partial}{\partial z_n}$$

on the ring of entire functions in $\mathbb{C}^n$. More generally, for S a positive integer, a *derivative operator D of order S* is a linear combination, with complex coefficients, of

$$\left(\frac{\partial}{\partial z_1}\right)^{\sigma_1} \cdots \left(\frac{\partial}{\partial z_n}\right)^{\sigma_n},$$

where $\underline{\sigma}$ runs over the set of elements in $\mathbb{N}^n$ satisfying $\|\underline{\sigma}\| = S$. This amounts to say that D is a linear combination, with complex coefficients, of products $\mathcal{D}_{\underline{w}_1} \cdots \mathcal{D}_{\underline{w}_S}$, where $(\underline{w}_1, \ldots, \underline{w}_S)$ ranges over a finite subset of $(\mathbb{C}^n)^S$.

Most often, tuples of numbers are underlined, like $\underline{w} = (w_1, \ldots, w_d) \in \mathbb{C}^d$; for $\underline{w}_1, \ldots, \underline{w}_{\ell_0}$ in $\mathbb{C}^d$ we write $\boldsymbol{w} = (\underline{w}_1, \ldots, \underline{w}_{\ell_0}) \in (\mathbb{C}^d)^{\ell_0}$. For $\underline{\sigma} \in \mathbb{N}^{\ell_0}$, $\underline{\tau} \in \mathbb{N}^{d_0}$, $\underline{t} \in \mathbb{Z}^{d_1}$ and $\underline{z} \in \mathbb{C}^d$ with $d = d_0 + d_1$, the function

$$\mathcal{D}_{\boldsymbol{w}}^{\underline{\sigma}}\left(\underline{z}^{\underline{\tau}} e^{\underline{t}\underline{z}}\right) = \mathcal{D}_{\underline{w}_1}^{\sigma_1} \cdots \mathcal{D}_{\underline{w}_{\ell_0}}^{\sigma_{\ell_0}} \left(z_1^{\tau_1} \cdots z_{d_0}^{\tau_{d_0}} e^{t_1 z_{d_0+1} + \cdots + t_{d_1} z_d}\right)$$

is an exponential polynomial for which explicit expressions will be given (see Lemmas 4.9 and 13.6).

For a complex function f which is continuous in a *polydisc*

$$\{\underline{z} \in \mathbb{C}^n \,;\, |\underline{z}| \le R\}$$

and analytic inside, we have again

$$\sup\{|f(\underline{z})| \,;\, |\underline{z}| \le R\} = \sup\{|f(\underline{z})| \,;\, |\underline{z}| = R\};$$

this number will be denoted $|f|_R$.

Our main tool will be Schwarz' Lemma, which is a sharp upper bound for the modulus of a complex function, taking into account its zeroes. See § 2.2.3 for one variable, § 6.2.1 for one point and several variables, § 4.3 for Cartesian products.

We shall use only very simple properties of analytic functions in $\mathbb{C}^n$ (see for instance [LelGru 1986], Chap. I, § 1). Cauchy's inequalities will occur in §§ 4.6 and 4.7: an entire function f in $\mathbb{C}^n$, whose Taylor expansion at the origin is

$$\sum_{\underline{\sigma} \in \mathbb{N}^n} a_{\underline{\sigma}} \underline{z}^{\underline{\sigma}} \quad \text{with} \quad a_{\underline{\sigma}} = \frac{1}{\underline{\sigma}!} \mathcal{D}^{\underline{\sigma}} f(0)$$

satisfies, for all $r > 0$:

$$|\mathcal{D}^{\underline{\sigma}} f(0)| \leq \frac{\underline{\sigma}!}{r^{\|\underline{\sigma}\|}} |f|_r.$$

One deduces, for $\underline{\zeta} \in \mathbb{C}^n$ and $r \geq 1 + |\underline{\zeta}|$,

$$|\mathcal{D}^{\underline{\sigma}} f(\underline{\zeta})| \leq \frac{\underline{\sigma}!}{(r - |\underline{\zeta}|)^{\|\underline{\sigma}\|}} |f|_r \leq \underline{\sigma}! |f|_r.$$

In § 4.3 we shall also use the fact that a continuous mapping $f : \mathbb{C}^n \to \mathbb{C}$ is analytic if and only if it is analytic in each z_j when the other variables are fixed. This is a consequence of Cauchy's integral formula for polydiscs; see for instance [Hö 1973], Th. 2.2.1.

Notation

Some notation has already been fixed in the prerequisites section. We complete it with the following ones which will be used throughout the book.

We shall use Kronecker's diagonal symbol:

$$\delta_{ij} = \begin{cases} 1 & \text{if } i = j, \\ 0 & \text{if } i \neq j. \end{cases}$$

For $x \in \mathbb{R}$, we set

$$\log_+ x = \log \max\{e, x\}$$

and we denote by $[x] \in \mathbb{N}$ the integral part of x, with $0 \le x - [x] < 1$.

The binomial coefficient

$$\binom{n}{k} = \frac{n!}{k!(n-k)!}$$

is 0 unless $0 \le k \le n$. More generally, an empty sum is equal to 0, while the value of an empty product is 1.

The number of elements in a finite set E will be denoted either by $\mathrm{Card}(E)$ or else by $|E|$.

- ◇ $\mathrm{Mat}_{d\times\ell}$ denotes the space of $d \times \ell$ matrices
- ◇ ${}^t\mathsf{M}$ is the transposed of a matrix M.
- ◇ $\mathsf{I}_d = \begin{pmatrix} 1 & 0 & \cdots & 0 \\ 0 & 1 & \cdots & 0 \\ \vdots & \vdots & \ddots & \vdots \\ 0 & 0 & \cdots & 1 \end{pmatrix}$ is the identity $d \times d$ matrix.
- ◇ For a positive integer d and a real number $S \ge 0$, the set of d-tuples

$$\mathbb{Z}^d[S] = \left\{ \underline{s} \in \mathbb{Z}^d \, ; \, |\underline{s}| \le S \right\}$$

has $(2[S]+1)^d$ elements.

- ◇ For $\underline{S} = (S_1, \ldots, S_d) \in \mathbb{R}^d_{>0}$, the set of d-tuples

$$\mathbb{Z}^d[\underline{S}] = \left\{ \underline{s} \in \mathbb{Z}^d \, ; \, |s_i| \le S_i \ \text{ for } 1 \le i \le d \right\}$$

has $(2[S_1]+1)\cdots(2[S_d]+1)$ elements.

◇ When $\mathcal{V}$ is a vector subspace of $\mathbb{C}^d$, we set

$$\mathcal{V}[S] = \mathcal{V} \cap \mathbb{Z}^d[S] \quad \text{and} \quad \mathcal{V}[\underline{S}] = \mathcal{V} \cap \mathbb{Z}^d[\underline{S}]$$

for $S \in \mathbb{R}_{>0}$ and $\underline{S} = (S_1, \dots, S_d) \in \mathbb{R}^d_{>0}$.

◇ For a finite subset E of an additive group G, and for m a positive integer,

$$E[m] = \{x_1 + \cdots + x_m \,;\, x_i \in E\} \subset G.$$

In the successive chapters we introduce further notation as follows.

In Chapter 1

The distance between two matrices of the same size M and M' is the maximum absolute value of the difference between the entries: for

$$\mathsf{M} = \left(x_{ij}\right)_{\substack{1\le i\le d\\ 1\le j\le \ell}} \quad \text{and} \quad \mathsf{M}' = \left(x'_{ij}\right)_{\substack{1\le i\le d\\ 1\le j\le \ell}},$$

in $\mathrm{Mat}_{d\times\ell}(\mathbb{C})$,

$$\mathrm{dist}(\mathsf{M}, \mathsf{M}') = \max_{\substack{1\le i\le d\\ 1\le j\le \ell}} |x_{ij} - x'_{ij}|.$$

In Chapter 2

◇ The degree of a polynomial f in one variable X is denoted by $\deg f$ or $\deg_X f$. For a polynomial f in several variables we denote by $\deg_X f$ and $\deg_{\underline{X}} f$ the partial degree with respect to one variable X and the total degree with respect to a set of variables $\underline{X} = (X_1, \dots, X_m)$.

◇ For $f \in \mathbb{C}[\mathrm{X}_1, \dots, \mathrm{X}_m]$, let $\mathrm{H}(f)$ be the maximum absolute value of the coefficients of f.

◇ $K[X_1^{\pm 1}, \dots, X_n^{\pm 1}]$ denotes the subring of $K(X_1, \dots, X_n)$ generated by K and

$$\{X_1, X_1^{-1}, \dots, X_n^{-1}, X_n\}$$

(see § 2.2.1).

In Chapter 3

◇ $v_p(\alpha)$ is the p-adic absolute value.

◇ M_k, M_k^∞ are the sets of normalized absolute values and of Archimedean normalized absolute values of a number field k.

◇ $d_v(k)$ is the local degree of k at v.

◇ $[k\colon\mathbb{Q}]$ is the degree of k over $\mathbb{Q}$.

◇ $\mathbb{Q}_p$ is the field of p-adic numbers.

◇ $\mathbb{P}_m$ denotes the projective space of dimension m.

◇ $\mathrm{H}(\alpha)$ is the usual height of an algebraic number.

◇ $\mathrm{L}(f)$, $\mathrm{L}(\alpha)$ are the length of a polynomial or of an algebraic number.

◇ $\mathrm{M}(f)$, $\mathrm{M}(\alpha)$ denote Mahler's measure of a polynomial or of an algebraic number.

◇ $\mathrm{den}(\gamma)$, $\lceil\overline{\gamma}\rceil$, $s(\alpha)$ denote the denominator, the house and the size of an algebraic number.

◇ $\mathrm{h}(\alpha)$, $\mathrm{h}(\gamma_0 : \cdots : \gamma_N)$ are the absolute logarithmic height of an algebraic number or of a projective point.

◇ $N_{K/k}$, $\mathrm{Tr}_{K/k}$ denote the norm and the trace attached to an extension K/k (see also § 4.2.3).

◇ $\mathrm{L}_2(f)$ is the Euclidean norm of $f \in \mathbb{C}[X]$.

In Chapter 4

◇ $\mathcal{A}_n$ is the space of entire functions in $\mathbb{C}^n$.

In Chapter 5

◇ $\mathbb{G}_{\mathrm{a}}$ and $\mathbb{G}_{\mathrm{m}}$ are the additive and multiplicative groups.

◇ res is the restriction map (§ 5.2.2).

◇ T_Φ is the algebraic subgroup of a torus $\mathbb{G}_{\mathrm{m}}^m$ associated with a subgroup Φ of $\mathbb{Z}^m$.

◇ $H(V; \underline{D})$ and $\mathcal{H}(V; \underline{D})$ are respectively an Hilbert function and the normalized homogeneous part of highest degree of an Hilbert-Samuel polynomial of an algebraic set V.

◇ τ_g is the translation by g in an abelian group.

In Chapter 6

◇ $\Theta_n(L)$ is defined in § 6.2.2.

◇ $\|\cdot\|_2$ is the Euclidean norm in Exercise 6.4.

In Chapter 7

◇ $\Theta(n; T_0, L)$ is defined in § 7.2.

◇ $\Delta(z; \tau)$ denote Fel'dman's polynomials introduced in § 7.7.

◇ Let K be a field, n a positive integer and $\mathcal{V}$ a vector subspace of K^n. We denote by $\pi_{\mathcal{V}}$ the canonical surjective linear map $K^n \longrightarrow K^n/\mathcal{V}$ with kernel $\mathcal{V}$.

In Chapter 8

- G^+ and G^- are algebraic subgroups of G.
- rank(I) is the rank of an ideal I (§ 8.2.1).
- $H(I;\, \underline{D})$ and $\mathcal{H}(I;\, \underline{D})$ are respectively an Hilbert function and the normalized homogeneous part of highest degree of an Hilbert-Samuel polynomial of I.
- $T_e(G)$ is the tangent space at the origin of an algebraic group G.
- In § 8.3.1, $\mathcal{V}^\perp$ is the subspace of $K[\underline{X}] = K[X_1, \ldots, X_{d_0}]$ consisting of the linear forms $a_1 X_1 + \cdots + a_{d_0} X_{d_0}$ which vanish identically on $\mathcal{V}$.

In Chapters 9 and 10

- *General case:* for a measure of linear independence of logarithms of algebraic numbers:

$$\Lambda = \beta_0 + \beta_1\lambda_1 + \cdots + \beta_m\lambda_m.$$

- *Homogeneous case:* $\beta_0 = 0$:

$$\Lambda = \beta_1\lambda_1 + \cdots + \beta_m\lambda_m.$$

- *Homogeneous rational case:* $\beta_0 = 0$ and $\beta_i = b_i \in \mathbb{Z}$:

$$\Lambda = b_1\lambda_1 + \cdots + b_m\lambda_m.$$

- $\delta_T(z;\tau)$, $\tau \in \mathbb{N}$ denote the polynomials of Fel'dman-Matveev (§ 9.2.1).
- In § 9.2.1 also we define

$$\delta(z;\sigma,\kappa) = \left(\frac{d}{dz}\right)^\kappa \delta(z;\sigma).$$

- $\mathcal{W}^\perp$ is the orthogonal of $\mathcal{W}$ in an Euclidean vector space (§ 10.2.4).

In Chapter 11

- $\exp_G\colon \mathbb{C}^d \to G(\mathbb{C})$ denotes the exponential map of an algebraic group G and Ω_G its kernel.
- For $d_0 \geq 0$ and $d_1 \geq 0$,

$$\mathcal{L}_G = \overline{\mathbb{Q}}^{d_0} \times \mathcal{L}^{d_1} = \exp^{-1}\bigl(G(\overline{\mathbb{Q}})\bigr)$$

(see § 11.1.2).

- $V_{\max}$, $V_{\min}$, $d_{\max}$, $d_{\min}$ are defined in § 11.1.2.

- In § 11.1.3 we introduce the $\overline{\mathbb{Q}}$-vector space $\widetilde{\mathcal{L}}$ spanned by 1 and $\mathcal{L}$; this is the set of linear combinations, with algebraic coefficients, of 1 and logarithms of algebraic numbers:

$$\widetilde{\mathcal{L}} = \Big\{\beta_0 + \beta_1\lambda_1 + \cdots + \beta_n\lambda_n\,;\\ n \geq 0,\ (\beta_0, \beta_1, \ldots, \beta_n) \in \overline{\mathbb{Q}}^{n+1},\ (\lambda_1, \ldots, \lambda_n) \in \mathcal{L}^n\Big\}.$$

- In § 11.6.2 we denote by $\widetilde{\mathcal{L}}_k$ the k-vector subspace of $\mathbb{C}$ spanned by 1 and $\mathcal{L}$.

In Chapter 12

- Property $\left(\begin{smallmatrix} A & B \\ C & 0 \end{smallmatrix}\right)$ is defined in § 12.1.3.
- $r_{\mathrm{str}}(\mathsf{M})$ is the structural rank of a matrix M (§ 12.1.4).

In Chapter 13

- $\mathcal{H}(G;\mathcal{T})$ is defined in § 13.1.

In Chapter 14

- $\varphi(D, h)$: simultaneous approximation measure
- $\mathsf{L}_{mn} = \Big(\lambda_{ij}\Big)_{\substack{1\leq i\leq m\\ 1\leq j\leq n}}$ denotes a $m \times n$ matrix with entries in $\mathcal{L}$.

In Chapter 15

- In § 15.1.1, $||\cdot||$ denotes the distance to the nearest integer:

$$||x|| = \min_{p\in\mathbb{Z}} |x - p|.$$

- $\Phi(D, H)$: transcendence measure for a complex number (§ 15.1.3) and measure of algebraic independence for a tuple (§ 15.1.5).
- $\psi(D, \mu)$: measure of algebraic approximation of a complex number (§ 15.1.3) and measure of simultaneous approximation of a tuple (§ 15.1.5).
- In § 15.3.3, for $K \subset \mathbb{C}$,

$$\mathcal{L}_K = \exp^{-1}(K^\times) = \big\{z \in \mathbb{C}\,;\ e^z \in K^\times\big\}.$$

1. Introduction and Historical Survey

In this first chapter we give some historical background on Baker's Theorem, both in the qualitative and in the quantitative form, in the homogeneous as well as in the nonhomogeneous version. We also describe the six exponentials Theorem, we present the state of the art on the problem of algebraic independence of logarithms of algebraic numbers. We conclude with a few comments on the Linear Subgroup Theorem.

1.1 Liouville, Hermite, Lindemann, Gel'fond, Baker

We start by quoting some of the oldest results of the theory.

The first example of a transcendental number was provided by J. Liouville in May 1844, in two notes in the Comptes Rendus de l'Académie des Sciences de Paris [Lio 1844a], [Lio 1844b]. He developed this subject seven years later in a well known paper [Lio 1851] in J. Math. Pures et Appl. (the so-called Liouville's Journal).

Theorem 1.1 (Liouville). *Let α be a complex number which is root of a nonzero polynomial in $\mathbb{Z}[X]$ of degree d. There exists a constant $c(\alpha) > 0$, which can be easily computed, such that, for any rational number p/q with $p/q \neq \alpha$ and $q > 0$,*

$$\left|\alpha - \frac{p}{q}\right| > \frac{c(\alpha)}{q^d}.$$

Liouville's Theorem is a result of diophantine approximation. It enabled J. Liouville to give the first example of transcendental numbers. It's a different matter to prove the transcendence of given numbers.

In 1873, Ch. Hermite published his four Notes in the Comptes Rendus de l'Académie des Sciences de Paris [He 1873] where he obtained the transcendence of e. One year later, G. Cantor [C 1874] gave a new proof that transcendental numbers are dense in the real line (using the result, due to R. Dedekind, that algebraic numbers are a countable set). In 1882, F. Lindemann [Li 1882a], [Li 1882b], [Li 1882c] proved the transcendence of π, thereby solving the famous greek problem of squaring the circle. Lindemann stated further results. One of them is now called the Hermite-Lindemann Theorem:

Theorem 1.2 (Hermite-Lindemann). *Let β be a nonzero complex number. Then one at least of the two numbers β and e^β is transcendental.*

Hence, if β is algebraic, then $\exp(\beta)$ is transcendental (for instance the number e itself is transcendental, as well as $e^{\sqrt{2}}$). If α is a nonzero algebraic number, and if λ is any nonzero determination of its logarithm, then λ is a transcendental number. For instance, $\log 2$ is a transcendental number. Taking $\alpha = 1$ and $\log 1 = 2i\pi$, one also deduces the transcendence of the number π.

The set $\mathcal{L}$ of logarithms of nonzero algebraic numbers, that is the inverse image of the multiplicative group $\overline{\mathbb{Q}}^\times$ by the exponential map:

$$\mathcal{L} = \exp^{-1}(\overline{\mathbb{Q}}^\times) = \{\lambda \in \mathbb{C}\,;\, e^\lambda \in \overline{\mathbb{Q}}^\times\}$$

will play an important role in these lectures. It is a $\mathbb{Q}$-vector subspace of $\mathbb{C}$, which contains $i\pi$ as well as all the usual logarithms of positive algebraic numbers.

The Theorem of Hermite-Lindemann can be written: $\overline{\mathbb{Q}} \cap \mathcal{L} = \{0\}$, which means:

Any nonzero element of $\mathcal{L}$ is transcendental.

Another important result, stated by F. Lindemann and proved by K. Weierstraß [We 1885], reads as follows:

Let $\beta_1, \ldots, \beta_n$ be pairwise distinct algebraic numbers. Then the numbers $e^{\beta_1}, \ldots, e^{\beta_n}$ are linearly independent over $\mathbb{Q}$.

This result is equivalent to the following algebraic independence statement.

Theorem 1.3* (Lindemann-Weierstraß[2]). *If $\beta_1, \ldots, \beta_n$ are algebraic numbers which are linearly independent over $\mathbb{Q}$, then the n numbers $e^{\beta_1}, \ldots, e^{\beta_n}$ are algebraically independent.*

This is one of the very few results on algebraic independence of numbers connected with the exponential function (see § 1.4 below).

After the contributions of J. Liouville, Ch. Hermite, F. Lindemann and K. Weierstraß, the next important step was provided by the work of C. L. Siegel [Si 1929], A. O. Gel'fond [G 1934] and Th. Schneider [Sch 1934], which led to the solution of Hilbert's seventh problem [Hi 1900]. The story of this problem is as follows. In his "Introductio in analysin infinitorum", L. Euler [Eu 1748] (Book I, Ch. VI, On Exponentials and Logarithms, N° 105, p.80) defined the exponential and logarithm functions, and said:

> From what we have seen, it follows that the logarithm of a number will not be a rational number unless the given number is a power of the base a. That is, unless the number b is a power of the base a, the logarithm of b cannot be expressed as a rational number. In case b is a power of the base a, then the logarithm of b cannot be an irrational number. If, indeed, $\log b = \sqrt{n}$, then $a^{\sqrt{n}} = b$, but this is impossible if both a and b are rational. It is especially

[2] The star * means that Theorem 1.3 will not be proved in the present volume.

desirable to know the logarithms of rational numbers, since from these it is possible to find the logarithms of fractions and also surds. Since the logarithms of numbers which are not the powers of the base are neither rational nor irrational, it is with justice that they are called transcendental quantities. For this reason, logarithms are said to be transcendental.

D. Hilbert [Hi 1900] proposed this question as the seventh of his problems:

> The expression α^β for an algebraic base α and an irrational algebraic exponent β, e.g. the number $2^{\sqrt{2}}$ or $e^\pi = i^{-2i}$, always represents a transcendental or at least an irrational number.

(See the biography [Re 1970] of Hilbert by C. Reid, Chap. XIX p. 164).

This problem was solved in 1934 by A. O. Gel'fond [G 1934] and Th. Schneider [Sch 1934], independently and simultaneously:

Theorem 1.4 (Gel'fond-Schneider). *If λ_1, λ_2 are $\mathbb{Q}$-linearly independent elements of $\mathcal{L}$, then they are $\overline{\mathbb{Q}}$-linearly independent.*

This shows that $\mathcal{L}$, which is a $\mathbb{Q}$-vector space, is not a $\overline{\mathbb{Q}}$-vector space. More precisely, the quotient λ_1/λ_2 of two nonzero elements of $\mathcal{L}$ is either a rational or a transcendental number. For instance $\log 2/\log 3$ is a transcendental number. Such a quotient cannot be an algebraic irrational number, like $i = \sqrt{-1}$ or like $\sqrt{2}$. The connection with Hilbert's problem is most easily seen by stating the Theorem of Gel'fond-Schneider as follows:

> *If λ and β are two complex numbers with $\lambda \neq 0$ and $\beta \notin \mathbb{Q}$, then one at least of the three numbers e^λ, β and $e^{\beta\lambda}$ is transcendental.*

Hence, if α is a nonzero algebraic number, λ any nonzero logarithm of α, and β an irrational algebraic number, then $\alpha^\beta = \exp(\beta\lambda)$ is a transcendental number. As an example, $2^{\sqrt{2}}$ is a transcendental number. The transcendence of e^π is obtained for instance by the choice $\alpha = 1$, $\lambda = 2i\pi$ and $\beta = -i/2$.

In his book [G 1952], A. O. Gel'fond emphasized the importance of getting a generalization of this statement to more than two logarithms (see § 1.2 below). Let $\lambda_1, \ldots, \lambda_n$ be n logarithms of algebraic numbers which are linearly independent over $\mathbb{Q}$. The question is to prove that they are also linearly independent over the field $\overline{\mathbb{Q}}$ of algebraic numbers. For $n = 2$, this is Theorem 1.4 of Gel'fond-Schneider. This problem was solved in 1966 by A. Baker [B 1966]:

Theorem 1.5 (Baker – Homogeneous Case). *If $\lambda_1, \ldots, \lambda_n$ are $\mathbb{Q}$-linearly independent elements of $\mathcal{L}$, then they are linearly independent over $\overline{\mathbb{Q}}$.*

Shortly later, A. Baker extended his result to a nonhomogeneous situation as follows:

Theorem 1.6 (Baker – General Case). *If $\lambda_1, \ldots, \lambda_n$ are $\mathbb{Q}$-linearly independent elements of $\mathcal{L}$, then the $n+1$ numbers $1, \lambda_1, \ldots, \lambda_n$ are linearly independent over $\overline{\mathbb{Q}}$.*

From Baker's Theorem 1.6, one easily deduces that if a number of the form

$$e^{\beta_0}\alpha_1^{\beta_1}\cdots\alpha_n^{\beta_n} = \exp\{\beta_0 + \beta_1\lambda_1 + \cdots + \beta_n\lambda_n\}$$

(with $\beta_i \in \overline{\mathbb{Q}}$, $\lambda_i \in \mathcal{L}$ and $\alpha_i = e^{\lambda_i} \in \overline{\mathbb{Q}}^\times$) is algebraic, then $\beta_0 = 0$, and moreover, either $\lambda_1, \ldots, \lambda_n$ are all zero, or else the numbers $1, \beta_1, \ldots, \beta_n$ are linearly dependent over $\mathbb{Q}$.

Also Theorem 1.6 shows that any nonzero element in the $\overline{\mathbb{Q}}$-vector space

$$\{\beta_1\lambda_1 + \cdots + \beta_n\lambda_n \ ; \ n \geq 0, \ \beta_i \in \overline{\mathbb{Q}}, \ \lambda_i \in \mathcal{L}\}$$

spanned by $\mathcal{L}$ is transcendental.

Theorem 1.6 includes not only the Theorem of Gel'fond-Schneider, but also the Theorem of Hermite-Lindemann (take $n = 1$). However it does not include all that is known on the transcendence of values of the exponential function, even if one does not mention results of algebraic independence (like Theorem 1.3 of Lindemann-Weierstraß). One such result, which is not included in Baker's Theorem, is the so-called *six exponentials Theorem* (see § 1.3, Th. 1.12 below).

It will be convenient to show that several statements are equivalent to Baker's homogeneous Theorem 1.5. As pointed out by J-P. Serre in his Bourbaki lecture on Baker's work [Ser 1970], it means that the natural map from the tensor product $\overline{\mathbb{Q}} \otimes_{\mathbb{Q}} \mathcal{L}$ in $\mathbb{C}$, which extends the injection from $\mathcal{L}$ to $\mathbb{C}$, is still injective (see Exercise 1.3. For a definition of the tensor product, see for instance [L 1993], Chap. XVI).

The only linear dependence relations, with algebraic coefficients, between logarithms of algebraic numbers, are the trivial ones, like

$$\log 24 = \sqrt{3}\log 9 + \left(1 - 2\sqrt{3}\right)\log 3 + \sqrt{2}\log 4 + (3 - 2\sqrt{2})\log 2.$$

Roughly speaking, if Theorem 1.5 is not true, then any vanishing nontrivial linear combination of elements of $\mathcal{L}$ with algebraic coefficients and *minimal length* would have the property that the coefficients are linearly independent over $\mathbb{Q}$, and at the same time the elements of $\mathcal{L}$ also are linearly independent over $\mathbb{Q}$.

Lemma 1.7. *Let $k \subset K$ be two fields, $\mathcal{E}$ be a K-vector space, and $\mathcal{M}$ be a k-vector subspace in $\mathcal{E}$. The three following statements are equivalent.*

(i) Let m be a positive integer and let $\lambda_1, \ldots, \lambda_m$ be elements of $\mathcal{M}$ which are linearly independent over k. Then these elements are also linearly independent over K in $\mathcal{E}$.

(ii) Let m be a positive integer. Let $\lambda_1, \ldots, \lambda_m$ be elements of $\mathcal{M}$, not all vanishing, and let $\beta_1, \ldots, \beta_m$ be k-linearly independent elements of K. Then

$$\beta_1\lambda_1 + \ldots + \beta_m\lambda_m \neq 0.$$

(iii) Let m be a positive integer. Let $\lambda_1, \ldots, \lambda_m$ be k-linearly independent elements of $\mathcal{M}$ and $\beta_1, \ldots, \beta_m$ be k-linearly independent elements of K. Then

$$\beta_1\lambda_1 + \ldots + \beta_m\lambda_m \neq 0.$$

Proof. We first remark that the implication *(i)* $\Rightarrow$ *(iii)* is trivial.
a) *Proof of (ii)* $\Rightarrow$ *(i).* Assume that for some $m \geq 1$ we have a relation $\beta_1\lambda_1 + \ldots + \beta_m\lambda_m = 0$ with $\beta_1, \ldots, \beta_m$ not all zero in K. Let $\beta'_1, \ldots, \beta'_s$ (with $0 < s \leq m$) be a basis of the k-vector space they span. We can write

$$\beta_i = \sum_{j=1}^{s} c_{ij}\beta'_j \qquad (1 \leq i \leq m),$$

with $c_{ij} \in k$, which are not all zero. Then

$$\sum_{j=1}^{s} \beta'_j \left(\sum_{i=1}^{m} c_{ij}\lambda_i \right) = 0.$$

Since $\beta'_1, \ldots, \beta'_s$ are k-linearly independent, we deduce from *(ii)*

$$\sum_{i=1}^{m} c_{ij}\lambda_i = 0 \qquad \text{for } 1 \leq j \leq s.$$

Therefore $\lambda_1, \ldots, \lambda_m$ are K-linearly dependent.
b) *Proof of (iii)* $\Rightarrow$ *(ii).* Assume $\beta_1\lambda_1 + \cdots + \beta_m\lambda_m = 0$ with $\beta_1, \ldots, \beta_m$ linearly independent over k in K and $\lambda_1, \ldots, \lambda_m$ in $\mathcal{M}$. We shall conclude $\lambda_1 = \cdots = \lambda_m = 0$. Renumbering $\lambda_1, \ldots, \lambda_m$ if necessary, we may assume that $\lambda_1, \ldots, \lambda_r$ (for some r with $0 \leq r \leq m$) is a basis of the k-vector space spanned by $\lambda_1, \ldots, \lambda_m$:

$$\lambda_i = \sum_{j=1}^{r} c_{ij}\lambda_j, \qquad (r+1 \leq i \leq m),$$

where c_{ij} are in k. We deduce

$$\sum_{j=1}^{r} \gamma_j\lambda_j = 0 \qquad \text{with} \qquad \gamma_j = \beta_j + \sum_{i=r+1}^{m} c_{ij}\beta_i, \quad (1 \leq j \leq r).$$

Using *(iii)* (with m replaced by r), we deduce from the linear independence of $\lambda_1, \ldots, \lambda_r$ over k that the r elements $\gamma_1, \ldots, \gamma_r$ are k-linearly dependent in K. However, since $\beta_1, \ldots, \beta_m$ are linearly independent over k, the only possibility is $r = 0$, which means $\lambda_1 = \cdots = \lambda_m = 0$. $\square$

When $k = \mathbb{Q}$, $K = \overline{\mathbb{Q}}$, $\mathcal{M} = \mathcal{L}$ and $\mathcal{E} = \mathbb{C}$, assertion *(i)* is nothing but Theorem 1.5 (see § 9.1.1 for an application of this Lemma 1.7 to the nonhomogeneous case). Other statements which are equivalent to Theorem 1.5 are given in Exercise 1.5.

1.2 Lower Bounds for $|a_1^{b_1} \cdots a_m^{b_m} - 1|$

Baker's Theorem 1.5 shows that expressions of the form

$$\beta_1\lambda_1 + \cdots + \beta_m\lambda_m$$

(where, for $1 \leq i \leq m$, β_i is an algebraic numbers and λ_i is a logarithm of an algebraic number) can vanish only in trivial cases. In fact, the proof yields a stronger result, giving an explicit lower bound for such nonzero numbers. We consider here these results in the easiest case to explain, namely $\beta_i \in \mathbb{Z}$, $\lambda_i = \log \alpha_i$ where $\alpha_i \in \mathbb{Z}$, $\alpha_i \geq 2$.

Let $a_1, \ldots, a_m, b_1, \ldots, b_m$ be rational integers with the a_i's all greater than one. We assume

$$a_1^{b_1} \cdots a_m^{b_m} \neq 1,$$

and we ask for a lower bound for the distance between these two numbers.

There is a trivial estimate: a nonzero rational number is at least as large as the inverse of a denominator:

$$\begin{aligned} \left|a_1^{b_1} \cdots a_m^{b_m} - 1\right| &\geq \prod_{b_i<0} a_i^{b_i} \\ &\geq \exp\left\{-\sum_{i=1}^{m} |b_i| \log a_i\right\} \\ &\geq \exp\{-mB \log A\}, \end{aligned}$$

where $B = \max\{|b_1|, \ldots, |b_m|\}$ and $A = \max\{a_1, \ldots, a_m\}$. This kind of estimate extends to algebraic α's. We shall call it *Liouville's inequality* (compare with Theorem 1.1; see also Chap. 2, Lemma 2.1, and Chap. 3, § 5).

The dependence in m and A in Liouville's inequality is sharp, but the main interest for applications is with the dependence in B. In order to see what can be expected, it is convenient to give a connection with measures of linear independence of logarithms of algebraic numbers. If

$$0 < \left|a_1^{b_1} \cdots a_m^{b_m} - 1\right| \leq \frac{1}{2},$$

then

$$\frac{1}{2}\left|b_1 \log a_1 + \cdots + b_m \log a_m\right| \leq \left|a_1^{b_1} \cdots a_m^{b_m} - 1\right| \leq 2\left|b_1 \log a_1 + \cdots + b_m \log a_m\right|$$

(see Exercise 1.1). Therefore the problem of obtaining a lower bound for the distance between 1 and the product $a_1^{b_1} \cdots a_m^{b_m}$ is equivalent to obtaining a lower bound for the nonzero number $b_1 \log a_1 + \cdots + b_m \log a_m$.

An easy application of Dirichlet's box principle (see Exercise 1.2) now yields:

Lemma 1.8. *Let $m, a_1, \ldots, a_m$ be rational integers, all of which are at least 2. Define $A = \max\{a_1, \ldots, a_m\}$. Then for every integer $B \geq 4 \log A$, there exist rational integers $b_1, \ldots, b_m$ with*

$$0 < \max_{1 \leq i \leq m} |b_i| < B$$

such that

$$\left| a_1^{b_1} \cdots a_m^{b_m} - 1 \right| \leq \frac{2m \log A}{B^{m-1}}.$$

If $a_1, \ldots, a_m$ are multiplicatively independent, then the left hand side is not zero. The upper bound is polynomial in $1/B$, while Liouville's inequality is exponential in $-B$. We shall see that, as far as the dependence in B is concerned, Lemma 1.8 is closer to the truth than Liouville's lower bound.

In 1935, one year after he had solved the seventh problem of D. Hilbert, A. O. Gel'fond used his transcendence method in order to derive a lower bound for a linear combination of two logarithms of algebraic numbers with algebraic coefficients (for references, see [G 1952], [FSh 1967], [B 1977] and [Sp 1982]). Let us give a simple example of such an estimate: for a_1, a_2 multiplicatively independent positive rational integers, and for $\epsilon > 0$, there exists a constant $C_1 = C_1(a_1, a_2, \epsilon)$, which can be explicitly computed, such that, for all $(b_1, b_2) \in \mathbb{Z}^2$ with $(b_1, b_2) \neq (0, 0)$, if we set $B = \max\{|b_1|, |b_2|, 2\}$, then

$$\left| a_1^{b_1} a_2^{b_2} - 1 \right| \geq C_1 \exp\left\{ -(\log B)^{5+\epsilon} \right\}.$$

In 1939, A. O. Gel'fond refined the estimate and replaced the exponent $5+\epsilon$ by $3+\epsilon$, and in 1949 he [3] reached $2 + \epsilon$. At the same time he gave an estimate which is valid for any $m \geq 2$ (see [G 1952], Th. III of Chap. 1, p.28):

Theorem 1.9. (Gel'fond's Ineffective Estimate). *Let $(a_1, \ldots, a_m)$ be a m-tuple of positive multiplicatively independent rational integers. For every $\delta > 0$, there exists a positive constant $C_2 = C_2(a_1, \ldots, a_m, \delta)$ such that, if $b_1, \ldots, b_m$ are rational integers, not all of which are zero, and if we set $B = \max\{|b_1|, \ldots, |b_m|, 2\}$, then*

$$\left| a_1^{b_1} \cdots a_m^{b_m} - 1 \right| \geq C_2 e^{-\delta B}.$$

For the proof of Theorem 1.9, the main tool is a result of diophantine approximation, which we shall take for granted. A. O. Gel'fond used a result of his own, which was a refinement of earlier results due to A. Thue, C. L. Siegel and F. Dyson. Here, for simplicity, we shall use the stronger result due to K. F. Roth, which we do not prove in these notes (see for instance [Ro 1955], [Sch 1957], [B 1975], [L 1983] or [Sc 1991]):

[3] Explicit estimates were provided later by A. Schinzel in [S 1967].

Theorem 1.10* (Thue-Siegel-Roth). *Let α be an algebraic number and let ϵ be a positive real number. There exists a number $C_0 = C_0(\alpha, \epsilon) > 0$ such that for any rational number p/q with $q > 0$ and $p/q \neq \alpha$,*

$$\left|\alpha - \frac{p}{q}\right| > \frac{C_0}{q^{2+\epsilon}}.$$

Proof of Theorem 1.9. We shall use Theorem 1.10 with $\epsilon = 1$:

$$\left|\alpha - \frac{p}{q}\right| > \frac{C_0(\alpha, 1)}{q^3}.$$

Let $\delta > 0$. Assume C_2 does not exist: for each real number $C > 0$ there exists $\underline{b} = (b_1, \ldots, b_m) \in \mathbb{Z}^m$ with

$$0 < \left|a_1^{b_1} \cdots a_m^{b_m} - 1\right| \leq C \exp\{-\delta B\}$$

where $B = \max\{2, |b_1|, \ldots, |b_m|\}$. Hence the set E_1 of $\underline{b} \in \mathbb{Z}^m$ for which

$$0 < \left|a_1^{b_1} \cdots a_m^{b_m} - 1\right| \leq \exp\{-\delta B\}$$

is infinite. Let N be a positive integer satisfying $N > (6m/\delta) \log A$, with $A = \max\{a_i\}$. Since the set $(\mathbb{Z}/N\mathbb{Z})^m$ is finite, there is an infinite subset E_2 of E_1 having all elements in the same class modulo N. This means that there exists $\underline{r} \in \mathbb{N}^m$ with $0 \leq r_i < N$ $(1 \leq i \leq m)$ such that, for all $\underline{b} \in E_2$,

$$b_i \equiv r_i \mod N \quad (1 \leq i \leq m).$$

Let E_3 be the set of $\underline{b} \in E_2$ with $B \geq N$. Once more this is an infinite set. For each $\underline{b} \in E_3$, there is a $\underline{x} \in \mathbb{Z}^m$ such that

$$b_i = r_i + N x_i \quad (1 \leq i \leq m).$$

We have $|x_i| \leq 1 + B/N \leq 2B/N$ $(1 \leq i \leq m)$. Let us define two rational numbers $s = a_1^{r_1} \cdots a_m^{r_m}$ and $t = a_1^{x_1} \cdots a_m^{x_m}$. Notice that s does not depend on $\underline{b} \in E_3$, while t depends on $\underline{b}$. From the construction of E_3 we deduce

$$0 < \left|st^N - 1\right| \leq e^{-\delta B}.$$

We now use the estimate $|x - 1| \leq |x^N - 1|$ which is valid for all $x > 0$ (the number $1 + x + \cdots + x^{N-1}$ is at least 1):

$$0 < \left|s^{1/N} t - 1\right| \leq e^{-\delta B}.$$

This shows that the rational number t is close to the algebraic number $\alpha = s^{-1/N}$ which is the real N-th root of $1/s$:

$$0 < |t - \alpha| \leq \alpha e^{-\delta B}.$$

Since the denominator of t is at most $A^{2mB/N}$, Theorem 1.10 yields:

$$|t - \alpha| \geq C_0(\alpha, 1)A^{-6mB/N}.$$

Combining the upper and lower bounds, we deduce the estimate

$$B\left(\delta - \frac{6m \log A}{N}\right) \leq -\log C_0(\alpha, 1) - \frac{1}{N} \log s,$$

which shows that the number B is bounded (the numbers δ, A, N, $C_0(\alpha, 1)$ and s do not depend on $\underline{b} \in E_3$), which is in contradiction with the fact that E_3 is an infinite set. □

This proof produces a lower bound for $\left|a_1^{b_1} \cdots a_m^{b_m} - 1\right|$ using a lower bound for $|\alpha - (p/q)|$. By means of similar arguments, one can go backwards and deduce nontrivial measures of rational approximation for algebraic numbers using measures of linear independence for logarithms of algebraic numbers (see § 10.4.1).

The proof of Theorem 1.9 does not enable one to compute the constant C_2, because one uses the Thue-Siegel-Roth Theorem which is not *effective*: the number C_0 in Theorem 1.10 depends on α and ϵ, but given α and ϵ we do not know how to compute it. The proof of Theorem 1.10 is by contradiction: if the result does not hold, there is a whole sequence of good rational approximations p_n/q_n to α, and this is the main point which makes the result ineffective.

A. O. Gel'fond applied his estimate to several number theoretic questions, in particular (with Y. V. Linnik) for Gauss' problem of determining all imaginary quadratic number fields with class number one. He also applied his lower bound to the study of several types of diophantine equations.

The question of effectivity here is a crucial one. To solve a diophantine equation is to give the complete list of solutions; for simplicity suppose we are looking for solutions in rational integers. A first question is to decide whether there are infinitely or only finitely many such solutions. Let us assume we are in the latter case. Sometimes it is possible to produce an upper bound for the *number* of solutions. Unless this upper bound is optimal, it will not be sufficient to derive an algorithm for completely solving the equation. On the other hand, if we know an upper bound for the maximum absolute value of the solutions themselves, then one deduces trivially such an algorithm (there are only finitely many integers below the bound). The last step is to produce an efficient algorithm which will complete the list of solutions, but we shall not address this issue here; our concern is to describe one basic tool which is efficient to produce an effective upper bound for the solutions.

We quote from Gel'fond's book [G 1952] (p.126 of the English edition):

> ... one can assume the fundamental problem in the analytic theory of transcendental numbers to be that of strengthening the analytic methods in the theory of transcendental numbers, so that it will be possible to apply them to the investigation of the behavior of linear forms in n logarithms of algebraic numbers.

Also, from p.177:

Nontrivial lower bounds for linear forms, with integral coefficients, of an arbitrary number of logarithms of algebraic numbers, obtained effectively by methods of the theory of transcendental numbers, will be of extraordinarily great significance in the solution of very difficult problems of modern number theory. Therefore, one may assume, as was already mentioned above, that the most pressing problem in the theory of transcendental numbers is the investigation of the measures of transcendence of finite sets of logarithms of algebraic numbers.

As we already know from § 1.1, this problem was solved in 1966 by A. Baker [B 1975]. The next refinement is due to N. I. Fel'dman [F 1968] two years later.

Let $a_1, \ldots, a_m$ be positive multiplicatively independent rational integers and $b_1, \ldots, b_m$ rational integers, not all of which are zero; then

$$\left|a_1^{b_1} \cdots a_m^{b_m} - 1\right| \geq \exp\left\{ - C_3 \log B\right\} = B^{-C_3},$$

where $C_3 = C_3(a_1, \ldots, a_m)$ is a positive effectively computable number.

Fel'dman's result is valid more generally when the integers $a_1, \ldots, a_m$ are replaced by algebraic numbers $\alpha_1, \ldots, \alpha_n$, (and also when $b_1, \ldots, b_m$ are replaced by algebraic numbers – in this case it is more convenient to state the result as a measure of linear independence for logarithms of algebraic numbers, that is a lower bound for a linear combination, with algebraic coefficients, of logarithms of algebraic numbers). Such estimates have many applications to various diophantine problems, including an effective improvement on Liouville's Theorem 1.1 due to Fel'dman (improving an earlier result of Baker – see § 10.4.1):

- *For each algebraic number α of degree $d \geq 3$, there exists two positive constants $c(\alpha)$ and $\eta(\alpha)$ such that, for $p/q \in \mathbb{Q}$,*

$$\left|\alpha - \frac{p}{q}\right| > \frac{c(\alpha)}{q^{d-\eta(\alpha)}}.$$

In 1993 E. Bombieri [Bo 1993] introduced a new method for obtaining effective irrationality measures for roots of high order of algebraic numbers and examined the applications to effective diophantine approximation in a number field by a finitely generated subgroup. A new effective solution of Thue's equation in number fields and the Baker-Feldman effective improvement to Liouville's Theorem resulted. The main tools were the Thue-Siegel Principle, Viola's version of Dyson's Lemma, and the geometry of numbers, there being no appeal to measure of linear independence of logarithms of algebraic numbers. E. Bombieri and P. B. Cohen [BoCoh 1997] extended this work to the nonarchimedean case and introduced, along the lines of ideas by P. Corvaja [Co 1997], the use of Laurent's determinantal method to replace Siegel's Lemma in this context.

It is stated in [BoCoh 1997] that Theorem 1 of that paper can be obtained directly from Baker's method, rather than from the Thue-Siegel method. The authors point out that this would lead to a sharper version of their Theorem 1 and that their Theorem 2, whose proof uses the geometry of numbers, could then be applied directly to this sharper result. This program has been carried out by Y. Bugeaud [Bu 1998b], both in the archimedean and nonarchimedean cases.

A detailed account of the history of measures of linear independence for logarithms of algebraic numbers until 1976 is given by A. Baker in [B 1977], and a more recent survey can be found in Chap. 4 § 1 of [FNe 1998] (see also [Mat 1998] and § 10.4). In this book we devote a lot of attention to this problem of giving explicit measures of linear independence for logarithms of algebraic numbers. Since the methods are usually not considered to be very simple, we try to make them easier to understand by introducing progressively some of the different tools which have been used so far for establishing the best known estimates. In Chap. 7 we prove a first explicit lower bound, which is not the best known one, but requires a minimum of technique. In particular the proof does not involve any derivation. One of the main tools is a zero estimate, which is proved in Chap. 5 by D. Roy, following P. Philippon [P 1986a]. A refinement of this zero estimate, involving derivations, is also due to P. Philippon [P 1986a], and is explained by D. Roy in Chap. 8.

Apart from the numerical value of the constant, the best known measures of linear independence for logarithms of algebraic numbers are proved twice, in Chapters 9 and 10, by means of *dual* methods.

A special case of the general measure of linear independence provided by Theorem 9.1 is the following:

- *Let $a_1, \ldots, a_m$, $b_1, \ldots, b_m$ be rational integers. Assume $a_i \geq 2$ for $1 \leq i \leq m$ and $a_1^{b_1} \cdots a_m^{b_m} \neq 1$. Define $B = \max\{2, |b_1|, \ldots, |b_m|\}$. Then*

$$\left|a_1^{b_1} \cdots a_m^{b_m} - 1\right| \geq \exp\bigl\{ - C(m)(\log B)(\log a_1) \cdots (\log a_m)\bigr\},$$

where $C(m)$ is a positive effectively computable number which depends only on m.

We describe the state of the art on this topic in § 10.4 for the results, in § 14.4 for the methods.

The second part of Lang's book [L 1978] deals with measures of linear independence for logarithms of algebraic numbers (not only for the usual exponential function, but also for elliptic functions). The introduction to Chap. X and XI of [L 1978] (pp.212–217) proposes far reaching conjectures. For instance:

Conjecture 1.11. *For any $\epsilon > 0$, there exists a constant $C_7(\epsilon) > 0$ such that, for any nonzero rational integers $a_1, \ldots, a_m$, $b_1, \ldots, b_m$ with $a_1^{b_1} \cdots a_m^{b_m} \neq 1$*

$$\left|a_1^{b_1} \cdots a_m^{b_m} - 1\right| \geq \frac{C_7(\epsilon)^m}{B^{m-1+\epsilon} A^{m+\epsilon}},$$

where $A = \max_{1 \leq i \leq m} |a_i|$ and $B = \max_{1 \leq i \leq m} |b_i|$.

A related open problem (see especially [B 1998], [P 1999b] and Exercise 1.11) is the *abc* Conjecture of D. W. Masser and J. Œsterlé:

(?) *abc***–Conjecture.** *For each $\epsilon > 0$ there exists a positive number $\kappa(\epsilon)$ which has the following property: if a, b and c are three positive rational integers which are relatively prime, with $a + b = c$, and if*

$$N = \prod_{p|abc} p$$

denotes the radical (or squarefree part) of the product abc, then

$$c < \kappa(\epsilon) N^{1+\epsilon}.$$

The example $a = 1$, $c = 3^{2^n}$, $b = c - a$ (where 2^n divides b) shows that the exponent $1 + \epsilon$ cannot be replaced by 1. This conjecture is closely related to a conjecture of L. Szpiro on the conductor of elliptic curves. An analog of the *abc* Conjecture for function fields is a theorem of W. W. Stothers (1981) and R. Mason (1984 — see [L 1993]). Using (ordinary as well as p-adic) measures of linear independence for logarithms of algebraic numbers, R. Tijdeman, C. L. Stewart and Yu Kunrui (see [StY 1991] and [Sc 1991] Epilogue) made a small step in the direction of the *abc* Conjecture:

$$\log c \le \kappa(\epsilon) N^{(1/3)+\epsilon}.$$

Applications of measures of linear independence of logarithms of algebraic numbers arise in many subjects: class number problems (A. Baker and H. Stark), p-adic L-functions (J. Ax and A. Brumer), knot theory (R. Riley), modular forms (R. W. K. Odoni), Ramanujan τ function (K. and M. Murty, T. N. Shorey), recurrent sequences (A. Schinzel, C. L. Stewart, J. H. Loxton, A. J. van der Poorten, M. Mignotte, R. Tijdeman, T. N. Shorey,...), diophantine equations (A. O. Gel'fond and many others, including A. Baker).

Improvement on measures of linear independence of logarithms of algebraic numbers are relevant for solving diophantine equations. There is a big industry on this topic with many achievements. When only two logarithms are involved, the numerical constants are so small that they enable one to solve completely whole families of diophantine equations.

Further applications (together with proofs and references) can be found in the following references: [G 1952], [Ser 1970], [Sho 1974], [B 1975], [B 1977], [V 1977], [L 1978], [Sp 1982], [Lox 1986], [ShoT 1986], [Ser 1989], [Ri 1994], [FNe 1998], Chap. 4 § 1.

Most applications involve only homogeneous linear combinations of logarithms with rational coefficients. However, an application to computer science of the lower bound in [NeW 1996] for $|\beta - \lambda|$ (see § 14.2.3) is given in [MuTi 1996].

Conjecture 14.25 in Chap. 14 proposes a sharp lower bound for

$$|\beta_0 + \beta_1\lambda_1 + \cdots + \beta_m\lambda_m|$$

(for $\beta_i \in \overline{\mathbb{Q}}$ and $\lambda_j \in \mathcal{L}$) when this number is nonzero.

In a few particular special cases, very sharp numerical constants are known: for instance, using completely different techniques, related with Padé approximation and Siegel's G-functions, E. Dubois, G. Rhin and Ph. Toffin [Rh 1987] proved: *for rational integers b_0, b_1 and b_2 with $B = \max\{|b_1|, |b_2|\} \geq 2$,*

$$|b_0 + b_1 \log 2 + b_2 \log 3| \geq B^{-13.3}.$$

Such an estimate has recently been extended to

$$|b_0 + b_1 \log 2 + b_2 \log 3 + b_3 \log 5 + b_4 \log 7|.$$

It should be pointed out that a similar method yields the irrationality of

$$\left(\log\left(1 + \frac{1}{q}\right)\right)\left(\log\left(1 - \frac{1}{q}\right)\right)$$

(work of A. I. Galochkin and M. Hata).

We just mention that there is also a rich related theory for elliptic logarithms [PW 1988c], [Hir 1991], with explicit estimates by S. David (for n complex logarithms) [D 1995] and G. Rémond and F. Urfels [RemU 1996] (for two elliptic p-adic logarithms). See also [FNe 1998], Chap. 4 § 3.

An important open question, related with the so-called *S-units equations*, is to give an effective lower bound for non-vanishing expressions of the form

$$\left|\sum_{i=1}^{n} \alpha_1^{b_{i1}} \cdots \alpha_m^{b_{im}}\right|$$

where $\alpha_1, \ldots, \alpha_m$ are nonzero algebraic numbers and b_{ij} $(1 \leq i \leq n, 1 \leq j \leq m)$ are rational integers. Nontrivial (but also noneffective) lower bounds are known from Schmidt's subspace theorem (a far reaching generalization of the Thue-Siegel-Roth theorem; see for instance [Sc 1980] and [Sc 1991]). Only the case $n = 2$ has been made effective so far, thanks to effective measures of linear independence of logarithms of algebraic numbers.

1.3 The Six Exponentials Theorem and the Four Exponentials Conjecture

Let us start with an easy question: *which are the real numbers t for which 2^t is a rational integer?* Of course all $t \in \mathbb{N}$ satisfy this requirement; but there are others: for $a \in \mathbb{N}$, $a \neq 0$, if we set $t = \log a / \log 2$, then $2^t = \exp(t \log 2) = a \in \mathbb{N}$. Hence

$$\{t \in \mathbb{R}; 2^t \in \mathbb{N}\} = \left\{\frac{\log a}{\log 2}; a \in \mathbb{N}, a > 0\right\}.$$

If we denote this set by E_1, then $E_1 \cap \mathbb{Q} = \mathbb{N}$.

We consider now the set

$$E_2 = \{t \in \mathbb{R}\,;\, 2^t \in \mathbb{N} \text{ and } 3^t \in \mathbb{N}\}.$$

This set contains $\mathbb{N}$ and is contained in E_1. In particular $E_2 \cap \mathbb{Q} = \mathbb{N}$. The following problem is still open: *is it true that* $E_2 = \mathbb{N}$? This means:

Problem. *Does there exist an irrational number which belongs to* E_2?

This question amounts to ask whether there exist two positive integers a and b such that

$$\frac{\log a}{\log 2} = \frac{\log b}{\log 3}$$

and at the same time this quotient is irrational. Another equivalent formulation is to ask whether a 2×2 matrix

$$\begin{pmatrix} \log a & \log b \\ \log 2 & \log 3 \end{pmatrix}$$

(with positive integers a and b) can be singular without a being a power of 2. We shall consider this question in a more general setting (the four exponentials Conjecture).

Finally we introduce a third set

$$E_3 = \{t \in \mathbb{R}\,;\, 2^t \in \mathbb{N},\ 3^t \in \mathbb{N} \text{ and } 5^t \in \mathbb{N}\}.$$

Of course we have $\mathbb{N} \subset E_3 \subset E_2 \subset E_1$. The six exponentials Theorem below implies $E_3 = \mathbb{N}$.

We may replace $\{2, 3, 5\}$ by any set of three distinct primes. More generally, if we consider three multiplicatively independent (complex) algebraic numbers, then there is no need to restrict the discussion to real values of t.

Theorem 1.12 (Six Exponentials). *Let $x_1, \ldots, x_d$ be complex numbers which are linearly independent over $\mathbb{Q}$ and let $y_1, \ldots, y_\ell$ also be complex numbers which are linearly independent over $\mathbb{Q}$. Assume $d\ell > d+\ell$. Then one at least of the $d\ell$ numbers*

$$\exp(x_i y_j), \qquad (1 \le i \le d,\ 1 \le j \le \ell)$$

is transcendental.

It is clear that the interesting case is $d = 3$, $\ell = 2$ (or $d = 2$, $\ell = 3$, which gives the same result because of the symmetry), and this explains the name of the result.

One conjectures that the conclusion is still valid under the weaker hypothesis $d\ell \ge d + \ell$:

Conjecture 1.13 (Four Exponentials Conjecture). *Let x_1, x_2 be two $\mathbb{Q}$-linearly independent complex numbers and y_1, y_2 also two $\mathbb{Q}$-linearly independent complex numbers. Then one at least of the 4 numbers*

$$\exp(x_i y_j), \qquad (i = 1, 2,\ j = 1, 2)$$

is transcendental.

The six exponentials Theorem 1.12 occurs for the first time in a paper by L. Alaoglu and P. Erdős [AEr 1944], when these authors try to prove Ramanujan's assertion that the quotient of two consecutive *superior highly composite numbers* [4] is a prime, they need to know that if x is a real number such that p_1^x and p_2^x are both rational numbers, with p_1 and p_2 distinct prime numbers, then x is an integer. However this statement (special case of the Conjecture 1.13) is yet unproven. They quote C. L. Siegel and claim that x indeed is an integer if one assumes p_i^x to be rational for *three* distinct primes p_i. This is just a special case of Theorem 1.12. They deduce that the quotient of two consecutive superior highly composite numbers is either a prime, or else a product of two primes.

Theorem 1.12 can be deduced from a very general result of Th. Schneider [Sch 1949]. Conjecture 1.13 is equivalent to the first of the eight problems at the end of Schneider's book [Sch 1957]. An explicit statement of the six exponentials Theorem, together with a proof, has been published independently and at about the same time by S. Lang [L 1965a], [L 1965b], [L 1966], Chap. 2 and K. Ramachandra [R 1968], [R 1969a], Chap. 2. They both formulated the four exponentials Conjecture 1.13 explicitly.

Remark. Taking $x_1 = 1$, $x_2 = |\lambda|/\lambda$, $y_1 = \lambda$ and $y_2 = |\lambda|$, one deduces from Conjecture 1.13:

(?) *For $\lambda \in \mathcal{L}$ with $\lambda \notin \mathbb{R}$, the number $e^{|\lambda|}$ is transcendental.*

Further similar open problems are proposed in § 11.6.1.

Baker's Theorem and the six exponentials Theorem do not cover all known transcendence results on the exponential function (without mentioning algebraic independence results). We give below (Theorems 1.16 and 1.17) stronger versions of the six exponentials Theorem.

1.4 Algebraic Independence of Logarithms

There are a few results of algebraic independence concerning the values of the exponential function which are not included in Theorem 1.3 of Lindemann-Weierstraß. For instance A. O. Gel'fond proved in 1949 that for a cubic number β and for $\lambda \in \mathcal{L} \setminus \{0\}$, if we set

$$\alpha = e^{\lambda}, \quad \alpha^{\beta} = e^{\beta\lambda} \quad \text{and} \quad \alpha^{\beta^2} = e^{\beta^2\lambda},$$

then the two numbers α^{β} and α^{β^2} are algebraically independent (see [G 1952]). This result has been extended, especially by G. V. Chudnovsky and P. Philippon. The best

[4] S. Ramanujan defines an integer n to be a *superior highly composite number* if there exists $\epsilon > 0$ such that the divisor function $d(n)$ (number of divisors of n) satisfies $d(m)m^{-\epsilon} < d(n)n^{-\epsilon}$ for $m \neq n$.

result known in this direction is due to G. Diaz: if β is algebraic of degree $d \geq 3$, then for any $\lambda \in \mathcal{L} \setminus \{0\}$, at least $\left[(d+1)/2\right]$ of the $d-1$ numbers $\alpha^{\beta}, \alpha^{\beta^2}, \ldots, \alpha^{\beta^{d-1}}$ are algebraically independent [Di 1989] (see § 15.4).

The most far reaching conjecture on the subject is due to S. Schanuel [L 1966]:

Conjecture 1.14 (Schanuel's Conjecture). *If $x_1, \ldots, x_n$ are $\mathbb{Q}$-linearly independent complex numbers, then, among the $2n$ numbers*

$$x_1, \ldots, x_n, e^{x_1}, \ldots, e^{x_n},$$

at least n are algebraically independent.

The conclusion may also be phrased: the *transcendence degree* over $\mathbb{Q}$ of the field

$$\mathbb{Q}\bigl(x_1, \ldots, x_n, e^{x_1}, \ldots, e^{x_n}\bigr)$$

is at least n.

This conjecture is believed to include all known transcendence results as well as all reasonable transcendence conjectures on the values of the exponential function. The special case where $x_1, \ldots, x_n$ are all algebraic is just Theorem 1.3 of Lindemann-Weierstraß. The other special case where $e^{x_1}, \ldots, e^{x_n}$ are algebraic would already have tremendous consequences:

Conjecture 1.15 (Algebraic Independence of Logarithms). *Let $\lambda_1, \ldots, \lambda_n$ be elements of $\mathcal{L}$ which are linearly independent over $\mathbb{Q}$. Then these numbers are algebraically independent.*

We are very far from this conjecture. Indeed, it is not yet even known that there exist at least two algebraically independent logarithms of algebraic numbers! In spite of this bad situation, interesting partial results are known, as we shall see. Instead of looking, for a fixed tuple $(\lambda_1, \ldots, \lambda_n) \in \mathcal{L}^n$, to the condition $P(\lambda_1, \ldots, \lambda_n) = 0$ for some $P \in \mathbb{Z}[X_1, \ldots, X_n]$, we fix $P \in \mathbb{Z}[X_1, \ldots, X_n]$ and we consider the set of zeros of P in $\mathcal{L}^n$.

Some such results are known in connection with lower bounds for the ranks of matrices whose entries are logarithms of algebraic numbers. We shall now look at this problem.

Conjecture 1.13 can be stated as follows. *Consider a 2×2 matrix whose entries are logarithms of algebraic numbers:*

$$\mathsf{M} = \begin{pmatrix} \lambda_{11} & \lambda_{12} \\ \lambda_{21} & \lambda_{22} \end{pmatrix}.$$

Assume that the two rows of M *are linearly independent over $\mathbb{Q}$ (in $\mathbb{C}^2$), and also that the two columns are linearly independent over $\mathbb{Q}$. Then the rank of* M *is* 2.

It is not difficult (Exercise 1.8) to deduce the four exponentials Conjecture 1.13 from Conjecture 1.15 on algebraic independence of logarithms.

Theorem 1.12 is equivalent to the following assertion:

Consider a $d \times \ell$ matrix whose entries are logarithms of algebraic numbers:

$$\mathsf{M} = \begin{pmatrix} \lambda_{11} & \cdots & \lambda_{1\ell} \\ \vdots & \ddots & \vdots \\ \lambda_{d1} & \cdots & \lambda_{d\ell} \end{pmatrix}.$$

Assume that the d rows of M are $\mathbb{Q}$-linearly independent (in $\mathbb{C}^\ell$), and also that the ℓ columns are $\mathbb{Q}$-linearly independent (in $\mathbb{C}^d$). If $d\ell > d + \ell$, then the rank of M is at least 2.

We obtain the equivalence with Theorem 1.12 by noticing that a $d \times \ell$ matrix has rank at most 1 if and only if it can be written $(x_i y_j)_{1\le i\le d, 1\le j\le \ell}$, for some complex numbers $x_1, \ldots, x_d, y_1, \ldots, y_\ell$ (see Exercise 1.9).

Assume now that $d\ell/(d+\ell)$ is *large. Is it possible to get a better lower bound for the rank of* M? We notice first that conditions on linear independence of rows and columns are no longer sufficient, as shown by matrices like the following one, which has rank 2:

$$\begin{pmatrix} \log 2 & \log 3 & \cdots & \log p_m \\ \log 3 & & & \\ \vdots & & 0 & \\ \log p_m & & & \end{pmatrix}$$

where $d = \ell = m$, and p_m is the m-th prime number. Here is a simple statement which extends Theorem 1.12 and will be proved in § 12.2.1.

Theorem 1.16. *Let $\mathsf{M} = \big(\lambda_{ij}\big)_{\substack{1\le i\le d\\ 1\le j\le \ell}}$ be a $d \times \ell$ matrix with entries in $\mathcal{L}$. Assume that for any $\underline{t} = (t_1, \ldots, t_d) \in \mathbb{Z}^d \setminus \{0\}$ and any $\underline{s} = (s_1, \ldots, s_\ell) \in \mathbb{Z}^\ell \setminus \{0\}$, we have*

$$\sum_{i=1}^{d} \sum_{j=1}^{\ell} t_i s_j \lambda_{ij} \neq 0.$$

Then the rank of M is at least $d\ell/(d+\ell)$.

Let us assume that Conjecture 1.14 is true.

- *How can one describe the rank of a matrix*

$$\mathsf{M} = \big(\lambda_{ij}\big)_{\substack{1\le i\le d\\ 1\le j\le \ell}}$$

with entries in $\mathcal{L}$?

This problem has been solved by D. Roy [Roy 1989] as follows: let $\lambda_1, \ldots, \lambda_r$ be a basis of the $\mathbb{Q}$-vector space spanned by the $d\ell$ entries of M: there exist rational integers a_{ijk} such that

$$\lambda_{ij} = \sum_{k=1}^{r} a_{ijk}\lambda_k, \qquad (1 \le i \le d, 1 \le j \le \ell).$$

Consider the matrix

$$\widetilde{\mathsf{M}} = \left(\sum_{k=1}^{r} a_{ijk}X_k\right)_{\substack{1\le i\le d\\1\le j\le \ell}} = \sum_{k=1}^{r} X_k\left(a_{ijk}\right)_{\substack{1\le i\le d\\1\le j\le \ell}},$$

with coefficients which are linear forms in unknowns $X_1, \dots, X_r$ and define the *structural rank of* M as the rank of the matrix $\widetilde{\mathsf{M}}$ whose entries are in the field $\mathbb{Q}(X_1, \dots, X_r)$ of rational functions in r variables:

$$r_{\mathrm{str}}(\mathsf{M}) = \operatorname{rank}(\widetilde{\mathsf{M}}).$$

This number does not depend on the choice of $\lambda_1, \dots, \lambda_r$, and Conjecture 1.15 plainly implies that the equality $\operatorname{rank}(\mathsf{M}) = r_{\mathrm{str}}(\mathsf{M})$ holds. One does not need the full force of Conjecture 1.15: one needs only the *homogeneous* special case of it: *if* $\lambda_1, \dots, \lambda_n$ *are elements of* $\mathcal{L}$ *which are linearly independent over* $\mathbb{Q}$*, and if* $P \in \mathbb{Q}[X_1, \dots, X_n]$ *is a nonzero* homogeneous *polynomial, then* $P(\lambda_1, \dots, \lambda_n)$ *is not zero.*

A quite remarkable fact, proved by D. Roy [Roy 1989], is that the converse holds, namely: *if* $\operatorname{rank}(\mathsf{M}) = r_{\mathrm{str}}(\mathsf{M})$ *holds for all matrices* M *with entries in* $\mathcal{L}$*, then the conjecture on homogeneous algebraic independence of logarithms is true.*

If one wishes to consider nonhomogeneous polynomials in logarithms of algebraic numbers, then it is sufficient to deal with matrices whose entries lie in the $\mathbb{Q}$-vector subspace of $\mathbb{C}$ spanned by 1 and $\mathcal{L}$.

More generally, denote by $\widetilde{\mathcal{L}}$ the $\overline{\mathbb{Q}}$-vector space spanned in $\mathbb{C}$ by 1 and $\mathcal{L}$:

$$\widetilde{\mathcal{L}} = \{\beta_0 + \beta_1\lambda_1 + \cdots + \beta_n\lambda_n;\ n \ge 0,\ \beta_i \in \overline{\mathbb{Q}},\ \lambda_i \in \mathcal{L}\}.$$

The *structural* rank of a matrix with entries in $\widetilde{\mathcal{L}}$ is defined as before, taking a basis of the $\overline{\mathbb{Q}}$-vector space spanned by the coefficients and considering matrices whose entries are linear forms. Again, it follows from [Roy 1989] that Conjecture 1.15 is equivalent to the fact that the rank equals the structural rank for matrices with coefficients in $\widetilde{\mathcal{L}}$. Moreover, the following partial result in the direction of Conjecture 1.15 is known [Roy 1992a]— see Chap. 12:

Theorem 1.17 (D. Roy). *Let* M *be a matrix whose entries are in* $\widetilde{\mathcal{L}}$*. Then*

$$\operatorname{rank}(\mathsf{M}) \ge \frac{1}{2} r_{\mathrm{str}}(\mathsf{M}).$$

Therefore, from this point of view, half of Conjecture 1.15 on algebraic independence of logarithms is now proved!

The proof of this result rests on the so-called *Linear Subgroup Theorem* (see § 1.5 below and Chap. 11). Further related results are given in Roy's papers, especially [Roy 1992c] where he answers a question of J-J. Sansuc on the density of finitely generated subgroups in the multiplicative group $k^\times$ of a number field k for the canonical embedding into $(k \otimes_{\mathbb{Q}} \mathbb{R})^\times$.

1.5 Diophantine Approximation on Linear Algebraic Groups

The *Linear Subgroup Theorem* 11.5 is a statement which provides a lower bound for the rank of matrices whose coefficients are either algebraic numbers or logarithms of algebraic numbers. We do not state the precise result here (all necessary information is provided in Chap. 11), but we only give some examples.

Let d_0, d_1, ℓ_0 and ℓ_1 be nonnegative integers. Define $d = d_0 + d_1$, $\ell = \ell_0 + \ell_1$, and assume $d > 0$ and $\ell > 0$. Consider the $d \times \ell$ matrix

$$\mathsf{M} = \begin{pmatrix} \mathsf{B}_0 & \mathsf{B}_1 \\ \mathsf{B}_2 & \mathsf{L} \end{pmatrix} = \begin{pmatrix} \beta_{11} & \cdots & \beta_{1\ell_0} & \beta_{1,\ell_0+1} & \cdots & \beta_{1\ell} \\ \vdots & \ddots & \vdots & \vdots & \ddots & \vdots \\ \beta_{d_0 1} & \cdots & \beta_{d_0\ell_0} & \beta_{d_0,\ell_0+1} & \cdots & \beta_{d_0\ell} \\ \beta_{d_0+1,1} & \cdots & \beta_{d_0+1,\ell_0} & \lambda_{11} & \cdots & \lambda_{1\ell_1} \\ \vdots & \ddots & \vdots & \vdots & \ddots & \vdots \\ \beta_{d1} & \cdots & \beta_{d\ell_0} & \lambda_{d_1 1} & \cdots & \lambda_{d_1\ell_1} \end{pmatrix}$$

where each of the three matrices B_0, B_1 and B_2 has algebraic entries, while the matrix L has entries in $\mathcal{L}$.

Under suitable assumptions, the following lower bound holds:

$$\operatorname{rank}(\mathsf{M}) \geq \frac{d_1\ell_1 + d_1\ell_0 + d_0\ell_1}{d_1 + \ell_1}.$$

This estimate is especially interesting when the right hand side is $> d - 1$, since in this case the conclusion can be written: $\operatorname{rank}(\mathsf{M}) = d$. This happens when

$$\ell_1 \geq d_1(d - \ell_0 - 1) + 1.$$

Two important examples, as we shall see shortly, are given by:
a) $\ell_0 = d - 1$ and $\ell_1 = 1$
and
b) $d_0 = d - 1$, $d_1 = 1$ and $\ell_0 + \ell_1 \geq d$.

A connection with Baker's Theorem 1.6 on non-vanishing of linear combinations of logarithms of algebraic numbers

$$\beta_0 + \beta_1\lambda_1 + \cdots + \beta_{n-1}\lambda_{n-1} - \lambda_n$$

arises from the following observation:

$$\det \begin{pmatrix} & & & X_1 \\ & \mathsf{I}_n & & \vdots \\ & & & X_n \\ Y_1 & \cdots & Y_n & T \end{pmatrix} = T - X_1Y_1 - \cdots - X_nY_n.$$

Hence, in case a), one can choose

$$\begin{pmatrix} & & & & 1 \\ & & & & \lambda_1 \\ & & \mathsf{I}_n & & \vdots \\ & & & & \lambda_{n-1} \\ \beta_0 & \beta_1 & \cdots & \beta_{n-1} & \lambda_n \end{pmatrix}$$

(with $d = n + 1$, $d_0 = 1$, $d_1 = d - 1$, $\ell_0 = d - 1$, $\ell_1 = 1$), while the following matrix belongs to case b):

$$\begin{pmatrix} & & & & \beta_0 \\ & & & & \beta_1 \\ & & \mathsf{I}_n & & \vdots \\ & & & & \beta_{n-1} \\ 1 & \lambda_1 & \cdots & \lambda_{n-1} & \lambda_n \end{pmatrix},$$

(with $d = n + 1$, $d_0 = d - 1$, $d_1 = 1$, $\ell_0 = 1$, $\ell_1 = d - 1$).

For a homogeneous linear combination of only two logarithms $\beta\lambda_1 - \lambda_2$, example a) with $\ell_1 = 1$ corresponds to Gel'fond's solution of Hilbert's seventh problem, while example b) with $d_1 = 1$ corresponds to Schneider's proof of Theorem 1.4. The relation between these two solutions is merely a transposition of the matrices; this *duality* will be introduced and studied in § 13.7, in connection with the Fourier-Borel transform.

This shows that transcendence results, like the theorems of Hermite-Lindemann, Gel'fond-Schneider, Baker, or the six exponentials Theorem, can be formulated as lower bounds for the rank of a matrix like M. A quantitative estimate of diophantine approximation is obtained from an effective version of the Linear Subgroup Theorem as follows. We start from a matrix as above

$$\mathsf{M} = \begin{pmatrix} \mathsf{B}_0 & \mathsf{B}_1 \\ \mathsf{B}_2 & \mathsf{L} \end{pmatrix} \begin{matrix} \}d_0 \\ \}d_1 \end{matrix}, \quad \underbrace{}_{\ell_0} \; \underbrace{}_{\ell_1}$$

where B_0, B_1, B_2 are matrices with algebraic entries, while the entries of L are in $\mathcal{L}$. A result of diophantine approximation (see Chap. 13) will be a lower bound for the *distance* between such a matrix M and a matrix with complex entries of low rank.

Therefore we consider another matrix, with complex entries, of the same size:

$$\mathsf{M}' = \begin{pmatrix} \mathsf{B}'_0 & \mathsf{B}'_1 \\ \mathsf{B}'_2 & \mathsf{L}' \end{pmatrix}$$

If the transcendence proof yields $\mathrm{rank}(\mathsf{M}) > r$, and if the matrix M' has rank at most r, then one can produce an explicit lower bound for the distance between the two matrices. Such a lower bound will depend on the parameters d_0, d_1, ℓ_0, ℓ_1, as well as on the rank of M'. The explicit estimate (Theorem 13.1) depends also on the heights of the algebraic numbers β_{ij} and $\alpha_{ij} = e^{\lambda_{ij}}$, on the degree of the number field generated by these $d\ell$ algebraic numbers, and on the absolute values of the λ_{ij}.

In the simplest case of square $d \times d$ matrices where M' is only supposed to be of rank at most $d - 1$, a lower bound for the distance between M and M' is equivalent to an estimate from below for the determinant of M. One obtains in this way effective versions of Baker's Theorem 1.6 (see § 14.4).

Finally, we point out that this type of result can be used to produce results of algebraic independence (see Chap. 15).

Exercises

Exercise 1.1.
a) Let z_1 and z_2 be two complex numbers. Denote by $x_i = \mathrm{Re}(z_i)$ the real part of z_i $(i = 1, 2)$. Check

$$|e^{z_1} - e^{z_2}| \le |z_1 - z_2| e^{\max\{x_1,x_2\}}.$$

Hint. *One solution is to check, for* $x = \mathrm{Re}(z)$,

$$\left| \int_0^1 e^{tz} dt \right| \le \int_0^1 e^{tx} dt.$$

Another solution starts from

$$|e^z - 1| \le e^{|z|} - 1.$$

Moreover, for $r > 0$ and $|z_1 - z_2| \le r$, check

$$|e^{z_1} - e^{z_2}| \le |z_1 - z_2| \frac{e^r - 1}{r} e^{\min\{x_1,x_2\}}.$$

b) For any $0 \le \theta < 1$, the condition $|z - 1| \le \theta$ implies, for the principal value of the complex logarithm,

$$|\log z| \le \frac{1}{1-\theta} |z - 1|.$$

Hint. *Check that, for any* t *and* ϑ *in* $\mathbb{R}$ *satisfying* $t \le \vartheta < 1$, *the following upper bound holds:*

$$|\log(1 - t)| \le \frac{|t|}{1 - \vartheta}.$$

c) Let $\vartheta \in \mathbb{R}$ and $v, w \in \mathbb{C}$ satisfy

$$|we^{-v} - 1| \le \vartheta \quad \text{and} \quad 0 \le \vartheta < 1.$$

Show that there exists $\lambda \in \mathbb{C}$ with $e^\lambda = w$ and

$$|\lambda - v| \le \frac{1}{1 - \vartheta} |we^{-v} - 1|.$$

Hint. *Define* $\lambda = v + \log(we^{-v})$ *where* log *is the principal value of the logarithm.*

d) For any $0 \le \theta \le \pi$, and $z \in \mathbb{C}$ satisfying $|z| \le \theta$, check

$$|z| \le \frac{\theta}{\sqrt{2 - 2\cos\theta}} |e^z - 1|.$$

Hint. *Check, for* $|z| = \theta \leq \pi$,

$$|e^z - 1| \geq |e^{i\theta} - 1|.$$

Exercise 1.2. Complete the proof of Lemma 1.8 by applying Dirichlet's pigeonhole principle to the points

$$b_1 \log a_1 + \cdots + b_m \log a_m, \qquad (0 \leq b_i < B,\ 1 \leq i \leq m)$$

which all lie in the interval $[0, mB\log A]$.

Hint. *Check* $B^{m-1}\log 2 \geq m\log A$ *and use Exercise 1.1.a. See also Lemma 4.11.*

Exercise 1.3. Show that the statements *(i)*, *(ii)* and *(iii)* in Lemma 1.7 are also equivalent to:

(iv) Let n be a nonnegative integer, $\lambda_1, \ldots, \lambda_{n+1}$ be elements of $\mathcal{M}$, and $\beta_1, \ldots, \beta_n$ elements of K. Assume $\lambda_1, \ldots, \lambda_n$ are K-linearly independent and

$$\beta_1\lambda_1 + \cdots + \beta_n\lambda_n = \lambda_{n+1}.$$

Then $\beta_1, \ldots, \beta_n$ are all in k.

(v) The natural map $\mathcal{M} \otimes_k K \to \mathcal{E}$, which extends the injection from $\mathcal{M}$ to $\mathcal{E}$, is still injective.

Hint. *Let* $(\mu_i)_{i\in I}$ *be a basis of the k-vector space* $\mathcal{M}$, *and let* $(\gamma_j)_{j\in J}$ *be a basis of the k-vector space* K. *Then* $\mu_i \otimes \gamma_j$ $(i \in I,\ j \in J)$ *is a basis of* $\mathcal{M} \otimes_k K$ *over* k:

$$\begin{aligned}
\mathcal{M} \otimes_k K &= \left\{ \sum_{i\in I} \mu_i \otimes \beta_i\,;\quad \beta_i \in K \quad \textit{with} \quad \mathrm{supp}(\beta_i)_{i\in I} \quad \textit{finite} \right\}\\
&= \left\{ \sum_{j\in J} \lambda_j \otimes \gamma_j\,;\quad \lambda_j \in \mathcal{M} \quad \textit{with} \quad \mathrm{supp}(\lambda_j)_{j\in J} \quad \textit{finite} \right\}\\
&= \left\{ \sum_{i\in I}\sum_{j\in J} c_{ij}\mu_i \otimes \gamma_j\,;\ c_{ij} \in k \ \textit{with}\ \mathrm{supp}(c_{ij})_{i\in I, j\in J}\ \textit{finite} \right\},
\end{aligned}$$

where finite support *means that all but finitely many elements vanish.*

The map $\mathcal{M} \otimes_k K \to \mathcal{E}$ *is nothing but*

$$\sum_{i\in I} \mu_i \otimes \beta_i \longmapsto \sum_{i\in I} \mu_i\beta_i, \qquad \sum_{j\in J} \lambda_j \otimes \gamma_j \longmapsto \sum_{j\in J} \lambda_j\gamma_j$$

as well as

$$\sum_{i\in I}\sum_{j\in J} c_{ij}\mu_i \otimes \gamma_j \longmapsto \sum_{i\in I}\sum_{j\in J} c_{ij}\mu_i\gamma_j.$$

Exercise 1.4. Let $k \subset K$ be two fields.

a) Let $\mathcal{V}$ be a K-vector subspace of K^d. Show that the following conditions are equivalent:

(i) $\mathcal{V}$ is intersection of hyperplanes which are defined by linear forms with coefficients in k.

(ii) $\mathcal{V}$ has a basis whose elements belong to k^d.

(iii) There exists a surjective linear map $K^d \longrightarrow K^r$ with kernel $\mathcal{V}$ whose matrix (in the canonical bases) has coefficients in k.

Such a subspace $\mathcal{V}$ is called *rational over* k.
b) Again let $\mathcal{V}$ be a vector subspace of K^d. Denote by $\pi_{\mathcal{V}}$ the canonical map $K^d \to K^d/\mathcal{V}$. Check $\dim_k\left(\pi_{\mathcal{V}}(k^d)\right) \geq \dim_K(K^d/\mathcal{V})$. Show that equality holds if and only if $\mathcal{V}$ is rational over k.
c) Let $\mathcal{V}$ be a K-vector space. A *k-structure on $\mathcal{V}$* is a k-vector subspace $\mathcal{V}'$ of $\mathcal{V}$ such that any basis of $\mathcal{V}'$ over k is a basis of $\mathcal{V}$ over K (see for instance [Roy 1995], § 1). In the case where $\mathcal{V}$ is a vector subspace of K^d, show that $\mathcal{V} \cap k^d$ is a k-structure on $\mathcal{V}$ if and only if $\mathcal{V}$ is rational over k.

Exercise 1.5. Show that Baker's homogeneous Theorem 1.5 is also equivalent to each of the following assertions:

(i) Let d be a positive integer. Let $\mathcal{W}$ be a subspace of $\mathbb{C}^d$ which is rational over $\overline{\mathbb{Q}}$ *(see Exercise 1.4)* such that $\mathcal{W} \cap \mathbb{Q}^d = 0$. Then $\mathcal{W} \cap \mathcal{L}^d = 0$.
(ii) Let n be a positive integer and E a subset of $\mathcal{L}^n$. The smallest (= intersection of all) subspace of $\mathbb{C}^n$ rational over $\overline{\mathbb{Q}}$ which contains E is rational over $\mathbb{Q}$.
(iii) Let ℓ, d be positive integers and $\underline{\lambda}_1, \ldots, \underline{\lambda}_\ell$ be $\mathbb{Q}$-linearly independent elements in $\mathcal{L}^d$. Then $\underline{\lambda}_1, \ldots, \underline{\lambda}_\ell$ are $\overline{\mathbb{Q}}$-linearly independent.
(iv) Let d be a positive integer and $\mathcal{W}$ a subspace of $\mathbb{C}^d$ which is rational over $\overline{\mathbb{Q}}$. Then

$$\mathcal{W} \cap \mathcal{L}^d = \bigcup_V V \cap \mathcal{L}^d,$$

where V ranges over the vector subspaces of $\mathbb{C}^d$ which are rational over $\mathbb{Q}$ and contained in $\mathcal{W}$.

Hint. *The implication (iii)* $\Rightarrow$ Theorem 1.5 *is clear (take $d = 1$), and (iv) $\Rightarrow$ (i) is easy (if $\mathcal{W} \cap \mathbb{Q}^d = 0$ then the only vector subspace of $\mathbb{C}^d$ which is rational over $\mathbb{Q}$ and contained in $\mathcal{W}$ is $\{0\}$).*

For the proof of Theorem 1.5 $\Rightarrow$ *(i), write $\mathcal{W}$ as intersection of $\overline{\mathbb{Q}}$-rational hyperplanes. For $(\lambda_1, \ldots, \lambda_d) \in \mathcal{W} \cap \mathcal{L}^d$, choose a basis of the $\mathbb{Q}$-vector subspace of $\mathbb{C}$ spanned by $\lambda_1, \ldots, \lambda_d$.*

For the proof of (i) $\Rightarrow$ (ii), consider a hyperplane Z in $\mathbb{C}^n$ which is rational over $\overline{\mathbb{Q}}$ and contains E. Let $\beta_1 z_1 + \cdots + \beta_n z_n = 0$ be an equation of Z. Select a basis of the $\mathbb{Q}$-vector subspace of $\mathbb{C}$ spanned by $\beta_1, \ldots, \beta_n$. Deduce from (i) that there exists a subspace W of $\mathbb{C}^n$, rational over $\mathbb{Q}$, with $E \subset W \subset Z$.

For (ii) $\Rightarrow$ (iii), transpose the matrix in $\mathrm{Mat}_{d\times\ell}(\mathcal{L})$ *whose columns vectors are $\underline{\lambda}_1, \ldots, \underline{\lambda}_\ell$ and apply (ii) with n replaced by ℓ.*

Finally, assuming (ii), we deduce that for any $\mathcal{W} \subset \mathbb{C}^d$ rational over $\overline{\mathbb{Q}}$, there exists $V \subset \mathbb{C}^d$ rational over $\mathbb{Q}$ and contained in $\mathcal{W}$ such that

$$\mathcal{W} \cap \mathcal{L}^d = V \cap \mathcal{L}^d,$$

which is more precise than (iv).

Exercise 1.6. A consequence of Baker's Theorem 1.6 is the transcendence of numbers like

$$\int_0^1 \frac{dt}{1+t^3} = \frac{1}{3}\left(\log 2 + \frac{\pi}{\sqrt{3}}\right)\cdot$$

Let P and Q be two nonzero polynomials with algebraic coefficients and $\deg P < \deg Q$. Assume Q has no multiple zero. Let γ be a contour in the complex plane, which is either closed, or has endpoints which are algebraic or infinite, and such that the definite integral

$$\int_\gamma \frac{P(z)}{Q(z)} dz$$

exists and is not zero. Then this integral is a transcendental number.

Hint. *See* [V 1971].

Exercise 1.7. Assume that the four exponentials Conjecture 1.13 is true. Deduce that if $z \in \mathbb{C}$ satisfies $|z| \in \mathbb{Q}$ and $e^{2i\pi z} \in \overline{\mathbb{Q}}$, then $z \in \mathbb{Q}$.

See fig. 1.18: Diaz' curve $e^{2i\pi z}$, $|z| = 1$. By the four exponentials Conjecture, apart from $z = \pm 1$, no point on this curve is algebraic; see [Di 1997a] for further comments on this topic.

Exercise 1.8.
a) Let $\lambda_1, \dots, \lambda_n$ be elements of $\mathcal{L}$ and let $P \in \overline{\mathbb{Q}}[X_1, \dots, X_n]$ be a nonzero polynomial with algebraic coefficients such that $P(\lambda_1, \dots, \lambda_n) = 0$. Assume that the Conjecture 1.15 on the algebraic independence of logarithms of algebraic numbers holds. Deduce that there is a vector subspace $\mathcal{V}$ of $\mathbb{C}^n$, rational over $\mathbb{Q}$, which is contained in the set of zeroes of P, and contains the point $(\lambda_1, \dots, \lambda_n)$.
b) Let C be a field with infinitely many elements, K a subfield of C and $\mathcal{V}$ a vector subspace of C^4, which is rational over K and contained in the hypersurface $z_1z_4 = z_2z_3$. Show that there exists $(a{:}b) \in \mathbb{P}_1(K)$ such that $\mathcal{V}$ is included either in the plane

$$\{(z_1, z_2, z_3, z_4) \in C^4; az_1 = bz_2,\ az_3 = bz_4\}$$

or in the plane

$$\{(z_1, z_2, z_3, z_4) \in C^4; az_1 = bz_3,\ az_2 = bz_4\}.$$

c) Deduce the four exponentials Conjecture 1.13 from the Conjecture 1.15 on algebraic independence of logarithms of algebraic numbers.

Exercise 1.9. Let K be a field, $\mathsf{M} \in \mathrm{Mat}_{d\times\ell}(K)$ a $d \times \ell$ matrix with entries in K and r a positive integer. Check that the two following properties are equivalent.
(i) $\mathrm{rank}(\mathsf{M}) \le r$.
(ii) There exist a $d \times r$ matrix $\mathsf{X} \in \mathrm{Mat}_{d\times r}(K)$ and a $r \times \ell$ matrix $\mathsf{Y} \in \mathrm{Mat}_{r\times\ell}(K)$ such that $\mathsf{M} = \mathsf{XY}$.

Hint. *Consider the linear map $K^\ell \to K^d$ associated to* M *in the canonical bases of K^ℓ and K^d respectively.*

Exercise 1.10. The following result, due to N. I. Feldman who improved a previous estimate of A. Baker, has been quoted in § 1.2.

Given positive integers $a_1, \dots, a_m$, there exists a positive constant C such that, for any tuple $(b_1, \dots, b_m)$ in $\mathbb{Z}^m$ for which

$$a_1^{b_1} \cdots a_m^{b_m} \neq 1,$$

we have

$$|a_1^{b_1} \cdots a_m^{b_m} - 1| \geq B^{-C}$$

with $B = \max\{2, |b_1|, \dots, |b_m|\}$.

Deduce from this estimate the following statement:

Given a finite set S of prime numbers, there exists a constant $c > 0$ such that, for any pair (x, y) of integers composed only of prime in S and satisfying $x > y \geq 2$, we have

$$x - y \geq x(\log x)^{-c}.$$

Exercise 1.11. (This is a refinement, due to P. Philippon, of [P 1999b], § 3). For relatively prime nonzero integers u, v with $v > 0$, define $\mathrm{h}(u/v) = \log \max\{|u|, v\}$ (see § 3.2). For a positive integer n, denote by $R(n)$ the radical of n:

$$R(n) = \prod_{p|n} p.$$

a) Let η be a real number in the range $0 < \eta < 1/2$ and let B be a positive integer. Assume the following property is true:

- *For any positive rational numbers a_1 and a_2, the inequality*

$$\prod_{p\in S} |a_1 a_2^B + 1|_p \geq \exp\Big\{-\eta B\big(\mathrm{h}(a_1) + \mathrm{h}(a_2) + \sum_{p\in S} \log p\big)\Big\}$$

holds with $S = \big\{p\ ;\ |a_1 a_2^B + 1|_p < 1\big\}$.

Deduce:

- *For any triple (a, b, c) of relatively prime positive integers with $a + b = c$, we have*

$$c \leq \big(R(abc)\big)^K \quad \textit{with} \quad K = \frac{\eta}{1 - 2\eta} B^2.$$

b) Let $K > 1$ be a real number. Assume:

- *For any triple (a, b, c) of relatively prime positive integers satisfying $a + b = c$, the inequality*

$$c \leq \big(R(abc)\big)^K$$

holds.

Deduce the following statement:

- *For any positive rational numbers a_1 and a_2, we have*

$$\prod_{p\in S} |a_1 a_2^B + 1|_p \geq \exp\Big\{-K\big(\mathrm{h}(a_1) + \mathrm{h}(a_2) + \sum_{p\in S} \log p\big)\Big\}$$

with $S = \big\{p\ ;\ |a_1 a_2^B + 1|_p < 1\big\}$.

Hint.
a) Write the decomposition of a/b into product of prime factors:

$$\frac{a}{b} = \prod_{p|ab} p^{e_p}$$

with $e_p = v_p(a/b)$ *and define*

$$\beta_p = \begin{cases} [e_p/B] & \text{if } e_p \geq 0, \\ -[-e_p/B] & \text{if } e_p \leq 0, \end{cases} \qquad a_1 = \prod_{p|ab} p^{e_p - B\beta_p} \quad \text{and} \quad a_2 = \prod_{p|ab} p^{\beta_p}$$

where $[\,\cdot\,]$ *denotes the integral part. Use the relation*

$$c = \prod_{p|c} |c|_p^{-1} = \prod_p \min\left\{1, |a_1 a_2^B + 1|_p\right\}^{-1}$$

and check

$$\mathrm{h}(a_1) \leq B \log R(ab), \quad \mathrm{h}(a_2) \leq \frac{1}{B} \log(ab) \leq \frac{2}{B} \log c$$

and

$$\sum_{p|c} \log p \leq \log R(c) \leq B \log R(c).$$

b) Define a *as the numerator of* $a_1 a_2^B$, b *as the denominator of* $a_1 a_2^B$, c *as the numerator of* $a_1 a_2^B + 1$ *and check*

$$\log R(ab) \leq \mathrm{h}(a_1) + \mathrm{h}(a_2).$$

Table 1.18. Diaz' curve $e^{2i\pi z}$, $|z| = 1$ (see Exercise 1.7).

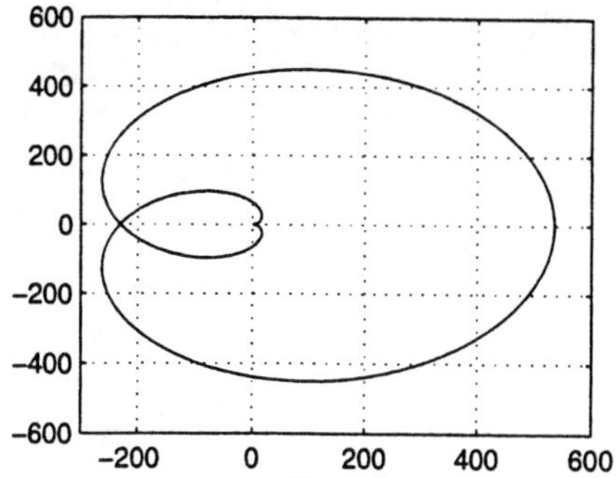

Figure 1

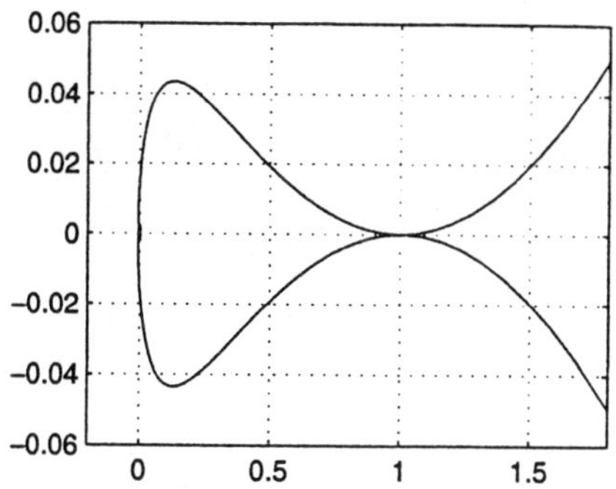

Figure 2

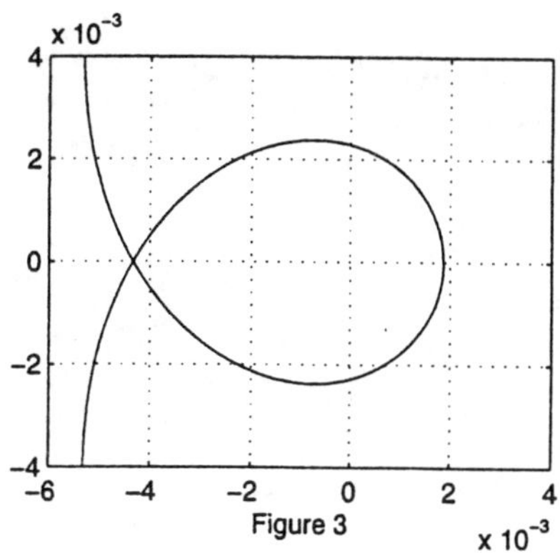

Figure 3

2. Transcendence Proofs in One Variable

The present chapter is an introduction to the method which will be developed in this book. However, we consider here only functions of a single variable. Our aim is to prove the theorems of Hermite-Lindemann and Gel'fond-Schneider by means of the alternants or interpolation determinants of M. Laurent [Lau 1989]. The real case of these two theorems (§§ 2.3 and 2.4) is easier, thanks to an estimate, due to G. Pólya (Lemma 2.2), for the number of real zeroes of real exponential polynomials. For the complex (i.e. general) case (§§ 2.5 and 2.6), another type of zero estimate, due to Y. V. Nesterenko, will be used. In the first section we explain the method, and in the second one we introduce a few auxiliary lemmas. It should be pointed out that the proof of our transcendence criterion (Lemma 2.1, which rests on Liouville's inequality) will be given only in the next chapter.

2.1 Introduction to Transcendence Proofs

A general transcendence problem is the following: consider two sequences, the first one is a sequence of entire functions of a single variable, say $\varphi_1, \varphi_2, \ldots$, while the second one is a sequence of complex numbers $\zeta_1, \zeta_2, \ldots$. We want to prove (under additional suitable assumptions!) that one at least of the numbers $\varphi_\lambda(\zeta_\mu)$ $(\lambda \geq 1, \mu \geq 1)$ is transcendental.

One can ask a related problem, which is also very important in transcendence theory, if the numbers ζ_μ are not all distinct. In this case, we introduce derivatives and replace $\varphi_\lambda(\zeta_\mu)$ by $(d/dz)^{\sigma_\mu}\varphi_\lambda(\zeta_\mu)$, where σ_μ is the number of n with $1 \leq n < \mu$ and $\zeta_n = \zeta_\mu$.

One main tool (Lemma 2.1) is a transcendence criterion which follows from Liouville's inequality. Roughly speaking, it says: *given complex numbers* $\theta_1, \ldots, \theta_m$, *if there exists a sequence of polynomials* $f_N \in \mathbb{Z}[X_1, \ldots, X_m]$ *such that* $f_N(\theta_1, \ldots, \theta_m)$ *is* very small *but not zero, then one at least of the numbers* $\theta_1, \ldots, \theta_m$ *is transcendental.*

Therefore our main goal will be to produce polynomials taking small nonzero values at a given point $(\theta_1, \ldots, \theta_m)$.

Let us give a few examples.

Example 1 (Theorem 1.2 of Hermite-Lindemann). Let β be a nonzero complex number. Put $\alpha = e^\beta$. We want to prove that one at least of the two numbers α, β is transcendental. We shall construct sequences of polynomials in $\mathbb{Z}[X, X', Y]$ having a small nonzero value at the point $(\alpha, \alpha^{-1}, \beta)$.

We start from the observation that the values of both functions z and e^z at the point β are polynomials in α and β. From the multiplication theorem which is satisfied by the exponential function, namely $e^{sz} = (e^z)^s$ for $s \in \mathbb{Z}$, we deduce that the values of these functions at the points $s\beta$, $s \in \mathbb{Z}$ are polynomials in α, α^{-1} and β. Further, the derivatives of any monomial in z and e^z have the same property, as shown by the differential equations they satisfy. For each set $\{\tau, t, \sigma, s\}$ of rational integers with $\tau \geq 0$ and $\sigma \geq 0$, we define a polynomial $f_{\tau t}^{(\sigma s)}(X, X', Y) \in \mathbb{Z}[X, X', Y]$ by

$$f_{\tau t}^{(\sigma s)}(X, X', Y) = \sum_{\kappa=0}^{\min\{\tau,\sigma\}} \frac{\sigma!\tau!}{\kappa!(\sigma-\kappa)!(\tau-\kappa)!} t^{\sigma-\kappa} s^{\tau-\kappa} Y^{\tau-\kappa} X^{\max\{ts,0\}} X'^{\max\{-ts,0\}},$$

so that

$$\left(\frac{d}{dz}\right)^{\sigma} \left(z^\tau e^{tz}\right)(s\beta) = f_{\tau t}^{(\sigma s)}(\alpha, \alpha^{-1}, \beta).$$

We wish to produce a sequence of functions, and a sequence of values. For this purpose we choose an ordering $(\tau_\lambda, t_\lambda)$, $\lambda \geq 1$, of $\mathbb{N}^2$, and we define

$$\varphi_\lambda(z) = z^{\tau_\lambda} e^{t_\lambda z}.$$

Also we consider the points $s\beta$, $s \in \mathbb{Z}$, but we repeat each of them infinitely often as follows: we choose an ordering (σ_μ, s_μ) $(\mu \geq 1)$ of $\mathbb{N} \times \mathbb{Z}$ and we define $\zeta_\mu = s_\mu \beta$. Hence

$$\left(\frac{d}{dz}\right)^{\sigma_\mu} \varphi_\lambda(\zeta_\mu) = \left(\frac{d}{dz}\right)^{\sigma_\mu} \left(z^{\tau_\lambda} e^{t_\lambda z}\right)(s_\mu \beta) \in \mathbb{Z}[\alpha, \alpha^{-1}, \beta].$$

We shall put these numbers into a square matrix and we shall prove that its determinant has a small nonzero absolute value. This will enable us to produce the required polynomial f as a determinant of a matrix whose entries are $f_{\tau t}^{(\sigma s)}$.

Example 2 (Theorem 1.4 of Gel'fond-Schneider with Gel'fond's method). For $\ell \in \mathbb{C}^\times$ and $\beta \in \mathbb{C} \setminus \mathbb{Q}$, define $\alpha_1 = e^\ell$, $\alpha_2 = e^{\ell\beta}$. The goal is to prove that one at least of the three numbers $\alpha_1, \alpha_2, \beta$ is transcendental. Hence we want to produce polynomials with small nonzero absolute value at the point $(\alpha_1, \alpha_1^{-1}, \alpha_2, \alpha_2^{-1}, \beta)$. We denote by $\mathbb{Z}[\alpha_1^{\pm 1}, \alpha_2^{\pm 1}, \beta]$ the ring generated by these five numbers.

We consider the functions e^z and $e^{\beta z}$, as well as monomials in these functions and their inverse, say $e^{t_1 z + t_2 \beta z}$, with $(t_1, t_2) \in \mathbb{Z}^2$. We take the derivatives of these functions at the points $s\ell$, with $s \in \mathbb{Z}$. All the values we get lie in $\mathbb{Z}[\alpha_1^{\pm 1}, \alpha_2^{\pm 1}, \beta]$, namely

$$\theta_{t_1 t_2}^{(\sigma s)} = \left(\frac{d}{dz}\right)^{\sigma} \left(e^{(t_1+t_2\beta)z}\right)(s\ell) = (t_1 + t_2\beta)^\sigma \alpha_1^{t_1 s} \alpha_2^{t_2 s}.$$

Choose an ordering $(t_{1\lambda}, t_{2\lambda})$ $(\lambda \geq 1)$, of $\mathbb{Z}^2$, as well as an ordering (σ_μ, s_μ) $(\mu \geq 1)$, of $\mathbb{N} \times \mathbb{Z}$, where the map $\mu \mapsto \sigma_\mu$ is non-decreasing. Define, for $\lambda \geq 1$ and $\mu \geq 1$,

$$\varphi_\lambda(z) = \exp\bigl((t_{1\lambda} + t_{2\lambda}\beta)z\bigr) \quad \text{and} \quad \zeta_\mu = s_\mu \ell.$$

The numbers $(d/dz)^{\sigma_\mu}\varphi_\lambda(\zeta_\mu) = \theta_{t_{1\lambda}t_{2\lambda}}^{(\sigma_\mu s_\mu)}$ belong to $\mathbb{Z}[\alpha_1^{\pm 1}, \alpha_1^{\pm 1}, \beta]$. Again we shall put these numbers into a square matrix whose determinant is small and nonzero, and this will give us a polynomial $f \in \mathbb{Z}[X_1, X_1^{-1}, X_2, X_2^{-1}, Y]$ such that $|f(\alpha_1, \alpha_2, \beta)|$ is small and nonzero.

Example 3 (Theorem of Gel'fond-Schneider with Schneider's method). As in Example 2 before, let ℓ be a nonzero complex number and β an irrational complex number. Define $\alpha_1 = e^\ell$, $\alpha_2 = e^{\ell\beta}$. We consider the values of the functions z and $e^{\ell z}$ (as well as monomials $z^\tau e^{t\ell z}$ with integers $(\tau, t) \in \mathbb{N} \times \mathbb{Z}$) at the points $s_1 + s_2\beta$, $(s_1, s_2) \in \mathbb{Z}^2$. For each $(\tau, t) \in \mathbb{N} \times \mathbb{Z}$, define $\phi_{\tau t} = z^\tau e^{t\ell z}$. Similarly, for each $(s_1, s_2) \in \mathbb{Z}^2$, define $\xi_{s_1 s_2} = s_1 + s_2\beta$. Since β is irrational, the points $\xi_{s_1 s_2}$ are pairwise distinct. For rational integers τ, t, s_1, s_2 with $\tau \geq 0$, the value of the function $\phi_{\tau t}$ at the point $\xi_{s_1 s_2}$ is nothing else than the number $\theta_{s_1 s_2}^{(\tau t)}$ of Example 2:

$$\phi_{\tau t}(\xi_{s_1 s_2}) = (s_1 + s_2\beta)^\tau \alpha_1^{s_1 t}\alpha_2^{s_2 t} = \theta_{s_1 s_2}^{(\tau t)}.$$

In order to reproduce the same notation as above, one needs to choose an ordering for the set of $(\tau, t) \in \mathbb{N} \times \mathbb{Z}$, as well as an ordering for the set of $(s_1, s_2) \in \mathbb{Z}^2$, and then one considers matrices $\Bigl(\varphi_\lambda(\zeta_\mu)\Bigr)_{1 \leq \lambda, \mu \leq L}$, where $\varphi_\lambda(z) = \phi_{\tau_\lambda t_\lambda}$ and $\zeta_\mu = s_{1\mu} + s_{2\mu}\beta$.

In all the examples which will be considered in this volume, the functions φ_λ will be exponential polynomials in one or several variables, i.e. linear combinations of functions of the form

$$z_1^{\tau_1} \cdots z_n^{\tau_n} \exp(x_1 z_1 + \cdots + x_n z_n),$$

where $\tau_1, \ldots, \tau_n$ are nonnegative integers, $x_1, \ldots, x_n$ complex numbers (these $2n$ numbers depend on λ). The points ζ_μ which will be considered will be of the form $s_1\underline{y}_1 + \cdots + s_m\underline{y}_m$, where $\underline{y}_1, \ldots, \underline{y}_m$ are fixed (i.e. independent on μ) elements, while $s_1, \ldots, s_m$ are rational integers which depend on μ.

Let us come back to the one dimensional case and consider all our numbers

$$\gamma_{\lambda\mu} = \left(\frac{d}{dz}\right)^{\sigma_\mu} \varphi_\lambda(\zeta_\mu) \qquad (\lambda \geq 1, \quad \mu \geq 1).$$

We express each of them as the value of a polynomial at a point $(\theta_1, \ldots, \theta_m)$. Our goal is to prove that one at least of $\theta_1, \ldots, \theta_m$ is transcendental.

Our method of proof shall involve putting these numbers into a matrix and then examining the determinant of submatrices of these matrices. We shall call the determinant of a matrix of the form

$$\Bigl(\varphi_\lambda(\zeta_\mu)\Bigr)_{1 \leq \lambda, \mu \leq L},$$

an *alternant* when no derivative is involved (the points ζ_μ are distinct, hence $\sigma_\mu = 0$ for all μ), and use the term *interpolation determinant* when the matrix is of the form

$$\left(\left(\frac{d}{dz}\right)^{\sigma_\mu} \varphi_\lambda(\zeta_\mu)\right)_{1\leq\lambda,\mu\leq L}.$$

We choose a large integer L (large enough to perform some computations which arise during the proof). The easiest case is when one knows that the square $L \times L$ matrix

$$\boldsymbol{M} = \begin{pmatrix} \gamma_{11} & \cdots & \gamma_{1L} \\ \vdots & \ddots & \vdots \\ \gamma_{L1} & \cdots & \gamma_{LL} \end{pmatrix}$$

is nonsingular. This will happen in each of the three above examples under the extra assumption that β and ℓ are real numbers (see Lemma 2.2, as well as §§ 2.3 and 2.4 below). In this situation, let Δ be the (nonzero) determinant of $\boldsymbol{M}$. As noticed by M. Laurent, a nontrivial upper bound for $|\Delta|$ (Lemmas 2.5 and 2.8) can be derived from *Schwarz' Lemma* which gives a sharp upper bound for the modulus on a disc of a function having a zero of high multiplicity at the origin (Lemma 2.4).

The number Δ is the value at the point $(\theta_1, \ldots, \theta_m)$ of a polynomial with integer coefficients, and Lemma 2.1 will lead to the desired transcendence result.

Unfortunately there are so far rather few results like Lemma 2.2 which enable us to show that the above square matrix is nonsingular. In the general situation, one only knows that the matrix with L rows and infinitely many columns:

$$\begin{pmatrix} \gamma_{11} & \gamma_{12} & \cdots & \gamma_{1\mu} & \cdots \\ \vdots & \vdots & \ddots & \vdots & \vdots \\ \gamma_{L1} & \gamma_{L2} & \cdots & \gamma_{L\mu} & \cdots \end{pmatrix}.$$

has maximal rank L. A *zero estimate* (which is sometimes called *multiplicity estimate* when derivatives are there) is a statement which shows that the matrix

$$\left(\gamma_{\lambda\mu}\right)_{\substack{1\leq\lambda\leq L \\ 1\leq\mu\leq L'}}$$

is of maximal rank L, for some L' bounded by cL, where c is some explicit (small) constant (this constant is of course at least one. If $c = 1$, we are in the first simpler situation). The estimates for the lower bound and the upper bound for the absolute value of the determinant are then more or less the same as in the case where $L' = L$. Zero estimates are known for exponential polynomials, and will be discussed later (especially in Chap. 5 and 8).

We conclude this section with some remarks on the set of ζ_μ's. We shall deal with sets of points (say in $\mathbb{C}^n$) of the form $s_1\underline{y}_1 + \cdots + s_m\underline{y}_m$, where $\underline{y}_1, \ldots, \underline{y}_m$ are fixed, and $s_1, \ldots, s_m$ are rational integers. One needs to choose the range for each s_j $(1 \leq j \leq m)$. The same will apply for the rational integers $(t_1, \ldots, t_d)$ related to the functions $e^{(t_1\underline{x}_1+\cdots+t_d\underline{x}_d)\underline{z}}$. There are several ways of choosing this range. All of them involve selecting positive numbers $S_1, \ldots, S_m$. We can then use the range $0 \leq s_j < S_j$, or $0 \leq s_j \leq S_j$, or else $|s_j| < S_j$, or finally $|s_j| \leq S_j$. Which one we choose does not really matter, and we shall select different options in different proofs. Here our favorite will be the last option.

2.2 Auxiliary Lemmas

In this section we state three auxiliary lemmas: the first one (a transcendence criterion which rests on Liouville's inequality) is arithmetic, the second (zero estimate for exponential polynomials) is a statement with an algebraic nature (even if Pólya's proof involves Rolle's Theorem), and the last one (Schwarz' Lemma) is analytic.

2.2.1 Transcendence Criterion

To begin with, we give a criterion for irrationality only. Let ϑ be a real number. The following conditions are equivalent

(i) *ϑ is irrational.*
(ii) *For any $\epsilon > 0$ there exists $p/q \in \mathbb{Q}$ such that*

$$0 < \left| \vartheta - \frac{p}{q} \right| < \frac{\epsilon}{q}.$$

(iii) *For any real number $Q > 1$ there exists an integer q in the range $1 \leq q < Q$ and a rational integer p such that*

$$0 < \left| \vartheta - \frac{p}{q} \right| < \frac{1}{qQ}.$$

(iv) *There exist infinitely many $p/q \in \mathbb{Q}$ such that*

$$0 < \left| \vartheta - \frac{p}{q} \right| < \frac{1}{\sqrt{5}q^2}.$$

Here, we are interested only in (ii)$\Rightarrow$(i) (see [Sc 1980] for further comments on these equivalences, and also § 15.1). Indeed if $\vartheta = a/b$ and $p/q \neq \vartheta$, then $aq - bp$ is a nonzero rational integer, hence has absolute value ≥ 1 and consequently

$$\left| \vartheta - \frac{p}{q} \right| \geq \frac{1}{bq}.$$

Therefore, in order to prove that some number is irrational, it is sufficient (and in fact also necessary) to produce good rational approximations. This criterion for irrationality extends into a transcendence criterion (cf. [FNe 1998], Chap. 2, § 1.5).

Instead of considering polynomials of degree 1 like $q\mathrm{X} - p$, one needs also to allow the degree to be large.

Lemma 2.1. *Let $\theta_1, \dots, \theta_m$ be complex numbers. Assume that for any $\kappa > 0$ there exists a polynomial $f \in \mathbb{Z}[\mathrm{X}_1, \dots, \mathrm{X}_m]$ and a positive integer T with*

$$\deg f + \log \mathrm{H}(f) \leq T$$

and

$$0 < |f(\theta_1, \dots, \theta_m)| \leq e^{-\kappa T}.$$

Then one at least of the numbers $\theta_1, \dots, \theta_m$ is transcendental.

As mentioned earlier, a proof of Lemma 2.1 will be given in § 3.5 (another proof is proposed in Exercise 2.2; see also § 15.1). Here is a sample of other references for a proof: [G 1952], Chap. I § 2, Lemma II; [L 1966], Chap. I; [W 1974], Chap. I § 2; [L 1978], Chap. 7 § 2; [W 1979a], Lemma 1.1.3; [F 1982], Lemma 9.2. See also [FNe 1998], Chap. I, § 1.7, Th. 1.5.

Remark. Let $\theta_1, \dots, \theta_k$ be nonzero complex numbers and $\theta_{k+1}, \dots, \theta_\ell$ be complex numbers. We shall apply Lemma 2.1 with $m = 2k + (\ell - k) = k + \ell$ to the m-tuple

$$\bigl(\theta_1, \dots, \theta_k, \theta_1^{-1}, \dots, \theta_k^{-1}, \theta_{k+1}, \dots, \theta_\ell\bigr).$$

Notation. The ring $\mathbb{Z}[\mathrm{X}_1^{\pm 1}, \dots, \mathrm{X}_k^{\pm 1}, \mathrm{Y}_1, \dots, \mathrm{Y}_{\ell-k}]$, generated by

$$\mathrm{X}_1, \dots, \mathrm{X}_k, \mathrm{X}_1^{-1}, \dots, \mathrm{X}_k^{-1}, \mathrm{Y}_1, \dots, \mathrm{Y}_{\ell-k},$$

is the image, in the field of rational functions $\mathbb{Q}(\mathrm{X}_1, \dots, \mathrm{X}_k, \mathrm{Y}_1, \dots, \mathrm{Y}_{\ell-k})$, of the ring

$$\mathbb{Z}[\mathrm{X}_1, \dots, \mathrm{X}_k, \mathrm{X}'_1, \dots, \mathrm{X}'_k, \mathrm{Y}_1, \dots, \mathrm{Y}_{\ell-k}],$$

under the obvious mapping $\mathrm{X}'_i \mapsto 1/\mathrm{X}_i$.

For $f \in \mathbb{Z}[\mathrm{X}_1^{\pm 1}, \dots, \mathrm{X}_k^{\pm 1}, \mathrm{Y}_1, \dots, \mathrm{Y}_{\ell-k}]$, we write

$$\deg f \leq D \quad \text{and} \quad \mathrm{H}(f) \leq H$$

if f is the image of a polynomial

$$F \in \mathbb{Z}[\mathrm{X}_1, \dots, \mathrm{X}_k, \mathrm{X}'_1, \dots, \mathrm{X}'_k, \mathrm{Y}_1, \dots, \mathrm{Y}_{\ell-k}]$$

for which $\deg F \leq D$ and $\mathrm{H}(F) \leq H$. We also denote by $f(\theta_1, \dots, \theta_\ell)$ the number $F\bigl(\theta_1, \dots, \theta_k, \theta_1^{-1}, \dots, \theta_k^{-1}, \theta_{k+1}, \dots, \theta_\ell\bigr)$.

2.2.2 Zero Estimate

The next lemma is our first and simplest example of a zero estimate ; it is due to G. Pólya. This result was already used in a similar context by A. O. Gel'fond and Yu. V. Linnik in Chap. 12 of [GLin 1962] (see also problem 75, Part V of Chap. 1 in [PoSz 1976]).

Lemma 2.2. *Let $a_1, \ldots, a_n$ be nonzero polynomials in $\mathbb{R}[X]$ of degrees $d_1, \ldots, d_n$, and let $w_1, \ldots, w_n$ be pairwise distinct real numbers. Then the real function of one real variable*

$$F(x) = \sum_{i=1}^{n} a_i(x)e^{w_i x}$$

has at most $d_1 + \cdots + d_n + n - 1$ real zeroes.

Remark. A set $\{f_1, \ldots, f_m\}$ of C^∞ real functions on a real interval $[a, b]$ is called *a Chebishev system on* $[a, b]$ if each nonzero element in the span over $\mathbb{R}$ has at most $m - 1$ zeroes. Therefore Lemma 2.2 states that the system

$$\left\{x^j e^{w_i x} \; ; \; 0 \le j \le d_i, \; 1 \le i \le n\right\}$$

is a Chebishev system on $\mathbb{R}$.

In Lemma 2.2 the zeroes are counted with multiplicities. For our application to Schneider's method in § 2.3, we need only an upper bound for the number of distinct real zeroes, but for Gel'fond's method in § 2.4, we have to take multiplicities into account. It is also interesting to remark that simple arguments from linear algebra show that the upper bound in Lemma 2.2 is best possible (see Exercise 2.3). Further related exercises are given in Chap. 6 of [W 1974] (in particular Exercise 6.1.c of [W 1974], where interpolation determinants are explicitly computed and further references are provided to N. I. Fel'dman's papers).

Proof. We first prove the following result. *Let N be a positive integer. If a continuously differentiable real function F of one real variable has at least N real zeroes (counting multiplicities), then its derivative F' has at least $N - 1$ real zeroes.*

Indeed, let $x_1, \ldots, x_k$ (with $k \ge 1$) be pairwise distinct real zeroes of F, in increasing order: $x_1 < x_2 < \cdots < x_k$. Let $n_1, \ldots, n_k$ be positive integers with $n_1 + \ldots + n_k \ge N$ and assume that, for each i, x_i is a zero of F of multiplicity at least n_i. Then x_i is a zero of F' with multiplicity at least $n_i - 1$ $(1 \le i \le k)$. Moreover, since $F(x_i) = F(x_{i+1})$ for $1 \le i \le k - 1$, it follows from Rolle's Theorem that F' has at least one zero in the open interval (x_i, x_{i+1}). Therefore F' has at least

$$(n_1 - 1) + \cdots + (n_k - 1) + (k - 1) \ge N - 1$$

real zeroes. This proves the preliminary statement.

We now prove Lemma 2.2 by induction on the integer $k := d_1 + \cdots + d_n + n - 1$. In the case $k = 0$, we have $n = 1$ and $d_1 = 0$, so $F(x) = a_1 e^{w_1 x}$ and the result is obvious. Assume $k \geq 1$. After multiplication of F by $e^{-w_n x}$, we may assume $w_n = 0$. Hence $w_i \neq 0$ for $1 \leq i < n$. Let N be a positive integer such that F has at least N real zeros. Then, as we have seen, its derivative F' has at least $N - 1$ real zeros. However, since $w_n = 0$, we have

$$F'(x) = \sum_{i=1}^{n-1} \tilde{a}_i(x) e^{w_i x} + \frac{d}{dx} a_n(x)$$

where

$$\tilde{a}_i = w_i a_i + \frac{d}{dx} a_i$$

is a polynomial of degree d_i for $1 \leq i < n$, while $(d/dx)a_n$ is of degree $d_n - 1$ (we consider here that the zero polynomial is of degree -1). One uses the induction hypothesis which yields $N - 1 \leq d_1 + \cdots + d_n + n - 2$, hence N is bounded as claimed. □

We will deduce from Lemma 2.2 that certain determinants are not zero.

Corollary 2.3. *Let $w_1, \ldots, w_n$ be pairwise distinct real numbers, $x_1, \ldots, x_m$ also pairwise distinct real numbers, and $\delta_1, \ldots, \delta_n, \kappa_1, \ldots, \kappa_m$ nonnegative integers, with $\delta_1 + \cdots + \delta_n = \kappa_1 + \cdots + \kappa_m$. Choose any ordering for the pairs (j, ν) with $0 \leq j < \delta_\nu$ and $1 \leq \nu \leq n$, and any ordering for the pairs (k, μ) with $0 \leq k < \kappa_\mu$ and $1 \leq \mu \leq m$. Then the determinant*

$$\det\left(\left(\frac{d}{dx}\right)^k \left(x^j e^{w_\nu x}\right)(x_\mu) \right)_{\substack{(j,\nu) \\ (k,\mu)}}$$

is not zero.

Proof. We have to show that if $a_{\nu j}$ are real numbers such that

$$\sum_{\nu=1}^{n} \sum_{j=0}^{\delta_\nu - 1} a_{\nu j} \left(\frac{d}{dx}\right)^k \left(x^j e^{w_\nu x}\right)(x_\mu) = 0$$

for $0 \leq k < \kappa_\mu$ and $1 \leq \mu \leq m$, then $a_{\nu j} = 0$ for all ν, j. This system of equations means that the function

$$F(x) = \sum_{\nu=1}^{n} \sum_{j=0}^{\delta_\nu - 1} a_{\nu j} x^j e^{w_\nu x}$$

has a zero at x_μ of multiplicity at least κ_μ, for $1 \leq \mu \leq m$, and hence the total number of zeroes of F is at least $\kappa_1 + \cdots + \kappa_m$. The polynomial

$$a_\nu(x) = \sum_{j=0}^{\delta_\nu - 1} a_{\nu j} x^j$$

either is 0 or is of degree $d_\nu \leq \delta_\nu - 1$, with $d_1 + \cdots + d_n + n - 1 \leq \delta_1 + \cdots + \delta_n - 1 < \kappa_1 + \cdots + \kappa_m$. From Lemma 2.2 one concludes $a_1 = \cdots = a_n = 0$, hence $a_{\nu j} = 0$ for all ν, j. □

Remark. For exponential polynomials in a single variable, one can use analytic arguments and also derive a zero estimate in the complex case (cf. Exercise 2.9).

2.2.3 Schwarz' Lemma

Our main tool from complex analysis will be Schwarz' Lemma. In this chapter we need only the easiest version of it, namely for analytic functions of a single variable with a single (multiple) zero.

Lemma 2.4. *Let T be a nonnegative integer, r and R real numbers satisfying $0 < r \leq R$ and Ψ a function of one complex variable which is an analytic in the disc $|z| \leq R$. Assume Ψ has a zero of multiplicity at least T at 0. Then*

$$|\Psi|_r \leq \left(\frac{R}{r}\right)^{-T} |\Psi|_R.$$

Proof. The function $\Phi(z) = z^{-T}\Psi(z)$ is analytic in the disc $|z| \leq R$. Since $r \leq R$, we have $|\Phi|_r \leq |\Phi|_R$. By the maximum modulus principle we deduce

$$|\Phi|_r = r^{-T}|\Psi|_r \quad \text{and} \quad |\Phi|_R = R^{-T}|\Psi|_R.$$

Lemma 2.4 follows. □

2.3 Schneider's Method with Alternants – Real Case

We give here the first proof of Theorem 1.4 of Gel'fond-Schneider in the real case (this is example 3 in § 2.1).

2.3.1 Upper Bound for an Alternant – One Variable

From Schwarz' Lemma 2.4 we deduce the following upper bound.

Lemma 2.5. *Let r and R be two real numbers with $0 < r \leq R$, $\varphi_1, \ldots, \varphi_L$ be functions of one complex variable, which are analytic in the disc $|z| \leq R$ of $\mathbb{C}$, and let $\zeta_1, \ldots, \zeta_L$ belong to the disc $|z| \leq r$. Then the absolute value of the determinant*

$$\Delta = \det \begin{pmatrix} \varphi_1(\zeta_1) & \dots & \varphi_L(\zeta_1) \\ \vdots & \ddots & \vdots \\ \varphi_1(\zeta_L) & \dots & \varphi_L(\zeta_L) \end{pmatrix}$$

is bounded from above by

$$|\Delta| \leq \left(\frac{R}{r}\right)^{-L(L-1)/2} L! \prod_{\lambda=1}^{L} |\varphi_\lambda|_R. \tag{2.6}$$

Notice also that the conclusion is trivial in the case $R = r$.

Note. This lemma will enable us to prove that certain determinants have a *small* absolute value. The main term on the right-hand side of (2.6) will be the first one: L will be *large* and R/r will be bounded away from 1 (say $R/r \geq e$), hence $(R/r)^{-L(L-1)/2}$ will be *small*. On the other hand, we shall check in the applications that the quantity $L!\prod_{\lambda=1}^{L} |\varphi_\lambda|_R$ is not *too big*, and in fact is *much smaller* than $(R/r)^{L(L-1)/2}$.

The left-hand side of (2.6) does not depend on R. The estimate is trivial in each of the following three cases:
1) for $R = r$,
2) if $\Delta = 0$,
3) if $\varphi_1, \dots, \varphi_L$ are all polynomials of degrees $0, 1, \dots, L-1$.

Otherwise, for $R \to \infty$ the right-hand side is unbounded. Hence there is at least one (and most often only one) optimal value for R which minimizes the right-hand side of (2.6).

Proof of Lemma 2.5. The determinant $\Psi(\zeta)$ of the matrix $\Big(\varphi_\lambda(\zeta_\mu\zeta)\Big)_{1\leq\lambda,\mu\leq L}$ is a function of one complex variable which is analytic in the disc $|\zeta| \leq R/r$.

a) We first prove that Ψ has a zero of multiplicity at least $L(L-1)/2$ at the origin.

Since the determinant is multi-linear and expressing each φ_λ in terms of its Taylor series, the problem can be reduced to the case where $\varphi_\lambda(z) = z^{n_\lambda}$ for some nonnegative integer n_λ and each $1 \leq \lambda \leq L$. In this case,

$$\Psi(\zeta) = \zeta^{n_1+\dots+n_L} \det\Big(\zeta_\mu^{n_\lambda}\Big)_{1\leq\lambda,\mu\leq L}.$$

If $\Psi(\zeta)$ does not vanish identically, then the n_λ are pairwise distinct, and the sum $n_1 + \dots + n_L$ is at least $0 + 1 + \dots + (L-1) = L(L-1)/2$. Hence we get the factor $\zeta^{L(L-1)/2}$ which we wanted.

The fact that $\Psi(\zeta)$ vanishes at 0 with multiplicity at least $L(L-1)/2$ can also be checked by taking derivatives of Ψ: applying Leibniz' rule on differentiating products yields by induction on k

$$\left(\frac{d}{d\zeta}\right)^k \Psi(\zeta) = \sum_{\kappa_1+\dots+\kappa_L=k} \frac{k!}{\kappa_1!\cdots\kappa_L!} \det\left(\left(\frac{d}{d\zeta}\right)^{\kappa_\mu} \varphi_\lambda(\zeta\zeta_\mu)\right)_{1\leq\lambda,\mu\leq L}.$$

At the point $\zeta = 0$, if one at least of the determinants at the right-hand side is not zero, then the numbers κ_μ are pairwise distinct, hence the integer $k = \kappa_1 + \cdots + \kappa_L$ is at least $L(L-1)/2$.

b) We now use Lemma 2.4 with $T = L(L-1)/2$, with r replaced by 1 and R by R/r:

$$|\Delta| = |\Psi(1)| \leq \left(\frac{R}{r}\right)^{-L(L-1)/2} |\Psi|_{R/r}.$$

The upper bound

$$|\Psi|_{R/r} \leq L! \prod_{\lambda=1}^{L} |\varphi_\lambda|_R,$$

which we get by expanding the determinant and by using a trivial upper bound for each of the $L!$ terms, yields the desired conclusion. □

Remark. Using Hadamard's inequality (see the proof of Lemma 3.25) one may replace $L!$ by $L^{L/2}$.

2.3.2 First Proof of the Real Case of Theorem 1.4 (Gel'fond-Schneider)

Proposition 2.7. *Let α_1 be a positive real number with $\alpha_1 \neq 1$, and β an irrational real number. Define $\alpha_2 = \alpha_1^\beta = \exp(\beta \log \alpha_1)$, where $\log \alpha_1$ is the real value of the logarithm of α_1, and*

$$c_1 = (2 + |\beta|)(1 + |\log \alpha_1|).$$

Then for any rational integers L, T_0, T_1, S and any real number E satisfying

$$T_0 \geq 2, \quad T_1 \geq 2, \quad S \geq 3, \quad E \geq e \quad \text{and} \quad L = (T_0+1)(2T_1+1) = (2S+1)^2,$$

there exists a polynomial $f \in \mathbb{Z}[X_1^{\pm 1}, X_2^{\pm 1}, Y]$ satisfying

$$\deg f \leq L(T_0 + 2T_1 S), \quad \mathrm{H}(f) \leq L!(2S)^{LT_0}$$

and

$$0 < |f(\alpha_1, \alpha_2, \beta)| \leq E^{-L^2/2} (SE)^{c_1 T_0 L} e^{c_1 T_1 SEL}.$$

Consequence. Given α_1, α_2 and β with $\alpha_2 = \alpha_1^\beta$ as in Proposition 2.7, we wish to deduce that one at least of these three numbers is transcendental. By Lemma 2.1, it is sufficient to check that for any $\kappa > 0$, we can choose the parameters in Proposition 2.7 so that

$$\kappa L\big(T_0 + 2T_1 S + \log L + T_0 \log(2S)\big) + c_1 L\big(T_0 \log(SE) + T_1 SE\big) \leq \frac{1}{2} L^2 \log E.$$

There are plenty of admissible values for these parameters L, T_0, T_1, S, E (and the explicit value of the constant c_1 is just irrelevant). For instance one can take a sufficiently large odd integer N, and choose

$$L = N^8, \quad T_0 = N^6 - 1,$$

$$T_1 = \frac{1}{2}(N^2 - 1) \quad S = \frac{1}{2}(N^4 - 1) \quad \text{and} \quad E = e.$$

With this choice of parameters, we have, for sufficiently large N,

$$\deg f + \log \mathrm{H}(f) \leq 5N^{14} \log N,$$

while

$$\log |f(\alpha_1, \alpha_2, \beta)| \leq -\frac{1}{3} N^{16}.$$

Given $\kappa > 0$, we take N sufficiently large so that these estimates are valid, and also so that $N^2 > 15\kappa \log N$. This completes the proof of Theorem 1.4 in the real case.

Another choice is the following: given $\kappa > 0$, take for T_1 a fixed integer $\geq \kappa + 2c_1$, and choose for L a much larger integer (which tends to infinity), which is an odd square divisible by $2T_1 + 1$. Then we put

$$T_0 = \frac{L}{2T_1 + 1} - 1, \quad S = \frac{1}{2}(\sqrt{L} - 1) \quad \text{and} \quad E = \sqrt{L}.$$

As $L \to \infty$, we have

$$\deg f + \log \mathrm{H}(f) \leq \frac{1}{2(2T_1 + 1)} L^2 \log L + O(L^2)$$

and

$$\log |f(\alpha_1, \alpha_2, \beta)| \leq \left(-\frac{1}{4} + \frac{c_1}{2T_1 + 1} \right) L^2 \log L + O(L^2).$$

□

Proof of Proposition 2.7.

Step 1. Construction of Δ

We choose any ordering $\{\varphi_1, \ldots, \varphi_L\}$ for the set of $(T_0 + 1)(2T_1 + 1)$ functions

$$\phi_{\tau t}(z) = z^\tau \alpha_1^{tz}, \qquad (0 \leq \tau \leq T_0, \ |t| \leq T_1),$$

and any ordering $\{\zeta_1, \ldots, \zeta_L\}$ for the set of $(2S + 1)^2$ real numbers

$$\xi_{s_1 s_2} = s_1 + s_2 \beta, \qquad \big((s_1, s_2) \in \mathbb{Z}^2, \ -S \leq s_1, s_2 \leq S\big).$$

Hence

$$\phi_{\tau t}(\xi_{s_1 s_2}) = (s_1 + s_2 \beta)^\tau \left(\alpha_1^{s_1} \alpha_2^{s_2}\right)^t.$$

We put the L^2 numbers $\varphi_\lambda(\zeta_\mu)$ into a square $L \times L$ matrix, and we take the determinant $\Delta = \det\Big(\varphi_\lambda(\zeta_\mu)\Big)_{1 \leq \lambda, \mu \leq L}$. It will also be convenient to write:

$$\Delta = \det\left((s_1 + s_2 \beta)^\tau \left(\alpha_1^{s_1} \alpha_2^{s_2}\right)^t \right)_{\substack{(\tau, t) \\ (s_1, s_2)}},$$

with $0 \le \tau \le T_0$, $|t| \le T_1$, and $-S \le s_1, s_2 \le S$. This means that we do not write explicitly the ordering which has been chosen for the rows and columns. A change in these orderings will introduce a factor ± 1, which has no effect on the absolute value of the determinant.

We shall show that Δ is not zero by means of the zero estimate Lemma 2.2 and produce an upper bound for $|\Delta|$ using Lemma 2.5.

Step 2. Consequence of the zero estimate

We prove that Δ does not vanish, which means that the matrix

$$\Big(\varphi_\lambda(\zeta_\mu)\Big)_{1\le\lambda,\mu\le L} = \Big((s_1+s_2\beta)^\tau\left(\alpha_1^{s_1+s_2\beta}\right)^t\Big)_{\substack{(\tau,t)\\(s_1,s_2)}},$$

with

$$0 \le \tau \le T_0, \quad -T_1 \le t \le T_1, \qquad -S \le s_1, s_2 \le S$$

has rank L. Indeed we may apply Corollary 2.3 with $n = 2T_1 + 1$, $m = (2S+1)^2$, $\delta_1 = \cdots = \delta_n = T_0 + 1$, $\kappa_1 = \cdots = \kappa_m = 1$,

$$\{w_1, \ldots, w_n\} = \{t \log \alpha_1\,;\, -T_1 \le t \le T_1\},$$

$$\{x_1, \ldots, x_m\} = \{s_1 + s_2\beta\,;\, -S \le s_1, s_2 \le S\},$$

Since $\log \alpha_1$ is not zero, the points $w_1, \ldots, w_n$ are pairwise distinct, and since β is irrational, the points $x_1, \ldots, x_m$ are pairwise distinct.

Step 3. Bound for the Degree and Height

For any quadruple (τ, t, s_1, s_2) of rational integers with

$$0 \le \tau \le T_0, \quad -T_1 \le t \le T_1, \qquad -S \le s_1, s_2 \le S$$

define

$$f_{\tau t}^{(s_1 s_2)} = (s_1 + s_2\mathrm{Y})^\tau \left(\mathrm{X}_1^{s_1}\mathrm{X}_2^{s_2}\right)^t \in \mathbb{Z}[\mathrm{X}_1^{\pm 1}, \mathrm{X}_2^{\pm 1}, \mathrm{Y}],$$

so that

$$\phi_{\tau t}(\xi_{s_1 s_2}) = f_{\tau t}^{(s_1 s_2)}(\alpha_1, \alpha_2, \beta).$$

We have

$$\deg f_{\tau t}^{(s_1 s_2)} \le T_0 + 2T_1 S, \quad \mathrm{H}(f_{\tau t}^{(s_1 s_2)}) \le (2S)^{T_0}.$$

Denote by f the determinant of the matrix

$$\Big(f_{\tau t}^{(s_1 s_2)}\Big)_{\substack{(\tau,t)\\(s_1,s_2)}},$$

with $0 \le \tau \le T_0$, $-T_1 \le t \le T_1$, and with $-S \le s_1, s_2 \le S$. Hence

$$\Delta = f(\alpha_1, \alpha_2, \beta).$$

and $f \in \mathbb{Z}[\mathrm{X}_1^{\pm 1}, \mathrm{X}_2^{\pm 1}, \mathrm{Y}]$ can be written

$$f = \sum_{\sigma} \epsilon(\sigma) \prod_{\tau=0}^{T_0} \prod_{t=-T_1}^{T_1} f_{\tau t}^{(\sigma(\tau,t))},$$

where σ runs over the set of bijective mappings from (τ, t) to (s_1, s_2) and $\epsilon(\sigma) \in \{-1, +1\}$. The degree of f is bounded by

$$\deg f \le \sum_{\tau=0}^{T_0} \sum_{t=-T_1}^{T_1} \max_{-S \le s_1, s_2 \le S} \deg f_{\tau t}^{(s_1 s_2)} \le L(T_0 + 2T_1 S).$$

We also need an upper bound for $\mathrm{H}(f)$. In fact it is easier to work with the *length* $\mathrm{L}(f)$, which is defined as the sum of the absolute values of the coefficients of f. Indeed the relations

$$\mathrm{L}(f+g) \le \mathrm{L}(f) + \mathrm{L}(g) \quad \text{and} \quad \mathrm{L}(fg) \le \mathrm{L}(f)\mathrm{L}(g)$$

show that the length $\mathrm{L}(\Delta)$ of a $L \times L$ determinant $\det\bigl(f_{\lambda\mu}\bigr)$ is bounded by

$$\mathrm{L}(\Delta) \le L! \max_{1 \le \lambda, \mu \le L} \mathrm{L}(f_{\lambda\mu})^L.$$

Here we have

$$\mathrm{L}(f_{\tau t}^{(s_1 s_2)}) \le (|s_1| + |s_2|)^\tau \le (2S)^{T_0},$$

hence

$$\mathrm{H}(f) \le \mathrm{L}(f) \le L!(2S)^{LT_0}.$$

Step 4. Upper bound for $|\Delta|$

We are going to use Lemma 2.5 with $r = S(1 + |\beta|)$ and $R = Er$. The choice of r is the obvious one: it is the radius of a disc containing all points $\zeta_1, \dots, \zeta_L$ (here, in the real case, it is the length of an interval centered at the origin which contains these points). We bound $|\phi_{\tau t}(z)|$ by $|z|^\tau e^{|tz \log \alpha_1|}$, hence

$$\max_{1 \le \lambda \le L} |\varphi_\lambda|_R \le R^{T_0} e^{T_1 R |\log \alpha_1|}$$

and from Lemma 2.5 we deduce

$$|\Delta| \le E^{-L(L-1)/2} L! R^{LT_0} e^{LT_1 R |\log \alpha_1|}.$$

We replace r and R by their values. From the crude inequalities

$$\log L < T_0 + T_1 < T_0 \log S + T_1 S E$$

$$\frac{1}{2} \log E < T_0 \log E \quad \text{and} \quad \log(1 + |\beta|) \le |\beta| \log(SE)$$

we deduce

$$LT_0 \log R + LT_1 R |\log \alpha_1| \le c_1 L\bigl(T_0 \log(SE) + T_1 SE\bigr).$$

This completes the proof of Proposition 2.7. □

Remark. A sharper estimate can easily be achieved for the degree: since

$$\deg f_{\tau t}^{(s_1 s_2)} \le T_0 + 2|t|S$$

and since

$$\sum_{t=-T_1}^{T_1} |t| = T_1(T_1+1) < \frac{1}{2}(T_1+1)(2T_1+1)$$

we can replace the estimate for the degree in Proposition 2.7 by

$$\deg f \le LT_0 + L(T_1+1)S.$$

We shall not use this remark in the present chapter but only later (see for instance Exercise 3.8).

2.4 Gel'fond's Method with Interpolation Determinants – Real Case

We first give a proof of the real case of Theorem 1.2 (Hermite-Lindemann), and then we give a second proof of Theorem 1.4 (Gel'fond-Schneider) in the real case.

2.4.1 Upper Bound for an Interpolation Determinant – One Variable

We generalize Lemma 2.5 by introducing derivatives.

Lemma 2.8. *Let $\varphi_1, \dots, \varphi_L$ be entire functions in $\mathbb{C}$, $\zeta_1, \dots, \zeta_L$ be elements of $\mathbb{C}$, $\sigma_1, \dots, \sigma_L$ nonnegative integers, and $0 < r \le R$ be real numbers, with $|\zeta_\mu| \le r$ $(1 \le \mu \le L)$. Then the absolute value of the determinant*

$$\Delta = \det\left(\left(\frac{d}{dz}\right)^{\sigma_\mu} \varphi_\lambda(\zeta_\mu)\right)_{1 \le \lambda, \mu \le L}$$

is bounded from above by

$$|\Delta| \le \left(\frac{R}{r}\right)^{-L(L-1)/2+\sigma_1+\cdots+\sigma_L} L! \prod_{\lambda=1}^{L} \max_{1\le\mu\le L} \sup_{|z|=R} \left|\left(\frac{d}{dz}\right)^{\sigma_\mu} \varphi_\lambda(z)\right|.$$

Proof. We claim that the function of one variable

$$\Psi(z) = \det\left(\left(\left(\frac{d}{dz}\right)^{\sigma_\mu} \varphi_\lambda\right)(z\zeta_\mu)\right)_{1 \le \lambda, \mu \le L}$$

has a zero at the origin of multiplicity at least

$$\frac{1}{2}L(L-1)-\sigma_1-\cdots-\sigma_L.$$

By multilinearity we reduce the proof of this claim to the special case $\varphi_\lambda(z)=z^{n_\lambda}$ for some $n_\lambda\in\mathbb{N}$ $(1\le\lambda\le L)$. In this special case we have

$$\Psi(z)z^{\sigma_1+\cdots+\sigma_L}=z^{n_1+\cdots+n_L}\det\left(\sigma_\mu!\binom{n_\lambda}{\sigma_\mu}\zeta_\mu^{n_\lambda}\right)_{1\le\lambda,\mu\le L}$$

where the binomial coefficient $\binom{n_\lambda}{\sigma_\mu}$ means 0 if $\sigma_\mu>n_\lambda$. If the right-hand side is not identically zero, then the numbers $n_1,\ldots,n_L$ are pairwise distinct, and then the right-hand side has a zero at the origin of multiplicity $n_1+\cdots+n_L\ge L(L-1)/2$. Our claim on the order of vanishing of Ψ at the origin easily follows.

Here again, like in the proof of Lemma 2.5, one can also check directly that the first $L(L-1)/2-\sigma_1-\cdots-\sigma_L$ derivatives of Ψ vanish at 0: if $(\kappa_1,\ldots,\kappa_L)\in\mathbb{N}^L$ is such that

$$\det\left(\left(\frac{d}{dz}\right)^{\sigma_\mu+\kappa_\mu}\varphi_\lambda(0)\right)_{1\le\lambda,\mu\le L}$$

is not zero, then $\sigma_1+\kappa_1+\cdots+\sigma_L+\kappa_L\ge L(L-1)/2$.

We conclude the proof of Lemma 2.8 by means of the Schwarz' Lemma 2.4, just as in Lemma 2.5. □

Remark. One could apply Cauchy's inequalities and bound the number

$$\sup_{|z|=R}\left|(d/dz)^{\sigma_\mu}\varphi_\lambda(z)\right|$$

by $\sigma_\mu!|\varphi_\lambda|_{R+1}$, for instance. In our applications a direct computation will be as easy.

2.4.2 Proof of the Real Case of the Theorem of Hermite-Lindemann

We develop here the first example of § 2.1, in the real case.

Proposition 2.9. *Let β be a nonzero real number. Define $\alpha=e^\beta$ and*

$$c_2=\max(|\beta|,|\beta|^{-1})+6.$$

Let T_0, T_1, S_0, S_1, L be rational integers which are all greater than 1, such that

$$L=(T_0+1)(2T_1+1)=(S_0+1)(2S_1+1).$$

Further let $E\ge e$ be a real number. Then there exists a polynomial $f\in\mathbb{Z}[X^{\pm1},Y]$ such that

$$\deg f\le L(T_0+T_1S_1),\quad \mathrm{H}(f)\le L!(T_0+T_1)^{LS_0}S_1^{LT_0}$$

and

$$0 < |f(\alpha,\beta)| \le E^{-L^2/2}\big((T_0+T_1)E\big)^{S_0L}(S_1E)^{c_2T_0L}e^{c_2T_1S_1EL}.$$

Consequence. Let us deduce from Lemma 2.1 that under the assumptions of Proposition 2.9 one at least of the two numbers α, β is transcendental. Let κ be a positive real number. By Lemma 2.1, it is sufficient to show that there exist parameters T_0, T_1, S_0, S_1, L, E satisfying the requirements of Proposition 2.9, and also such that

$$\frac{1}{2}L\log E > \kappa\big(T_0+T_1S_1+\log L+S_0\log(T_0+T_1)+T_0\log S_1\big)$$
$$+c_2\big(S_0\log\big((T_0+T_1)E\big)+T_0\log(S_1E)+T_1S_1E\big).$$

We give two sets of admissible choices for these parameters.

a) Let N be a sufficiently large odd integer. Choose

$$T_0=S_0=N^2-1,\quad T_1=S_1=\frac{1}{2}(N-1),\quad L=N^3\quad\text{and}\quad E=e.$$

In this case

$$\deg f+\log \mathrm{H}(f)\le 4N^5\log N$$

while

$$\log|f(\alpha,\beta)|\le -\frac{1}{3}N^6.$$

b) Choose for S_1 any integer with $S_1 > 2c_2+\kappa$. Next take for N a sufficiently large odd integer, which is a multiple of $2S_1+1$. Define

$$T_0=N-1,\quad T_1=\frac{1}{2}(N-1),\quad L=N^2,$$

$$S_0+1=\frac{L}{2S_1+1}\quad\text{and}\quad E=\sqrt{L}.$$

Now

$$\deg f+\log \mathrm{H}(f)\le \frac{1}{2S_1+1}N^4\log N+O(N^4)$$

and

$$\log|f(\alpha,\beta)|\le\left(-\frac{1}{2}+\frac{2c_2}{2S_1+1}\right)N^4\log N+O(N^4).$$

□

Proof of Proposition 2.9. We start from the fact that the values, at the points $s\beta$ ($s\in\mathbf{Z}$), of the derivatives of any monomial in the two functions z and e^z belong to the ring $\mathbb{Z}[\alpha^{\pm1},\beta]$: for τ, t, σ and s rational integers with $\tau\ge 0$ and $\sigma\ge 0$, we have

$$\gamma_{\tau t}^{(\sigma s)}=\left(\frac{d}{dz}\right)^{\sigma}\left(z^{\tau}e^{tz}\right)(s\beta)\in\mathbb{Z}[\alpha^{\pm1},\beta].$$

We build a matrix out of these numbers:

$$M = \left(\gamma_{\tau t}^{(\sigma s)}\right)_{\substack{(\tau,t)\\(\sigma,s)}}$$

where the index of rows is, say, (τ, t), while the index of columns is (σ, s) (any ordering of these pairs will do). We want to give an upper bound for the rank of this matrix. Here, in the real case, the matrix M will be square, we shall just prove that the determinant of M is not zero, and then apply Lemma 2.2 to get the conclusion.

We consider the $L \times L$ determinant

$$\Delta = \det\left(\left(\frac{d}{dz}\right)^{\sigma}\left(z^{\tau}e^{tz}\right)(s\beta)\right)_{\substack{(\tau,t)\\(\sigma,s)}}$$

where (τ, t) is the index of rows $(0 \le \tau \le T_0,\ -T_1 \le t \le T_1)$, while (σ, s) is the index of columns $(0 \le \sigma \le S_0,\ -S_1 \le s \le S_1)$.

We use Corollary 2.3 with

$$n = 2T_1 + 1, \quad \delta_1 = \cdots = \delta_n = T_0 + 1,$$

$$\{w_1, \ldots, w_n\} = \{-T_1, -T_1 + 1, \ldots, -1, 0, 1, \ldots, T_1\},$$

$$\{x_1, \ldots, x_m\} = \{0, \pm\beta, \ldots, \pm S_1\beta\}, \quad \kappa_1 = \cdots = \kappa_m = S_0 + 1.$$

Since $\beta \neq 0$, the x_i's are pairwise distinct and hence the determinant

$$\Delta = \det\left(\left(\frac{d}{dz}\right)^{\sigma}\left(z^{\tau}e^{tz}\right)(s\beta)\right)_{\substack{(\tau,t)\\(\sigma,s)}} \quad \text{with} \quad \begin{cases} 0 \le \tau \le T_0, & |t| \le T_1,\\ 0 \le \sigma \le S_0, & |s| \le S_1 \end{cases}$$

is not zero.

We write $\Delta = f(\alpha, \beta)$, where f is a polynomial in $\mathbb{Z}[\mathrm{X}^{\pm 1}, \mathrm{Y}]$, which can be explicitly written as

$$f = \det\left(f_{\tau t}^{(\sigma s)}\right)_{\substack{(\tau,t)\\(\sigma,s)}},$$

with

$$f_{\tau t}^{(\sigma s)}(\mathrm{X}^{\pm 1}, \mathrm{Y}) = \sum_{\kappa=0}^{\min\{\tau,\sigma\}} \frac{\sigma!\tau!}{\kappa!(\sigma-\kappa)!(\tau-\kappa)!} t^{\sigma-\kappa} s^{\tau-\kappa} \mathrm{X}^{ts}\mathrm{Y}^{\tau-\kappa}.$$

For each polynomial $f_{\tau t}^{(\sigma s)} \in \mathbb{Z}[\mathrm{X}^{\pm 1}, \mathrm{Y}]$ we have

$$\deg f_{\tau t}^{(\sigma s)} \le T_0 + T_1 S_1 \quad \text{and} \quad \mathrm{H}(f_{\tau t}^{(\sigma s)}) \le (T_0 + T_1)^{S_0} S_1^{T_0}.$$

We deduce

$$\deg f \le L(T_0 + T_1 S_1) \quad \text{and} \quad \mathrm{H}(f) \le L!(T_0 + T_1)^{LS_0} S_1^{LT_0}.$$

We use Lemma 2.8 with $r = \max\{1, S_1|\beta|\}$ and $R = Er$, with λ replaced by (τ, t), with μ replaced by (σ, s), with the following functions $\{\varphi_1, \ldots, \varphi_L\}$:

$$\phi_{\tau t}(z) = z^{\tau}e^{tz}, \qquad (0 \le \tau \le T_0, \quad |t| \le T_1),$$

and finally with the points

$$\{\zeta_1, \ldots, \zeta_L\} = \{s\beta\,;\ -S_1 \le s \le S_1\},$$

each of them being repeated $S_0 + 1$ times. We obtain

$$\log|\Delta| \le -\left(\frac{L(L-1)}{2} - LS_0\right)\log E + \log(L!) + \sum_{\tau=0}^{T_0}\sum_{t=-T_1}^{T_1} \max_{0\le\sigma\le S_0} \log \sup_{|z|=R}\left|\left(\frac{d}{dz}\right)^{\sigma}\phi_{\tau t}(z)\right|.$$

We bound $\sup_{|z|=R}\left|(d/dz)^{\sigma}\phi_{\tau t}(z)\right|$ as follows. Using Leibniz' formula for derivatives of products, we have

$$\left(\frac{d}{dz}\right)^{\sigma}\phi_{\tau t}(z) = \sum_{\kappa=0}^{\min\{\tau,\sigma\}} \kappa!\binom{\tau}{\kappa}\binom{\sigma}{\kappa} t^{\sigma-\kappa} z^{\tau-\kappa} e^{tz}.$$

Bounding $\kappa!\binom{\tau}{\kappa}$ from above by τ^{κ}, we find that

$$\sum_{\kappa=0}^{\min\{\tau,\sigma\}} \kappa!\binom{\tau}{\kappa}\binom{\sigma}{\kappa}|t|^{\sigma-\kappa} \le \sum_{\kappa=0}^{\sigma}\binom{\sigma}{\kappa}\tau^{\kappa}|t|^{\sigma-\kappa} = (\tau + |t|)^{\sigma} \le (T_0 + T_1)^{S_0}.$$

Combining these two results, we obtain

$$\max_{0\le\sigma<S_0} \sup_{|z|=R}\left|\left(\frac{d}{dz}\right)^{\sigma}\phi_{\tau t}(z)\right| \le (T_0 + T_1)^{S_0} R^{T_0} e^{T_1 R}.$$

This inequality holds for any pair (τ, t). Notice that we have defined r so that $r \ge 1$ in such a way that $R \ge 1$. Otherwise the term R^{T_0} should be replaced by $\max\{1, R^{T_0}\}$. Anyway, it turns out that S_1 will always be chosen $> 1/|\beta|$. We finally use the following estimate:

$$\log L + T_0 \log R + T_1 R \le c_2\bigl(T_0 \log(S_1 E) + T_1 S_1 E\bigr).$$

□

2.4.3 Second Proof of the Real Case of Theorem 1.4

Proposition 2.10. *Let α_1 and β be two real numbers, with $\alpha_1 > 0$, $\alpha_1 \ne 1$ and $\beta \notin \mathbb{Q}$. Define $\alpha_2 = \alpha_1^{\beta} = \exp(\beta \log \alpha_1)$, where $\log \alpha_1$ is the real valued logarithm of α_1. Let $E \ge e$ be a real number and T, S_0, S_1, L be four integers, all greater than one, with*

$$L = (2T + 1)^2 = (S_0 + 1)(2S_1 + 1).$$

Then there exists a polynomial $f \in \mathbb{Z}[X_1^{\pm 1}, X_2^{\pm 1}, Y]$ satisfying

$$\deg f \le LS_0 + 2LTS_1, \quad H(f) \le L!(2T)^{LS_0},$$

and

$$0 < |f(\alpha_1, \alpha_2, \beta)| \le E^{-L^2/2}(TE)^{c_3 S_0 L} e^{c_3 T S_1 E L}$$

with $c_3 = 5(1 + |\beta|)(1 + |\log \alpha_1|)$.

Consequence. The real case of Gel'fond-Schneider Theorem follows. The conditions on the parameters are the same as for Proposition 2.7, provided that we replace the parameters T_0, T_1 and S of Schneider's method respectively by S_0, S_1 and T (replace also c_1 by c_3). This relationship between the parameters is related to the fact that the matrix associated with one method is just the transpose of the matrix which belongs to the other method (see § 13.7 for this *duality*). This completes the second proof of Theorem 1.4 in the real case.

Proof of Proposition 2.10.

We consider the following $L \times L$ determinant:

$$\Delta = \det\Big((t_1 + t_2\beta)^\sigma \alpha_1^{t_1 s} \alpha_2^{t_2 s}\Big)_{\substack{(t_1,t_2)\\(\sigma,s)}}.$$

with $-T \le t_1, t_2 \le T$ on one hand, $0 \le \sigma \le S_0$ and $-S_1 \le s \le S_1$ on the other. This number Δ is well defined only up to a multiplicative factor ± 1 (depending on the orderings of the rows and of the columns which is implicit). The zero estimate (Corollary 2.3) implies $\Delta \neq 0$. We repeat the estimate which was made in § 2.3: by expanding the determinant, we find that Δ is the value, at the point $\alpha_1, \alpha_2, \beta$, of a polynomial $f \in \mathbb{Z}[X_1^{\pm 1}, X_2^{\pm 1}, Y]$ which satisfies the given bounds for $\deg f$ and $H(f)$.

We estimate $|\Delta|$ from above as follows: for $\max\{|t_1|, |t_2|\} \le T$, define $\phi_{t_1 t_2}(z) = e^{(t_1+t_2\beta)z}$. Then

$$\Delta = \det\left(\left(\frac{d}{dz}\right)^\sigma \phi_{t_1 t_2}(s \log \alpha_1)\right)_{\substack{(t_1,t_2)\\(\sigma,s)}}.$$

with $|\underline{t}| \le T$ and $0 \le \sigma \le S_0$, $|s| \le S_1$. We deduce from Lemma 2.8:

$$\frac{1}{L}\log|\Delta| \le -\frac{L-1}{2}\log E + S_0 \log E + \log L + \max_{\sigma, t_1, t_2} \log \sup_{|z|=R} \left|\left(\frac{d}{dz}\right)^\sigma \phi_{t_1 t_2}(z)\right|$$

where $r = \max\{1, S_1 |\log \alpha_1|\}$ and $R = Er$. From

$$\left(\frac{d}{dz}\right)^\sigma \phi_{t_1 t_2}(z) = (t_1 + t_2\beta)^\sigma e^{(t_1+t_2\beta)z},$$

we deduce

$$\log \sup_{|z|=R} \left|\left(\frac{d}{dz}\right)^\sigma \phi_{t_1 t_2}(z)\right| \le S_0 \log\big(T(1+|\beta|)\big) + TR(1+|\beta|),$$

hence

$$\begin{aligned}\frac{1}{L}\log|\Delta| &\le \\ -\frac{L-1}{2}&\log E+\log L+S_0\log\bigl(TE(1+|\beta|)\bigr)+TS_1E(1+|\beta|)(1+|\log\alpha_1|)\\ &\le -\frac{1}{2}L\log E+c_3\bigl(S_0\log(TE)+TS_1E\bigr).\end{aligned}$$

□

2.5 Gel'fond-Schneider's Theorem in the Complex Case

We complete now the proof of Theorem 1.4 in the complex case. The transcendence of the number e^{π} will follow.

2.5.1 Statement of Proposition 2.11

Proposition 2.11. *Let λ be a nonzero complex number and β an irrational complex number. Define $\alpha_1=e^{\lambda}$ and $\alpha_2=e^{\beta\lambda}$. Let L, T_0, T_1 and S be positive rational integers and E a real number satisfying*

$$T_0\ge 2,\quad T_1\ge 2,\quad S\ge 3,\quad E\ge e,\quad L=(T_0+1)(T_1+1)$$
$$S^2>T_0(T_1+1)\quad\textit{and}\quad S>T_1.$$

Then there exists a polynomial $f\in\mathbb{Z}[X_1^{\pm1},X_2^{\pm1},Y]$ such that

$$\deg f\le L(T_0+2T_1S),\quad \mathrm{H}(f)\le L!S^{LT_0}$$

and

$$0<|f(\alpha_1,\alpha_2,\beta)|\le E^{-L^2/2}(SE)^{c_4T_0L}e^{c_4T_1SLE}$$

with a constant c_4 which depends only on λ and β.

2.5.2 Proof of Theorem 1.4 in the Complex Case

Given $c>0$, we choose the parameters in such a way that

$$c\bigl(T_0+T_1S+T_0\log S\bigr)+c_4\bigl(T_0\log(SE)+T_1SE\bigr)<\frac{1}{2}L\log E.$$

Here are two sets of admissible choices for the parameters (there are many further possibilities).

(i) Take a sufficiently large integer N, and choose

$$L=N^8,\quad T_0=N^6-1,\quad T_1=N^2-1,\quad S=2N^4\quad\text{and}\quad E=e.$$

(ii) Take for T_1 a sufficiently large integer, choose for S a large integer such that S^2 is a multiple of $2T_1$. Then put $T_0=S^2/2T_1$ and $E=S$.

2.5.3 Sketch of the Proof of Proposition 2.11.

Consider a matrix

$$M = \left((s_1 + s_2\beta)^\tau \left(\alpha_1^{s_1}\alpha_2^{s_2}\right)^t \right)_{\substack{(\tau,t)\\(s_1,s_2)}},$$

where (τ, t) is the index of rows, (s_1, s_2) the index of columns. The parameters τ, t, s_1 and s_2 will run over nonnegative integers with $0 \le \tau \le T_0$, $0 \le t \le T_1$ and $0 \le s_1, s_2 \le S$.

We first want to prove that the rank of M is maximal, i.e. equal to the minimum between the number $(T_0 + 1)(T_1 + 1)$ of rows and the number $(S + 1)^2$ of columns, under suitable assumptions on the parameters. Here we shall assume that $(S + 1)^2$ is larger than $(T_0+1)(T_1+1)$ and use a zero estimate from Y. V. Nesterenko (Proposition 2.12 below) to achieve this goal. Another solution to the same problem deals with the case where $(T_0 + 1)(T_1 + 1)$ is bigger than $(S + 1)^2$ and involves a multiplicity estimate (cf. Chap. 8).

The idea of proof of the zero estimate is the following. We introduce the points $s_1\gamma_1 + s_2\gamma_2 \in \mathbb{C} \times \mathbb{C}^\times$, where[5]

$$\gamma_1 = (1, \alpha_1), \quad \gamma_2 = (\beta, \alpha_2), \quad s_1\gamma_1 + s_2\gamma_2 = (s_1 + s_2\beta, \alpha_1^{s_1}\alpha_2^{s_2}).$$

If the rank of M is less than $(T_0+1)(T_1+1)$, then there exist complex (in fact algebraic) numbers $p_{\tau t}$, not all zero, such that

$$\sum_{\tau=0}^{T_0}\sum_{t=0}^{T_1} p_{\tau t}(s_1 + s_2\beta)^\tau \alpha_1^{s_1 t}\alpha_2^{s_2 t} = 0$$

for all $(s_1, s_2) \in \mathbb{Z}^2$ with $0 \le s_1, s_2 \le S$. Therefore we get a nonzero polynomial

$$P(X, Y) = \sum_{\tau=0}^{T_0}\sum_{t=0}^{T_1} p_{\tau t}X^\tau Y^t \in \mathbb{C}[X, Y]$$

such that $P(s_1\gamma_1 + s_2\gamma_2) = 0$ for all (s_1, s_2) with $0 \le s_1, s_2 \le S$. Here comes the main argument: if S' and S'' are positive integers with $S' + S'' = S$, then all polynomials

$$P\left(s_1' + s_2'\beta + X, \alpha_1^{s_1'}\alpha_2^{s_2'}Y\right), \qquad (s_1', s_2') \in \mathbb{Z}^2 \quad 0 \le s_1', s_2' \le S'$$

vanish at $s_1''\gamma_1 + s_2''\gamma_2$ for $(s_1'', s_2'') \in \mathbb{Z}^2$ with $0 \le s_1'', s_2'' \le S''$. If we can *eliminate* the variable Y between these $(S'+1)^2$ polynomials, then we obtain a nonzero polynomial in X only, for which the number of zeroes is bounded by the degree.

Once we know the rank of M, we extract a nonsingular matrix of maximal size and we consider its determinant $\Delta \ne 0$. There are two ways of getting an upper bound for $|\Delta|$.

---

[5] We write additively the law on the abelian group $\mathbb{C} \times \mathbb{C}^\times$.

- Either one uses Schneider's method like in § 2.3.b: apply Lemma 2.5 to the alternant involving the functions

$$\phi_{\tau t}(z) = z^\tau \alpha_1^{tz}, \qquad 0 \le \tau \le T_0, \quad 0 \le t \le T_1,$$

and the points

$$s_1 + s_2\beta, \qquad 0 \le s_1, s_2 \le S.$$

- Or else (Gel'fond's method) one considers the interpolation determinant constructed with the functions

$$\psi_{s_1 s_2}(z) = e^{(s_1+s_2\beta)z} \qquad 0 \le s_1, s_2 \le S$$

with the derivatives $(d/dz)^\tau$, and with the points $t \log \alpha_1$ $(0 \le \tau \le T_0,$ $0 \le t \le T_1)$. The desired estimate then follows from Lemma 2.8.

2.5.4 Zero Estimate

One important tool in the proofs of zero estimates is *elimination*. Let L be a field and $f_1, \ldots, f_m$ be polynomials in $L[T]$. To *eliminate* T between $f_1, \ldots, f_m$ is to find polynomials $A_1, \ldots, A_m$ in $L[T]$ such that $A_1 f_1 + \cdots + A_m f_m$ is a nonzero constant polynomial. A necessary condition to achieve this goal is that the polynomials $f_1, \ldots, f_m$ have no common factor in the factorial ring $L[T]$. The resultant (see for instance § 8, Chap. 4 of [L 1993]) is a convenient tool for studying the converse.

One of the first appearances of the resultant in transcendental number theory is in the paper [Bor 1899], where É. Borel proved a transcendence measure for the number e. In 1932, J. F. Koksma and J. Popken [KoPop 1932] used resultants when they established a transcendence measure for e^π. Next A. O. Gel'fond [G 1952] used similar arguments in his proof of a transcendence criterion; this criterion is one of the basic tools he introduced for proving results of algebraic independence (see also [FNe 1998] as well as § 15.5.1). Another fundamental tool in this work of Gel'fond's is a zero estimate. Gel'fond proved his zero estimate by analytic means. Such analytic arguments have been developed after, but turned out not to be sufficient for solving some other problems of diophantine approximation.

Later, W. D. Brownawell and D. W. Masser [BrMa 1980] used resultants in order to prove zero estimates. Further algebraic arguments have been introduced by W. D. Brownawell and D. W. Masser [BrMa 1980], then refined by D. W. Masser [Ma 1981b] (see also [Mo 1983]), G. Wüstholz [Wü 1989], P. Philippon [P 1986a], [P 1996] and Y. V. Nesterenko (see Chap. 5 and 8).

Here we shall deal with polynomials in two variables: $f_1, \ldots, f_m$ belong to $K[X, Y]$ where K is a field. We eliminate (if possible) Y in $L[Y]$ where $L = K(X)$. Multiplying by a denominator in $K[X]$ yields polynomials $A_1, \ldots, A_m$ in $K[X, Y]$ such that

$$A_1(X, Y) f_1(X, Y) + \cdots + A_m(X, Y) f_m(X, Y) = R(X)$$

is a nonzero polynomial in $K[X]$. Therefore the number of $x \in K$ such that $f_1, \ldots, f_m$ have a common zero $(x, y) \in K^2$ is at most the degree of R.

Remark. In the proof of Proposition 2.11 (as well as in many other transcendence proofs), two rings of polynomials occur: one for the zero estimate, here it is $K[X, Y]$, and one for applying the transcendence Criterion 2.1, which has been denoted by $\mathbb{Z}[\mathrm{X}_1^{\pm 1}, \mathrm{X}_2^{\pm 1}, \mathrm{Y}]$.

The variables (X, Y) will be specialized in the functions z, e^z, and we shall consider the values at the points $z = s_1 + s_2\beta$. Therefore (X, Y) will be specialized in $(s_1 + s_2\beta, \alpha_1^{s_1}\alpha_2^{s_2}) = s_1\gamma_1 + s_2\gamma_2$.

On the other hand the variables $(\mathrm{X}_1^{\pm 1}, \mathrm{X}_2^{\pm 1}, \mathrm{Y})$ related to Liouville's inequality will be specialized in $(\alpha_1, \alpha_2, \beta)$.

The following zero estimate is due to Y. V. Nesterenko [LauMN 1995].

Proposition 2.12. *Let K be a field of zero characteristic, T_0 and T_1 be two positive integers, $(x_1, y_1), \dots, (x_N, y_N)$, $(\xi_1, \eta_1), \dots, (\xi_M, \eta_M)$ elements of $K \times K^\times$, with $y_1, \dots, y_N$ pairwise distinct in $K^\times$ and $\xi_1, \dots, \xi_M$ pairwise distinct in K. Assume*

$$N > T_1 \quad \textit{and} \quad M > T_0(T_1 + 1).$$

Then there is no nonzero polynomial $P \in K[X, Y]$, with degree at most T_0 in X and degree at most T_1 in Y, which satisfies

$$P(x_\nu + \xi_\mu, y_\nu\eta_\mu) = 0 \quad \textit{for } 1 \le \nu \le N \textit{ and } 1 \le \mu \le M.$$

Proof. The proof involves an elimination procedure as follows: a nonzero polynomial $P \in K[X, Y]$ of degree at most T_0 in X and at most T_1 in Y can be written

$$P(X, Y) = \sum_{j=1}^{r} Q_j(X) Y^{k_j},$$

where r is an integer with $1 \le r \le T_1 + 1$, where $0 \le k_1 < k_2 < \cdots < k_r \le T_1$ are integers, and where $Q_1, \dots, Q_r$ are nonzero polynomials in $K[X]$ (of degree at most T_0). We shall assume $k_1 = 0$, since there is no loss of generality to assume that Y does not divide P. We want to eliminate Y between the polynomials

$$P(x_\nu + X, y_\nu Y) = \sum_{j=1}^{r} Q_j(x_\nu + X) y_\nu^{k_j} Y^{k_j}, \qquad (1 \le \nu \le N).$$

We shall find $\nu_1, \dots, \nu_r$ in $\{1, \dots, N\}$ such that the polynomial

$$\Delta(X) = \det\Big(Q_j(x_{\nu_i} + X) y_{\nu_i}^{k_j} \Big)_{1 \le i, j \le r}$$

is not identically zero. Next, we show that there exist $S_1, \dots, S_r$ in $K[X]$, with

$$\Delta(X) = \sum_{i=1}^{r} P(x_{\nu_i} + X, y_{\nu_i} Y) S_i(X).$$

Indeed, if $\mathcal{C}_j$ denotes the column vector $\left(Q_j(x_{\nu_i} + X)y_{\nu_i}^{k_j}\right)_{1\le i\le r}$ of $\Delta(X)$, then $\mathcal{C}_1 + Y^{k_2}\mathcal{C}_2 + \cdots + Y^{k_r}\mathcal{C}_r$ is the column vector $\left(P(x_{\nu_i} + X, y_{\nu_i}Y)\right)_{1\le i\le r}$. We find the polynomials S_i by expanding the determinant with the first column replaced by this linear combination.

In order to construct $\Delta \neq 0$, we notice that if $q_jX^{d_j}$ is the leading term of $Q_j(X)$, then the coefficient of $X^{d_1+\cdots+d_r}$ in $\Delta(X)$ is

$$q_1\cdots q_r \det\left(y_{\nu_i}^{k_j}\right)_{1\le i,j\le r}.$$

The key point of the proof is provided by the following lemma.

Lemma 2.13. *Let* $0 \le k_1 < k_2 < \cdots < k_r$ *be integers and* $\mathcal{M}$ *a subset of* $K^\times$ *with* $\operatorname{Card}\mathcal{M} > k_r$. *Then there exist* $a_1, \ldots, a_r$ *in* $\mathcal{M}$ *such that*

$$\det\left(a_i^{k_j}\right)_{1\le i,j\le r} \neq 0.$$

Proof. The proof is by induction on r. For $r = 1$, the result is trivial. Assume $a_1, \ldots, a_{r-1}$ are such that

$$\det\left(a_i^{k_j}\right)_{1\le i,j\le r-1} \neq 0.$$

Introduce the polynomial

$$P(z) = \det\begin{pmatrix} a_1^{k_1} & \cdots & a_1^{k_r} \\ \vdots & \ddots & \vdots \\ a_{r-1}^{k_1} & \cdots & a_{r-1}^{k_r} \\ z^{k_1} & \cdots & z^{k_r} \end{pmatrix}.$$

Then P is of exact degree k_r. Since $\operatorname{Card}\mathcal{M} > k_r$, there exists $a_r \in \mathcal{M}$ such that $P(a_r) \neq 0$. □

Proof of Proposition 2.12. We are now able to complete the proof of Proposition 2.12. Since $k_r \le T_1 < N$, and since $y_1, \ldots, y_N$ are pairwise distinct, we can use Lemma 2.13 with $\mathcal{M} = \{y_1, \ldots, y_N\}$. We deduce that there exist $\nu_1, \ldots, \nu_r$ in $\{1, \ldots, N\}$ such that

$$\det\left(y_{\nu_i}^{k_j}\right)_{1\le i,j\le r} \neq 0.$$

Then

$$\Delta(X) = \det\left(Q_j(x_{\nu_i} + X)y_{\nu_i}^{k_j}\right)_{1\le i,j\le r}$$

is a nonzero polynomial of exact degree $\deg Q_1 + \cdots + \deg Q_r$, hence

$$\deg \Delta \le rT_0 \le T_0(T_1 + 1).$$

Further, if

$$P(x_\nu + \xi_\mu, y_\nu \eta_\mu) = 0 \quad \text{for } 1 \le \nu \le N \text{ and } 1 \le \mu \le M,$$

then

$$\Delta(\xi_\mu) = 0 \quad \text{for } 1 \le \mu \le M.$$

However $\xi_1, \dots, \xi_M$ are pairwise distinct, and $M > T_0(T_1 + 1) \ge \deg \Delta$. This completes the proof of Proposition 2.12. □

2.5.5 Proof of Proposition 2.11

Step 1. The Matrix $\boldsymbol{M}$

We consider the $L \times (2S+1)^2$ matrix

$$\boldsymbol{M} = \left((s_1 + s_2\beta)^\tau \left(\alpha_1^{s_1}\alpha_2^{s_2}\right)^t \right)_{\substack{(\tau,t) \\ (s_1,s_2)}},$$

with $0 \le \tau \le T_0$, $0 \le t \le T_1$, and with $\underline{s} = (s_1, s_2) \in \mathbb{Z}^2$, $\max\{|s_1|, |s_2|\} \le S$.

Step 2. The Determinant Δ

Our first goal is to show that this matrix $\boldsymbol{M}$ has rank L. We use Proposition 2.12 with $N = S$, $M = S^2$, (ξ_μ, η_μ) are M points in the set

$$(s_1 + s_2\beta, \alpha_1^{s_1}\alpha_2^{s_2}) \quad (\max\{|s_1|, |s_2]\} \le \frac{S}{2}),$$

while (x_ν, y_ν) are N of these points with $y_1, \dots, y_N$ pairwise distinct. Notice that the numbers $s_1 + s_2\beta$ ($\underline{s} \in \mathbb{Z}^2$) are pairwise distinct, and that α_1 and α_2 are not both roots of unity. From the conditions

$$S^2 > T_0(T_1 + 1) \quad \text{and} \quad S > T_1$$

one deduces that the assumptions of Proposition 2.12 are satisfied.

Therefore there exist L elements $\underline{s}^{(1)}, \dots, \underline{s}^{(L)}$ in $\mathbb{Z}^2$ with

$$\max\{|s_1^{(\mu)}|, |s_2^{(\mu)}|\} \le S \quad (1 \le \mu \le L)$$

such that the determinant

$$\Delta = \det\left((s_1^{(\mu)} + s_2^{(\mu)}\beta)^\tau \left(\alpha_1^{s_1^{(\mu)}} \alpha_2^{s_2^{(\mu)}}\right)^t \right)_{\substack{(\tau,t) \\ 1 \le \mu \le L}}$$

is not zero.

Step 3. Estimates for Degree and Height

Define $f \in \mathbb{Z}[\mathrm{X}_1^{\pm 1}, \mathrm{X}_2^{\pm 1}, \mathrm{Y}]$ by

$$f = \det\left((s_1^{(\mu)} + s_2^{(\mu)}\mathrm{Y})^\tau \left(\mathrm{X}_1^{s_1^{(\mu)}}\mathrm{X}_2^{s_2^{(\mu)}}\right)^t \right)_{\substack{(\tau,t)\\ 1\le\mu\le L}},$$

so that

$$\Delta = f(\alpha_1, \alpha_2, \beta).$$

Then

$$\deg f \le L(T_0 + 2T_1S) \quad \text{and} \quad \mathrm{H}(f) \le L!S^{LT_0}.$$

Step 4. Upper Bound for $|\Delta|$

We claim that the absolute value of the determinant Δ is bounded from above by

$$\frac{1}{L}\log|\Delta| \le -\frac{1}{2}L\log E + c_4\big(T_0\log(SE) + T_1SE\big).$$

As we have seen, there are two proofs of this fact.

- Either one applies Lemma 2.5 to the alternant involving the functions

$$\phi_{\tau t}(z) = z^\tau e^{t\lambda z}, \qquad (0 \le \tau \le T_0,\ 0 \le t \le T_1)$$

  and the points

$$s_1^{(\mu)} + s_2^{(\mu)}\beta, \qquad (1 \le \mu \le L).$$

  The estimates of § 2.3.b are still valid and produce the desired upper bound.
- Or else one applies Lemma 2.8 to the interpolation determinant constructed with the functions $\psi_{s_1^{(\mu)}s_2^{(\mu)}}$ $(1 \le \mu \le L)$, where

$$\psi_{s_1s_2}(z) = e^{(s_1+s_2\beta)z} \quad \text{for} \quad \underline{s} = (s_1, s_2) \in \mathbb{Z}^2, \quad 0 \le s_1, s_2 \le S$$

  with the derivatives $(d/dz)^\tau$ $(0 \le \tau \le T_0)$, and with the points $t\lambda$ $(0 \le t \le T_1)$. The estimates are the same as in § 2.4.c. There is only a change of notation: the parameters S_0, S_1 and T which occur there are now replaced respectively by T_0, T_1 and S.

□

2.6 Hermite-Lindemann's Theorem in the Complex Case

We complete the proof of the complex case of Theorem 1.2 and deduce in particular the transcendence of π.

2.6.1 Multiplicity Estimate

The proof of the Theorem of Hermite-Lindemann involves the complex analytic functions z and e^z. For $P \in \mathbb{C}[X, Y]$, the derivative $(d/dz)F$ of the function

$$F(z) = P(z, e^z)$$

is a polynomial in z and e^z, which we call $\mathcal{D}P$:

$$\frac{d}{dz}P(z, e^z) = \mathcal{D}P(z, e^z).$$

It is plain that $\mathcal{D}$ is the derivative operator $(\partial/\partial X) + Y(\partial/\partial Y)$. Hence when K is any field of zero characteristic we can define $\mathcal{D}$ on $K[X, Y]$ by

$$\mathcal{D} = \frac{\partial}{\partial X} + Y\frac{\partial}{\partial Y}.$$

Here is the multiplicity estimate of Y. V. Nesterenko [NeW 1996], Lemma 2.

Proposition 2.14. *Let K be a field of zero characteristic and T_0, T_1, S_0 and M be positive integers satisfying*

$$(S_0 + 1)M > (T_0 + M)(T_1 + 1).$$

Let $(\xi_1, \eta_1), \dots, (\xi_M, \eta_M)$ be elements in $K \times K^\times$ with $\xi_1, \dots, \xi_M$ pairwise distinct. Then there is no nonzero polynomial $P \in K[X, Y]$, of degree at most T_0 in X and of degree at most T_1 in Y which satisfies

$$\mathcal{D}^\sigma P(\xi_\mu, \eta_\mu) = 0 \quad \text{for } 1 \le \mu \le M \text{ and } 0 \le \sigma \le S_0. \tag{2.15}$$

The proof is essentially the same as the proof of Proposition 2.12: we shall eliminate Y using $T_1 + 1$ derivatives, and get a polynomial in X which vanishes at ξ_j with multiplicity at least $S_0 + 1 - T_1$.

Proof. Let us suppose that a polynomial P satisfies all the conditions of the lemma, equalities (2.15) and $P \neq 0$. We assume, as we may without loss of generality, that Y does not divide the polynomial P, and also that P has degree at least 1 with respect to Y. Let us define the numbers $k_0 = 0 < k_1 < \cdots < k_n \le T_1$ by the conditions

$$P(X, Y) = \sum_{i=0}^{n} Q_i(X)Y^{k_i}, \quad Q_i(X) = b_i X^{m_i} + \cdots \in K[X],$$
$$b_i \neq 0, \quad i = 0, \dots, n.$$

For $0 \le \sigma \le n$, we consider the polynomials

$$\mathcal{D}^\sigma P(X, Y) = \sum_{i=0}^{n} Q_{\sigma i}(X) \cdot Y^{k_i}, \tag{2.16}$$

where

$$Q_{\sigma i}(X) = \sum_{j=0}^{\sigma} \binom{\sigma}{j} Q_i^{(\sigma - j)}(X) k_i^j = b_i k_i^\sigma \cdot X^{m_i} + \cdots.$$

It follows from this representation that the determinant

$$\Delta(X) = \det\bigl(Q_{\sigma i}(X)\bigr)_{0\le i,\sigma\le n}$$

can be written

$$\Delta(X) = \det\bigl(b_i k_i^\sigma \cdot X^{m_i} + \cdots\bigr)_{0\le i,\sigma\le n} = b_0\cdots b_n B X^{m_0+\cdots+m_n} + \cdots,$$

where B is a Vandermonde determinant constructed from the numbers $k_0,\ldots,k_n$, hence $B \neq 0$. Now from (2.16) we derive

$$\Delta(X) = \sum_{\sigma=0}^{n} \Delta_\sigma(X)\cdot \mathcal{D}^\sigma P(X,Y), \quad \Delta_\sigma(X)\in K[X],$$

and for any $\tau\in\mathbb{Z}$, $0\le\tau\le S_0-n$, with some $c_{\tau j\sigma}\in K$,

$$\Delta^{(\tau)}(\xi_j) = \sum_{\sigma=0}^{n+\tau} c_{\tau j\sigma}\cdot \mathcal{D}^\sigma P(\xi_j,\eta_j) = 0, \quad j=1,\ldots,M.$$

Since $n\le T_1$ and $\deg\Delta(X) = m_0+\cdots+m_n \le (n+1)T_0 \le T_0(T_1+1)$, we deduce

$$(S_0+1-n)M \le \deg\Delta(X) \le T_0(T_1+1),$$

and $(S_0+1)M \le T_0(T_1+1)+nM \le (T_0+M)(T_1+1)$. This contradicts the assumption and completes the proof of Proposition 2.14. □

2.6.2 Proof of Theorem 1.2

Proposition 2.17. *Let β be a nonzero complex number. There exists $c_5>0$ with the following property. Define $\alpha = e^\beta$. For any integers T_0, T_1, S_0, S_1, L, all of which are at least 2, and any real number $E\ge e$, satisfying*

$$L=(T_0+1)(T_1+1), \quad (S_0+1)(2S_1+1) > (T_0+2S_1+1)(T_1+1),$$

there exists a polynomial $f\in\mathbb{Z}[X^{\pm1},Y]$ such that

$$\deg f \le L(T_0+T_1S_1), \quad \mathrm{H}(f)\le L!(T_0+T_1)^{LS_0}S_1^{LT_0}$$

and

$$0<|f(\alpha,\beta)| \le E^{-L^2/2}\bigl((T_0+T_1)E\bigr)^{S_0L}(S_1E)^{c_5T_0L}e^{c_5T_1S_1EL}.$$

The proof is essentially the same as in § 2.4 (which dealt only with the case where β is real), apart from the zero estimate.

We consider the $L\times(S_0+1)(2S_1+1)$ matrix

$$M = \left(\left(\frac{d}{dz}\right)^\sigma \bigl(z^\tau e^{tz}\bigr)(s\beta)\right)_{\substack{(\tau,t)\\(\sigma,s)}},$$

where (τ, t) is the index of rows $(0 \le \tau \le T_0, 0 \le t \le T_1)$, while (σ, s) is the index of columns $(0 \le \sigma \le S_0, -S_1 \le s \le S_1)$. Notice that

$$\left(\frac{d}{dz}\right)^{\sigma} (z^{\tau} e^{tz})(s\beta) = \mathcal{D}^{\sigma}\left(X^{\tau} Y^{t}\right)(s\beta, \alpha^{s}).$$

Here, the variables (X, Y) related with the zero estimate are specialized in (z, e^z), and for $z = s\beta$ we specialize (X, Y) in $(s\beta, \alpha^s)$, while the variables $(\mathrm{X}^{\pm 1}, \mathrm{Y})$ in the statement of Proposition 2.17 are specialized in (α, β).

Given the conditions on the parameters, we can apply Proposition 2.14 with the points (ξ_μ, η_μ) $(1 \le \mu \le M)$ as $(s\beta, \alpha^s)$ $(|s| \le S_1)$, hence $M = 2S_1 + 1$ and conclude that the matrix $\boldsymbol{M}$ has rank L.

We select L elements $(\sigma^{(\mu)}, s^{(\mu)})$ $(1 \le \mu \le L)$, in $\mathbb{Z}^2$, with $0 \le \sigma^{(\mu)} \le S_0$ and $|s^{(\mu)}| \le S_1$, in such a way that the $L \times L$ determinant

$$\Delta = \det\left(\left(\frac{d}{dz}\right)^{\sigma^{(\mu)}} \left(z^{\tau} e^{tz}\right)(s^{(\mu)}\beta)\right)_{\substack{(\tau,t)\\ 1\le\mu\le L}}$$

is not zero. We derive the upper bound

$$\frac{1}{L}\log|\Delta| \le -\frac{1}{2}L\log E + \Big(S_0\log\big((T_0+T_1)E\big) + c_5 T_0\log(S_1 E) + c_5 T_1 S_1 E\Big),$$

exactly like in § 2.4 above.

In order to complete the proof of Theorem 1.2 it remains to choose our parameters T_0, T_1, S_0, S_1 and E so that one can apply Lemma 2.1. Hence we require that

$$S_0\log\big((T_0+T_1)E\big) + T_0\log(S_1 E) + T_1 S_1 E$$

is small compared with $L \log E$. The main difference with the proof in § 2.4.b lies in the requirement

$$(S_0+1)(2S_1+1) > (T_0+2S_1+1)(T_1+1),$$

which replaces the equality between L and $(S_0 + 1)(2S_1 + 1)$ which took place in § 2.4.

Here are two sets of admissible choices for these parameters.

(i) Let N be a sufficiently large integer. Choose

$$T_1 = S_1 = N, \quad T_0 = S_0 = N^2 \quad \text{and} \quad E = e.$$

(ii) Choose for S_1 any sufficiently large integer, for T_1 a large multiple of S_1, and define

$$T_0 = E = T_1, \quad S_0 = \frac{T_1^2}{S_1}.$$

□

Exercises

Exercise 2.1. Let ϑ be a real number. Show that the following properties are equivalent.

(i) $\vartheta \notin \mathbb{Q}$

(ii) There exists an integer $k \geq 1$ and a sequence $(P_N)_{N \geq N_0}$ of polynomials in $\mathbb{Z}[X]$ of degree $\leq k$ such that $P_N(\vartheta) \neq 0$ and $|P_N(\vartheta)| \to 0$ as $N \to \infty$.

(iii) For any $k \geq 1$ there exist infinitely many $P \in \mathbb{Z}[X]$ of degree $\leq k$ such that

$$0 < |P(\vartheta)| \leq \frac{1}{\mathrm{H}(P)}.$$

Exercise 2.2.
a) Check that the following statement is equivalent with Lemma 2.1:

> *Given any algebraic numbers $\gamma_1, \ldots, \gamma_m$, there exists a positive constant $c = c(\gamma_1, \ldots, \gamma_m)$ which satisfies the following property.*
> *Let $f \in \mathbb{Z}[X_1, \ldots, X_m]$ and T be a positive number such that the absolute values of the coefficients of f are all at most e^T and the total degree of f is at most T. If the number $f(\gamma_1, \ldots, \gamma_m)$ is nonzero, then*
>
> $$|f(\gamma_1, \ldots, \gamma_m)| \geq e^{-cT}.$$

b) If $\alpha \in \mathbb{C}^\times$ is a root of a polynomial with rational integer coefficients whose absolute values are bounded above by some number H, then $|\alpha| \geq (1 + H)^{-1}$.

Hint. *See* *§ 3.5.1.*

c) For $1 \leq i \leq m$, let $\gamma_i \in \mathbb{C}$ be root of a polynomial in $\mathbb{Z}[X]$ of degree d_i, leading coefficient a_i, and complex roots γ_{ij} $(1 \leq j \leq d_i)$. Let $f \in \mathbb{Z}[X_1, \ldots, X_m]$ be a polynomial in m variables of total degree at most L with integer coefficients. Define

$$\Delta = (a_1 \cdots a_m)^{d_1 \cdots d_m L}.$$

Check that the polynomial

$$F(X) = \Delta \prod_{j_1=1}^{d_1} \cdots \prod_{j_m=1}^{d_m} \bigl(X - f(\gamma_{1,j_1}, \ldots, \gamma_{m,j_m})\bigr)$$

has rational integer coefficients.

Hint. *Use the Theorem on symmetric polynomials* ([L 1993], *Chap. IV).*

d) Deduce the statement in a).

Hint. *Use b) for the number $\alpha = f(\gamma_1, \ldots, \gamma_m)$.*

e) Show also that the statement in a) provides a necessary and sufficient condition for the numbers $\gamma_1, \ldots, \gamma_m$ to be all algebraic.

Hint. *Let θ be a complex transcendental number. Using Dirichlet's box principle (see for instance Lemma 4.11), show that there exists a positive real number $c = c(\theta)$ such that, for each $T \geq 2$, there is a nonzero polynomial $P \in \mathbb{Z}[X]$ of total degree at most T and of coefficients bounded in absolute value by e^T, such that $|P(\theta)| \leq e^{-cT^2}$. See also Proposition 15.2.*

Exercise 2.3.
a) Let $w_1, \ldots, w_n$ distinct real numbers, $d_1, \ldots, d_n$ nonnegative rational integers, and $u_1, \ldots, u_N$ distinct real numbers, with $N = d_1 + \cdots + d_n + n - 1$. Show that there exist polynomials $a_1, \ldots, a_n$ in $\mathbb{R}[t]$, of degrees $d_1, \ldots, d_n$ respectively, such that the function

$$F(t) = \sum_{i=1}^{n} a_i(t)e^{w_i t}$$

has a simple zero at each point $u_1, \ldots, u_N$ and no other zero.

Hint. *Use linear algebra as well as Lemma 2.2.*

b) Give also a generalization where the u_j are no more distinct, but multiplicities are required.

Exercise 2.4 (*Algebraic version of Lemma* 2.2: upper bound for the number of consecutive integral zeroes of an exponential polynomial; see [MiW 1994].)
a) Let K be a field, $\alpha_1, \ldots, \alpha_n$ nonzero elements of K which are pairwise distinct, and $a_1, \ldots, a_n$ nonzero polynomials in $K[X]$, of degrees say $d_1, \ldots, d_n$. Then the function $\mathbb{Z} \longrightarrow K$ which is defined by

$$F(m) = \sum_{i=1}^{n} a_i(m)\alpha_i^m \tag{2.18}$$

cannot vanish on a set of $d_1 + \cdots + d_n + n$ consecutive integers.
b) Let $d_1, \ldots, d_n$ be nonnegative integers and let $E \subset \mathbb{Z}$ be a set of $d_1 + \cdots + d_n + n - 1$ consecutive integers. Show that there exist nonzero polynomials $a_1, \ldots, a_n$ in $K[X]$, of degrees respectively $d_1, \ldots, d_n$, such that the function $\mathbb{Z} \longrightarrow K$ which is defined by (2.18) vanishes on E.

Exercise 2.5. Let $\underline{w}_1, \ldots, \underline{w}_t$ be $\mathbb{Q}$-linearly independent elements in $\mathbb{C}^n$. Show that the t functions $e^{\underline{w}_1 \cdot \underline{z}}, \ldots, e^{\underline{w}_t \cdot \underline{z}}$ are algebraically independent over the field $\mathbb{Q}(z_1, \ldots, z_n)$.

Hint. *Use induction like in the proof of Lemma 2.2.*

Exercise 2.6. Under the assumptions of Lemma 2.5, let $E \geq 1$ be a real number. For $1 \leq \mu \leq L$, define $R_\mu = E|\zeta_\mu|$. Further, denote by $\mathfrak{S}_L$ the symmetric group on $\{1 \ldots, L\}$ of order $L!$. Check

$$|\Delta| \leq E^{-L(L-1)/2} L! \max_{\sigma \in \mathfrak{S}_L} \prod_{\lambda=1}^{L} |\varphi_\lambda|_{R_{\sigma(\lambda)}}.$$

Deduce

$$|\Delta| \leq E^{-L(L-1)/2} L! \prod_{\mu=1}^{L} \max_{1 \leq \lambda \leq L} |\varphi_\lambda|_{R_\mu}.$$

Exercise 2.7.
a) Let $L \geq 14$ be an integer, $f: \mathbb{C}^2 \to \mathbb{C}$ an analytic function in $\mathbb{C}^2$ (entire function of two variables), $x_1, \dots, x_L, y_1, \dots, y_L$ complex numbers and r_1, r_2, R_1, R_2, E real numbers satisfying

$$R_1 \geq r_1 \geq \max_{1 \leq \lambda \leq L} |x_\lambda|, \quad R_2 \geq r_2 \geq \max_{1 \leq \mu \leq L} |y_\mu|, \quad \max\left\{\frac{R_1}{r_1}, \frac{R_2}{r_2}\right\} \geq E \geq e.$$

For $1 \leq \lambda \leq L$ and $1 \leq \mu \leq L$, assume that the number

$$u_{\lambda\mu} = f(x_\lambda, y_\mu)$$

is in $\mathbb{Z}$. Assume also

$$\log \sup \left\{|f(z, w)|\,;\; |z| \leq R_1,\ |w| \leq R_2\right\} \leq \frac{1}{3} L \log E.$$

Show that the determinant of the matrix $\left(u_{\lambda\mu}\right)_{1 \leq \lambda, \mu \leq L}$ is zero.

Hint. *For $L \geq 14$, check $3L + 6\log(L!) < L^2$.*

b) Let d and ℓ be positive integers satisfying $d\ell > d + \ell$, let $u_1, \dots, u_d$ be $\mathbb{Q}$-linearly independent real numbers, and let $v_1, \dots, v_\ell$ be also $\mathbb{Q}$-linearly independent real numbers. Using question a), show that at least one of the $d\ell$ numbers

$$e^{u_i v_j}, \qquad (1 \leq i \leq d,\ 1 \leq j \leq \ell)$$

is irrational.

Hint. *Choose a large integer N, define $T = N^\ell$, $S = N^d$, $L = N^{d\ell}$, so that $T^d = S^\ell = L$. Consider the two sets, each consisting of L distinct complex numbers:*

$$\mathcal{X} = \left\{t_1 u_1 + \cdots + t_d u_d\,;\; (t_1, \dots, t_d) \in \mathbb{Z}^d,\ 0 \leq t_i < T,\ (1 \leq i \leq d)\right\}$$

and

$$\mathcal{Y} = \left\{s_1 v_1 + \cdots + s_\ell v_\ell,\,;\; (s_1, \dots, s_\ell) \in \mathbb{Z}^\ell,\ 0 \leq s_j < S,\ (1 \leq j \leq \ell)\right\}.$$

Assume $e^{uv} \in \mathbb{Q}$ for any $u \in \mathcal{X}$ and any $v \in \mathcal{Y}$. Let D be a positive integer such that $De^{u_i v_j} \in \mathbb{Z}$ for $1 \leq i \leq d$ and $1 \leq j \leq \ell$. Use question a) with $E = e$ for the entire function of two complex variables $(z, w) \mapsto D^{TS} e^{zw}$ and get a contradiction.

c) *Application.* Deduce from b) the following statement: *let x be a real number and let p_1, p_2, p_3 be three pairwise distinct prime numbers. If the three numbers p_1^x, p_2^x and p_3^x are all rational, then x is a nonnegative integer.*

Exercise 2.8.
a) Let d and ℓ be positive integers, $x_1, \dots, x_d$ real numbers which are linearly independent over $\mathbb{Q}$ and $y_1, \dots, y_\ell$ also $\mathbb{Q}$-linearly independent real numbers. Further let L, T and S be positive integers and E a real number satisfying

$$L = T^d = S^\ell \quad \text{and} \quad E \geq e.$$

Show that there exists a polynomial f in $d\ell$ variables with integer coefficients satisfying

$$\deg f \leq LTS, \quad \mathrm{H}(f) \leq L!$$

and

$$0 < |f(\underline{\zeta})| \leq E^{-L^2/2} e^{cTSEL}$$

where c depends only on $x_1, \ldots, x_d, y_1, \ldots, y_\ell$, while $\underline{\zeta} \in \mathbb{C}^{d\ell}$ denotes the point with coordinates $e^{x_i y_j}$ $(1 \leq i \leq d, 1 \leq j \leq \ell)$.

b) Deduce the real case of the six exponentials Theorem 1.12.

Hint. *See also* [Lau 1989] *and* [Pi 1993].

Exercise 2.9. Complete the proof of the Theorems of Hermite-Lindemann and Gel'fond-Schneider as well as of the six exponentials Theorem in the complex cases by means of the following result of R. Tijdeman (refining an earlier estimate of Gel'fond's [G 1952], Chap. III, § 4, Lemma III; see also [F 1982], Chap. 9, § 4, Lemma 8.9 and [W 1974], Chap. 6), in place of Nesterenko's zero and multiplicity estimates.

Let $a_1, \ldots, a_n$ be polynomials in $\mathbb{C}[z]$ of degrees $d_1, \ldots, d_n$, and let $w_1, \ldots, w_n$ pairwise distinct complex numbers. Define

$$\Omega = \max\{|w_1|, \ldots, |w_n|\}.$$

Then the number of zeroes (counting multiplicities) of the function

$$F(z) = \sum_{i=1}^{n} a_i(z) e^{w_i z}$$

in the disc $|z| \leq R$ of $\mathbb{C}$ is at most $2(d_1 + \cdots + d_n + n - 1) + 5R\Omega$.

Exercise 2.10.

a) Let L' and L be two integers, $1 \leq L' \leq L$. Let $\delta_{\lambda\mu}$ $(1 \leq \lambda, \mu \leq L)$ be L^2 complex numbers, $E, M_1, \ldots, M_L$ positive real numbers, $\zeta_1, \ldots, \zeta_L$ complex numbers and $\varphi_1, \ldots, \varphi_{L'}$ entire functions in $\mathbb{C}$. Assume

$$\delta_{\lambda\mu} = \varphi_\lambda(\zeta_\mu) \quad \text{for } 1 \leq \lambda \leq L', 1 \leq \mu \leq L.$$

Assume further, $E \geq e$, and, for $1 \leq \mu \leq L$,

$$\sup_{|z|=E} |\varphi_\lambda(z\zeta_\mu)| \leq M_\lambda \quad \text{for } 1 \leq \lambda \leq L'$$

and

$$|\delta_{\lambda\mu}| \leq M_\lambda \quad \text{for } L' < \lambda \leq L.$$

Show that the absolute value of the determinant Δ of the $L \times L$ matrix $\left(\delta_{\lambda\mu}\right)_{1 \leq \lambda, \mu \leq L}$ is bounded by

$$|\Delta| \leq E^{-L'(L'-1)/2} L! M_1 \cdots M_L.$$

Hint. *See § 7.2.*

b) Check that for any $K \leq L$ the polynomial $f \in \mathbb{Z}[X_1^{\pm 1}, X_2^{\pm 1}, Y]$, given by the proof of Proposition 2.11, satisfies

$$\max_{\|\kappa\| \leq K} |\mathcal{D}^\kappa f(\alpha_1, \alpha_2, \beta)| \leq E^{-(L-K)^2/2} (SE)^{c_4' T_0 SL} e^{c_4' T_1 SLE}$$

with a constant c_4' which depends only on λ and β. Here, for $\kappa = (\kappa_1, \kappa_2, \kappa_3) \in \mathbb{N}^3$, $\mathcal{D}^\kappa$ denotes the derivative $(\partial/\partial X_1)^{\kappa_1}(\partial/\partial X_2)^{\kappa_2}(\partial/\partial Y)^{\kappa_3}$.

Hint. *See* [LauRoy 1999a].

c) Prove a similar result for Proposition 2.17.

Exercise 2.11. Let $x_1, \ldots, x_d$ be complex numbers which are linearly independent over $\mathbb{Q}$, and $y_1, \ldots, y_\ell$ complex numbers which are also linearly independent over $\mathbb{Q}$. Let K_0 be the field generated by the $d\ell$ numbers $e^{x_i y_j}$ $(1 \le i \le d, 1 \le j \le \ell)$. Define also

$$K_1 = K_0(x_1, \ldots, x_d), \; K_2 = K_0(y_1, \ldots, y_\ell) \text{ and } K_3 = K_0(x_1, \ldots, x_d, y_1, \ldots, y_\ell).$$

For $i = 0, 1, 2, 3$, show that the field K_i has transcendence degree at least 1 under the assumption which is provided by the second column of the table 2.19. The functions which are involved are displayed in the third column, the points are $\mathbb{Z}y_1 + \cdots + \mathbb{Z}y_\ell$. The fourth column tells whether the proof involves or not derivatives. The fifth (and last) column gives the reference of the corresponding theorem.

Hint. *Compare with § 11.2.*

Table 2.19.

Field	Assumption	Functions	$\frac{d}{dz}$	Theorem
K_0	$d\ell > d + \ell$	$e^{x_1 z}, \ldots, e^{x_d z}$	No	1.12
K_1	$d \ge 1, \ell \ge 2$	$z, e^{x_1 z}, \ldots, e^{x_d z}$	No	1.4
K_2	$d \ge 2, \ell \ge 1$	$e^{x_1 z}, \ldots, e^{x_d z}$	Yes	1.4
K_3	$d \ge 1, \ell \ge 1$	$z, e^{x_1 z}, \ldots, e^{x_d z}$	Yes	1.2

3. Heights of Algebraic Numbers

A nonzero rational integer has absolute value at least 1. A nonzero rational number has absolute value at least the inverse of any denominator. Liouville's inequality (§ 3.5) is an extension of these estimates and provides a lower bound for the absolute value of any nonzero algebraic number. More specifically, if we are given finitely many (fixed) algebraic numbers $\gamma_1, \ldots, \gamma_t$, and a polynomial $P \in \mathbb{Z}[X_1, \ldots, X_t]$ which does not vanish at the point $(\gamma_1, \ldots, \gamma_t)$ then we can estimate from below $|P(\gamma_1, \ldots, \gamma_t)|$. The lower bound will depend upon the degrees of P with respect to each of the X_i's, the absolute values of its coefficients as well as some measure of the γ_i's.

In order to obtain such lower bounds, we introduce a notion of *height* for an algebraic number (§ 3.2). There are several such heights (§ 3.4) and they all satisfy the fundamental property that for each fixed d and H, the set of algebraic numbers of degree at most d and height at most H is finite. It follows that there exists a function depending on d and H which bounds from below the absolute value of a nonzero algebraic number of degree at most d and height at most H. Now the problem is to compute explicitly such a function, and also to give an upper bound for the height of $P(\gamma_1, \ldots, \gamma_t)$ in terms of $P \in \mathbb{Z}[X_1, \ldots, X_t]$ and the heights of the γ_j's. From this point of view the so-called *absolute logarithmic height* is more convenient than the others, because it has several equivalent definitions:

- The first one is the integral, on the unit circle, of the logarithm of the modulus of the minimal polynomial of the given algebraic number (§ 3.3),
- The second one involves the absolute values (see § 3.1) of the conjugates and the leading coefficient of the minimal polynomial of the algebraic number,
- The third one is phrased in terms of the absolute values — Archimedean and ultrametric — of the algebraic number.

We study this height with somewhat more details than are strictly necessary, because it is an important tool in many situations. We conclude this chapter with Lehmer's problem and related questions (§ 3.6).

3.1 Absolute Values on a Number Field

We need a little bit of algebraic number theory. There are plenty of references on this subject (see, for example, [Ar 1967]; [Bou 1985] Chap. 6; [FrTa 1991] Chap. 1,2,3; [L 1970]; [L 1978]Chap. 4 § 1, pp. 77–84 and Chap. 7 § 1, pp. 159–162; [L 1983] Chap. 3 § 1, pp. 50–54; [L 1993], Chap. 12; [Neu 1999], Chap. 2; [Sc 1999]; [Ser 1989], Chap. 2 § 1–3, pp. 7–16; [Sil 1986], Chap. VIII § 5).

We explain briefly the basic facts we shall need, detailed proofs can be found in [L 1993], (especially Chap. 12) which we take as basic reference.

3.1.1 p-adic Valuation and p-adic Absolute Values over $\mathbb{Q}$

For $x \in \mathbb{Q}$, $x \neq 0$, we write the decomposition of x into a product of prime factors as follows

$$x = \pm \prod_p p^{v_p(x)}.$$

This defines, for each prime number p, a map v_p from $\mathbb{Q}^\times$ to $\mathbb{Z}$, which we extend by $v_p(0) = \infty$. The map $v_p : \mathbb{Q} \longrightarrow \mathbb{Z} \cup \{\infty\}$ thus obtained is the *p-adic valuation* over $\mathbb{Q}$. One can easily prove that it satisfies the following properties:

(i) for $x \in \mathbb{Q}$, $v_p(x) = \infty$ is equivalent to $x = 0$
(ii) for $(x, y) \in \mathbb{Q}^2$, $v_p(xy) = v_p(x) + v_p(y)$
(iii) for $(x, y) \in \mathbb{Q}^2$, $v_p(x + y) \geq \min\{v_p(x), v_p(y)\}$.

To v_p is associated an absolute value $|\cdot|_p$, which is the map from $\mathbb{Q}$ to $\mathbb{Q}$ defined by

$$|x|_p = p^{-v_p(x)}$$

for $x \neq 0$ and $|0|_p = 0$.

The *p-adic absolute value* satisfies the following properties:

(i) for $x \in \mathbb{Q}$, $|x|_p = 0$ is equivalent to $x = 0$
(ii) for $(x, y) \in \mathbb{Q}^2$, $|xy|_p = |x|_p |y|_p$
(iii) for $(x, y) \in \mathbb{Q}^2$, $|x + y|_p \leq \max\{|x|_p, |y|_p\}$.

Such an absolute value is called an *ultrametric absolute value*. It is a *nonarchimedean* absolute value: the set $\{|n|_p ; n \in \mathbb{Z}\}$ is bounded. An important property of an ultrametric absolute value $|\cdot|$ is the fact that $|x| < |y|$ implies $|x + y| = \max\{|x|, |y|\} = |y|$. This property will be used several times.

This p-adic absolute value defines a distance on $\mathbb{Q}$, hence a topology. The ball of radius p^{-r} (with $r \in \mathbb{Z}$) with $a \in \mathbb{Q}$ as its center:

$$\mathcal{D}(a, r) = \{x \in \mathbb{Q}\,;\; |x - a|_p \leq p^{-r}\} = \{x \in \mathbb{Q}\,;\; v_p(x - a) \geq r\}$$

is the set of rational numbers x such that the difference $x - a$ is divisible by p^r, i.e. such that $x - a$ is the product of p^r by a rational number with denominator not divisible by p. For $r \geq 1$, this means that the numerator of $x - a$ (written as a quotient of two coprime integers) is congruent to 0 modulo p^r.

The completion of $\mathbb{Q}$ for the p-adic valuation is *the field $\mathbb{Q}_p$ of p-adic numbers.* Each element x of $\mathbb{Q}_p$ can be written as

$$x = \frac{a_{-N}}{p^N} + \frac{a_{-N+1}}{p^{N-1}} + \cdots + a_0 + a_1 p + \cdots + a_n p^n + \cdots,$$

with $a_i \in \{0, \ldots, p-1\}$. Such a series is called the *Hensel's expansion* of x. For $x \neq 0$, the least $n \in \mathbb{Z}$ for which $a_n \neq 0$ is nothing else than $v_p(x)$.

Two absolute values on a field are said to be *equivalent* if they define the same topology on that field.

One can show (Ostrowski's Theorem, see for instance [K 1980] or [Neu 1999], Chap. 2 § 4) that any nontrivial absolute value on $\mathbb{Q}$ is equivalent to either a p-adic absolute value or to the usual absolute value on $\mathbb{Q}$.

If one fixes a nonzero rational number x and takes the product of all these absolute values of x, then something quite interesting occurs. A property known as the *product formula* holds:

$$|x| \prod_p |x|_p = 1 \quad \text{for all } x \in \mathbb{Q}^\times,$$

which can also be written additively:

$$\sum_p v_p(x) \log p = \log |x| \quad \text{for all } x \in \mathbb{Q}^\times.$$

The fact that this property holds in many common types of fields is of great importance in algebraic number theory as well as in the study of diophantine and transcendence problems. We shall return to the product formula later in this chapter.

3.1.2 Number Fields

Let α be an algebraic number. The image of the homomorphism $\mathbb{Q}[X] \longrightarrow \mathbb{C}$, which maps $f \in \mathbb{Q}[X]$ onto $f(\alpha)$, is the field $\mathbb{Q}(\alpha)$ generated by α over $\mathbb{Q}$. The kernel of the same homomorphism is a prime (hence maximal) ideal of $\mathbb{Q}[X]$, which has a uniquely defined monic generator. This generator f is called the *irreducible polynomial of α over $\mathbb{Q}$*. The degree of f is called *the degree* of the algebraic number α. Two algebraic numbers are called *conjugate* if they have the same irreducible polynomial over $\mathbb{Q}$.

Let a_0 be the least positive integer such that $g = a_0 f$ has integer coefficients. The product $g = a_0 f$, say

$$g(X) = a_0 X^d + \cdots + a_d \in \mathbb{Z}[X],$$

is the *minimal polynomial of α over $\mathbb{Z}$*. This polynomial g is irreducible in the factorial ring $\mathbb{Z}[X]$ (see [L 1993], Chap. 4 § 2), which means that g is irreducible in $\mathbb{Q}[X]$ and the rational integers $a_0, \ldots, a_d$ are relatively prime. The number α is called an *algebraic integer* if $a_0 = 1$, a *unit* if $a_0 = 1$ and $a_d = \pm 1$. The set of algebraic integers is a ring in the field $\overline{\mathbb{Q}}$, and the units are the invertible elements of this ring.

A *number field* is a subfield k of $\mathbb{C}$ which, considered as a vector space over $\mathbb{Q}$, is of finite dimension. We denote this dimension by $[k : \mathbb{Q}]$ and we call it the *degree* of k (over $\mathbb{Q}$). For instance, when γ is an algebraic number, then $k = \mathbb{Q}(\gamma)$ is a number field of degree $[\mathbb{Q}(\gamma) : \mathbb{Q}]$ equal to the degree of γ and such a γ is called a *generator* of the number field k.

When k is a number field, each $\gamma \in k$ is algebraic over $\mathbb{Q}$. On the other hand, using the fact that $[k_3 : k_2][k_2 : k_1] = [k_3 : k_1]$ when $k_1 \subset k_2 \subset k_3$ are finite extensions (see [L 1993], Chap. 5 § 1), it follows easily that for each number field k there exist $\alpha_1, \dots, \alpha_n$ in k such that $k = \mathbb{Q}(\alpha_1, \dots, \alpha_n)$.

Let k be a number field of degree d. If $k = \mathbb{Q}(\gamma)$ for some $\gamma \in k$, then there are exactly d distinct embeddings of k into $\mathbb{C}$. Indeed, if $\gamma_1, \dots, \gamma_d$ are the roots of f in $\mathbb{C}$ (these are *the conjugates of* γ), then the d embeddings of k into $\mathbb{C}$ are given by

$$\begin{array}{ccc} k & \longrightarrow & \mathbb{C} \\ \gamma & \longmapsto & \gamma_i \end{array}$$

$(1 \le i \le d)$. By induction one deduces that any number field k of degree d has exactly d embeddings into $\mathbb{C}$. Moreover, a number $\gamma \in k$ is a generator of k over $\mathbb{Q}$ if and only if the d images of γ under these embeddings are distinct. From this follows the *Theorem of the primitive element* (see Exercise 3.1): *for each number field k there exists an algebraic number $\gamma \in k$ such that $k = \mathbb{Q}(\gamma)$.*

We shall now study the set of absolute values of a number field. To do this, we have to study how an absolute value can be extended.

We can deal with the trivial absolute value (defined by $|0| = 0$ and $|x| = 1$ for $x \neq 0$) as follows: *if K/k is a finite extension, the unique extension to K of the trivial absolute value on k is the trivial absolute value on K.* Indeed, for $\alpha \in K^\times$, there exist a positive integer d and $a_0, \dots, a_d$ in k with $a_0 a_d \neq 0$ such that $a_0\alpha^d + a_1\alpha^{d-1} + \cdots + a_d = 0$. Since v is trivial on k, we have

$$|a_i\alpha^{d-i}| < \begin{cases} |a_0\alpha^d| & \text{for} \quad 1 \le i \le d \qquad \text{if } |\alpha| > 1, \\ |a_d| = 1 & \text{for} \quad 0 \le i \le d-1 \quad \text{if } |\alpha| < 1. \end{cases}$$

Notice as well that since v is trivial, it is ultrametric and thus if $|x| < |y|$ then $|x + y| = \max\{|x|, |y|\} = |y|$. Therefore, we conclude that $|\alpha| = 1$.

Let $|\cdot|$ be a nontrivial absolute value on a number field k. The restriction of this absolute value to $\mathbb{Q}$ is equivalent either to the usual absolute value on $\mathbb{Q}$ (in this case the absolute value is *Archimedean*), or else to a p-adic absolute value (in this case the absolute value is said to be *ultrametric*).

In each equivalence class v of nontrivial absolute values, we choose the representative $|\cdot|_v$ which is *normalized* by

$$\begin{cases} |x|_v = x & \text{if } x \in \mathbb{Q},\ x > 0, \text{ and } v \text{ is Archimedean}, \\ |p|_v = \dfrac{1}{p} & \text{if } v \text{ extends the } p\text{-adic valuation of } \mathbb{Q}. \end{cases}$$

We write $v \mid \infty$ if v is Archimedean, and $v \mid p$ if v extends the p-adic valuation. We denote by M_k (resp. M_k^∞) the set of normalized absolute values (resp. Archimedean normalized absolute values) of k. For $v \in M_k$, the completion of k at v will be denoted by k_v.

3.1.3 Archimedean Absolute Values over a Number Field

Let k be a number field, γ a generator of k over $\mathbb{Q}$, and f the irreducible polynomial of γ over $\mathbb{Q}$.

To each complex embedding $\sigma\colon k \longrightarrow \mathbb{C}$ we associate a normalized Archimedean absolute value v_σ defined by $|x|_{v_\sigma} = |\sigma(x)|$ for $x \in k$. Conversely, let v be a normalized Archimedean absolute value on k. The completion k_v of k is an extension of the completion $\mathbb{R}$ of $\mathbb{Q}$. We denote by γ_v the image of γ in k_v. Then $\mathbb{R}(\gamma_v)$ is a finite extension of $\mathbb{R}$ (because γ_v is a root of f), hence is either $\mathbb{R}$ or $\mathbb{C}$. We know which one it is by writing the decomposition of $f \in \mathbb{Q}[X]$ into irreducible factors in $\mathbb{R}[X]$: $f = f_1 \cdots f_r$, where $r = r_1 + r_2$, $f_1, \dots, f_{r_1}$ are of degree $d_1 = \dots = d_{r_1} = 1$, while $f_{r_1+1}, \dots, f_r$ are of degree $d_{r_1+1} = \dots = d_r = 2$. If γ_v is root of one of the f_i's of degree 1, then $\mathbb{R}(\gamma_v) = \mathbb{R}$, while if γ_v is root of one of the f_i's of degree 2, then $\mathbb{R}(\gamma_v) = \mathbb{C}$. In any case, we have $k_v = \mathbb{R}(\gamma_v)$, since $\mathbb{R}$ and $\mathbb{C}$ are complete, and we get a complex embedding σ_v of k into $\mathbb{C}$ such that $v_{\sigma_v} = v$. Hence the mapping $\sigma \mapsto v_\sigma$ is surjective.

If $\sigma(\gamma) \in \mathbb{R}$, then $\sigma(k) \subset \mathbb{R}$ and $k_v = \mathbb{R}$. The embedding σ and the absolute value v are called *real*. If $\sigma(\gamma) \notin \mathbb{R}$, then $k_v = \mathbb{C}$. Here the embedding σ and the absolute value v are called *complex*. We denote by d_v the degree $[k_v : \mathbb{R}]$:

$$d_v = \begin{cases} 1 & \text{if } v \text{ is real,} \\ 2 & \text{if } v \text{ is complex.} \end{cases}$$

Let σ_1 and σ_2 be two distinct embeddings of k into $\mathbb{C}$ which give rise to the same Archimedean absolute value v. For any $\alpha \in k$ we have

$$|\sigma_1(\alpha)|_v = |\sigma_2(\alpha)|_v \quad \text{and} \quad |1 - \sigma_1(\alpha)|_v = |1 - \sigma_2(\alpha)|_v,$$

which implies that the complex numbers $\sigma_1(\alpha)$ and $\sigma_2(\alpha)$ are conjugate. Therefore, to a real absolute value v corresponds one and only one real embedding of k, while to a complex absolute value v correspond two (complex conjugate) embeddings of k into $\mathbb{C}$. We deduce that the number of elements in M_k^∞ (i.e. the number of nonequivalent Archimedean absolute values of k) is $r = r_1 + r_2$, where (as before) r_1 is the number of real roots of f, while r_2 is the number of pairs of conjugate complex roots of f, with $d = r_1 + 2r_2$. We can index the irreducible factors of f over $\mathbb{R}$ by $v \in M_k^\infty$ (instead of $1 \le i \le r$):

$$f = \prod_{v \in M_k^\infty} f_v \quad \text{and} \quad d = \sum_{v \in M_k^\infty} d_v \quad \text{with} \quad d_v = \deg f_v.$$

The d-tuple $(|\gamma_1|,\dots,|\gamma_d|)$ consists of the elements $|\gamma|_v$ $(v\in M_k^\infty)$, where each $|\gamma|_v$ is repeated d_v times. For instance, supposing that the minimal polynomial of γ over $\mathbb{Q}$ is $f(X)=a_0X^d+\cdots+a_d$, we get

$$\prod_{v\in M_k^\infty}|\gamma|_v^{d_v}=\prod_{i=1}^d|\gamma_i|=\left|\frac{a_d}{a_0}\right|$$

and

$$\prod_{v\in M_k^\infty}\max\{1,|\gamma|_v\}^{d_v}=\prod_{i=1}^d\max\{1,|\gamma_i|\}.$$

3.1.4 Ultrametric Absolute Values over a Number Field

Let p be a prime number. The absolute value $|\cdot|_p$ on $\mathbb{Q}_p$ has a unique extension to any finite extension K of $\mathbb{Q}_p$. This is due to the fact that $\mathbb{Q}_p$ is complete (see [L 1993], Chap. 12, Prop. 2.5 or [Neu 1999], Chap. 2 Th. 4.8). This extension is given as follows. For $\alpha\in K$, let $\mathrm{N}_{K/\mathbb{Q}_p}(\alpha)$ denote the norm of the $\mathbb{Q}_p$-endomorphism of K which maps x onto αx. If n is the degree of K over $\mathbb{Q}_p$, the extension $|\cdot|_p$ of the p-adic absolute value of $\mathbb{Q}_p$ to K is defined by

$$|\alpha|_p=|\mathrm{N}_{K/\mathbb{Q}_p}(\alpha)|_p^{1/n}.$$

Denote by $\overline{\mathbb{Q}}_p$ the algebraic closure of $\mathbb{Q}_p$, equipped with this absolute value. Then $\overline{\mathbb{Q}}_p$ is not complete (which makes a difference with the Archimedean situation). This is not a serious drawback, and we could take $\overline{\mathbb{Q}}_p$ as the analog of the field of complex number. But we shall prefer to denote by $\mathbb{C}_p$ the completion of $\overline{\mathbb{Q}}_p$ for the absolute value $|\cdot|_p$. This is a complete field in which $\overline{\mathbb{Q}}_p$ is dense, and moreover $\mathbb{C}_p$ is algebraically closed (we shall need only that it contains an algebraic closure of $\mathbb{Q}_p$, hence it also contains an algebraic closure of $\mathbb{Q}$).

Again let $k=\mathbb{Q}(\gamma)$ be a number field of degree d and f the irreducible polynomial of γ over $\mathbb{Q}$. Denote by $\gamma_1^{(p)},\dots,\gamma_d^{(p)}$ the roots of f in $\mathbb{C}_p$. There are d distinct embeddings of k into $\mathbb{C}_p$ (each embedding maps a root of f onto another root of the same). These embeddings are given by

$$\begin{array}{ccc} k & \longrightarrow & \mathbb{C}_p\\ \gamma & \longmapsto & \gamma_i^{(p)}\end{array}$$

$(1\le i\le d)$. To each such embedding $\sigma\colon k\longrightarrow\mathbb{C}_p$ we associate an ultrametric absolute value $v_\sigma\mid p$ defined by $|x|_{v_\sigma}=|\sigma(x)|_p$.

Let v be an absolute value on k which extends the p-adic absolute value of $\mathbb{Q}$. We view the completion k_v of k as an extension of $\mathbb{Q}_p$ and denote by γ_v the image of γ into k_v. Then $\mathbb{Q}_p(\gamma_v)$ is a finite extension of $\mathbb{Q}_p$. But we can say more about the degree of this extension. Consider the decomposition of $f\in\mathbb{Q}[X]$ into irreducible factors [6] in $\mathbb{Q}_p[X]$: $f = f_1 \cdots f_r$ (notice that the number r of irreducible factors varies with p). Since γ_v is a root of f into $\mathbb{C}_p$, there is a unique i, $1 \le i \le r$, such that γ_v is a root of f_i. Therefore f_i has a root in the field $\mathbb{Q}_p(\gamma_v)$, which is an extension of $\mathbb{Q}_p$ of degree $d_v = \deg(f_i)$, and $k_v = \mathbb{Q}_p(\gamma_v)$. This number d_v is called *the local degree* at v. From this it follows that k_v is (isomorphic to) a subfield of $\mathbb{C}_p$, and we get an embedding σ_v of k into $\mathbb{C}_p$ such that $v_{\sigma_v} = v$. Hence the mapping $\sigma \mapsto v_\sigma$ is surjective.

Let σ_1 and σ_2 be two distinct embeddings of k into $\mathbb{C}_p$. They give rise to the same ultrametric absolute value v if and only if $\sigma_1(\gamma)$ and $\sigma_2(\gamma)$ are conjugate over $\mathbb{Q}_p$, which means that they are roots of the same irreducible factor f_i (cf. [L 1993], Chap. 12, Prop. 3.2 or [Neu 1999], Chap. 2, Prop. 8.2). Therefore the number of distinct embeddings σ into $\mathbb{C}_p$ associated to a given absolute value $v \mid p$ is the local degree $d_v = [k_v : \mathbb{Q}_p]$ of v, and the number of elements $v \in M_k$ with $v \mid p$ is the number r of irreducible factors of f over $\mathbb{Q}_p$. This enables us to write

$$f = \prod_{v \in M_k, v \mid p} f_v \quad \text{and} \quad d = \sum_{v \in M_k, v \mid p} d_v \quad \text{with} \quad d_v = \deg f_v.$$

The d-tuple $\left(|\gamma_1^{(p)}|_p, \ldots, |\gamma_d^{(p)}|_p\right)$ consists of the elements $|\gamma|_v$ $(v \in M_k, v \mid p)$, where each $|\gamma|_v$ is repeated d_v times. For instance

$$\prod_{v \in M_k, v \mid p} |\gamma|_v^{d_v} = \prod_{i=1}^{d} |\gamma_i^{(p)}|_p = \left|\frac{a_d}{a_0}\right|_p$$

and

$$\prod_{v \in M_k, v \mid p} \max\{1, |\gamma|_v\}^{d_v} = \prod_{i=1}^{d} \max\{1, |\gamma_i^{(p)}|_p\}.$$

The next lemma shows that this last number is $1/|a_0|_p$.

Lemma 3.1. *Let p be a prime number. Let*

$$f(X) = a_0 X^d + a_1 X^{d-1} + \cdots + a_d$$

be a polynomial in $\mathbb{Z}[X]$ with degree d and $\gcd(a_0, \ldots, a_d) = 1$. *Denote the roots of f in $\mathbb{C}_p$ by $\alpha_1, \ldots, \alpha_d$:*

$$f(X) = a_0 \prod_{i=1}^{d} (X - \alpha_i).$$

Then

$$|a_0|_p \prod_{i=1}^{d} \max\{1, |\alpha_i|_p\} = 1.$$

It follows from Lemma 3.1 that for each subset I of $\{1, \ldots, d\}$, the number

[6] Since f is irreducible in $\mathbb{Q}[X]$ the polynomials $f_1, \ldots, f_r$ in $\mathbb{Q}_p[X]$ are pairwise distinct.

$$a_0 \prod_{i \in I} \alpha_i$$

is an algebraic integer.

Proof. We may assume $|\alpha_1|_p \leq \cdots \leq |\alpha_d|_p$. Since the numbers a_i are relatively prime, $\max\{|a_0|_p, \ldots, |a_d|_p\} = 1$. Let us write a_i/a_0 as a symmetric function of the α_i:

$$\frac{a_i}{a_0} = (-1)^i \sum_{1 \leq s_1 < \cdots < s_i \leq d} \alpha_{s_1} \cdots \alpha_{s_i} \qquad (1 \leq i \leq d).$$

If $|\alpha_i|_p \leq 1$ for all $i = 1, \ldots, d$, then $|a_i|_p \leq |a_0|_p$ and $\max\{|a_0|_p, \ldots, |a_d|_p\} = |a_0|_p = 1$, which gives the desired result. Otherwise let j $(1 \leq j \leq d)$, be such that

$$|\alpha_1|_p \leq \cdots \leq |\alpha_{j-1}|_p \leq 1 < |\alpha_j|_p \leq \cdots \leq |\alpha_d|_p.$$

Then, using the main property of ultrametric absolute values, we obtain

$$\max\left\{\left|\frac{a_i}{a_0}\right|_p ; 1 \leq i \leq d\right\} = \left|\frac{a_{d-j+1}}{a_0}\right|_p = |\alpha_j \cdots \alpha_d|_p = \prod_{i=1}^{d} \max\{1, |\alpha_i|_p\},$$

hence

$$\max\{|a_1|_p, \ldots, |a_d|_p\} = |a_0|_p \prod_{i=1}^{d} \max\{1, |\alpha_i|_p\}.$$

Since this number is at least $|a_0|_p$, we deduce

$$\max\{|a_0|_p, \ldots, |a_d|_p\} = |a_0|_p \prod_{i=1}^{d} \max\{1, |\alpha_i|_p\},$$

hence the result. □

We have already defined an algebraic integer as an algebraic number whose irreducible polynomial over $\mathbb{Q}$ has coefficients in $\mathbb{Z}$ (which means that the minimal polynomial of α over $\mathbb{Z}$ is monic). From Lemma 3.1 we deduce at once:

Corollary 3.2. *Let α be an algebraic number. The following conditions are equivalent:*
(i) *α is an algebraic integer.*
(ii) *There exists a monic polynomial in $\mathbb{Z}[X]$ which vanishes at α.*
(iii) *For each number field k containing α, and for each ultrametric absolute value v of k, we have $|\alpha|_v \leq 1$.*
(iv) *There exists a number field k containing α such that, for each ultrametric absolute value v of k, we have $|\alpha|_v \leq 1$.*

Remark 1. Let α be an algebraic number with conjugates $\alpha_1, \ldots, \alpha_d$ (with $d = [\mathbb{Q}(\alpha) \colon \mathbb{Q}]$ and, say, $\alpha_1 = \alpha$). If $D \in \mathbb{Z}$ is such that

$$\left| D \prod_{i \in I} \alpha_i \right|_v \leq 1$$

for all subsets I of $\{1, \dots, d\}$ and all ultrametric absolute values v, then

$$|D|_p \prod_{i=1}^{d} \max\{1, |\alpha_i|_p\} \leq 1$$

for each prime number p and each embedding of $\mathbb{Q}(\alpha_1, \dots, \alpha_d)$ into $\mathbb{C}_p$. Hence $|D|_p \leq |a_0|_p$ for each p, which means that a_0 divides D. This shows that a_0 is the positive generator of the ideal of $D \in \mathbb{Z}$ for which, for any subset $\{i_1, \dots, i_t\}$ of $\{1, \dots, d\}$, the number $D\alpha_{i_1} \cdots \alpha_{i_t}$ is an algebraic integer.

Remark 2. (M. Laurent). Even if we do not need it, the relation between valuations and prime ideals in a number field is worth mentioning. We just quote one result involving ideals. Let α be a nonzero algebraic number. The ring of integers $\mathcal{O}_k$ of the number field $k = \mathbb{Q}(\alpha)$ is a Dedekind domain. The principal fractional ideal (α) can be written $\mathcal{B}/\mathcal{C}$, where $\mathcal{B}$ and $\mathcal{C}$ are nonzero relatively prime integral ideals of k. Let us show that

$$\mathcal{C} = \{\gamma \in \mathcal{O}_k \,;\, \gamma\alpha \in \mathcal{O}_k\} \quad \text{and} \quad \mathrm{N}\mathcal{C} = a_0,$$

where $\mathrm{N}\mathcal{C}$ is the absolute norm of the ideal $\mathcal{C}$.

We write

$$(\alpha) = \prod_{\mathcal{P}} \mathcal{P}^{m_{\mathcal{P}}(\alpha)},$$

where $\mathcal{P}$ runs over the set of prime ideals of $\mathcal{O}_k$. Hence

$$\mathcal{B} = \prod_{\mathcal{P}} \mathcal{P}^{\max\{0, m_{\mathcal{P}}(\alpha)\}}, \quad \mathcal{C} = \prod_{\mathcal{P}} \mathcal{P}^{\max\{0, -m_{\mathcal{P}}(\alpha)\}}.$$

Recall that the *absolute norm* of $\mathcal{P}$ is defined by $\mathrm{N}\mathcal{P} = \mathrm{Card}(\mathcal{O}_k/\mathcal{P})$. If $v \in M_k$ is the ultrametric absolute value associated to $\mathcal{P}$ and d_v the local degree, then

$$|\alpha|_v^{d_v} = \mathrm{N}\mathcal{P}^{-m_{\mathcal{P}}(\alpha)}$$

(in view of the product formula below, the product of the left hand side for all ultrametric v, as well as the product of the left hand side for all prime ideals $\mathcal{P}$, is $1/|\mathrm{N}(\alpha)|$, where $\mathrm{N}(\alpha)$ is the absolute norm of α). Indeed, for $\gamma \in \mathcal{O}_k$ and $m \geq 1$, we have

$$\gamma \in \mathcal{P}^m \iff |\gamma|_v^{d_v} \leq \mathrm{N}\mathcal{P}^{-m}.$$

Using Corollary 3.2, we conclude

$$\begin{aligned} \mathcal{C} &= \{\gamma \in \mathcal{O}_k \,;\, |\gamma|_v \leq |\alpha|_v^{-1} \text{ for all ultrametric } v \in M_k\} \\ &= \{\gamma \in \mathcal{O}_k \,;\, \gamma\alpha \in \mathcal{O}_k\} \end{aligned}$$

and

$$\mathcal{B} = \{\gamma\alpha \,;\, \gamma \in \mathcal{C}\}.$$

Further, by the multiplicativity property of N, we deduce from Lemma 3.1:

$$\mathrm{N}\mathcal{C} = \prod_{\mathcal{P}} \mathrm{N}\mathcal{P}^{\max\{0, -m_{\mathcal{P}}(\alpha)\}} = \prod_{v \text{ ultrametric}} \max\{1, |\alpha|_v^{d_v}\} = a_0.$$

3.1.5 The Product Formula

Again let k be a number field of degree d. Let $\alpha \in k$ have minimal polynomial $a_0X^d + \cdots + a_d$ over $\mathbb{Z}$. If v is an ultrametric absolute value of k, say $v \mid p$, with $|\alpha|_v > 1$, then, by Lemma 3.1, we deduce that p divides a_0. On the other hand, if $\alpha \neq 0$, then the minimal polynomial of α^{-1} is $a_dX^d + \cdots + a_0$. Hence, if $|\alpha|_v < 1$, then p divides a_d. As a consequence, for each $\alpha \neq 0$ in k, the set of v in M_k for which $|\alpha|_v \neq 1$ is finite.

The *product formula* reads

$$\prod_{v \in M_k} |\alpha|_v^{d_v} = 1 \quad \text{for} \quad \alpha \in k,\ \alpha \neq 0.$$

We already know this formula holds in the rational case $k = \mathbb{Q}$. The general case readily follows by considering a_d/a_0, which is the *absolute norm* of α, namely $N_{k/\mathbb{Q}}(\alpha)$ with $k = \mathbb{Q}(\alpha)$.

We shall need a generalization of the relations $d = \sum_{v \in M_k^\infty} d_v = \sum_{v|p} d_v$ when the basis field $\mathbb{Q}$ is replaced by a finite extension. For this purpose it will be convenient to write $d_v = d_v(k)$. Let K be a finite extension of k. One can define a map from M_K onto M_k by mapping w onto the restriction v of w on k, in which case one writes $w \mid v$. We claim that for each $v \in M_k$,

$$\sum_{w|v} d_w(K) = [K : k] d_v(k)$$

(see [L 1993], Chap. 12 Prop. 3.3 and [Neu 1999], Chap. 2 § 8). Indeed, for $\gamma \in K$ such that $K = \mathbb{Q}(\gamma)$, we also have $K = k(\gamma)$, and the irreducible polynomial g of γ over k (which is of degree $[K : k]$) can be decomposed into irreducible factors in $k_v[X]$, say $g = \prod_{w|v} g_w$, where g_w is of degree $[K_w : k_v]$. Therefore, for each $v \in M_k$,

$$\sum_{w|v} [K_w : k_v] = [K : k].$$

Since $d_w(K) = [K_w : k_v] d_v(k)$, our claim follows.

An alternate proof of this relation (suggested by Dong Ping Ping) is as follows. For $\alpha \in k$ and $\gamma \in K$ such that $k = \mathbb{Q}(\alpha)$ and $K = \mathbb{Q}(\gamma)$, there exists a polynomial $Q \in \mathbb{Q}[X]$ such that $\alpha = Q(\gamma)$. Let f be the minimal polynomial of γ over $\mathbb{Q}$ and denote by $\alpha_1, \ldots, \alpha_{d_v(k)}$ the conjugates of α (in $\mathbb{C}$ if v is Archimedean, in $\mathbb{C}_p$ if v is ultrametric) which induce the absolute value v on k. Among the $[K : \mathbb{Q}]$ roots of f in $\mathbb{C}$ (resp. in $\mathbb{C}_p$), there are exactly $[K : k] d_v(k)$ roots whose images by Q belong to the set $\{\alpha_1, \ldots, \alpha_{d_v(k)}\}$. These roots are all the roots of f in $\mathbb{C}$ (resp. in $\mathbb{C}_p$) which induce the absolute values w in K over v. Therefore there are precisely $\sum_{w|v} d_w(K)$ such roots.

3.2 The Absolute Logarithmic Height (Weil)

Let k be a number field. For $\alpha \in k$ we define

$$H_k(\alpha) = \prod_{v \in M_k} \max\{1, |\alpha|_v\}^{d_v}.$$

This is a finite product (all but finitely many factors in the right hand side are equal to 1). Let K be a finite extension of k. For $\alpha \in k$ we obtain

$$\begin{aligned} H_K(\alpha) &= \prod_{w \in M_K} \max\{1, |\alpha|_w\}^{d_w(K)} \\ &= \prod_{v \in M_k} \max\{1, |\alpha|_v\}^{\sum_{w|v} d_w(K)} \\ &= H_k(\alpha)^{[K:k]}. \end{aligned}$$

This shows that the number $H_k(\alpha)^{1/[k:\mathbb{Q}]}$ does not depend on the number field k containing α. The logarithm of this number will play an important role. When α is an algebraic number and K a number field which contains α, we define

$$\mathrm{h}(\alpha) = \frac{1}{[K:\mathbb{Q}]} \log H_K(\alpha) = \frac{1}{[K:\mathbb{Q}]} \sum_{v \in M_K} D_v \log \max\{1, |\alpha|_v\},$$

where D_v denotes the local degree at $v \in M_K$. This is the (Weil) *absolute logarithmic height* of the number α. It does not depend on the choice of the number field K containing α, but only on α.

Example. For two rational integers a, b which are relatively prime,

$$\mathrm{h}\left(\frac{a}{b}\right) = \log \max\{|a|, |b|\}.$$

Property 3.3. *For algebraic numbers α_1 and α_2,*

$$\mathrm{h}(\alpha_1\alpha_2) \le \mathrm{h}(\alpha_1) + \mathrm{h}(\alpha_2) \tag{3.4}$$

and

$$\mathrm{h}(\alpha_1 + \alpha_2) \le \log 2 + \mathrm{h}(\alpha_1) + \mathrm{h}(\alpha_2). \tag{3.5}$$

Moreover, for any algebraic number $\alpha \neq 0$ and for any $n \in \mathbb{Z}$,

$$\mathrm{h}(\alpha^n) = |n|\mathrm{h}(\alpha). \tag{3.6}$$

Proof. The upper bound (3.4) is a consequence of the estimate

$$\max\{1, xy\} \le \max\{1, x\} \max\{1, y\} \quad \text{for all } x \ge 0,\ y \ge 0,$$

while (3.5) follows from the inequality

$$\max\{1, x+y\} \le 2\max\{1,x\}\max\{1,y\} \quad \text{for all } x \ge 0,\ y \ge 0.$$

Since

$$\max\{1, x^n\} = \max\{1,x\}^n \quad \text{for all } x > 0,\ n \in \mathbb{Z},\ n \ge 0,$$

property (3.6) reduces to $\mathrm{h}(\alpha) = \mathrm{h}(1/\alpha)$ for $\alpha \neq 0$, which follows from the product formula, since $\max\{1,x\} = x\max\{1, 1/x\}$ for $x > 0$. □

Remark. The term $\log 2$ in the right hand side of the estimate (3.5) cannot be replaced by a smaller absolute constant, as shown by the following example: $\alpha_1 = q/(q-1)$, $\alpha_2 = q/(q+1)$ with q an even integer. Another example is $\alpha_1 = \alpha_2 = 1$.

The next lemma provides an upper bound for the absolute logarithmic height of an algebraic number which is given as the value of a polynomial in algebraic numbers $\gamma_1, \ldots, \gamma_t$.

When $f \in \mathbb{C}[\mathrm{X}_1, \ldots, \mathrm{X}_t]$ is a polynomial in t variables, with complex coefficients, we denote by $\mathrm{L}(f)$ its *length*, which is the sum of the modulus of its complex coefficients. The length is very convenient because it satisfies the inequalities

$$\mathrm{L}(f+g) \le \mathrm{L}(f) + \mathrm{L}(g) \quad \text{and} \quad \mathrm{L}(fg) \le \mathrm{L}(f)\mathrm{L}(g)$$

which will be used repeatedly in the transcendence proofs. Indeed, if we write

$$f = \sum_{\underline{\lambda}} p_{\underline{\lambda}} \underline{\mathrm{X}}^{\underline{\lambda}} \quad \text{and} \quad g = \sum_{\underline{\mu}} q_{\underline{\mu}} \underline{\mathrm{X}}^{\underline{\mu}},$$

where $p_{\underline{\lambda}}$ and $q_{\underline{\mu}}$ are complex numbers, $\underline{\lambda} = (\lambda_i)$ and $\underline{\mu} = (\mu_i)$ run over finite subsets of $\mathbb{N}^t$, while $\underline{\mathrm{X}}^{\underline{\lambda}}$ stands for $\prod_{i=1}^t \mathrm{X}_i^{\lambda_i}$, then the length of

$$fg = \sum_{\underline{\nu}} \sum_{\underline{\lambda}+\underline{\mu}=\underline{\nu}} p_{\underline{\lambda}} q_{\underline{\mu}} \underline{\mathrm{X}}^{\underline{\nu}}$$

satisfies

$$\mathrm{L}(fg) = \sum_{\underline{\nu}} \Big| \sum_{\underline{\lambda}+\underline{\mu}=\underline{\nu}} p_{\underline{\lambda}} q_{\underline{\mu}} \Big| \le \sum_{\underline{\nu}} \sum_{\underline{\lambda}+\underline{\mu}=\underline{\nu}} |p_{\underline{\lambda}} q_{\underline{\mu}}| = \mathrm{L}(f)\mathrm{L}(g).$$

We shall prove (as a consequence of Lemma 3.8 below) the following estimate:

Lemma 3.7. *Let $f \in \mathbb{Z}[\mathrm{X}_1, \ldots, \mathrm{X}_t]$ be a nonzero polynomial in t variables with rational integer coefficients. Let $\gamma_1, \ldots, \gamma_t$ be algebraic numbers. Then*

$$\mathrm{h}\big(f(\gamma_1, \ldots, \gamma_t)\big) \le \log \mathrm{L}(f) + \sum_{i=1}^t (\deg_{\mathrm{X}_i} f)\mathrm{h}(\gamma_i).$$

Applying Lemma 3.7 with $f(\mathrm{X}_1, \ldots, \mathrm{X}_n) = \mathrm{X}_1 + \cdots + \mathrm{X}_n$, one can deduce a generalization of (3.5):

$$\mathrm{h}(\alpha_1 + \cdots + \alpha_n) \le \log n + \mathrm{h}(\alpha_1) + \cdots + \mathrm{h}(\alpha_n).$$

When p_1/q_1 and p_2/q_2 are two rational numbers with $(p_1, q_1) = (p_2, q_2) = 1$ and $q_i > 0$, then (3.5) (as well as Lemma 3.7) yields

$$\mathrm{h}\left(\frac{p_1}{q_1} + \frac{p_2}{q_2}\right) \le \log 2 + \log \max\{|p_1|, q_1\} + \log \max\{|p_2|, q_2\}.$$

However, it is sometimes more efficient to write $p_1/q_1 = a/c$ and $p_2/q_2 = b/c$ with $\gcd(a, b, c) = 1$ and $c > 0$:

$$\begin{aligned}\mathrm{h}\left(\frac{a}{c} + \frac{b}{c}\right) &\le \log \max\{|a+b|, c\}\\ &\le \log 2 + \log \max\{|a|, |b|, c\}.\end{aligned}$$

This example suggests a refinement of Lemma 3.7, using a notion of simultaneous height for several numbers. Let K be a number field of degree D. Let $\gamma_0, \ldots, \gamma_\nu$ and λ be elements of K with $(\gamma_0, \ldots, \gamma_\nu) \neq (0, \ldots, 0)$ and $\lambda \neq 0$. From the product formula, it follows that the number

$$\frac{1}{D} \sum_{v \in M_K} D_v \log \max\{|\gamma_0|_v, \ldots, |\gamma_\nu|_v\},$$

which is attached to the $(\nu+1)$-tuple $(\gamma_0, \ldots, \gamma_\nu) \in K^{\nu+1}$, is the same as the number

$$\frac{1}{D} \sum_{v \in M_K} D_v \log \max\{|\lambda\gamma_0|_v, \ldots, |\lambda\gamma_\nu|_v\},$$

which is attached to the $(\nu+1)$-tuple $(\lambda\gamma_0, \ldots, \lambda\gamma_\nu) \in K^{\nu+1}$. Therefore this number, which we will denote by $\mathrm{h}(\gamma_0 : \cdots : \gamma_\nu)$, depends only on the class $(\gamma_0 : \cdots : \gamma_\nu)$ of $(\gamma_0, \ldots, \gamma_\nu)$ in the projective space $\mathbb{P}_\nu(K)$. For instance $\mathrm{h}(\alpha) = \mathrm{h}(1 : \alpha)$.

Lemma 3.8. *Let K be a number field and $\nu_1, \ldots, \nu_\ell$ be positive integers. For $1 \le i \le \ell$, let $\gamma_{i1}, \ldots, \gamma_{i\nu_i}$ be elements of K and denote by $\underline{\gamma}$ the point $(\gamma_{ij})_{1 \le j \le \nu_i, 1 \le i \le \ell}$ in $K^{\nu_1 + \cdots + \nu_\ell}$. Further, let f be a nonzero polynomial in $\nu_1 + \cdots + \nu_\ell$ variables, with coefficients in $\mathbb{Z}$, of total degree at most N_i with respect to the ν_i variables corresponding to $\gamma_{i1}, \ldots, \gamma_{i\nu_i}$. Then*

$$\mathrm{h}\big(f(\underline{\gamma})\big) \le \log \mathrm{L}(f) + \sum_{i=1}^{\ell} N_i \mathrm{h}(1 : \gamma_{i1} : \cdots : \gamma_{i\nu_i}).$$

Recall that $\mathrm{L}(f)$ denotes the length of f (sum of the absolute values of the coefficients). We deduce Lemma 3.7 from Lemma 3.8 by taking $\nu_i = 1$ for $1 \le i \le \ell$.

Proof. Write

$$f = \sum_{\underline{\lambda}} p_{\underline{\lambda}} \prod_{i=1}^{\ell} \prod_{j=1}^{\nu_i} X_{ij}^{\lambda_{ij}},$$

where $p_{\underline{\lambda}}$ are rational integers and $\underline{\lambda} = (\lambda_{ij})$ runs over a finite subset of $\mathbb{N}^{\nu_1+\cdots+\nu_\ell}$. Let v be an absolute value of k. If v is ultrametric, then

$$\begin{aligned} \log \max\{1, |f(\underline{\gamma})|_v\} &\le \log \max \left\{1, \max_{\underline{\lambda}} \prod_{i=1}^{\ell} \prod_{j=1}^{\nu_i} |\gamma_{ij}|_v^{\lambda_{ij}}\right\} \\ &\le \sum_{i=1}^{\ell} N_i \log \max\{1, |\gamma_{i1}|_v, \ldots, |\gamma_{i\nu_i}|_v\}. \end{aligned}$$

If v is Archimedean, then

$$\begin{aligned} \log \max\{1, |f(\underline{\gamma})|_v\} &\le \log \mathrm{L}(f) + \log \max \left\{1, \max_{\underline{\lambda}} \prod_{i=1}^{\ell} \prod_{j=1}^{\nu_i} |\gamma_{ij}|_v^{\lambda_{ij}}\right\} \\ &\le \log \mathrm{L}(f) + \sum_{i=1}^{\ell} N_i \log \max\{1, |\gamma_{i1}|_v, \ldots, |\gamma_{i\nu_i}|_v\}. \end{aligned}$$

Using the relation $\sum_{v \in M_k^\infty} D_v = D$, we easily deduce the conclusion. □

3.3 Mahler's Measure

Lemma 3.9. *Let $f \in \mathbb{C}[X]$ be a nonzero polynomial of degree d:*

$$f(X) = a_0 X^d + a_1 X^{d-1} + \cdots + a_{d-1} X + a_d = a_0 \prod_{i=1}^{d} (X - \alpha_i).$$

Then

$$|a_0| \prod_{i=1}^{d} \max\{1, |\alpha_i|\} = \exp\left(\int_0^1 \log |f(e^{2i\pi t})| dt\right).$$

Proof. This is a special case of Jensen's formula for analytic functions. Since both sides of the conclusion of Lemma 3.9 are multiplicative functions of f, it is sufficient to consider the case where f is either a_0 or else $X - \alpha$. In the first case the left hand side is $|a_0|$ and the desired equality plainly holds. In the latter case, the left hand side is $\max\{1, |\alpha|\}$. Therefore Lemma 3.9 is equivalent to the fact that, for any complex number α,

$$\int_0^1 \log |e^{2i\pi t} - \alpha| dt = \log \max\{1, |\alpha|\}.$$

(See for instance [M 1976], pp. 5–6). □

Under the notation of Lemma 3.9, we define *Mahler's measure* of f by

$$\mathrm{M}(f) = |a_0| \prod_{i=1}^{d} \max\{1, |\alpha_i|\}.$$

This is a multiplicative function:

$$\mathrm{M}(f_1 f_2) = \mathrm{M}(f_1)\mathrm{M}(f_2)$$

for f_1 and f_2 in $\mathbb{C}[X]$, a fact which follows immediately from the definition of M.

When α is an algebraic number with minimal polynomial $f \in \mathbb{Z}[X]$ over $\mathbb{Z}$, we define its *Mahler's measure* by $\mathrm{M}(\alpha) = \mathrm{M}(f)$.

Lemma 3.10. *Let α be an algebraic complex number of degree d. Then*

$$\mathrm{h}(\alpha) = \frac{1}{d} \log \mathrm{M}(\alpha).$$

Proof. Denote, as before, by $a_0 > 0$ the leading coefficient of the minimal polynomial of α, by k the number field $\mathbb{Q}(\alpha)$, and, for $v \in M_k$, by d_v the local degree at v. From the definition of $\mathrm{M}(\alpha)$ follows

$$\mathrm{M}(\alpha) = a_0 \prod_{v \in M_k^\infty} \max\{1, |\alpha|_v\}^{d_v}.$$

In Lemma 3.1 we have proved

$$|a_0|_p^{-1} = \prod_{v|p} \max\{1, |\alpha|_v\}^{d_v}.$$

Therefore the product formula

$$a_0 = \prod_p |a_0|_p^{-1}$$

implies

$$a_0 = \prod_{v \notin M_k^\infty} \max\{1, |\alpha|_v\}^{d_v},$$

which provides the desired conclusion. □

3.4 Usual Height and Size

There are several other notions of heights or size (in French: *taille*) for algebraic numbers. We shall give a few examples (see also the appendix to this Chap. 3). One main property of a height is that the set of algebraic numbers of bounded height and degree should be finite. For instance, for any $\nu \geq 1$, $D \geq 1$ and $h \geq 1$, the set of projective points $\underline{\gamma} \in \mathbb{P}_\nu(\overline{\mathbb{Q}})$ with $\mathrm{h}(\underline{\gamma}) \leq h$, and for which there exists a system of projective coordinates $\underline{\gamma} = (\gamma_0 : \cdots : \gamma_\nu)$ satisfying

$$\left[\mathbb{Q}(\gamma_0, \ldots, \gamma_\nu) : \mathbb{Q}\right] \leq D,$$

is a finite subset of $\mathbb{P}_\nu(\overline{\mathbb{Q}})$. This is *a completely elementary result due to Northcott* ([L 1991], Chap. II, Th. 2.2; see also [Sc 1991], Lemma 7C). To give estimates for the number of elements of such sets is also an interesting question (see the reference to Schanuel's work in [L 1983], [L 1991] and [Sc 1999], Th. 3B).

The *usual height* $\mathrm{H}(f)$ of a polynomial $f(X) = a_0X^d + a_1X^{d-1} + \cdots + a_d \in \mathbb{Z}[X]$ is the maximum of the complex modulus of its coefficients:

$$\mathrm{H}(f) = \max\{|a_0|, \ldots, |a_d|\}.$$

The *usual height* $\mathrm{H}(\alpha)$ of an algebraic number α is the usual height of its minimal polynomial over $\mathbb{Z}$.

The *house* of an algebraic number is the maximum of the modulus of its conjugates in $\mathbb{C}$:

$$\overline{|\alpha|} = \max\{|\alpha_1|, \ldots, |\alpha_d|\}$$

when the minimal polynomial of α is written in $\mathbb{C}[X]$ as

$$f(X) = a_0X^d + \cdots + a_d = a_0\prod_{i=1}^{d}(X - \alpha_i).$$

The *denominator* $\mathrm{den}(\alpha)$ of α is the positive generator of the ideal of $D \in \mathbb{Z}$ for which $D\alpha$ is an algebraic integer. It is a divisor of a_0.

Among several notions of *size*, one of the most frequently used is

$$s(\alpha) = \log\max\{\mathrm{den}(\alpha)\,;\,\overline{|\alpha|}\}.$$

Lemma 3.11. *For $\alpha \in \overline{\mathbb{Q}}$ of degree d, we have*

$$\frac{1}{d}\log\mathrm{H}(\alpha) - \log 2 \leq \mathrm{h}(\alpha) \leq \frac{1}{d}\log\mathrm{H}(\alpha) + \frac{1}{2d}\log(d+1)$$

and

$$\frac{1}{d}s(\alpha) \leq \mathrm{h}(\alpha) \leq \log\mathrm{den}(\alpha) + \log\max\left\{1, \overline{|\alpha|}\right\} \leq 2s(\alpha).$$

Proof. The first part of the conclusion can be written

$$2^{-d}\mathrm{H}(\alpha) \le \mathrm{M}(\alpha) \le \mathrm{H}(\alpha)\sqrt{d+1}.$$

The left inequality follows from the identity which relates the coefficients of a polynomial with the roots of this polynomial:

$$a_j = (-1)^j a_0 \sum_{1\le s_1<\cdots<s_j\le d} \alpha_{s_1}\cdots\alpha_{s_j}, \qquad (1\le j\le d).$$

The number of terms in the sum is $\binom{d}{j} \le 2^d$, and each of these terms is bounded from above by $\mathrm{M}(\alpha)/a_0$.

The right inequality follows from the arithmetico-geometric inequality:

$$\exp\left(\int_0^1 \log\left|f\left(e^{2i\pi t}\right)\right|dt\right) \le \int_0^1 \left|f\left(e^{2i\pi t}\right)\right|dt.$$

Using this bound for f^p, with p positive real, we deduce

$$\mathrm{M}(f) \le \left(\int_0^1 \left|f\left(e^{2i\pi t}\right)\right|^p dt\right)^{1/p}.$$

For $p = 2$ we obtain the desired estimate.

The proof of the second series of inequalities does not involve any difficulty and is left as an exercise. □

Remark 1. Some authors (for instance W. M. Schmidt in [Sc 1991], Ch. I, § 7) prefer another normalization of the absolute height, using the Euclidean norm at the Archimedean places; so the modified logarithmic height of a rational number a/b (with a, b relatively prime) is then $\log\sqrt{a^2+b^2}$ (see Exercise 3.2.b).

Remark 2. The fact that M is a multiplicative function on the ring $\mathbb{C}[X]$:

$$\mathrm{M}(f_1 f_2) = \mathrm{M}(f_1)\mathrm{M}(f_2),$$

combined with the estimates

$$2^{-d}\mathrm{H}(f) \le \mathrm{M}(f) \le \sqrt{d+1}\,\mathrm{H}(f) \tag{3.12}$$

for $d = \deg f$, yields

$$\mathrm{H}(f_1)\mathrm{H}(f_2) \le 2^d\sqrt{d+1}\,\mathrm{H}(f_1 f_2)$$

where $d = \deg(f_1 f_2)$.

Such an upper bound for the product $\mathrm{H}(f_1)\mathrm{H}(f_2)$ in terms of $\mathrm{H}(f_1 f_2)$ already appears in the seminal paper [KoPop 1932] of J. F. Koksma and J. Popken, where the authors give a transcendence measure for e^π (see [FNe 1998], Chap. 2, § 4.2, Th. 27 p. 102). In their paper Koksma and Popken introduce some of the main tools which will enable A. O. Gel'fond, at the end of the 40's, to create his method of algebraic independence (see [G 1952]). Gel'fond established sharp estimates concerning the height of polynomials and extended his investigations to polynomials in several variables. Related results also occur in the book *Diophantine Geometry* of S. Lang in 1962 (see also [L 1983]). The above simple proof, which rests on the multiplicativity of the measure M, is due to K. Mahler [M 1962]. Further references on this topic are given in [Ev 1998].

3.5 Liouville's Inequalities

3.5.1 Introduction

One characteristic of transcendence proofs, and more generally of results of diophantine approximation, is that some variant of the following fact is needed: *if a rational integer is nonzero, then its absolute value is at least* 1. One of the variants of this fact asserts that *if a rational number* p/q (where p and q are relatively prime rational integers and $q > 0$) *is nonzero, then* $|p/q| \geq 1/q$. Now the bound depends on the number considered. In terms of the logarithmic height $\mathrm{h}(p/q) = \log\max\{|p|, q\}$ of p/q (with $(p, q) = 1$ and $q > 0$), the previous inequality yields:

$$\log|x| \geq -\mathrm{h}(x) \quad \text{for all } x \in \mathbb{Q}^\times.$$

Liouville's inequality is a generic name for similar lower bounds for nonzero algebraic numbers α.

There is a simple lower bound for the modulus of a nonzero complex algebraic number α in terms of the usual height $\mathrm{H}(\alpha)$:

$$|\alpha| \geq \frac{1}{\mathrm{H}(\alpha)+1}\cdot$$

Since $\mathrm{H}(\alpha^{-1}) = \mathrm{H}(\alpha)$, this lower bound is equivalent to an upper bound $|\alpha| \leq \mathrm{H}(\alpha)+1$ (which plainly holds also for $\alpha = 0$). More generally, if α is a complex number which is root of a nonzero polynomial $f(X) = a_0X^n + \cdots + a_n \in \mathbb{Z}[X]$ of degree n with $\max\limits_{0\leq i\leq n} |a_i| \leq H$ (f need not be the minimal polynomial of α), then $|\alpha| \leq H + 1$. Indeed, this estimate holds trivially if $|\alpha| \leq 1$, while if $|\alpha| > 1$, then

$$\begin{aligned}|\alpha| \leq |a_0\alpha| &= |a_1 + a_2\alpha^{-1} + \cdots + a_n\alpha^{-n+1}|\\ &\leq H\bigl(1 + |\alpha|^{-1} + \cdots + |\alpha|^{-n+1}\bigr) < H\bigl(1 - |\alpha|^{-1}\bigr)^{-1}.\end{aligned}$$

One of the most useful inequalities of Liouville's type is

$$\log|\alpha|_v \geq -[\mathbb{Q}(\alpha)\colon \mathbb{Q}]\mathrm{h}(\alpha) \tag{3.13}$$

for all $\alpha \in \overline{\mathbb{Q}}$, $\alpha \neq 0$, and all absolute values v of $\mathbb{Q}(\alpha)$. For the proof, we first remark that for all $\alpha \in \overline{\mathbb{Q}}$ (including $\alpha = 0$), we have

$$\log|\alpha|_v \leq [\mathbb{Q}(\alpha)\colon \mathbb{Q}]\mathrm{h}(\alpha).$$

Further, if $\alpha \neq 0$, then $\mathrm{h}(\alpha) = \mathrm{h}(\alpha^{-1})$ (see (3.6)).

From Lemma 3.8 we now deduce the following statement: *under the hypotheses of Lemma 3.8, if the number $f(\underline{\gamma})$ is nonzero, then for all absolute values v of the number field k, we have*

$$\log|f(\underline{\gamma})|_v \geq -D\log \mathrm{L}(f) - D\sum_{i=1}^{\ell} N_i \mathrm{h}(1:\gamma_{i1}:\cdots:\gamma_{i\nu_i}),$$

where $D = [K:\mathbb{Q}]$.

In the next section we give a slight refinement, where $D\log \mathrm{L}(f)$ is replaced by $(D-1)\log \mathrm{L}(f)$ when v is an Archimedean absolute value.

3.5.2 The Main Lower Bound

Proposition 3.14 *(Liouville's inequality). Let K be a number field of degree D, v be an Archimedean absolute value of K and $\nu_1,\ldots,\nu_\ell$ be positive integers. For $1 \leq i \leq \ell$, let $\gamma_{i1},\ldots,\gamma_{i\nu_i}$ be elements of K. Further, let f be a polynomial in $\nu_1+\cdots+\nu_\ell$ variables, with coefficients in $\mathbb{Z}$, which does not vanish at the point $\underline{\gamma} = \left(\gamma_{ij}\right)_{1\leq j\leq \nu_i, 1\leq i\leq \ell}$. Assume f is of total degree at most N_i with respect to the ν_i variables corresponding to $\gamma_{i1},\ldots,\gamma_{i\nu_i}$. Then*

$$\log|f(\underline{\gamma})|_v \geq -(D-1)\log \mathrm{L}(f) - D\sum_{i=1}^{\ell} N_i \mathrm{h}(1:\gamma_{i1}:\cdots:\gamma_{i\nu_i}).$$

The simplest case $\ell = 1$, $\nu_1 = 1$ can be written as follows: *for a polynomial $f \in \mathbb{Z}[X]$ of degree at most N and an algebraic number $\alpha \in \mathbb{C}$ of degree d which is not a root of f, we have*

$$|f(\alpha)| \geq \mathrm{L}(f)^{1-d} e^{-dN\mathrm{h}(\alpha)}.$$

(We deduce this estimate from Proposition 3.14 by taking for v the Archimedean absolute value associated with the given embedding of $\mathbb{Q}(\alpha)$ in $\mathbb{C}$).

Proof. We write the product formula for $f(\underline{\gamma}) \neq 0$:

$$D_v \log|f(\underline{\gamma})|_v = -\sum_{w\neq v} D_w \log|f(\underline{\gamma})|_w,$$

where w runs over the absolute values of K distinct from v. If w is Archimedean we have

$$\log|f(\underline{\gamma})|_w \leq \sum_{i=1}^{\ell} N_i \log\max\left\{1, |\gamma_{i1}|_w,\ldots,|\gamma_{i\nu_i}|_w\right\} + \log \mathrm{L}(f).$$

The sum of D_w for w Archimedean and $w \neq v$ is $D - D_v \leq D - 1$. If w is ultrametric, the same estimate holds without the term $\log \mathrm{L}(f)$. We conclude the proof by using the bound

$$\sum_{w \neq v} D_w \sum_{i=1}^{\ell} N_i \log \max\{1, |\gamma_{i1}|_w, \ldots, |\gamma_{i\nu_i}|_w\} \leq D \sum_{i=1}^{\ell} N_i \mathrm{h}(1 : \gamma_{i1} : \cdots : \gamma_{i\nu_i}).$$

□

3.5.3 Further Lower Bounds

Using inequality (3.13) for $\alpha = \beta - (p/q)$ (or, alternatively, if v is Archimedean, using Proposition 3.14 for the polynomial in a single variable $f(\mathrm{X}) = q\mathrm{X} - p$), we deduce that for each algebraic number β, there exists a constant $c(\beta) > 0$ such that for all $p/q \in \mathbb{Q}$ with $q > 0$ and $p/q \neq \beta$, and for any absolute value v of $\mathbb{Q}(\beta)$, we have

$$\left| \beta - \frac{p}{q} \right|_v \geq \frac{c(\beta)}{\max\{|p|, q\}^d}$$

with $d = [\mathbb{Q}(\beta) : \mathbb{Q}]$ (and $|p|$ is the usual absolute value of p). An admissible value for $c(\beta)$ is $2^{-d} e^{-dh(\beta)}$.

Finally, the *size inequality*

$$\begin{cases} \log |\alpha|_v \geq -(d-1) \log \lceil \overline{\alpha} \rceil - d \log \operatorname{den} \alpha & \text{if } v \text{ is Archimedean} \\ \log |\alpha|_v \geq -d \log \lceil \overline{\alpha} \rceil - d \log \operatorname{den} \alpha & \text{if } v \text{ is ultrametric} \end{cases}$$

for all $\alpha \in \overline{\mathbb{Q}}$, $\alpha \neq 0$ is proved by writing

- that the norm over $\mathbb{Q}$ of the product $\alpha \cdot \operatorname{den}(\alpha)$ is a nonzero rational integer if v is Archimedean,
- the product formula for $\alpha \cdot \operatorname{den}(\alpha)$ if v is ultrametric.

A *Liouville number* is a real number ϑ such that, for any $\kappa > 0$, there exists $p/q \in \mathbb{Q}$ with $q \geq 2$ and

$$0 < \left| \vartheta - \frac{p}{q} \right| \leq \frac{1}{q^\kappa}.$$

From Liouville's inequality one deduces that a Liouville number is transcendental.

3.5.4 Proof of Lemma 2.1

From Proposition 3.14 one deduces the following result:

Given algebraic numbers $\gamma_1, \ldots, \gamma_m$ and a polynomial $f \in \mathbb{Z}[\mathrm{X}_1, \ldots, \mathrm{X}_m]$ which does not vanish at the point $(\gamma_1, \ldots, \gamma_m)$, we have

$$|f(\gamma_1, \ldots, \gamma_m)| \geq e^{-cT}$$

where $T = \deg f + \log \mathrm{H}(f)$,

$$c = D\big(2 + \mathrm{h}(\gamma_1) + \cdots + \mathrm{h}(\gamma_m)\big) \quad \text{and} \quad D = [\mathbb{Q}(\gamma_1, \dots, \gamma_m) : \mathbb{Q}].$$

Lemma 2.1 easily follows.

3.5.5 Estimates for Determinants

Most often we shall use Proposition 3.14 for a polynomial given by a determinant. We need to produce upper bounds for the degrees and heights of this polynomial; such estimates are given by the following simple lemma:

Lemma 3.15. *Let* L *be a positive integer and* $p_{\lambda\mu}$ $(1 \le \lambda, \mu \le L)$ *be* L^2 *polynomials in* $\nu_1 + \cdots + \nu_\ell$ *variables* X_{ij} $(1 \le j \le \nu_i,\ 1 \le i \le \ell)$, *with coefficients in* $\mathbb{Z}$. *Define, for* $1 \le \lambda \le L$,

$$M_\lambda = \max_{1 \le \mu \le L} \mathrm{L}(p_{\lambda\mu})$$

and

$$N_{i\lambda} = \max_{1 \le \mu \le L} \deg_{\underline{\mathrm{X}}_i} p_{\lambda\mu} \quad (1 \le i \le \ell),$$

where $\deg_{\underline{\mathrm{X}}_i}$ *denotes the total degree with respect to the set of variables* $\mathrm{X}_{i1}, \dots, \mathrm{X}_{i,\nu_i}$. *Then*

$$\Delta = \det\Big(p_{\lambda\mu}\Big)_{1 \le \lambda, \mu \le L}$$

is a polynomial in $\mathbb{Z}[\underline{\mathrm{X}}_1, \dots, \underline{\mathrm{X}}_\ell]$ *of length bounded by*

$$\mathrm{L}(\Delta) \le L! \prod_{\lambda=1}^{L} M_\lambda$$

and degrees bounded by

$$\deg_{\underline{\mathrm{X}}_i} \Delta \le \sum_{\lambda=1}^{L} N_{i\lambda} \quad (1 \le i \le \ell).$$

Consequently if γ_{ij} $(1 \le j \le \nu_i,\ 1 \le i \le \ell)$ are algebraic numbers in a number field of degree $\le D$ such that the polynomial Δ does not vanish at the point

$$\underline{\gamma} = \big(\gamma_{ij}\big)_{\substack{1 \le j \le \nu_i \\ 1 \le i \le \ell}} \in \mathbb{C}^{\nu_1 + \cdots + \nu_\ell},$$

then

$$\log|\Delta(\underline{\gamma})| \ge$$
$$-(D-1)\left(\log(L!) + \sum_{\lambda=1}^{L} \log M_\lambda\right) - DL \sum_{i=1}^{\ell} \left(\mathrm{h}(1 : \gamma_{i1} : \cdots : \gamma_{i\nu_i}) \sum_{\lambda=1}^{L} N_{i\lambda}\right).$$

Remark. In § 2.2.1 we introduced the ring $\mathbb{Z}[\mathrm{X}_1^{\pm 1}, \dots, \mathrm{X}_k^{\pm 1}, \mathrm{Y}_1, \dots, \mathrm{Y}_{\ell-k}]$. Let $\Delta = \det\bigl(p_{\lambda\mu}\bigr)$ be the determinant of a $L \times L$ matrix with coefficients in this ring. Assume

$$\max_{1 \le \mu \le L} \deg_{\mathrm{X}_i^{\pm 1}} p_{\lambda\mu} \le N_{i\lambda} \quad (1 \le i \le k),$$

$$\max_{1 \le \mu \le L} \deg_{\mathrm{Y}_j} p_{\lambda\mu} \le N'_{j\lambda} \quad (1 \le j \le \ell - k)$$

and

$$\mathrm{L}(p_{\lambda\mu}) \le M_\lambda$$

for $1 \le \lambda \le L$. We can apply Lemma 3.15 with ℓ replaced by $\ell + k$, with $\nu_j = 1$ for all j and with

$$\mathrm{X}_{i1} = \begin{cases} \mathrm{X}_i & \text{for } 1 \le i \le k, \\ \mathrm{X}_{i-k}^{-1} & \text{for } k < i \le 2k, \\ \mathrm{Y}_{i-2k} & \text{for } 2k < i \le \ell + k. \end{cases}$$

We deduce

$$\deg_{\mathrm{X}_i^{\pm 1}} \Delta \le \sum_{\lambda=1}^{L} N_{i\lambda} \quad (1 \le i \le k),$$

$$\deg_{\mathrm{Y}_j} \Delta \le \sum_{\lambda=1}^{L} N'_{j\lambda} \quad (1 \le j \le \ell - k).$$

Further, let $\underline{\gamma} = (\gamma_1, \dots, \gamma_\ell)$ be a ℓ-tuple of algebraic numbers in a number field of degree D with $\gamma_j \neq 0$ for $1 \le j \le k$. Assume $\Delta(\underline{\gamma}) \neq 0$. We can apply Proposition 3.14, but one must carefully add the contributions of $\deg_{\mathrm{X}_i}$ and $\deg_{\mathrm{X}_i^{-1}}$:

$$\log|\Delta(\underline{\gamma})| \ge$$
$$(D-1)\sum_{\lambda=1}^{L} \log M_\lambda - (D-1)\log(L!) - 2D\sum_{i=1}^{k}\sum_{\lambda=1}^{L} N_{i\lambda} - D\sum_{j=1}^{\ell-k}\sum_{\lambda=1}^{L} N'_{j\lambda}.$$

See for instance Exercise 3.8.

3.6 Lower Bound for the Height

We quoted Northcott's Theorem in § 3.4 as a fundamental property of any height. Another important property of the absolute logarithmic height (which distinguishes this height from most other ones) is that for $\alpha \in \overline{\mathbb{Q}}^\times$, $\mathrm{h}(\alpha) = 0$ if and only if α is a root of unity (i.e. a torsion point in the multiplicative group $\mathbb{G}_{\mathrm{m}}(\overline{\mathbb{Q}}) = \overline{\mathbb{Q}}^\times$). This raises the important problem of estimating $\mathrm{h}(\alpha)$ from below when it does not vanish.

3.6.1 Kronecker's Theorem

By definition the values of Mahler's measure M are ≥ 1.

If a nonzero algebraic number α satisfies $\mathrm{M}(\alpha) < 2$, then α is an algebraic integer, and α^{-1} also, which means that α is a unit. In other terms the (absolute logarithmic) height of an algebraic number which is not a unit is at least $(\log 2)/d$.

Let α be nonzero algebraic integer. Assume $\mathrm{M}(\alpha) = 1$, which means that all conjugates of α have modulus at most 1. *Then α is a root of unity.* Indeed, if α has degree d, then each α^ℓ with $\ell \geq 1$ is a root of a monic polynomial, with rational integer coefficients, of degree d, whose coefficients have usual absolute values at most 2^d. The set of such polynomials is finite, hence so is the set of α^ℓ $(\ell \geq 1)$. The conclusion plainly follows.

Using Corollary 3.2, one deduces the following statement, due to L. Kronecker [Kr 1857]: *if k is a number field and α a nonzero element of k such that $|\alpha|_v \leq 1$ for all $v \in M_k$, then α is a root of unity.*

Therefore the only algebraic numbers α which satisfy $\mathrm{h}(\alpha) = 0$ are 0 and the roots of unity. The other ones satisfy $\mathrm{h}(\alpha) > 0$. To give a sharp lower bound for $\mathrm{h}(\alpha)$, when α is a unit but not a root of unity, in terms of the degree of α is an interesting and difficult problem (see [L 1991], Chap. IX, § 7).

Remark. For an algebraic number α of degree d, since $\mathrm{h}(\alpha) = (1/d)\log \mathrm{M}(\alpha)$, the conditions $\mathrm{h}(\alpha) > 0$ and $\mathrm{M}(\alpha) > 1$ are plainly equivalent.

If $\kappa > 0$ and $\alpha \in \overline{\mathbb{Q}}$ satisfy $\mathrm{h}(\alpha) \geq \kappa$, then from the inequality

$$e^{\kappa d} > 1 + \kappa d$$

with $d = [\mathbb{Q}(\alpha) : \mathbb{Q}]$ we deduce $\mathrm{M}(\alpha) > 1 + \kappa d$.

Conversely, if $\alpha \in \overline{\mathbb{Q}}$ has degree d and if $\kappa > 0$, $\epsilon > 0$ satisfy

$$\mathrm{M}(\alpha) \geq 1 + \kappa d \quad \text{and} \quad \kappa d < \epsilon,$$

then (using Exercise 1.1.a) we deduce

$$\mathrm{h}(\alpha) > c\kappa \quad \text{with} \quad c(\epsilon) = \frac{1}{\epsilon}\log(1+\epsilon).$$

Notice that $c(\epsilon) \to 1$ as $\epsilon \to 0$. More precisely we have

$$1 < c(\epsilon) < 1 + \epsilon.$$

3.6.2 Lehmer's Problem

In 1933 D. H. Lehmer [Le 1933] asked whether it is true that for every positive ϵ there exists an algebraic integer α for which $1 < \mathrm{M}(\alpha) < 1+\epsilon$. The answer is not yet known but it is easy to see (Exercise 3.9) that for each positive integer d there exists a positive number $c(d)$ such that, for any nonzero algebraic number α which is not a root of unity and is of degree at most d, the inequality $\mathrm{h}(\alpha) \geq c(d)$ is valid. The example $\alpha = 2^{1/d}$ shows that such a function $c(d)$ must satisfy $c(d) \leq (\log 2)/d$. It is widely believed that there exists a positive absolute constant c_0 such that $c(d) \geq c_0/d$. This problem is known as *Lehmer's problem* (see Chap. 7 of [BerDGPS 1992]) and an answer would have various applications. The first one is due to D. H. Lehmer himself in [Le 1933]: he introduced the subject while looking for large prime numbers. Next, following A. Schinzel, C. Pinner and J. Vaaler related the Mahler measure of a polynomial and the number of its irreducible non-cyclotomic factors. Polynomials with small measure also occur in ergodic theory and dynamical systems (works of Ia. Sinaï, W. Lawton, E. Bombieri and J. E. Taylor). We refer to M. J. Mossinghoff's thesis [Mos 1995] for further references (see also [Ev 1998]).

The smallest known value $\mathrm{M}(\alpha) > 1$ for an algebraic number α is the root $1.1762808183\ldots$ of the reciprocal[7] polynomial of degree 10 :

$$X^{10}+X^9-X^7-X^6-X^5-X^4-X^3+X+1=X^5Q\left(X+\frac{1}{X}\right)$$

with

$$Q(Y)=(Y+1)^2(Y-1)(Y+2)(Y-2)-1$$

This example is due to D. H. Lehmer [Le 1933].

In 1980 and then in 1989, D. Boyd developed an algorithm for searching polynomials with small Mahler's measure. He found all polynomials of degree at most 20 and Mahler's measure at most 1.3. In his thesis M. J. Mossinghoff [Mos 1995] listed 1560 irreducible non-cyclotomic polynomials with Mahler's measure less than 1.3 and degree at most 64. None of these has Mahler's measure less than Lehmer's degree 10 example reported in 1933.

The first result in the direction of Lehmer's problem is due to A. Schinzel and H. Zassenhaus [SZa 1965]: when $\alpha \neq 0$ is an algebraic integer of degree $d \geq 2$ which is not a root of unity, then

$$\overline{|\alpha|} > 1+4^{-s-2}$$

where $2s$ is the number of nonreal conjugates of α. Therefore

$$\mathrm{M}(\alpha) > 1+\frac{c}{2^d}$$

for some absolute constant $c > 0$.

In 1971, by means of an averaging technique in Fourier analysis, P. E. Blanksby and H. L. Montgomery [BlMon 1971] refined this result and proved, for an algebraic integer of degree $d > 1$ which is not a root of unity,

$$\mathrm{M}(\alpha) > 1+\frac{1}{52d\log(6d)}.$$

A consequence is the estimate

$$\overline{|\alpha|} > 1+\frac{1}{30d^2\log(6d)}.$$

[7] A polynomial $f \in \mathbb{C}[X]$ of degree d is *reciprocal* if $f(X)=X^d f(1/X)$.

Also in that year Smyth [Sm 1971] used Parseval's formula to prove, under the same assumptions, that if α^{-1} is not a conjugate of α, then $\mathrm{M}(\alpha) \geq 1.3247179572\ldots$, this number being the real root of $X^3 - X - 1$ and the smallest PV-number [8]. An interesting consequence of his result is that it solves Lehmer's problem when d is *odd* (one can use $c_0 = 0.2811\ldots$ in this case). In 1978, C. L. Stewart [Ste 1978] introduced a method from transcendental number theory to prove

$$\mathrm{M}(\alpha) > 1 + \frac{1}{10^4 d \log d}$$

for $d \geq 2$. This is marginally weaker than the previous result of Blanksby-Montgomery, but the interest lies in the method.

3.6.3 Dobrowolski's Theorem

In 1979, E. Dobrowolski [Do 1979] succeeded to extend Stewart's argument and to obtain the following statement: for each $\epsilon > 0$, there exists an integer $d_0(\epsilon)$ such that, for any $d > d_0(\epsilon)$ and any nonzero algebraic number α of degree $\leq d$ which is not a root of unity,

$$\mathrm{h}(\alpha) > \frac{1-\epsilon}{d}\left(\frac{\log\log d}{\log d}\right)^3,$$

which can be written

$$\mathrm{M}(\alpha) > 1 + (1-\epsilon)\left(\frac{\log\log d}{\log d}\right)^3.$$

In 1981, independently, D. C. Cantor and E. G. Straus [CaStr 1982] and U. Rausch [Ra 1985] simplified Dobrowolski's proof by introducing a determinant and replaced $1-\epsilon$ by $2-\epsilon$. Finally R. Louboutin [Lo 1983] reached $(9/4)-\epsilon$ by a modification of this determinant. The same result with $(9/4)-\epsilon$ has been also obtained by M. Meyer [Me 1988], using a construction of an auxiliary function (like Dobrowolski), but with Thue-Siegel lemma replaced by a refinement due to E. Bombieri and J. Vaaler [BoVa 1983].

Dobrowolski's result is effective: by [Do 1979], for all $d \geq 2$,

$$\mathrm{M}(\alpha) > 1 + \frac{1}{1200}\left(\frac{\log\log d}{\log d}\right)^3.$$

P. Voutier [Vou 1996] improved this bound: for $d \geq 2$,

$$\mathrm{h}(\alpha) > \frac{1}{4d}\left(\frac{\log\log d}{\log d}\right)^3.$$

---

[8] A *Pisot-Vijayaraghavan number*, or *PV-number*, is a real algebraic integer > 1 all of whose other conjugates lie inside the open unit disc. A *Salem number* is a real algebraic integer > 1 all of whose other conjugates lie inside the closed unit disc, with at least one conjugate on the unit circle. See [BerDGPS 1992].

Let α be a nonzero algebraic integer of degree $\leq d$ with $d \geq 2$. Since $\log\lceil\alpha\rceil \geq \mathrm{h}(\alpha)$, we deduce, if α is not a root of unity,

$$\log\lceil\alpha\rceil > \frac{1}{4d}\left(\frac{\log\log d}{\log d}\right)^3.$$

The estimate (see [Du 1993])

$$\lceil\alpha\rceil > 1 + \left(\frac{64}{\pi^2} - \epsilon\right)\frac{1}{d}\left(\frac{\log\log d}{\log d}\right)^3 \quad \text{for} \quad d > d_0(\epsilon)$$

is sharper for large d, while

$$\log\lceil\alpha\rceil > \frac{\log\bigl(d + (1/2)\bigr)}{d^2} \quad \text{for} \quad d \geq 1$$

(see [Mat 1991]) is stronger for small d. From the latter one deduces that a nonzero algebraic integer α satisfying $\mathrm{h}(\alpha) < 2/(3d^3)$ is a root of unity (compare with Theorem 3.16). Another uniform estimate is [Vou 1996]:

$$\lceil\alpha\rceil > 1 + \frac{1}{2d}\left(\frac{\log\log d}{\log d}\right)^3 \quad \text{for} \quad d \geq 2.$$

Our aim in the rest of this section is twofold. On one hand we wish to establish a lower bound which will be useful later (namely in Chap. 7, proof of Lemma 7.19):

Theorem 3.16. *Let d be a positive integer and α be a nonzero algebraic number of degree $\leq d$ which is not a root of unity. Then*

$$\mathrm{h}(\alpha) > \frac{1}{11d^3}.$$

On the other hand we wish to give a further example of a *transcendence proof* using an interpolation determinant. This will produce a sharpening of Theorem 3.16, but only for sufficiently large d:

Theorem 3.17. *There exists a positive integer d_0 such that, for any integer $d \geq d_0$ and a nonzero algebraic number α of degree $\leq d$ which is not a root of unity,*

$$\mathrm{h}(\alpha) \geq \frac{1}{250d}\left(\frac{\log\log d}{\log d}\right)^3.$$

From the previous discussion it is clear that $d_0 = 2$ is an admissible value for the constant in Theorem 3.17, and in fact this would follow from the argument given below. But assuming that d is sufficiently large will simplify the estimates (we insist that d is only an *upper bound* for the degree of α, and not the actual degree).

It may be useful for the reader if we repeat that there is no loss of generality, in the proofs of Theorems 3.16 and 3.17, to assume that α is an algebraic integer. Indeed, we have seen that the result is obvious unless α is a unit.

3.6.4 Fermat's Little Theorem

One main tool is Fermat's little theorem, which is used as follows:

Lemma 3.18. *Let p be a prime number and $f \in \mathbb{Z}[X_1, \ldots, X_k]$ a polynomial in k variables with integer coefficients. Then there exists $g \in \mathbb{Z}[X_1, \ldots, X_k]$ such that*

$$f(X_1^p, \ldots, X_k^p) - f(X_1, \ldots, X_k)^p = pg(X_1, \ldots, X_k).$$

Proof. For simplicity write $\underline{X}$ for $(X_1, \ldots, X_k)$ and $\underline{X}^p$ for $(X_1^p, \ldots, X_k^p)$, so that the conclusion is just

$$f(\underline{X}^p) - f(\underline{X})^p = pg(\underline{X}).$$

The result holds for a monomial $f(\underline{X}) = aX_1^{i_1} \cdots X_k^{i_k}$:

$$f(\underline{X}^p) - f(\underline{X})^p = (a - a^p)X_1^{pi_1} \cdots X_k^{pi_k}$$

and p divides the integer $a - a^p$. If the result holds for f_1 and for f_2, namely if

$$f_1(\underline{X}^p) - f_1(\underline{X})^p = pg_1(\underline{X}) \quad \text{and} \quad f_2(\underline{X}^p) - f_2(\underline{X})^p = pg_2(\underline{X}),$$

then it holds for $f = f_1 + f_2$, because the coefficients of the polynomial

$$(f_1 + f_2)^p - f_1^p - f_2^p = \sum_{h=1}^{p-1} \binom{p}{h} f_1^h f_2^{p-h}$$

are rational integers which are all divisible by p. Therefore one may choose

$$g = g_1 + g_2 - \sum_{h=1}^{p-1} \frac{(p-1)!}{h!(p-h)!} f_1^h f_2^{p-h}.$$

Lemma 3.18 plainly follows. □

Remark. An explicit expression for g can be given. For instance for $k = 1$ if the given polynomial f is $f(X) = a_0X^d + a_1X^{d-1} + \cdots + a_d$, then one can write:

$$g(X) = \sum_{i=0}^{d} \frac{a_{d-i} - a_{d-i}^p}{p} X^{pi} - \sum_{\substack{i_0+\cdots+i_d=p \\ 0 \le i_h < p, (0 \le h \le d)}} \frac{(p-1)!}{i_0! \cdots i_d!} a_d^{i_0} \cdots a_0^{i_d} X^{i_1+2i_2+\cdots+di_d}.$$

We now give a very simple proof of the following estimate, once more due to E. Dobrowolski [Do 1978]: *if a nonzero algebraic integer α of degree $\le d$ is not a root of unity, then*

$$\lceil\overline{\alpha}\rceil > 1 + \frac{1}{4ed^2}$$

Let α be a nonzero algebraic integer of degree d. Denote by Σ the set of embeddings of $\mathbb{Q}(\alpha)$ into $\mathbb{C}$, so that

$$\{\sigma\alpha\,;\,\sigma\in\Sigma\}=\{\alpha_1,\ldots,\alpha_d\}$$

is the set of conjugates of α. For any positive integer h, the value of the *Newton sum*

$$S_h=\sum_{\sigma\in\Sigma}\sigma\alpha^h$$

(which is the *trace* of α^h from $\mathbb{Q}(\alpha)$ to $\mathbb{Q}$) is a rational integer. Let p be a prime number. Fermat's little Theorem gives the congruence $S_h\equiv S_h^p \pmod p$. On the other hand, using Lemma 3.18 with $k=d$ for the polynomial $f=X_1^h+\cdots+X_d^h$, we can write $S_{hp}-S_h^p=pg(\alpha_1,\ldots,\alpha_d)$, for some $g\in\mathbb{Z}[X_1,\ldots,X_d]$. Now $g(\alpha_1,\ldots,\alpha_d)$ is an algebraic integer, and since it is a rational number, we get $S_{hp}\equiv S_h^p \pmod p$. This shows that the three numbers S_{hp}, S_h and S_h^p are congruent modulo p. For any $h\geq 1$ we have

$$|S_h|\leq d\,\overline{|\alpha|}^{\,h}.$$

We now assume $\overline{|\alpha|}\leq 1+1/(4ed^2)$. By the so-called *Bertrand's Postulate* (which was proved by Chebishev in 1850 – see [HaWr 1938], Chap. 22 and [GLin 1962], Th. 3.5.1), there exists a prime number p in the range $2ed<p<4ed$. For $1\leq h\leq d$, the estimates

$$|S_h|\leq d\left(1+\frac{1}{4ed^2}\right)^d\leq de \quad\text{and}\quad |S_{hp}|\leq d\left(1+\frac{1}{4ed^2}\right)^{4ed^2}\leq de$$

hold. Therefore $|S_h-S_{hp}|\leq 2de<p$, which implies $S_h=S_{hp}$ for $1\leq h\leq d$. This means that α and α^p have the same minimal polynomial, i.e. that they are conjugates. One deduces from the following lemma that α is a root of unity.

Lemma 3.19. *Let α be a nonzero algebraic number. Assume that there exist two distinct positive rational integers h and ℓ such that α^h and α^ℓ are conjugate. Then α is a root of unity.*

Proof. Let K be the splitting field of α over $\mathbb{Q}$: if Σ denotes the set of embeddings of the field $\mathbb{Q}(\alpha)$ in $\mathbb{C}$, then K is the field generated over $\mathbb{Q}$ by $\{\sigma\alpha\,;\,\sigma\in\Sigma\}$. From the assumption that α^h and α^ℓ are conjugate, we deduce that there exists an element φ in the Galois group of K over $\mathbb{Q}$ such that $\varphi(\alpha^h)=\alpha^\ell$. By induction, for any $n\geq 1$, we deduce $\varphi^n(\alpha^{h^n})=\alpha^{\ell^n}$. Let m be the order of φ in the Galois group. Then $\alpha^{h^m}=\alpha^{\ell^m}$. Since $h\neq\ell$, we conclude that α is a root of unity. □

Proof of Theorem 3.16. We first notice that the inequality

$$\left(1+\frac{1}{4ed^2}\right)^{11d^2}>e$$

holds for $d\geq 2$. Hence for $d\geq 2$ we deduce

$$\mathrm{h}(\alpha) \geq \frac{1}{d} \log \lceil\overline{\alpha}\rceil \geq \frac{1}{d} \log\left(1 + \frac{1}{4ed^2}\right) > \frac{1}{11d^3}.$$

□

Remark. Using the same arguments, E. Dobrowolski [Do 1978] also proves that a nonzero algebraic integer α which is a not root of unity satisfies

$$\lceil\overline{\alpha}\rceil \geq 1 + \frac{\log d}{6d^2}.$$

As we have seen, sharper results [Mat 1991], [Du 1993], [Vou 1996] are now available.

In order to prove Theorem 3.17, we need some preparation.

The proof of the next lemma involves the *norm* $\mathrm{N}_{k/\mathbb{Q}}: k \to \mathbb{Q}$ of a number field k: for $\alpha \in k$, the norm of α with respect to the extension $k/\mathbb{Q}$ is the determinant of the endomorphism $x \mapsto \alpha x$ of the $\mathbb{Q}$-vector space k. If we denote again by Σ the set of embeddings of k into $\mathbb{C}$, then

$$\mathrm{N}_{k/\mathbb{Q}}(\alpha) = \prod_{\sigma\in\Sigma} \sigma\alpha.$$

The *absolute norm* of an algebraic number α is $\mathrm{N}_{\mathbb{Q}(\alpha)/\mathbb{Q}}(\alpha) \in \mathbb{Q}$. When α is an algebraic integer, we have $\mathrm{N}_{k/\mathbb{Q}}(\alpha) \in \mathbb{Z}$ (the absolute value of this integer is nothing else than the absolute norm of the principal ideal (α) in the ring of integers of k). The following relation holds for any element α in a number field k:

$$\mathrm{N}_{k/\mathbb{Q}}(\alpha) = \left(\mathrm{N}_{\mathbb{Q}(\alpha)/\mathbb{Q}}(\alpha)\right)^{[k:\mathbb{Q}(\alpha)]}.$$

Lemma 3.20. *Let p be a prime number and α an algebraic integer of degree d. Denote by Σ the set of embeddings of $\mathbb{Q}(\alpha)$ into $\mathbb{C}$. Then the number*

$$\prod_{\sigma\in\Sigma}\prod_{\tau\in\Sigma}(\tau\alpha^p - \sigma\alpha)$$

is a rational integer which is divisible by p^d.

Proof. The minimal polynomial $f \in \mathbb{Z}[X]$ of α over $\mathbb{Z}$ can be written

$$f(X) = \prod_{\sigma\in\Sigma}(X - \sigma\alpha).$$

Hence

$$\prod_{\sigma\in\Sigma}\prod_{\tau\in\Sigma}(\tau\alpha^p - \sigma\alpha) = \prod_{\tau\in\Sigma} f(\tau\alpha^p) = \mathrm{N}_{\mathbb{Q}(\alpha)/\mathbb{Q}}\big(f(\alpha^p)\big).$$

Since α is an algebraic integer and $f \in \mathbb{Z}[X]$, the number $f(\alpha^p)$ is also an algebraic integer, hence its norm is a rational integer.

Using Lemma 3.18 with $k=1$, we write $f(X^p)-f(X)^p = pg(X)$ for some $g\in\mathbb{Z}[X]$. Since $f(\tau\alpha)=0$ for all $\tau\in\Sigma$, we deduce $f(\tau\alpha^p)=pg(\tau\alpha)$ and

$$\prod_{\tau\in\Sigma} f(\tau\alpha^p) = p^d\prod_{\tau\in\Sigma} g(\tau\alpha).$$

Finally, we observe that the number

$$\prod_{\tau\in\Sigma} g(\tau\alpha) = \mathrm{N}_{\mathbb{Q}(\alpha)/\mathbb{Q}}\big(g(\alpha)\big)$$

is a rational integer. □

Remark. Using the product formula in place of the norm would enable us to deal with algebraic numbers in place of algebraic integers. Compare with Lemma 3.23 below.

Lemma 3.21. *Let α be an algebraic integer of degree d, measure $\mathrm{M}(\alpha)$ and length $\mathrm{L}(\alpha)$. Let p be a prime number. If α is neither 0 nor a root of unity, then*

$$\mathrm{M}(\alpha)\ge\left(\frac{p}{\mathrm{L}(\alpha)}\right)^{1/p}.$$

Proof. From Lemma 3.19 (with $h=1$ and $\ell=p$) we deduce that α and α^p are not conjugate. If $f(X)=X^d+a_1X^{d-1}+\cdots+a_d$ is the minimal polynomial of α, then the number $f(\alpha^p)$ is a nonzero algebraic integer. Hence its norm $N_{\mathbb{Q}(\alpha)/\mathbb{Q}}\big(f(\alpha^p)\big)$ over $\mathbb{Q}$ is a nonzero rational integer which, by Lemma 3.20, is divisible by p^d. A trivial upper bound for the absolute value of this number

$$N_{\mathbb{Q}(\alpha)/\mathbb{Q}}\big(f(\alpha^p)\big)=\prod_{\sigma\in\Sigma}\sum_{j=0}^{d} a_{d-j}\sigma\alpha^{pj}$$

(where $a_0=1$) is obtained as follows:

$$\left|N_{\mathbb{Q}(\alpha)/\mathbb{Q}}\big(f(\alpha^p)\big)\right|\le\prod_{\sigma\in\Sigma}\sum_{j=0}^{d}|a_{d-j}|\max\{1,|\sigma\alpha|^{pj}\}\le\mathrm{L}(\alpha)^d\mathrm{M}(\alpha)^{pd}.$$

Therefore

$$p^d\le\mathrm{L}(\alpha)^d\mathrm{M}(\alpha)^{pd}.$$

□

Remark. If we use the fact that there is a prime number p in the interval $[2\mathrm{L}(\alpha),4\mathrm{L}(\alpha)]$, together with the estimate $2\log x\ge x\log 2$ which holds in the range $2\le x\le 4$, we deduce

$$\mathrm{M}(\alpha)\ge\left(\frac{p}{\mathrm{L}(\alpha)}\right)^{1/p}\ge 2^{1/2\mathrm{L}(\alpha)}\ge 1+\frac{\log 2}{2\mathrm{L}(\alpha)}\cdot$$

3.6.5 Dobrowolski's Proof of Theorem 3.17

In order to obtain a lower bound for $\mathrm{M}(\alpha)$ which depends only on the degree of α, one main idea of Dobrowolski's is to use the same outline, not for the minimal polynomial f of α itself, but for a suitable multiple of f^T, where T is a large integer.

The argument is as follows (see [Do 1979] and [Sc 1999], § 6). All throughout the proof we assume d is sufficiently large. Select two parameters L and T (depending on d) with $L > dT$. The first step establishes the existence of a nonzero polynomial F in $\mathbb{Z}[X]$, of degree $\le L$ and length

$$\mathrm{L}(F) \le L^{2dT^2/L},$$

which satisfies

$$\left(\frac{d}{dX}\right)^t F(\alpha) = 0 \quad \text{for} \quad 0 \le t < T.$$

The second step is the zero estimate: there exists a prime number p in the range

$$2 \le p \le p_0 \quad \text{where} \quad p_0 := \frac{3}{2} \cdot \frac{L}{d} \log \frac{L}{d}$$

such that $F(\alpha^p) \neq 0$.

Assuming for the moment these two preliminary steps, we complete the proof of Theorem 3.17.

We deduce from Lemma 3.20 that the number

$$N = N_{\mathbb{Q}(\alpha)/\mathbb{Q}} F(\alpha^p) = \prod_{\tau\in\Sigma} F(\tau\alpha^p)$$

is a nonzero rational integer which is a multiple of p^{dT}. Hence

$$|N| \ge p^{dT}.$$

On the other hand the estimate

$$\prod_{\tau\in\Sigma} |F(\tau\alpha^p)| \le \mathrm{L}(F)^d \prod_{\tau\in\Sigma} \max\{1, |\tau\alpha|\}^{pL}$$

gives

$$|N| \le \mathrm{L}(F)^d \mathrm{M}(\alpha)^{pL}.$$

Therefore

$$\mathrm{M}(\alpha) \ge p^{dT/(pL)} \mathrm{L}(F)^{-d/(pL)} \ge p^{dT/(pL)} L^{-2d^2T^2/(pL^2)}.$$

We wish now to choose the parameters T and L such that the quantity

$$\min_{p\le p_0} \left\{ \frac{dT}{pL} \log p - \frac{2d^2T^2}{pL^2} \log L \right\}$$

is *large*: its value will provide a lower bound for $\log \mathrm{M}(\alpha)$. Choose for instance the parameters as follows:

$$T = \left[5\frac{\log d}{\log\log d}\right] \quad \text{and} \quad L = dT^2.$$

Since

$$\frac{dT^2}{L} > \frac{p}{2T^2\log p} + \frac{2d^2T^3\log L}{L^2\log p},$$

one deduces the lower bound

$$\log \mathrm{M}(\alpha) \geq \frac{1}{2T^3},$$

which yields the conclusion of Theorem 3.17.

Let us come back to the first step: the construction of F. Using Dirichlet's box principle (see Exercise 3.12), Dobrowolski ([Do 1979], Lemma 1) shows the existence of $F \in \mathbb{Z}[X]$ with a zero of multiplicity $\geq T$ at α and with the following upper bound for its height:

$$\mathrm{H}(F) \leq \left(\left(2^{3/2}(L+1)L^{(T-1)/2}\right)^{dT}\mathrm{M}(\alpha)^{TL}\right)^{1/(L-dT)}.$$

In order to deduce the desired upper bound for the length of F, it suffices to check

$$2^{3dT/2}(L+1)^L\mathrm{M}(\alpha)^{TL} \leq L^{(3dT^2/2)-(2d^2T^3/L)+(dT/2)}.$$

Given our choice of parameters (recall that d is sufficiently large), this estimate is satisfied as soon as $\log \mathrm{M}(\alpha) \leq (1/11)\log\log d$, an assumption which of course does not involve any loss of generality for the proof of Theorem 3.17.

In order to complete the proof of Theorem 3.17, we only need to prove the zero estimate of the second step: *one at least of the numbers $F(\alpha^p)$, with p prime in the range $2 \leq p \leq p_0$, is not zero.* The number of primes in this range is $> L/d$. We are going to check that the set

$$\left\{\sigma\alpha^p\,;\, \sigma \in \Sigma,\ p \leq p_0\right\}.$$

has more than L elements. It will follow that the nonzero polynomial $F \in \mathbb{Z}[X]$ cannot vanish at all points in this set, which is what we want.

By Lemma 3.19, for $p_1 \neq p_2$ and for any σ and τ in Σ, we have $\sigma\alpha^{p_1} \neq \tau\alpha^{p_2}$. If there is a prime p for which the elements $\sigma\alpha^p$ $(\sigma \in \Sigma)$, are not pairwise distinct, then α^p is of degree $< d$, and we complete the proof of our claim by means of an inductive argument, thanks to the following lemma (compare with Lemma 3 of [Ra 1985]):

Lemma 3.22. *Let α be a nonzero algebraic integer which is not a root of unity. Define $d = [\mathbb{Q}(\alpha):\mathbb{Q}]$. Assume that there exists a positive integer n such that $[\mathbb{Q}(\alpha^n):\mathbb{Q}] < d$. Then there exists an algebraic integer β such that*

$$[\mathbb{Q}(\beta):\mathbb{Q}] \leq \frac{d}{2} \quad \textit{and} \quad \mathrm{M}(\beta) \leq \mathrm{M}(\alpha).$$

Proof. Define $k = \mathbb{Q}(\alpha^n)$ and notice that α is a root of $X^n - \alpha^n \in k[X]$. Hence the irreducible polynomial g of α over k is a divisor of $X^n - \alpha^n$ in $k[X]$. It follows that the constant term, say $\beta \in k$, of g, can be written $\zeta\alpha^r$, where $r = [\mathbb{Q}(\alpha) : k]$ is the degree of g, while ζ is a n-th root of unity. Therefore we have

$$\mathrm{h}(\beta) = r\mathrm{h}(\alpha)$$

and

$$[\mathbb{Q}(\beta) : \mathbb{Q}] \leq [k : \mathbb{Q}] = \frac{d}{r} = \frac{1}{r}[\mathbb{Q}(\alpha) : \mathbb{Q}],$$

hence

$$\mathrm{M}(\beta) \leq \mathrm{M}(\alpha).$$

□

It is interesting to compare the previous sketch of proof with the usual one in transcendental number theory: a nonzero number is constructed and its absolute value is estimated from above and from below. But here, in place of a sharp analytic upper bound (Schwarz' Lemma) and a weak arithmetic lower bound (Liouville's inequality), we have a sharp arithmetic lower bound (coming from Fermat's little Theorem) and a trivial upper bound. However it is possible to give the proof in a way which is closer to the usual one, involving a sharp (ultrametric) upper bound together with the product formula. Here is the needed p-adic estimate.

Lemma 3.23. *Let α be an algebraic number of degree $d = [\mathbb{Q}(\alpha) : \mathbb{Q}]$. Let p be a prime number, $F \in \mathbb{Z}[X]$ a polynomial of degree L which vanishes at α with multiplicity at least T and v a place of $\mathbb{Q}(\alpha)$ which extends the p-adic valuation of $\mathbb{Q}$. Then*

$$|F(\alpha^p)|_v \leq p^{-T} \max\{1, |\alpha|_v\}^{pL}.$$

Proof. Consider first the case where F is the minimal polynomial f of α over $\mathbb{Z}$. Using Lemma 3.18 with $k = 1$, we can write $f(X^p) = f(X)^p + pg(X)$ for some $g \in \mathbb{Z}[X]$ of degree $\leq pd$. Hence

$$|g(\alpha)|_v \leq \max\{1, |\alpha|_v\}^{pd}$$

and

$$|f(\alpha^p)|_v = |pg(\alpha)|_v \leq p^{-1} \max\{1, |\alpha|_v\}^{pd},$$

which is what we wanted.

In the general case, F is divisible by f^T: let $G \in \mathbb{Z}[X]$ satisfy $F = f^T G$. Hence G has degree $L - dT$ and

$$|G(\alpha^p)|_v \leq \max\{1, |\alpha|_v\}^{p(L-dT)}.$$

Therefore

$$|F(\alpha^p)|_v = |f(\alpha^p)|_v^T |G(\alpha^p)|_v \leq p^{-T} \max\{1, |\alpha|_v\}^{pL}.$$

□

Remark 1. From Lemma 3.23 one deduces the following lower bound, which could be used in the proof of Theorem 3.17 in place of the one which we derived from Lemma 3.20 : *Under the assumptions of Lemma 3.23, assume that α is an algebraic integer and $F(\alpha^p) \neq 0$. Let Σ the set of embeddings of $\mathbb{Q}(\alpha)$ into $\mathbb{C}$. Then*

$$\prod_{\tau\in\Sigma} |F(\tau\alpha^p)| \geq p^{dT}.$$

This follows from the product formula

$$\left(\prod_{\tau\in\Sigma} |F(\tau\alpha^p)|\right)\left(\prod_{v} |F(\alpha^p)|_v\right) = 1$$

where v runs over the set of ultrametric absolute values of $\mathbb{Q}(\alpha)$, using the upper bound $|F(\alpha^p)|_v \leq 1$ for any ultrametric absolute value v of $\mathbb{Q}(\alpha)$ such that $|p|_v = 1$.

Remark 2. Under the hypotheses of Lemma 3.23, assume further that α is an integer. Then one can derive the conclusion in the general case from the special case $F = f$ by means of an ultrametric Schwarz' lemma as follows.

Let Σ_v be the set of embeddings of $\mathbb{Q}(\alpha)$ into an algebraically closed field $\mathbb{C}_v$ containing the completion of $\mathbb{Q}(\alpha)$ at v. The analytic function $z \mapsto F(z)$ on $\mathbb{C}_v$ vanishes at the points $z = \sigma\alpha$ $(\sigma \in \Sigma_v)$ with multiplicity $\geq T$. Since $|\sigma\alpha|_v \leq 1$ for any $\sigma \in \Sigma_v$, we deduce

$$\left|\frac{F(w)}{\prod_{\sigma\in\Sigma_v}(w-\sigma\alpha)^T}\right|_v \leq R^{-dT} \sup_{|z|_v=R} |F(z)|_v$$

for any $w \in \mathbb{C}_v$ with $|w|_v \leq 1$ and any $R > 1$. Let $R \to 1$: for the same $w \in \mathbb{C}_v$, we obtain

$$|F(w)|_v \leq \prod_{\sigma\in\Sigma_v} |w-\sigma\alpha|_v^T = |f(w)|_v^T.$$

Remark 3. In [AmD 1999], Th. 3.1, F. Amoroso and S. David prove a multidimensional generalization of Lemma 3.23. They first prove the corresponding estimate when α is an integer, and deduce the general case by means of the strong approximation Theorem.

3.6.6 Proof of Theorem 3.17 Following Cantor-Straus and Rausch

We shall now provide the details of the proof of Theorem 3.17 by means of the idea of Cantor, Straus and Rausch which does not use Dirichlet's box principle, but replaces the auxiliary function by a determinant.

We proceed by induction on the degree $d = [\mathbb{Q}(\alpha) : \mathbb{Q}]$. We may assume that d is sufficiently large, and that the conclusion of Theorem 3.17 holds for any algebraic integer of degree $< d$. Let α be an algebraic integer of degree d. By Lemma 3.22, we may assume that for any prime number p, the number α^p has degree d over $\mathbb{Q}$.

Let $P = \{p_1, \dots, p_r\}$ be a set of r distinct primes. Define $L = d(T + r)$. As we have seen, Dobrowolski's original proof involved the construction of a nonzero auxiliary polynomial f of degree $< L$ which vanishes at the points $\sigma_1\alpha, \dots, \sigma_d\alpha$ with multiplicity $\geq T$. The statement that *not all of the numbers* $f(\sigma_i\alpha^{p_j})$ *are zero* will be called *the zero estimate*.

In place of this construction, Cantor, Straus and Rausch consider the system of equations which occurs in the zero estimate, namely

$$\begin{cases} \dfrac{1}{t!}\left(\dfrac{d}{dX}\right)^t f(\sigma_i\alpha) = 0, & (0 \leq t < T, \quad 1 \leq i \leq d) \\ f(\sigma_i\alpha^{p_j}) = 0, & (1 \leq j \leq r, \quad 1 \leq i \leq d), \end{cases}$$

where the unknowns are the coefficients of $f \in \mathbb{Z}[X]$ with $\deg f < L$. The number of unknowns (the coefficients of f) is L, which is also the number of equations. Let Δ be the determinant of this system. We are going to write down Δ explicitly.

We consider the set $\{\zeta_1, \zeta_2, \dots, \zeta_L\}$ of complex numbers defined by

$$\begin{cases} \zeta_{(i-1)T+j} = \sigma_i\alpha & \text{for } 1 \leq i \leq d \text{ and } 1 \leq j \leq T, \\ \zeta_{dT+d(j-1)+i} = \sigma_i\alpha^{p_j} & \text{for } 1 \leq i \leq d \text{ and } 1 \leq j \leq r. \end{cases}$$

This means that

- each of the d numbers $\sigma_1\alpha, \dots, \sigma_d\alpha$ is repeated T times,
- each of the dr numbers $\sigma_i\alpha^{p_j}$ $(1 \leq i \leq d, 1 \leq j \leq r)$ occurs just once.

Next define nonnegative integers $\{t_1, \dots, t_L\}$ by

$$\begin{cases} t_{(i-1)T+j} = j - 1 & \text{for } 1 \leq i \leq d \text{ and } 1 \leq j \leq T, \\ t_{dT+i} = 0 & \text{for } 1 \leq i \leq dr. \end{cases}$$

Therefore, for $1 \leq \lambda \leq L$,

$$t_\lambda = \mathrm{Card}\{\mu ;\ 1 \leq \mu < \lambda,\ \zeta_\mu = \zeta_\lambda\} < \begin{cases} T & \text{if } \zeta_\lambda \text{ is of the form } \sigma_i(\alpha), \\ 1 & \text{if } \zeta_\lambda \text{ is of the form } \sigma_i\alpha^{p_j} \end{cases}$$

and we have

$$\Delta = \det\left(\binom{\mu - 1}{t_\lambda}\zeta_\lambda^{\mu-1-t_\lambda}\right)_{1 \leq \lambda, \mu \leq L},$$

where the binomial coefficient $\binom{m}{n}$ is defined as 0 for $n > m$.

We first check (zero estimate) $\Delta \neq 0$. Indeed, otherwise, there would exist a nonzero polynomial of degree $< L$ vanishing at $\sigma_1\alpha, \dots, \sigma_d\alpha$ with multiplicity $\geq T$, and with a root at each point $\sigma_i\alpha^{p_j}$ $(1 \leq i \leq d, 1 \leq j \leq r)$. Since the $d(r+1)$ numbers

$$\sigma_1\alpha, \dots, \sigma_d\alpha \quad \text{and} \quad \sigma_i\alpha^{p_j}, \quad (1 \leq i \leq d,\ 1 \leq j \leq r)$$

are pairwise distinct (recall Lemmas 3.19 and 3.22), and since a nonzero polynomial of degree $< L$ has not more than $L - 1$ roots (counting multiplicities), that is not possible.

We now invoke Fermat's little Theorem (Lemma 3.20) in order to get a lower bound for the absolute value of the interpolation determinant Δ.

Lemma 3.24. *We have*

$$|\Delta| \geq \prod_{j=1}^{r} p_j^{dT}.$$

Proof. For $T_1, \dots, T_m$ positive integers with $T_1 + \cdots + T_m = L$, consider the determinant $D \in \mathbb{Z}[X_1, \dots, X_m]$ of the following $L \times L$ matrix

$$\boldsymbol{M} = (\, \boldsymbol{M}_1 \quad \boldsymbol{M}_2 \quad \cdots \quad \boldsymbol{M}_m \,)$$

where, for $1 \leq j \leq m$, $\boldsymbol{M}_j$ denotes the $L \times T_j$ block

$$\boldsymbol{M}_j = \begin{pmatrix} 1 & 0 & \cdots & 0 \\ X_j & 1 & \cdots & 0 \\ X_j^2 & 2X_j & \cdots & 0 \\ \vdots & \vdots & \ddots & \vdots \\ X_j^{\mu-1} & (\mu-1)X_j^{\mu-2} & \cdots & \binom{\mu-1}{T_j-1}X_j^{\mu-T_j} \\ \vdots & \vdots & \ddots & \vdots \\ X_j^{L-1} & (L-1)X_j^{L-2} & \cdots & \binom{L-1}{T_j-1}X_j^{L-T_j} \end{pmatrix}.$$

The square of D is a symmetric polynomial in $X_1, \dots, X_m$. Moreover for $1 \leq i < j \leq m$ the polynomial D is divisible, in the ring $\mathbb{Z}[X_1, \dots, X_m]$, by $(X_i - X_j)^{T_iT_j}$.

Choose $m = d(r+1)$, $T_1 = \cdots = T_d = T$, $T_{d+1} = \cdots = T_m = 1$. If we define, for $0 \leq j \leq r$ and $1 \leq i \leq d$,

$$\xi_{jd+i} = \sigma_i\alpha^{p_j}$$

where $p_0 = 1$, then we have $\Delta = \pm D(\xi_1, \dots, \xi_m)$. It follows from Lemma 3.20 that Δ^2 is a rational integer, which is divisible by p_j^{2dT} for $1 \leq j \leq r$. □

Next we produce an upper bound for the absolute value of Δ:

Lemma 3.25. *We have*

$$|\Delta| \le L^{d(T^2+r)/2}\mathrm{M}(\alpha)^{L(T+p_1+\cdots+p_r)}.$$

Proof. We use the so-called *Hadamard's inequality* (see for instance [F 1982], Appendix C):

- *the determinant* Δ *of a* $L \times L$ *matrix* $(a_{\lambda\mu})_{1\le\lambda,\mu\le L}$ *satisfies the inequality*

$$|\Delta|^2 \le \prod_{\lambda=1}^{L}\sum_{\mu=1}^{L}|a_{\lambda\mu}|^2.$$

Here we get

$$|\Delta|^2 \le \prod_{\lambda=1}^{L}\sum_{\mu=1}^{L}\binom{\mu-1}{t_\lambda}^2 \max\{1,|\zeta_\lambda|\}^{2(\mu-1)}.$$

We split the product on λ in two parts: the first one is

$$\begin{aligned}\prod_{\lambda=1}^{dT}\sum_{\mu=1}^{L}\binom{\mu-1}{t_\lambda}^2 \max\{1,|\zeta_\lambda|\}^{2(\mu-1)} &= \prod_{i=1}^{d}\prod_{j=1}^{T}\sum_{\mu=1}^{L}\binom{\mu-1}{j-1}^2 \max\{1,|\sigma_i\alpha|\}^{2L}\\ &\le \prod_{i=1}^{d}\prod_{t=0}^{T-1} L^{2t+1}\max\{1,|\sigma_i\alpha|\}^{2L}\\ &\le L^{dT^2}\mathrm{M}(\alpha)^{2LT}\end{aligned}$$

and the second

$$\begin{aligned}\prod_{\lambda=dT+1}^{L}\sum_{\mu=1}^{L}\binom{\mu-1}{t_\lambda}^2 \max\{1,|\zeta_\lambda|\}^{2(\mu-1)} &= \prod_{i=1}^{d}\prod_{j=1}^{r}\sum_{\mu=1}^{L}\max\{1,|\sigma_i\alpha^{p_j}|\}^{2L}\\ &\le L^{dr}\mathrm{M}(\alpha)^{2L(p_1+\cdots+p_r)}.\end{aligned}$$

□

We now complete the proof of Theorem 3.17. From Lemmas 3.24 and 3.25 we derive

$$\prod_{j=1}^{r} p_j^{dT} \le L^{d(T^2+r)/2}\mathrm{M}(\alpha)^{L(T+p_1+\cdots+p_r)}.$$

Recall that $L = d(T+r)$. Take for $p_1,\ldots,p_r$ the first r primes with $r = T^2 - T$, and define

$$T = \left[5\frac{\log d}{\log\log d}\right].$$

From the prime number Theorem ([HaWr 1938], Th. 6 and Chap. 22) one deduces

$$\sum_{j=1}^{r} \log p_j \simeq r \log r \quad \text{and} \quad \sum_{j=1}^{r} p_j \simeq \frac{1}{2} r^2 \log r$$

as $r \to \infty$. Since, as soon as d is sufficiently large, we have

$$\frac{dT^2}{L} > \frac{dT \log L}{2L \log T} + \frac{4}{5},$$

we deduce

$$\log \mathrm{M}(\alpha) > \frac{1}{T^3}.$$

□

3.6.7 Further Related Questions and Results

Several related results are worth mentioning.

A. Schinzel, E. Dobrowolski and W. Lawton gave lower bounds for the height for an algebraic number α (which is neither zero nor a root of unity) in terms of the number of non-vanishing coefficients of a polynomial $f \in \mathbb{Z}[X]$ such that $f(\alpha) = 0$.

In [Mi 1979], M. Mignotte gave a lower bound for $|\alpha - 1|$ which is stronger than Liouville's one in terms of the degree; this is specially interesting when α has a small Mahler's measure. The proof involved an auxiliary polynomial. This estimate was improved in [MiW 1994] by means of an interpolation determinant. A p-adic analogue has been obtained by Y. Bugeaud [Bu 1998a]. Further refinements are due to E. M. Matveev [Mat 1996b], and F. Amoroso [Am 1996] and [Am 1998]. A remarkable connection with Riemann's hypothesis is described in [Am 1996], while [Am 1998] contains a survey of such results.

Higher dimensional generalizations of Kronecker's Theorem, Lehmer's problem and Dobrowolski's estimate have been considered from different points of view.

In [AmD 1999], F. Amoroso and S. David extend Dobrowolski's result to simultaneous approximation. Lehmer's Problem is related to the multiplicative group $\mathbb{G}_{\mathrm{m}}$. Here is a generalization to $\mathbb{G}_{\mathrm{m}}^n$ suggested in [AmD 1999].

Conjecture 3.26. *For each positive integer $n \geq 1$ there exists a positive number $c_1(n)$ such that, if $\underline{\alpha} = (\alpha_1, \ldots, \alpha_n)$ is a n-tuple of multiplicatively independent algebraic numbers and if $\omega(\underline{\alpha})$ denotes the minimum degree of a nonzero polynomial in $\mathbb{Q}[X_1, \ldots, X_n]$ which vanishes at $\underline{\alpha}$, then*

$$\mathrm{h}(1 : \alpha_1 : \cdots : \alpha_n) \geq \frac{c_1(n)}{\omega(\underline{\alpha})}. \tag{3.27}$$

A partial result is proved in [AmD 1999]:

$$\mathrm{h}(1 : \alpha_1 : \cdots : \alpha_n) \geq \frac{c_2(n)}{\omega(\underline{\alpha})\bigl(1 + \log \omega(\underline{\alpha})\bigr)^{\kappa(n)}}$$

for some positive constants $c_2(n)$ and $\kappa(n)$ which depend only on n.

A consequence (see Exercise 3.13) of Conjecture 3.26 is the following open problem:

(?) *For each positive integer $n \geq 1$ there exist a positive number $c_3(n)$ having the following property. Let $\alpha_1, \ldots, \alpha_n$ be multiplicatively independent algebraic numbers. Define $D = [\mathbb{Q}(\alpha_1, \ldots, \alpha_n) : \mathbb{Q}]$. Then*

$$\prod_{i=1}^{n} \mathrm{h}(\alpha_i) \geq \frac{c_3(n)}{D}. \tag{3.28}$$

A weaker estimate of the form

$$\prod_{i=1}^{n} \mathrm{h}(\alpha_i) \geq \frac{c_4(n)}{D(1+\log D)^{n\kappa(n)}} \tag{3.29}$$

is proved in [AmD 1999].

Another kind of higher dimensional *Lehmer type* problem arose from a work of S. Zhang in 1992 on positive line bundles on arithmetic varieties. He showed that *if $\mathcal{V}$ is a curve of a linear torus which is not a translate of a subtorus of positive dimension by a torsion point, then there exists a positive constant c such that $\mathcal{V}$ has only finitely many algebraic points of height $\leq c$.* Hence for sufficiently small c these points have a vanishing height. A more elementary proof of this result for the special case of the curve $x + y = 1$ in the torus $\mathbb{G}_{\mathrm{m}}^2$ was given by D. Zagier, with the best possible value for the constant: *any solution $(x, y) \in \overline{\mathbb{Q}}^2$ of the equation $x + y = 1$ with $x \neq 0$ and $x^6 \neq 1$ satisfies*

$$\mathrm{h}(x) + \mathrm{h}(y) \geq \frac{1}{2} \log\left(\frac{1+\sqrt{5}}{2}\right).$$

F. Beukers and D. Zagier gave sharp explicit results on this question and mentioned a number of applications. W. M. Schmidt, then E. Bombieri and U. Zannier, and then W. M. Schmidt again, extended Zagier's elementary argument to higher dimensional subvarieties of $\mathbb{G}_{\mathrm{m}}^n$. A survey on this topic is given by W. M. Schmidt in [Sc 1999].

Algebraic units u such that $1 - u$ is also a unit are sometimes called *exceptional units*. See [Sil 1996] for a survey of this topic.

One main tool is the notion of height for subvarieties of an affine or projective space. So far we have considered only the height of an algebraic point, which has dimension 0.

Subvarieties of codimension 1 are hypersurfaces . Lemma 3.9 suggested to Mahler a natural extension of his measure to polynomials in several variables (see [S 1999]):

$$\log \mathrm{M}(f) = \int_0^1 \cdots \int_0^1 \log |f(e^{2i\pi t_1}, \ldots e^{2i\pi t_n})| dt_1 \cdots dt_n.$$

The name *generalized cyclotomic polynomial* is sometimes used for a polynomial f (in several variables) which defines a hypersurface $\mathcal{V}$ of a torus containing a translate of a subtorus by a torsion point. By a result of D. Boyd, W. Lawton and C. Smyth, an irreducible polynomial f which is not a generalized cyclotomic polynomial has $\mathrm{M}(f) > 1$.

For a subvariety of any dimension, H. Gillet, C. Soulé, G. Faltings and P. Philippon introduced closely related notions of height[9]. For instance (see [DP 1999]) the canonical height of a hypersurface defined by $F = 0$ (where F is an irreducible polynomials with coefficients in $\mathbb{Z}$) is nothing else than $\log \mathrm{M}(F)$ where M is Mahler's measure (in several variables).

In the case (which we are interested in) of a subvariety of a torus, a *canonical height* can be defined (à la Néron-Tate), which vanishes exactly for the subvarieties containing a translate of a subtorus by a torsion point. For a hypersurface V of $\mathbb{G}_{\mathrm{m}}^n$ defined over $\mathbb{Q}$, say $f = 0$, which is not an algebraic subgroup of $\mathbb{G}_{\mathrm{m}}$, a lower bound for $\mathrm{h}(V)$ (that is for $\log \mathrm{M}(f)$) has been given in [AmD 2000].

An interesting related topic (see for instance [DP 1999]) is then to compare the height of the variety $\mathcal{V}$ with the minimum height of algebraic points on $\mathcal{V}$. The limit distribution of small points on algebraic tori has been studied by Y. Bilu.

These problems are the multiplicative analogues of a conjecture of F. A. Bogomolov about the discreteness of algebraic points on an algebraic curve of genus at least 2 with respect to the distance induced by the Néron-Tate height on the Jacobian. We do not deal here with Abelian varieties, and we shall only refer to work by L. Szpiro, J-F. Burnol, S. Zhang, E. Bombieri and U. Zannier, E. Ullmo, S. David and P. Philippon (extensions to semi-abelian varieties have also been considered by B. Poonen and A. Chambert-Loir).

There are further lower bounds for the height of algebraic numbers. For instance A. Schinzel and E. Dobrowolski got estimates which depend on the number of nonzero coefficients of the minimal polynomial. In [Mat 1996a], E. M. Matveev proves, for some classes of algebraic integers, a sharper estimate than Dobrowolski's one including the discriminant Δ of α: he replaces the degree d of α by $d/\Delta^{1/d}$.

Other estimates are due to M. Langevin; on one hand he proves [La 1986]: *Let $\mathcal{V}$ be a neighborhood of a point of the unit circle. There exists two effectively computable constants $c > 1$ and $D_0 > 0$ such that for any nonzero algebraic number α of degree $D \geq D_0$, all of whose conjugates are outside $\mathcal{V}$, the inequality $\mathrm{M}(\alpha) > c^D$ holds.* On the other hand, after a joint work with E. Reyssat and G. Rhin, he answered two questions of Favard by proving lower bounds for the *diameter* of an algebraic integer α, which is defined as

$$\operatorname{diam}(\alpha) = \max_{1 \leq i \neq j \leq d} |\alpha_i - \alpha_j|,$$

where $\{\alpha_1, \ldots, \alpha_d\}$ is the set of conjugates of α. These lower bounds are

$$\operatorname{diam}(\alpha) \geq \sqrt{3} \quad \text{for } d = [\mathbb{Q}(\alpha) : \mathbb{Q}] \geq 2$$

[9] A special case was already considered by W. M. Schmidt (see [Sc 1980]) who defined the height of a vector subspace by considering the Plücker coordinates of the corresponding point in a Grassmanian.

and, for any $\epsilon > 0$,

$$\operatorname{diam}(\alpha) \geq 2 - \epsilon \qquad \text{for } d \geq d_0(\epsilon).$$

For further results and references on the height of algebraic numbers, see Chap. 7 of [BerDGPS 1992], [S 1999] and [Sc 1999].

Open Problems

1. (Lehmer's problem [Le 1933] — see § 3.6). Does there exist an absolute constant $c_0 > 0$ such that, for any nonzero algebraic number which is not a root of unity, $d\mathrm{h}(\alpha) \geq c_0$?

2. (Conjecture of Schinzel and Zassenhaus [SZa 1965]). Does there exist an absolute constant $c > 0$ such that, for any nonzero algebraic integer of degree d which is not a root of unity, $\overline{|\alpha|} \geq 1 + (c/d)$?

Since, for any algebraic integer α of degree d, we have $\mathrm{h}(\alpha) \leq \log \overline{|\alpha|} \leq d\mathrm{h}(\alpha)$, the conjecture of Schinzel and Zassenhaus would follow from a positive answer to Lehmer's problem 1 above.

3. (A. Dubickas) Check that for any nonzero algebraic integer α of degree d which is not a root of unity,

$$\log \max\left\{\overline{|\alpha|}\ ;\ \overline{|\alpha^{-1}|}\right\} \geq \frac{1}{d}\log 2.$$

4. (D. Boyd [Boy 1980]) The minimal value for $\overline{|\alpha|}$ when α is a nonzero algebraic integer of degree d which is not a root of unity should be reached for the roots of

$$X^d + X^{2d/3} - 1$$

with d multiple of 3. An example is $X^3 + X^2 - 1$.

5. In the case $f(X) = qX - p$ with p and q rational integers, Liouville's inequality (Theorem 1.1) gives an estimate for the approximation of algebraic numbers by rational numbers. In this special case this lower bound is not the best known (Theorem 1.10 of Thue-Siegel-Roth-Schmidt; see [Sc 1980]). Is it possible to improve the estimate in the general case of Proposition 3.14? Even an ineffective result might be useful.

Exercises

Exercise 3.1. Let $\alpha_1, \ldots, \alpha_s$ be algebraic numbers. Define $k = \mathbb{Q}(\alpha_1, \ldots, \alpha_s)$ and $d = [k : \mathbb{Q}]$. Show that there exist rational integers $a_2, \ldots, a_s$ with $0 \leq a_i \leq d(d-1)/2$ such that the number $\gamma = \alpha_1 + a_2\alpha_2 + \cdots + a_s\alpha_s$ satisfies $k = \mathbb{Q}(\gamma)$.

Hint. *See* [MiW 1977] *Lemme 3.*

Exercise 3.2.
a) For $f \in \mathbb{C}[X_1, \ldots, X_t]$, we denote by $|f|_1$ the upper bound of $|f(\underline{z})|$ on the unit polydisc:

$$|f|_1 = \sup\left\{|f(z_1, \ldots, z_t)|\,;\ \underline{z} \in \mathbb{C}^t,\ |z_i| = 1,\ 1 \leq i \leq t\right\}.$$

Hence $|f|_1 \leq \mathrm{L}(f)$. Show that in Lemmas 3.7, 3.8 and Proposition 3.14, one can replace $\log \mathrm{L}(f)$ by $\log |f|_1$.

Hint. *Start by proving the following statement: if $a_0, \ldots, a_N$, y are complex numbers, then*

$$\left|\sum_{i=0}^{N} a_i y^i\right| \leq \sup_{|z|=1} \left|\sum_{i=0}^{N} a_i z^i\right| \cdot \max\left(1, |y|\right)^N.$$

When $|y| \leq 1$, this inequality follows from the maximum modulus principle for $a_0 + a_1 z + \cdots + a_N z^N$. When $|y| > 1$, perform the change of variables $z' = 1/z$.

Deduce by induction: for a polynomial $f \in \mathbb{C}[X_1, \ldots, X_t]$, when $y_1, \ldots, y_t$ are complex numbers,

$$|f(y_1, \ldots, y_t)| \leq |f|_1 \prod_{i=1}^{t} \max(1, |y_i|)^{\deg_{X_i} f}.$$

b) For an algebraic number γ of degree d and minimal polynomial

$$a_0 X^d + \cdots + a_d = a_0 \prod_{i=1}^{d} (X - \gamma_i),$$

define a modified Mahler's measure by

$$\widetilde{\mathrm{M}}(\gamma) = a_0 \prod_{i=1}^{d} \sqrt{1 + |\gamma_i|}$$

and a modified absolute logarithmic height by

$$\widetilde{\mathrm{h}}(\gamma) = \frac{1}{d} \log \widetilde{\mathrm{M}}(\gamma).$$

Check, under the assumptions of Lemma 3.7,

$$\widetilde{\mathrm{h}}\big(f(\gamma_1, \ldots, \gamma_t)\big) \leq \log \mathrm{H}(f) + \sum_{i=1}^{t} \left(\deg_{X_i} f\right) \widetilde{\mathrm{h}}(\gamma_i).$$

Hint. *Compare with* [Sc 1991], *Chap. I, § 7, Lemma 7D.*

c) Let k be a number field of degree d. For $\underline{\gamma} = (\gamma_0 : \cdots : \gamma_\nu) \in \mathbb{P}_\nu(k)$, define

$$\widetilde{\mathrm{h}}(\underline{\gamma}) = \frac{1}{d} \sum_{v \in M_k} d_v \log \|\underline{\gamma}\|_v,$$

where

$$\|\underline{\gamma}\|_v = \begin{cases} \max\{|\gamma_0|_v, \ldots, |\gamma_\nu|_v\} & \text{for } v \text{ ultrametric,} \\ \sqrt{|\gamma_0|_v^2 + \cdots + |\gamma_\nu|_v^2} & \text{for } v \text{ Archimedean.} \end{cases}$$

Check that one can replace the height h by this modified height $\widetilde{\mathrm{h}}$ and at the same time the length L by the usual height H in Lemmas 3.7, 3.8 and 3.14.

Exercise 3.3.
a) Let N and M be positive integers and $\vartheta_1, \ldots, \vartheta_N, \theta_1, \ldots, \theta_M$ algebraic numbers. Check that

$$\mathrm{h}(1 : \vartheta_1 : \cdots : \vartheta_N : \theta_1 : \cdots : \theta_M) \le \mathrm{h}(1 : \vartheta_1 : \cdots : \vartheta_N) + \mathrm{h}(1 : \theta_1 : \cdots : \theta_M).$$

Deduce, for algebraic numbers $\vartheta_0, \ldots, \vartheta_s$, not all of which are zero,

$$\mathrm{h}(\vartheta_0 : \cdots : \vartheta_s) \le \sum_{i=0}^{s} \mathrm{h}(\vartheta_i).$$

b) Let $a_1, \ldots, a_n$ be rational integers, $b_1, \ldots, b_n$ be non-vanishing integers and $\beta_1, \ldots, \beta_n$ algebraic numbers. Define

$$N = \max\{|a_1|, |b_1|, \ldots, |a_n|, |b_n|\}$$

and

$$\gamma = \frac{a_1}{b_1}\beta_1 + \cdots + \frac{a_n}{b_n}\beta_n.$$

Then

$$\mathrm{h}(\gamma) \le n(n+1)\log N + \log n + \sum_{i=1}^{n} \mathrm{h}(\beta_i).$$

Hint. *This is Lemma 2.7 of* [W 1980].

c) Let $L_1, \ldots, L_k, N_1, \ldots, N_k$ and M be positive integers. For $1 \le i \le k$, let $\gamma_{0i}, \ldots, \gamma_{N_i i}$ be algebraic numbers. Assume that for each $i = 1, \ldots, k$, at least one of the numbers $\gamma_{0i}, \ldots, \gamma_{N_i i}$ is nonzero and denote by γ_i the point in $\mathbb{P}_{N_i}(\overline{\mathbb{Q}})$ with projective coordinates $(\gamma_{0i} : \cdots : \gamma_{N_i i})$. We will also write $\underline{\gamma}$ for the point $(\gamma_{vi})_{0 \le v \le N_i, 1 \le i \le k}$ in $\overline{\mathbb{Q}}^{N_1 + \cdots + N_k + k}$. Furthermore, let $F_0, \ldots, F_M$ be polynomials in $N_1 + \cdots + N_k + k$ variables, with coefficients in $\mathbb{Z}$, each of which is homogeneous of degree L_i with respect to the $N_i + 1$ variables $X_{0i}, \ldots, X_{N_i i}$. Assume that one at least of the $M+1$ numbers $\theta_\mu = F_\mu(\underline{\gamma})$ $(0 \le \mu \le M)$ is nonzero, and define θ as the point in $\mathbb{P}_M(\overline{\mathbb{Q}})$ with projective coordinates $(\theta_0 : \cdots : \theta_M)$. Then

$$\mathrm{h}(\theta) \le \log \max_{0 \le \mu \le M} \mathrm{L}(F_\mu) + \sum_{i=1}^{k} L_i \mathrm{h}(\gamma_i).$$

d) Let $P \in \overline{\mathbb{Q}}[X_0, \ldots, X_n, Y]$ be a homogeneous polynomial in $n+2$ variables such that $P(0, \ldots, 0, 1) \neq 0$. Let $(\alpha_0 : \cdots : \alpha_n : \beta) \in \mathbb{P}_{n+1}(\overline{\mathbb{Q}})$ satisfy $P(\alpha_0 : \cdots : \alpha_n : \beta) = 0$. Then

$$\mathrm{h}(\alpha_0 : \cdots : \alpha_n : \beta) \le \mathrm{h}(\alpha_0 : \cdots : \alpha_n) + \mathrm{h}(p) + \log N,$$

where $N+1$ is the number of monomials in P and p is the projective point which is defined by the sequence of coefficients of P.
(Compare with [Ser 1989], § 2.3, N°4, Prop. 14.)
e) For any polynomial $F \in \mathbb{Z}[X, T]$, there exists a constant $c > 0$ such that, if α and β are algebraic numbers with $F(\alpha, \beta) = 0$, and if the polynomial $F(\alpha, T) \in \mathbb{Q}(\alpha)[T]$ is not zero, then $\mathrm{h}(\beta) \leq c \max\{1, \mathrm{h}(\alpha)\}$.

Exercise 3.4. For $f \in \mathbb{Z}[X]$ a nonzero polynomial, define $t(f) = \deg f + \log \mathrm{H}(f)$. For an algebraic number α with minimal polynomial $f_\alpha \in \mathbb{Z}[X]$, define $t(\alpha) = t(f_\alpha)$. Check the following *Liouville's inequality*:

If $f \in \mathbb{Z}[X]$ and $\alpha \in \overline{\mathbb{Q}}$ satisfy $f(\alpha) \neq 0$, then

$$|f(\alpha)| \geq e^{-t(f)t(\alpha)}.$$

Hint. *One may use Proposition 3.14 together with the estimates*

$$(N+1)^{d-1}(d+1)^{N/2} \leq e^{dN} \quad \textit{for integers } d \geq 1 \textit{ and } N \geq 0.$$

Remark. Another way of proving a lower bound for $|f(\alpha)|$ is to use the fact that the resultant of f and f_α is a nonzero rational integer - see [Bor 1899].

Exercise 3.5. Show that in Proposition 3.14, if the Archimedean absolute value v is not real, then the conclusion can be refined as

$$\log |f(\underline{\gamma})|_v \geq -\left(\frac{d}{2} - 1\right) \log |f|_1 - \frac{d}{2} \sum_{i=1}^{t} N_i \mathrm{h}(1 : \gamma_{i1} : \cdots : \gamma_{i\nu_i}).$$

Show also that if v is an ultrametric absolute value of k, then

$$\log |f(\underline{\gamma})|_v \geq -\frac{d}{d_v} \left(\log |f|_1 + \sum_{i=1}^{t} N_i \mathrm{h}(1 : \gamma_{i1} : \cdots : \gamma_{i\nu_i}) \right).$$

where d_v is, as usual, the local degree at v.

Hint. *Use Exercise 3.2.a.*

Exercise 3.6. Let $f \in \mathbb{Z}[X]$ be a nonzero polynomial of degree d with leading coefficient $a_0 > 0$ and let $\alpha \in \mathbb{C}$ be a zero of f.
a) Let p/q be a rational number with $q > 0$ such that $f(p/q) \neq 0$. Show that

$$\left| \alpha - \frac{p}{q} \right| \geq \frac{\max\{1, |\alpha|\}}{q(|p| + q)^{d-1} \mathrm{M}(f)}.$$

b) Deduce that for an algebraic number α of degree d, if we set

$$c(\alpha) = \begin{cases} \dfrac{1}{2^{d-1} \mathrm{M}(\alpha)} & \text{if } |\alpha| \leq 1, \\[2ex] \dfrac{|\alpha|}{(2 + |\alpha|)^{d-1} \mathrm{M}(\alpha)} & \text{if } |\alpha| > 1, \end{cases}$$

then for all $p/q \in \mathbb{Q}$ with $p/q \neq \alpha$ we have

$$\left|\alpha - \frac{p}{q}\right| \geq \frac{c(\alpha)}{q^d}.$$

c) Show that, for each $\kappa > |f'(\alpha)|$, there are only finitely many $p/q \in \mathbb{Q}$ with

$$\left|\alpha - \frac{p}{q}\right| \leq \frac{1}{\kappa q^d}.$$

Example: Let α be a real quadratic number, which is root of a polynomial $aX^2 + bX + c$ of discriminant $\Delta = b^2 - 4ac > 0$. Then for each $\kappa > \sqrt{\Delta}$ there exist $q_0 > 0$ such that, for $p/q \in \mathbb{Q}$ with $q > q_0$,

$$\left|\alpha - \frac{p}{q}\right| > \frac{1}{\kappa q^2}.$$

Exercise 3.7.
a) Let β be a nonzero algebraic number and λ a nonzero logarithm of an algebraic number. Define $\alpha = e^\lambda$ and $D = [\mathbb{Q}(\alpha, \beta) : \mathbb{Q}]$. Then

$$|\beta\lambda| > \left(2e^{\mathrm{h}(\alpha)+\mathrm{h}(\beta)}\right)^{-D}.$$

Hint. *Using (3.13), deduce* $|\beta| \geq e^{-D\mathrm{h}(\beta)}$. *Using Proposition 3.14, show that* $|\alpha - 1| \geq 2\left(2e^{\mathrm{h}(\alpha)}\right)^{-D}$ *if* $\alpha \neq 1$. *From Exercise 1.1, deduce* $\min\{|\alpha - 1|, 1\} < 2|\lambda|\}$.

b) Let $\lambda_1, \ldots, \lambda_m$ be logarithms of algebraic numbers and $b_1, \ldots, b_m$ rational integers. Let D be the degree of a number field containing the m algebraic numbers $\alpha_j = \exp(\lambda_j)$ $(1 \leq j \leq m)$. If the number

$$\Lambda = b_1\lambda_1 + \cdots + b_m\lambda_m$$

is nonzero, then

$$|\Lambda| \geq 2^{-D} \exp\left\{-D\sum_{j=1}^{m} |b_j|\mathrm{h}(\alpha_j)\right\}.$$

Exercise 3.8. Let $\alpha_1, \ldots, \alpha_{n+1}$ be nonzero algebraic numbers and $\beta_1, \ldots, \beta_n$ be algebraic numbers. Denote by D the degree of the number field

$$\mathbb{Q}(\alpha_1, \ldots, \alpha_{n+1}, \beta_1, \ldots, \beta_n).$$

Let $T_0, T_1, S_1, \ldots, S_{n+1}$ be positive rational integers. Define $L = \binom{T_0+n}{n}(2T_1 + 1)$ and $S^* = \max\{S_1, \ldots, S_{n+1}\}$. Further let $\underline{s}^{(1)}, \ldots, \underline{s}^{(L)}$ be any elements in the set $\mathbb{Z}^{n+1}(\underline{S})$ of $\underline{s} = (s_1, \ldots, s_{n+1}) \in \mathbb{Z}^{n+1}$ which satisfy $|s_i| \leq S_i$ $(1 \leq i \leq n+1)$. Let Δ be the determinant of the $L \times L$ matrix

$$\left((s_1^{(\mu)} + s_{n+1}^{(\mu)}\beta_1)^{\tau_1} \cdots (s_n^{(\mu)} + s_{n+1}^{(\mu)}\beta_n)^{\tau_n}\left(\alpha_1^{s_1^{(\mu)}} \cdots \alpha_{n+1}^{s_{n+1}^{(\mu)}}\right)^t\right)_{\substack{(\tau,t)\\ 1\leq\mu\leq L}},$$

where (τ, t) ranges over the set of elements $(\tau_1, \ldots, \tau_n, t) \in \mathbb{N}^n \times \mathbb{Z}$ for which $\tau_1 + \cdots + \tau_n \leq T_0$ and $|t| \leq T_1$. Assume $\Delta \neq 0$. Prove

$$\frac{1}{L}\log|\Delta| \geq -(D-1)\big(T_0\log(2S^*) + \log L\big) - D(T_1+1)\sum_{i=1}^{n+1} S_i\mathrm{h}(\alpha_i) - DT_0\mathrm{h}(1:\beta_1:\cdots:\beta_n).$$

Hint. *Use Lemma 3.15 with* $\ell = 2n + 3$,

$$\nu_1 = \cdots = \nu_{2n+2} = 1, \quad \nu_{2n+3} = n,$$

$$p_{\lambda\mu} = \prod_{j=1}^{n} \left(s_j^{(\mu)} + s_{n+1}^{(\mu)} X_{2n+3,j}\right)^{\tau_j} \cdot \prod_{i=1}^{n+1} \left(X_{i1}^{\max\{ts_i^{(\mu)},0\}} X_{n+1+i,1}^{\max\{-ts_i^{(\mu)},0\}}\right)$$

where (τ, t) *corresponds to the index* λ,

$$N_{i\lambda} = N_{n+1+i,\lambda} = t \max |s_i^{(\mu)}| \leq tS_i \quad (1 \leq i \leq n+1)$$

and $N_{2n+3,\lambda} \leq T_0$, *so that* $L(p_{\lambda\mu}) \leq (2S^*)^{T_0}$,

$$\sum_{\lambda=1}^{L} N_{i\lambda} = \sum_{\lambda=1}^{L} N_{n+1+i,\lambda} \leq \frac{1}{2} L(T_1 + 1) S_i \quad (1 \leq i \leq n+1)$$

and

$$\sum_{\lambda=1}^{L} N_{2n+3,\lambda} \leq LT_0$$

Next apply Proposition 3.14 with $\gamma_{i1} = \alpha_i$ *for* $1 \leq i \leq n+1$, $\gamma_{i1} = \alpha_{i-n-1}^{-1}$ *for* $n+2 \leq i \leq 2n+2$ *and* $\gamma_{2n+3,j} = \beta_j$ *for* $1 \leq j \leq n$.

Exercise 3.9.
a) Check that for a nonzero algebraic number α, of degree $d \in \{1, 2, 3, 4, 5\}$, which is not a root of unity, the number $d\mathrm{h}(\alpha) = \log \mathrm{M}(\alpha)$ is bounded from below by the value given in table 3.30 (the last column provides a polynomial which yields the minimum).

Table 3.30

$d =$	$d\mathrm{h}(\alpha) \geq$	minimum for
1	$\log 2 = 0.6931\ldots$	$X - 2$
2	$\log\left((1+\sqrt{5})/2\right) = 0.4812\ldots$	$X^2 - X - 1$
3	$0.2811\ldots$	$X^3 - X - 1$
4	$0.3223\ldots$	$X^4 - X - 1$
5	$0.2998\ldots$	$X^5 - X^4 + X^3 - X + 1$

b) Show that the proof (see § 3.6) of Kronecker's result is effective: if d is a positive integer, there exists a positive number $c(d)$ such that, for any nonzero algebraic numbers α which is not a root of unity and is of degree at most d, the inequality $\mathrm{h}(\alpha) \geq c(d)$ is valid.

Hint. *Let α be an algebraic number of degree at most d. Assume that there exists a positive integer ℓ such that*

$$\mathrm{M}(\alpha)^{\ell} < 1 + 2^{-d} \quad \textit{and} \quad \ell \geq d(2^{d+1}+1)^{d+1}.$$

Check $\mathrm{H}(\alpha^j) \leq 2^d$ for $0 \leq j \leq \ell$ and deduce that the numbers $1, \alpha, \ldots, \alpha^{\ell}$ are not pairwise distinct.

c) Let A and d be two positive integers, H and C two positive real numbers, and α a nonzero algebraic number of degree d. Assume

$$d\mathrm{h}(\alpha) \leq \frac{1}{H}, \quad \frac{1}{C^2} = \left(\frac{\pi}{A}\right)^2 + \left(\frac{2A-1}{H}\right)^2$$

and

$$C > 2^d e^{(2A-1)/H}.$$

Show that α is a root of unity of order $< 2A$.

Hint. *Show that there exists an integer r in the range $1 \leq r \leq 2A-1$ such that $|\log(\alpha^r)| \leq 1/C$. Deduce $|\alpha^r - 1| < 2/C$. Use Liouville's inequality (3.14) for $f(X) = X^r - 1$ and conclude.*

d) Deduce from c) that a suitable value for $c(d)$ in question b) above is 2^{-2d-4} (compare with [SZa 1965]).

Hint. *Choose $A = 2^{d+2}$, $H = A^2$.*

Exercise 3.10. (see [CaStr 1982] and [Ra 1985]). Let a, b, c be positive real numbers and α an algebraic integer of degree $\leq d$ which is not a root of unity. Assume that there is a prime p in the range $ad < p \leq bd$. Assume also

$$\left(1 + \frac{c}{d^2}\right)^{bd^2} \leq \frac{a}{2}.$$

Deduce

$$\overline{|\alpha|} > 1 + \frac{c}{d^2}.$$

See also [Do 1978] for a much stronger estimate.

Exercise 3.11. Show that the polynomial D which occurs in the proof of Lemma 3.24 (namely the so-called *confluent Vandermonde determinant*) is equal to

$$\pm \prod_{1 \leq i < j \leq m} (X_i - X_j)^{T_i T_j}$$

Exercise 3.12.
a) Let ν, μ, ℓ be positive integers, p_{ij} $(1 \leq i \leq \nu, 1 \leq j \leq \mu)$ polynomials in $\mathbb{Z}[X_1, \ldots, X_{\ell}]$

and $\underline{\gamma} = (\gamma_1, \dots, \gamma_\ell)$ a tuple of algebraic numbers in a number field of degree D. Define, for $1 \leq j \leq \mu$ and $1 \leq k \leq \ell$,

$$N_{kj} = \max_{1 \leq i \leq \nu} \deg_{X_k} p_{ij}$$

and

$$V_j = \left(\sum_{i=1}^{\nu} \mathrm{L}(p_{ij}) \right) \prod_{k=1}^{\ell} e^{N_{kj} \mathrm{h}(\gamma_k)}.$$

Assume $\nu > D\mu$. Show that there exist $x_1 \dots, x_\nu$ in $\mathbb{Z}$ satisfying

$$\sum_{i=1}^{\nu} x_i p_{ij}(\underline{\gamma}) = 0 \qquad (1 \leq j \leq \mu)$$

and

$$0 < \max_{1 \leq i \leq \nu} |x_i| \leq 2 + \left(2^{\mu} \left(V_1 \cdots V_\mu \right)^D \right)^{1/(\nu - \mu D)}.$$

Hint. *Use Dirichlet's box principle as in the proof of Lemmas 4.11 and 4.12. See also Lemma 4 of* [MiW 1978] *and Lemma 1 of* [Do 1979].

b) Deduce the existence of F for the first step of the proof of Dobrowolski's Theorem given in § 3.6.5.

Exercise 3.13. Let $\alpha_1, \dots, \alpha_n$ be nonzero algebraic numbers. Denote by $[\mathbb{Q}(\underline{\alpha}) : \mathbb{Q}]$ the degree of the number field they generate.
a) Recall the notation $\omega(\underline{\alpha})$ in Conjecture 3.26. Check

$$\omega(\underline{\alpha}) \leq n [\mathbb{Q}(\underline{\alpha}) : \mathbb{Q}]^{1/n}.$$

Hint. *Using linear algebra, for any integer δ satisfying*

$$\binom{\delta + n}{n} > [\mathbb{Q}(\underline{\alpha}) : \mathbb{Q}],$$

show that there exists a nonzero polynomial $P \in \mathbb{Q}[\underline{X}]$ of total degree $\leq \delta$ such that $P(\underline{\alpha}) = 0$.

b) Assume $\alpha_1, \dots, \alpha_n$ are multiplicatively independent. Let $\epsilon > 0$. Show that there exist multiplicatively independent algebraic numbers $\gamma_1, \dots, \gamma_n$ such that

$$[\mathbb{Q}(\underline{\alpha}) : \mathbb{Q}] \mathrm{h}(\alpha_1) \cdots \mathrm{h}(\alpha_n) \geq (1 - \epsilon) [\mathbb{Q}(\underline{\gamma}) : \mathbb{Q}] \mathrm{h}(1 : \gamma_1 : \cdots : \gamma_n)^n.$$

Hint. *Since $\alpha_1, \dots, \alpha_n$ are multiplicatively independent, we have $\mathrm{h}(\alpha_i) > 0$ for $1 \leq i \leq n$. Let N be a sufficiently large integer. Define $A_i = [N \mathrm{h}(\alpha_i)]$ and select $\gamma_i = \alpha_i^{1/A_i}$.*

c) Deduce that Conjecture 3.28 is a consequence of Conjecture 3.26
d) Check also that (3.29) follows from (3.27).

Hint. *Use Theorem 3.16.*

Exercise 3.14. Let $P \in \mathbb{C}[X]$ be a nonzero polynomial of degree $\leq d$ and let p be a prime number. Check

$$\prod_{\zeta} |P(\zeta)| \leq p^d \mathrm{M}(P)^{p-1},$$

where ζ (in the product of the left hand side) ranges over the set with $p - 1$ elements of primitive p-th roots of unity.

Annex to Chapter 3. Inequalities Between Different Heights of a Polynomial - From a Manuscript by Alain Durand

Let $f \in \mathbb{C}[X]$ be a nonzero polynomial with complex coefficients of degree d:

$$f = a_0X^d + a_1X^{d-1} + \cdots + a_d = a_0\prod_{i=1}^{d}(X - \alpha_i).$$

There are several notions of *height* for f. For instance we have Mahler's measure of f (see § 3.3):

$$\mathrm{M}(f) = |a_0|\prod_{i=1}^{d}\max\{1, |\alpha_i|\},$$

the usual height of f (see § 3.4):

$$\mathrm{H}(f) = \max\{|a_0|, |a_1|, \ldots, |a_d|\},$$

the Euclidean norm of f:

$$\mathrm{L}_2(f) = \left(|a_0|^2 + |a_1|^2 + \cdots + |a_d|^2\right)^{1/2} = \left(\int_0^1 |f(e^{2i\pi t})|^2 dt\right)^{1/2},$$

the sup norm on the unit disc (or on the unit circle, which is the same by the maximum modulus principle):

$$|f|_1 = \sup_{|z|\le 1}|f(z)| = \sup_{|z|=1}|f(z)|,$$

and finally the length of f (see § 3.2):

$$\mathrm{L}(f) = |a_0| + |a_1| + \cdots + |a_d|.$$

Inequalities like

$$\boxed{(d+1)^{-1/2}\mathrm{M}(f) \le \mathrm{H}(f) \le \mathrm{L}_2(f) \le |f|_1 \le \mathrm{L}(f) \le 2^d\mathrm{M}(f)}$$

relate these functions. Table 3.31 below (due to the late Alain Durand) provides an upper bound for the quotient of one of the norms (left column) by another one (first row). In each case but two, below the upper bound is displayed one polynomial for which the estimate is optimal (where f_d denotes the polynomial $1 + X + \cdots + X^d$). There are two exceptions where the optimal result is not known:
(1) By (3.12),

$$\mathrm{M}(f) \le \sqrt{d+1}\,\mathrm{H}(f).$$

There are examples which show that for d sufficiently large, there exist polynomials f of degree d with $\mathrm{H}(f) = 1$ and $\mathrm{M}(f) \ge \sqrt{d+1} - (\log d)/2$ (see [Dur 1990], p.56).
(2) One can also prove that

$$\mathrm{L}(f) \le \sqrt{d}\,|f|_1$$

Table 3.31

	$M(f)$	$H(f)$	$L_2(f)$	$\lvert f\rvert_1$	$L(f)$
$M(f) \leq$	1	$\sqrt{d+1}$ (1)	1 X^d	1 X^d	1 X^d
$H(f) \leq$	$\binom{d}{[d/2]}$ $(X+1)^d$	1	1 X^d	1 X^d	1 X^d
$L_2(f) \leq$	$\binom{2d}{d}^{1/2}$ $(X+1)^d$	$\sqrt{d+1}$ f_d	1	1 X^d	1 X^d
$\lvert f\rvert_1 \leq$	2^d $(X+1)^d$	$d+1$ f_d	$\sqrt{d+1}$ f_d	1	1 X^d
$L(f) \leq$	2^d $(X+1)^d$	$d+1$ f_d	$\sqrt{d+1}$ f_d	$\sqrt{d}$ (2)	1

and give examples of polynomials f of degree d with $\lvert f\rvert_1 = 1$ and $L(f) \geq \sqrt{d} - 3d^{1/6}$ for d sufficiently large (again see A. Durand, op. cit., 64–65, or J-P. Kahane, Sur les polynômes à coefficients unimodulaires, Bull. London Math. Soc., **12** (1980), 321–342).

4. The Criterion of Schneider-Lang

Baker's Theorem 1.6 was proved in 1966. In 1980, D. Bertrand and D. W. Masser realized that it was a consequence of a result which was known earlier, namely *the criterion of Schneider-Lang for Cartesian products* [BertMa 1980], [Ma 1981a]. The main purpose of this chapter is to explain this argument (§ 4.2).

We shall state the criterion of Schneider-Lang (§ 4.1) only for entire functions (there is an extension to meromorphic functions, which is relevant for the study of analytic subgroups of commutative algebraic groups). We shall give a proof (§ 4.6) only of the special case which we use for the proof of Baker's Theorem, namely when one considers exponential polynomials.

It is possible to prove the main result of this chapter (Corollary 4.2) by means of interpolation determinants, using the multiplicity estimate of P. Philippon (see Chap. 8). It is possible also to prove it with interpolation determinants and without zero estimate (see [W 1997b]). Here, we shall use the old fashioned argument which rests on Thue-Siegel's Lemma and the construction of an auxiliary function (§ 4.5). This classical method deserves a place in these lectures: we remind the reader that, until recently, all known transcendence proofs of the theorems of Gel'fond-Schneider, Baker, or the six exponentials Theorem, involved a construction of an auxiliary function by means of Thue-Siegel's Lemma.

We need also a Schwarz' Lemma for entire functions which vanish with a high multiplicity on a Cartesian product (§ 4.3). This lemma is usually proved by means of integral formulae. Here we shall prove it by induction.

4.1 Algebraic Values of Entire Functions Satisfying Differential Equations

The criterion of Schneider-Lang is a general statement dealing with values of meromorphic functions of one or several complex variables, satisfying differential equations.

The first general result dealing with analytic or meromorphic functions of one variable and containing the solution to Hilbert's seventh problem appears in [Sch 1949]. In fact one can deduce the transcendence of α^β (Gel'fond-Schneider Theorem 1.4) from this criterion, either by using the two functions z and α^z without derivatives (Schneider's method), or else e^z and $e^{\beta z}$ with derivatives (Gel'fond's method). The statement is rather complicated, and Th. Schneider made successful attempts to simplify it [Sch 1957]. Schneider's criteria in [Sch 1957], Chap. II, § 3, Th.12 and 13 deal only with Gel'fond's method, i.e. involve derivatives. Further simplifications have been introduced by S. Lang later: either for Schneider's method (see [L 1966], Chap. III, § 1, Th.1), or else for Gel'fond's method and functions satisfying differential equations (see [L 1965b], [L 1966], Chap. III, § 1, Th.1 and [L 1993], Appendix 1). This last result is known as the *criterion of Schneider-Lang*. It has been extended by S. Lang to functions of several variables [L 1965b], [L 1966], Chap. IV, § 1, Th.1. Incidentally, the first transcendence proof involving functions of several variables goes back to [Sch 1941], where Th. Schneider proved the transcendence of the values $B(a, b)$ of the Beta function at rational points. Until 1970, all transcendence results in several variables involved only Cartesian products. The introduction of deeper tools from complex function theory occurred in the joint work [BoL 1970] of E. Bombieri and S. Lang, where Lelong's measure of the zeroes of an analytic function plays an important role (see [LelGru 1986], Chap. VI, §. 2 and [W 1979b], § 7.4). Then E. Bombieri [Bo 1970], introducing Hörmander's L^2-estimates into the subject (see also [LelGru 1986], Chap. VI, §. 2, Th. 1 and [W 1979b], Th. 5.1.1), answered a question of Nagata raised in [L 1966]: in this context it turns out that the convenient generalization of a finite set of $\mathbb{C}$ to higher dimension is the notion of hypersurface, and the number of elements is replaced by the degree of the hypersurface. However the special case (dealing with Cartesian products) which was considered by Th. Schneider and S. Lang is sufficient for our purpose.

The criterion of Schneider-Lang has many consequences. The one variable case already contains the theorems of Hermite-Lindemann and Gel'fond-Schneider (see [L 1993], Appendix 1: *the transcendence of e and π*). Here for simplicity we state the criterion only for entire functions in $\mathbb{C}^n$, and we prove it only for exponentials and polynomials.

We denote by $\mathcal{A}_n$ the ring of entire functions in $\mathbb{C}^n$. We introduce the following definition: an entire function f in $\mathbb{C}^n$ is *of finite order of growth* if

$$\limsup_{R\to\infty} \frac{\log\log |f|_R}{\log R} < \infty, \quad \text{where} \quad |f|_R = \sup_{|\underline{z}|\le R} |f(\underline{z})|.$$

For instance a polynomial, or an exponential function $e^{t_1z_1+\cdots+t_nz_n}$, satisfy this property (the left hand side is 0 if f is a polynomial, 1 for an exponential function).

Theorem 4.1* *(Criterion of Schneider-Lang for entire functions[10]). Let d and n be two integers with $d > n \ge 1$, K be a number field, and $f_1, \dots, f_d$ be algebraically independent entire functions of finite order of growth. Assume, for $1 \le \nu \le n$ and $1 \le i \le d$, that the partial derivative $(\partial/\partial z_\nu) f_i$ of f_i belongs to the ring $K[f_1, \dots, f_d]$. Further, let $(\underline{y}_1, \dots, \underline{y}_n)$ be a basis of $\mathbb{C}^n$ over $\mathbb{C}$. Then the numbers*

[10] A more general statement is valid for meromorphic functions in n variables, of finite order of growth, assuming that at least $n + 1$ of them are algebraically independent; see [L 1966], Chap. IV, § 1, Th. 1 and [W 1979b], Corollaire 5.1.2.

$$f_i(s_1\underline{y}_1 + \cdots + s_n\underline{y}_n), \qquad (1 \le i \le d,\ (s_1, \ldots, s_n) \in \mathbb{Z}^n)$$

do not all belong to K.

An important point in this criterion is that two bases of $\mathbb{C}^n$ are concerned: the first one, implicit, is the canonical basis. The $\overline{\mathbb{Q}}$-structure of $\mathbb{C}^n$ is involved by the fact that the partial derivatives of the functions are supposed to be polynomials in $f_1, \ldots, f_d$ with algebraic coefficients. The second basis $(\underline{y}_1, \ldots, \underline{y}_n)$ is not supposed to be defined over $\overline{\mathbb{Q}}$.

We shall prove (and use) only the following corollary.

Corollary 4.2 *(Criterion of Schneider-Lang for $\mathbb{C}^{d_0} \times (\mathbb{C}^\times)^{d_1}$). Let d_0, d_1 and n be three integers with $0 \le d_0 \le n < d_0 + d_1$. Let $\underline{x}_1, \ldots, \underline{x}_{d_1}$ be $\mathbb{Q}$-linearly independent elements of $\overline{\mathbb{Q}}^n$, and $(\underline{y}_1, \ldots, \underline{y}_n)$ be a basis of $\mathbb{C}^n$ over $\mathbb{C}$. Write $\underline{y}_j = (y_{1j}, \ldots, y_{nj})$ $(1 \le j \le n)$. Then one at least of the following $(d_0 + d_1)n$ numbers*

$$y_{hj}, \quad e^{\underline{x}_i \underline{y}_j}, \qquad (1 \le h \le d_0,\ 1 \le i \le d_1,\ 1 \le j \le n)$$

is transcendental.

Proof of Corollary 4.2 as a consequence of Theorem 4.1. Define

$$d = d_0 + d_1, \qquad f_h(\underline{z}) = z_h, \quad (1 \le h \le d_0)$$

and

$$f_{d_0+i}(\underline{z}) = e^{\underline{x}_i \underline{z}}, \quad (1 \le i \le d_1).$$

From the assumption of linear independence of $\underline{x}_1, \ldots, \underline{x}_{d_1}$ over $\mathbb{Q}$, it follows easily (see Exercise 2.4) that the functions $f_1, \ldots, f_d$ are algebraically independent. These functions satisfy differential equations, for $1 \le j \le n$,

$$\frac{\partial}{\partial z_j} f_h = \delta_{hj} = \begin{cases} 0 & \text{if } h \ne j, \\ 1 & \text{if } h = j, \end{cases} \qquad (1 \le h \le d_0),$$

and

$$\frac{\partial}{\partial z_j} f_{d_0+i} = x_{ji} f_{d_0+i}, \qquad (1 \le i \le d_1),$$

where $\underline{x}_i = (x_{1i}, \ldots, x_{ni})$ $(1 \le i \le d_1)$. Let K be the field generated over $\mathbb{Q}$ by the $(d_0 + 2d_1)n$ numbers

$$x_{ji}, \quad f_h(\underline{y}_j) = y_{hj} \quad \text{and} \quad f_{d_0+i}(\underline{y}_j) = e^{\underline{x}_i \underline{y}_j},$$
$$(1 \le h \le d_0, \quad 1 \le i \le d_1, \quad 1 \le j \le n).$$

From the addition theorem which is satisfied by the exponential function, it follows that the values of $f_1, \ldots, f_d$ at the points of $s_1\underline{y}_1 + \cdots + s_n\underline{y}_n$ $(\underline{s} = (s_1, \ldots, s_n) \in \mathbb{Z}^n)$ all belong to K. Hence we deduce from Theorem 4.1 that K is not a number field. □

We shall use only two special cases of Corollary 4.2 (however, see Exercise 4.2). Here is the case $d_0 = 0$:

Corollary 4.3. *Let $\underline{x}_1, \ldots, \underline{x}_d$ be elements of $\overline{\mathbb{Q}}^n$ which generate a subgroup of rank at least $n+1$, and let $\{\underline{y}_1, \ldots, \underline{y}_\ell\}$ be a subset of $\mathbb{C}^n$ which contains a basis of $\mathbb{C}^n$ over $\mathbb{C}$. Then one at least of the $d\ell$ numbers $\underline{x}_i\underline{y}_j$ $(1 \leq i \leq d,\ 1 \leq j \leq \ell)$ does not belong to $\mathcal{L}$.*

For $n = 1$, Corollary 4.3 is clearly equivalent to Theorem 1.4 (Gel'fond-Schneider).

Next the case $d_0 = 1$, $d_1 = n$.

Corollary 4.4. *Let $\underline{x}_1, \ldots, \underline{x}_d$ be $\mathbb{Q}$-linearly independent elements in $\overline{\mathbb{Q}}^d$, and let $(\underline{y}_1, \ldots, \underline{y}_d)$ be a basis of $\mathbb{C}^d$ over $\mathbb{C}$. Write $\underline{y}_j = (y_{1j}, \ldots, y_{dj}) \in \mathbb{C}^d$ and assume that the d numbers y_{1j} $(1 \leq j \leq d)$ are algebraic. Then one at least of the d^2 numbers $\underline{x}_i\underline{y}_j$ $(1 \leq i \leq d,\ 1 \leq j \leq d)$ does not belong to $\mathcal{L}$.*

For $d = 1$ this is clearly equivalent to Theorem 1.2 (Hermite-Lindemann).

In § 4.2 we show that Theorem 1.5 (homogeneous case of Baker's Theorem) follows from Corollary 4.3, and Theorem 1.6 (nonhomogeneous case) from Corollary 4.4.

4.2 First Proof of Baker's Theorem

4.2.1 A Special Case

We explain the idea of D. Bertrand and D. W. Masser [BertMa 1980] in a special case, before considering the general case. Assume

$$\ell_1 + \sqrt[3]{2}\ell_2 + \sqrt[3]{4}\ell_3 = 0,$$

where ℓ_1, ℓ_2, ℓ_3 are nonzero elements of $\mathcal{L}$. Using Theorem 1.4 of Gel'fond-Schneider (i.e. the case $n = 1$ of Corollary 4.3), one deduces that the three numbers ℓ_1, ℓ_2, ℓ_3 are linearly independent over $\mathbb{Q}$.

We multiply the given relation by $\sqrt[3]{2}$ and by $\sqrt[3]{4}$:

$$2\ell_3 + \sqrt[3]{2}\ell_1 + \sqrt[3]{4}\ell_2 = 0 \quad \text{and} \quad 2\ell_2 + 2\sqrt[3]{2}\ell_3 + \sqrt[3]{4}\ell_1 = 0.$$

Therefore the three functions e^{z_1}, e^{z_2} and $e^{\sqrt[3]{2}z_1 + \sqrt[3]{4}z_2}$ take algebraic values at the points $\underline{y}_1 = (\ell_2, \ell_3)$, $\underline{y}_2 = (\ell_1, \ell_2)$ and $\underline{y}_3 = (2\ell_3, \ell_1)$. We define $\underline{x}_1 = (1, 0)$, $\underline{x}_2 = (0, 1)$ and $\underline{x}_3 = (\sqrt[3]{2}, \sqrt[3]{4})$, so that

$$\begin{array}{lll} \underline{x}_1\underline{y}_1 = \ell_2, & \underline{x}_1\underline{y}_2 = \ell_1, & \underline{x}_1\underline{y}_3 = 2\ell_3, \\ \underline{x}_2\underline{y}_1 = \ell_3, & \underline{x}_2\underline{y}_2 = \ell_2, & \underline{x}_2\underline{y}_3 = \ell_1, \\ \underline{x}_3\underline{y}_1 = -\ell_1, & \underline{x}_3\underline{y}_2 = -2\ell_3, & \underline{x}_3\underline{y}_3 = -2\ell_2. \end{array}$$

By Theorem 1.12 (six exponentials Theorem), the matrix

$$\begin{pmatrix} \ell_2 & \ell_1 & 2\ell_3 \\ \ell_3 & \ell_2 & \ell_1 \end{pmatrix}$$

has rank 2, hence $\{\underline{y}_1, \underline{y}_2, \underline{y}_3\}$ contains a basis of $\mathbb{C}^2$ over $\mathbb{C}$. This is in contradiction with Corollary 4.3. Hence a relation $\ell_1 + \sqrt[3]{2}\ell_2 + \sqrt[3]{4}\ell_3 = 0$ with $\ell_i \in \mathcal{L}$ implies $\ell_1 = \ell_2 = \ell_3 = 0$.

There is a better way of proving the same result, where Theorem 1.12 is not needed: instead of multiplying by $\sqrt[3]{2}$ and $\sqrt[3]{4}$, we make use of the three embeddings of the field $K = \mathbb{Q}(\sqrt[3]{2})$ into $\mathbb{C}$. We deal with the general case using this way.

4.2.2 The Main Result

Theorem 4.5. *Let K be a number field of degree d over $\mathbb{Q}$, let $(\beta_1, \ldots, \beta_d)$ be a basis of the $\mathbb{Q}$-vector space K, and let $\ell_1, \ldots, \ell_d$ be elements of $\mathcal{L}$. Assume $\beta_1\ell_1 + \cdots + \beta_d\ell_d \in \overline{\mathbb{Q}}$. Then $\ell_1 = \ldots = \ell_d = 0$.*

We first deduce (in § 4.2.4) Theorem 4.5 from Corollaries 4.3 and 4.4. Next (§ 4.2.5) we shall show that Baker's Theorem 1.6 follows from Theorem 4.5 (compare with Lemma 1.7).

4.2.3 Trace

We shall need a well-known result from algebraic number theory which we now recall:

Lemma 4.6. *Let K be a number field. Then $(x, y) \mapsto \mathrm{Tr}(xy)$ is a nondegenerate bilinear form on K.*

The *trace* $K \to \mathbb{Q}$ is a $\mathbb{Q}$-linear map $\mathrm{Tr} = \mathrm{Tr}_{K/\mathbb{Q}}$ which can be defined as follows: $\mathrm{Tr}(\gamma) = \sum_\sigma \gamma^\sigma$, where σ runs over the set of complex embeddings of K (this is a finite set with $[K : \mathbb{Q}]$ elements; see Chap. 3), while $\gamma^\sigma \in \mathbb{C}$ stands for the image of $\gamma \in K$ under σ. If $a_0X^n + a_1X^{n-1} + \cdots + a_n$ is the minimal polynomial of γ over $\mathbb{Z}$, then the sum of the complex conjugates of γ is $-a_1/a_0$, and $\mathrm{Tr}(\gamma) = -[K : \mathbb{Q}(\gamma)]a_1/a_0 \in \mathbb{Q}$.

Let $(\beta_1, \ldots, \beta_d)$ is a basis of the $\mathbb{Q}$-vector space K (with $d = [K : \mathbb{Q}]$). The non-degeneracy of the trace means that the determinant $\det\bigl(\mathrm{Tr}(\beta_i\beta_j)\bigr)_{1\le i,j\le d}$ is not zero.

Proof. Define $B = \bigl(\beta_k^{\sigma_i}\bigr)_{1\le i,k\le d}$. A simple computation shows that the product of B by its transpose is $\bigl(\mathrm{Tr}(\beta_i\beta_j)\bigr)_{1\le i,j\le d}$, hence

$$\det\bigl(\mathrm{Tr}(\beta_i\beta_j)\bigr)_{1\le i,j\le d} = (\det B)^2.$$

The fact we need is that $\det B$ is not zero. If c_σ are complex numbers such that

$$\sum_\sigma c_\sigma \beta_i^\sigma = 0 \quad \text{for} \quad 1 \le i \le d,$$

since $(\beta_1, \ldots, \beta_d)$ is a basis of K over $\mathbb{Q}$, it follows by linearity

$$\sum_\sigma c_\sigma \beta^\sigma = 0 \quad \text{for all} \quad \beta \in K.$$

This implies $c_\sigma = 0$ for all σ, as shown by the theorem of linear independence of characters (the restrictions of σ to $K^\times$ are distinct characters of the multiplicative group $K^\times$ in $\mathbb{C}^\times$, hence they are linearly independent; see for instance [L 1993], Chap. VI, Theorem 4.1). □

4.2.4 Proof of Theorem 4.5 Using Corollaries 4.3 and 4.4

Denote by $\{\sigma_1, \ldots, \sigma_d\}$ the embeddings of K into $\mathbb{C}$. For $1 \le i \le d$, define the complex number λ_i by

$$\lambda_i = \sum_{k=1}^d \beta_k^{\sigma_i} \ell_k.$$

We consider three cases.

First case: One at least (but not all) of $\lambda_1, \ldots, \lambda_d$ vanishes. We choose an ordering of $\{\sigma_1, \ldots, \sigma_d\}$ so that

$$\lambda_1 \neq 0, \ldots, \lambda_n \neq 0, \quad \lambda_{n+1} = \cdots = \lambda_d = 0.$$

By assumption we have $1 \le n < d$.

For $1 \le i \le d$, define $\underline{x}_i \in \mathbb{C}^n$ by

$$\underline{x}_i = (\beta_i^{\sigma_1}, \ldots, \beta_i^{\sigma_n}).$$

If $a_1, \ldots, a_d$ are rational numbers such that $a_1\underline{x}_1 + \cdots + a_d\underline{x}_d = 0$, then $\sum_{i=1}^d a_i\beta_i = 0$, hence $a_1 = \cdots = a_d = 0$. This shows that $\underline{x}_1, \ldots, \underline{x}_d$ are $\mathbb{Q}$-linearly independent.

For $1 \le j \le d$, we define $\underline{y}_j \in \mathbb{C}^n$ by

$$\underline{y}_j = (\beta_j^{\sigma_1}\lambda_1, \ldots, \beta_j^{\sigma_n}\lambda_n).$$

From Lemma 4.6 we deduce that the matrix B has rank d. It follows that the $d \times n$ matrix $B_n = \left(\beta_k^{\sigma_i}\right)_{1 \le k \le d, 1 \le i \le n}$ has rank n (its n columns are independent in K^d). The product of B_n by the $n \times n$ diagonal matrix $\operatorname{diag}(\lambda_1, \ldots, \lambda_n)$ is the $d \times n$ matrix whose row vectors are $\underline{y}_1, \ldots, \underline{y}_d$:

$$\begin{pmatrix} \beta_1^{\sigma_1}\lambda_1 & \cdots & \beta_1^{\sigma_n}\lambda_n \\ \vdots & \ddots & \vdots \\ \beta_d^{\sigma_1}\lambda_1 & \cdots & \beta_d^{\sigma_n}\lambda_n \end{pmatrix} = \begin{pmatrix} \beta_1^{\sigma_1} & \cdots & \beta_1^{\sigma_n} \\ \vdots & \ddots & \vdots \\ \beta_d^{\sigma_1} & \cdots & \beta_d^{\sigma_n} \end{pmatrix} \begin{pmatrix} \lambda_1 & \cdots & 0 \\ \vdots & \ddots & \vdots \\ 0 & \cdots & \lambda_n \end{pmatrix}.$$

Therefore $\{\underline{y}_1, \ldots, \underline{y}_d\}$ contains a basis of $\mathbb{C}^n$.

Finally, we check $\underline{x}_i\underline{y}_j \in \mathcal{L}$ as follows: from the assumption $\lambda_{n+1} = \cdots = \lambda_d = 0$, we deduce

$$\underline{x}_i\underline{y}_j = \sum_{\nu=1}^{n} \beta_i^{\sigma_\nu}\beta_j^{\sigma_\nu}\lambda_\nu = \sum_{\nu=1}^{d} \beta_i^{\sigma_\nu}\beta_j^{\sigma_\nu}\lambda_\nu = \sum_{\nu=1}^{d} \beta_i^{\sigma_\nu}\beta_j^{\sigma_\nu}\sum_{k=1}^{d}\beta_k^{\sigma_\nu}\ell_k = \sum_{k=1}^{d} c_{ijk}\ell_k$$

with

$$c_{ijk} = \sum_{\nu=1}^{d} \beta_i^{\sigma_\nu}\beta_j^{\sigma_\nu}\beta_k^{\sigma_\nu} = \mathrm{Tr}(\beta_i\beta_j\beta_k) \in \mathbb{Q}.$$

Using Corollary 4.3 we conclude that this case is impossible.

Second case: None of the numbers $\lambda_1, \ldots, \lambda_d$ is zero. One of the embeddings of K into $\mathbb{C}$, say σ_1, is the natural embedding given by the fact that $\beta_1, \ldots, \beta_d$ are complex numbers. Therefore $\lambda_1 \in \overline{\mathbb{Q}}$. For $1 \le k \le d$, define $\underline{x}_k \in \mathbb{C}^d$ by $\underline{x}_k = (\beta_k^{\sigma_1}, \ldots, \beta_k^{\sigma_d})$. By Lemma 4.6, the matrix $B = \left(\beta_k^{\sigma_i}\right)_{1\le i,k\le d}$ is regular. Hence these d elements $\underline{x}_1, \ldots, \underline{x}_d$ of $\overline{\mathbb{Q}}^d$ are linearly independent over $\overline{\mathbb{Q}}$.

For $1 \le j \le d$, define $\underline{y}_j \in \mathbb{C}^d$ by $\underline{y}_j = (\beta_j^{\sigma_1}\lambda_1, \ldots, \beta_j^{\sigma_d}\lambda_d)$. Since B has rank d and since

$$\begin{pmatrix} \beta_1^{\sigma_1}\lambda_1 & \cdots & \beta_1^{\sigma_d}\lambda_d \\ \vdots & \ddots & \vdots \\ \beta_d^{\sigma_1}\lambda_1 & \cdots & \beta_d^{\sigma_d}\lambda_d \end{pmatrix} = \begin{pmatrix} \beta_1^{\sigma_1} & \cdots & \beta_1^{\sigma_d} \\ \vdots & \ddots & \vdots \\ \beta_d^{\sigma_1} & \cdots & \beta_d^{\sigma_d} \end{pmatrix} \begin{pmatrix} \lambda_1 & \cdots & 0 \\ \vdots & \ddots & \vdots \\ 0 & \cdots & \lambda_d \end{pmatrix},$$

it follows that $(\underline{y}_1, \ldots, \underline{y}_d)$ is a basis of $\mathbb{C}^d$.

Finally, we have $y_{1j} \in \overline{\mathbb{Q}}$ for $1 \le j \le d$ and, as in the first case,

$$\underline{x}_i\underline{y}_j = \sum_{k=1}^{d} \mathrm{Tr}(\beta_i\beta_j\beta_k)\ell_k \in \mathcal{L} \quad \text{for } 1 \le i \le d \text{ and } 1 \le j \le d.$$

From Corollary 4.4 we deduce again that this case is impossible.

Third case: $\lambda_1 = \cdots = \lambda_d = 0$. Since the first and second case are excluded, we have $\lambda_1 = \cdots = \lambda_d = 0$, which is what we wanted to prove. □

4.2.5 Proof of Baker's Theorem 1.6 as a Consequence of Theorem 4.5

Let $\ell_1, \ldots, \ell_m$ be $\mathbb{Q}$-linearly independent elements in $\mathcal{L}$ and $\gamma_0, \gamma_1, \ldots, \gamma_m$ be algebraic numbers. Assume

$$\gamma_0 + \gamma_1\ell_1 + \cdots + \gamma_m\ell_m = 0,$$

We shall deduce $\gamma_0 = \cdots = \gamma_m = 0$.

Define $K = \mathbb{Q}(\gamma_1, \ldots, \gamma_m)$, choose a basis $(\beta_1, \ldots, \beta_d)$ of the $\mathbb{Q}$-vector space K, and write the elements γ_j (for $1 \le j \le m$) as linear combinations of the β_i with rational coefficients c_{ji}:

$$\gamma_j = \sum_{i=1}^{d} c_{ji}\beta_i, \qquad (1 \le j \le m).$$

Then we have the relation

$$\sum_{i=1}^{d} \beta_i \ell_i' \in \overline{\mathbb{Q}}$$

where

$$\ell_i' = \sum_{j=1}^{m} c_{ji}\ell_j \in \mathcal{L},$$

and Theorem 4.5 implies $\ell_1' = \cdots = \ell_d' = 0$. However the linear independence of $\ell_1, \ldots, \ell_m$ over $\mathbb{Q}$ shows that the relations

$$\sum_{j=1}^{m} c_{ji}\ell_j = 0, \qquad (1 \le i \le d)$$

imply $c_{ji} = 0$ for $1 \le i \le d$ and $1 \le j \le m$, hence $\gamma_1 = \cdots = \gamma_m = 0$, and finally $\gamma_0 = 0$. □

4.3 Schwarz' Lemma for Cartesian Products

Schwarz' Lemma for analytic functions of a single variable provides a sharp upper bound for the values of a function having lot of zeroes. It is a difficult (and interesting) problem to extend this lemma to functions of several variables (see Chap. 7 of [W 1979b]). We shall deal with Cartesian products. After the work of Th. Schneider [Sch 1941] and S. Lang [L 1966] connected with the proof of the criterion of Schneider-Lang, F. Gross [Gr 1969] used interpolation formulae for Cartesian products and extended Pólya's Theorem (cf. Exercise 4.5) to several variables. Here we use an inductive argument in place of such integral formulae.

Here is the main result of this section.

Proposition 4.7. *Let $\mathcal{E}_1, \ldots, \mathcal{E}_n$ be subsets of $\mathbb{C}$, each with S_1 elements. Define $\mathcal{E} = \mathcal{E}_1 \times \cdots \times \mathcal{E}_n \subset \mathbb{C}^n$. Let $r > 0$ satisfy $r \ge \max_{1\le i\le n} \max_{\zeta\in\mathcal{E}_i} |\zeta|$. Let R be a positive real number such that $R \ge 18^n r$. Let f be an entire function of n variables such that*

$$\mathcal{D}^{\underline{\sigma}} f(\underline{\xi}) = 0 \quad \textit{for all } \underline{\xi} \in \mathcal{E} \textit{ and all } \underline{\sigma} \in \mathbb{N}^n \textit{ with } |\underline{\sigma}| < S_0.$$

Then

$$|f|_r \le |f|_R \left(\frac{R}{18^n r}\right)^{-S_0 S_1}.$$

The proof of Proposition 4.7 will rest on the study of the ideal in $\mathcal{A}_n$ generated by n polynomial functions $P_1(z_1), \ldots, P_n(z_n)$.

When $P \in \mathbb{C}[X]$ is a polynomial in one variable and ζ a complex zero of P, we denote by $\mathrm{m}_P(\zeta)$ the multiplicity of ζ as a zero of P. We also consider that the zero polynomial has degree -1.

For a function $f \in \mathcal{A}_n$, an index j with $1 \leq j \leq n$ and an integer p, we say that *f is a polynomial in z_j of degree $< p$* if the Taylor expansion at the origin of f involves only z_j^h with $0 \leq h < p$. We insist that f is not necessarily assumed to be a polynomial in z_i for $i \neq j$.

Lemma 4.8. *Let m and n be rational integers with $1 \leq m \leq n$. For $m \leq i \leq n$, let $P_i \in \mathbb{C}[X]$ be a nonzero monic polynomial of degree p_i and let $\mathcal{E}_i = P_i^{-1}(0) \subset \mathbb{C}$ be the set of zeroes of P_i. Denote by $\mathcal{I}$ the ideal, in the ring $\mathcal{A}_n$ of entire functions of n complex variables, generated by the $n - m + 1$ functions $P_m(z_m), \ldots, P_n(z_n)$. Further define $\mathcal{E} = \mathcal{E}_m \times \cdots \times \mathcal{E}_n \subset \mathbb{C}^{n-m+1}$.*
a) If $f \in \mathcal{I}$ is a polynomial of degree $< p_i$ in z_i for $m \leq i \leq n$, then $f = 0$.
b) The ideal $\mathcal{I}$ consists of the elements $f \in \mathcal{A}_n$ which satisfy

$$\mathcal{D}^{\underline{\kappa}} f(z_1, \ldots, z_{m-1}, \zeta_m, \ldots, \zeta_n) = 0$$

for all $(\zeta_m, \ldots, \zeta_n) \in \mathcal{E}$, and for all $(\kappa_1, \ldots, \kappa_n) \in \mathbb{N}^n$ with $\kappa_i = 0$ for $1 \leq i < m$ and $0 \leq \kappa_i < \mathrm{m}_{P_i}(\zeta_i)$ for $m \leq i \leq n$.
c) For each function f in $\mathcal{A}_n$, there is a unique family $(f_0, f_m, \ldots, f_n)$ of $n - m + 2$ functions in $\mathcal{A}_n$ satisfying the following properties:

(i)

$$f(\underline{z}) = f_0(\underline{z}) + \sum_{i=m}^{n} f_i(\underline{z}) P_i(z_i).$$

(ii) *For $m \leq j \leq n$, f_0 is a polynomial in z_j of degree $< p_j$.*
(iii) *For $m \leq i < j \leq n$, f_i is a polynomial in z_j of degree $< p_j$.*

d) Let r and R be positive real numbers with $R \geq 5r$. Assume each $\mathcal{E}_i$ is contained in the disc of the complex plane of radius r. Then for $i \in \{0, m, m+1, \ldots, n\}$, we have

$$|f_i|_R \leq 9^{(n-m+1)p} R^{-p_i} |f|_R,$$

where $p = \max\{p_m, \ldots, p_n\}$ and $p_0 = 0$.
e) Let $k \in \{1, \ldots, n\}$. If f is a polynomial in z_k of degree $\leq d$, then so are $f_0, f_m, \ldots, f_n$.

Remark 1. The decomposition given in c) is unique, but not canonical: for $n > m$ it depends on the ordering of $P_m(z_m), \ldots, P_n(z_n)$. In any case a choice should be made if we want unicity in the decomposition of, say, $P_1(z_1)P_2(z_2)$ for $m = 1$ and $n = 2$. For instance when $m = 1, n = 2$, $p_1 \geq 1$ and $p_2 = 0$, the decomposition which arises from Lemma 4.8 is trivial: $f_0 = f_1 = 0$, $f_2 = f$, while for $p_1 = 0$ and $p_2 \geq 1$, the decomposition $f(\underline{z}) = f_1(\underline{z}) + f_2(\underline{z}) P_2(z_2)$ is not trivial.

Remark 2. From a) and c) we deduce

$$f \in \mathcal{I} \quad \text{if and only if} \quad f_0 = 0.$$

Remark 3. For the proof of Proposition 4.7, it will be sufficient to use the case $m = 1$ of Lemma 4.8. To a certain extent, this amounts to take $P_1 = \cdots = P_{m-1} = 0$, and this is very useful for the inductive argument in the proof of Lemma 4.8.

Proof of Lemma 4.8. We split the proof into several steps.

Step 1. We first notice that one inclusion in b) is obvious: for $m \le i \le n$, the polynomial P_i obviously satisfies

$$\left(\frac{d}{dX}\right)^{\kappa_i} P_i(\zeta_i) = 0$$

for all $\zeta_i \in \mathcal{E}_i$ and for all $\kappa_i \in \mathbb{N}$ with $0 \le \kappa_i < \mathrm{m}_{P_i}(\zeta_i)$. Therefore each $f \in \mathcal{I}$ satisfies the vanishing condition stated in b).

Step 2. The proof of Lemma 4.8 will use induction on n. In this second step (which is split into five substeps) we prove the case $m = n$ together with a slight refinement of d), namely with

$$|f_0|_R \le 3^p |f|_R \quad \text{and} \quad |f_n|_R \le \left(\frac{3}{R}\right)^p |f|_R.$$

This refinement will be useful for the induction hypothesis in step 3.6.

We shall sometimes write z for z_n, P in place of P_n, $\mathcal{E}$ for $\mathcal{E}_n$, p for p_n, and also $f(z)$ for $f(z_1, \ldots, z_{n-1}, z)$ when $z_1, \ldots, z_{n-1}$ are fixed in $\mathbb{C}^{n-1}$.

Step 2.1. We prove property b). The ideal $\mathcal{I}$ is the principal ideal of $\mathcal{A}_n$ generated by the polynomial $P(z_n)$. Let $f \in \mathcal{A}_n$ satisfy

$$\left(\frac{\partial}{\partial z_n}\right)^{\kappa} f(z_1, \ldots, z_{n-1}, \zeta) = 0$$

for all $\zeta \in \mathcal{E}$ and for all $\kappa \in \mathbb{N}$ with $0 \le \kappa < \mathrm{m}_P(\zeta)$.

The function

$$g(z_1, \ldots, z_n) = \frac{f(z_1, \ldots, z_{n-1}, z_n)}{P(z_n)}$$

is continuous on $\mathbb{C}^n$, is an entire function of $z_n \in \mathbb{C}$ for each $(z_1, \ldots, z_{n-1}) \in \mathbb{C}^{n-1}$ and is an entire function of $(z_1, \ldots, z_{n-1}) \in \mathbb{C}^n$ for each $z_n \in \mathbb{C}$ (one could restrict to $z_n \in \mathbb{C} \setminus \mathcal{E}$ but it is true also for all $z_n \in \mathbb{C}$). Hence $g \in \mathcal{A}_n$ and therefore $f \in \mathcal{I}$.

Step 2.2. Property a) readily follows: if $f = gP$ is a polynomial in z_n of degree $< p$, then $g = f/P$ is a polynomial in z_n of degree < 0, hence $g = 0$.

Step 2.3. Unicity in c) is a consequence of a): if $f_0 + f_n P = 0$ then $f_0 \in \mathcal{I}$; the condition $\deg_{z_n} f_0 < p$ yields $f_0 = 0$, hence $f_n P = 0$ and finally $f_n = 0$.

Step 2.4. We now prove the existence of a decomposition in c):

$$f(z) = f_0(z) + f_n(z)P(z)$$

with

$$\deg_{z_n} f_0 < p, \quad |f_0|_R \leq 3^p |f|_R \quad \text{and} \quad |f_n|_R \leq \left(\frac{3}{R}\right)^p |f|_R,$$

which includes the announced refinement of the estimate for d).

For $p = 0$ we have $P = 1$ and we take $f_0 = 0$, $f_n = f$.

For $p = 1$, we write $P(X) = X - \zeta$ and we define

$$f_0(z_1, \ldots, z_n) = f(z_1, \ldots, z_{n-1}, \zeta),$$

$$f_n(z_1, \ldots, z_n) = \frac{f(z_1, \ldots, z_n) - f(z_1, \ldots, z_{n-1}, \zeta)}{z_n - \zeta},$$

so that f_0 and f_n are in $\mathcal{A}_n$.

We deduce the estimate (valid for $p = 1$):

$$|f_0|_R \leq |f|_R \quad \text{and} \quad |f_n|_R \leq \frac{2}{R - r}|f|_R.$$

Assume $p \geq 2$. We prove the existence of f_0 and f_n together with an estimate

$$|f_0|_R \leq A_p |f|_R \quad \text{and} \quad |f_n|_R \leq \left(\frac{2}{R - r}\right)^p |f|_R$$

with some number $A_p \geq 1$, by induction on p. Since $R \geq 3r$ (indeed we assumed $R \geq 5r$), we have $2/(R - r) \leq 3/R$, and the desired estimate for $|f_n|_R$ will follow. At the end of this step 2.4 we shall check the inequality $A_p \leq 3^p$.

Let ζ be a root of P. Define $Q(X) = P(X)/(X - \zeta)$. We first write, as before,

$$f(z) = f(\zeta) + (z - \zeta)g(z),$$

where $g \in \mathcal{A}_n$ satisfies

$$|g|_R \leq \frac{2}{R - r}|f|_R.$$

We use the induction hypothesis: there exist g_0 and f_n in $\mathcal{A}_n$, where g_0 is a polynomial in $z = z_n$ of degree $< p - 1$, such that $g(z) = g_0(z) + Q(z) f_n(z)$ where

$$|g_0|_R \leq A_{p-1} |g|_R \quad \text{and} \quad |f_n|_R \leq \left(\frac{2}{R - r}\right)^{p-1} |g|_R.$$

The last inequality yields

$$|f_n|_R \leq \left(\frac{2}{R - r}\right)^p |f|_R,$$

as wanted. Define $f_0(z) = f(\zeta) + (z - \zeta)g_0(z)$, so that $f(z) = f_0(z) + P(z)f_n(z)$, and $f_0 \in \mathcal{A}_n$ is a polynomial in $z = z_n$ of degree $< p$. We have

$$|f_0|_R \le |f|_R + (R+r)|g_0|_R \le |f|_R \left(1 + 2A_{p-1}\frac{R+r}{R-r}\right).$$

This proves the desired estimate for f_0, with $A_p = 1 + 2A_{p-1}(R+r)/(R-r)$. Since the estimate for $p = 1$ holds with $A_1 = 1$, we deduce that it holds for $p \ge 2$ with

$$\begin{aligned} A_p &= 1 + \frac{2(R+r)}{R-r} + \cdots + \left(\frac{2(R+r)}{R-r}\right)^{p-1} \\ &= \frac{R-r}{R+3r}\left(\left(\frac{2(R+r)}{R-r}\right)^p - 1\right) \\ &< 2^p \frac{R+r}{R+3r}\left(\frac{R+r}{R-r}\right)^{p-1} \\ &< 2 \cdot 3^{p-1} < 3^p, \end{aligned}$$

because $R \ge 5r$.

Step 2.5. We now check property e) by induction on p. Let $k \in \{1, \dots, n\}$ and d satisfy $\deg_{z_k} f \le d$. Recall the construction in step 2.4.

For $p = 0$ since $f_0 = 0$ and $f_n = f$ we have $\deg_{z_k} f_0 \le d$ and $\deg_{z_k} f_n \le d$.

For $p = 1$ again $\deg_{z_k} f_0 \le d$ and $\deg_{z_k} f_n \le d$. Moreover, if $k = n$, then $\deg_{z_n} f_n \le d - 1$.

For $p \ge 2$ and $k < n$ we have (with the notation of step 2.4)

$$f(z) = f(\zeta) + (z - \zeta)g(z),\ \ g(z) = g_0(z) + Q(z)f_n(z),\ \ f_0(z) = f(\zeta) + (z - \zeta)g_0(z)$$

and we find successively, using the induction hypothesis on p,

$$\deg_{z_k} g \le d, \quad \deg_{z_k} g_0 \le d, \quad \deg_{z_k} f_n \le d, \quad \deg_{z_k} f_0 \le d.$$

For $p \ge 2$ and $k = n$ we have

$$\deg_{z_k} g \le d - 1, \quad \deg_{z_k} g_0 \le d - 1, \quad \deg_{z_k} f_n \le d - 1, \quad \deg_{z_k} f_0 \le d.$$

This completes the proof of Lemma 4.8 in the special case $m = n$.

Step 3. We now prove Lemma 4.8 by induction on m, for decreasing values $m = n, n-1, \dots, 1$. The start of the induction is $m = n$ which has been carried out in step 2. Let m satisfy $1 \le m < n$. Assume the result holds for $m + 1$.

Step 3.1. We prove the existence of a decomposition in c). Let $f \in \mathcal{A}_n$. Write $\underline{z}$ for $(z_1, \dots, z_n)$. Using step 2 (with a renumbering of $z_m, \dots, z_n$), we get a decomposition

$$f(\underline{z}) = \varphi_0(\underline{z}) + \varphi_1(\underline{z})P_m(z_m)$$

with φ_0 and φ_1 in $\mathcal{A}_n$ and φ_0 a polynomial in z_m of degree $< p_m$. The induction hypothesis (with $m+1$) yields

$$\varphi_0(\underline{z}) = \psi_0(\underline{z}) + \psi_{m+1}(\underline{z})P_{m+1}(z_{m+1}) + \cdots + \psi_n(\underline{z})P_n(z_n)$$

and

$$\varphi_1(\underline{z}) = \theta_0(\underline{z}) + \theta_{m+1}(\underline{z})P_{m+1}(z_{m+1}) + \cdots + \theta_n(\underline{z})P_n(z_n)$$

with ψ_0 and θ_0 in $\mathcal{A}_n$ polynomials of degree $< p_j$ in z_j for $m+1 \le j \le n$. Further, for $m+1 \le i < j \le n$, ψ_i and θ_i in $\mathcal{A}_n$ are polynomials of degree $< p_j$ in z_j. Furthermore, by step 2.5, since φ_0 is a polynomial of degree $< p_m$ in z_m, so are $\psi_0, \psi_{m+1}, \ldots, \psi_n$. The functions

$$f_0(\underline{z}) = \psi_0(\underline{z}), \quad f_m(\underline{z}) = \theta_0(\underline{z})$$

and

$$f_i(\underline{z}) = \psi_i(\underline{z}) + \theta_i(\underline{z})P_m(z_m) \quad (m+1 \le i \le n)$$

are in $\mathcal{A}_n$ and satisfy

$$f(\underline{z}) = f_0(\underline{z}) + f_m(\underline{z})P_m(z_m) + \cdots + f_n(\underline{z})P_n(z_n).$$

We already know that $f_0 = \psi_0$ is a polynomial of degree $< p_j$ in z_j for $m \le j \le n$ and that $f_m = \theta_0$ is a polynomial of degree $< p_j$ in z_j for $m+1 \le j \le n$. Finally, for $m+1 \le i < j \le n$, $f_i(\underline{z}) = \psi_i(\underline{z}) + \theta_i(\underline{z})P_m(z_m)$ is a polynomial of degree $< p_j$ in z_j.

Step 3.2. We prove property e). Recall the construction and notation in step 3.1.

If f is a polynomial in z_k of degree $\le d$, it follows from step 2.5 that the same holds for φ_0 and φ_1, hence (by induction hypothesis) for each of ψ_i and θ_i, and finally also for each f_i if $k \ne m$.

For $k = m$ we distinguish the case $p_m = 0$ and $p_m \ge 1$. In the former case (where $P_m = 1$) there is no difficulty. In the latter, since f and φ_0 have degree $\le d$ in z_m, we have $\deg_{z_m} \varphi_1 \le d - p_m$, hence $\deg_{z_m} \theta_i \le d - p_m$ and the inequality $\deg_{z_m} f_i \le d$ follows.

Step 3.3. We prove property a). Let $f \in \mathcal{I}$ be a polynomial in z_i of degree $< p_i$ for $i \ge m$. Let $(\zeta_{m+1}, \ldots, \zeta_n) \in \mathcal{E}_{m+1} \times \cdots \times \mathcal{E}_n$ (the case where this set is empty is trivial) and let $\underline{\kappa} \in \mathbb{N}^n$ satisfy $\kappa_i = 0$ for $1 \le i \le m$ and $0 \le \kappa_i < \mathrm{m}_{P_i}(\zeta_i)$ for $m+1 \le i \le n$. For any $(z_1, \ldots, z_{m-1}) \in \mathbb{C}^{m-1}$,

$$Q = \mathcal{D}^{\underline{\kappa}} f(z_1, \ldots, z_{m-1}, X, \zeta_{m+1}, \ldots, \zeta_n)$$

is a polynomial in $\mathbb{C}[X]$ of degree $< p_m$. Since $f \in \mathcal{I}$, we deduce from step 1

$$\left(\frac{d}{dX}\right)^{\kappa_m} Q(\zeta_m) = 0$$

for any $\zeta_m \in \mathcal{E}_m$ and any $\kappa_m \in \mathbb{N}$ in the range $0 \le \kappa_m < \mathrm{m}_{P_m}(\zeta_m)$. The sum of multiplicities of Q is

$$\geq \sum_{\zeta_m \in \mathcal{E}_m} \mathrm{m}_{P_m}(\zeta_m) = p_m.$$

Since $\deg Q < p_m$, we deduce $Q = 0$.

We expand $f(\underline{z})$:

$$f(\underline{z}) = \sum_{h=0}^{p_m-1} z_m^h \varphi_h(z_1, \ldots, z_{m-1}, z_{m+1}, \ldots, z_n)$$

where φ_h is a polynomial in z_j of degree $< p_j$ for $m+1 \leq j \leq n$. Since $Q = 0$ we have

$$\mathcal{D}^{\underline{\kappa}} \varphi_h(z_1, \ldots, z_{m-1}, \zeta_{m+1}, \ldots, \zeta_n) = 0 \quad \text{for} \quad 0 \leq h < p_m.$$

Each φ_h is an entire functions of $n-1$ variables. We use assertion b) of the induction hypothesis (with n, m replaced by $n-1, m+1$ respectively): each φ_h lies in the ideal of the ring of entire functions in $z_1, \ldots, z_{m-1}, z_{m+1}, \ldots, z_n$ generated by $P_{m+1}(z_{m+1}), \ldots, P_n(z_n)$. Using assertion a) of the inductive hypothesis with $m+1$, we obtain $\varphi_h = 0$ for any $h = 0, \ldots, p_m - 1$, hence $f = 0$.

It will be useful to point out the following result: if f_0 in the decomposition c) satisfies

$$\mathcal{D}^{\underline{\kappa}} f_0(z_1, \ldots, z_{m-1}, \zeta_m, \ldots, \zeta_n) = 0$$

for any $\underline{\zeta} \in \mathcal{E}$ and any $\underline{\kappa} \in \mathbb{N}^n$ with $\kappa_1 = \cdots = \kappa_{m-1} = 0$ and $0 \leq \kappa_i < \mathrm{m}_{P_i}(\zeta_i)$ for $m \leq i \leq n$, then $f_0 = 0$.

Step 3.4. We prove unicity of the decomposition in c). Assume

$$f_i(\underline{z}) P_i(z_i) + \cdots + f_n(\underline{z}) P_n(z_n) = 0$$

for some i in the range $\{0, m, m+1, \ldots, n\}$, where $P_0(z_0) = 1$ if $i = 0$. It suffices to check $f_i = 0$.

If $i = 0$ then

$$f_0(\underline{z}) = -f_m(\underline{z}) P_m(z_m) - \cdots - f_n(\underline{z}) P_n(z_n),$$

hence $f_0 \in \mathcal{I}$. However f_0 is a polynomial in z_j of degree $< p_j$ for $j \geq m$. From step 3.3 we infer $f_0 = 0$.

Assume now $m \leq i \leq n$ and consider the function $f_i(\underline{z}) P_i(z_i) \in \mathcal{A}_n$. On one hand it lies in the ideal of $\mathcal{A}_n$ generated by the polynomials $P_{i+1}(z_{i+1}), \ldots, P_n(z_n)$. On the other hand it is a polynomial in z_j of degree $< p_j$ for $i+1 \leq j \leq n$. Assertion a) of the inductive hypothesis (with $m+1$) yields $f_i(\underline{z}) P_i(z_i) = 0$. Since $P_i \not\equiv 0$, we conclude $f_i = 0$.

Step 3.5. We prove property b). Let $f \in \mathcal{A}_n$ satisfy the vanishing conditions of b):

$$\mathcal{D}^{\underline{\kappa}} f(z_1, \ldots, z_{m-1}, \zeta_m, \ldots, \zeta_n) = 0$$

for all $\underline{\zeta} \in \mathcal{E}$ and all $\underline{\kappa} \in \mathbb{N}^n$ with $\kappa_i = 0$ for $1 \leq i < m$ and $0 \leq \kappa_i < \mathrm{m}_{P_i}(\zeta_i)$ for $m \leq i \leq n$.

Using c), write

$$f(\underline{z}) = f_0(\underline{z}) + f_m(\underline{z})P_m(z_m) + \cdots + f_n(\underline{z})P_n(z_n).$$

Using step 1 we deduce that f_0 satisfies the same vanishing conditions $\mathcal{D}^{\underline{\kappa}} f_0(\underline{z}, \underline{\zeta}) = 0$. From the last remark of step 3.3 we deduce $f_0 = 0$, hence $f \in \mathcal{I}$.

Step 3.6. For the proof of property d), we come back to the construction of f_i in step 3.1. From step 2.4 we deduce

$$|\varphi_0|_R \le 3^{p_m}|f|_R \quad \text{and} \quad |\varphi_1|_R \le \left(\frac{3}{R}\right)^{p_m} |f|_R.$$

By induction assume, for $j = 0$ and for $m + 1 \le j \le n$,

$$|\psi_j|_R \le C_{m+1}R^{-p_j}|\varphi_0|_R \quad \text{and} \quad |\theta_j|_R \le C_{m+1}R^{-p_j}|\varphi_1|_R$$

with $C_{m+1} = 3^{p(n-m)}(1 + 2^p)^{n-m-1}$ (so that $C_n = 3^p$). Using the inequalities

$$C_{m+1}(3^{p_m} + 2^{p_m}3^{p_m}) \le C_m$$

and

$$|P_j|_R \le (R + r)^{p_j} \le (2R)^{p_j},$$

we obtain, for $j = 0, m + 1, \ldots, n$,

$$\begin{aligned} |f_j|_R &\le |\psi_j|_R + (2R)^{p_m}|\theta_j|_R \\ &\le C_{m+1}R^{-p_j}|f|_R(3^{p_m} + 2^{p_m}3^{p_m}) \\ &\le C_m R^{-p_j}|f|_R. \end{aligned}$$

Since $f_m = \theta_0$, these estimates also hold for $j = m$. Finally the inequality $C_m \le 9^{(n-m+1)p}$ concludes the proof of Lemma 4.8. □

Proof of Proposition 4.7. Let $f \in \mathcal{A}_n$ satisfy the hypotheses of Proposition 4.7. From Lemma 4.8 with $m = n$, we deduce that f belongs to the ideal $\mathcal{I}$ generated in $\mathcal{A}_n$ by the functions $P_1(z_1), \ldots, P_n(z_n)$, where

$$P_i(X) = \prod_{\xi \in \mathcal{E}_i} (X - \xi)^{S_0}, \qquad (1 \le i \le n),$$

and that there exist entire functions $f_1, \ldots, f_n$ in $\mathcal{A}_n$ with

$$f(\underline{z}) = f_1(\underline{z})P_1(z_1) + \cdots + f_n(\underline{z})P_n(z_n)$$

and

$$|f_i|_R \le 9^{np}R^{-p_i}|f|_R, \qquad (1 \le i \le n).$$

Since $|P_i|_r \le (2r)^{S_0S_1}$ and $p = p_1 = \cdots = p_n = S_1S_0$, using the maximum modulus principle $|f_i|_r \le |f_i|_R$, we deduce

$$|f|_r \le \sum_{i=1}^{n} |f_i|_r (2r)^{S_0 S_1} \le n2^p 3^{2np} \left(\frac{r}{R}\right)^p |f|_R.$$

Finally, we bound $n2^p3^{2np}$ by 18^{np}. □

4.4 Exponential Polynomials

Let d_1, ℓ_1 and n be positive integers, $\underline{x}_1, \dots, \underline{x}_{d_1}$ and $\underline{y}_1, \dots, \underline{y}_{\ell_1}$ be elements of $\mathbb{C}^n$. For $\underline{\tau} \in \mathbb{N}^n$ and $\underline{t} \in \mathbb{Z}^{d_1}$ we write $\underline{z}^{\underline{\tau}}$ in place of $z_1^{\tau_1} \cdots z_n^{\tau_n}$ and $\underline{t}\underline{x}$ in place of $t_1\underline{x}_1 + \cdots + t_{d_1}\underline{x}_{d_1}$. Hence $\underline{t}\underline{x}\underline{z}$ denotes the complex number $\sum_{i=1}^{d_1}\sum_{\nu=1}^{n} t_i x_{\nu i} z_\nu$ (scalar product of $\underline{t}\underline{x}$ and $\underline{z}$ in $\mathbb{C}^n$). Consider the function

$$\underline{z}^{\underline{\tau}} e^{\underline{t}\underline{x}\underline{z}} = z_1^{\tau_1} \cdots z_n^{\tau_n} \exp\bigl((t_1\underline{x}_1 + \cdots + t_{d_1}\underline{x}_{d_1})\underline{z}\bigr).$$

We are interested in the derivative $\mathcal{D}^{\underline{\sigma}}$, for $\underline{\sigma} \in \mathbb{N}^n$, of this function at the point $\underline{s}\underline{y} = s_1\underline{y}_1 + \cdots + s_{\ell_1}\underline{y}_{\ell_1}$ where $\underline{s} \in \mathbb{Z}^{\ell_1}$. In the next lemma we write explicitly this number $\mathcal{D}^{\underline{\sigma}}\bigl(\underline{z}^{\underline{\tau}} e^{\underline{t}\underline{x}\underline{z}}\bigr)(\underline{s}\underline{y})$. We state the result for values of complex exponential functions, but an equivalent algebraic statement holds over a field of zero characteristic.

Lemma 4.9. *For $\underline{\tau} \in \mathbb{N}^n$, $\underline{t} \in \mathbb{Z}^{d_1}$, $\underline{\sigma} \in \mathbb{N}^n$ and $\underline{s} \in \mathbb{Z}^{\ell_1}$, define a polynomial $P_{\underline{\tau}\underline{t}}^{(\underline{\sigma}\underline{s})}$ in $n(d_1 + \ell_1)$ variables with coefficients in $\mathbb{Z}$ as follows:*

$$P_{\underline{\tau}\underline{t}}^{(\underline{\sigma}\underline{s})}(X, Y) =$$

$$\sum_{\underline{\kappa}} \prod_{\nu=1}^{n} \left(\frac{\sigma_\nu!\tau_\nu!}{\kappa_\nu!(\sigma_\nu - \kappa_\nu)!(\tau_\nu - \kappa_\nu)!} \left(\sum_{i=1}^{d_1} t_i X_{\nu i}\right)^{\sigma_\nu - \kappa_\nu} \left(\sum_{j=1}^{\ell_1} s_j Y_{\nu j}\right)^{\tau_\nu - \kappa_\nu} \right),$$

where $\underline{\kappa} = (\kappa_1, \dots, \kappa_n)$ runs over the set of elements in $\mathbb{N}^n$ for which

$$0 \le \kappa_\nu \le \min\{\sigma_\nu, \tau_\nu\} \quad \text{for} \quad 1 \le \nu \le n.$$

Then we have

$$\mathcal{D}^{\underline{\sigma}}\bigl(\underline{z}^{\underline{\tau}} e^{\underline{t}\underline{x}\underline{z}}\bigr)(\underline{s}\underline{y}) = P_{\underline{\tau}\underline{t}}^{(\underline{\sigma}\underline{s})}(\underline{x}, \underline{y}) \cdot \prod_{i=1}^{d_1}\prod_{j=1}^{\ell_1} e^{\underline{x}_i \underline{y}_j t_i s_j},$$

where $\underline{x}$ stands for $(x_{\nu i})_{1 \le \nu \le n, 1 \le i \le d_1}$ and $\underline{y}$ for $(y_{\nu j})_{1 \le \nu \le n, 1 \le j \le \ell_1}$. The total degree of $P_{\underline{\tau}\underline{t}}^{(\underline{\sigma}\underline{s})}$ with respect to the nd_1 variables $X_{\nu i}$ (resp. to the $n\ell_1$ variables $Y_{\nu j}$) is bounded by $\|\underline{\sigma}\|$ (resp. $\|\underline{\tau}\|$). If T and S are positive real numbers with $T \ge \max\{\|\underline{t}\|, 1\}$ and $S \ge \max\{\|\underline{s}\|, 1\}$, then the length of $P_{\underline{\tau}\underline{t}}^{(\underline{\sigma}\underline{s})}$ is at most

$$T^{\|\underline{\sigma}\|}S^{\|\underline{\tau}\|}\min\left\{\left(1+\frac{|\underline{\tau}|}{TS}\right)^{\|\underline{\sigma}\|};\left(1+\frac{|\underline{\sigma}|}{TS}\right)^{\|\underline{\tau}\|}\right\}.$$

Proof. A simple computation based on Leibniz' formula yields the relation between $\mathcal{D}^{\underline{\sigma}}\left(\underline{z}^{\underline{\tau}}e^{\underline{t}\underline{x}\underline{z}}\right)(s\underline{y})$ and $P_{\underline{\tau}\underline{t}}^{(\underline{\sigma}s)}(\underline{x},\underline{y})$. The length of $P_{\underline{\tau}\underline{t}}^{(\underline{\sigma}s)}$ is bounded as follows:

$$\begin{aligned} L(P_{\underline{\tau}\underline{t}}^{(\underline{\sigma}s)}) &\le \sum_{\underline{\kappa}}\prod_{\nu=1}^{n}\left(\frac{\sigma_\nu!\tau_\nu!}{\kappa_\nu!(\sigma_\nu-\kappa_\nu)!(\tau_\nu-\kappa_\nu)!}\|\underline{t}\|^{\sigma_\nu-\kappa_\nu}\|\underline{s}\|^{\tau_\nu-\kappa_\nu}\right)\\ &\le \prod_{\nu=1}^{n}\sum_{\kappa_\nu=0}^{\sigma_\nu}\frac{\sigma_\nu!}{\kappa_\nu!(\sigma_\nu-\kappa_\nu)!}\tau_\nu^{\kappa_\nu}\|\underline{t}\|^{\sigma_\nu-\kappa_\nu}\|\underline{s}\|^{\tau_\nu-\kappa_\nu}\\ &\le \prod_{\nu=1}^{n}\sum_{\kappa_\nu=0}^{\sigma_\nu}\frac{\sigma_\nu!}{\kappa_\nu!(\sigma_\nu-\kappa_\nu)!}\tau_\nu^{\kappa_\nu}T^{\sigma_\nu-\kappa_\nu}S^{\tau_\nu-\kappa_\nu}\\ &\le T^{\|\underline{\sigma}\|}S^{\|\underline{\tau}\|}\left(1+\frac{|\underline{\tau}|}{TS}\right)^{\|\underline{\sigma}\|}. \end{aligned}$$

Lemma 4.9 easily follows (by symmetry – see also § 13.7). □

4.5 Construction of an Auxiliary Function

There are mainly two variants for the construction of an auxiliary function: either one shows that there is a nontrivial solution to a system of homogeneous linear equations, which forces the function to vanish at a finite collection of points; or else one builds a function which is small on a large disc. In either cases the construction of the auxiliary function involves Dirichlet's box principle, either for a system of equations (Thue-Siegel's Lemma), of for a system of inequalities. This alternative does not make too much difference: with both approaches one then proves by induction that the constructed function at the same time satisfies the vanishing conditions at the given points, and has a small maximum modulus on large discs. We choose here the second option for two reasons: firstly the other one is already explained at several other places ([L 1966], Chap. IV, [Bo 1970], [W 1979b], § 5.4, [LelGru 1986], Chap. 6). Secondly the construction we are going to work out would enable us to use Schwarz' Lemma only with a single point (in the construction of the auxiliary function) and to use instead a multiplicity estimate (see step 4 in § 4.6 below).

The following auxiliary function arises from [W 1981].

Proposition 4.10. *Let L and n be positive integers, N, U, V, R, r positive real numbers and $\varphi_1,\dots,\varphi_L$ entire functions in $\mathbb{C}^n$. Define $W=N+U+V$ and assume*

$$W\ge 12n^2,\quad e\le\frac{R}{r}\le e^{W/6},\quad \sum_{\lambda=1}^{L}|\varphi_\lambda|_R\le e^U$$

and

$$(2W)^{n+1} \leq LN\big(\log(R/r)\big)^n.$$

Then there exist rational integers $p_1, \ldots, p_L$, with

$$0 < \max_{1 \leq \lambda \leq L} |p_\lambda| \leq e^N,$$

such that the function $F = p_1\varphi_1 + \cdots + p_L\varphi_L$ satisfies

$$|F|_r \leq e^{-V}.$$

A variant of Proposition 4.10 is given in Exercise 4.7.

The proof involves an application of Dirichlet's box principle which goes back to the work of A. Thue and C. L. Siegel. We also use a simple interpolation formula obtained by truncating the Taylor expansion at the origin of an entire function.

We start with the following version of Thue-Siegel's Lemma (cf. [W 1974], lemme 1.3.2):

Lemma 4.11. *Let v_{ij} $(1 \leq i \leq \nu, 1 \leq j \leq \mu)$ be real numbers, U a positive integer satisfying*

$$U \geq \max_{1 \leq j \leq \mu} \sum_{i=1}^{\nu} |v_{ij}|,$$

and X, ℓ positive integers such that

$$\ell^\mu < (X+1)^\nu.$$

Then there exist rational integers $\xi_1, \ldots, \xi_\nu$ with

$$0 < \max_{1 \leq i \leq \nu} |\xi_i| \leq X,$$

and

$$\max_{1 \leq j \leq \mu} \left| \sum_{i=1}^{\nu} v_{ij}\xi_i \right| \leq \frac{UX}{\ell}.$$

Proof. We consider the mapping φ from the set

$$\mathcal{E} = \big\{\underline{\xi} = (\xi_1, \ldots, \xi_\nu) \in \mathbb{Z}^\nu\,;\, 0 \leq \xi_i \leq X, (1 \leq i \leq \nu)\big\}$$

to $\mathbb{R}^\mu$ which maps $\underline{\xi}$ to $\underline{\eta} = (\eta_1, \ldots, \eta_\mu)$ with

$$\eta_j = \sum_{i=1}^{\nu} v_{ij}\xi_i, \qquad (1 \leq j \leq \mu).$$

For $1 \leq j \leq \mu$, denote by $-V_j$ (resp. W_j) the sum of negative (resp. positive) elements among $\{v_{1j}, \ldots, v_{\nu j}\}$:

$$V_j = \sum_{i=1}^{\nu} \max\{0, -v_{ij}\}, \quad W_j = \sum_{i=1}^{\nu} \max\{0, v_{ij}\} \qquad (1 \le j \le \mu),$$

so that $V_j + W_j \le U$ for $1 \le j \le \mu$. For $\underline{\xi} \in \mathcal{E}$, the image $\underline{\eta} = \varphi(\underline{\xi})$ belongs to the set

$$\mathcal{F} = \{\underline{\eta} = (\eta_1, \ldots, \eta_\mu) \in \mathbb{R}^\mu\,;\, -XV_j \le \eta_j \le XW_j,\ (1 \le j \le \mu)\}.$$

For $1 \le j \le \mu$, we decompose the interval $[-XV_j, XW_j]$ into ℓ intervals, each of length at most UX/ℓ, so that $\mathcal{F}$ is cut into ℓ^μ pieces $\mathcal{F}_k$ $(1 \le k \le \ell^\mu)$. Since

$$\ell^\mu < (1+X)^\nu = \operatorname{Card}\mathcal{E},$$

Dirichlet's box principle shows that there exist two distinct elements $\underline{\xi}'$ and $\underline{\xi}''$ in $\mathcal{E}$ whose images under φ belong to the same $\mathcal{F}_k$. Denote by $\underline{\xi}$ the difference $\underline{\xi}' - \underline{\xi}''$ in $\mathbb{Z}^\nu$, and by $\underline{\eta}$ the image $\varphi(\underline{\xi})$ of $\underline{\xi}$. We deduce

$$\underline{\xi} = (\xi_1, \ldots, \xi_\nu) \in \mathbb{Z}^\nu, \quad 0 < \max_{1 \le i \le \nu} |\xi_i| \le X,$$

and

$$\underline{\eta} = (\eta_1, \ldots, \eta_\mu) \in \mathbb{R}^\mu, \quad \max_{1 \le j \le \mu} |\eta_j| \le \frac{UX}{\ell},$$

which concludes the proof of Lemma 4.11. □

We apply Lemma 4.11 as follows:

Lemma 4.12. *Let X be a positive integer, U, V be positive real numbers and u_{ij} $(1 \le i \le \nu, 1 \le j \le \mu)$ be complex numbers. Assume*

$$\sum_{i=1}^{\nu} |u_{ij}| \le e^U, \qquad (1 \le j \le \mu)$$

and

$$\left(\sqrt{2}Xe^{U+V} + 1\right)^{2\mu} \le (X+1)^\nu.$$

Then there exists $(\xi_1, \ldots, \xi_\nu) \in \mathbb{Z}^\nu$ satisfying

$$0 < \max_{1 \le i \le \nu} |\xi_i| \le X$$

and

$$\max_{1 \le j \le \mu} \left| \sum_{i=1}^{\nu} u_{ij}\xi_i \right| \le e^{-V}.$$

It is important to notice that the numbers u_{ij} are not supposed to be algebraic, there are just complex numbers.

Proof. By homogeneity, replacing if necessary u_{ij} by $u_{ij}e^{-U}$ and V by $U+V$, we may assume $U = 0$. From the hypothesis one deduces that there exists an integer ℓ which satisfies

$$\ell^{2\mu} < (X+1)^{\nu} \quad \text{and} \quad \sqrt{2}\frac{X}{\ell} \le e^{-V}.$$

We solve the system of linear inequalities

$$\begin{cases} \max_{1\le j\le\mu} \left| \sum_{i=1}^{\nu} \mathrm{Re}(u_{ij})\xi_i \right| \le \dfrac{X}{\ell}, \\ \max_{1\le j\le\mu} \left| \sum_{i=1}^{\nu} \mathrm{Im}(u_{ij})\xi_i \right| \le \dfrac{X}{\ell}, \end{cases}$$

by means of Lemma 4.11. □

Here is the simple interpolation formula we need:

Lemma 4.13. *Let r and R be real numbers satisfying $0 < r < R$, T a positive integer, and F an entire function in $\mathbb{C}^n$. Then*

$$|F|_r \le (1+\sqrt{T})\left(\frac{r}{R}\right)^T |F|_R + \sum_{\|\underline{\tau}\|<T} |\mathcal{D}^{\underline{\tau}}F(0)|\frac{r^{\|\underline{\tau}\|}}{\underline{\tau}!}.$$

Exercise 4.6 gives a variant of Lemma 4.13.

Proof. We truncate the Taylor expansion of F at the origin: define

$$G(\underline{z}) = F(\underline{z}) - \sum_{\|\underline{\tau}\|<T} \mathcal{D}^{\underline{\tau}}F(0)\frac{\underline{z}^{\underline{\tau}}}{\underline{\tau}!}.$$

Let $\underline{z}_0 \in \mathbb{C}^n$ satisfy $|\underline{z}_0| = r$ and $|F(\underline{z}_0)| = |F|_r$. Define two entire functions f and g of a single variable w by

$$f(w) = F(\underline{z}_0 w), \quad g(w) = G(\underline{z}_0 w).$$

Since g has a zero of multiplicity at least T at the origin, Schwarz' Lemma yields

$$|f(\underline{z}_0)| = |g(1)| \le \left(\frac{r}{R}\right)^T |g|_{R/r}.$$

Using Cauchy-Schwarz' inequality

$$\left|\sum_{t=0}^{T-1} x_t y_t\right|^2 \le \left(\sum_{t=0}^{T-1} |x_t|^2\right)\left(\sum_{t=0}^{T-1} |y_t|^2\right)$$

and Parseval's formula

$$\sum_{t=0}^{\infty} |a_t|^2 \varrho^{2t} = \frac{1}{2\pi\varrho} \int_{|w|=\varrho} |f(w)|^2 dw \leq |f|_\varrho^2$$

for

$$f(w) = \sum_{t=0}^{\infty} a_t w^t \quad \text{and} \quad \varrho > 0,$$

we deduce that the polynomial

$$P(w) = f(w) - g(w) = \sum_{t=0}^{T-1} a_t w^t$$

satisfies

$$|P|_{R/r}^2 \leq \left(\sum_{t=0}^{T-1} |a_t| \left(\frac{R}{r} \right)^t \right)^2 \leq T|f|_{R/r}^2.$$

Hence

$$|g|_{R/r} \leq (1 + T^{1/2})|f|_{R/r} \quad \text{and} \quad |G(\underline{z}_0)| \leq (1 + T^{1/2}) \left(\frac{r}{R} \right)^T |F|_R.$$

We conclude

$$|F|_r = |F(\underline{z}_0)| \leq |G(\underline{z}_0)| + \sum_{\|\underline{\tau}\| < T} |\mathcal{D}^{\underline{\tau}} F(0)| \frac{|\underline{z}_0|^{\|\underline{\tau}\|}}{\underline{\tau}!}.$$

□

Proof of Proposition 4.10. Define the integer T by

$$\frac{4}{3} W \leq T \log \frac{R}{r} < \frac{4}{3} W + \log \frac{R}{r}.$$

For $1 \leq \lambda \leq L$ and $\underline{\tau} \in \mathbb{N}^n$, define

$$u_{\lambda\underline{\tau}} = 2T^n \mathcal{D}^{\underline{\tau}} \varphi_\lambda(0) \frac{r^{\|\underline{\tau}\|}}{\underline{\tau}!}.$$

We consider the linear system of inequalities:

$$\left| \sum_{\lambda=1}^{L} p_\lambda u_{\lambda\underline{\tau}} \right| \leq e^{-V}, \qquad (\underline{\tau} \in \mathbb{N}^n, \;\; \|\underline{\tau}\| < T).$$

The unknowns are $p_1, \ldots, p_L$ in $\mathbb{Z}$, and the number μ of inequalities is $\binom{T+n-1}{n} \leq T^n$. For $\|\underline{\tau}\| < T$ we have

$$\sum_{\lambda=1}^{L} |\mathcal{D}^{\underline{\tau}} \varphi_\lambda(0)| \frac{r^{\|\underline{\tau}\|}}{\underline{\tau}!} \leq \sum_{\lambda=1}^{L} |\varphi_\lambda|_r \leq e^U,$$

hence

$$\sum_{\lambda=1}^{L} |u_{\lambda\underline{\tau}}| \leq 2T^n e^U.$$

From the assumptions $R/r \geq e$ and $W \geq 12n^2$ we deduce

$$T \leq \frac{4}{3}W + 1 \quad \text{and} \quad 3\left(\frac{4}{3}W + 1\right)^n < e^{W/3}.$$

Therefore $3T^n < e^{W/3}$, which gives

$$2\sqrt{2}T^n e^W + 1 \leq e^{4W/3}.$$

Since $R/r \leq e^{W/6}$ we have $T \log(R/r) < (3/2)W$. We use the upper bound for $(2W)^{n+1}$ in the hypothesis of Proposition 4.10 and we bound $(3/2)^{n-1}$ by 2^{n-1} in order to get $(8/3)WT^n \leq LN$. Therefore we have

$$\left(2\sqrt{2}T^n e^W + 1\right)^{2T^n} < e^{LN},$$

which enables us to use Lemma 4.12 with $X = [e^N]$ and with e^U replaced by $2T^n e^U$). We deduce that there exists a nontrivial solution $(p_1, \ldots, p_L) \in \mathbb{Z}^L$ with $\max_{1\leq\lambda\leq L} |p_\lambda| \leq e^N$. The function $F = p_1\varphi_1 + \cdots + p_L\varphi_L$ then satisfies

$$\sum_{\|\underline{\tau}\|<T} |\mathcal{D}^{\underline{\tau}} F(0)| \frac{r^{\|\underline{\tau}\|}}{\underline{\tau}!} \leq \frac{1}{2}e^{-V}.$$

From the estimates

$$|F|_R \leq \sum_{\lambda=1}^{L} |p_\lambda|\, |\varphi_\lambda|_R \leq e^{N+U}$$

and $1 + \sqrt{T} \leq (1/2)e^{W/3}$, we deduce

$$\left(1 + \sqrt{T}\right)\left(\frac{r}{R}\right)^T |F|_R \leq \frac{1}{2}e^{W/3}\left(\frac{R}{r}\right)^{-T} e^{N+U} \leq \frac{1}{2}e^{-V}.$$

Lemma 4.13 provides the conclusion

$$|F|_r \leq e^{-V}.$$

□

4.6 Direct Proof of Corollary 4.2

We already deduced Corollary 4.2 from Theorem 4.1, but since we did not include a proof of Theorem 4.1, we produce a direct proof of Corollary 4.2. We decompose it into several steps.

Step 1. Introducing the parameters

Assume that the numbers y_{hj} and $e^{\underline{x}_i\underline{y}_j}$ which occur in the conclusion of Corollary 4.2 all belong to a number field K. Choose rational integers T_0, T_1, S_0 and S_1, all of which are at least 2. We shall see that an admissible choice is to take for S_1 a large, fixed, positive integer, to take $T_0 = T_1$ ($= T$, say) and S_0 integers which tend to infinity and are related by

$$S_0S_1 = c_0T^{d/n},$$

where $c_0 > 1$ is a suitable integer which depends only on the $\underline{x}_i$'s and the $\underline{y}_j$'s. All numbers $c_1, \ldots, c_{13}$ below are positive real numbers which can be explicitly computed in terms of $\underline{y}_1, \ldots, \underline{y}_n, \underline{x}_1, \ldots, \underline{x}_{d_1}$.

Define $L = (T_0+1)^{d_0}(T_1+1)^{d_1}$, and denote by $\{\varphi_1, \ldots, \varphi_L\}$ the set of L functions of n complex variables:

$$\left\{\underline{z}^{\underline{\tau}}e^{\underline{txz}}\,; 0 \le \tau_h \le T_0, (1 \le h \le d_0),\ 0 \le t_i \le T_1, (1 \le i \le d_1)\right\},$$

where

$$\underline{z}^{\underline{\tau}}e^{\underline{txz}} = z_1^{\tau_1}\cdots z_{d_0}^{\tau_{d_0}}\exp\big((t_1\underline{x}_1 + \cdots + t_{d_1}\underline{x}_{d_1})\underline{z}\big).$$

We embed $\mathbb{N}^{d_0}$ into $\mathbb{N}^n$ by $(\tau_1, \ldots, \tau_{d_0}) \mapsto (\tau_1, \ldots, \tau_n)$ with $\tau_{d_0+1} = \cdots = \tau_n = 0$.

We shall consider the derivatives of these functions at the points $\underline{sy} = s_1\underline{y}_1 + \cdots + s_n\underline{y}_n$ for $\underline{s} \in \mathbb{N}^n$ with $0 \le s_j < S_1$.

Step 2. A lower bound: Liouville's estimate

The numbers $\mathcal{D}^{\underline{\sigma}}\big(\underline{z}^{\underline{\tau}}e^{\underline{txz}}\big)(\underline{sy})$ belong to the number field K. In order to use Liouville's inequality, we apply Lemma 4.9 with

$$\underline{\tau} \in \mathbb{N}^{d_0}, \quad |\underline{\tau}| \le T_0; \qquad \underline{t} \in \mathbb{N}^{d_1}, \quad |\underline{t}| \le T_1; \qquad \ell_1 = n, \quad \underline{s} \in \mathbb{N}^n, \quad |\underline{s}| < S_1,$$

so that the total degree of the polynomial $P_{\underline{\tau t}}^{(\sigma s)}$ in Lemma 4.9 is at most

$$d_0T_0 + \|\underline{\sigma}\|,$$

and the length of the same polynomial is bounded by

$$(d_1T_1)^{\|\underline{\sigma}\|}(nS_1 + |\underline{\sigma}|)^{d_0T_0}.$$

It follows that if $F = p_1\varphi_1 + \cdots + p_L\varphi_L$ is a linear combination of $\varphi_1, \ldots, \varphi_L$ with integer coefficients p_λ, with $\max_{1\le\lambda\le L}|p_\lambda| \le e^N$ (for some $N > 0$), and if $\underline{\sigma} \in \mathbb{N}^n$ and $\underline{s} \in \mathbb{N}^n$ with $|\underline{s}| < S_1$ are such that $\mathcal{D}^{\underline{\sigma}}F(\underline{sy})$ does not vanish, then

$$\log|\mathcal{D}^{\underline{\sigma}}F(\underline{sy})| \ge -c_1\big(N + \|\underline{\sigma}\|\log T_1 + T_0\log(S_1 + \|\underline{\sigma}\|) + T_1S_1\big), \tag{4.14}$$

where $c_1 \ge 1$ can be explicitly computed by means of Liouville's inequality (Lemma 2.1 or 3.14).

Step 3. The auxiliary function

Our goal now is to choose the coefficients $p_\lambda \in \mathbb{Z}$ in a suitable way. Since $\underline{x}_1, \ldots, \underline{x}_{d_1}$ are linearly independent over $\mathbb{Q}$, the functions $z_1, \ldots, z_n, e^{\underline{x}_1\underline{z}}, \ldots, e^{\underline{x}_{d_1}\underline{z}}$ are algebraically independent (see Exercise 2.5), hence the functions $\varphi_1, \ldots, \varphi_L$ are linearly independent over $\mathbb{C}$.

We introduce a new parameter $E \geq e$ (which we shall choose at the end of the proof). We are going to define four constants $c_2, \ldots, c_5$, and to check the hypotheses of Proposition 4.10 with

$$r = c_2 S_1, \quad R = Er, \quad U = V = c_3 L^{1/n} \log E, \quad N = c_4 U,$$

under the following assumption:

$$T_0 \log(S_1 E) + T_1 S_1 E \leq c_5 L^{1/n} \log E. \tag{4.15}$$

We start with the definition of c_2. We shall need $1 + |\underline{s}\underline{y}| \leq r$ for $\underline{s} \in \mathbb{N}^n$ with $|\underline{s}| < S_1$. We choose

$$c_2 = |\underline{y}_1| + \cdots + |\underline{y}_n| + 2;$$

(the condition $c_2 \geq 2$ will be useful). Next consider c_4: the quantity $c_1 N$, which occurs in the lower bound of step 2, should be smaller than U. We take $c_4 = 1/(2c_1)$ (since $c_1 \geq 1$, we have $c_4 < 1$, hence $N < U$).

The main assumption in Proposition 4.10 is

$$(2W)^{n+1} \leq LN\big(\log(R/r)\big)^n.$$

Here, $R/r = E$, $N = c_4 U$, $U = c_3 L^{1/n} \log E$, $W \leq 3U$, hence this condition is satisfied if $6^{n+1} c_3^n \leq c_4$. We define $c_3 = c_4^{1/n} 6^{-1-(1/n)}$.

The conditions $W \geq 12n^2$ and $W \geq 6\log(R/r)$ can now be written $c_3 L^{1/n} \log E \geq \max\{6n^2, 3\log E\}$, or equivalently

$$L \geq 3^n c_3^{-n} \max\left\{1, \left(\frac{2n^2}{\log E}\right)^n\right\}.$$

Since $E \geq e$ and $c_3^{-n} = 6^{n+1} c_4^{-1} = 2^{n+2} 3^{n+1} c_1$, it is sufficient to assume $L \geq 6^{2n+2} n^{2n} c_1$.

Finally, we estimate $|\varphi_\lambda|_R$:

$$\begin{aligned}\log|\varphi_\lambda|_R &\leq d_0 T_0 \log R + (\|\underline{x}_1\| + \cdots + \|\underline{x}_{d_1}\|) T_1 R \\ &\leq d_0 T_0 \log(c_2 S_1 E) + c_2(\|\underline{x}_1\| + \cdots + \|\underline{x}_{d_1}\|) T_1 S_1 E.\end{aligned}$$

We use a very crude bound for the number L:

$$\log L = d_0 \log(T_0 + 1) + d_1 \log(T_1 + 1) \leq d_0 T_0 + d_1 T_1 \leq d_0 T_0 + c_2 d_1 T_1 S_1 E.$$

Therefore

$$\log \sum_{\lambda=1}^{L} |\varphi_\lambda|_R \leq d_0 T_0 \log(e c_2 S_1 E) + c_2(d_1 + \|\underline{x}_1\| + \cdots + \|\underline{x}_{d_1}\|) T_1 S_1 E.$$

Since $S_1 \geq 2$, $E \geq e$ and $c_2 \geq 2$, we have $\log(ec_2 S_1 E) \leq c_2 \log(S_1 E)$, and we conclude

$$\log \sum_{\lambda=1}^{L} |\varphi_\lambda|_R \leq c_6\bigl(T_0 \log(S_1 E) + T_1 S_1 E\bigr),$$

with

$$c_6 = c_2(d_0 + d_1 + \|\underline{x}_1\| + \cdots + \|\underline{x}_{d_1}\|).$$

The condition (4.15) will guarantee $\log \sum_{\lambda=1}^{L} |\varphi_\lambda|_R \leq U$ if we take $c_5 = c_3/c_6$.

Now c_2, c_4, c_3, c_6, c_5 have been successively defined, and the hypotheses of Proposition 4.10 have been checked. Let F be the function which is constructed in this proposition: $\log|F|_r \leq -U$. Since $1 + |\underline{s}\underline{y}| \leq r$, we deduce from Cauchy's inequalities $|\mathcal{D}^{\underline{\sigma}} F(\underline{s}\underline{y})| \leq \underline{\sigma}!|F|_r$:

$$\log|\mathcal{D}^{\underline{\sigma}} F(\underline{s}\underline{y})| \leq -c_3 L^{1/n} \log E + \log(\underline{\sigma}!) \tag{4.16}$$

for all $\underline{\sigma} \in \mathbb{N}^n$ and $\underline{s} \in \mathbb{N}^n$ with $|\underline{s}| < S_1$.

Step 4. Lower bound for the order of vanishing of F at sy

We compare the upper bound (4.14) with the lower bound (4.16) for the number $|\mathcal{D}^{\underline{\sigma}} F(\underline{s}\underline{y})|$, with $\underline{\sigma} \in \mathbb{N}^n$, $|\underline{\sigma}| < S_0$, and $\underline{s} \in \mathbb{N}^n$, $|\underline{s}| < S_1$. These two estimates are not consistent if the parameters satisfy

$$c_3 L^{1/n} \log E > nS_0 \log S_0 + c_1\bigl(N + nS_0 \log T_1 + T_0 \log(S_1 + nS_0) + T_1 S_1\bigr).$$

Recall that $2c_1 N \leq c_3 L^{1/n} \log E$. We impose the following condition[11] on the parameters (which includes all preceding ones)

$$L^{1/n} \log E > c_7\bigl(S_0 \log(S_0 T_1) + T_0 \log(S_0 S_1 E) + T_1 S_1 E\bigr), \tag{4.17}$$

where $c_7 \geq \max\{c_5^{-1}, 2nc_1/c_3\}$. We deduce that for $\underline{\sigma} \in \mathbb{N}^n$ with $|\underline{\sigma}| < S_0$, the function $\mathcal{D}^{\underline{\sigma}} F$ has a zero at each point $\underline{s}\underline{y}$ ($\underline{s} \in \mathbb{N}^n$, $|\underline{s}| < S_1$).

We shall also choose the parameters so that $(S_0 S_1)^n$ is large compared with $L = (T_0 + 1)^{d_0}(T_1 + 1)^{d_1}$:

$$S_0 S_1 \geq c_8 L^{1/n}, \tag{4.18}$$

where c_8 is a sufficiently large number, namely $c_8 \geq 1 + 2c_3$. Hence, at this stage of the proof, we get a system of relations

$$|\mathcal{D}^{\underline{\sigma}} F(\underline{s}\underline{y})| = 0, \quad \text{for } \underline{\sigma} \in \mathbb{N}^n \text{ with } |\underline{\sigma}| < S_0 \text{ and for } \underline{s} \in \mathbb{N}^n \text{ with } |\underline{s}| < S_1,$$

where the number L of coefficients of the constructed polynomial is smaller than the number of relations. A good zero estimate (which takes care of the derivatives – this will be called a *multiplicity estimate*) would enable us to conclude the proof right now (without using the Schwarz' Lemma of § 4.3). Such a result will be proved in Chap. 8. Here, we shall avoid such a multiplicity estimate, but use Proposition 4.7 instead.

[11] Another lower bound for c_7 will occur later in step 6.

Step 5. Upper bound for the first non-vanishing coefficient

We want to estimate the *first* non-vanishing coefficient in the Taylor expansion of F at one of the $\underline{s}\underline{y}$. Let S_0' be the largest integer such that

$$\mathcal{D}^{\underline{\sigma}}F(\underline{s}\underline{y}) = 0 \quad \text{for all } \underline{s} \in \mathbb{N}^n \text{ with } |\underline{s}| < S_1 \text{ and all } \underline{\sigma} \in \mathbb{N}^n \text{ with } |\underline{\sigma}| < S_0'.$$

The existence of S_0' follows from the fact that the function F does not vanish identically, as noticed before. We have already computed the lower bounds $S_0' \geq S_0$. From the definition of S_0' we deduce that there exist $\underline{\sigma}^0 \in \mathbb{N}^n$ and $\underline{s}^0 \in \mathbb{N}^n$ with

$$\mathcal{D}^{\underline{\sigma}^0} F(\underline{s}^0\underline{y}) \neq 0 \quad \text{and } |\underline{\sigma}^0| = S_0',\ |\underline{s}^0| < S_1.$$

Since $\underline{y}_1, \dots, \underline{y}_n$ is a basis of $\mathbb{C}^n$, there exists a positive constant c_9 such that, for $\underline{w} \in \mathbb{C}^n$ with $|\underline{w}| < |\underline{y}_1| + \cdots + |\underline{y}_n| + 1$, there exists $\underline{z} \in \mathbb{C}^n$ with $\underline{w} = z_1\underline{y}_1 + \cdots + z_n\underline{y}_n$ and $|z_i| \leq c_9$. Now we use Proposition 4.7 for the function $f(\underline{z}) = F(z_1\underline{y}_1 + \cdots + z_n\underline{y}_n)$, with a new parameter $E' \geq E$, and with

$$\mathcal{E}_1 = \cdots = \mathcal{E}_n = \{0, 1, \dots, S_1 - 1\}, \quad r = c_9 S_1, \quad R = 18^n E' r,$$

(and also with S_0 and E replaced by S_0' and E' respectively). We deduce the upper bound

$$\log |f|_r \leq -S_0' S_1 \log E' + \log |f|_R.$$

The same computation as before yields a constant c_{10} such that

$$\log |f|_R \leq N + c_{10}\big(T_0 \log(S_1 E') + T_1 S_1 E'\big).$$

On the other hand from the choice of c_9 and Cauchy's inequalities we deduce

$$\begin{aligned} \log |\mathcal{D}^{\sigma^0} F(\underline{s}^0\underline{y})| &\leq n S_0' \log S_0' + \log |f|_r \\ &\leq -S_0' S_1 \log E' + n S_0' \log S_0' + N + c_{10}\big(T_0 \log(S_1 E') + T_1 S_1 E'\big). \end{aligned}$$

It is useful to notice that

$$S_0' S_1 \log E' \geq S_0 S_1 \log E > 2 c_3 L^{1/n} \log E = 2U > 2N.$$

This enables us to deduce from the previous upper bound:

$$\log |\mathcal{D}^{\sigma^0} F(\underline{s}^0\underline{y})| \leq -\frac{1}{2} S_0' S_1 \log E' + n S_0' \log S_0' + c_{10}\big(T_0 \log(S_1 E') + T_1 S_1 E'\big).$$

Step 6. Conclusion and choice of parameters

We shall be able to conclude the proof if the lower bound from step 2 does not match with the upper bound from step 5. This means that the parameters should satisfy

$$\begin{aligned} \frac{1}{2} S_0' S_1 \log E' > c_1\big(N + n S_0' \log T_1 + T_0 \log(S_1 + n S_0') + T_1 S_1\big) + \\ + n S_0' \log S_0' + c_{10}\big(T_0 \log(S_1 E') + T_1 S_1 E'\big). \end{aligned}$$

We already know $S_0S_1 \geq c_8L^{1/n}$ from (4.18). By (4.17), we deduce that each of the two terms c_1N and $T_0 \log S_1$ is small compared with $(1/2)S_0'S_1 \log E'$. Therefore it is sufficient to add the following constraint[12] on the parameters:

$$S_0'S_1 \log E' > c_{11}\Big(S_0' \log(S_0'T_1) + T_0 \log(S_0'E') + T_1S_1E'\Big). \tag{4.19}$$

The conditions on the parameters are (4.17), (4.18) and (4.19), with $E' \geq E \geq e$, $L = (T_0+1)^{d_0}(T_1+1)^{d_1}$. In loose terms, that means: each of the numbers

$$S_0 \log(S_0T_1), \quad T_0 \log(S_0S_1E), \quad T_1S_1E,$$

is small compared to $L^{1/n} \log E$, next each of the numbers

$$S_0' \log(S_0'T_1), \quad T_0 \log(S_0'E'), \quad T_1S_1E',$$

is small compared to $S_0'S_1 \log E'$, and finally $L^{1/n}$ is small compared to S_0S_1.

We are free to choose T_0, T_1, S_0, S_1 and E. The number S_0' cannot be chosen (we just know that it is at least S_0), but we can choose E' as a function of S_0'.

Here is an admissible choice for the parameters: we start with a large integer S_1. How large it should be is easy to state explicitly in terms of d, n, $\underline{x}_1, \ldots, \underline{x}_{d_1}$, $\underline{y}_1, \ldots, \underline{y}_n$ and the degree D of the number field K. Next we choose an integer T (at the end we get the conclusion by letting T tend to infinity, while S_1 is fixed) which is a n-th power and also a multiple of S_1^n (both conditions are not important: they just enable us to avoid taking integral parts!). We define

$$T_0 = T_1 = T, \quad S_0 = \frac{c_0T^{d/n}}{S_1}, \quad E = S_0^{(d-n)/d}, \quad E' = S_0'^{(d-n)/d}.$$

The left hand side of (4.17) is equivalent to $\big((d/n)-1\big)T^{d/n}\log T$, while the right hand side is at most $(c_{12}/S_1)T^{d/n}\log T$. In the same way the left hand side of (4.19) is $\big(1-(n/d)\big)S_1S_0'\log S_0'$, while the right hand side is at most $c_{13}S_0'\log S_0'$.

With this choice of parameters the conditions (4.17) and (4.19) are satisfied, and the proof of Corollary 4.2 is therefore complete. □

Exercises

Exercise 4.1. Prove Theorem 4.1 using the method of § 4.6.

Hint. *In Liouville's estimate, derivatives must be estimated for functions satisfying differential equations; see* [L 1966], *Chap. 4, § 2, Lemma 1,* [Bo 1970], *Lemma 1, and* [W 1979b], *Lemme 5.4.2.*

---

[12] provided that c_7 is sufficiently large. This is the extra condition on c_7 which has been announced in step 4

Exercise 4.2. Deduce Corollary 4.2 from Baker's Theorem 1.6.

Remark. It follows that Corollary 4.1 can be deduced from Corollary 4.2 and Corollary 4.3 together. In particular the special case $d_0 \le 1$ of Corollary 4.1 implies the general case.

Hint. *Using Theorem 1.5, prove the following statement (compare with Exercise 1.5):*

Let $\mathcal{V}$ be a vector subspace of $\mathbb{C}^m$ defined by homogeneous linear equations $\sum_{\nu=1}^{m} z_\nu \ell_{\nu j} = 0$ $(1 \le j \le \ell)$, with $\ell_{\nu j} \in \mathcal{L}$. If $\mathcal{V} \cap \mathbb{Q}^m = 0$, then $\mathcal{V} \cap \overline{\mathbb{Q}}^m = 0$

Deduce that Corollary 4.2 is a consequence of Theorem 1.5 (take $m = dr + 1$, where r is the dimension of the $\overline{\mathbb{Q}}$-vector space spanned by $\underline{x}_1, \ldots, \underline{x}_d$).

Exercise 4.3.
a) (Blaschke products.) Let z and ζ be two complex numbers. Define $r = |z|$ and $\varrho = |\zeta|$. Assume R satisfies $r \le R$, $\varrho \le R$ and $r\varrho < R^2$. Check

$$\frac{|r-\varrho|}{R^2 - r\varrho} \le \left|\frac{z-\zeta}{R^2 - z\overline{\zeta}}\right| \le \frac{r+\varrho}{R^2 + r\varrho}.$$

b) Let $\zeta \in \mathbb{C}$ and $R > 0$ satisfy $|\zeta| \le R$. Check that the function

$$z \longmapsto \frac{z-\zeta}{R^2 - z\overline{\zeta}}$$

can be continued as an analytic function on an open neighborhood of the closed disc $|z| \le R$ with

$$\left|\frac{z-\zeta}{R^2 - z\overline{\zeta}}\right| = \frac{1}{R} \quad \text{for} \quad |z| = R.$$

Hint. *For $|z| = R$ we have $R^2 - z\overline{\zeta} = z(\overline{z} - \overline{\zeta})$.*

c) Let $N, \kappa_1, \ldots, \kappa_N$ be positive integers and r, R positive real numbers with $r \le R$. Let $\zeta_1, \ldots, \zeta_N$ be distinct elements in the disc $|\zeta| \le R$ and let f be an analytic function in an open neighborhood of the closed disc $|z| \le R$ which vanishes at each ζ_i with multiplicity $\ge \kappa_i$ $(1 \le i \le N)$. Check

$$|f|_r \le |f|_R \prod_{i=1}^{N} \left(\frac{R^2 + r|\zeta_i|}{R(r + |\zeta_i|)}\right)^{-\kappa_i}.$$

d) Prove the following variant of Proposition 4.7.

Let $n, N_1, \ldots, N_n$ be positive integers and E, Θ positive real numbers with $E \ge 1$. For $1 \le i \le n$, let r_i, ϱ_i, R_i be positive real numbers satisfying

$$r_i \le R_i, \quad \varrho_i \le R_i \quad \text{and} \quad \frac{R_i^2 + r_i\varrho_i}{R_i(r_i + \varrho_i)} \ge E$$

and let $\mathcal{E}_i$ be a subset with N_i elements of the disc $|z| \le \varrho_i$ of $\mathbb{C}$. Define $\mathcal{E} = \mathcal{E}_1 \times \cdots \times \mathcal{E}_n \subset \mathbb{C}^n$. Let f be a complex function of n variables which is analytic in an open neighborhood of the closed polydisc

$$\{\underline{z} \in \mathbb{C}^n \,;\, |z_i| \le R_i \ (1 \le i \le n)\}.$$

Assume

$$\mathcal{D}^{\underline{\sigma}} f(\underline{\xi}) = 0 \quad \textit{for any } \underline{\xi} \in \mathcal{E} \textit{ and any } \underline{\sigma} \in \mathbb{N}^n \textit{ with } \sigma_1 N_1 + \cdots + \sigma_n N_n < \Theta.$$

Then

$$|f|_r \leq E^{-\Theta} |f|_R.$$

Hint. *See F. Gramain, Lemmes de Schwarz pour des produits Cartésiens, Publ. Math. St Etienne, to appear.*

Exercise 4.4 (*Schwarz' Lemma for complete intersections of hyperplanes without multiplicities*).

Let $P_1, \ldots, P_n$ be n polynomials in $\mathbb{C}[z_1, \ldots, z_n]$. Assume, for $1 \leq i \leq n$, that P_i is a product of polynomials of degree 1, and define $p_j = \deg P_j$. Assume that the set

$$S = \{\underline{s} \in \mathbb{C}^n \,;\ P_1(\underline{s}) = \cdots = P_n(\underline{s}) = 0\}$$

has exactly $p_1 \cdots p_n$ elements. Define $p = \min\{p_1, \ldots, p_n\}$ and $r = \max\{|\underline{s}| \,;\ \underline{s} \in S\}$. Show that there exists a constant $c > 0$, which depends only on $P_1, \ldots, P_n$, and satisfies the following property: for all $R \geq \max\{1, 2r\}$ and all $f \in \mathcal{A}_n$ which vanishes at each point of S, there exist $f_1, \ldots, f_n$ in $\mathcal{A}_n$ such that

$$|f_j|_R \leq c \left(\frac{r}{R}\right)^{-p} |f|_R, \qquad (1 \leq j \leq n).$$

Deduce the following Schwarz' Lemma for a function f which vanishes on S:

$$|f|_r \leq c' \left(\frac{r}{R}\right)^{-p} |f|_R,$$

where c' depends only on $P_1, \ldots, P_n$.

Remark. In the case $n = 2$, the assumption that P_1 and P_2 are products of linear polynomials can be slightly relaxed: one of them is any polynomial, the other is product of polynomials of degree 1 or 2; see [W 1983].

Exercise 4.5. Let n be a positive integer. Show that there exists a positive number $c = c(n)$ with the following property: let f be an entire function in $\mathbb{C}^n$ which satisfies $f(\mathbb{N}^n) \subset \mathbb{Z}$ and

$$\log |f|_R \leq cR \quad \text{for all sufficiently large } R.$$

Then f is a polynomial.
Compare with [Gr 1969].

Remark. By Pólya's Theorem (see for instance [F 1982], Chap. 2, § 3, Th. 1.2), an entire function f of a single variable such that

$$f(\mathbb{N}) \subset \mathbb{Z} \quad \text{and} \quad \limsup_{R \to \infty} \frac{1}{R} \log |f|_R < \log 2$$

is a polynomial.

Exercise 4.6. Let $R_1, \dots, R_n$ be positive real numbers. Denote by $D(0, R)$ the polydisc

$$\{\underline{z} \in \mathbb{C}^n \,;\, |z_\nu| \le R_\nu,\ (1 \le \nu \le n)\}.$$

Let F be an analytic function in an open neighborhood of this polydisc. Write the Taylor expansion of F at the origin

$$F(\underline{z}) = \sum_{\underline{\tau} \in \mathbb{N}^n} a_{\underline{\tau}} \underline{z}^{\underline{\tau}}.$$

Let $r_1, \dots, r_n$ be positive real numbers with $r_\nu \le R_\nu$ $(1 \le \nu \le n)$ and let $T_1, \dots, T_n$ be positive integers.
a) Assume $a_{\underline{\tau}} = 0$ for $0 \le \tau_\nu < T_\nu$ $(1 \le \nu \le n)$. Check

$$\sup_{\underline{z} \in D(0,r)} |F(\underline{z})| \le \sup_{\underline{z} \in D(0,R)} |F(\underline{z})| \max_{1 \le \nu \le n} \left(\frac{R_\nu}{r_\nu}\right)^{-T_\nu}.$$

b) Check

$$\sup_{\underline{z} \in D(0,r)} |F(\underline{z})| \le \left(1 + \sqrt{T_1 \cdots T_n}\right) \sup_{\underline{z} \in D(0,R)} |F(\underline{z})| \max_{1 \le \nu \le n} \left(\frac{R_\nu}{r_\nu}\right)^{-T_\nu} + \sum_{\substack{0 \le \tau_\nu < T_\nu \\ 1 \le \nu \le n}} |a_{\underline{\tau}}| r^{\underline{\tau}}$$

and also

$$\sup_{\underline{z} \in D(0,r)} |F(\underline{z})| \le \sup_{\underline{z} \in D(0,R)} |F(\underline{z})| \max_{1 \le \nu \le n} \left(\frac{R_\nu}{r_\nu}\right)^{-T_\nu} + 2 \sum_{\substack{0 \le \tau_\nu < T_\nu \\ 1 \le \nu \le n}} |a_{\underline{\tau}}| r^{\underline{\tau}}.$$

Exercise 4.7. Let L, n be positive integers and $N, U, V, R_1, \dots, R_n, r_1, \dots, r_n$ be positive real numbers. Define $W = N + U + V$ and assume

$$W \ge 20n^3, \quad e \le \frac{R_\nu}{r_\nu} \le e^{W/9n}, \quad (1 \le \nu \le n)$$

and

$$3W^{n+1} \le LN \prod_{\nu=1}^{n} \log \frac{R_\nu}{r_\nu}.$$

Let $\varphi_1, \dots, \varphi_L$ be entire functions in $\mathbb{C}^n$ satisfying

$$\sum_{\lambda=1}^{L} \sup_{\substack{|z_\nu| \le R_\nu \\ 1 \le \nu \le n}} |\varphi_\lambda(\underline{z})| \le e^U.$$

Show that there exist rational integers $p_1, \dots, p_L$, with

$$0 < \max_{1 \le \lambda \le L} |p_\lambda| \le e^N,$$

such that the function $F = p_1\varphi_1 + \cdots + p_L\varphi_L$ satisfies

$$|F|_r \le e^{-V}.$$

Exercise 4.8. Let $n_0, n_1, \ldots, n_L$, N be nonnegative integers, $\zeta_0, \ldots, \zeta_N$ pairwise distinct complex numbers and $r_1, \ldots, r_L$, $R_1, \ldots, R_L$ positive real numbers satisfying

$$0 = n_0 \leq n_1 < n_2 < \cdots < n_L = N$$

and

$$R_\mu \geq r_\mu \geq \max_{0 \leq i \leq n_\mu} |\zeta_i| \quad (1 \leq \mu \leq L).$$

Let $f_1, \ldots, f_L$ be complex functions of a single variable which are analytic in a neighborhood of the closed disc $|z| \leq \max_{1 \leq \mu \leq L} R_\mu$. Define $F = (f_1, \ldots, f_L)$. Assume that for $0 \leq \nu < L$ and $n_\nu \leq i < n_{\nu+1}$, the $L \times (\nu + 1)$ matrix

$$\Big(F(\zeta_{n_1}), \ldots, F(\zeta_{n_\nu}), F(\zeta_i)\Big)$$

has rank $\leq \nu$. Show that the determinant of the $L \times L$ matrix

$$A = \Big(F(\zeta_{n_1}), \ldots, F(\zeta_{n_L})\Big)$$

has absolute value bounded by

$$|\det A| \leq L! \left(\prod_{\mu=1}^{L} \left(\frac{R_\mu^2 + r_\mu^2}{2R_\mu r_\mu} \right)^{-n_\mu} \right) \max_{\tau \in \mathfrak{S}_L} \prod_{\lambda=1}^{L} |f_\lambda|_{R_{\tau(\lambda)}}$$

where $\mathfrak{S}_L$ denotes the symmetric group on $\{1, \ldots, L\}$.

5. Zero Estimate, by Damien Roy

In this chapter, we present the zero estimate of P. Philippon [P 1986a] in the context of linear commutative algebraic groups when no order of vanishing is prescribed. This result takes into account the multidegrees of the obstruction subgroup and improves in this way the earlier zero estimates of D. W. Masser [Ma 1981b] and D. W. Masser and G. Wüstholz [MaWü 1981]. A refinement will be given in Chap. 8 when multiplicities are introduced.

5.1 The Main Result

Let K be an algebraically closed field of characteristic zero and let d_0, d_1 be nonnegative integers with $d = d_0 + d_1 > 0$. We denote by $\mathbb{G}_{\mathrm{a}}^{d_0} \times \mathbb{G}_{\mathrm{m}}^{d_1}$, or more simply by G, the group $K^{d_0} \times (K^\times)^{d_1}$ with its group law written additively.

The basic functions on G are the polynomials in

$$X_1, \ldots, X_{d_0}, Y_1, \ldots, Y_{d_1}, Y_1^{-1}, \ldots, Y_{d_1}^{-1}.$$

They form a subring

$$K[G] = K[X_1, \ldots, X_{d_0}, Y_1, \ldots, Y_{d_1}, Y_1^{-1}, \ldots, Y_{d_1}^{-1}]$$

of the field $K(\underline{X}, \underline{Y})$ of rational functions in $\underline{X}, \underline{Y}$. Given nonnegative integers $D_0, D_1, \ldots, D_{d_1}$, we say that an element P of $K[G]$ is of *multidegree* $\le (D_0, D_1, \ldots, D_{d_1})$ if its total degree in $X_1, \ldots, X_{d_0}$ is $\le D_0$ and if, for $j = 1, \ldots, d_1$, its degree in Y_j and its degree in Y_j^{-1} are $\le D_j$.

The object of a zero estimate is to give constraints between the multidegrees of a nonzero element of $K[G]$ vanishing at each point of a given subset of G. Here, we use the group structure on G by assuming that this subset has the form

$$\Sigma[d] = \{\sigma_1 + \cdots + \sigma_d \,;\, (\sigma_1, \ldots, \sigma_d) \in \Sigma^d\}$$

for some positive integer d and some other subset Σ of G containing the neutral element e of G.

In § 3, we recall the notion of *algebraic* subgroup G^* of G and show that these are products of the form $G^* = \mathcal{V} \times T_\Phi$ where $\mathcal{V}$ is a subspace of K^{d_0}, Φ is a subgroup of $\mathbb{Z}^{d_1}$ and T_Φ is given by

$$T_\Phi = \left\{(y_1, \ldots, y_{d_1}) \in (K^\times)^{d_1}\,;\ y_1^{\varphi_1} \cdots y_{d_1}^{\varphi_{d_1}} = 1 \text{ for all } \underline{\varphi} = (\varphi_1, \ldots, \varphi_{d_1}) \in \Phi\right\}.$$

We show that the *dimension* of G^* is $d^* = d_0^* + d_1^*$ where $d_0^* = \dim(\mathcal{V})$ and $d_1^* = d_1 - \operatorname{rank}(\Phi)$, and that G^* is *connected* if and only if Φ is a direct factor of $\mathbb{Z}^{d_1}$. Let $r = \operatorname{rank}(\Phi)$. We also attach to G^* the polynomial

$$\mathcal{H}(G^*; D_0, D_1, \ldots, D_{d_1}) = \frac{d^*!}{d_0^*!} 2^{d_1^*} D_0^{d_0^*} \sum |\det \mathsf{M}_{i_1,\ldots,i_r}| D_{j_1} \cdots D_{j_{d_1^*}}$$

which is closely related to an Hilbert function of G^*. In this expression, the sum extends to all partitions of $\{1, \ldots, d_1\}$ into disjoint subsets $\{i_1, \ldots, i_r\}$ and $\{j_1, \ldots, j_{d_1^*}\}$ with $i_1 < \ldots < i_r$ and $j_1 < \ldots < j_{d_1^*}$, and the symbol $\mathsf{M}_{i_1,\ldots,i_r}$ denotes the $r \times r$ matrix formed by the columns of indices $i_1, \ldots, i_r$ of a fixed $r \times d_1$ matrix M whose rows generate Φ. We also make the convention that the empty matrix has determinant 1. Thus the sum reduces to $D_1 \cdots D_{d_1}$ when $r = 0$. In particular, for the group G, we have

$$\mathcal{H}(G; D_0, D_1, \ldots, D_{d_1}) = \frac{d!}{d_0!} 2^{d_1} D_0^{d_0} D_1 \cdots D_{d_1}$$

Our aim is to show:

Theorem 5.1 *(P. Philippon). Let Σ be a subset of G containing e. Assume that, for given integers $D_0, D_1, \ldots, D_{d_1} \geq 0$, there exists a nonzero element P of $K[G]$ of multidegree $\leq (D_0, D_1, \ldots, D_{d_1})$ which vanishes on $\Sigma[d]$. Then there exists a connected algebraic subgroup G^* of G of dimension $< d$ such that*

$$\operatorname{Card}\left(\frac{\Sigma + G^*}{G^*}\right) \mathcal{H}(G^*; D_0, D_1, \ldots, D_{d_1}) \leq \mathcal{H}(G; D_0, D_1, \ldots, D_{d_1}).$$

In the above inequality, the expression $(\Sigma + G^*)/G^*$ stands for the image of Σ under the canonical map from G to G/G^*. It consists of all translates of G^* of the form $\sigma + G^*$ with $\sigma \in \Sigma$. The conclusion of the theorem implies that it is a finite set even though Σ may be infinite. The group G^* produced by Theorem 5.1 is often called an *obstruction subgroup*.

Exercise 5.3 provides a partial converse to this result. In the applications where Σ often has the form $\Sigma'[S]$ for some fixed Σ' and a large integer S, the conclusion of the theorem appears to be optimal up to a constant factor. The example below illustrates this.

5.1.1 An Example of Application

Theorem 5.1 will be applied in chapters 6 and 7 with a group G of the form $G = \mathbb{G}_{\mathrm{a}}^{d_0} \times \mathbb{G}_{\mathrm{m}}$. Since $\mathbb{Z}$ has only two direct factors, $\{0\}$ and $\mathbb{Z}$, the connected algebraic subgroups G^* of such a group G are of the form $G^* = \mathcal{V} \times G_1^*$ where $\mathcal{V}$ is a subspace of K^{d_0} and G_1^* is either $\mathbb{G}_{\mathrm{m}} = K^\times$ or $\{1\}$. Moreover, if $\mathcal{V}$ has dimension

n, then G^* has dimension $n+1$ in the first case and dimension n in the second case. For integers $D_0, D_1 \geq 0$, this gives:

$$\mathcal{H}(G^*;\, D_0, D_1) = \begin{cases} 2(n+1)D_0^n D_1 & \text{if } G_1^* = K^\times, \\ \\ D_0^n & \text{if } G_1^* = \{1\}. \end{cases}$$

The next example provides a simple application of the theorem with the group $G = \mathbb{G}_a \times \mathbb{G}_m$. Another example, with the group $G = \mathbb{G}_a^2$, is given in Exercise 5.1.

Example. Let $\beta \in K$ and $\alpha_1, \alpha_2 \in K^\times$. Assume that $\beta \notin \mathbb{Q}$ and that α_1, α_2 are multiplicatively independent (i.e. that they generate a subgroup of rank two of $K^\times$). Fix a positive integer S and consider the subset Σ of the group $G = \mathbb{G}_a \times \mathbb{G}_m$ given by

$$\Sigma = \big\{(s_1 + s_2\beta, \alpha_1^{s_1}\alpha_2^{s_2});\ \underline{s} = (s_1, s_2) \in \mathbb{Z}^2, |\underline{s}| \leq S\big\}.$$

Suppose that, for some positive integers D_0, D_1, there exists a nonzero polynomial $P \in K[G] = K[X, Y, Y^{-1}]$ of bidegree $\leq (D_0, D_1)$ which vanishes at each point of

$$\Sigma[2] = \big\{(s_1 + s_2\beta, \alpha_1^{s_1}\alpha_2^{s_2});\ \underline{s} = (s_1, s_2) \in \mathbb{Z}^2, |\underline{s}| \leq 2S\big\}.$$

Then, according to Theorem 5.1, there exists a connected algebraic subgroup G^* of G of dimension < 2 such that

$$\operatorname{Card}\left(\frac{\Sigma + G^*}{G^*}\right) \mathcal{H}(G^*;\, D_0, D_1) \leq 4 D_0 D_1.$$

There are only three possibilities for G^*. If G^* has dimension 0, it is reduced to the neutral element $\{e\}$ of G, where $e = (0, 1)$. If G^* has dimension 1, it is either $K \times \{1\}$ or $\{0\} \times K^\times$. In all cases, the set $(\Sigma + G^*)/G^*$ has cardinality $(2S+1)^2$. This follows from our assumptions that $\beta \notin \mathbb{Q}$ and that α_1, α_2 are multiplicatively independent. Moreover, the quantity $\mathcal{H}(G^*;\, D_0, D_1)$ is 1 in the first case, D_0 in the second and $2D_1$ in the third. Since D_0 and D_1 are assumed to be positive, we deduce that, in all cases, we must have:

$$D_0 D_1 \geq \frac{1}{4}(2S+1)^2.$$

One appreciates this result by observing that conversely, for any pair of positive integers D_0, D_1 with $D_0 D_1 \geq 2(2S+1)^2$, there exists a nonzero polynomial $P \in K[G]$ of bidegree $\leq (D_0, D_1)$ which vanishes at each point of $\Sigma[2]$. To see why this is true, denote by E the subspace of $K[G]$ consisting of polynomials of bidegree $\leq (D_0, D_1)$, denote by F the vector space of K-valued functions on the set $\Sigma[2]$, and consider the linear map $\varphi\colon E \to F$ which sends a polynomial $P \in E$ to the K-valued function on $\Sigma[2]$ induced by P. Since E has dimension $(D_0+1)(2D_1+1) > 2D_0D_1$ while F has dimension $(4S+1)^2 < 4(2S+1)^2$, the map φ cannot be injective when $D_0 D_1 \geq 2(2S+1)^2$, and then, its kernel contains a nonzero element.

5.2 Some Algebraic Geometry

As above, K denotes an algebraically closed field of characteristic zero. We also fix a positive integer n.

5.2.1 Algebraic Subsets of K^n

An *algebraic subset of* K^n is a subset of K^n which is the set of common zeros of a family of polynomials in $K[X_1, \ldots, X_n]$.

From this definition, it follows that the intersection of a family of algebraic subsets of K^n is again an algebraic subset of K^n and that a finite union of algebraic subsets of K^n is also an algebraic subset of K^n.

An algebraic subset of K^n is called *irreducible* if it is not empty and cannot be written as the union of two algebraic subsets of K^n properly contained in it. An irreducible algebraic subset of K^n is also called an *algebraic subvariety* of K^n or simply a *subvariety* of K^n. Thus, if an algebraic subvariety of K^n is contained in a finite union of algebraic subsets of K^n, then it is contained in one of them. The space K^n and the points of K^n are examples of subvarieties of K^n, but the empty set is excluded. One can show that each algebraic subset V of K^n is a finite (possibly empty) union of subvarieties $V_1, \ldots, V_s$ of K^n:

$$V = V_1 \cup \cdots \cup V_s.$$

In this decomposition of V, one can impose the condition $V_i \not\subseteq V_j$ for $i \neq j$. In this case the subvarieties V_i are uniquely determined: they are the maximal subvarieties of K^n contained in V, and they are called the *irreducible components of* V.

An algebraic subset of K^n is said to be *connected* if it cannot be written as the union of two disjoint non empty algebraic sets. Thus, an irreducible algebraic set is always connected but the converse is false.

The *dimension* of an algebraic subvariety V of K^n is the largest integer d for which there exists a strictly increasing chain

$$V_0 \subset V_1 \subset \ldots \subset V_d = V$$

of subvarieties of K^n ending with V. It can be shown that this integer always exists and is $\leq n$. In fact, any chain like the above can be refined to a maximal one by inserting subvarieties before V_0 or between two consecutive ones until this becomes impossible without introducing repetitions, and the number of subvarieties in any such maximal chain ending with V is $d + 1$ where d is the dimension of V. It follows from the definition that if $V_1 \subset V_2$ are two distinct subvarieties of K^n, then $\dim(V_1) < \dim(V_2)$. The dimension of a point is 0, and for K^n it is n.

The *dimension* of a nonempty algebraic subset of K^d is defined as the maximum of the dimensions of its irreducible components. Thus, if V, V' are two non empty algebraic subsets of K^n with $V' \subseteq V$, then we have $\dim(V') \leq \dim(V)$, *with equality if and only if V and V' have a common irreducible component of dimension* $\dim(V)$.

This follows from the fact that each irreducible component of V' is contained in one of V. *In particular, an algebraic subset V of K^n of dimension d contains only finitely many subvarieties of K^n of dimension d.*

An algebraic subset of K^n is said to be *equidimensional* if all its irreducible components have the same dimension.

5.2.2 Hilbert–Samuel Polynomial

Let V be a nonempty algebraic subset of K^n and let $K[V]$ be the set of all maps from V to K induced by polynomials in $K[X_1, \ldots, X_n]$. Then $K[V]$ is a subring of the ring of all functions from V to K and the restriction map

$$\mathrm{res}_V : K[\underline{X}] \longrightarrow K[V]$$

is a surjective homomorphism of rings. Its kernel is thus an ideal of $K[\underline{X}]$. It consists of all polynomials of $K[\underline{X}]$ vanishing identically on V. It is called the *ideal* of V and denoted $I(V)$. The above map therefore induces an isomorphism of rings

$$K[\underline{X}]/I(V) \simeq K[V] .$$

Observe that if V_1 and V_2 are two algebraic subsets of K^n, then we have $V_1 \subseteq V_2$ if and only if $I(V_2) \subseteq I(V_1)$. This follows from the fact that V_i is the set of common zeros of the elements of $I(V_i)$. In particular, V_1 and V_2 are equal if and only if $I(V_1) = I(V_2)$. Another property that we will need is that an algebraic subset V of K^n is irreducible if and only if $I(V)$ is a prime ideal of $K[\underline{X}]$.

For each integer $D \geq 0$, we denote by $K[\underline{X}]_{\leq D}$ the vector space over K consisting of all polynomials in $X_1, \ldots, X_n$ of total degree $\leq D$. The *Hilbert function* of an algebraic subset $V \neq \emptyset$ of K^n is the map $H(V; -) : \mathbb{N} \to \mathbb{N}$ given by

$$H(V; D) = \dim_K \left(\mathrm{res}_V K[\underline{X}]_{\leq D}\right)$$

for each integer $D \in \mathbb{N}$. This is also given by

$$H(V; D) = \dim_K \left((K[\underline{X}]_{\leq D} + I)/I\right)$$

where $I = I(V)$ is the ideal of V. This function carries many information about V. First of all, it can be shown that for all sufficiently large $D \in \mathbb{N}$, its value at D is given by a polynomial in D whose degree is the dimension of V:

$$H(V; D) = \sum_{i=0}^{d} a_i D^i \qquad \text{with} \quad d = \dim(V).$$

This polynomial is called the *Hilbert–Samuel polynomial* of V. It can be shown that all its coefficients are rational numbers and that $d!$ is a common denominator for them. In particular, the product $d!a_d$ is a positive integer. It is called the *degree* of V and denoted $\deg(V)$. Geometrically, the intersection of V with a linear affine subvariety L of K^n of dimension $n - d$ is *in general* finite and its cardinality is equal to $\deg(V)$. In fact, when $V \cap L$ is finite, its cardinality is at most equal to $\deg(V)$. A reference for this is, for example, § 12, Chap. VII of [ZSa 1958], or § 7, Chap. I of [Har 1977].

Example. For $V = K^n$, we have

$$\begin{aligned} H(K^n; D) &= \dim_K \left(K[\underline{X}]_{\leq D} \right) \\ &= \binom{D+n}{n} \\ &= \frac{1}{n!} D^n + \cdots , \end{aligned}$$

hence the degree of K^n is 1.

5.2.3 Multihomogeneous Hilbert–Samuel Polynomial

One can get more information about a nonempty algebraic set V of K^n by partitioning the set of variables $\underline{X} = (X_1, \ldots, X_n)$ into subsets

$$\underline{X}^{(1)} = (X_1^{(1)}, \ldots, X_{n_1}^{(1)}) , \quad \ldots , \quad \underline{X}^{(k)} = (X_1^{(k)}, \ldots, X_{n_k}^{(k)})$$

with $n_1 + \ldots + n_k = n$ and by considering the function $H(V; -)\colon \mathbb{N}^k \to \mathbb{N}$ given for any $\underline{D} = (D_1, \ldots, D_k) \in \mathbb{N}^k$ by

$$H(V; \underline{D}) = \dim_K \left(\mathrm{res}_V K[\underline{X}^{(1)}, \ldots, \underline{X}^{(k)}]_{\leq \underline{D}} \right)$$

where $K[\underline{X}^{(1)}, \ldots, \underline{X}^{(k)}]_{\leq \underline{D}}$ denotes the K-vector space of all polynomials in the ring $K[\underline{X}^{(1)}, \ldots, \underline{X}^{(k)}]$ which have total degree $\leq D_i$ in each set of variables $\underline{X}^{(i)}$. An element of $K[\underline{X}^{(1)}, \ldots, \underline{X}^{(k)}]_{\leq \underline{D}}$ is said to be a polynomial of *multidegree* $\leq (D_1, \ldots, D_k)$ *with respect to the sets of variables* $\underline{X}^{(1)}, \ldots, \underline{X}^{(k)}$ and the function $H(V; \underline{D})$ so defined is called a *multihomogeneous Hilbert function* of V. This function is also given by

$$H(V; \underline{D}) = \dim_K \left((K[\underline{X}^{(1)}, \ldots, \underline{X}^{(k)}]_{\leq \underline{D}} + I)/I \right)$$

where $I = I(V)$ denotes the ideal of V. As before it can be shown that for all sufficiently large integers $D_1, \ldots, D_k$, its value at the point $\underline{D}$ is given by a polynomial in $D_1, \ldots, D_k$ of total degree equal to the dimension of V

$$H(V; \underline{D}) = \sum_{|\underline{i}| \leq d} a_{\underline{i}} \underline{D}^{\underline{i}} \qquad \text{with} \quad d = \dim(V).$$

This polynomial is called the *Hilbert–Samuel multihomogeneous polynomial* of V corresponding to our partition of $\underline{X}$. The reason why it is said "multihomogeneous" is that if we add to each set $\underline{X}^{(i)}$ a new variable $X_0^{(i)}$ to create a new set $\underline{\widetilde{X}}^{(i)} = (X_0^{(i)}, X_1^{(i)}, \ldots, X_{n_i}^{(i)})$, and if we consider the ideal $\tilde{I}$ of $K[\underline{\widetilde{X}}^{(1)}, \ldots, \underline{\widetilde{X}}^{(k)}]$ generated by all polynomials which are separately homogeneous in each set of variables $\underline{\widetilde{X}}^{(i)}$ and whose image under the specializations $X_0^{(1)} \to 1, \ldots, X_0^{(k)} \to 1$ fall into $I(V)$, then $H(V;\underline{D})$ becomes the dimension over K of the quotient $(K[\underline{\widetilde{X}}^{(1)}, \ldots, \underline{\widetilde{X}}^{(k)}]_{\underline{D}} + \widetilde{I})/\widetilde{I}$ where $K[\underline{\widetilde{X}}^{(1)}, \ldots, \underline{\widetilde{X}}^{(k)}]_{\underline{D}}$ stands for the space consisting of 0 and of all polynomials which are separately homogeneous of degree D_i in each set of variables $\underline{\widetilde{X}}^{(i)}$.

We denote by

$$\mathcal{H}(V;\underline{D}) = d! \sum_{|\underline{i}|=d} a_{\underline{i}} \underline{D}^{\underline{i}}$$

the product by $d!$ of the homogeneous part of degree $d = \dim(V)$ of this polynomial. It can be proven that the coefficients of $\mathcal{H}(V;\underline{D})$ are integral and nonnegative. More generally, the numbers

$$c_{\underline{i}} = i_1! \cdots i_k! a_{\underline{i}} \qquad (\text{with } |\underline{i}| = d)$$

are also integral. They are called the *multidegrees* of V.

To present the geometric interpretation of these numbers, fix a k-tuple $\underline{i} = (i_1, \ldots, i_k) \in \mathbb{N}^k$ with $|\underline{i}| = d$. If one index i_j is $> n_j$, then $c_{\underline{i}} = 0$. Otherwise, choose a linear affine subvariety L of K^n of dimension d defined by i_j inhomogeneous linear forms in $\underline{X}^{(j)}$ for $j = 1, \ldots, k$. Then, *in general*, L will meet V in a finite set of points of cardinality equal to $c_{\underline{i}}$. In fact, if $V \cap L$ is finite, its cardinality is $\le c_{\underline{i}}$. A reference for this is [Vd 1928].

Example. For $V = K^n$, we have

$$\begin{aligned} H(K^n;\underline{D}) &= \dim_K\left(K[\underline{X}^{(1)}, \ldots, \underline{X}^{(k)}]_{\le \underline{D}}\right) \\ &= \binom{D_1+n_1}{n_1} \cdots \binom{D_k+n_k}{n_k} \\ &= \frac{1}{n_1!} \cdots \frac{1}{n_k!} D_1^{n_1} \cdots D_k^{n_k} + \cdots, \end{aligned}$$

hence

$$\mathcal{H}(K^n;\underline{D}) = \frac{n!}{n_1! \cdots n_k!} D_1^{n_1} \cdots D_k^{n_k}.$$

We state without proof the following important fact (see Theorem 8 of [Vd 1928], or Exercises 8.4 and 8.5 in Chap. 8):

Proposition 5.2*. *If V is an algebraic subset of K^n of dimension d and if $V_1, \ldots, V_r$ are its irreducible components of dimension d, then*

$$\mathcal{H}(V;\underline{D}) = \sum_{i=1}^{r} \mathcal{H}(V_i;\underline{D}).$$

5.2.4 Philippon's Upper Bound

The following result is a special case of P. Philippon's general upper bound for the function $\mathcal{H}$ (Proposition 3.3 of [P 1986a]); it plays a central role in the proof of zero estimates:

Theorem 5.3. *Let U be an equidimensional algebraic subset of K^n and let V be the set of common zeros in U of a family $\mathcal{F}$ of polynomials of multidegree $\leq \underline{D}$. Assume that V is not empty. Then, we have*

$$\mathcal{H}(V;\underline{D}) \leq \mathcal{H}(U;\underline{D}).$$

For the proof, we will need the following lemma:

Lemma 5.4. *Let W be an algebraic subvariety of K^n of dimension $d \geq 1$ and let Z be the set of zeros in K^n of a polynomial P of multidegree $\leq \underline{D}$. Assume that $W \cap Z$ is not empty and distinct from W. Then, $W \cap Z$ is an equidimensional algebraic subset of K^n of dimension $d-1$, and we have*

$$\mathcal{H}(W \cap Z;\underline{D}) \leq \mathcal{H}(W;\underline{D}).$$

Proof. The fact that $W \cap Z$ is equidimensional of dimension $d-1$ follows from Theorem 1.11A and Exercise 1.8 in Chap. I of [Har 1977]. The inequality involving the function $\mathcal{H}$ follows from Lemma 3.1 of [P 1986a]. It is however simple to give a direct proof of this inequality. For any k-tuple of integers $\underline{T} = (T_1, \dots, T_k) \in \mathbb{N}^k$, the restriction map

$$\operatorname{res}_W\left(K[\underline{X}]_{\leq \underline{T}+\underline{D}}\right) \longrightarrow \operatorname{res}_{Z\cap W}\left(K[\underline{X}]_{\leq \underline{T}+\underline{D}}\right)$$

is linear, surjective and, since P vanishes identically on Z, its kernel contains the image of $\operatorname{res}_W\left(K[\underline{X}]_{\leq \underline{T}}\right)$ under multiplication by P. Moreover, the multiplication by P is injective on $K[W] = \operatorname{res}_W\left(K[\underline{X}]\right)$ because W is irreducible and not contained in Z. Comparing dimensions, this implies

$$\begin{aligned} H\left(Z \cap W;\underline{T}+\underline{D}\right) &= \dim_K\left(\operatorname{res}_{Z\cap W}\left(K[\underline{X}]_{\leq \underline{T}+\underline{D}}\right)\right) \\ &\leq \dim_K\left(\operatorname{res}_W\left(K[\underline{X}]_{\leq \underline{T}+\underline{D}}\right)\right) - \dim_K\left(\operatorname{res}_W\left(K[\underline{X}]_{\leq \underline{T}}\right)\right) \\ &= H\left(W;\underline{T}+\underline{D}\right) - H\left(W;\underline{T}\right). \end{aligned}$$

Now, fix a point $\underline{C} = (C_1, \dots, C_k) \in \mathbb{N}^k$ and, for each integer $t \geq 0$, define

$$p(t) = H\left(W;\underline{C}+t\underline{D}\right) \quad \text{and} \quad q(t) = H\left(Z \cap W;\underline{C}+t\underline{D}\right).$$

Then, assuming that $\underline{C}$ has sufficiently large positive coordinates, the integers $p(t)$ and $q(t)$ are given by polynomials in t of degree d and $d-1$ respectively and the preceding observation applied with $\underline{T} = \underline{C} + (t-1)\underline{D}$ gives

$$q(t) \leq p(t) - p(t-1).$$

Moreover, the definition of the function $\mathcal{H}$ implies that

$$\mathcal{H}\big(W; \underline{D}\big) = d! \lim_{t\to\infty} \frac{p(t)}{t^d} \quad \text{and} \quad \mathcal{H}\big(Z \cap W; \underline{D}\big) = (d-1)! \lim_{t\to\infty} \frac{q(t)}{t^{d-1}}.$$

Combining these formulas with the above upper bound for $q(t)$ gives finally

$$\mathcal{H}\big(Z \cap W; \underline{D}\big) \leq (d-1)! \lim_{t\to\infty} \frac{p(t) - p(t-1)}{t^{d-1}} = \mathcal{H}\big(W; \underline{D}\big). \qquad \square$$

Proof of Theorem 5.3. Let $d = \dim(U)$ and $r = d - \dim(V)$. By induction on the integer $i = 0, \ldots, r$, we shall construct an equidimensional algebraic subset $V_i \subseteq U$ of dimension $d - i$ which contains V and satisfies

$$\mathcal{H}(V_i; \underline{D}) \leq \mathcal{H}(U; \underline{D}). \tag{5.5}$$

For $i = 0$, we set $V_0 = U$. Assume that V_i is constructed for an integer $i \geq 0$ with $i < r$, and let $W_1, \ldots, W_s$ be its irreducible components. We have

$$V \subseteq W_1 \cup \cdots \cup W_s.$$

Since $\dim(W_j) = d - i > \dim(V)$, there exists for each j a polynomial P_j in the family $\mathcal{F}$ which does not vanish everywhere on W_j; let Z_j be the set of zeros of P_j in K^n. We define

$$V_{i+1} = (W_1 \cap Z_1) \cup \cdots \cup (W_s \cap Z_s).$$

By construction, V_{i+1} contains V, therefore $V_{i+1} \neq \emptyset$. Without loss of generality, we may assume that there exists an integer $t \geq 1$ such that $W_j \cap Z_j \neq \emptyset$ for $j = 1, \ldots, t$, and $W_j \cap Z_j = \emptyset$ for $j > t$. Then, Lemma 5.4 shows that $W_j \cap Z_j$ is an equidimensional algebraic subset of K^n of dimension $d - i - 1$ for $j = 1, \ldots, t$. Therefore, V_{i+1} is also equidimensional of dimension $d - i - 1$ and its irreducible components are the union of those of $W_j \cap Z_j$ for $j = 1, \ldots, t$. By virtue of Proposition 5.2, this gives

$$\mathcal{H}(V_{i+1}; \underline{D}) \leq \sum_{j=1}^{t} \mathcal{H}(W_j \cap Z_j; \underline{D}).$$

Since each P_j is of multidegree $\leq \underline{D}$, we also have, by Lemma 5.4,

$$\sum_{j=1}^{t} \mathcal{H}(W_j \cap Z_j; \underline{D}) \leq \sum_{j=1}^{t} \mathcal{H}(W_j; \underline{D}).$$

Since $W_1, \ldots, W_t$ are among the irreducible components of V_i of dimension $d - i$, Proposition 5.2 gives

$$\sum_{j=1}^{t} \mathcal{H}(W_j; \underline{D}) \leq \mathcal{H}(V_i; \underline{D}).$$

Combining these inequalities with (5.5), we get $\mathcal{H}(V_{i+1};\underline{D}) \le \mathcal{H}(U;\underline{D})$ as required. This shows the existence of $V_0, \dots, V_r$. Since V and V_r have the same dimension $d-r$, the inclusion $V \subseteq V_r$ implies that the irreducible components of V of dimension $d-r$ are among those of V_r; therefore applying Proposition 5.2 and using the relation (5.5) with $i = r$, we get

$$\mathcal{H}(V;\underline{D}) \le \mathcal{H}(V_r;\underline{D}) \le \mathcal{H}(U;\underline{D}).$$

The proof is complete. □

5.3 The Group G and its Algebraic Subgroups

An *affine algebraic group* is an algebraic subset U of K^n which also has a group structure with the group law $(\underline{x}, \underline{y}) \mapsto \underline{x} \cdot \underline{y}$ and the map $\underline{x} \mapsto \underline{x}^{-1}$ given by polynomials:

$$\underline{x} \cdot \underline{y} = (P_1(\underline{x}, \underline{y}), \dots, P_n(\underline{x}, \underline{y})) \qquad \underline{x}^{-1} = (Q_1(\underline{x}), \dots, Q_n(\underline{x}))$$

with $P_1, \dots, P_n \in K[\underline{X}, \underline{Y}]$ and $Q_1, \dots, Q_n \in K[\underline{X}]$. An *algebraic subgroup* of U is a subgroup of U which, as a set, is also an algebraic subset of K^n. When U is a commutative group, it is common to denote the group law by $+$ and the inverse of an element g by $-g$.

Let $d_0, d_1 \ge 0$ be integers not both zero, and let G be the product $K^{d_0} \times (K^\times)^{d_1}$ of d_0 copies of the additive group of K with d_1 copies of the multiplicative group $K^\times$ of all nonzero elements of K. Then, G is a commutative group for the product structure. Its group law written additively is given by

$$(\underline{x}, \underline{y}) + (\underline{x}', \underline{y}') = (x_1 + x_1', \dots, x_{d_0} + x_{d_0}', y_1 y_1', \dots, y_{d_1} y_{d_1}').$$

However, G is not an affine algebraic group in our sense because it is not an algebraic subset of $K^{d_0} \times K^{d_1}$ and the map $g \mapsto -g$ is not given by polynomials.

To correct this situation, we put $n = d_0 + 2d_1$ and consider the algebraic subset U of K^n given by

$$U = \{(\underline{x}, \underline{y}, \underline{z}) \in K^{d_0} \times K^{d_1} \times K^{d_1}\,;\; y_1 z_1 = \cdots = y_{d_1} z_{d_1} = 1\}.$$

This is a subgroup of $K^{d_0} \times (K^\times)^{2d_1}$ and since the inverse of an element $(\underline{x}, \underline{y}, \underline{z})$ of U is given by $(-\underline{x}, \underline{z}, \underline{y})$, we see that U is an affine algebraic group. Moreover, the projection map π from $K^{d_0} \times K^{d_1} \times K^{d_1}$ to $K^{d_0} \times K^{d_1}$ which sends a point $(\underline{x}, \underline{y}, \underline{z})$ to $(\underline{x}, \underline{y})$ induces a group isomorphism $\pi: U \to G$. We will use it to carry on G the structures that apply to U.

(i) A *polynomial map* or simply a *polynomial* on G will be a map $f: G \to K$ such that $f \circ \pi: U \to K$ is induced by a polynomial on K^n. These maps form a ring which we will denote $K[G]$. Since for all $(\underline{x}, \underline{y}, \underline{z}) \in U$ we have $z_i = y_i^{-1}$ $(i = 1, \dots, d_1)$, this ring is simply the ring of all maps from G to K induced by polynomials in $X_1, \dots, X_{d_0}, Y_1, \dots, Y_{d_1}, Y_1^{-1}, \dots, Y_{d_1}^{-1}$. Thus $K[G]$ is isomorphic to $K[\underline{X}, \underline{Y}^{\pm 1}]$ and we will identify both rings.

(ii) An *algebraic subset* of G will be a subset E of G such that $\pi^{-1}(E)$ is an algebraic subset of K^n. By virtue of (i), an algebraic subset of G is therefore the set of common zeros in G of a family of polynomials in $K[G] = K[\underline{X}, \underline{Y}^{\pm 1}]$. When $E \neq \emptyset$, we will denote by $K[E]$ the ring of all maps from E to K induced by elements of $K[G]$. It is isomorphic to $K[G]/I$ where I is the ideal of all elements of $K[G]$ vanishing identically on E. This ideal I will be called the *ideal* of E, and denoted $I(E)$.

A subgroup E of G will be said to be *algebraic* if it is an algebraic subset of G. For example, for each vector subspace $\mathcal{V}$ of K^{d_0} and each subgroup Φ of $\mathbb{Z}^{d_1}$, the set $\mathcal{V} \times T_\Phi$ is an algebraic subgroup of G because it is defined by linear equations in $\underline{X}$ and by the equations $\underline{Y}^{\underline{\varphi}} - 1 = 0$ with $\underline{\varphi} \in \Phi$ (see the definition of T_Φ at the beginning of the chapter). We will show that each algebraic subgroup of G is of this form.

(iii) Let $D_0 \in \mathbb{N}$ and $\underline{D} = (D_1, \dots, D_{d_1}) \in \mathbb{N}^{d_1}$. We introduce new variables $\underline{Z} = (Z_1, \dots, Z_{d_1})$ and say that a polynomial $Q \in K[\underline{X}, \underline{Y}, \underline{Z}]$ has *multidegree* $\leq (D_0, \underline{D})$ if its total degree in $\underline{X}$ is $\leq D_0$ and if, for $j = 1, \dots, d_1$, its total degree in Y_j and Z_j is $\leq D_j$. Then, in accordance with § 5.1, a polynomial $P \in K[G] = k[\underline{X}, \underline{Y}^{\pm 1}]$ has *multidegree* $\leq (D_0, \underline{D})$ if there exists $Q \in K[\underline{X}, \underline{Y}, \underline{Z}]$ of multidegree $\leq (D_0, \underline{D})$ such that

$$P = Q(X_1, \dots, X_{d_0}, Y_1, \dots, Y_{d_1}, Y_1^{-1}, \dots, Y_{d_1}^{-1}).$$

Given an algebraic subset $E \neq \emptyset$ of G, we define its *Hilbert function* $H(E; D_0, \underline{D})$ in two equivalent ways either as $H(\pi^{-1}(E); D_0, \underline{D})$ or as the dimension over K of the space of maps $f: E \to K$ induced by elements of $K[G]$ of multidegree $\leq (D_0, \underline{D})$. We also define the *dimension* of E as the dimension of $\pi^{-1}(E)$. When this dimension is m, we define $\mathcal{H}(E; D_0, \underline{D})$ as the product by $m!$ of the homogeneous part of degree m of the polynomial in $D_0, D_1, \dots, D_{d_1}$ which coincides with $H(E; D_0, \underline{D})$ for large integral values of the parameters.

5.3.1 Structure of the Algebraic Subgroups

Proposition 5.6. *Let G^* be an algebraic subgroup of G. Then there exist a subspace $\mathcal{V}$ of K^{d_0} and a subgroup Φ of $\mathbb{Z}^{d_1}$ such that*

$$G^* = \mathcal{V} \times T_\Phi.$$

The ideal of G^* is generated by the homogeneous linear forms $a_1 X_1 + \cdots + a_{d_0} X_{d_0}$ vanishing identically on $\mathcal{V}$ and by the elements $\underline{Y}^{\underline{\varphi}} - 1$ with $\underline{\varphi} \in \Phi$.

Proof. Let Φ be the set of all $\underline{\varphi} \in \mathbb{Z}^{d_1}$ such that $\underline{Y}^{\underline{\varphi}} - 1$ vanishes identically on G^* and let M be the set of all homogeneous linear forms in $K[\underline{X}]_1 = KX_1 + \ldots + KX_{d_0}$

which vanish identically on G^*. Then, Φ is a subgroup of $\mathbb{Z}^{d_1}$ and M is a subspace of $K[\underline{X}]_1$. Consider the ideal I of $K[G]$ generated by all elements of M and by all polynomials $\underline{Y}^{\underline{\varphi}} - 1$ with $\underline{\varphi} \in \Phi$. As we previously observed, the zero set of I is an algebraic subgroup of G in the form of a product

$$\mathcal{V} \times \boldsymbol{T}_\Phi$$

where $\mathcal{V}$ is the subspace of K^{d_0} defined by the linear forms of M. Since $I \subseteq I(G^*)$, we have $G^* \subseteq \mathcal{V} \times \boldsymbol{T}_\Phi$. We will show that the two groups are equal.

Choose a subspace N of $K[\underline{X}]_1$ such that $K[\underline{X}]_1 = M \oplus N$. Choose also a basis $L_1, \dots, L_s$ of N and a set S of representatives of the classes of $\mathbb{Z}^{d_1}$ modulo Φ. Consider the subspace F of $K[G]$ consisting of all expressions of the form

$$\sum_{\underline{\alpha} \in S} A_{\underline{\alpha}}(L_1(\underline{X}), \dots, L_s(\underline{X}))\underline{Y}^{\underline{\alpha}}$$

where $A_{\underline{\alpha}} \in K[T_1, \dots, T_s]$ are polynomials. Since $K[\underline{X}]_1 = M \oplus N$, every element of $K[\underline{X}]_1$ is congruent modulo I to an element of N and so, every element of $K[\underline{X}]$ is congruent modulo I to a polynomial in $L_1, \dots, L_s$. Moreover, for every $\underline{\beta} \in \mathbb{Z}^{d_1}$, there is an $\underline{\alpha} \in S$ such that $\underline{\varphi} = \underline{\beta} - \underline{\alpha} \in \Phi$ and so, $\underline{Y}^{\underline{\beta}} = \underline{Y}^{\underline{\alpha}} + \underline{Y}^{\underline{\alpha}}(\underline{Y}^{\underline{\varphi}} - 1)$ is congruent to $\underline{Y}^{\underline{\alpha}}$ modulo I. This shows that every element of $K[G]$ is congruent modulo I to an element of F. Now, consider the restriction map from $K[G]$ to $K[G^*]$. We will prove that it is injective on F. If we take this for granted, we get $I = I(G^*)$ because $K[G] = I + F$ and I is contained in the kernel $I(G^*)$ of the restriction map. Since G^* is the zero set of $I(G^*)$, this will show $G^* = \mathcal{V} \times \boldsymbol{T}_\Phi$.

Assume on the contrary that the restriction map from $K[G]$ to $K[G^*]$ is not injective on F. Then, there exist distinct elements $\underline{\alpha}_1, \dots, \underline{\alpha}_r$ of S and nonzero polynomials $A_1, \dots, A_r \in K[T_1, \dots, T_s]$ such that

$$P(\underline{X}, \underline{Y}) = \sum_{i=1}^{r} A_i(L_1(\underline{X}), \dots, L_s(\underline{X}))\underline{Y}^{\underline{\alpha}_i}$$

vanishes identically on G^*. Choose them such that

$$\sum_{i=1}^{r} (\deg(A_i) + 1)$$

is minimal. Take any point $(\underline{x}, \underline{y}) \in G^*$ and consider the polynomial

$$\underline{y}^{\underline{\alpha}_1} P(\underline{X}, \underline{Y}) - P(\underline{X} + \underline{x}, \underline{y}\underline{Y}).$$

By construction, it vanishes identically on G^*. On the other hand, it can be written in the form

$$\sum_{i=1}^{r} B_i(L_1(\underline{X}), \dots, L_s(\underline{X}))\underline{Y}^{\underline{\alpha}_i}$$

with

$$B_i(T_1, \ldots, T_s) = \underline{y}^{\underline{\alpha}_1} A_i(T_1, \ldots, T_s) - \underline{y}^{\underline{\alpha}_i} A_i(T_1 + L_1(\underline{x}), \ldots, T_s + L_s(\underline{x}))$$

for $i = 1, \ldots r$. If $B_1, \ldots, B_r$ are not all zero, we get the expected contradiction because $\deg(B_i) = \deg(A_i)$ when $\underline{y}^{\underline{\alpha}_i} \neq \underline{y}^{\underline{\alpha}_1}$ and because either $B_i = 0$ or $\deg(B_i) < \deg(A_i)$ when $\underline{y}^{\underline{\alpha}_i} = \underline{y}^{\underline{\alpha}_1}$. It thus remains to show that $(\underline{x}, \underline{y})$ can be chosen so that at least one B_i is $\neq 0$. If $r \geq 2$, the polynomial $\underline{Y}^{\underline{\alpha}_2} - \underline{Y}^{\underline{\alpha}_1}$ does not vanish identically on G^* because $\underline{\alpha}_2 - \underline{\alpha}_1 \notin \Phi$. Then, we get $B_2 \neq 0$ by choosing $(\underline{x}, \underline{y}) \in G^*$ such that $\underline{y}^{\underline{\alpha}_2} \neq \underline{y}^{\underline{\alpha}_1}$. If $r = 1$, the polynomial A_1 has a positive degree m and the homogeneous part of B_1 of degree $m - 1$ is

$$-\underline{y}^{\underline{\alpha}_1} \sum_{i=1}^{s} L_i(\underline{x}) \frac{\partial A}{\partial T_i}(T_1, \ldots, T_s)$$

where A denotes the homogeneous part of A_1 of degree m. Since the characteristic of K is zero, the derivatives $\partial A/\partial T_1, \ldots, \partial A/\partial T_s$ are not all zero. So, there is at least one coefficient of B_1 of the form $L(\underline{x})$ where L is a nonzero element of N. Since L does not vanish identically on G^*, we get $B_1 \neq 0$ by choosing $(\underline{x}, \underline{y}) \in G^*$ such that $L(\underline{x}) \neq 0$. □

Looking more closely at the above argument, one gets a formula for the Hilbert function of $G^* = \mathcal{V} \times T_\Phi$ in terms of $\mathcal{V}$ and Φ:

Proposition 5.7. *Let $\mathcal{V}$ be a subspace of K^{d_0}, let Φ be a subgroup of $\mathbb{Z}^{d_1}$ and let $G^* = \mathcal{V} \times T_\Phi$. Then, the Hilbert function of G^* is*

$$H(G^*; D_0, \underline{D}) = \binom{D_0 + d_0^*}{d_0^*} \chi(\underline{D})$$

where $d_0^* = \dim_K \mathcal{V}$ and where $\chi(\underline{D})$ denotes the number of cosets of Φ of the form $\underline{\alpha} + \Phi$ with $\underline{\alpha} = (\alpha_1, \ldots, \alpha_{d_1}) \in \mathbb{Z}^{d_1}$ satisfying $|\alpha_i| \leq D_i$ for $i = 1, \ldots, d_1$.

The details of the proof are left to the reader. In the case where $\mathcal{V} = K^{d_0}$ and $\Phi = \{0\}$, this result gives

$$H(G; D_0, \underline{D}) = \binom{D_0 + d_0}{d_0} (2D_1 + 1) \cdots (2D_{d_1} + 1),$$

so G has dimension d and

$$\mathcal{H}(G; D_0, \underline{D}) = \frac{d!}{d_0!} 2^{d_1} D_0^{d_0} D_1 \cdots D_{d_1}. \tag{5.8}$$

5.3.2 Translations and Automorphisms

For each $g \in G$, we denote by $\tau_g: G \to G$, the operator of translation by g in G:

$$\tau_g(x) = g + x \qquad \text{for all } x \in G.$$

Looking at the addition law in G, we see that each τ_g is given in coordinates by polynomials of degree 1. We will need these operators in the proofs of the next three lemmas.

Lemma 5.9. *Let V be a nonempty algebraic subset of G and let $g \in G$. Then, $g + V$ is an algebraic subset of G with the same dimension as V and we have*

$$\mathcal{H}(g + V;\, D_0, \underline{D}) = \mathcal{H}(V;\, D_0, \underline{D})$$

for all $(D_0, \underline{D}) \in \mathbb{N}^{1+d_1}$. Moreover, $g + V$ is irreducible if V is irreducible.

Proof. By hypothesis, V is the set of common zeros in G of a family of polynomials $\{P_j\}_{j \in J}$. Therefore, $g+V = \tau_g(V)$ is the set of common zeros in G of the polynomials $P_j \circ \tau_{-g}$ with $j \in J$. This proves that $g + V$ is an algebraic subset of G.

The vector space of functions from $g + V$ to K is isomorphic to the vector space of functions from V to K under the map which sends a function $f: g + V \to K$ to the composite $f \circ \tau_g: V \to K$. If f is induced by a polynomial of multidegree $\leq (D_0, \underline{D})$, then $f \circ \tau_g$ is also induced by a polynomial of multidegree $\leq (D_0, \underline{D})$, and conversely. We therefore have

$$H(g + V;\, D_0, \underline{D}) = H(V;\, D_0, \underline{D})$$

for all $(D_0, \underline{D}) \in \mathbb{N}^{1+d_1}$. This shows that V and $g + V$ have the same Hilbert-Samuel polynomial. Consequently, they have the same dimension, and the polynomials $\mathcal{H}(g + V;\, D_0, \underline{D})$ and $\mathcal{H}(V;\, D_0, \underline{D})$ coincide.

Finally, assume that V is irreducible. If $g + V$ were not irreducible, it could be written as the union of two algebraic subsets V_1, V_2 of G both distinct from $g + V$; then V would be the union of $-g + V_1$ and $-g + V_2$, and this is a contradiction since both are algebraic subsets of G which are distinct from V. Therefore $g + V$ is irreducible. □

Lemma 5.10. *Let G^* be an algebraic subgroup of G, and let E be a finite and nonempty union of translates of G^* in G. Then, E is an algebraic subset of G and, for all $(D_0, \underline{D}) \in \mathbb{N}^{1+d_1}$, we have*

$$\mathcal{H}(E;\, D_0, \underline{D}) = \mathrm{Card}(E/G^*)\mathcal{H}(G^*;\, D_0, \underline{D}).$$

Proof. Let d^* be the dimension of G^*. Lemma 5.9 shows that each translate $g + G^*$ of G^* is an algebraic subset of G of dimension d^* and that the polynomials $\mathcal{H}(g+G^*;\, D_0, \underline{D})$ and $\mathcal{H}(G^*;\, D_0, \underline{D})$ coincide. Since E is a finite union of translates of G^*, E is therefore an algebraic subset of G of dimension d^*, and the conclusion follows from Proposition 5.2. □

Lemma 5.11. *Let V and X be algebraic subsets of G. Define*

$$E = \{g \in G\,;\, g + V \subseteq X\}.$$

Then E is an algebraic subset of G. Moreover, if X is defined in G by polynomials of multidegree $\leq (D_0, \underline{D})$, then E is also defined in G by polynomials of multidegree $\leq (D_0, \underline{D})$.

Proof. Let $\{P_j\}_{j\in J}$ be a family of polynomials whose set of common zeros in G is X. We have

$$\begin{aligned} E &= \{g \in G\,;\, g + v \in X \text{ for all } v \in V\} \\ &= \{g \in G\,;\, P_j(g + v) = 0 \text{ for all } j \in J \text{ and } v \in V\}. \end{aligned}$$

This shows that E is the set of common zeros in G of the polynomials $P_j \circ \tau_v$ with $j \in J$ and $v \in V$. Therefore E is an algebraic subset of G. Furthermore, if the polynomials P_j are of multidegree $\leq (D_0, \underline{D})$, then the same holds for the polynomials $P_j \circ \tau_v$. This proves the second part of the lemma. □

We define an *endomorphism* of G as a group homomorphism $\psi\colon G \to G$ which satisfies $f \circ \psi \in K[G]$ for all $f \in K[G]$. It can be shown that such a map has the form

$$\psi(\underline{x}, \underline{y}) = (L(\underline{x}), \underline{y}^{\underline{\alpha}_1}, \ldots, \underline{y}^{\underline{\alpha}_{d_1}}) \tag{5.12}$$

for a linear map $L\colon K^{d_0} \to K^{d_0}$ and elements $\underline{\alpha}_1, \ldots, \underline{\alpha}_{d_1}$ of $\mathbb{Z}^{d_1}$. Moreover, ψ is invertible if and only if L is invertible and $\{\underline{\alpha}_1, \ldots, \underline{\alpha}_{d_1}\}$ is a basis of $\mathbb{Z}^{d_1}$. In this case, ψ^{-1} is also an endomorphism of G; we say that ψ is an *automorphism* of G.

An automorphism ψ of G maps an algebraic subset E of G into an algebraic subset $\psi(E)$ of G. It also preserves irreducibility and maps the irreducible components of E into those of $\psi(E)$. This uses the same arguments as in the proof of Lemma 5.9. Finally, it preserves dimension because if $V_0 \subset \ldots \subset V_d \subseteq E$ is a maximal chain of subvarieties of G contained in E then $\psi(V_0) \subset \ldots \subset \psi(V_d) \subseteq \psi(E)$ is another one contained in $\psi(E)$, so E and $\psi(E)$ have the same dimension d.

Theorem 5.13. *Let G^* be an algebraic subgroup of G. Then, there is one and only one pair $(\mathcal{V}, \Phi)$ consisting of a subspace $\mathcal{V}$ of K^{d_0} and a subgroup Φ of $\mathbb{Z}^{d_1}$ such that*

$$G^* = \mathcal{V} \times T_\Phi.$$

Let $\overline{\Phi}$ be the largest subgroup of $\mathbb{Z}^{d_1}$ containing Φ with the same rank as Φ. Then, $G_0^* = \mathcal{V} \times T_{\overline{\Phi}}$ is an irreducible component of G^* and the other ones are translates of G_0^*. Their number is

$$[G^* : G_0^*] = [\overline{\Phi} : \Phi].$$

Moreover, G^* is equidimensional and its dimension is

$$\dim(G^*) = \dim(\mathcal{V}) + (d_1 - \operatorname{rank}(\Phi)).$$

The group G_0^* is called the *neutral component* of G^*. The theorem shows that the irreducible components of an algebraic subgroup of G are disjoint. Therefore, it is equivalent to say that an algebraic subgroup of G is irreducible or connected (see also Exercise 5.5).

Proof. By Proposition 5.6, we know that there exist a subspace $\mathcal{V}$ of K^{d_0} and a subgroup Φ of $\mathbb{Z}^{d_1}$ such that $G^* = \mathcal{V} \times T_\Phi$. Let $r = d_0 - \dim(\mathcal{V})$ and $s = \operatorname{rank}(\Phi)$. The theorem of elementary divisors shows that there are a basis $\{\underline{\alpha}_1, \dots, \underline{\alpha}_{d_1}\}$ of $\mathbb{Z}^{d_1}$ and nonzero integers $m_1, \dots, m_s$ such that $\{m_1\underline{\alpha}_1, \dots, m_s\underline{\alpha}_s\}$ is a basis of Φ (see Theorem 7.8, Chap. III of [L 1993]). Then, $\overline{\Phi}$ is the subgroup of $\mathbb{Z}^{d_1}$ generated by $\underline{\alpha}_1, \dots, \underline{\alpha}_s$. Choose an invertible linear map $L\colon K^{d_0} \to K^{d_0}$ such that $L(\mathcal{V})$ is the subspace of K^{d_0} defined by $X_1 = \ldots = X_r = 0$ and consider the automorphism ψ of G given by (5.12). By construction, ψ maps G^* to the subgroup of G defined by

$$X_1 = \ldots = X_r = 0 \quad \text{and} \quad Y_1^{m_1} = \ldots = Y_s^{m_s} = 1.$$

It also maps $G_0^* = \mathcal{V} \times T_{\overline{\Phi}}$ to the subgroup of G defined by

$$X_1 = \ldots = X_r = 0 \quad \text{and} \quad Y_1 = \ldots = Y_s = 1.$$

This shows that $\psi(G_0^*)$ is a subgroup of $\psi(G^*)$ of index $m_1 \cdots m_s = [\overline{\Phi} : \Phi]$ and therefore $[G^* : G_0^*] = [\overline{\Phi} : \Phi]$. It also implies that the ideal of $\psi(G_0^*)$ is the kernel of the surjective ring homomorphism

$$\begin{aligned} K[X_1, \dots, X_{d_0}, Y_1^{\pm 1}, \dots, Y_{d_1}^{\pm 1}] &\longrightarrow K[X_{r+1}, \dots, X_{d_0}, Y_{s+1}^{\pm 1}, \dots, Y_{d_1}^{\pm 1}] \\ P(X_1, \dots, X_{d_0}, Y_1, \dots, Y_{d_1}) &\longmapsto \\ &P(0, \dots, 0, X_{r+1}, \dots, X_{d_0}, 1, \dots, 1, Y_{s+1}, \dots, Y_{d_1}). \end{aligned}$$

Since the image ring has no zero divisor other than 0, the ideal of $\psi(G_0^*)$ is prime. Therefore, $\psi(G_0^*)$ is irreducible and so G_0^* is irreducible. Since G^* is a finite union of translates of G_0^* and since these are disjoint and irreducible, we conclude that the translates of G_0^* in G^* are all the irreducible components of G^*. In particular, G^* is equidimensional. To compute $\dim G^* = \dim G_0^* = \dim \psi(G_0^*)$, we use Proposition 5.7. It gives

$$H(\psi(G_0^*); D_0, \underline{D}) = \frac{1}{(d_0 - r)!}(D_0 + 1) \cdots (D_0 + d_0 - r)(2D_{s+1} + 1) \cdots (2D_{d_1} + 1),$$

and so, $\psi(G_0^*)$ has dimension $(d_0 - r) + (d_1 - s) = \dim(\mathcal{V}) + (d_1 - \operatorname{rank}(\Phi))$

It remains to show that if $\mathcal{V}'$ is a subspace of K^{d_0} and if Φ' is a subgroup of $\mathbb{Z}^{d_1}$ such that $G^* = \mathcal{V}' \times T_{\Phi'}$, then $\mathcal{V}' = \mathcal{V}$ and $\Phi' = \Phi$. In this case, we get $G^* = \mathcal{V}'' \times T_{\Phi''}$ where $\mathcal{V}'' = \mathcal{V} \cap \mathcal{V}'$ and $\Phi'' = \Phi + \Phi'$. By the formula for $\dim G^*$, this implies

$$\dim(\mathcal{V}) = \dim(\mathcal{V}') = \dim(\mathcal{V}'') \quad \text{and} \quad \operatorname{rank}(\Phi) = \operatorname{rank}(\Phi') = \operatorname{rank}(\Phi'').$$

So, $\mathcal{V} = \mathcal{V}'$ and $\overline{\Phi} = \overline{\Phi}' = \overline{\Phi}''$. The indices of Φ, Φ' and Φ'' in $\overline{\Phi}''$ are thus finite and equal to the number of irreducible components of G^*. This gives $[\Phi'' : \Phi] = [\Phi'' : \Phi'] = 1$, so $\Phi = \Phi' = \Phi''$. □

5.3.3 An Explicit Formula for $\mathcal{H}(G^*; D_0, \underline{D})$

Proposition 5.14. *Let $\mathcal{V}$ be a subspace of K^{d_0} of dimension d_0^* and let Φ be a subgroup of $\mathbb{Z}^{d_1}$ of rank r generated by the rows of an $r \times d_1$ matrix M with coefficients in $\mathbb{Z}$. Put $d^* = d_0^* + d_1^*$ where $d_1^* = d_1 - r$ and denote by G^* the group $\mathcal{V} \times T_\Phi$. Then, we have*

$$\mathcal{H}(G^*, D_0, \underline{D}) = \frac{d^*!}{d_0^*!} 2^{d_1^*} D_0^{d_0^*} \sum |\det(\mathsf{M}_{i_1,\dots,i_r})| D_{j_1} \cdots D_{j_{d_1^*}}$$

where the summation extends to all partitions of $\{1, \dots, d_1\}$ into disjoint subsets $\{i_1, \dots, i_r\}$ and $\{j_1, \dots, j_{d_1^*}\}$ with $i_1 < \dots < i_r$ and $j_1 < \dots < j_{d_1^*}$ and where $\mathsf{M}_{i_1,\dots,i_r}$ denotes the $r \times r$ matrix formed by the columns of M of indices $i_1, \dots, i_r$.

Proof. When $r = 0$, the formula reduces to (5.8) because of our convention that an empty matrix has determinant 1 (see § 5.1). Hence, we may assume $r \geq 1$.

Since $G^* \subseteq G$, we have $H(G^*; D_0, \underline{D}) \leq H(G; D_0, \underline{D})$ for all $(D_0, \underline{D}) \in \mathbb{N}^{d_1+1}$. Therefore, (5.8) shows that the degree of $\mathcal{H}(G^*; D_0, \underline{D})$ in D_j is ≤ 1 for $j = 1, \dots, d_1$. On the other hand, Proposition 5.7 implies that $\mathcal{H}(G^*; D_0, \underline{D})$ is the product of $D_0^{d_0^*}$ by a homogeneous polynomial in $\underline{D}$. Since $\dim G^* = d^*$, this gives

$$\mathcal{H}(G^*; D_0, \underline{D}) = \frac{d^*!}{d_0^*!} D_0^{d_0^*} \sum_{1 \leq j_1 < \dots < j_{d_1^*} \leq d_1} c_{j_1,\dots,j_{d_1^*}} D_{j_1} \cdots D_{j_{d_1^*}}$$

where $c_{j_1,\dots,j_{d_1^*}}$ denotes the largest integer c for which there exists an affine linear subvariety L of K^n defined by d_0^* inhomogeneous linear forms in $\underline{X}$ and by one inhomogeneous linear form in Y_{j_k} and Z_{j_k} for each $k = 1, \dots, d_1^*$ such that $\pi^{-1}(G^*) \cap L$ is finite with cardinality c.

For simplicity, let us compute $c_{j_1,\dots,j_{d_1^*}}$ when $j_1 = r+1, \dots, j_{d_1^*} = d_1$. It is always possible to reduce to that case by permuting the coordinates in K^n. We have

$$\operatorname{Card}(\pi^{-1}(G^*) \cap L) = \operatorname{Card}(G^* \cap \pi(L \cap U)).$$

The set $\pi(L \cap U)$ is defined in G by d_0^* inhomogeneous linear forms of $K[\underline{X}]$ and by polynomials of the form

$$a_j Y_j + b_j Y_j^{-1} + c_j$$

for $j = r+1, \dots, d_1$. It is therefore a product

$$\pi(L \cap U) = L_0 \times L_1 \times \cdots \times L_{d_1}$$

where L_0 is an affine linear subvariety of K^{d_0} of codimension $\le d_0^*$, where $L_1 = \ldots = L_r = K^\times$ and where, for $j = r+1, \ldots, d_1$, L_j is either $K^\times$ or a subset of $K^\times$ of cardinality at most 2. We get

$$\mathrm{Card}(G^* \cap \pi(L \cap U)) = \mathrm{Card}(\mathcal{V} \cap L_0)\mathrm{Card}(T_\Phi \cap (L_1 \times \ldots \times L_{d_1})).$$

The set $\mathcal{V} \cap L_0$ either contains one point or is empty or infinite. Assume $\mathsf{M} = (m_{i,j})$. For each $(a_{r+1}, \ldots, a_{d_1}) \in L_{r+1} \times \ldots \times L_{d_1}$, the number of points of T_Φ in $(K^\times)^r \times \{a_{r+1}\} \times \ldots \times \{a_{d_1}\}$ is equal to the number of solutions $(y_1, \ldots, y_r) \in (K^\times)^r$ of the system

$$\begin{cases} y_1^{m_{1,1}} \cdots y_r^{m_{1,r}} = a_{r+1}^{-m_{1,r+1}} \cdots a_{d_1}^{-m_{1,d_1}} \\ \qquad \cdots \\ y_1^{m_{r,1}} \cdots y_r^{m_{r,r}} = a_{r+1}^{-m_{r,r+1}} \cdots a_{d_1}^{-m_{r,d_1}} \end{cases}$$

If $\det \mathsf{M}_{1,\ldots,r} \neq 0$, this is a set of cardinality $|\det \mathsf{M}_{1,\ldots,r}|$ independent of the choice of $(a_{r+1}, \ldots, a_{d_1})$ (see Exercise 5.6). In this case, we get that $G^* \cap \pi(L \cap U)$ either is infinite or consists of at most $2^{d_1^*} |\det \mathsf{M}_{1,\ldots,r}|$ points. Since this upper bound is achieved for a suitable choice of equations defining L, we obtain

$$c_{r+1,\ldots,d_1} = 2^{d_1^*} |\det \mathsf{M}_{1,\ldots,r}|.$$

If $\det \mathsf{M}_{1,\ldots,r} = 0$, the above system has either no solutions or an infinite number of solutions (see Exercise 5.6). So, $G^* \cap \pi(L \cap U)$ is either empty or infinite and the above formula for $c_{r+1,\ldots,d_1}$ still holds. □

5.4 Proof of the Main Result

Let the notation be as in the statement of Theorem 5.1 and let X_1 be the set of zeros of P in G. For each integer $r \ge 2$, we define

$$X_r = \bigcap_{(\sigma_1,\ldots,\sigma_{r-1}) \in \Sigma^{r-1}} \big(-\sigma_1 - \cdots - \sigma_{r-1} + X_1\big).$$

Alternatively, X_r is the set of common zeros in G of the polynomials $P \circ \tau_{\sigma_1+\cdots+\sigma_{r-1}}$ with $(\sigma_1, \ldots, \sigma_{r-1}) \in \Sigma^{r-1}$. Therefore, it is defined in G by polynomials of multidegree $\le (D_0, \underline{D})$.

The sets $X_1, X_2, \ldots$ are related by the formulas

$$X_{r+1} = \bigcap_{\sigma \in \Sigma} \big(-\sigma + X_r\big), \qquad (r \ge 1). \tag{5.15}$$

Since $e \in \Sigma$, this implies

$$X_1 \supseteq X_2 \supseteq \cdots \supseteq X_{d+1} \supseteq \cdots$$

Since P vanishes on $\Sigma[d]$, X_{d+1} contains e ; therefore this set is not empty. On the other hand, since P is not identically zero on G, Lemma 5.4 gives $\dim(X_1) = d - 1$. Consequently, there exists a positive integer $r \leq d$ such that

$$\dim(X_r) = \dim(X_{r+1}).$$

Let m be the common dimension of X_r and X_{r+1}, and let V be an irreducible component of dimension m of X_{r+1}. Using (5.15), we get

$$V \subseteq \bigcap_{\sigma \in \Sigma} (-\sigma + X_r);$$

hence for all $\sigma \in \Sigma$, $\sigma + V$ is contained in X_r. We set

$$E = \{g \in G\,;\, g + V \subseteq X_r\}.$$

We just showed $\Sigma \subseteq E$. We also set

$$G^* = \{g \in G\,;\, g + V = V\}$$

and

$$R = \{g + V\,;\, g \in E\}.$$

From Lemma 5.9 we deduce that the elements in the set R are, like V, algebraic subvarieties of G of dimension m. Since they are contained in X_r, and since X_r has dimension m, R is a finite set. We also notice that G^* is a subgroup of G, that E is stable under translation by the elements of G^*, and that there is a bijection

$$E/G^* \longrightarrow R.$$

Therefore E is a finite union of translates of G^*. Now recall that, by Lemma 5.9, the translates of V are irreducible subsets of G of the same dimension as V. So, for any $g \in G$, the condition $g + V \subseteq V$ is equivalent to $g + V = V$. Then Lemma 5.11, with $X = V$, shows that G^* is an algebraic subset of G. Hence G^* is an algebraic subgroup of G. Since it is contained in $-v + V$ for any $v \in V$, its dimension is $\leq m < d$. Applying again Lemma 5.11, but with $X = X_r$, shows that E is an algebraic subset of G which is defined, like X_r, by polynomials of multidegree $\leq (D_0, \underline{D})$. Since E is a finite union of translates of G^*, Lemma 5.10 gives

$$\mathcal{H}(E;\, D_0, \underline{D}) = \mathrm{Card}(E/G^*)\mathcal{H}(G^*;\, D_0, \underline{D}).$$

Since E is defined in G by polynomials of multidegree $\leq (D_0, \underline{D})$, Theorem 5.3 provides an upper bound for the left hand side of the previous equality:

$$\mathcal{H}(E;\, D_0, \underline{D}) \leq \mathcal{H}(G;\, D_0, \underline{D}).$$

Finally, since $\Sigma \subseteq E$, we have

$$\mathrm{Card}(E/G^*) \geq \mathrm{Card}\left(\frac{\Sigma + G^*}{G^*}\right).$$

This implies

$$\mathrm{Card}\left(\frac{\Sigma + G^*}{G^*}\right) \mathcal{H}(G^*;\, D_0, \underline{D}) \le \mathcal{H}(G;\, D_0, \underline{D}).$$

This inequality also holds with the neutral component G_0^* of G^* instead of G^* because

$$\mathrm{Card}\left(\frac{\Sigma + G_0^*}{G_0^*}\right) \le [G^* : G_0^*]\, \mathrm{Card}\left(\frac{\Sigma + G^*}{G^*}\right)$$

and because by Lemma 5.10 and Theorem 5.13 we have

$$\mathcal{H}(G^*;\, D_0, \underline{D}) = [G^* : G_0^*]\, \mathcal{H}(G_0^*;\, D_0, \underline{D}).$$

□

Exercises

Exercise 5.1. Let β_1, β_2 be elements of K such that $1, \beta_1, \beta_2$ are linearly independent over $\mathbb{Q}$. Let S be a positive integer, and consider the subset Σ of $G = \mathbb{G}_{\mathrm{a}}^2$ given by

$$\Sigma = \left\{(s_1 + s_3\beta_1, s_2 + s_3\beta_2)\, ;\; \underline{s} = (s_1, s_2, s_3) \in \mathbb{Z}^3,\;\; |\underline{s}| \le S\right\}.$$

Show that if a nonzero polynomial $P \in K[G] = K[X_1, X_2]$ vanishes at each point of $\Sigma[2]$, then its degree D satisfies $D \ge (2S+1)^{3/2}$. Conversely, show that, if an integer $D \ge 1$ satisfies $D \ge 4(2S+1)^{3/2}$, then there exists a nonzero polynomial $P \in K[G]$ of degree $\le D$ which vanishes at each point of $\Sigma[2]$.

Exercise 5.2. Let k be a subfield of K. An algebraic subgroup of G is said to be *defined over* k if it is of the form $\mathcal{V} \times T_\Phi$ for a subspace $\mathcal{V}$ of K^{d_0} defined over k.

(a) Show that if an algebraic subgroup G^* of G is defined over k, then it is the set of common zeros in G of a family of elements of $k[\underline{X}, \underline{Y}^{\pm 1}]$.
(b) Show that if the set Σ of Theorem 5.1 is contained in $k^{d_0} \times (k^\times)^{d_1}$ then, in the conclusion of the theorem, one can assume that the group G^* is defined over k.

Hint. *Let $G^* = \mathcal{V} \times T_\Phi$ and let $\mathcal{V}_k$ be the largest subspace of $\mathcal{V}$ defined over k. Show that the group $G_k^* = \mathcal{V}_k \times T_\Phi$ is connected and that one has*

$$\mathrm{Card}\left(\frac{\Sigma + G_k^*}{G_k^*}\right) = \mathrm{Card}\left(\frac{\Sigma + G^*}{G^*}\right)$$

and

$$\mathcal{H}(G_k^*;\, D_0, \underline{D}) \le \mathcal{H}(G^*;\, D_0, \underline{D}).$$

Exercise 5.3. (A converse to Theorem 5.1) Let G^* be a connected algebraic subgroup of G with $G^* \neq G$ and let Σ be a finite subset of G. Show that for any $(D_0, \underline{D}) \in \mathbb{N} \times \mathbb{N}^{d_1}$ satisfying

$$\operatorname{Card}\left(\frac{\Sigma + G^*}{G^*}\right) H(G^*; D_0, \underline{D}) < H(G; D_0, \underline{D}) \tag{5.16}$$

there exists a nonzero polynomial $P \in K[\underline{X}, \underline{Y}^{\pm 1}]$ of multidegree $\leq (D_0, \underline{D})$ which vanishes at each point of Σ.

Hint. *Let $E = \bigcup_{\sigma \in \Sigma}(\sigma + G^*)$. Show that the left hand side of (5.16) is $\geq H(E; D_0, \underline{D})$.*

Exercise 5.4. Use Theorem 5.3 with $d_0 = n$ and $d_1 = 0$ to prove that if F is a finite algebraic subset of K^n defined by polynomials of $K[X_1, \ldots, X_n]$ of degree $\leq D$, then the cardinality of F is $\leq D^n$.

Hint. *Show that for a finite algebraic subset F of K^n, the polynomial $\mathcal{H}(F; X)$ is constant, equal to the cardinality of F.*

Exercise 5.5. Let H be an algebraic subgroup of G of dimension m and let V be an irreducible component of H of the same dimension. Define

$$H_0 = \{g \in G\,;\, g + V = V\} \quad \text{and} \quad R = \{g + V\,;\, g \in H\}.$$

(a) Show that H_0 is an algebraic subgroup of H, that R is the set of all irreducible components of H and that the quotient H/H_0 is in bijection with R.

(b) Deduce from (a) that H is equidimensional, that its irreducible components are disjoint and that the one which contains e is an algebraic subgroup of H.

Exercise 5.6. Let $\mathsf{M} = (m_{ij})$ be an $r \times r$ matrix with coefficients in $\mathbb{Z}$. Consider the endomorphism ψ of $(K^\times)^r$ given by

$$\psi(y_1, \ldots, y_r) = (y_1^{m_{11}} \cdots y_r^{m_{1r}}, \ldots, y_1^{m_{r1}} \cdots y_r^{m_{rr}}).$$

(a) Show that the kernel of ψ is infinite if $\det \mathsf{M} = 0$ and that otherwise its cardinality is $|\det \mathsf{M}|$.

(b) Show that ψ is surjective if and only if $\det \mathsf{M} \neq 0$.

6. Linear Independence of Logarithms of Algebraic Numbers

In Chap. 4, we proved Baker's homogeneous Theorem 1.5: *if logarithms of algebraic numbers are linearly independent over* $\mathbb{Q}$*, then they are linearly independent over* $\overline{\mathbb{Q}}$. The proof was an extension of Gel'fond's solution to Hilbert's seventh problem. Here we give a second proof of the same theorem, using an extension of Schneider's method. The two main tools are an upper bound for the absolute value of an alternant in several variables (Proposition 6.6) and the zero estimate (namely Theorem 5.1).

The idea of proof is as follows. Assume there is a relation

$$\lambda_{n+1} = \beta_1\lambda_1 + \cdots + \beta_n\lambda_n$$

with algebraic coefficients $\beta_1, \ldots, \beta_n$ between logarithms of algebraic numbers $\lambda_1, \ldots, \lambda_{n+1}$. Set $\alpha_j = e^{\lambda_j}$ $(1 \le j \le n+1)$. We first introduce analytic functions which take algebraic values in the number field

$$K = \mathbb{Q}\bigl(\beta_1, \ldots, \beta_n, \alpha_1, \ldots, \alpha_{n+1}\bigr)$$

at many points. Consider the $n+1$ functions in $n+1$ variables

$$z_1, \ldots, z_n,\ e^{z_{n+1}},$$

as well as monomials in these functions:

$$\underline{z}^{\underline{\tau}} e^{tz_{n+1}} = z_1^{\tau_1} \cdots z_n^{\tau_n} e^{tz_{n+1}}$$

for $\underline{\tau} = (\tau_1, \ldots, \tau_n) \in \mathbb{N}^n$ and $t \in \mathbb{Z}$. The corresponding algebraic group (needed for using the zero estimate of Chap. 5) is $G = \mathbb{G}_{\mathrm{a}}^n \times \mathbb{G}_{\mathrm{m}}$. We estimate these functions at the points

$$\underline{\eta}_j = \bigl(\delta_{j1}, \ldots, \delta_{jn}, \lambda_j\bigr) \quad (1 \le j \le n) \quad \text{and} \quad \underline{\eta}_{n+1} = \bigl(\beta_1, \ldots, \beta_n, \lambda_{n+1}\bigr)$$

in $\mathbb{C}^{n+1}$ (where δ_{ji} is Kronecker's diagonal symbol) as well as at linear combinations of these points:

$$\begin{aligned} \underline{s}\eta &= s_1\underline{\eta}_1 + \cdots + s_{n+1}\underline{\eta}_{n+1} \\ &= \bigl(s_1 + s_{n+1}\beta_1, \ldots, s_n + s_{n+1}\beta_n, s_1\lambda_1 + \cdots + s_{n+1}\lambda_{n+1}\bigr) \end{aligned}$$

for $\underline{s} = (s_1, \ldots, s_{n+1}) \in \mathbb{Z}^{n+1}$. We have

$$\left(\underline{z}^{\underline{\tau}} e^{t z_{n+1}}\right)(\underline{s}\eta) = \prod_{i=1}^{n} (s_i + s_{n+1}\beta_i)^{\tau_i} \prod_{j=1}^{n+1} \alpha_j^{s_j t} \in K.$$

The given linear dependence relation between $\lambda_1, \ldots, \lambda_{n+1}$ means that these points $\underline{s}\eta$ all belong to the hyperplane W in $\mathbb{C}^{n+1}$ of equation

$$\lambda_1 z_1 + \cdots + \lambda_n z_n = z_{n+1}.$$

If we prefer we can also consider only functions of n variables, replacing z_{n+1} by $\lambda_1 z_1 + \cdots + \lambda_n z_n$.

We needed *many* functions and *many* points, and we found infinitely many of them, which is too much! We select finite subsets of these: we introduce *large* parameters T_0, T_1 and S, so that we can deal with only finitely many functions and points, namely with $\underline{\tau}$, t and $\underline{s}$ restricted to[13]

$$\|\underline{\tau}\| \le T_0, \quad |t| \le T_1, \quad |\underline{s}| \le S.$$

With the values of these functions at the given points we build a matrix

$$\boldsymbol{M} = \left(\left(\underline{z}^{\underline{\tau}} e^{t z_{n+1}}\right)(\underline{s}\eta) \right)_{\substack{(\underline{\tau}, t) \\ \underline{s}}}$$

like in Chap. 2. The proof now decomposes into three steps, which may be introduced in any order. The idea is to consider a $L \times L$ minor Δ of maximal size, namely with

$$L = \binom{T_0 + n}{n}(2T_1 + 1).$$

An analytic estimate yields an upper bound for $|\Delta|$ - this is the step of the proof where the hyperplane W is required. Since Δ is an algebraic number, Liouville's estimate provides a lower bound for $|\Delta|$ provided that it is not zero. Under suitable conditions for the parameters T_0, T_1, S, the zero estimate shows that either $\boldsymbol{M}$ has maximal rank L, or else there is a linear dependence relation over $\mathbb{Q}$ between either $\lambda_1, \ldots, \lambda_{n+1}$ or $1, \beta_1, \ldots, \beta_n$. Putting all this information together enables one to deduce Baker's Theorem.

6.1 Applying the Zero Estimate

In this section, K is an algebraically closed field of zero characteristic. Our aim is to prove the following result:

Proposition 6.1. *Let n be a positive integer, $\alpha_1, \ldots, \alpha_{n+1}$ be nonzero elements of K and $\beta_1, \ldots, \beta_n$ be elements of K. Assume the numbers $1, \beta_1, \ldots, \beta_n$ are linearly independent over $\mathbb{Q}$. Let T_0, T_1 and S be positive integers satisfying*

---

[13] As a matter of fact we shall need to consider all points $\underline{s}\eta$ with $|\underline{s}| \le (n+1)S$.

$$(2S+1)^{n+1} > 2(n+1)T_0^n T_1, \quad T_0 > 2S \quad \textit{and} \quad (2S+1)^2 > \frac{n+1}{n} T_0.$$

Assume further:

- *either* $\alpha_1, \ldots, \alpha_{n+1}$ *are multiplicatively independent,*
- *or else* $\alpha_1, \ldots, \alpha_{n+1}$ *generate a multiplicative subgroup of* $K^\times$ *of rank* n *and* $(2S+1)^n > 2(n+1)T_0^{n-1}T_1$.

Consider the following matrix

$$M = \left((s_1 + s_{n+1}\beta_1)^{\tau_1} \cdots (s_n + s_{n+1}\beta_n)^{\tau_n} \left(\alpha_1^{s_1} \cdots \alpha_{n+1}^{s_{n+1}}\right)^t \right)_{\substack{(\underline{\tau},t) \\ \underline{s}}},$$

where the index of rows is

$$(\underline{\tau}, t) = (\tau_1, \ldots, \tau_n, t) \in \mathbb{N}^n \times \mathbb{Z}, \qquad \|\underline{\tau}\| \le T_0, \qquad |t| \le T_1,$$

while the index of columns is $\underline{s} = (s_1, \ldots, s_{n+1}) \in \mathbb{Z}^{n+1}$, $|\underline{s}| \le (n+1)S$.

Then the rank of M *is*

$$\binom{T_0+n}{n}(2T_1+1).$$

In the special case $n = 1$, Nesterenko's zero estimate (Proposition 2.12 of Chap. 2) provides a slightly stronger result. Such an improvement is useful for sharpening measures of linear independence of two logarithms by means of Schneider's method [LauMiNe 1993].

For the proof of Proposition 6.1, we need one more lemma which explains how the hypothesis on the linear independence of $1, \beta_1, \ldots, \beta_n$ is related with the conclusion of Theorem 5.1.

For $\underline{\beta} = (\beta_1, \ldots, \beta_n) \in K^n$, the condition that $1, \beta_1, \ldots, \beta_n$ are linearly independent over $\mathbb{Q}$ is equivalent to say that for each nonzero linear form $\varphi\colon K^n \to K$ which maps $\mathbb{Z}^n$ into $\mathbb{Z}$, we have $\varphi(\underline{\beta}) \notin \mathbb{Z}$. We reformulate this condition by replacing $\ker\varphi$ by any hyperplane of K^n (see Exercise 6.1), and then we extend the property to any subspace of K^n distinct from K^n.

Notation. Let M be a finitely generated $\mathbb{Z}$-module given with a set of generators $\{x_1, \ldots, x_n\}$. For any nonnegative integer S we set

$$M[S] = \left\{ s_1x_1 + \cdots + s_nx_n \;;\; (s_1, \ldots, s_n) = \underline{s} \in \mathbb{Z}^n, \; |\underline{s}| \le S \right\}.$$

For instance

$$\mathbb{Z}^n[S] = \left\{ (s_1, \ldots, s_n) \in \mathbb{Z}^n \;;\; |\underline{s}| \le S \right\}$$

has $(2S+1)^n$ elements. Another example is $Y = \mathbb{Z}^n + \mathbb{Z}(\beta_1, \ldots, \beta_n) \subset K^n$: then

$$Y[S] = \left\{ (s_1 + s_{n+1}\beta_1, \ldots, s_n + s_{n+1}\beta_n) \;;\; (s_1, \ldots, s_{n+1}) \in \mathbb{Z}^{n+1}[S] \right\}.$$

Lemma 6.2. *Let $\beta_1, \dots, \beta_n$ be elements of K. Define*

$$\begin{aligned} Y &= \mathbb{Z}^n + \mathbb{Z}(\beta_1, \dots, \beta_n) \subset K^n \\ &= \left\{(s_1 + s_{n+1}\beta_1, \dots, s_n + s_{n+1}\beta_n)\,;\, (s_1, \dots, s_{n+1}) \in \mathbb{Z}^{n+1}\right\}. \end{aligned}$$

Then the following conditions are equivalent.
(i) The numbers $1, \beta_1, \dots, \beta_n$ are linearly independent over $\mathbb{Q}$.
(ii) For any vector subspace $V \subset K^n$ of codimension $r \geq 1$, we have

$$\mathrm{rk}_{\mathbb{Z}}\left(\frac{Y+V}{V}\right) \geq r+1.$$

(iii) For any $S \geq 1$ and any vector subspace $V \subset K^n$ of codimension $r \geq 1$, we have

$$\mathrm{Card}\left(\frac{Y[S]+V}{V}\right) \geq (2S+1)^{r+1}.$$

(ii)′ For any vector subspace $W \subset K^{n+1}$ of codimension $r \geq 1$, containing $(\beta_1, \dots, \beta_n, -1)$, we have

$$\mathrm{rk}_{\mathbb{Z}}\left(\frac{\mathbb{Z}^{n+1}+W}{W}\right) \geq r+1.$$

(iii)′ For any $S \geq 1$ and any vector subspace $W \subset K^{n+1}$ of codimension $r \geq 1$, containing $(\beta_1, \dots, \beta_n, -1)$, we have

$$\mathrm{Card}\left(\frac{\mathbb{Z}^{n+1}[S]+W}{W}\right) \geq (2S+1)^{r+1}.$$

Proof of Lemma 6.2. The proofs of *(ii)* ⇔ *(ii)′* and of *(iii)* ⇔ *(iii)′* are easily obtained by considering the linear surjective map

$$\begin{array}{ccc} K^{n+1} & \longrightarrow & K^n \\ (z_1, \dots, z_{n+1}) & \longmapsto & (z_1 + z_{n+1}\beta_1, \dots, z_n + z_{n+1}\beta_n) \end{array}$$

whose kernel is the line $K(\beta_1, \dots, \beta_n, -1)$.

We prove the implication *(ii)′* ⇔ *(i)*, using Exercise 1.4, as follows: condition *(i)* means that the point $(\beta_1, \dots, \beta_n, -1)$ is not contained in a hyperplane which is rational over $\mathbb{Q}$. This is equivalent to say that this point is not contained in a subspace of K^{n+1}, of positive codimension, which is rational over $\mathbb{Q}$.

On the other hand, since $\mathbb{Z}^{n+1}$ contains a basis of K^{n+1}, the inequality

$$\mathrm{rk}_{\mathbb{Z}}\left(\frac{\mathbb{Z}^{n+1}+W}{W}\right) \geq r$$

always holds. Equality holds if and only if W is intersection of hyperplanes of K^{n+1} which are rational over $\mathbb{Q}$ (which means that W is rational over $\mathbb{Q}$).

The fact that *(iii)'* $\Rightarrow$ *(ii)'* is easy: denote by $\underline{e}_1, \ldots, \underline{e}_{n+1}$ the canonical basis of K^{n+1}, and put

$$\varrho = \mathrm{rk}_{\mathbb{Z}}\left(\frac{\mathbb{Z}^{n+1} + W}{W}\right).$$

Let $\underline{\eta}_1, \ldots, \underline{\eta}_\varrho$ be elements in $\mathbb{Z}^{n+1}$ whose classes modulo W give a basis of $(\mathbb{Z}^{n+1} + W)/W$. Define $k_{ij} \in \mathbb{Z}$ by

$$\underline{e}_i - \sum_{j=1}^{\varrho} k_{ij}\underline{\eta}_j \in W, \qquad (1 \le i \le n+1).$$

Then

$$\mathrm{Card}\left(\frac{\mathbb{Z}^{n+1}[S] + W}{W}\right) \le c(2S+1)^\varrho, \qquad \text{with} \quad c = \sum_{j=1}^{\varrho}\sum_{i=1}^{n+1} |k_{ij}|.$$

In particular, if $\varrho \le r$, then $c(2S+1)^\varrho < (2S+1)^{r+1}$ as soon as $S \ge c$.

Finally we check *(ii)'* $\Rightarrow$ *(iii)'*. Denote again by $\underline{e}_1, \ldots, \underline{e}_{n+1}$ the canonical basis of K^{n+1}, and by ϱ the rank over $\mathbb{Z}$ of $(\mathbb{Z}^{n+1} + W)/W$. Let $\{i_1, \ldots, i_\varrho\}$ be indices in $\{1, \ldots, n+1\}$ such that the classes modulo W of $\underline{e}_{i_1}, \ldots, \underline{e}_{i_\varrho}$ are linearly independent over $\mathbb{Z}$. Then the $(2S+1)^\varrho$ classes of

$$s_{i_1}\underline{e}_{i_1} + \cdots + s_{i_\varrho}\underline{e}_{i_\varrho} \in \mathbb{Z}^{n+1}[S], \qquad (s_{i_1}, \ldots, s_{i_\varrho}) \in \mathbb{Z}^\varrho[S],$$

are pairwise distinct. □

Proof of Proposition 6.1. We apply Theorem 5.1 with

$$d_0 = n,\ d_1 = 1,\ d = n+1,\ G = \mathbb{G}_{\mathrm{a}}^n \times \mathbb{G}_{\mathrm{m}},\ D_0 = T_0,\ D_1 = T_1$$

and

$$\Sigma = \left\{\left(s_1 + s_{n+1}\beta_1, \ldots, s_n + s_{n+1}\beta_n, \alpha_1^{s_1} \cdots \alpha_{n+1}^{s_{n+1}}\right)\ ;\ \underline{s} \in \mathbb{Z}^{n+1}[S]\right\}.$$

If the conclusion of Proposition 6.1 does not hold, then there exists a nonzero polynomial $P \in K[X_1, \ldots, X_n, Y^{\pm 1}]$, of total degree at most T_0 in the variables $X_1, \ldots, X_n$ and of degree at most T_1 in Y, which vanishes at all the points of the subset $\Sigma[n+1]$ of $K^n \times K^\times$. We deduce that there exists a connected algebraic subgroup G^* of G such that $G^* \neq G$ and

$$\mathrm{Card}\left(\frac{\Sigma + G^*}{G^*}\right) \mathcal{H}(G^*; T_0, T_1) \le \mathcal{H}(G; T_0, T_1).$$

As explained in § 5.1.1, we have

$$\mathcal{H}(G; T_0, T_1) = 2(n+1)T_0^n T_1,$$

and there exists a vector subspace V of K^n, of dimension say δ, such that $G^* = V \times G_1^*$, where G_1^* is either $\{1\}$ or $\mathbb{G}_{\mathrm{m}}$ and

$$\mathcal{H}(G^*;\, T_0, T_1) = \begin{cases} T_0^\delta & \text{if } G_1^* = \{1\},\\ 2(\delta+1)T_0^\delta T_1 & \text{if } G_1^* = \mathbb{G}_{\mathrm{m}}. \end{cases}$$

Hence the conclusion of Theorem 5.1 is

$$\mathrm{Card}\left(\frac{\Sigma + (V \times G_1^*)}{V \times G_1^*}\right) \le \begin{cases} 2(n+1)T_0^{n-\delta}T_1 & \text{if } G_1^* = \{1\},\\ \dfrac{n+1}{\delta+1}T_0^{n-\delta} & \text{if } G_1^* = \mathbb{G}_{\mathrm{m}}. \end{cases}$$

Since each β_i is irrational, the elements

$$\bigl(s_1 + s_{n+1}\beta_1, \ldots, s_n + s_{n+1}\beta_n\bigr) \in K^n \qquad \underline{s} \in \mathbb{Z}^{n+1}$$

are pairwise distinct. From the assumption

$$(2S+1)^{n+1} > 2(n+1)T_0^n T_1$$

we deduce $V \ne \{0\}$, so that we have $1 \le \delta \le n$.

Consider firstly the case $G_1^* = \{1\}$. In this case we plainly have

$$\mathrm{Card}\left(\frac{\Sigma + (V \times \{1\})}{V \times \{1\}}\right) \ge \mathrm{Card}\Bigl\{\alpha_1^{s_1}\cdots\alpha_{n+1}^{s_{n+1}} \in K^\times \; ; \; \underline{s} \in \mathbb{Z}^{n+1}[S]\Bigr\}.$$

If $\alpha_1, \ldots, \alpha_{n+1}$ are multiplicatively independent, the right hand side has $(2S+1)^{n+1}$ elements, and we get a contradiction as before. If $\alpha_1, \ldots, \alpha_{n+1}$ generate a multiplicative subgroup of $K^\times$ of rank n, then

$$\mathrm{Card}\Bigl\{\alpha_1^{s_1}\cdots\alpha_{n+1}^{s_{n+1}} \in K^\times \; ; \; \underline{s} \in \mathbb{Z}^{n+1}[S]\Bigr\} \ge (2S+1)^n.$$

Now from the assumption $(2S+1)^n > 2(n+1)T_0^{n-1}T_1$ and the inequality $\delta \ge 1$ we again derive a contradiction.

Therefore $G_1^* = \mathbb{G}_{\mathrm{m}}$. Since $G^* \ne G$, we have $\delta \le n-1$. Define $Y = \mathbb{Z}^n + \mathbb{Z}(\beta_1, \ldots, \beta_n) \subset K^n$. The conclusion of Theorem 5.1 becomes

$$\mathrm{Card}\left(\frac{Y[S]+V}{V}\right) \le \frac{n+1}{\delta+1}T_0^{n-\delta}.$$

Using both assumptions

$$(2S+1)^{n+1} > 2(n+1)T_0^n T_1 \quad \text{and} \quad T_0 \ge 2S+1$$

we deduce

$$(2S+1)^{n-\delta+1} > 2(n+1)T_0^{n-\delta}.$$

Again we derive a contradiction, which completes the proof of Proposition 6.1. □

Remark. For $m \ge 1$ and $\lambda_1, \ldots, \lambda_m$ in $\mathbb{C}$, define $\alpha_j = e^{\lambda_j}$ $(1 \le j \le m)$. If $\lambda_1, \ldots, \lambda_m$ are linearly independent over $\mathbb{Q}$, then $\alpha_1, \ldots, \alpha_m$ generate a multiplicative subgroup of $\mathbb{C}^\times$ of rank m or $m-1$. The rank is m (which means that $\alpha_1, \ldots, \alpha_m$ are multiplicatively independent) if and only if the $m+1$ numbers $\lambda_1, \ldots, \lambda_m, 2i\pi$ are $\mathbb{Q}$-linearly independent. Otherwise, when the rank is $m-1$, the set of $\underline{k} \in \mathbb{Z}^m$

such that $\alpha_1^{k_1}\cdots\alpha_m^{k_m}=1$ is a rank one $\mathbb{Z}$-module: there is an element $\underline{k}^0$ in $\mathbb{Z}^m\setminus\{0\}$, which is unique up to a multiplicative factor ± 1, such that, for $\underline{k}\in\mathbb{Z}^m$,

$$\alpha_1^{k_1}\cdots\alpha_m^{k_m}=1 \iff \underline{k}\in\underline{k}^0\mathbb{Z}.$$

For instance when

$$m=1,\ \lambda=\frac{2i\pi}{k^0},\ \alpha=e^{2i\pi/k^0}$$

with $k^0\in\mathbb{Z}\setminus\{0\}$, the ideal of $k\in\mathbb{Z}$ such that $\alpha^k=1$ is $k^0\mathbb{Z}$.

6.2 Upper Bounds for Alternants in Several Variables

In this section we denote by $f_1,\ldots,f_L$ analytic functions in $\mathbb{C}^n$, and by $\underline{\zeta}_1,\ldots,\underline{\zeta}_L$ elements of $\mathbb{C}^n$. Our aim is to give an upper bound for the absolute value of the determinant

$$\Delta=\det\Bigl(f_\lambda(\underline{\zeta}_\mu)\Bigr)_{1\le\lambda,\mu\le L},$$

following Michel Laurent [Lau 1989], [Lau 1992], [Lau 1994]. The case $n=1$ has been considered in Lemma 2.5. The estimate for the general case is stated in Proposition 6.6 below. For the proof, we show that the function of a single complex variable z

$$\Psi(z)=\det\Bigl(f_\lambda(z\underline{\zeta}_\mu)\Bigr)_{1\le\lambda,\mu\le L}$$

has a zero of high multiplicity at the origin. Then Schwarz' Lemma provides the desired upper bound. At the end of this section, we remark that the proofs can be considered as *elementary* so far as no complex analysis is required.

6.2.1 Schwarz' Lemma

We first apply Schwarz' Lemma in one variable (Lemma 2.4).

Lemma 6.3. *Let $r>0$ and $R>0$ be positive real numbers such that*

$$\max_{1\le\mu\le L}|\underline{\zeta}_\mu|\le r \quad \textit{and} \quad R\ge r.$$

Let T be the multiplicity of the zero of the function Ψ at the origin. Then

$$|\Delta|\le\left(\frac{R}{r}\right)^{-T}L!\prod_{\lambda=1}^{L}|f_\lambda|_R.$$

Proof. Define $E=R/r$. Since $E\ge 1$, we deduce from Lemma 2.4 an upper bound for the absolute value of the number $\Delta=\Psi(1)$:

$$|\Psi(1)| \leq E^{-T} |\Psi|_E.$$

From $|z\underline{\zeta}_\mu| \leq Er = R$ for $|z| \leq E$, we deduce

$$|\Psi|_E \leq L! \prod_{\lambda=1}^{L} |f_\lambda|_R.$$

This completes the proof of Lemma 6.3. □

6.2.2 Estimate for the Multiplicity of Ψ at the Origin

The proof (in § 2.3.1) of Lemma 2.5 (dealing with the one dimensional case) involves the number

$$\Theta_1(L) = \min\{\kappa_1 + \cdots + \kappa_L\} = 0 + 1 + \cdots + (L-1) = \frac{L(L-1)}{2},$$

where the minimum runs over the L-tuples $(\kappa_1, \ldots, \kappa_L)$ of nonnegative integers which are pairwise distinct. In the general case $n \geq 1$, we define

$$\Theta_n(L) = \min\left\{\|\underline{\kappa}_1\| + \cdots + \|\underline{\kappa}_L\|\right\}$$

where the minimum runs over the L-tuples $(\underline{\kappa}_1, \ldots, \underline{\kappa}_L)$ of elements in $\mathbb{N}^n$ which are pairwise distinct.

Lemma 6.4. *The function Ψ has a zero at $z = 0$ of multiplicity at least $\Theta_n(L)$.*

Proof. Since the determinant is multilinear, by expanding each f_λ in Taylor series at the origin, we may assume that each f_λ is a monomial $f_\lambda(\underline{\zeta}) = \underline{\zeta}^{\underline{\kappa}_\lambda}$, with $\underline{\kappa}_\lambda \in \mathbb{N}^n$. In this case $f_\lambda(z\underline{\zeta}) = \underline{\zeta}^{\underline{\kappa}_\lambda} z^{\|\underline{\kappa}_\lambda\|}$.

In the row indexed by λ, we have a common factor $z^{\|\underline{\kappa}_\lambda\|}$:

$$\Psi(z) = \det\left(\underline{\zeta}_\mu^{\underline{\kappa}_\lambda}\right)_{1\leq\lambda,\mu\leq L} \cdot z^{\|\underline{\kappa}_1\|+\cdots+\|\underline{\kappa}_L\|}.$$

If Ψ is not the zero function, then the elements $\underline{\kappa}_1, \ldots, \underline{\kappa}_L$ in $\mathbb{N}^n$ are pairwise distinct, and Ψ has a zero at 0 of multiplicity $\|\underline{\kappa}_1\| + \cdots + \|\underline{\kappa}_L\|$, which proves our claim. □

6.2.3 Lower Bound for Θ_n

Here is a lower bound for the number $\Theta_n(L)$:

Lemma 6.5. *For any $L \geq 1$ and $n \geq 1$, we have*

$$\Theta_n(L) > \frac{n}{n+1}(n!)^{1/n} L^{(n+1)/n} - n(n+1)L.$$

Moreover for $L \geq (4n)^{2n}$, we have

$$\Theta_n(L) > \frac{n}{e} L^{(n+1)/n}.$$

Proof. Let us check the first estimate. One may assume $L > (n+1)^{2n}/n!$, otherwise the result is trivial.

The smallest value for the sum $\|\underline{\kappa}_1\| + \cdots + \|\underline{\kappa}_L\|$ is reached by choosing for $\underline{\kappa}_\mu$ successively:

- $(0, \ldots, 0)$;
- the n elements of $\mathbb{N}^n$ of length 1:

$$(1, 0, \ldots, 0), \ldots, (0, \ldots, 0, 1);$$

- the $\binom{n+1}{2} = \binom{n+1}{n-1}$ elements of length 2:

$$(2, 0, \ldots, 0),\ (1, 1, 0, \ldots, 0), \ldots,\ (0, \ldots, 0, 1, 1),\ (0, \ldots, 0, 2);$$

- and so on.

In general, for a a nonnegative integer, the number of elements $\underline{\kappa} \in \mathbb{N}^n$ of length $\|\underline{\kappa}\| = a$ is the coefficient of z^a in the series

$$\sum_{\underline{\kappa} \in \mathbb{N}^n} z^{\|\underline{\kappa}\|} = \left(\sum_{k=0}^{\infty} z^k \right)^n = \frac{1}{(1-z)^n} = \sum_{a \geq 0} \binom{n+a-1}{a} z^a,$$

hence this number is

$$\binom{n+a-1}{a} = \binom{n+a-1}{n-1}.$$

For any positive integer A we have

$$\sum_{k=0}^{A-1} \binom{n+k}{n} = \binom{n+A}{n+1}.$$

This is an easy consequence (by induction) of the formula

$$\binom{n+k-1}{n+1} + \binom{n+k-1}{n} = \binom{n+k}{n+1}.$$

Let A be the positive integer such that

$$\sum_{a=0}^{A} \binom{n+a-1}{n-1} = \binom{n+A}{n} \leq L < \binom{n+A+1}{n}.$$

We have

$$\|\underline{\kappa}_1\| + \cdots + \|\underline{\kappa}_L\| \geq \sum_{a=0}^{A} a\binom{n+a-1}{n-1} = n\sum_{a=1}^{A}\binom{n+a-1}{n}$$
$$= n\sum_{a=0}^{A-1}\binom{n+a}{n} = n\binom{n+A}{n+1},$$

hence

$$\Theta_n(L) \geq n\binom{n+A}{n+1}.$$

We use the estimates

$$\binom{n+A+1}{n} \leq \frac{(n+A+1)^n}{n!} \qquad \text{and} \qquad \binom{n+A}{n+1} \geq \frac{A^{n+1}}{(n+1)!}.$$

We get

$$\Theta_n(L) \geq \frac{n}{(n+1)!}\cdot A^{n+1} \geq \frac{n}{n+1}(n!)^{1/n}\cdot\left(\frac{A}{A+n+1}\right)^{n+1}\cdot L^{1+(1/n)}.$$

Define ϵ by the condition

$$\epsilon^n n! L = (n+1)^{2n}.$$

Then the number $c = (n+1)/\epsilon$ satisfies

$$\left(1-\frac{1}{c}\right)^{n+1} > 1-\epsilon.$$

The inequalities

$$\frac{(A+n+1)^n}{n!} \geq \binom{A+n+1}{n} > L \geq \frac{c^n}{n!}(n+1)^n$$

yield

$$A+n+1 > c(n+1) \qquad \text{and} \qquad \frac{A}{A+n+1} > 1-\frac{1}{c}.$$

The first estimate in Lemma 6.5 follows. The second inequality is a consequence of the first one, since

$$(n!)^{1/n} \geq \frac{n+1}{e} + \left(\frac{n+1}{4n}\right)^2$$

for any $n \geq 1$. □

6.2.4 Conclusion

We now combine the preceding lemmas as follows:

Proposition 6.6. *Let $f_1, \dots, f_L$ be analytic functions in $\mathbb{C}^n$, $\underline{\zeta}_1, \dots, \underline{\zeta}_L$ elements of $\mathbb{C}^n$, r, R and E positive real numbers such that*

$$\max_{1 \le \mu \le L} |\underline{\zeta}_\mu| \le r \qquad \text{and} \qquad E = \frac{R}{r} \ge 1.$$

Then the determinant

$$\Delta = \det\Big(f_\lambda(\underline{\zeta}_\mu) \Big)_{1 \le \lambda, \mu \le L}$$

is bounded from above by

$$\log |\Delta| \le -\Theta_n(L) \log E + \log(L!) + \sum_{\lambda=1}^{L} \log |f_\lambda|_R.$$

Here we shall use only the following consequence of Proposition 6.6:

Corollary 6.7. *Let $\lambda_1, \dots, \lambda_n$, $\beta_1, \dots, \beta_n$ be complex numbers. For $1 \le i \le n$ define $\alpha_i = \exp(\lambda_i)$. There exists a positive constant c, which depends only on $n, \lambda_1, \dots, \lambda_n, \beta_1, \dots, \beta_n$ and which satisfies the following property: let T_0, T_1, S be rational integers at least 2 and E a real number at least e. Assume the number $L = \binom{T_0+n}{n}(2T_1+1)$ satisfies $L \ge (4n)^{2n}$. Let $\underline{s}^{(1)}, \dots, \underline{s}^{(L)}$ be any elements in $\mathbb{Z}^{n+1}[S]$. Consider the $L \times L$ determinant Δ of the matrix*

$$\left((s_1^{(\mu)} + s_{n+1}^{(\mu)}\beta_1)^{\tau_1} \cdots (s_n^{(\mu)} + s_{n+1}^{(\mu)}\beta_n)^{\tau_n} \left(\alpha_1^{s_1^{(\mu)}+s_{n+1}^{(\mu)}\beta_1} \cdots \alpha_n^{s_n^{(\mu)}+s_{n+1}^{(\mu)}\beta_n} \right)^t \right)$$

with $(\underline{\tau}, t) = (\tau_1, \dots, \tau_n, t) \in \mathbb{N}^{n+1}$, $\tau_1 + \cdots + \tau_n \le T_0$ and $t \le T_1$, and with $1 \le \mu \le L$. Then

$$|\Delta| \le \exp\left\{ -\frac{n}{e} L^{1+(1/n)} \log E + cL\big(T_0 \log(SE) + T_1 SE\big) \right\}.$$

Proof. We consider the functions

$$f_{\underline{\tau}t}(\underline{z}) = z_1^{\tau_1} \cdots z_n^{\tau_n} \left(\alpha_1^{z_1} \cdots \alpha_n^{z_n} \right)^t$$

and the points

$$\underline{\zeta}_\mu = \left(s_1^{(\mu)} + s_{n+1}^{(\mu)}\beta_1, \dots, s_n^{(\mu)} + s_{n+1}^{(\mu)}\beta_n \right) \in \mathbb{C}^n.$$

For any $R > 0$ we plainly have

$$\log |f_{\underline{\tau}t}|_R \le T_0 \log R + T_1 R \sum_{i=1}^{n} |\lambda_i|.$$

We choose

$$r = S\left(1 + \max_{1\leq j\leq n} |\beta_j|\right), \qquad R = Er.$$

From Proposition 6.6 one deduces

$$\log|\Delta| \leq -\Theta_n(L)\log E + \log L! + LT_0\log R + LT_1 R\sum_{i=1}^{n}|\lambda_i|.$$

Finally we conclude thanks to Lemma 6.5. □

6.2.5 Avoiding the Use of Complex Analysis

We conclude this section with the following remark: the proofs we give in these lectures do not require any complex analysis. From this point of view they are *elementary* in the sense of [GLin 1962]. The only point where analysis played any role so far was in the use of Schwarz' Lemma 2.4, in the proof of Lemma 6.3. But we use it only for exponential polynomials in one variable, and in this case the estimate is quite easy:

Lemma 6.8. *Let a_{ij} (for $0 \leq i \leq s$, $1 \leq j \leq t$) and w_j (for $1 \leq j \leq t$) be complex numbers. Assume that the exponential polynomial*

$$F(z) = \sum_{i=0}^{s}\sum_{j=1}^{t} a_{ij} z^i e^{w_j z}$$

has a zero of multiplicity at least T at the origin. Then for $z_0 \in \mathbb{C}$ and $R \geq |z_0|$ we have

$$|F(z_0)| \leq \left(\frac{R}{|z_0|}\right)^{-T} \sum_{i=0}^{s}\sum_{j=1}^{t} |a_{ij}| R^i e^{|w_j|R}.$$

Proof. We consider the Taylor expansion of F at the origin:

$$F(z) = \sum_{n\geq 0}\alpha_n z^n \qquad \text{where} \qquad \alpha_n = \sum_{i=0}^{\min\{s,n\}}\sum_{j=1}^{t} a_{ij}\frac{w_j^{n-i}}{(n-i)!}.$$

By assumption $\alpha_0 = \cdots = \alpha_{T-1} = 0$. For $n \geq T$ we have $(R/|z_0|)^T \leq (R/|z_0|)^n$ (because $R \geq |z_0|$), hence

$$|F(z_0)| = \left|\sum_{n\geq T}\alpha_n z_0^n\right| \leq \sum_{n\geq T}|\alpha_n|\,|z_0|^n \leq \left(\frac{R}{|z_0|}\right)^{-T}\sum_{n\geq T}|\alpha_n|\,R^n.$$

We now use the trivial bound

$$\sum_{n\geq T}|\alpha_n|\,R^n \leq \sum_{i=0}^{\min\{s,n\}}\sum_{j=1}^{t}|a_{ij}|\sum_{n\geq i}\frac{|w_j|^{n-i}}{(n-i)!}R^n,$$

where the right hand side is nothing else than

$$\sum_{i=0}^{s}\sum_{j=1}^{t}|a_{ij}|R^i e^{|w_j|R}.$$

□

6.3 A Second Proof of Baker's Homogeneous Theorem

We already gave a proof of Baker's Theorem (both in the homogeneous and nonhomogeneous cases) in Chap. 4. The proof which we shall give now is quite different, and will enable us (in Chap. 7) to get quantitative estimates (measures of linear independence).

Proposition 6.9. *Let $\lambda_1, \ldots, \lambda_{n+1}$ be $\mathbb{Q}$-linearly independent complex numbers and $\beta_1, \ldots, \beta_n$ be complex numbers such that $1, \beta_1, \ldots, \beta_n$ are $\mathbb{Q}$-linearly independent and*

$$\lambda_{n+1} = \beta_1\lambda_1 + \cdots + \beta_n\lambda_n.$$

For $1 \le j \le n+1$ define $\alpha_j = e^{\lambda_j}$. There exists a positive constant c_0 with the following property. Let T_0, T_1, S, L be rational integers and E a real number satisfying

$$T_0 \ge 2, \qquad T_1 \ge 2, \qquad S \ge 2, \quad E \ge e, \quad L = \binom{T_0+n}{n}(2T_1+1),$$

$$T_0 > 2S \quad \text{and} \quad (2S+1)^{n+1} > 2(n+1)T_0^n T_1.$$

Then there exists a polynomial $f \in \mathbb{Z}[X_1^{\pm 1}, \ldots, X_{n+1}^{\pm 1}, Y_1, \ldots, Y_n]$, of degree and height bounded by

$$\deg f \le c_0 L\big(T_0\log(SE) + T_1 SE\big) \quad \text{and} \quad \log \mathrm{H}(f) \le c_0 L\big(T_0\log(SE) + T_1 SE\big)$$

such that

$$0 < |f(\alpha_1, \ldots, \alpha_{n+1}, \beta_1, \ldots, \beta_n)| \le \exp\left\{-\frac{n}{e}L^{1+1/n}\log E + c_0 L\big(T_0\log(SE) + T_1 SE\big)\right\}.$$

Proof We shall work with the following $n+1$ functions of n variables:

$$z_1, \ldots, z_n, \alpha_1^{z_1}\cdots\alpha_n^{z_n},$$

where $\alpha_1^{z_1}\cdots\alpha_n^{z_n}$ stands for $\exp(z_1\lambda_1 + \cdots + z_n\lambda_n)$. The main fact is that, at all the points of the form

$$(s_1 + s_{n+1}\beta_1, \ldots, s_n + s_{n+1}\beta_n), \qquad (s_1, \ldots, s_{n+1}) \in \mathbb{Z}^{n+1},$$

these functions take values in the ring

$$\mathbb{Z}[\alpha_1^{\pm 1}, \ldots, \alpha_{n+1}^{\pm 1}, \beta_1, \ldots, \beta_n].$$

Another important property of our functions (which is implicit in the proof - see step 1) is that they are algebraically independent: if P is a nonzero polynomial in $n+1$ variables (with, say, complex coefficients), then the function

$$F(z_1, \ldots, z_n) = P\big(z_1, \ldots, z_n, \alpha_1^{z_1} \cdots \alpha_n^{z_n}\big)$$

does not vanish identically (Exercise 2.5).

Step 1. Using the zero estimate

Consider the $L \times (2S+1)^{n+1}$ matrix

$$M = \Big((s_1 + s_{n+1}\beta_1)^{\tau_1} \cdots (s_n + s_{n+1}\beta_n)^{\tau_n} \alpha_1^{ts_1} \cdots \alpha_{n+1}^{ts_{n+1}} \Big)_{\substack{(\underline{\tau}, t) \\ \underline{s}}},$$

where the index of row is $(\underline{\tau}, t)$ and the index of column is $\underline{s}$; $(\underline{\tau}, t)$ runs over the $(n+1)$-tuples $(\tau_1, \ldots, \tau_n, t)$ of elements in $\mathbb{N}^{n+1}$ satisfying $\tau_1 + \cdots + \tau_n \le T_0$ and $t \le T_1$. Hence the number of rows is L. On the other hand $\underline{s}$ runs over the $(n+1)$-tuples in $\mathbb{Z}^{n+1}[S]$, hence there are $(2S+1)^{n+1}$ columns. The ordering of the rows or columns will be irrelevant: we shall be interested only in the rank of M.

Our hypothesis that $\lambda_1, \ldots, \lambda_{n+1}$ are linearly independent over $\mathbb{Q}$ implies that the rank of the multiplicative subgroup of $\mathbb{C}^\times$ generated by $\alpha_1, \ldots, \alpha_{n+1}$ is at least n. The assumptions

$$T_0 \ge 2S+1 \quad \text{and} \quad (2S+1)^{n+1} > 2(n+1)T_0^n T_1$$

imply

$$(2S+1)^2 > \frac{n+1}{n} T_0 \quad \text{and} \quad (2S+1)^n > 2(n+1)T_0^{n-1} T_1.$$

This allows us to use Proposition 6.1 (with $K = \mathbb{C}$), and we deduce that the matrix M has rank L.

Step 2. Definition of Δ.

Let $\underline{s}^{(1)}, \ldots, \underline{s}^{(L)}$ be elements in $\mathbb{Z}^{n+1}[S]$ such that the following $L \times L$ determinant

$$\Delta = \det\Big((s_1^{(\mu)} + s_{n+1}^{(\mu)}\beta_1)^{\tau_1} \cdots (s_n^{(\mu)} + s_{n+1}^{(\mu)}\beta_n)^{\tau_n} \alpha_1^{ts_1^{(\mu)}} \cdots \alpha_{n+1}^{ts_{n+1}^{(\mu)}} \Big)_{\substack{(\underline{\tau}, t) \\ 1 \le \mu \le L}}$$

is not zero.

Step 3. Upper bound for $|\Delta|$

For $(\underline{\tau}, t) = (\tau_1, \ldots, \tau_n, t) \in \mathbb{Z}^{n+1}$, we define

$$f_{\underline{\tau}t} = z_1^{\tau_1} \cdots z_n^{\tau_n} \big(\alpha_1^{z_1} \cdots \alpha_n^{z_n}\big)^t.$$

Our assumption $\lambda_{n+1} = \beta_1\lambda_1 + \cdots + \beta_n\lambda_n$ (this is the only place in the proof where it is used) enables us to write, for $(s_1, \ldots, s_{n+1}) \in \mathbb{Z}^{n+1}$,

$$\alpha_1^{s_1+s_{n+1}\beta_1} \cdots \alpha_n^{s_n+s_{n+1}\beta_n} = \alpha_1^{s_1} \cdots \alpha_{n+1}^{s_{n+1}}.$$

Therefore

$$\Delta = \det\Big(f_{\underline{\tau}t}\big(s_1^{(\mu)} + s_{n+1}^{(\mu)}\beta_1, \ldots, s_n^{(\mu)} + s_{n+1}^{(\mu)}\beta_n\big)\Big)_{\substack{(\underline{\tau},t) \\ 1\le\mu\le L}},$$

and we can use Corollary 6.7:

$$\frac{1}{L}\log|\Delta| \le -\frac{n}{e}L^{1/n}\log E + c_0\big(T_0\log(SE) + T_1SE\big).$$

Step 4. Estimates for degree and height

From Lemma 3.15 we deduce that there exists a polynomial

$$f \in \mathbb{Z}[\mathrm{X}_1^{\pm1}, \ldots, \mathrm{X}_{n+1}^{\pm1}, \mathrm{Y}_1, \ldots, \mathrm{Y}_n]$$

such that

$$\Delta = f(\alpha_1, \ldots, \alpha_{n+1}, \beta_1, \ldots, \beta_n),$$

$$\deg f \le c_0L\big(T_0\log(SE) + T_1SE\big) \quad \text{and} \quad \log \mathrm{H}(f) \le c_0L\big(T_0\log(SE) + T_1SE\big)$$

This completes the proof of Proposition 6.9. □

Proof of Theorem 1.5. Let us deduce that one at least of the numbers $\alpha_1, \ldots, \alpha_{n+1}$, $\beta_1, \ldots, \beta_n$ is transcendental. Thanks to Lemma 1.7, this will imply Baker's Theorem 1.5.

By Lemma 2.1, it suffices to show that for any $\kappa > 0$, there exist parameters T_0, T_1, S and E satisfying the assumptions of Proposition 6.9 as well as

$$L^{1/n}\log E > \kappa\big(T_0\log(SE) + T_1SE\big).$$

For instance one can take

$$T_1 = [\log S]^{2n}, \qquad T_0 = [S^{1+1/n}(\log S)^{-3}] \qquad \text{and} \qquad E = e,$$

with S sufficiently large. □

Exercises

Exercise 6.1. With the notation of Lemma 6.2, prove that the five conditions (*i*) to (*iii*)′ are also equivalent to the following ones:

(*iv*) For any hyperplane V of K^n which is rational over $\mathbb{Q}$,

$$\mathrm{rk}_{\mathbb{Z}}\left(\frac{Y+V}{V}\right) \geq 2.$$

(*v*) For any $S \geq 1$ and any hyperplane $V \subset K^n$ which is rational over $\mathbb{Q}$, we have

$$\mathrm{Card}\left(\frac{Y[S]+V}{V}\right) \geq (2S+1)^2.$$

Exercise 6.2. Let S be a positive integer and $v_1, \ldots, v_{m-r}$ be $m-r$ linearly independent elements in $\mathbb{Z}^m[S]$. Show that the vector space V they span in $\mathbb{C}^m$ is contained in a hyperplane of equation $b_1 z_1 + \cdots + b_m z_m = 0$ with $(b_1, \ldots, b_m) \in \mathbb{Z}^m[S'] \setminus \{0\}$ and

$$S' = (m-r)!S^{m-r}.$$

Hint. *The absolute value of a determinant of a $N \times N$ matrix with entries of absolute values $\leq X$ is at most $N!X^N$.*

Exercise 6.3. (Quantitative version of the implication (*i*) ⇒ (*iii*)′ in Lemma 6.2)

Let K be a field of characteristic zero, S be a positive integer and V be a subspace of K^m of codimension $r \geq 1$, such that

$$\mathrm{Card}\left(\frac{\mathbb{Z}^m[S]+V}{V}\right) < (2S+1)^{r+1}.$$

Show that
1) There exists a basis $(\underline{v}_1, \ldots, \underline{v}_{m-r})$ of V with $\underline{v}_j \in \mathbb{Z}^m[2S+1]$ for $1 \leq j \leq m-r$.
2) The vector space V is intersection of r hyperplanes of equations

$$b_{i1}z_1 + \cdots + b_{im}z_m = 0 \qquad (1 \leq i \leq r),$$

where, for $1 \leq i \leq r$, $\underline{b}_i = (b_{i1}, \ldots, b_{im})$ is in $\mathbb{Z}^m[2S+1]$.

Exercise 6.4. (With the collaboration of D. Roy, W.M. Schmidt and J. Thunder).
Let $S \geq 2$ be a positive integer and V a subspace of $\mathbb{R}^m$ of codimension $r \geq 1$ satisfying the following condition (the same as in Exercise 6.3)

$$\mathrm{Card}\left(\frac{\mathbb{Z}^m[S]+V}{V}\right) < (2S+1)^{r+1}.$$

a) Show that the intersection $V \cap \mathbb{Z}^m(2S+1)$ contains more than $(2S+1)^{m-r-1}$ points and contains a basis of V.

Hint. *See Exercise 7.4.*

Hence $\Lambda = V \cap \mathbb{Z}^m$ is a *lattice* in V of dimension $m - r$.

b) Denote by B^m the unit ball in $\mathbb{R}^m$ for the Euclidean norm. Define, for $X > 0$,

$$\mathbb{R}^m(X) = \{(x_1, \ldots, x_m) \in \mathbb{R}^m \ ; \ |x_i| \le X, \ (1 \le i \le m)\}.$$

Hence $\mathbb{R}^m(X) \subset \sqrt{m}XB^m$. Finally, recall (see § 10.2.4 and [Sc 1991]) that the *determinant* $\det\Lambda$ of the lattice Λ in V is the volume of V/Λ, i.e. the volume of a fundamental domain of Λ in V. Check

$$\det\Lambda \le \left(2\sqrt{m}\right)^{m-r}\mathrm{vol}(B^{m-r})(2S+1),$$

where vol is the Euclidean volume.

Hint. *Choose a basis of V belonging to $\mathbb{Z}^m[2S+1]$, and denote by P the corresponding parallelepiped. Check that P contains a fundamental domain of V/Λ.*

Define $K = V \cap \mathbb{R}^m(2S+1)$. Check

$$\mathrm{Card}(K \cap \Lambda)\det\Lambda \le \mathrm{vol}(K+P).$$

Here, $\mathrm{vol}(K+P)$ *is the volume of $K+P$ in V (for the metric induced by the metric of $\mathbb{R}^m$). Check also*

$$\mathrm{vol}(K+P) \le \left(2\sqrt{m}(2S+1)\right)^{m-r}\mathrm{vol}(B^{m-r}).$$

c) Denote by $\|\cdot\|_2$ the Euclidean norm in $\mathbb{R}^m$. Show that there is a basis $\underline{v}_1, \ldots, \underline{v}_{m-r}$ of V, where $\underline{v}_i \in \Lambda$ satisfy

$$\|\underline{v}_1\|_2 \cdots \|\underline{v}_{m-r}\|_2 \le 2^{m-r}\left(2\sqrt{m}\right)^{m-r}(2S+1).$$

Hint. *By Minkowski's Theorem, if $\lambda_1 \le \lambda_2 \le \cdots \le \lambda_{m-r}$ are the successive minima of Λ with respect to $B^m \cap V$, then*

$$\lambda_1 \cdots \lambda_{m-r} \le \frac{2^{m-r}\det\Lambda}{\mathrm{vol}\left(B^m \cap V\right)}.$$

d) Let $V^\perp$ be the *orthogonal complement* of V in $\mathbb{R}^m$:

$$V^\perp = \{\underline{x} \in \mathbb{R}^m \ ; \ \langle \underline{x}, \underline{y}\rangle = 0 \text{ for all } \underline{y} \in V\},$$

where $\langle\ ,\ \rangle$ denotes the usual scalar product in $\mathbb{R}^m$. Then $\Lambda^\perp = \mathbb{Z}^m \cap V^\perp$ is a lattice in $V^\perp$ of dimension r, with $\det\Lambda^\perp = \det\Lambda$ (see [Sc 1991], Chap. 1). Deduce that V is intersection of r hyperplanes in $\mathbb{R}^m$ of equations

$$\langle \underline{b}_i, \underline{z}\rangle = 0, \qquad (1 \le i \le r),$$

where $\underline{b}_i = (b_{i1}, \ldots, b_{im})$ are in $\mathbb{Z}^m$ and satisfy

$$\|\underline{b}_1\|_2 \cdots \|\underline{b}_r\|_2 \le 4^{m-r}m^{(m-r)/2}(2S+1).$$

e) Give better estimates in the special case $m = 2$ and $r = 1$.

Exercise 6.5. Let E be a normed vector space of dimension n over $\mathbb{R}$, L a lattice in E, and λ a positive real number such that there exists a basis for E which consists of vectors of L of norm $\le \lambda$. Show that there exists a basis of L which consists of vectors of L of norm $\le n\lambda$.

Exercise 6.6. Let K be a field, m a positive integer, and V a vector subspace of K^m of dimension d. The following properties are equivalent.
(i) If $\pi_V\colon K^m \longrightarrow K^m/V$ is the canonical projection, then $\big(\pi_V(\underline{e}_1),\dots,\pi_V(\underline{e}_{m-d})\big)$ is a basis of K^m/V.
(ii) For $\underline{z}=(z_1,\dots,z_m)\in V$, the conditions $z_{m-d+1}=\cdots=z_m=0$ imply $\underline{z}=0$.
(iii) The restriction to V of the projection $K^m \longrightarrow K^d$ on the last d coordinates is injective.
(iv) V is intersection of $m-d$ hyperplanes of equations

$$z_j=\sum_{i=m-d+1}^{m} a_{ij}z_i,\qquad (1\le j\le m-d).$$

Exercise 6.7.
a) Let n be a positive integer, L a sufficiently large integer, $f\colon\mathbb{C}^{2n}\to\mathbb{C}$ an entire function of $2n$ complex variables, $\underline{x}_1,\dots,\underline{x}_L,\underline{y}_1,\dots,\underline{y}_L$ elements of $\mathbb{C}^n$ and r_1,r_2,R_1,R_2,E real numbers which satisfy

$$R_1\ge r_1\ge \max_{1\le\lambda\le L}|\underline{x}_\lambda|,\quad R_2\ge r_2\ge \max_{1\le\mu\le L}|\underline{y}_\mu|,\quad \max\left\{\frac{R_1}{r_1},\frac{R_2}{r_2}\right\}\ge E\ge e.$$

For $1\le\lambda\le L$ and $1\le\mu\le L$, assume that the number

$$u_{\lambda\mu}=f(\underline{x}_\lambda,\underline{y}_\mu)$$

is in $\mathbb{Z}$. Further assume

$$\log\sup\big\{|f(\underline{z},\underline{w})|\,;\,|\underline{z}|\le R_1,\ |\underline{w}|\le R_2\big\}\le \frac{n}{3}L^{1/n}\log E.$$

Show that the determinant of the matrix $\big(u_{\lambda\mu}\big)_{1\le\lambda,\mu\le L}$ is zero.

b) Let d,ℓ be positive integers and a_{ij} $(1\le i\le d,\ 1\le j\le\ell)$ positive rational numbers. Assume, for any $\underline{t}=(t_1,\dots,t_d)\in\mathbb{Z}^d\setminus\{0\}$ and any $\underline{s}=(s_1,\dots,s_\ell)\in\mathbb{Z}^\ell\setminus\{0\}$,

$$\prod_{i=1}^{d}\prod_{j=1}^{\ell}a_{ij}^{t_is_j}\ne 1.$$

Show that the rank n of the matrix $\big(\log a_{ij}\big)_{\substack{1\le i\le d\\ 1\le j\le\ell}}$ is bounded from below by

$$n\ge\frac{d\ell}{d+\ell}\cdot$$

(Compare with Theorems 1.16 and 12.17, where the numbers $e^{\lambda_{ij}}$ are algebraic, while here, a_{ij} are positive rational numbers).

Hint. *Use Exercise 1.9, Theorem 5.1 together with part a) of the present exercise for the function* $e^{\underline{z}\underline{w}}$.

7. Homogeneous Measures of Linear Independence

Three chapters (7, 9 and 10) will be devoted to measures of linear independence of logarithms of algebraic numbers. Here we prove the first of such estimates by using the method of Chap. 6. The proof is much simpler than the ones in the next chapters, and nevertheless the estimate is rather sharp.

We state the main result (Theorem 7.1) in § 7.1. In § 7.2 we prove a lower bound for the order of vanishing of an interpolation determinant which will enable us in § 7.3 to give an upper bound for the absolute value of this determinant. Using the zero estimate, a nonzero determinant is constructed in § 7.4. The transcendence argument is given in § 7.5, where a more precise result than the general case of Theorem 7.1 is established. We deduce Theorem 7.1 in § 7.6 for the general case (measure of linear independence of logarithms of algebraic numbers over the field algebraic numbers), and in § 7.7 for the rational case (measure of linear independence over the field rational numbers), introducing Fel'dman's polynomials. Finally in § 7.2 we remove the hypothesis (occurring in Theorem 7.1) that the logarithms are linearly independent over $\mathbb{Q}$.

7.1 Statement of the Measure

7.1.1 The Main Result

This chapter is devoted to the proof of the following measure of linear independence for logarithms of algebraic numbers.

Theorem 7.1. *Let $\lambda_1, \ldots, \lambda_m$ be $\mathbb{Q}$-linearly independent logarithms of algebraic numbers. For $1 \leq i \leq m$ define $\alpha_i = \exp(\lambda_i)$. Let $\beta_1, \ldots, \beta_m$ be algebraic numbers, not all of which are zero. Denote by D the degree of the number field $\mathbb{Q}(\alpha_1, \ldots, \alpha_m, \beta_1, \ldots, \beta_m)$ over $\mathbb{Q}$. Further, let $A_1, \ldots, A_m$, B and E be positive real numbers, which satisfy $E \geq e$,*

$$\log A_i \geq \max\left\{\mathrm{h}(\alpha_i), \frac{E|\lambda_i|}{D}, \frac{\log E}{D}\right\} \qquad (1 \leq i \leq m)$$

and

$$\log B \geq \max\left\{\mathrm{h}(\beta_1), \ldots, \mathrm{h}(\beta_m)\right\},$$

Furthermore, assume
(i) Either (general case)

$$B \geq \max\left\{E^{1/D},\ \frac{D}{\log E},\ 2^6 m^4 \cdot \frac{D \log A}{\log E}\right\}$$

where $A = \max\{A_1, \ldots, A_m\}$,
or else
(ii) (rational case) $(\beta_1, \ldots, \beta_m) \in \mathbb{Q}^m$, *and*

$$B \geq \max\{e,\ E^{1/D}\}.$$

Then the absolute value of the number

$$\Lambda = \beta_1\lambda_1 + \cdots + \beta_m\lambda_m$$

is bounded from below by

$$|\Lambda| > \exp\{-C(m)D^{m+2}(\log B)^2(\log A_1)\cdots(\log A_m)(\log E)^{-m-1}\},$$

with

$$C(m) = 2^{2m+7}m^{3m+8}$$

Remark 1. The conclusion of Theorem 7.1 is sharp in terms of each A_i separately: in the rational case it has the form A_i^{-C}, which is best possible (see Exercise 10.5). It can be sharpened in terms of B, replacing $(\log B)^2$ by $\log B$ (see Theorem 9.1).

For small values of m the method of this chapter yields the sharpest known numerical estimates for $C(m)$. In the special case $m = 2$, our method is closely related with Schneider's solution of Hilbert's seventh problem (§ 2.3), which has been developed in [MiW 1978] for producing quantitative measures of linear independence for two logarithms of algebraic numbers. The sharpest know estimates for two logarithms are given in [LauMN 1995]; the method of the present chapter in case $m = 2$ is very close to the proof given in [LauMN 1995]; the main difference is that we do not pay so much attention to the numerical value of $C(2)$. Our constant $C(2)$ is $> 8 \cdot 10^6$, while the corresponding constant in [LauMN 1995] (for the homogeneous rational case) is < 100. Also for $m = 3$, according to P. Voutier, the conclusion holds in the homogeneous rational case with $C(3)$ replaced by $1.1 \cdot 10^7$.

Further comments on the different available methods are postponed to §§ 10.4 and 14.4.

Remark 2. Assume that the numbers $\beta_1, \ldots, \beta_m$ are rational integers, say $\beta_i = b_i$ $(1 \leq i \leq m)$. The number

$$\Lambda = b_1\lambda_1 + \cdots + b_m\lambda_m$$

satisfies

$$|\Lambda| \geq \left|\alpha_1^{b_1}\cdots\alpha_m^{b_m} - 1\right| \cdot e^{|\Lambda|}$$

(see Exercise 1.1.a and § 9.4.4). From Liouville's inequality (Exercise 3.7.b), we deduce

$$|\Lambda| \geq 2^{-D} A^{-mDB_0},$$

where

$$A = \max_{1 \leq i \leq m} A_i \quad \text{and} \quad B_0 = \max\{|b_1|, \dots, |b_m|\}.$$

For simplicity we order $\lambda_1, \dots, \lambda_m$ so that $A_1 \leq \cdots \leq A_m$. Hence $A = A_m$, and it follows that we may assume, without loss of generality,

$$B_0 \geq \frac{C(m)}{m} \left(\frac{D \log B}{\log E} \right)^2 \cdot \frac{D \log A_1}{\log E} \cdots \frac{D \log A_{m-1}}{\log E} - \frac{\log 2}{m}.$$

In particular one may assume that the number

$$B = \max\{e, \ E^{1/D}, \ B_0\}$$

satisfies

$$B \geq \frac{C(m)}{2m}, \quad B \geq \frac{D}{\log E} \quad \text{and} \quad B \geq \frac{D \log A_{m-1}}{\log E},$$

but we may not assume that B is greater than $D(\log A_m)/\log E$.

Notice also that in the case $E = e$, the assumption $e|\lambda_i| \leq D \log A_i$ cannot be omitted in Theorem 7.1. For instance take a rational approximation a/b to $\sqrt{2}$, and choose $m = 2$, $\lambda_1 = 2i\pi a$, $\lambda_2 = 2i\pi b$, $\beta_1 = 1$, $\beta_2 = -\sqrt{2}$, $D = 2$, $A_1 = A_2 = e$, so that $|\Lambda| = 2\pi |a - b\sqrt{2}|$.

7.1.2 Sketch of Proof

A nontrivial (but rather weak) measure of linear independence can be deduced from Proposition 6.9 (where the constant c_0 can be explicitly computed). The idea is the following. We first assume that $\lambda_1, \dots, \lambda_m$ are $\mathbb{Q}$-linearly independent, that $1, \beta_1, \dots, \beta_{m-1}$ are $\mathbb{Q}$-linearly independent and that $\beta_m = -1$. Apply Proposition 6.9 with $n = m - 1$, with $\lambda_1, \dots, \lambda_n, \beta_1, \dots, \beta_n$ having the same meaning as in Theorem 7.1, but with λ_{n+1} replaced by $\lambda_m + \Lambda$, so that the hypothesis of Proposition 6.9 is satisfied. The number α_{n+1} in Proposition 6.9 is replaced by $\alpha_m e^{\Lambda}$. From Proposition 6.9 we obtain a polynomial $f \in \mathbb{Z}[X_1^{\pm 1}, \dots, X_m^{\pm 1}, Y_1, \dots, Y_{m-1}]$, of degree and height explicitly bounded, such that the number

$$f(\alpha_1, \dots, \alpha_{m-1}, \alpha_m e^{\Lambda}, \beta_1, \dots, \beta_{m-1})$$

is nonzero and has a small absolute value. From Liouville's inequality, since $\alpha_1, \dots, \alpha_m, \beta_1, \dots, \beta_{m-1}$ are algebraic, we deduce not only that $\Lambda \neq 0$, but also that $|\Lambda|$ cannot be too small (see Proposition 15.3).

The estimate which can be reached with this argument is rather weak (compared with Theorem 7.1 for instance), but is certainly not trivial and would be quite sufficient for solving several diophantine problems.

A refinement arises from the observation that not only the values of the polynomial f, but also the values of its first derivatives at the point $(\alpha_1, \ldots, \alpha_{m-1}, \alpha_m e^\Lambda, \beta_1, \ldots, \beta_{m-1})$ have small absolute values. This observation arises in the work of M. Laurent and D. Roy [LauRoy 1999a] (see Exercise 15.4).

Our approach will be slightly different. We repeat the proof of Chap. 6, but we introduce two matrices: an arithmetic one, involving algebraic numbers, and an analytic one, involving values of functions of $m-1$ variables. In Chap. 6 we had $\Lambda = 0$, and the two matrices were the same. Here, the difference between the two matrices is controlled by $|\Lambda|$.

The entries of the arithmetic matrix are the numbers

$$\gamma_{\underline{\tau}t}^{(\underline{s})} = \prod_{i=1}^{m-1} (s_i + s_m\beta_i)^{\tau_i} \prod_{j=1}^{m} \alpha_j^{s_j t}$$

which already occurred in Chap. 6. The entries of the analytic matrix are the values of the exponential monomials in $m-1$ variables

$$f_{\underline{\tau}t}(\underline{z}) = \underline{z}^{\underline{\tau}} e^{t(\beta_1 z_1 + \cdots + \beta_{m-1} z_{m-1})}$$

at the points

$$\underline{s}\eta = \bigl(s_1 + s_m\beta_1, \ldots, s_{m-1} + s_m\beta_{m-1}\bigr) \in \mathbb{C}^{m-1}.$$

The connection between both is

$$f_{\underline{\tau}t}(\underline{s}\eta) = \gamma_{\underline{\tau}t}^{(\underline{s})} e^{t s_m \Lambda};$$

this amounts to replace α_m by $\alpha_m e^\Lambda$ in the definition of $\gamma_{\underline{\tau}t}^{(\underline{s})}$.

The zero estimate shows that the arithmetic matrix has maximal rank; we extract a nonzero determinant, which is the arithmetic determinant Δ_{an}, and we also get an analytic determinant Δ_{ar} which is extracted in the same way from the analytic matrix.

From Liouville's inequality we deduce a lower bound for $|\Delta_{\mathrm{ar}}|$, while Schwarz's Lemma yields an upper bound for $|\Delta_{\mathrm{an}}|$. The difference $|\Delta_{\mathrm{ar}} - \Delta_{\mathrm{an}}|$ can easily be bounded by a multiple of $|\Lambda|$. A crude estimate for this difference yields a nontrivial but weaker measure than our Theorem 7.1, namely:

- *Under the assumptions of Theorem 7.1 with $E = e$, define*

$$H = \max\{e, B, A_1, \ldots, A_m\}.$$

Then

$$|\Lambda| \geq \exp\bigl\{-(10^3 m^3 D \log H)^{\kappa(m)}\bigr\}$$

with $\kappa(m) = 2^m (m!)^2$.

A complete proof of this estimate is worked out in [W 1992], Chap. 7.

A refinement is due to M. Laurent [Lau 1994]: he writes the expansion of this difference in powers of $|\Lambda|$ and finds that the first coefficients are pretty small, essentially as small as $|\Delta_{\rm an}|$ (they are almost interpolation polynomials, again). As a consequence, a sharp upper bound for $|\Delta_{\rm ar}|$ can be deduced.

A further refinement is the following: in the upper bound for $|\Delta_{\rm an}|$, we replace the function $\Theta_n(L)$ (where $n = m - 1$ is the number of variables) arising in § 6.2.2 by a larger function, which takes into account the fact that our alternant $\left(f_\lambda(\underline{\zeta}_\mu)\right)$ involves functions f_λ of the form

$$z_1^{\tau_1} \cdots z_n^{\tau_n} \varphi_t(\theta_1 z_1 + \cdots + \theta_n z_n)$$

where $\underline{\tau} \in \mathbb{N}^n$ and φ_t are analytic functions of a single variable (see § 7.2), while $\theta_1, \ldots, \theta_n$ are fixed complex numbers. The point is that, apart from monomials in $z_1, \ldots, z_n$, the functions f_λ depend only of one variable. There is a connection (see §§ 13.7 and 14.4) with the main idea of Baker's extrapolation argument, where derivatives are taken (in an n-dimensional space) of an auxiliary function, at several points which all lie on a complex line (of dimension one).

In the first sketch of proof we gave involving Proposition 6.9, it was necessary to assume that the numbers $\lambda_1, \ldots, \lambda_m$ are $\mathbb{Q}$-linearly independent and also that the numbers $1, \beta_1, \ldots, \beta_{m-1}$ are $\mathbb{Q}$-linearly independent. The first assumption on the λ's is not a serious restriction: we explain in § 7.8 how an induction can get rid of it (with a minimal cost). The assumption on the β's is certainly a more serious one: for instance it rules out the rational case, which is the most important for diophantine applications! There is a simple way of dealing with this issue: one does not require that $1, \beta_1, \ldots, \beta_{m-1}$ are $\mathbb{Q}$-linearly independent, but only that there is no *small* linear dependence relations between these numbers (see Exercise 7.1).

However if we want to deduce an estimate in the general case from the special case by means of an induction process, then the estimate we reach at the end is not so sharp (again the details are given in [W 1992], Lemma 7.4). Here we follow another way.

The transcendence argument shows that if a nonzero number of the shape

$$\beta_1\lambda_1 + \cdots + \beta_{m-1}\lambda_{m-1} - \lambda_m$$

has a sufficiently small absolute value, then the arithmetic matrix has not maximal rank. We deduce from the zero estimate that there exists a vector subspace $\mathcal{V}$ of $\mathbb{C}^m$, of codimension $r \geq 1$, which contains the point $(\beta_1, \ldots, \beta_{m-1}, -1)$, such that for some positive integer S the number of elements in $\left(\mathbb{Z}^m[S]+\mathcal{V}\right)/\mathcal{V}$ is relatively small. In particular we can make it smaller than $(2S + 1)^{r+1}$. By Lemma 6.2, this implies that $1, \beta_1, \ldots, \beta_{m-1}$ satisfy a linear dependence condition over $\mathbb{Q}$, and an explicit bound for the coefficients can be given (Exercise 6.3).

However this is not very efficient. It is much better to use directly the information on the upper bound for $\mathrm{Card}\Big(\left(\mathbb{Z}^m[S]+\mathcal{V}\right)/\mathcal{V}\Big)$. Instead of constructing a determinant with analytic functions in $\mathbb{C}^{m-1}$ (which we view as $\mathbb{C}^m/\mathbb{C}(\beta_1, \ldots, \beta_{m-1}, -1)$), we take analytic functions in

$$\mathcal{V}/\mathbb{C}(\beta_1, \ldots, \beta_{m-1}, -1),$$

which involve only $d = m - 1 - r$ complex variables. This is explained in § 7.4.

7.2 Lower Bound for a Zero Multiplicity

In this section we show how to improve the analytic upper bound of Proposition 6.6 for the absolute value of an alternant when functions are of a special form, namely a product of a polynomial by a function of a single variable. At the same time, we introduce another refinement, which is motivated by the next section: we derive an upper bound for the absolute value of a determinant which is not exactly an alternant, but where some of the rows have constant entries. The entries of the other rows are values of analytic functions.

Let n, T_0 and L be positive integers, $\underline{\zeta}_1, \ldots, \underline{\zeta}_L$ elements of $\mathbb{C}^n$, $\varphi_1, \ldots, \varphi_L$ analytic functions in $\mathbb{C}$, $\theta_1, \ldots, \theta_n$ complex numbers, and $p_1, \ldots, p_L$ polynomials in $\mathbb{C}[z_1, \ldots, z_n]$ of total degree $\leq T_0$. We define, for $1 \leq \lambda \leq L$,

$$f_\lambda(z_1, \ldots, z_n) = p_\lambda(z_1, \ldots, z_n)\varphi_\lambda(\theta_1 z_1 + \cdots + \theta_n z_n).$$

Let I be a subset of $\{1, \ldots, L\}$, and let $\delta_{\lambda\mu}$ $(1 \leq \lambda \leq L$ with $\lambda \notin I$, and $1 \leq \mu \leq L)$ be complex numbers. For $\lambda \in I$ and $1 \leq \mu \leq L$, we define $\delta_{\lambda\mu} = f_\lambda(\underline{\zeta}_\mu)$. We consider the matrix

$$\Big(\delta_{\lambda\mu}\Big)_{1 \leq \lambda, \mu \leq L},$$

whose determinant we denote by Δ_I.

We shall determine an upper bound for $|\Delta_{\mathrm{ar}}|$ by expressing Δ_{ar} as a sum of terms involving these Δ_I's. For this purpose we need to bound each $|\Delta_I|$ from above too. To do this we consider the following function $D_I(z)$ (of a single variable z) which is closely related to Δ_I. Let

$$d_{\lambda\mu}(z) = \begin{cases} f_\lambda(\underline{\zeta}_\mu z) & \text{for } \lambda \in I, \\ \delta_{\lambda\mu} & \text{for } \lambda \notin I. \end{cases}$$

We define

$$D_I(z) = \det\Big(d_{\lambda\mu}(z)\Big)_{1 \leq \lambda, \mu \leq L}.$$

Our first step will be to bound from below the multiplicity of the zero of $D_I(z)$ at $z = 0$. A simple application of Schwarz' Lemma will then give us an upper bound for $|D_I(z)|$, and hence for $|\Delta_I| = |D_I(1)|$.

The upper bound we shall produce depends on the following quantity:

$$\Theta(n; T_0, L) = \min\left\{\|\underline{\kappa}_1\| + \cdots + \|\underline{\kappa}_L\|\right\}$$

where the minimum runs over the L-tuples $(\underline{\kappa}_1, \ldots, \underline{\kappa}_L)$ of elements of $\mathbb{N}^n$ which are pairwise distinct and satisfy $\kappa_{\lambda 2} + \cdots + \kappa_{\lambda n} \leq T_0$ for $1 \leq \lambda \leq L$. The point is that this sum does not involve the first coordinate $\kappa_{\lambda 1}$ of

$$\underline{\kappa}_\lambda = (\kappa_{\lambda 1}, \ldots, \kappa_{\lambda n}).$$

For $n = 1$ we plainly have $\Theta(1; T_0, L) = L(L-1)/2$ (compare with Lemma 2.5).

Lemma 7.2. *The function $D_I(z)$ has a zero at $z = 0$ of multiplicity at least $\Theta(n; T_0, |I|)$, where $|I|$ is the number of elements in I.*

Proof. For $n = 1$ the argument is the same as in the proof of Lemma 2.5.

For $n \geq 2$, we note that the multiplicity of the zero of $D_I(z)$ at the origin is not affected by a change of variables in $\mathbb{C}^n$. Also such a change of variables will not modify the total degree of polynomials in $z_1, \ldots, z_n$. Therefore we may assume $\theta_2 = \cdots = \theta_n = 0$ as well as

$$p_\lambda(z_1, \ldots, z_n) = z_1^{a_{\lambda 1}} \cdots z_n^{a_{\lambda n}} \qquad (1 \leq \lambda \leq L)$$

for some $a_\lambda \in \mathbb{N}^n$.

Since the determinant is multilinear, by expanding each φ_λ in Taylor series centered at $z_1 = 0$, we can write $D_I(z)$ as a sum of determinants each of which is a constant times $z^{\|\underline{\kappa}_1\| + \cdots + \|\underline{\kappa}_{|I|}\|}$ times the determinant of the matrix $M_{\underline{\kappa}} = \big(m_{ij}\big)$ with

$$m_{ij} = \begin{cases} \zeta_{\mu 1}^{\kappa_{\lambda 1}} \cdots \zeta_{\mu n}^{\kappa_{\lambda n}} & \text{if } \lambda \in I, \\ \delta_{\lambda\mu} & \text{otherwise} \end{cases}$$

and $\underline{\kappa}_\lambda = (\kappa_{\lambda 1}, \ldots, \kappa_{\lambda n}) \in \mathbb{N}^n$. If the elements $\underline{\kappa}_1, \ldots, \underline{\kappa}_{|I|}$ are not pairwise distinct, then $\det M_{\underline{\kappa}} = 0$. In our expression for $f_\lambda(\underline{\zeta}_\mu)$, we had $a_{\lambda 1} + \cdots + a_{\lambda n} \leq T_0$ and so $a_{\lambda 2} + \cdots + a_{\lambda n} \leq T_0$. Since, by our preliminary reduction, $\varphi_\lambda(\theta_1 z_1 + \cdots + \theta_n z_n)$ depends only on z_1, only the $\kappa_{\lambda 1}$'s can increase, and $\kappa_{\lambda 2} + \cdots + \kappa_{\lambda n} = a_{\lambda 2} + \cdots + a_{\lambda n} \leq T_0$ remains valid. Therefore the smallest value of $\|\underline{\kappa}_1\| + \cdots + \|\underline{\kappa}_{|I|}\|$ for which the determinant of $M_{\underline{\kappa}}$ could be nonzero, is $\Theta(n; T_0, |I|)$. This proves Lemma 7.2. □

Here is a lower bound for the number $\Theta(n; T_0, L)$:

Lemma 7.3. *For n, T_0 and L positive integers with $n \geq 2$ we have*

$$\Theta(n; T_0, L) \geq \frac{L}{2}\left(\frac{L+1}{\binom{T_0+n-1}{n-1}} - \frac{T_0}{n} - 1\right).$$

Notice that this estimate yields a stronger result than Lemma 6.3 only if T_0 is small compared with $L^{1/n}$. In our case we shall have $L = \binom{T_0+n}{n}(2T_1 + 1)$, the main term in the final lower bound for $(1/L)\Theta(n; T_0, L)$ involves

$$\frac{L}{2\binom{T_0+n-1}{n-1}} = \frac{(T_0+n)(2T_1+1)}{2n}$$

in place of

$$\frac{n}{e} \cdot L^{1/n} \leq \frac{n}{e} \cdot (T_0 + n)(2T_1 + 1)^{1/n}.$$

In our application, T_1 will be large compared with n.

Proof. Without loss of generality we may assume that the right hand side of the conclusion is nonnegative, which means $L \geq \binom{T_0+n}{n}$.

The smallest value for the sum $\|\underline{\kappa}_1\| + \cdots + \|\underline{\kappa}_L\|$ is reached when we choose successively, for each integer $a = 0, 1, \ldots$, all points in the domain

$$\mathcal{D}_a = \left\{(x_1, \ldots, x_n) \in \mathbb{N}^n\,;\; x_2 + \cdots + x_n \leq T_0,\; x_1 + \cdots + x_n = a\right\}$$

and we stop when the total number of points reaches L. For $a \geq T_0$ the number of points in $\mathcal{D}_a$ is $\binom{T_0+n-1}{n-1}$ (once $(x_2, \ldots, x_n)$ is chosen, there is exactly one value for x_1). For $a < T_0$ the number of points in $\mathcal{D}_a$ is $\binom{a+n-1}{n-1}$ (we just forget the condition involving T_0), hence the number of points we get by varying a between 0 and, say, A (with $A \geq T_0$), is

$$(A - T_0 + 1)\binom{T_0 + n - 1}{n - 1} + \sum_{a=0}^{T_0-1}\binom{a + n - 1}{n - 1} = \left(A - T_0 + 1 + \frac{T_0}{n}\right)\binom{T_0 + n - 1}{n - 1}.$$

Therefore, if A is such that the above quantity is at most L, then

$$\Theta(n; T_0, L) \geq \sum_{a=T_0}^{A}\binom{T_0 + n - 1}{n - 1}a = \frac{1}{2}\binom{T_0 + n - 1}{n - 1}(A - T_0 + 1)(A + T_0).$$

We choose for A the largest integer such that the required condition

$$\left(A - T_0 + 1 + \frac{T_0}{n}\right)\binom{T_0 + n - 1}{n - 1} \leq L$$

is satisfied, namely

$$A = \left[\frac{L}{\binom{T_0+n-1}{n-1}} + T_0 - \frac{T_0}{n} - 1\right].$$

Then

$$\left(A - T_0 + \frac{T_0}{n} + 2\right)\binom{T_0 + n - 1}{n - 1} \geq L + 1.$$

We use this last inequality twice: on one hand we deduce

$$A - T_0 + 1 \geq \frac{L + 1}{\binom{T_0+n-1}{n-1}} - \frac{T_0}{n} - 1.$$

On the other hand, since, for $T_0 \geq 1$ and $n \geq 2$, we have

$$2T_0 - \frac{T_0}{n} - 2 + \frac{1}{\binom{T_0+n-1}{n-1}} \geq 0,$$

we deduce

$$A + T_0 \geq \frac{L}{\binom{T_0+n-1}{n-1}}.$$

Therefore

$$\Theta(n; T_0, L) \geq \frac{L}{2}(A - T_0 + 1) \geq \frac{L}{2}\left(\frac{L+1}{\binom{T_0+n-1}{n-1}} - \frac{T_0}{n} - 1\right).$$

This completes the proof of Lemma 7.3. □

7.3 Upper Bound for the Arithmetic Determinant

The basic idea, due to M. Laurent [Lau 1994], is to expand the arithmetic determinant in order to improve the upper bound.

Lemma 7.4. *Let*

$$A = \left(a_{\lambda\mu}\right)_{1\leq\lambda,\mu\leq L} \qquad \textit{and} \qquad B = \left(b_{\lambda\mu}\right)_{1\leq\lambda,\mu\leq L}$$

be two $L \times L$ matrices with complex coefficients, and let ϵ be a complex number. Define

$$\Delta = \det(A + \epsilon B).$$

Moreover, for each subset I of $\{1, \ldots, L\}$, define

$$\Delta_I = \det\left(c_{\lambda\mu}^{(I)}\right)_{1\leq\lambda,\mu\leq L} \qquad \textit{with} \qquad c_{\lambda\mu}^{(I)} = \begin{cases} a_{\lambda\mu} & \textit{if } \lambda \in I \\ b_{\lambda\mu} & \textit{if } \lambda \notin I. \end{cases}$$

Let r, χ_0, χ_1, χ_2, V be positive real numbers. Assume, for each I,

$$\log|\Delta_I| \leq -\chi_0|I|^{1+1/r} + (\chi_1 - V)|I| + \chi_2.$$

Assume also

$$|\epsilon| \leq e^{-V}.$$

Then

$$\log|\Delta| \leq -LV + \left(\frac{r}{\chi_0}\right)^r \left(\frac{\chi_1}{r+1}\right)^{r+1} + \chi_2 + L\log 2.$$

Proof. From the multilinearity of the determinant we have

$$\Delta = \sum_{I\subset\{1,\ldots,L\}} \epsilon^{L-|I|}\Delta_I.$$

Hence

$$\log|\Delta| \leq L\log 2 + \max_I\left\{\log|\Delta_I| - (L-|I|)V\right\}$$
$$\leq -LV + L\log 2 + \max_I\left\{\log|\Delta_I| + |I|V\right\}.$$

The function

$$y = x(-\chi_0 x^{1/r} + \chi_1)$$

reaches its maximum at $x = \Big(r\chi_1/\big((r+1)\chi_0\big)\Big)^r$ and the value of this maximum is

$$\left(\frac{r}{\chi_0}\right)^r \cdot \left(\frac{\chi_1}{r+1}\right)^{r+1}.$$

This completes the proof of Lemma 7.4. □

Remark. If

$$\frac{r^r\chi_1^{r+1}}{(r+1)^{r+1}\chi_0^r} \leq \frac{1}{2}LV,$$

then the conclusion can be written

$$\log|\Delta| \leq -\frac{1}{2}LV + \chi_2 + L\log 2.$$

Lemma 7.5. *Let n, T_0, L and L' be positive integers with $L' \leq L$, $\varphi_1, \dots, \varphi_{L'}$ be entire functions in $\mathbb{C}$, $\theta_1, \dots, \theta_n$ be complex numbers, and $a_{\lambda i}$ (for $1 \leq i \leq n$, $1 \leq \lambda \leq L'$) be nonnegative rational integers with $a_{\lambda 1} + \dots + a_{\lambda n} \leq T_0$. We define, for $1 \leq \lambda \leq L'$,*

$$f_\lambda(z_1, \dots, z_n) = z_1^{a_{\lambda 1}} \cdots z_n^{a_{\lambda n}} \varphi_\lambda(\theta_1 z_1 + \dots + \theta_n z_n).$$

Further let $\underline{\zeta}_1, \dots, \underline{\zeta}_L$ be elements of $\mathbb{C}^n$. Furthermore, for $L'+1 \leq \lambda \leq L$ and $1 \leq \mu \leq L$ let $\delta_{\lambda\mu}$ be a complex number. For $1 \leq \lambda \leq L'$ and $1 \leq \mu \leq L$ we define $\delta_{\lambda\mu} = f_\lambda(\underline{\zeta}_\mu)$. Finally, let $E > 1$ and $M_1, \dots, M_L$ be positive real numbers satisfying

$$M_\lambda \geq \log\sup_{|z|=E}\max_{1\leq\mu\leq L}|f_\lambda(z\underline{\zeta}_\mu)| \qquad 1 \leq \lambda \leq L',$$

$$M_\lambda \geq \log\max_{1\leq\mu\leq L}|\delta_{\lambda,\mu}| \qquad L'+1 \leq \lambda \leq L.$$

We consider the determinant

$$\Delta = \det\Big(\delta_{\lambda\mu}\Big)_{1\leq\lambda,\mu\leq L}.$$

Then we have

$$\log|\Delta| \leq -\Theta(n; T_0, L')\log E + \log(L!) + M_1 + \dots + M_L.$$

Proof. In the case $L' = L$, the result follows from Lemmas 6.1 and 7.2. The general case involves the same arguments. For $1 \leq \mu \leq L$, we define functions $d_{1\mu}(z), \dots, d_{L\mu}(z)$ of a single variable $z \in \mathbb{C}$ by

$$d_{\lambda\mu}(z) = \begin{cases} f_\lambda(\underline{\zeta}_\mu z) & \text{for } 1 \le \lambda \le L', \\ \delta_{\lambda\mu} & \text{for } L' < \lambda \le L. \end{cases}$$

This means that for $\lambda > L'$ the function $d_{\lambda\mu}$ is constant. From Lemma 7.2 we deduce that the function

$$D(z) = \det\Big(d_{\lambda\mu}(z)\Big)_{1\le\lambda,\mu\le L}$$

has a zero at the origin of multiplicity $\ge \Theta(n; T_0, L')$. We conclude the proof of Lemma 7.5 by using Schwarz Lemma like in the proof of Lemma 6.1:

$$\log|\Delta| = \log|D(1)| \le -\Theta(n; T_0, L')\log E + \log \sup_{|z|=E} |D(z)|.$$

For $|z| = E$, we plainly have

$$\log|D(z)| \le \log(L!) + M_1 + \cdots + M_L.$$

□

Here is a consequence of Lemmas 7.4 and 7.5, which will give a sharp upper bound for the absolute value of the determinant Δ_{ar} in the transcendence proof.

Proposition 7.6. *Let $T_0 \ge 0$, $T_1 > 0$ be integers and $E > 1$ a real number. Define $L = \binom{T_0+n}{n}(2T_1+1)$. Let φ_t $(t \in \mathbb{Z}, |t| \le T_1)$ be analytic functions of one variable, let $\theta_1, \ldots, \theta_n$ be complex numbers and, for $(\underline{\tau}, t) = (\tau_1, \ldots, \tau_n, t) \in \mathbb{N}^n \times \mathbb{Z}$ with $\tau_1 + \cdots + \tau_n \le T_0$ and $|t| \le T_1$, define*

$$f_{\underline{\tau}t}(z_1, \ldots, z_n) = z_1^{\tau_1} \cdots z_n^{\tau_n} \varphi_t(\theta_1 z_1 + \cdots + \theta_n z_n).$$

For the same $(\underline{\tau}, t)$, let $b_{\underline{\tau}t1}, \ldots, b_{\underline{\tau}tL}$ be complex numbers. Further, let $\underline{\zeta}_1, \ldots, \underline{\zeta}_L$ be elements in $\mathbb{C}^n$. Define

$$V = \frac{1}{2n}(T_0+n)(2T_1+1)\log E.$$

Assume that, for each $(\underline{\tau}, t)$ as above, we have a positive real number $M_{\underline{\tau}t}$ for which

$$M_{\underline{\tau}t} \ge \log \sup_{|z|=E} \max_{1\le\mu\le L} |f_{\underline{\tau}t}(z\underline{\zeta}_\mu)| \qquad M_{\underline{\tau}t} \ge \log \max_{1\le\mu\le L} |b_{\underline{\tau}t\mu}|$$

and

$$\log(2L) + M_{\underline{\tau}t} \le \frac{V}{4}.$$

Finally, let ϵ be a complex number with

$$|\epsilon| \le e^{-V}.$$

Then the determinant

$$\Delta = \det\Big(f_{\underline{\tau}t}(\underline{\zeta}_\mu) + \epsilon b_{\underline{\tau}t\mu}\Big)_{\substack{(\underline{\tau},t) \\ 1\le\mu\le L}}$$

has absolute value bounded by

$$|\Delta| \le e^{-LV/4}.$$

Proof. The set of $(\underline{\tau}, t) \in \mathbb{N}^n \times \mathbb{Z}$ satisfying $\|\underline{\tau}\| \le T_0$ and $|t| \le T_1$ has L elements. For each subset I, we define $\Delta_I = \det\big(c^{(I)}_{\underline{\tau}t\mu}\big)_{\substack{(\underline{\tau},t) \\ 1\le\mu\le L}}$ where

$$c^{(I)}_{\underline{\tau}t\mu} = \begin{cases} f_{\underline{\tau}t}(\underline{\zeta}_\mu) & \text{for } (\underline{\tau}, t) \in I, \\ b_{\underline{\tau}t\mu} & \text{for } (\underline{\tau}, t) \notin I. \end{cases}$$

From Lemmas 7.3 and 7.5, we see that the hypotheses of Lemma 7.4 are satisfied with $r = 1$ and

$$\chi_0 = \frac{1}{2}(\log E)\binom{T_0+n-1}{n-1}^{-1},$$

$$\chi_1 = V - \chi_0 + \frac{1}{2n}(T_0+n)\log E, \quad \chi_2 = \log(L!) + \sum_{\|\underline{\tau}\|\le T_0}\sum_{t=-T_1}^{T_1} M_{\underline{\tau}t}.$$

The assumption $\log(2L) + M_{\underline{\tau}t} \le V/4$ implies $\chi_2 + L\log 2 \le LV/4$. Since $T_1 \ge 1$ we have

$$\chi_1 - V < \frac{1}{2n}T_0\log E + \frac{1}{2}\log E \le \frac{V}{3}.$$

Finally, from $\chi_1 < 4V/3$ one deduce

$$\begin{aligned}\left(\frac{r}{\chi_0}\right)^r\left(\frac{\chi_1}{r+1}\right)^{r+1} &< \frac{2}{\log E}\binom{T_0+n-1}{n-1}\cdot\left(\frac{2V}{3}\right)^2 \\ &\le \frac{8nV^2L}{9(T_0+n)(2T_1+1)\log E} \\ &\le \frac{4LV}{9} < \frac{LV}{2}.\end{aligned}$$

The desired result plainly follows from Lemma 7.4. □

7.4 Construction of a Nonzero Determinant

Let K be an algebraically closed field of zero characteristic, $\alpha_1, \ldots, \alpha_{n+1}$ be nonzero elements of K, $\beta_1, \ldots, \beta_n$ be elements of K, and $T_0, T_1, S_1, \ldots, S_{n+1}$ be positive integers.

Let $\mathcal{V}$ be a vector subspace of K^{n+1} over K which contains the point $(\beta_1, \ldots, \beta_n, -1)$. We denote by $d+1$ the dimension of $\mathcal{V}$, by $\pi_{\mathcal{V}}$ the canonical map from K^{n+1} onto $K^{n+1}/\mathcal{V}$, by $(\underline{e}_1, \ldots, \underline{e}_{n+1})$ the canonical basis of K^{n+1} and we assume that $\pi_{\mathcal{V}}(\underline{e}_1), \ldots, \pi_{\mathcal{V}}(\underline{e}_{n-d})$ is a basis of $K^{n+1}/\mathcal{V}$. This means that if $\underline{z} = (z_1, \ldots, z_{n+1}) \in \mathcal{V}$ satisfies $z_{n-d+1} = \cdots = z_{n+1} = 0$, then $\underline{z} = 0$. We use the notation

$$\mathcal{V}[\underline{S}] = \mathcal{V} \cap \mathbb{Z}^{n+1}[\underline{S}].$$

We consider the following matrix

$$M_{\mathrm{ar}} = \left(\prod_{j=n-d+1}^{n} \left(s_j + s_{n+1}\beta_j\right)^{\tau_j} \prod_{i=1}^{n+1} \alpha_i^{s_i t} \right)_{\substack{(\underline{\tau},t)\\ \underline{s}}},$$

where the index of rows is $(\underline{\tau}, t) = (\tau_{n-d+1}, \ldots, \tau_n, t) \in \mathbb{N}^d \times \mathbb{Z}$ with $\tau_{n-d+1} + \cdots + \tau_n \leq T_0$ and $|t| \leq T_1$, while the index of columns is $\underline{s} \in \mathcal{V}[2(d+1)\underline{S}]$. The number of rows is $L_d = \binom{T_0+d}{d}(2T_1+1)$.

Our goal is to deduce from the zero estimate of Chap. 5 the following result:

Proposition 7.7. *Assume that $\alpha_1, \ldots, \alpha_{n+1}$ generate a multiplicative subgroup of $K^\times$ of rank $\geq n$ and that the parameters T_0, T_1 and $S_1, \ldots, S_{n+1}$ satisfy*

$$T_0 > 4S_i \qquad (1 \leq i \leq n+1)$$

and

$$(2S_1+1)\cdots(2S_{n+1}+1) > 2(n+1)T_0^n T_1.$$

Assume further that for all $s \in \mathbb{Z}$ with $0 < s \leq 4S_{n+1}$, we have $(s\beta_1, \ldots, s\beta_n, -s) \notin \mathbb{Z}^{n+1}[4\underline{S}]$. Finally, assume that

$$\mathrm{Card}\left(\frac{\mathbb{Z}^{n+1}[\underline{S}] + \mathcal{V}}{\mathcal{V}} \right) \leq \frac{n+1}{d+1} T_0^{n-d}$$

and that there is no subspace in K^{n+1}, of dimension $d'+1$ with $d' < d$, containing $(\beta_1, \ldots, \beta_n, -1)$, which satisfies this inequality with d replaced by d'. Then the matrix M_{ar} has rank L_d.

The proof of this Proposition requires some preparation. The first auxiliary lemma is a *counting argument* which will be used several times later also. It is a substitute, for the category of sets, of the relation $\dim_K(V/W) = \dim_K(V) - \dim_K(W)$ for the category of K-vector spaces.

Lemma 7.8. *Let $\mathcal{C}$ be a finite set and $f\colon \mathcal{C} \longrightarrow \mathcal{C}'$ be a mapping. Then*

$$\operatorname{Card}\mathcal{C} = \sum_{u\in f(\mathcal{C})} \operatorname{Card} f^{-1}(u).$$

Proof. The map f induces on $\mathcal{C}$ an equivalence relation with $\operatorname{Card} f(\mathcal{C})$ classes, namely

$$\{f^{-1}(u)\,;\, u \in f(\mathcal{C})\}.$$

□

From Lemma 7.8 one deduces

$$\operatorname{Card} f(\mathcal{C}) \min_{u\in f(\mathcal{C})} \operatorname{Card} f^{-1}(u) \le \operatorname{Card}\mathcal{C} \le \operatorname{Card} f(\mathcal{C}) \max_{u\in f(\mathcal{C})} \operatorname{Card} f^{-1}(u). \qquad (7.9)$$

When $\psi\colon G_1 \longrightarrow G_2$ is a homomorphism of $\mathbb{Z}$-modules and $\mathcal{C}$ a finite subset of G_1, if we define $\widetilde{\mathcal{C}} = \{\underline{\lambda} - \underline{\lambda}'\,;\, \underline{\lambda} \in \mathcal{C},\, \underline{\lambda}' \in \mathcal{C}\}$, then

$$\operatorname{Card}\psi(\mathcal{C}) \cdot \operatorname{Card}\big(\widetilde{\mathcal{C}} \cap \ker\psi\big) \ge \operatorname{Card}\mathcal{C}.$$

Indeed, one applies Lemma 7.8 to the restriction $f\colon \mathcal{C} \longrightarrow \psi(\mathcal{C})$ of ψ to $\mathcal{C}$. If $\lambda^{(1)}, \dots, \lambda^{(t)}$ are distinct elements in the same class $f^{-1}(u)$, then $0, \lambda^{(2)} - \lambda^{(1)}, \dots, \lambda^{(t)} - \lambda^{(1)}$ are distinct elements in $\widetilde{\mathcal{C}} \cap \ker\psi$.

For instance take $G_1 = \mathbb{Z}^{n+1}$, $\mathcal{C} = \mathbb{Z}^{n+1}[\underline{S}]$, and ψ is the restriction to $\mathbb{Z}^{n+1}$ of the canonical map $K^{n+1} \longrightarrow K^{n+1}/\mathcal{V}$. Since $\widetilde{\mathcal{C}}$ is contained in $\mathbb{Z}^{n+1}[2\underline{S}]$, we deduce

$$\operatorname{Card}\left(\frac{\mathbb{Z}^{n+1}[\underline{S}] + \mathcal{V}}{\mathcal{V}}\right) \operatorname{Card}\big(\mathcal{V}[2\underline{S}]\big) \ge (2S_1 + 1)\cdots(2S_{n+1} + 1).$$

Proof of Proposition 7.7. Assume that the rank of M_{ar} is less than L_d: there exists a nonzero polynomial in $K[X_{n-d+1}, \dots, X_n, Y^{\pm 1}]$, of total degree $\le T_0$ in $X_{n-d+1}, \dots, X_n$ and of degree $\le T_1$ in $Y^{\pm 1}$ which vanishes on the set $\Sigma[d+1]$, where

$$\Sigma = \left\{\left(s_{n-d+1} + s_{n+1}\beta_{n-d+1}, \dots, s_n + s_{n+1}\beta_n, \alpha_1^{s_1} \cdots \alpha_{n+1}^{s_{n+1}}\right); \underline{s} \in \mathcal{V}[2\underline{S}]\right\}.$$

We use Theorem 5.1 for the algebraic group $G = \mathbb{G}_{\mathrm{a}}^d \times \mathbb{G}_{\mathrm{m}}$, with d_0 replaced by d, $d_1 = 1$ and $D_0 = T_0$, $D_1 = T_1$. We deduce the existence of a connected algebraic subgroup G^* of G, $G^* \ne G$, satisfying the conclusion of Theorem 5.1. We can write $G^* = V \times G_1^*$, where V is a vector subspace of K^d, of dimension say δ, while G_1^* is either $\{1\}$ or $\mathbb{G}_{\mathrm{m}}$. Since

$$\mathcal{H}(G;\underline{T}) = 2(d+1)T_0^d T_1 \quad\text{and}\quad \mathcal{H}(G^*;\underline{T}) = \begin{cases} T_0^\delta & \text{if } G_1^* = \{1\},\\ 2(\delta+1)T_0^\delta T_1 & \text{if } G_1^* = \mathbb{G}_{\mathrm{m}},\end{cases}$$

(see § 5.1.1), we have

$$\mathrm{Card}\left(\frac{\Sigma+(V\times K^\times)}{V\times K^\times}\right)\le\begin{cases}2(d+1)T_0^{d-\delta}T_1 & \text{if } G_1^*=\{1\},\\ \dfrac{d+1}{\delta+1}T_0^{d-\delta} & \text{if } G_1^*=\mathbb{G}_\mathrm{m}.\end{cases}$$

We are going to prove firstly $V\neq 0$, secondly $G_1^*=\mathbb{G}_\mathrm{m}$.

We claim that the elements

$$\left\{(s_{n-d+1}+s_{n+1}\beta_{n-d+1},\dots,s_n+s_{n+1}\beta_n)\,;\ \underline{s}\in\mathcal{V}[2\underline{S}]\right\}$$

are pairwise distinct. Indeed, if this is not true, then there exists $\underline{s}\in\mathcal{V}[4\underline{S}]$ with $\underline{s}\neq 0$, $s_{n+1}\ge 0$ and

$$s_i+s_{n+1}\beta_i=0\qquad\text{for}\qquad n-d+1\le i\le n.$$

Therefore the point

$$(s_1,\dots,s_{n+1})+(s_{n+1}\beta_1,\dots,s_{n+1}\beta_n,-s_{n+1})$$

belongs to $\mathcal{V}$ and has its $d+1$ last components which vanish. Hence the first $n-d$ components also are zero, and $(s_{n+1}\beta_1,\dots,s_{n+1}\beta_n,-s_{n+1})\in\mathcal{V}[4\underline{S}]$, contrary to our assumption. This proves our claim.

From this claim we deduce, whether G_1^* is $\{1\}$ or $\mathbb{G}_\mathrm{m}$,

$$\mathrm{Card}\left(\frac{\Sigma+(0\times G_1^*)}{0\times G_1^*}\right)=\mathrm{Card}\big(\mathcal{V}[2\underline{S}]\big).$$

We derive from Lemma 7.8

$$\mathrm{Card}\left(\frac{\mathbb{Z}^{n+1}[\underline{S}]+\mathcal{V}}{\mathcal{V}}\right)\mathrm{Card}\big(\mathcal{V}[2\underline{S}]\big)\ge(2S_1+1)\cdots(2S_{n+1}+1).$$

From our choice of $\mathcal{V}$ and our hypothesis on $S_1,\dots,S_{n+1}$, we deduce

$$\frac{n+1}{d+1}T_0^{n-d}\mathrm{Card}\big(\mathcal{V}[2\underline{S}]\big)\ge(2S_1+1)\cdots(2S_{n+1}+1)>2(n+1)T_0^nT_1,$$

hence

$$\mathrm{Card}\big(\mathcal{V}[2\underline{S}]\big)>2(d+1)T_0^dT_1.$$

Therefore

$$\mathrm{Card}\left(\frac{\Sigma+(0\times G_1^*)}{0\times G_1^*}\right)>2(d+1)T_0^dT_1,$$

which implies $V\neq\{0\}$.

Assume now $G_1^*=\{1\}$. We use the assumption that $\alpha_1,\dots,\alpha_{n+1}$ generate a multiplicative group of rank $\ge n$ and we apply the counting argument of Lemma 7.8 (see Exercise 7.5.a): the number of distinct points in

$$\left\{\alpha_1^{s_1}\cdots\alpha_{n+1}^{s_{n+1}}\,;\ \underline{s}\in\mathcal{V}[2\underline{S}]\right\}$$

is at least $(4\max\{S_i\}+1)^{-1}\mathrm{Card}\big(\mathcal{V}[2\underline{S}]\big)$. The hypothesis $T_0\ge 4S_i+1$ shows that this number is greater than

$$2(d+1)T_0^{d-1}T_1.$$

Hence

$$\mathrm{Card}\left(\frac{\Sigma + (V \times \{1\})}{V \times \{1\}}\right) > 2(d+1)T_0^{d-1}T_1.$$

Since we already know that δ is at least 1, we get a contradiction. From the condition $G^* \neq G$ we conclude $G_1^* = \mathbb{G}_{\mathrm{m}}$ and $\delta < d$.

Let $\theta\colon \mathcal{V} \to K^d$ be the linear map which sends $(z_1, \ldots, z_{n+1})$ onto the point $(z_{n-d+1} + z_{n+1}\beta_{n-d+1}, \ldots, z_n + z_{n+1}\beta_n)$. Using once more the assumption that $\underline{e}_1, \ldots, \underline{e}_{n-d}$ are linearly independent modulo $\mathcal{V}$, we deduce that θ is surjective with kernel $K(\beta_1, \ldots, \beta_n, -1)$. We define $W = \theta^{-1}(V)$. Hence W is a vector subspace of $\mathcal{V}$, of dimension $\delta + 1 < d + 1$, containing $(\beta_1, \ldots, \beta_n, -1)$, such that

$$\mathrm{Card}\left(\frac{\mathcal{V}[2\underline{S}] + W}{W}\right) \le \frac{d+1}{\delta+1}T_0^{d-\delta}.$$

We apply Lemma 7.8 to the canonical map

$$\psi\colon \frac{K^{n+1}}{W} \longrightarrow \frac{K^{n+1}}{\mathcal{V}}$$

with

$$\mathcal{C} = \pi_W\left(\mathbb{Z}^{n+1}[\underline{S}]\right) = \frac{\mathbb{Z}^{n+1}[\underline{S}] + W}{W},$$

$$\psi(\mathcal{C}) = \pi_{\mathcal{V}}\left(\mathbb{Z}^{n+1}[\underline{S}]\right) = \frac{\mathbb{Z}^{n+1}[\underline{S}] + \mathcal{V}}{\mathcal{V}}$$

and

$$\widetilde{\mathcal{C}} \cap \ker\psi = \pi_W\left(\mathbb{Z}^{n+1}[2\underline{S}]\right) \cap \ker\psi = \frac{\mathcal{V}[2\underline{S}] + W}{W}.$$

We get

$$\begin{aligned} \mathrm{Card}\left(\frac{\mathbb{Z}^{n+1}[\underline{S}] + W}{W}\right) &\le \mathrm{Card}\left(\frac{\mathbb{Z}^{n+1}[\underline{S}] + \mathcal{V}}{\mathcal{V}}\right) \cdot \mathrm{Card}\left(\frac{\mathcal{V}[2\underline{S}] + W}{W}\right) \\ &\le \frac{d+1}{\delta+1}T_0^{d-\delta} \cdot \frac{n+1}{d+1}T_0^{n-d} \\ &\le \frac{n+1}{\delta+1}T_0^{n-\delta}. \end{aligned}$$

Since $\dim_K(W) = \delta + 1 < d + 1$, this contradicts the hypothesis on $\mathcal{V}$. □

7.5 The Transcendence Argument — General Case

In this section as well as in § 7.6 we use the following notation.

Let $\lambda_1, \ldots, \lambda_{n+1}$ be logarithms of nonzero algebraic numbers $\alpha_i = \exp(\lambda_i)$ $(1 \le i \le n+1)$ and $\beta_1, \ldots, \beta_n$ be algebraic numbers with $0 < \max\{|\beta_1|, \ldots, |\beta_n|\} \le 1$. Assume that the numbers $\lambda_1, \ldots, \lambda_{n+1}$ are $\mathbb{Q}$-linearly independent. By Baker's Theorem 1.5, the number

$$\Lambda = \beta_1\lambda_1 + \cdots + \beta_n\lambda_n - \lambda_{n+1}$$

is nonzero.

Let D be the degree over $\mathbb{Q}$ of the number field $\mathbb{Q}(\alpha_1, \ldots, \alpha_{n+1}, \beta_1, \ldots, \beta_n)$ and let $A_1, \ldots, A_{n+1}$, B_1 and E be positive real numbers which satisfy

$$\log A_i \ge \max\left\{\mathrm{h}(\alpha_i), \frac{E|\lambda_i|}{D}\right\} \qquad (1 \le i \le n+1)$$

and

$$\mathrm{h}(\beta_1 : \cdots : \beta_n : 1) \le \log B_1, \qquad e \le E \le B_1^D.$$

Theorem 7.10. *Let T_0, T_1 and $S_1, \ldots, S_{n+1}$ be $n+3$ positive rational integers satisfying the following conditions:*

$$T_0 > 4 \max_{1 \le i \le n+1} S_i$$

and

$$(2S_1 + 1) \cdots (2S_{n+1} + 1) > 2(n+1)T_0^n T_1. \tag{7.11}$$

Define

$$V = \frac{1}{2n}(T_0 + 1)(2T_1 + 1)\log E$$

and assume

$$\frac{V}{4} \ge DT_0 \log\big(4e(n+1)B_1 S\big) + \tag{7.12}$$

$$2(n+1)D(T_1 + 1)\sum_{i=1}^{n+1} S_i \log A_i + D\log L + \log T_1$$

where $L := \binom{T_0+n}{n}(2T_1 + 1)$ and $S = \max_{1 \le i \le n+1} S_i$. Then

$$|\Lambda| > e^{-nV}.$$

Proof.

Step 1. Liouville's inequality

We begin with an easy case, when there exists such a rational integer $s \in \mathbb{Z}$ with $0 < s \le 4S_{n+1}$ and

$$(s\beta_1, \ldots, s\beta_n, -s) \in \mathbb{Z}^{n+1}[4\underline{S}].$$

In this case we write $b_i = s\beta_i$, $(1 \le i \le n)$ and $b_{n+1} = -s$. Hence $s\Lambda = b_1\lambda_1 + \cdots + b_{n+1}\lambda_{n+1}$ and $b_i \in \mathbb{Z}$ with $|b_i| \le 4S_i$. We use Liouville's estimate (Exercise 3.7.b):

$$s|\Lambda| \ge 2^{-D} \exp\left\{-4D\sum_{i=1}^{n+1} S_i \mathrm{h}(\alpha_i)\right\}.$$

This gives

$$\begin{aligned}\log|\Lambda| &\ge -D\log 2 - 4D\sum_{i=1}^{n+1} S_i \log A_i - \log(4S_{n+1})\\ &> -nV,\end{aligned}$$

which is much stronger than our ultimate goal.

Therefore we shall now assume

$$(s\beta_1, \ldots, s\beta_n, -s) \notin \mathbb{Z}^{n+1}[4\underline{S}]$$

for $0 < s \le 4S_{n+1}$.

This condition will be needed twice. Firstly it occurs in the assumption of Proposition 7.7. Secondly it enables us to check

$$\begin{aligned}\mathrm{Card}\Big\{\big(s_1 + s_{n+1}\beta_1, \ldots, s_n + s_{n+1}\beta_n\big);\ \underline{s} \in \mathbb{Z}^{n+1}[\underline{S}]\Big\} &> 2(n+1)T_0^n T_1\\ &> (n+1)T_0^n.\end{aligned}$$

Indeed from Dirichlet's box principle, we deduce that the points

$$\big(s_1 + s_{n+1}\beta_1, \ldots, s_n + s_{n+1}\beta_n\big) \in \mathbb{C}^n \qquad \big(\underline{s} \in \mathbb{Z}^{n+1}[\underline{S}]\big)$$

are pairwise distinct. Using the lower bound (7.11) we obtain

$$\mathrm{Card}\big(\mathbb{Z}^{n+1}[\underline{S}]\big) = (2S_1 + 1)\cdots(2S_{n+1} + 1) > 2(n+1)T_0^n T_1.$$

This fact will be needed to check that some integer d (introduced in step 2) is at least 1. The analytic argument (step 4) will involve complex functions of d variables.

Step 2. Choice of $\mathcal{V}$

We remark that there exist vector subspaces $\mathcal{V}$ of $\mathbb{C}^{n+1}$ which contain the point $(\beta_1, \ldots, \beta_n, -1)$ and also satisfy

$$\mathrm{Card}\left(\frac{\mathbb{Z}^{n+1}[\underline{S}] + \mathcal{V}}{\mathcal{V}}\right) \le \frac{n+1}{d+1} T_0^{n-d}$$

with $d = \dim_{\mathbb{C}}(\mathcal{V}) - 1$. Indeed $\mathcal{V} = \mathbb{C}^{n+1}$ is such a space. Among them, we choose one (which we call $\mathcal{V}$) of minimal dimension $d+1$. By step 1, the image of $\mathbb{Z}^{n+1}[\underline{S}]$ under the mapping

$$\begin{array}{ccc} \mathbb{C}^{n+1} & \longrightarrow & \mathbb{C}^n \\ \underline{z} & \longmapsto & \big(z_1 + z_{n+1}\beta_1, \dots, z_n + z_{n+1}\beta_n\big) \end{array}$$

has more than $(n+1)T_0^n$ elements, hence $\mathcal{V} \neq \mathbb{C}(\beta_1, \dots, \beta_n, -1)$, which means $d \geq 1$.

Let $\pi_{\mathcal{V}}$ denote the canonical map from $\mathbb{C}^{n+1}$ onto $\mathbb{C}^{n+1}/\mathcal{V}$ and let $\underline{e}_1, \dots, \underline{e}_{n+1}$ be the canonical basis of $\mathbb{C}^{n+1}$. Since $\mathcal{V} \ni (\beta_1, \dots, \beta_n, -1)$, we have

$$\pi_{\mathcal{V}}(\underline{e}_{n+1}) = \beta_1 \pi_{\mathcal{V}}(\underline{e}_1) + \cdots + \beta_n \pi_{\mathcal{V}}(\underline{e}_n),$$

hence there exists a basis of $\mathbb{C}^{n+1}/\mathcal{V}$ of the form $\big(\pi_{\mathcal{V}}(\underline{e}_{i_1}), \dots, \pi_{\mathcal{V}}(\underline{e}_{i_{n-d}})\big)$, with $1 \leq i_1 < \cdots < i_{n-d} \leq n$. For ease of notation we shall assume $\{i_1, \dots, i_{n-d}\} = \{1, \dots, n-d\}$.

Writing $\pi_{\mathcal{V}}(\underline{e}_i)$ in terms of $\pi_{\mathcal{V}}(\underline{e}_1), \dots, \pi_{\mathcal{V}}(\underline{e}_{n-d})$, we see that there exist $(n-d)(d+1)$ complex numbers $u_i^{(j)}$ such that

$$\underline{e}_i + \sum_{j=1}^{n-d} u_i^{(j)} \underline{e}_j \in \mathcal{V} \qquad \text{for} \quad n-d+1 \leq i \leq n+1.$$

These $d+1$ elements of $\mathcal{V}$ can be written

$$\big(u_i^{(1)}, \dots, u_i^{(n-d)}, 0, \dots, 0, 1, 0, \dots, 0\big) \qquad (n-d+1 \leq i \leq n+1)$$

and they form a basis of $\mathcal{V}$. One deduces that $\mathcal{V}$ is intersection of $n-d$ hyperplanes

$$z_j = \sum_{i=n-d+1}^{n+1} u_i^{(j)} z_i \qquad (1 \leq j \leq n-d).$$

We define

$$\theta_i = \lambda_i + \sum_{j=1}^{n-d} u_i^{(j)} \lambda_j \qquad (n-d+1 \leq i \leq n+1).$$

Then, for $\underline{z} \in \mathcal{V}$, we have

$$\sum_{i=n-d+1}^{n+1} z_i \theta_i = \sum_{j=1}^{n+1} z_j \lambda_j.$$

In particular, since $(\beta_1, \dots, \beta_n, -1)$ is in $\mathcal{V}$,

$$\sum_{i=n-d+1}^{n} \beta_i \theta_i = \theta_{n+1} + \Lambda.$$

Let $V = \mathcal{V} \cap \mathbb{Z}^{n+1}$.

Step 3. Lower bound for $|\Delta_{\mathrm{ar}}|$

Thanks to Proposition 7.7, we know that the matrix M_{ar} has rank $L_d = \binom{T_0+d}{d}(2T_1+1)$. Therefore there exist L_d elements $\underline{s}^{(1)}, \dots, \underline{s}^{(L_d)}$ in $V[2(n+1)\underline{S}]$ such that, if we define

$$\gamma_{\underline{\tau}t}^{(\mu)} = \prod_{j=n-d+1}^{n} \left(s_j^{(\mu)} + s_{n+1}^{(\mu)}\beta_j\right)^{\tau_j} \prod_{i=1}^{n+1} \alpha_i^{s_i^{(\mu)} t} \qquad (1 \le \mu \le L_d),$$

then the $L_d \times L_d$ determinant

$$\Delta_{\mathrm{ar}} = \det\left(\gamma_{\underline{\tau}t}^{(\mu)}\right)_{\substack{(\underline{\tau},t)\\ \mu}}$$

is not zero. As in § 7.4 above, $(\underline{\tau}, t)$ runs over the elements $(\tau_{n-d+1}, \ldots, \tau_n, t)$ in $\mathbb{N}^d \times \mathbb{Z}$ with $\|\underline{\tau}\| \le T_0$ and $|t| \le T_1$, while μ ranges over $\{1, \ldots, L_d\}$.

From Liouville's inequality we deduce (see Exercise 3.8, but replace S_i by $2(n+1)S_i$):

$$\frac{1}{L_d} \log |\Delta_{\mathrm{ar}}| > -U_1$$

with

$$U_1 = (D-1)\big(T_0 \log(4(n+1)S) + \log L_d\big) + DT_0 \log B_1 + 2(n+1)D(T_1+1)\sum_{i=1}^{n+1} S_i \log A_i.$$

Step 4. Analytic argument

For each $(\underline{\tau}, t) = (\tau_{n-d+1}, \ldots, \tau_n, t)$ in $\mathbb{N}^d \times \mathbb{Z}$ with $\|\underline{\tau}\| \le T_0$ and $|t| \le T_1$, we define a function $f_{\underline{\tau}t}$ of d complex variables:

$$f_{\underline{\tau}t}(z_{n-d+1}, \ldots, z_n) = \prod_{i=n-d+1}^{n} \left(z_i^{\tau_i} e^{t\theta_i z_i}\right).$$

For $\underline{s} \in \mathcal{V} \cap \mathbb{Z}^{n+1}$, if we set

$$\underline{\xi}_{\underline{s}} = (z_{n-d+1}, \ldots, z_n) = \left(s_{n-d+1} + s_{n+1}\beta_{n-d+1}, \ldots, s_n + s_{n+1}\beta_n\right) \in \mathbb{C}^d,$$

we have

$$\sum_{i=n-d+1}^{n} \theta_i z_i = \sum_{j=1}^{n+1} s_j \lambda_j + s_{n+1}\Lambda.$$

Hence for $z \in \mathbb{C}$ we have

$$f_{\underline{\tau}t}(z\underline{\xi}_{\underline{s}}) = z^{\|\underline{\tau}\|} \left(\prod_{i=n-d+1}^{n} \left(s_i + s_{n+1}\beta_i\right)^{\tau_i}\right) \cdot \left(\prod_{j=1}^{n+1} e^{ts_j\lambda_j z}\right) \cdot e^{ts_{n+1}\Lambda z}.$$

We define $\underline{\zeta}_1, \ldots, \underline{\zeta}_{L_d}$ in $\mathbb{C}^d$ by $\underline{\zeta}_\mu = \underline{\xi}_{\underline{s}^{(\mu)}}$:

$$\underline{\zeta}_\mu = \left(s_{n-d+1}^{(\mu)} + s_{n+1}^{(\mu)}\beta_{n-d+1}, \ldots, s_n^{(\mu)} + s_{n+1}^{(\mu)}\beta_n\right) \qquad (1 \le \mu \le L_d),$$

so that

$$f_{\underline{\tau}t}(\underline{\zeta}_\mu) = \gamma_{\underline{\tau}t}^{(\mu)} e^{ts_{n+1}^{(\mu)}\Lambda}.$$

Let us check the hypotheses of Proposition 7.6 with $\epsilon = \Lambda$, n replaced by d, L by L_d, V by

$$V_d = \frac{1}{2d}(T_0 + d)(2T_1 + 1)\log E,$$

with

$$b_{\underline{\tau}t\mu} = \gamma_{\underline{\tau}t}^{(\mu)}\left(1 - e^{t\Lambda s_{n+1}^{(\mu)}}\right)\Lambda^{-1}$$

and finally with

$$M_{\underline{\tau}t} = T_0 \log(4(n+1)ES) + 2(n+1)D(T_1+1)\sum_{i=1}^{n+1} S_i \log A_i + \log\frac{T_1}{2}.$$

Since $V > \log\bigl(2(n+1)T_1 S_{n+1} E\bigr)$, without loss of generality we may assume

$$2(n+1)|\Lambda| T_1 S_{n+1} E < 1.$$

Recall the assumptions $|\beta_i| \le 1$ and $E|\lambda_i| \le D\log A_i$. For $z \in \mathbb{C}$ with $|z| \le E$, we have

$$\begin{aligned}\log|f_{\underline{\tau}t}(z\underline{\zeta}_\mu)| &\le T_0\log\bigl(4(n+1)ES\bigr) + 2(n+1)T_1 E\left(\sum_{i=1}^{n+1} S_i|\lambda_i| + S_{n+1}|\Lambda|\right)\\ &\le T_0\log\bigl(4(n+1)ES\bigr) + 2(n+1)T_1 E\sum_{i=1}^{n+1} S_i\log A_i + 1\\ &\le M_{\underline{\tau}t}.\end{aligned}$$

We use the estimate in Exercise 1.1.a, with $r = 1$, $z_1 = ts_{n+1}^{(\mu)}\Lambda$, $z_2 = 0$. Since $|z_1| < 1/7$ we have

$$(e-1)e^{|z_1|} < 2,$$

and we get

$$|e^{z_1} - 1| \le 2|z_1|,$$

which yields

$$|b_{\underline{\tau}t\mu}| \le 2\bigl|\gamma_{\underline{\tau}t}^{(\mu)} ts_{n+1}^{(\mu)}\bigr|.$$

This enables us to bound $|b_{\underline{\tau}t\mu}|$ in the same way as $|f_{\underline{\tau}t}(z\underline{\zeta}_\mu)|$, but with E replaced by 1. More precisely, using the estimate

$$\log\bigl(8(n+1)S_{n+1}\bigr) \le T_0\log E + 2(n+1)S_{n+1}$$

as well as the hypothesis $\bigl|\log|\alpha_i|\bigr| \le D\log A_i$ we deduce

$$\begin{aligned}\log|b_{\underline{\tau}t\mu}| &\le T_0\log\bigl(4(n+1)S\bigr) + 2(n+1)DT_1\sum_{i=1}^{n+1} S_i\log A_i + \log\bigl(4(n+1)T_1 S_{n+1}\bigr)\\ &\le M_{\underline{\tau}t}.\end{aligned}$$

Since $D \log B_1 \geq \log E$ and $V \leq V_d$, from (7.12) we conclude

$$\log(2L_d) + M_{\underline{\tau t}} \leq \frac{1}{4} V_d \qquad \text{and} \qquad U_1 < \frac{1}{4} V_d.$$

Step 5. Conclusion of the proof

From step 3 we deduce that the conclusion of Proposition 7.6 is not satisfied. Therefore

$$|\Lambda| > e^{-V_d} \geq e^{-nV}.$$

This completes the proof of Theorem 7.10.

□

7.6 Proof of Theorem 7.1 — General Case

We shall deduce from Theorem 7.10 an intermediate result, from which we shall then deduce the so-called *general case* of Theorem 7.1.

7.6.1 A Consequence of Theorem 7.10

Recall the assumptions at the beginning of § 7.5.

Corollary 7.13. *Assume further*

$$E \leq A_i^D \quad \textit{for} \quad 1 \leq i \leq n+1.$$

Let N_0 be a positive integer, N and C_0 positive real numbers satisfying the following conditions:

$$2 + \frac{2(n+1)^2}{N}\left(1 + \frac{1}{N_0}\right) + \frac{1}{200} + \frac{1}{8n} \leq \frac{N_0}{4n}$$

and

$$C_0 \geq 2^{-n}(n+1)N^{n+1}N_0^{n+2}\left(1 + \frac{1}{8N_0}\right).$$

Assume also

$$\frac{B_1}{\log B_1} \geq \frac{4e(n+1)C_0}{(N_0-1)N} \cdot \frac{D^{n+1}(\log A)^n}{(\log E)^{n+1}}.$$

Then

$$|\Lambda| > \exp\left\{-(N_0+1)C_0 D^{n+3}(\log B_1)^2(\log A_1)\cdots(\log A_{n+1})(\log E)^{-n-2}\right\}.$$

Proof.

Step 1. The parameters are not too small

As a preliminary remark we deduce from the hypotheses of Corollary 7.13

$$N_0 > 8n, \quad N_0 N > 8n(n+1)^2,$$

hence

$$\frac{C_0}{N_0 N} > 2^{4n+1} n^n N_0 \quad \text{and} \quad C_0 > 2^{4n+6} n^n N_0.$$

Step 2. Choice of parameters

Define a real number U by

$$U = C_0 D^{n+3} (\log B_1)^2 (\log A_1) \cdots (\log A_{n+1}) (\log E)^{-n-2}$$

and rational integers $T_0, T_1, S_1, \ldots, S_{n+1}$ by

$$T_0 = \left[\frac{U}{D \log B_1}\right] \qquad T_1 = \left[\frac{N_0 D \log B_1}{\log E}\right],$$

$$S_i = \left[\frac{U}{N D T_1 \log A_i}\right] \qquad (1 \le i \le n+1).$$

From the assumptions $D \log B_1 \ge \log E$ and $D \log A_j \ge \log E$ we deduce

$$T_0 \ge C_0, \quad T_1 \ge N_0 \quad \text{and} \quad S_j \ge \frac{C_0}{N_0 N} - 1.$$

Step 3. The conditions $T_0 > 4S_j$ and $T_1 > 6$ of Theorem 7.11 are clearly satisfied. Moreover from $T_1 \ge N_0$ we deduce

$$T_1 + 1 \le \left(1 + \frac{1}{N_0}\right) T_1,$$

$$T_1 > \left(1 - \frac{1}{N_0}\right)(T_1 + 1) > (N_0 - 1) \frac{D \log B_1}{\log E}$$

and

$$2T_1 + 1 > \left(2 - \frac{1}{N_0}\right)(T_1 + 1) > \frac{(2N_0 - 1) D \log B_1}{\log E}.$$

In particular the number

$$V = \frac{1}{2n}(T_0 + 1)(2T_1 + 1) \log E$$

satisfies

$$V > \frac{2N_0 - 1}{2n} \cdot U.$$

Step 4. We check (7.12)

For $1 \le j \le n+1$ we have

$$\begin{aligned} S_j &\le \frac{U}{NDT_1 \log A_j} \\ &\le \frac{U \log E}{(N_0-1)ND^2(\log B_1)(\log A_j)} \\ &\le \frac{C_0}{(N_0-1)N} \cdot D^{n+1}(\log A)^n(\log B_1)(\log E)^{-n-1}, \end{aligned}$$

hence

$$4e(n+1)S_j \le \frac{4e(n+1)C_0}{(N_0-1)N} \cdot D^{n+1}(\log A)^n(\log B_1)(\log E)^{-n-1} \le B_1$$

and

$$DT_0 \log\bigl(4e(n+1)B_1 S\bigr) \le 2U.$$

Next we have

$$\begin{aligned} D(T_1+1)\sum_{i=1}^{n+1} S_i \log A_i &\le (n+1)\left(1+\frac{1}{N_0}\right) \max_{1\le i\le n+1} DT_1 S_i \log A_i \\ &\le (n+1)\left(1+\frac{1}{N_0}\right) \cdot \frac{U}{N}, \end{aligned}$$

hence

$$2(n+1)D(T_1+1)\sum_{i=1}^{n+1} S_i \log A_i \le 2(n+1)^2\left(1+\frac{1}{N_0}\right) \cdot \frac{U}{N}.$$

We finally need to estimate $D\log L + \log T_1$. One easily checks (with $A = \max_{1\le j\le n+1} A_j$)

$$\begin{aligned} L &\le (T_0+n)^n(2T_1+1) \\ &\le 2(n+1)^{n+1}T_0^n T_1 \\ &\le 2(n+1)^{n+1}N_0 \cdot \frac{U^n}{(D\log B_1)^{n-1}\log E} \\ &\le 2(n+1)^{n+1}N_0C_0^n \left(\frac{D\log B_1}{\log E}\right)^{n+1}\left(\frac{D\log A}{\log E}\right)^n. \end{aligned}$$

On the other hand we have

$$\frac{U}{D} \ge C_0\left(\frac{D\log B_1}{\log E}\right) \quad \text{and} \quad \frac{U}{D} \ge \left(\frac{D\log A}{\log E}\right).$$

One deduces

$$D\log L + \log T_1 < \frac{U}{200}.$$

Step 5. We check (7.11)

We need a lower bound for $(2S_1+1)\cdots(2S_{n+1}+1)$. Since

$$S_i \geq \frac{C_0}{N_0 N} - 1$$

we have

$$2S_i + 1 > 2(1-\eta)(S_i+1) > \frac{2(1-\eta)U}{NDT_1 \log A_i}$$

where

$$\eta = \frac{N_0 N}{C_0 - 1} \leq \frac{1}{2^5 n^n N_0 - 1}.$$

We obtain

$$(2S_1+1)\cdots(2S_{n+1}+1) > \frac{2^{n+1}(1-\eta)^{n+1}U^{n+1}}{\left(NDT_1\right)^{n+1}(\log A_1)\cdots(\log A_{n+1})}.$$

Notice the estimate

$$(1-\eta)^{n+1}\left(1+\frac{1}{8N_0}\right) > 1.$$

On the other hand we have the upper bounds

$$T_0^n T_1 \leq \frac{U^n T_1}{(D\log B_1)^n}$$

and

$$T_1 \leq N_0 D(\log B_1)(\log E)^{-1}.$$

From our condition on C_0 together with the definition of U we deduce

$$U \geq \frac{(n+1)N^{n+1}}{2^n}\cdot\left(1+\frac{1}{8N_0}\right)\cdot D(\log A_1)\cdots(\log A_{n+1})T_1^{n+2}(\log B_1)^{-n}$$

and

$$\begin{aligned}(2S_1+1)\cdots(2S_{n+1}+1) &> \frac{2^{n+1}U^{n+1}}{\left(1+\frac{1}{8N_0}\right)\left(NDT_1\right)^{n+1}(\log A_1)\cdots(\log A_{n+1})}\\ &\geq \frac{2(n+1)U^n T_1}{(D\log B_1)^n}\\ &\geq 2(n+1)T_0^n T_1.\end{aligned}$$

Obviously (7.11) follows.

Step 6. End of the proof

The conclusion of Corollary 7.13 follows from the estimates

$$(T_0+1)(2T_1+1) \leq 2\left(1+\frac{1}{C_0}\right)\left(1+\frac{1}{2N_0}\right)T_0T_1 \leq 2\left(1+\frac{1}{N_0}\right)T_0T_1$$

which imply

$$nV < \left(1 + \frac{1}{N_0}\right) T_0 T_1 \log E \leq (N_0 + 1)U.$$

□

7.6.2 End of the Proof in the General Case

We complete the proof of the general case of Theorem 7.1. We assume the hypotheses of that statement are satisfied.

Step 1. Further assumptions

Until step 4 we assume that $\beta_m = -1$ and that $|\beta_i| \leq 1$ for $1 \leq i \leq m-1$. We write $n = m - 1$ so that

$$\Lambda = \beta_1 \lambda_1 + \cdots + \beta_n \lambda_n - \lambda_{n+1}.$$

We shall prove the desired result with a slightly better value for the constant, namely with the constant $C(m)$ replaced by

$$C'(n) = (4n^2 + 13n + 1) 2^{2n+5} (n+1)^{3n+9}.$$

Step 2. Choice of parameters

There are many possibilities for the choice of N_0 and N. We give an example without trying to optimize:

$$N = 2(n+1), \qquad N_0 = n(4n+13), \qquad C_0 = 2^{2n+5}(n+1)^{3n+7}.$$

In this case we have

$$2 + \frac{2(n+1)^2}{N}\left(1 + \frac{1}{N_0}\right) + \frac{1}{200} + \frac{1}{8n} = n + 3 + \frac{n+1}{n(4n+13)} + \frac{1}{200} + \frac{1}{8n}$$

and

$$\frac{N_0}{4n} = n + 3 + \frac{1}{4}\cdot$$

One easily checks

$$\frac{n+1}{n(4n+13)} + \frac{1}{200} + \frac{1}{8n} < \frac{1}{4}\cdot$$

On the other hand

$$2^{-n}(n+1)N^{n+1}N_0^{n+2} = 2n^{n+2}(n+1)^{n+2}(4n+13)^{n+2}$$

and

$$n^{n+2}(4n+13)^{n+2}\left(1 + \frac{1}{8N_0}\right) \leq 2^{2n+4}(n+1)^{2n+5}.$$

Step 3. Choice of B_1

We set $B_1 = B^{n+2}$ (a larger exponent would increase $C'(n)$ but would enable us to reduce the value of the number $2^6(n+1)^4$ occurring in the conditions on B in the general case of Theorem 7.1). The condition $B_1^D \geq E$ is clearly satisfied. Since

$$\log B \geq \max_{1\leq i\leq n} \mathrm{h}(\beta_i) \quad \text{we have} \quad \log B_1 \geq \mathrm{h}(1\colon \beta_1\colon \cdots \colon \beta_n)$$

(see Exercise 3.3.a). Next using

$$B \geq \frac{D}{\log E} \quad \text{and} \quad B \geq 2^6(n+1)^4 \frac{D \log A}{\log E}$$

we deduce

$$B^{n+1} \geq 2^{6n}(n+1)^{4n} \left(\frac{D}{\log E}\right) \left(\frac{D \log A}{\log E}\right)^n.$$

On the other hand from $N_0 - 1 = 4n^2 + 13n - 1 \geq 4(n+1)^2$ one deduces

$$\frac{4e(n+1)C_0}{(N_0-1)N} \leq e \cdot 2^{2n+4}(n+1)^{3n+5}.$$

Hence in order to check the condition

$$\frac{B_1}{\log B_1} \geq \frac{4e(n+1)C_0}{(N_0-1)N} \cdot \frac{D^{n+1}(\log A)^n}{(\log E)^{n+1}},$$

it remains to check

$$2^{4n-4}(n+1)^{n-5} B \geq e(n+2) \log B.$$

Since $B \geq 2^6(n+1)^4$ it suffices to use

$$2^{4n+2}(n+1)^{n-1} \geq e(n+2)\big(6 \log 2 + 4 \log(n+1)\big).$$

Step 4. Conclusion of the proof in the case $\beta_m = -1$, $|\beta_i| \leq 1$

Notice that

$$(n+2)^2 C_0(N_0+1) = (4n^2+13n+1)2^{2n+5}(n+1)^{3n+7}(n+2)^2.$$

Denote this number by $C'(n)$. Applying Corollary 7.13, we get the conclusion of Theorem 7.1 with $C(m)$ replaced by $C'(m-1)$ (recall $m = n+1$). It is useful for the last step to notice that we have not used the full force of the hypothesis $\max \mathrm{h}(\beta_i) \leq \log B$, but only the weaker condition $\mathrm{h}(1\colon \beta_1\colon \cdots \colon \beta_n) \leq n \log B$.

Step 5 Removing the Extra Assumption of Step 1

The proof of Theorem 7.1 is complete under the extra assumptions $\beta_m = -1$ and $|\beta_i| \leq 1$ for $1 \leq i \leq m-1$ of Step 1, and in this case the constant $C(m)$ can be replaced by $C'(m)$. We now remove these assumptions.

Since the result is symmetric in $\beta_1, \ldots, \beta_m$, without loss of generality we may assume $|\beta_m| \geq \max_{1\leq i<m} |\beta_j|$. Define $\beta_j' = -\beta_j/\beta_m$ $(1 \leq j \leq m)$ and

$$\Lambda' = \beta_1'\lambda_1 + \cdots + \beta_{m-1}'\lambda_{m-1} - \lambda_m.$$

The assumption $\max_{1\le i\le m} \mathrm{h}(\beta_i) \le \log B$ implies $\mathrm{h}(1:\beta_1':\cdots:\beta_{m-1}') \le (m-1)\log B$, and since $\beta_m \neq 0$, Liouville's inequality (3.13) gives

$$|\beta_m| \ge B^{-D}.$$

Since $\Lambda = -\beta_m\Lambda'$, and since Λ' satisfies the conditions $\beta_m' = -1$ and $|\beta_i'| \le 1$, we deduce from step 4

$$|\Lambda| \ge |\beta_m\Lambda'| \ge B^{-D}\exp\bigl\{-C'(m-1)D^{m+2}(\log B)^2(\log A_1)\cdots(\log A_m)\bigr\}.$$

Since

$$1 + C'(m-1) \le 2^{2m+7}m^{3m+8} = C(m),$$

the conclusion of Theorem 7.1 follows. □

7.7 The Rational Case: Fel'dman's Polynomials

Our goal in this section is to complete the rational case of Theorem 7.1.

We need a variant of Theorem 7.10 related to the rational case. We introduce the following notation which will be valid for Theorem 7.14 and Corollary 7.17.

Let $\lambda_1,\ldots,\lambda_{n+1}$ be $\mathbb{Q}$-linearly independent logarithms of nonzero algebraic numbers and $b_1,\ldots,b_{n+1}$ be rational integers with $b_{n+1}\neq 0$. Define $\alpha_i = \exp(\lambda_i)$ $(1\le i\le n+1)$ and

$$\Lambda = b_1\lambda_1 + \cdots + b_{n+1}\lambda_{n+1}.$$

Denote by D the degree of the number field $\mathbb{Q}(\alpha_1,\ldots,\alpha_{n+1})$ over $\mathbb{Q}$. Let $A_1,\ldots,A_{n+1}$, B_1 and E be positive real numbers, which satisfy

$$\log A_i \ge \max\left\{\mathrm{h}(\alpha_i),\ \frac{E|\lambda_i|}{D}\right\} \qquad (1\le i\le n+1) \quad\text{and}\quad E\ge e.$$

Theorem 7.14. *Let T_0, T_1 and $S_1,\ldots,S_{n+1}$ be positive rational integers satisfying the following conditions:*

$$T_0 > 4S_i \qquad (1\le i\le n+1)$$

and

$$(2S_1+1)\cdots(2S_{n+1}+1) > 2(n+1)T_0^nT_1.$$

Define

$$V = \frac{1}{2n}(T_0+1)(2T_1+1)\log E,$$

$$L = \binom{T_0+n}{n}(2T_1+1),$$

$$B_1 = E^{1/D} \max_{1 \le j \le n} \left\{ 1 + \frac{2n(n+1)}{T_0} \big(|b_{n+1}| S_j + |b_j| S_{n+1} \big) \right\}$$

and assume

$$\frac{V}{4} \ge DT_0 \log(eB_1) + 2(n+1)D(T_1+1) \sum_{i=1}^{n+1} S_i \log A_i + D \log L + \log T_1 .$$

Then

$$|\Lambda| > e^{-nV}.$$

In order to omit the condition

$$B \ge 2^6 m^4 \cdot \frac{D \log A}{\log E}$$

which occurred in the general case, the idea, arising in the work of N. I. Fel'dman (see § 10.4.1), is to replace, in the transcendence part of the proof, the numbers

$$(s_j + s_{n+1}\beta_j)^{\tau_j}$$

by binomial coefficients like

$$\binom{|s_j b_{n+1} - s_{n+1} b_j|}{\tau_j}.$$

More precisely, when τ is a nonnegative integer and z a complex number, we define

$$\Delta(z;\tau) = \frac{1}{\tau!}(z+1)\cdots(z+\tau),$$

with $\Delta(z;0) = 1$. For any $m \in \mathbb{Z}$, the number $\Delta(m;\tau)$ is a rational integer, and for $m \ge 0$ we have

$$\Delta(m;\tau) = \binom{m+\tau}{\tau}.$$

We get a basis of the space of polynomials in $\mathbb{C}[z_1, \ldots, z_d]$ of degree $\le T_0$ by taking

$$\prod_{i=1}^{d} \Delta(z_i;\tau_i), \qquad (\underline{\tau} \in \mathbb{N}^d, \quad \|\underline{\tau}\| \le T_0).$$

We shall use the simple estimate (see [Y 1989], I, Lemma 2.4 p.128 and [W 1993], Lemma 3.3).

Lemma 7.15. *Let R, T_0 be positive real numbers. For $\underline{z} \in \mathbb{C}^d$ and $\underline{\tau} \in \mathbb{N}^d$ with $|\underline{z}| \le R$ and $\|\underline{\tau}\| \le T_0$, we have*

$$\prod_{i=1}^{d} \left| \Delta(z_i;\tau_i) \right| \le \left(\frac{dR}{T_0} + 1 \right)^{T_0} e^{\|\underline{\tau}\|}$$

and

$$\prod_{i=1}^{d} \left|\Delta(z_i;\tau_i)\right| \le \left(\frac{dR}{T_0}+1\right)^{\|\underline{\tau}\|} e^{T_0}.$$

Hence for $R > 0$ and $T_0 > 0$, we have

$$\max_{\|\underline{\tau}\|\le T_0} \sup_{|\underline{z}|\le R} \prod_{i=1}^{d} \left|\Delta(z_i;\tau_i)\right| \le \left(\frac{dR}{T_0}+1\right)^{T_0} e^{T_0}.$$

Proof. Since $\Delta(z;0)=1$ we may assume that the number $t=\|\underline{\tau}\|$ is not 0.

For $z\in\mathbb{C}$ and $\tau\in\mathbb{N}$, we have

$$|\,\Delta\,(z;\tau)| \le \frac{1}{\tau!}(|z|+\tau)^{\tau}.$$

Therefore, for $\underline{z}\in\mathbb{C}^d$ and $\underline{\tau}\in\mathbb{N}^d$,

$$\prod_{i=1}^{d} \left|\Delta(z_i;\tau_i)\right| \le \frac{1}{\|\underline{\tau}\|!}\binom{\|\underline{\tau}\|}{\underline{\tau}} \prod_{i=1}^{d} (|z_i|+\tau_i)^{\tau_i}.$$

Since

$$\sum_{\|\underline{\tau}\|=t} \binom{\|\underline{\tau}\|}{\underline{\tau}} \prod_{i=1}^{d} (|z_i|+\tau_i)^{\tau_i} = (\|\underline{z}\|+\|\underline{\tau}\|)^{\|\underline{\tau}\|},$$

we deduce

$$\sum_{\|\underline{\tau}\|=t} \prod_{i=1}^{d} \left|\Delta(z_i;\tau_i)\right| \le \frac{1}{t!}(\|\underline{z}\|+t)^{t}.$$

For any positive integer t, we have

$$\left(1+\frac{1}{t}\right)^{t} \le e$$

and by induction we deduce

$$t^t \le t!e^t.$$

Since $\|\underline{z}\| \le d|\underline{z}|$, for $|\underline{z}|\le R$ and $\|\underline{\tau}\|=t\ge 1$ we have

$$\prod_{i=1}^{d} \left|\Delta(z_i;\tau_i)\right| \le \left(\frac{dR}{t}+1\right)^{t} e^{t}.$$

Finally, for $1\le t\le T$,

$$\left(1+\frac{1}{t}\right)^{t} \le \left(1+\frac{1}{T}\right)^{T} \quad \text{and} \quad \left(1+\frac{1}{t}\right)^{t} e^{t} \le \left(1+\frac{1}{T}\right)^{t} e^{T}$$

(the right hand sides are increasing functions of T). □

Lemma 7.16. *Given positive integers d, T_0, T_1 and L with*

$$L = \binom{T_0+d}{d}(2T_1+1),$$

analytic functions φ_t ($t \in \mathbb{Z}$, $|t| \le T_1$) in $\mathbb{C}^d$, points $\underline{\zeta}_1, \ldots, \underline{\zeta}_L$ in $\mathbb{C}^d$ and a nonzero complex number b, define, for $\underline{\tau} \in \mathbb{N}^d$ with $\|\underline{\tau}\| \le T_0$, and for $t \in \mathbb{Z}$ with $|t| \le T_1$,

$$f_{\underline{\tau}t}(\underline{z}) = \underline{z}^{\underline{\tau}}\varphi_t(\underline{z}) \qquad \textit{and} \qquad \tilde{f}_{\underline{\tau}t}(\underline{z}) = \left(\prod_{i=1}^{d}\Delta(z_i;\tau_i)\right)\varphi_t(\underline{z}).$$

Then

$$\det\Bigl(\tilde{f}_{\underline{\tau}t}(b\underline{\zeta}_\mu)\Bigr)_{\substack{(\underline{\tau},t)\\ 1\le\mu\le L}} = \left(\prod_{\|\underline{\tau}\|\le T_0}\prod_{j=1}^{d}\frac{b^{\tau_j}}{\tau_j!}\right)^{T_1+1}\cdot\det\Bigl(f_{\underline{\tau}t}(\underline{\zeta}_\mu)\Bigr)_{\substack{(\underline{\tau},t)\\ 1\le\mu\le L}}.$$

Proof. Since $\Delta(bX;\tau)$ is the product by $b^\tau/\tau!$ of a monic polynomial of degree τ in X, we deduce that $\tilde{f}_{\underline{\tau}t}(b\underline{z})$ is the sum of $\bigl(\prod_{1\le j\le d} b^{\tau_j}/\tau_j!\bigr)f_{\underline{\tau}t}(\underline{z})$ with a linear combination of $f_{\underline{\tau}'t}(\underline{z})$ for $\underline{\tau}' \in \mathbb{N}^d$ with $\|\underline{\tau}'\| < \|\underline{\tau}\|$. The desired result follows by multilinearity. □

Proof. We repeat the proof of Theorem 7.10, with a few modifications.

Step 1. Using Liouville's inequality

Without loss of generality we may assume that $b_1, \ldots, b_{n+1}$ are relatively prime, and also

$$(b_1, \ldots, b_{n+1}) \notin \mathbb{Z}^{n+1}[4\underline{S}].$$

Step 2. Choice of $\mathcal{V}$

Let $\mathcal{V}$ be a vector subspace of $\mathbb{C}^{n+1}$, containing $(b_1, \ldots, b_{n+1})$, and of minimal dimension say $d+1$, for which

$$\mathrm{Card}\left(\frac{\mathbb{Z}^{n+1}[\underline{S}]+\mathcal{V}}{\mathcal{V}}\right) \le \frac{n+1}{d+1}T_0^{n-d}.$$

After a permutation of the coordinates if necessary, we may assume that the first $n-d$ elements of the canonical basis of $\mathbb{C}^n$, viz. $\underline{e}_1, \ldots, \underline{e}_{n-d}$, taken modulo $\mathcal{V}$, give a basis of $\mathbb{C}^n/\mathcal{V}$.

We define θ_i, ($n-d+1 \le i \le n+1$) in the same way as in step 2 of the proof of Theorem 7.10: let $u_i^{(j)}$ be complex numbers such that

$$\underline{e}_i - \sum_{j=1}^{n-d} u_i^{(j)}\underline{e}_j \in \mathcal{V}.$$

Let

$$\theta_i = \lambda_i + \sum_{j=1}^{n-d} u_i^{(j)} \lambda_j .$$

Then

$$b_{n-d+1}\theta_{n-d+1} + \cdots + b_{n+1}\theta_{n+1} = \Lambda .$$

Step 3. Lower bound for $|\widetilde{\Delta}_{\mathrm{ar}}|$

We define

$$\widetilde{\gamma}_{\underline{\tau}t}^{(\mu)} = \prod_{j=n-d+1}^{n} \Delta\big(s_j^{(\mu)} b_{n+1} - s_{n+1}^{(\mu)} b_j ; \tau_j\big) \prod_{i=1}^{n+1} \alpha_i^{s_i^{(\mu)} t} \qquad (1 \le \mu \le L_d)$$

and

$$\widetilde{\Delta}_{\mathrm{ar}} = \det\left(\widetilde{\gamma}_{\underline{\tau}t}^{(\mu)} \right)_{\substack{(\underline{\tau},t) \\ \mu}} .$$

By Lemma 7.16,

$$\widetilde{\Delta}_{\mathrm{ar}} = \Delta_{\mathrm{ar}} \prod_{\|\underline{\tau}\| \le T_0} \left(\prod_{j=n-d+1}^{n} \frac{b_{n+1}^{\tau_j}}{\tau_j !} \right)^{T_1+1} \neq 0 .$$

We bound the product of $b_{n+1}^{\tau_j}/\tau_j!$ by

$$\prod_{j=n-d+1}^{n} \frac{b_{n+1}^{\tau_j}}{\tau_j!} \le \left(\frac{|b_{n+1}|}{T_0} + 1 \right)^{T_0} e^{T_0} .$$

Since

$$B_1 \ge e \left(\frac{|b_{n+1}|}{T_0} + 1 \right)$$

we deduce

$$|\widetilde{\Delta}_{\mathrm{ar}}| \le \Delta_{\mathrm{ar}} \left(\frac{|b_{n+1}|}{T_0} + 1 \right)^{L_d T_0} e^{L_d T_0} \le \Delta_{\mathrm{ar}} B_1^{L_d T_0} .$$

The number $\widetilde{\Delta}_{\mathrm{ar}}$ is the value at the point $\alpha_1, \ldots, \alpha_{n+1}, \alpha_1^{-1}, \ldots, \alpha_{n+1}^{-1}$, of a polynomial with coefficients in $\mathbb{Z}$ of length at most

$$L_d!(eB_1)^{T_0 L_d} .$$

This estimate follows from the upper bound

$$1 + \frac{2n(n+1)}{T_0} \big(|b_{n+1}| S_j + |b_j| S_{n+1} \big) \le B_1 .$$

We use again Liouville's inequality: we deduce from Proposition 3.15

$$\frac{1}{L_d} \log |\widetilde{\Delta}_{\mathrm{ar}}| \ge -U_1$$

with

$$U_1 = (D-1)T_0 \log(eB_1) + (D-1)\log L_d + 2(n+1)D(T_1+1)\sum_{i=1}^{n+1} S_i \log A_i.$$

Therefore

$$\frac{1}{L_d}\log|\Delta_{\rm ar}| \geq -\widetilde{U}_1$$

with

$$\begin{aligned}\widetilde{U}_1 &= U_1 + T_0 \log B_1\\ &< DT_0\log(eB_1) + (D-1)\log L_d + 2(n+1)D(T_1+1)\sum_{i=1}^{n+1} S_i \log A_i.\end{aligned}$$

Step 4. Conclusion of the proof

For each $(\underline{\tau}, t)$ as in step 4 of the proof of Theorem 7.10, we define a function $\tilde{f}_{\underline{\tau}t}$ of d complex variables:

$$\tilde{f}_{\underline{\tau}t}(z_{n-d+1}, \ldots, z_n) = \prod_{i=n-d+1}^{n} \left(\Delta(z_i; \tau_i) e^{t\theta_i z_i}\right).$$

For $\underline{s} \in \mathcal{V} \cap \mathbb{Z}^{n+1}$, put

$$\underline{\xi}_{\underline{s}} = \left(s_{n-d+1}b_{n+1} - s_{n+1}b_{n-d+1}, \ldots, s_n b_{n+1} - s_{n+1}b_n\right) \in \mathbb{C}^d.$$

Since

$$\sum_{i=n-d+1}^{n} \theta_i(s_i b_{n+1} - s_{n+1}b_i) = -s_{n+1}\Lambda + b_{n+1}\sum_{j=1}^{n+1} s_j\lambda_j,$$

for $z \in \mathbb{C}$ we have

$$\tilde{f}_{\underline{\tau}t}(z\underline{\xi}_{\underline{s}}) = \prod_{i=n-d+1}^{n} \Delta\big(z(s_i b_{n+1} - s_{n+1}b_i); \tau_i\big) \cdot \prod_{j=1}^{n+1} e^{ts_j\lambda_j z} \cdot e^{-ts_{n+1}\Lambda z}.$$

For $1 \leq \mu \leq L_d$ let $\underline{\zeta}_{\mu} = \underline{\xi}_{\underline{s}^{(\mu)}}$. Then

$$\tilde{f}_{\underline{\tau}t}(\underline{\zeta}_{\mu}) = \widetilde{\gamma}_{\underline{\tau}t}^{(\mu)} e^{-ts_{n+1}^{(\mu)}\Lambda}.$$

We use Proposition 7.6 with

$$M_{\underline{\tau}t} = DT_0\log(eB_1) + 2(n+1)D(T_1+1)\sum_{i=1}^{n+1} S_i \log A_i + \log\frac{T_1}{2}.$$

For

$$R = 2(n+1)E \max_{1\leq j\leq n+1}\{|b_{n+1}|S_j + |b_j|S_{n+1}\}$$

we have $(dR/T_0) + 1 \le B_1$, because

$$1 + \frac{2n(n+1)E}{T_0}\bigl(|b_{n+1}|S_j + |b_j|S_{n+1}\bigr) \le B_1^D.$$

The conclusion of Proposition 7.6 is not satisfied, therefore

$$|\Lambda| > e^{-nV}.$$

This completes the proof of Theorem 7.14. □

We shall deduce from Theorem 7.14 the following result, which (as we shall see) is sharper than the rational case of Theorem 7.1.

Corollary 7.17. *Under the assumptions stated before Theorem 7.14, assume*

$$E \le A_i^D \quad \text{for} \quad 1 \le i \le n+1 \qquad \text{and} \qquad B_1 \ge \max\bigl\{B_0,\ E^{1/D},\ e\bigr\},$$

where

$$B_0 = \max\{|b_1|, \dots, |b_{n+1}|\}.$$

Let N, N_0 and C_0 be positive real numbers satisfying the following conditions:

$$2 + \frac{2(n+1)^2}{N}\left(1 + \frac{1}{N_0}\right) + \frac{1}{200} + \frac{1}{8n} \le \frac{N_0}{4n}$$

and

$$C_0 \ge 2^{-n}(n+1)N^{n+1}N_0^{n+2}\left(1 + \frac{1}{8N_0}\right).$$

Then

$$|\Lambda| > \exp\bigl\{-(N_0+1)C_0D^{n+3}(\log B_1)^2(\log A_1)\cdots(\log A_{n+1})(\log E)^{-n-2}\bigr\}.$$

Proof. The differences between Theorem 7.14 and Theorem 7.10 are that the definition of B_1 is not the same, and that

$$DT_0\log(eB_1) \quad \text{replaces} \quad DT_0\log\bigl(4e(n+1)B_1S\bigr)\bigr).$$

Also the only difference between Corollary 7.17 and Corollary 7.13 lies in the conditions involving B_1. Therefore we just repeat the proof of Corollary 7.13 (we define U, T_0, S_j in the same way) and we only need to check $DT_0\log(eB_1) \le 2U$.

We claim that the number

$$E^{1/D}\max_{1\le j\le n}\left\{1 + \frac{2n(n+1)}{T_0}\bigl(|b_{n+1}|S_j + |b_j|S_{n+1}\bigr)\right\},$$

is bounded by B_1^2. Indeed, from the inequalities

$$T_0 > \left(1 - \frac{1}{C_0}\right) \frac{U}{D \log B}$$

and

$$S_j \leq \frac{U}{NDT_1 \log A_j} < \frac{U \log E}{N(N_0 - 1)D^2(\log A_j)(\log B)}$$

one deduces

$$\frac{S_j}{T_0} < \frac{C_0}{N(N_0 - 1)(C_0 - 1)}$$

and

$$\frac{2n(n+1)}{T_0}\big(|b_{n+1}|S_j + |b_j|S_{n+1}\big) < \frac{C_0 B}{2(n+1)(C_0 - 1)}.$$

Finally the number

$$\frac{1}{e} + \frac{C_0}{2(n+1)(C_0 - 1)}$$

is not greater than 1. □

Proof of Theorem 7.1 in the rational case.

We now complete the proof of the rational case of Theorem 7.1. We repeat the proof of § 7.6.2, until step 3, where we take $B_1 = B$. We deduce that the rational case of Theorem 7.1 holds in the special case where $\beta_1, \dots, \beta_m$ are relatively prime rational integers, say $\beta_i = b_i$, with $|b_i| \leq B$. Moreover in that case we can replace the constant $C(m)$ by $C'(m-1)/(m+1)^2$, where C' is the constant of step 4 in § 7.6.2. We consider now the general case.

Let $(\beta_1, \dots, \beta_m)$ be a tuple of rational numbers with $\mathrm{h}(\beta_i) \leq \log B$. Define p/q as the positive rational number (with relatively prime (p, q)) such that the numbers $b_i = (p/q)\beta_i$ are relatively prime rational integers. So q is the least positive integer such that $q\beta_i \in \mathbb{Z}$ for $1 \leq i \leq m$, and p is the least common denominator to these integers. We have

$$\mathrm{h}(p : qb_1 : \cdots : qb_m) = \mathrm{h}(1 : \beta_1 : \cdots : \beta_m) \leq m \log B,$$

hence

$$p \leq B^m \quad \text{and} \quad \max_{1 \leq i \leq m} |b_i| \leq B^m.$$

Define

$$\Lambda' = \frac{p}{q}\Lambda = b_1\lambda_1 + \cdots + b_m\lambda_m.$$

We have

$$|\Lambda| \geq p^{-1}|\Lambda'| \geq B^{-m}|\Lambda'|$$

and

$$B^{-m} > \exp\{-mD^{m+2}(\log B)^2(\log A_1)\cdots(\log A_m)(\log E)^{-m-1}\}.$$

Since

$$C'(m-1)\cdot\frac{m^2}{(m+1)^2} + m < C(m),$$

the conclusion of Theorem 7.1 in the rational case follows. □

7.8 Linear Dependence Relations between Logarithms

In this section we prove the following result:

Proposition 7.18. *In the statement of Theorem 7.1, we may replace the hypothesis that the numbers* $\lambda_1, \ldots, \lambda_m$ *are linearly independent over* $\mathbb{Q}$ *by the extra hypotheses* $\Lambda \neq 0$ *and*

$$D^3(\log B)^2(\log A_i) \geq (\log D)(\log E)^2$$

for $1 \leq i \leq m$.

For instance, under the assumptions of Theorem 7.1, the extra hypothesis of Proposition 7.18 is satisfied as soon as $B \geq D$.

We need the following lemma:

Lemma 7.19. *Let* $\lambda_1, \ldots, \lambda_m$ *(with* $m \geq 2$*) be* $\mathbb{Q}$*-linearly dependent logarithms of algebraic numbers. Define* $\alpha_j = e^{\lambda_j}$ $(1 \leq j \leq m)$. *For* $1 \leq j \leq m$, *let* $\log A_j \geq 1$ *be an upper bound for* $\max\{\mathrm{h}(\alpha_j), |\lambda_j|/D\}$. *Further D be the degree of the number field* $K = \mathbb{Q}(\alpha_1, \ldots, \alpha_m)$ *over* $\mathbb{Q}$. *Then there exist rational integers* $n_1, \ldots, n_m$, *not all of which are zero, such that*

$$n_1\lambda_1 + \cdots + n_m\lambda_m = 0$$

and

$$|n_k| \leq \bigl(11(m-1)D^3\bigr)^{m-1} \frac{(\log A_1)\cdots(\log A_m)}{\log A_k}$$

for $1 \leq k \leq m$.

Remark 1. The need for such a result appeared at an early stage of Baker's theory. Baker himself ([B 1966], IV) used his transcendence arguments to establish such an estimate. Then Stark [St 1973] obtained sharper estimates by means of geometry of numbers (see also [V 1977], Lemma 9). Since trancendence methods provide weaker results for this particular problem, they are no more used for it now, and indeed we shall follow Stark's approach. However one should mention that for the similar problem related to elliptic curves or abelian varieties in place of the multiplicative group, Baker's transcendence method is still a powerful tool for estimating small dependence relations between periods (see for instance the work of Masser and Wüstholz [MaWü 1990] on Faltings' isogeny Theorem, and [D 1995] for explicit estimates dealing with elliptic curves).

Remark 2. Let $\alpha_1, \ldots, \alpha_m$ be multiplicatively dependent algebraic numbers. The set G of $\underline{n} \in \mathbb{Z}^m$ such that $\alpha_1^{n_1} \cdots, \alpha_m^{n_m} = 1$ is a nonzero subgroup of $\mathbb{Z}^m$, and from Lemma 7.19 one deduces an upper bound for $|\underline{n}|$, where $\underline{n}$ is some nonzero element in G (i.e. an upper bound for the first minimum of the discrete subgroup G in $\mathbb{R}^m$). Much more information on this $\mathbb{Z}$-module G is available: see for instance [Mat 1993b] and [Bert 1997].

Remark 3. According to E. M. Matveev, the factor $\bigl(11(m-1)D^3\bigr)^{m-1}$ can be replaced by $C^m D \log D$ with some absolute constant $C > 0$.

Proof of Lemma 7.19 (cf. [W 1980], Lemma 4.1, [L 1991] Chap. 9 § 7; see also [Ma 1988]). We assume, as we may without loss of generality, that $m \geq 2$, and that any $m-1$ elements among $\lambda_1, \ldots, \lambda_m$ are linearly independent over $\mathbb{Q}$. Thus there exists a unique (up to a factor ± 1) set of relatively prime integers $n_1, \ldots, n_m$ such that

$$n_1\lambda_1 + \cdots + n_m\lambda_m = 0.$$

Hence

$$\alpha_1^{n_1} \cdots \alpha_m^{n_m} = 1.$$

Fix an integer k in the range $1 \leq k \leq m$. Define $c_1, \ldots, c_m$ by

$$c_j = \bigl(11(m-1)D^3 \log A_j\bigr)^{-1} \quad (1 \leq j \leq m,\ j \neq k)$$

and

$$c_k = \bigl(11(m-1)D^3\bigr)^{m-1} \prod_{\substack{1\leq j\leq m \\ j\neq k}} \log A_j,$$

so that $c_1 \cdots c_m = 1$. Using Minkowski's linear form Theorem (e.g. [Sc 1980], p.33 Th. 2C) we deduce that there exist integers $\nu_1, \ldots, \nu_m$, not all of which are zero, such that

$$\left|\nu_j - \frac{\nu_k n_j}{n_k}\right| \leq c_j, \qquad (1 \leq j \leq m,\ j \neq k) \qquad \text{and} \qquad |\nu_k| \leq c_k.$$

(one could even ask for strict inequalities $|\nu_j - \nu_k n_j/n_k| < c_j$, but this will not be necessary). We want to prove the relation $\nu_1\lambda_1 + \cdots + \nu_m\lambda_m = 0$. We first show that the number $\alpha = \alpha_1^{\nu_1} \cdots \alpha_m^{\nu_m}$ is a root of unity. Using (3.4) and (3.6) for the number

$$\alpha^{n_k} = \prod_{j=1}^{m} \alpha_j^{\nu_j n_k} = \prod_{1\leq j\leq m} \alpha_j^{\nu_j n_k - \nu_k n_j},$$

we get

$$|n_k| \mathrm{h}(\alpha) \leq \sum_{1\leq j\leq m, j\neq k} |\nu_j n_k - \nu_k n_j| \mathrm{h}(\alpha_j),$$

hence

$$\mathrm{h}(\alpha) \leq \sum_{j\neq k} c_j \mathrm{h}(\alpha_j) \leq \frac{1}{11D^3}.$$

So by Theorem 3.16 it follows that α is a root of unity. Let M be the order of α. Then α is a M-th primitive root of unity, hence of degree $[\mathbb{Q}(\alpha) : \mathbb{Q}] = \varphi(M) \leq [K : \mathbb{Q}] = D$ over $\mathbb{Q}$ (where φ is Euler's function), therefore $M \leq 2D^2$ (sharper estimates are valid - see Exercise 7.2.a) and

$$M \sum_{j=1}^{m} \nu_j\lambda_j \in 2i\pi\mathbb{Z}.$$

We observe that

$$\left|M\sum_{j=1}^{m}\nu_j\lambda_j\right| = \left|M\sum_{j=1}^{m}\left(\nu_j-\frac{\nu_k n_j}{n_k}\right)\lambda_j\right| \le M\sum_{1\le j\le m,\, j\ne k} c_j|\lambda_j| < 2\pi$$

and we conclude

$$\sum_{i=1}^{m}\nu_j\lambda_j = 0.$$

Therefore there exists a nonzero integer $\ell\in\mathbb{Z}$ with $(\nu_1,\dots,\nu_m)=(\ell n_1,\dots,\ell n_m)$ We deduce the inequality $|n_k|\le|\nu_k|$, from which the desired upper bound $|n_k|\le c_k$ readily follows. □

Remark. The coefficient 11 in the conclusion of Lemma 7.19 can be replaced by 9 (see [W 1980]). In fact, as pointed out in § 3.6.3, $11D^3$ can be replaced by a smaller function. But for our purpose a much weaker result would already be sufficient.

We deduce from Lemma 7.19 the following result:

Lemma 7.20. *Let K be a number field of degree D and $\lambda_1,\dots,\lambda_m$ elements of $\mathcal{L}$ such that $\alpha_i = e^{\lambda_i}$ is in K for $1\le i\le m$. For $1\le i\le m$ let $A_i\ge e^{1/D}$ be a real number such that*

$$\log A_i \ge \mathrm{h}(\alpha_i) \quad \textit{and} \quad \log A_i \ge \frac{1}{D}|\lambda_i|.$$

For each nonempty subset I of $\{1,\dots,m\}$, define

$$A_I=\max\bigl\{e,\ \max_{i\in I}A_i\bigr\},\qquad N_I=\bigl[(11nD^3\log A_I)^{n-1}\bigr]$$

and let Φ_I be a nondecreasing positive valued real function satisfying the following three conditions.

(1) For $\emptyset\ne I'\subset I\subset\{1,\dots,m\}$ and for any $B_0>0$ we have

$$\Phi_{I'}(B_0)\le\Phi_I(B_0).$$

(2) For any nonempty subset I of $\{1,\dots,m\}$ for which the numbers $\{\lambda_i\}_{i\in I}$ are $\mathbb{Q}$-linearly independent and for any $(\beta_i)_{i\in I}\in K^I\setminus\{0\}$, the inequality

$$\left|\sum_{i\in I}\beta_i\lambda_i\right|\ge\exp\{-\Phi_I(B_0)\}$$

holds with

$$\log B_0=\max_{i\in I}\mathrm{h}(\beta_i).$$

(3) For any nonempty subset I of $\{1,\dots,m\}$, the inequality

$$\Phi_{I\setminus\{k\}}(2N_IB_0^2)+\log N_I\le\Phi_I(B_0)$$

holds any $k\in I$ satisfying $A_I=\max\bigl\{e,\ A_k\bigr\}$.

Then, for any $(\beta_1, \dots, \beta_m) \in K^m$ *for which the number*

$$\Lambda = \beta_1\lambda_1 + \cdots + \beta_m\lambda_m$$

is nonzero, the inequality

$$|\Lambda| \geq \exp\{-\Phi_{\{1,\dots,m\}}(B_0)\}$$

holds

Proof of Lemma 7.20. The proof is by induction on the number n of elements in the set

$$I = \{i \in \{1, \dots, m\}\;;\; \beta_i \neq 0\}.$$

We start with the case $n = 1$. Write $I = \{i_1\}$ and $\Lambda = \beta_{i_1}\lambda_{i_1}$. Notice that $N_{\{i_1\}} = 1$. Since $\lambda_{i_1} \neq 0$ we deduce from assumptions (1) and (2):

$$|\Lambda| \geq \exp\{-\Phi_{\{i_1\}}(B_0)\} \geq \exp\{-\Phi_{\{1,\dots,m\}}(B_0)\}.$$

Using the same argument we deduce that the result holds if the numbers λ_i $(i \in I)$ are $\mathbb{Q}$-linearly independent. Assume $n \geq 2$ and assume there is a nontrivial relation

$$\sum_{i\in I} a_i\lambda_i = 0$$

where $(a_i)_{i\in I} \in \mathbb{Z}^I \setminus \{0\}$. Using Lemma 7.19, we deduce that there exists such a relation, and there exists an index $k \in I$, for which

$$\max_{i\in I} |a_i| \leq N_I, \quad a_k \neq 0 \quad \text{and} \quad A_I = \max\{e,\; A_k\}.$$

Using Lemma 3.7 with

$$f(\mathrm{X}_1, \mathrm{X}_2) = a_k\mathrm{X}_1 - a_i\mathrm{X}_2,$$

we deduce, for $i \in I$,

$$\mathrm{h}(a_k\beta_i - a_i\beta_k) \leq \log B_0' \quad \text{with} \quad B_0' = 2N_I B_0^2.$$

From the induction hypothesis for

$$a_k\Lambda = \sum_{\substack{i\in I\\ i\neq k}} (a_k\beta_i - a_i\beta_k)\lambda_i,$$

we conclude

$$|a_k\Lambda| \geq \exp\{-\Phi_{I\setminus\{k\}}(B_0')\}.$$

Lemma 7.20 easily follows from assumptions (1) and (3). □

Proof of Proposition 7.18. For $I \subset \{1, \dots, m\}$, define

$$\Phi_I(B_0) = C(n)D^{n+2}(\log B_I)^2\Big(\prod_{i\in I} \log A_i\Big)(\log E)^{-n-1},$$

where $n = |I|$ and where B_I is defined as follows: in the general case,

$$B_I = \max\left\{B_0,\ E^{1/D},\ \frac{D}{\log E},\ 2^6 n^4 \cdot \frac{D \log A_I}{\log E}\right\}$$

and in the rational case,

$$B_I = \max\{B_0,\ e,\ E^{1/D}\}.$$

Define also B'_I in the same way, but with B_0 replaced by $B'_0 = 2N_I B_0^2$.

We check the hypotheses of Lemma 7.20. We start with (1). For $I' \subset I$ define $n' = |I'|$, $n = |I|$. Since $B_{I'} \le B_I$, $C(n') \le C(n)$ and $D \log A_i \ge \log E$, we have

$$C(n')(\log B_{I'})^2 \le C(n)(\log B_I)^2 \prod_{i \in I \setminus I'} \frac{D \log A_i}{\log E},$$

hence $\Phi_{I'}(B_0) \le \Phi_I(B_0)$.

The assumption (2) follows from Theorem 7.1.

It remains to check (3). Using the inequalities $D \log B_I \ge \log E$, $D \log A_i \ge \log E$ and

$$D^3(\log B)^2(\log A_i) \ge (\log D)(\log E)^2$$

of Proposition 7.18, we deduce

$$\begin{aligned}\Phi_I(B_0) &\ge C(n) \cdot \frac{D^3(\log B_I)^2(\log A_I)}{(\log E)^2}\\ &\ge C(n)\max\{D \log B_I,\ D \log A_I,\ \log D\},\end{aligned}$$

hence

$$\begin{aligned}\log N_I &\le (n-1)\log(11 n D^3 \log A_I)\\ &\le \frac{1}{2} C(n) \cdot \frac{D^3(\log B_I)^2(\log A_I)}{(\log E)^2}\\ &\le \frac{1}{2}\Phi_I(B_0).\end{aligned}$$

Similarly

$$\Phi_{I\setminus\{k\}}(B'_0) \le \frac{1}{2}\Phi_I(B_0).$$

The hypotheses of Lemma 7.20 are satisfied, and therefore Proposition 7.18 follows. Moreover we insist that a loose upper bound for N_I suffices: as already pointed out in the preceding remark, a weaker coefficient than $11D^3$ in Lemma 7.19 would have been sufficient. □

Open Problem

Does there exist an absolute constant $C > 0$ such that, for all $p/q \in \mathbb{Q}$ with $q > 1$,

$$\left| e^{\pi} - \frac{p}{q} \right| > q^{-C} \, ?$$

This would mean that the number e^{π} is not a Liouville number. This problem is related with the case $m = 2$ of Theorem 7.1: take $\lambda_1 = i\pi$, $\lambda_2 = \log(p/q)$, $\beta_1 = i$, $\beta_2 = 1$, $D = 2$, $\log A_1 = e\pi$, $\log A_2 = \log p$ (without loss of generality we may assume $p > q$). The trouble is that β_1 is not rational, hence we are in the general case and we need a condition $B \geq c \log p$, where c is a positive constant.

Exercises

Exercise 7.1. Let $\lambda_1, \ldots, \lambda_m$ be logarithms of nonzero algebraic numbers, not all of which are zero, and $\beta_1, \ldots, \beta_m$ algebraic numbers. Assume

$$\beta_1\lambda_1 + \cdots + \beta_m\lambda_m = 0.$$

Deduce that there is a linear dependence relation $k_1\beta_1 + \cdots + k_m\beta_m = 0$, with $(k_1, \ldots, k_m) \in \mathbb{Z}^m \setminus \{0\}$ and with an explicit upper bound for $|\underline{k}|$.

Hint. *Repeat the proof of Theorem 7.1. The upper bound for $|\underline{k}|$ arises from the zero estimate.*

Exercise 7.2. Let K be a number field of degree D, α be an element in $K^\times$, $\lambda \in \mathcal{L}$ a logarithm of α and m a positive integer such that $e^{\lambda/m} \in K^\times$. Check

$$m \leq 11 D^3 \max\left\{ \mathrm{h}(\alpha), \frac{|\lambda|}{D} \right\}.$$

Hint. *Use Exercise 7.1 and compare with* [W 1980], *Lemma 4.2.*

More precisely,
a) If α is a root of unity, then

$$|m| \leq \frac{1}{\pi} D^2 |\lambda|.$$

Hint. *Define $N = \max\{n \geq 1 \,;\, \varphi(n) \leq D\}$. Check $N \leq 2D^2$ and $|m| \leq (N/2\pi)|\lambda|$.*

Remark. Stronger upper bounds for N hold:

$$N \leq 4D \log\log(D+7) \quad \text{for} \quad D \geq 2$$

and

$$N \leq (e^{\gamma} + \epsilon) D \log\log D \quad \text{for} \quad D \geq D_0(\epsilon)$$

where γ is Euler's constant ($e^{\gamma} = 1.78107\ldots$). See [MiW 1978], III, Proposition A3 (Appendix) p.74.

b) If α is a unit but not a root of unity, then

$$|m| \leq 11D^3\mathrm{h}(\alpha).$$

Hint. *Use Theorem 3.16.*

c) If α is not a unit, then

$$|m| \leq \frac{1}{\log 2} \log |N_{K/\mathbb{Q}}(\alpha \cdot \mathrm{den}(\alpha)|$$

where $\mathrm{den}(\alpha) \in \mathbb{Z}$ is the denominator of α (see § 3.4) and $N_{K/\mathbb{Q}}$ is the norm $K^\times \to \mathbb{Q}^\times$.

Exercise 7.3.
a) Check the following formula for the number $\Theta(n; T_0, L)$: define A and $\varrho \in \mathbb{N}$ by the conditions $A \geq T_0$ and

$$L = (A - T_0 + 1)\binom{T_0 + n - 1}{n - 1} + \binom{T_0 + n - 1}{n} + \varrho, \quad \text{with} \quad 0 \leq \varrho < \binom{T_0 + n - 1}{n - 1}.$$

Then

$$\Theta(n; T_0, L) = \sum_{a=1}^{T_0 - 1} \binom{a + n - 1}{n - 1} a + \frac{1}{2}\binom{T_0 + n - 1}{n - 1}(A - T_0 + 1)(A + T_0) + (A + 1)\varrho.$$

b) Show that in the conclusion of Lemma 7.3, strict inequality holds for $n \geq 2$:

$$\Theta(n; T_0, L) > \frac{L}{2}\left(\frac{L + 1}{\binom{T_0+n-1}{n-1}} - \frac{T_0}{n} - 1\right).$$

Hint. *For $T_0 = 1$ check*

$$\Theta(n; 1, L) = \frac{1}{2}(A + 1)(nA + 2\varrho) \geq \frac{1}{2n}(L + n - 1)(L - 1),$$

where $L = nA + 1 + \varrho$ with $0 \leq \varrho < n$.

c) Assuming $L \geq \binom{T_0+n}{n}$, deduce

$$\Theta(n; T_0, L) > \frac{L^2}{2\binom{T_0+n-1}{n-1}}.$$

Exercise 7.4. Let K be a field of characteristic zero, m and S positive integers, and $\mathcal{V}$ a vector subspace of K^m.
a) Show that there exists $\underline{x} \in \mathbb{Z}^m[S]$ such that

$$\mathrm{Card}\left(\frac{\mathbb{Z}^m[S] + \mathcal{V}}{\mathcal{V}}\right) \mathrm{Card}\big((\underline{x} + \mathcal{V}) \cap \mathbb{Z}^m[S]\big) \geq (2S + 1)^m.$$

Hint. *Use Lemma 7.8.*

b) Let W be a vector subspace of K^m of dimension d. Check the inequality

$$\mathrm{Card}\big((\underline{x} + W) \cap \mathbb{Z}^m[S]\big) \le (2S+1)^d$$

for each $\underline{x} \in K^m$.

Hint. *Show first that there is no loss of generality to assume $\underline{x} \in \mathbb{Z}^m$. After a permutation of coordinates, one may also assume that the projection $K^m \longrightarrow K^d$ on the first d coordinates maps W isomorphically onto K^d. Then the image of $(\underline{x} + W) \cap \mathbb{Z}^m[S]$ under this projection has at most $(2S+1)^d$ elements.*

c) Assume

$$\mathrm{Card}\left(\frac{\mathbb{Z}^m[S] + \mathcal{V}}{\mathcal{V}}\right) < (2S+1)^{r+1}$$

where $r \ge 1$ is the codimension of $\mathcal{V}$. Show that $\mathcal{V} \cap \mathbb{Z}^m[2S+1]$ contains more than $(2S+1)^{m-r-1}$ points, and that these points span $\mathcal{V}$ as a vector space.

Exercise 7.5. Let K be a field, $\alpha_1, \dots, \alpha_m$ be elements in $K^\times$ which generate a multiplicative subgroup of rank $\ge m-1$. Let $\mathfrak{S}$ be a finite subset of $\mathbb{Z}^m$ and S a positive real number such that $|s_j| \le S$ for any $\underline{s} = (s_1, \dots, s_m) \in \mathfrak{S}$. For $\underline{s} \in \mathbb{Z}^m$ write $\underline{\alpha}^{\underline{s}}$ for $\alpha_1^{s_1} \cdots \alpha_m^{s_m}$.

a) Check

$$\mathrm{Card}\big\{\underline{\alpha}^{\underline{s}};\ \underline{s} \in \mathfrak{S}\big\} \ge \frac{\mathrm{Card}(\mathfrak{S})}{2S+1}.$$

Hint. *Check that for any $\underline{s}^0 \in \mathfrak{S}$, the number of $\underline{s} \in \mathfrak{S}$ such that $\underline{\alpha}^{\underline{s}} = \underline{\alpha}^{\underline{s}^0}$ is $\le 2S+1$. Next apply Lemma 7.8.*

b) More precisely, show that the number of elements in the image in $K^\times / K^\times_{\mathrm{tors}}$ of the set

$$\big\{\underline{\alpha}^{\underline{s}};\ \underline{s} \in \mathfrak{S}\big\}$$

is at least

$$\frac{\mathrm{Card}(\mathfrak{S})}{2S+1}.$$

Hint. *If $\alpha_1, \dots, \alpha_m$ are multiplicatively independent, then the number of elements in this image is just $\mathrm{Card}(\mathfrak{S})$. Otherwise the map*

$$\begin{array}{cccc} \psi: & \mathbb{Z}^m & \longrightarrow & \dfrac{K^\times}{K^\times_{\mathrm{tors}}} \\ & (s_1, \dots, s_m) & \longmapsto & \text{class of } \underline{\alpha}^{\underline{s}} \end{array}$$

has a kernel which is a subgroup of $\mathbb{Z}^m$ of rank 1. Check, for $\underline{a} \in \mathbb{Z}^m \setminus \{0\}$,

$$\mathrm{Card}\big(\mathbb{Z}^m[S] \cap \mathbb{Z}\underline{a}\big) \le 2S+1.$$

Apply Lemma 7.8 with $\mathcal{C} = \mathcal{S}$, $\mathcal{C}' = K^\times / K^\times_{\mathrm{tors}}$, and f is the restriction of ψ to $\mathcal{C}$.

8. Multiplicity Estimate by Damien Roy

This chapter refines the zero estimate of Chapter 5 by taking multiplicities into account. The main result that we shall present here is again essentially due to P. Philippon (see [P 1986a]) and again we restrict to commutative linear algebraic groups. This allows us to be more concrete and brings simplifications in the proof of the result. For an outline of the zero estimate of P. Philippon on a general commutative algebraic group, the reader may consult the expository papers [Bert 1987] and [Roy 2000b].

8.1 The Main Result

The notation is the same as in Chapter 5. In particular, we work with the group $G = \mathbb{G}_a^{d_0} \times \mathbb{G}_m^{d_1} = K^{d_0} \times (K^\times)^{d_1}$ and the corresponding ring

$$K[G] = K[X_1, \ldots, X_{d_0}, Y_1, \ldots, Y_{d_1}, Y_1^{-1}, \ldots, Y_{d_1}^{-1}],$$

where K denotes an algebraically closed field of characteristic 0. The new feature is that, for each $w = (\xi_1, \ldots, \xi_{d_0}, \eta_1, \ldots, \eta_{d_1}) \in K^d$, we introduce a derivation $\mathcal{D}_w$ of $K[G]$ by putting

$$\mathcal{D}_w = \sum_{h=1}^{d_0} \xi_h \frac{\partial}{\partial X_h} + \sum_{i=1}^{d_1} \eta_i Y_i \frac{\partial}{\partial Y_i}.$$

We will discuss these derivations in more details in § 8.3.2. At this point, we simply need the following definitions.

Given a point $g \in G$, a vector subspace $\mathcal{W}$ of K^d and an integer $N \geq 0$, we say that an element P of $K[G]$ *vanishes to order* $> N$ *at* g *with respect to* $\mathcal{W}$ if

$$\mathcal{D}_{w_1} \cdots \mathcal{D}_{w_s} P(g) = 0$$

for any integer s with $0 \leq s \leq N$ and any $w_1, \ldots, w_s \in \mathcal{W}$, this condition being interpreted as $P(g) = 0$ when $s = 0$.

Given an algebraic subgroup G^* of G, we define the *tangent space* of G^* at the neutral element e to be the subspace $T_e(G^*)$ of K^d consisting of all points $w \in K^d$ such that $\mathcal{D}_w$ maps into itself the ideal $I(G^*)$ of all elements of $K[G]$ vanishing identically on G^*. We will show in § 8.3.2 that, when G^* is written in the form of a product $\mathcal{V} \times T_\Phi$ where $\mathcal{V}$ is a subspace of K^{d_0} and Φ a finitely generated subgroup of $\mathbb{Z}^{d_1}$, then we have $T_e(G^*) = \mathcal{V} \times L$ where L denotes the subspace of K^{d_1} consisting of the common zeros of the linear forms $\varphi_1 Y_1 + \cdots + \varphi_{d_1} Y_{d_1}$ with $(\varphi_1, \ldots, \varphi_{d_1}) \in \Phi$. In particular, we have $T_e(G) = K^d$.

The following statement deals with a relative situation where we have a pair of algebraic subgroups G^-, G^+ of G with $G^- \subset G^+$, and a polynomial P which vanishes identically, with multiplicities, on a family of translates of G^- inside G^+. It can be thought as a multiplicity estimate on the quotient G^+/G^- although it does not require that the polynomial map induced by P on the group G^+ factors through the quotient. It also avoids embedding the quotient as an algebraic subset of some affine space.

Theorem 8.1. *Let G^- and G^+ be connected algebraic subgroups of G with $G^- \subset G^+$, let Σ be a subset of G^+ containing e, let $S_0 \geq 0$ be an integer, and let $\mathcal{W}$ be a vector subspace of $T_e(G^+)$. Denote by d^+ the dimension of G^+ and assume that, for given integers $D_0, D_1, \ldots, D_{d_1} \geq 0$ and $S_0 \geq 0$, there exists a nonzero element P of $K[G]$ of multidegree $\leq (D_0, \underline{D}) = (D_0, D_1, \ldots, D_{d_1})$ which does not vanish identically on G^+ but vanishes to order $> (d^+)S_0$ with respect to $\mathcal{W}$ at each point of $\Sigma[d^+]+G^-$. Then there exists a connected algebraic subgroup G^* of G^+ of dimension $< d^+$, containing G^- such that, if we set*

$$\ell_0' = \dim_K \left(\frac{\mathcal{W} + T_e(G^*)}{T_e(G^*)} \right),$$

then

$$\binom{S_0 + \ell_0'}{\ell_0'} \operatorname{Card} \left(\frac{\Sigma + G^*}{G^*} \right) \mathcal{H}(G^*; D_0, \underline{D}) \leq \mathcal{H}(G^+; D_0, \underline{D}).$$

Moreover, we may assume that G^* is an irreducible component of the set of zeros in G^+ of a family of polynomials of $K[G]$ of multidegree $\leq (D_0, \underline{D})$.

The last assertion of the theorem is often expressed in short by saying that G^* is *incompletely defined* in G^+ by polynomials of multidegree $\leq (D_0, \underline{D})$. If we write $T_e(G^+) = \mathcal{V}^+ \times L^+$, this is equivalent to saying that $T_e(G^*)$ has the form $\mathcal{V}^* \times L^*$ where $\mathcal{V}^*$ is a subspace of $\mathcal{V}^+$ and where L^* is the intersection of L^+ with a subspace of K^{d_1} defined by linear forms $\varphi_1 Y_1 + \cdots + \varphi_{d_1} Y_{d_1}$ with $(\varphi_1, \ldots, \varphi_{d_1}) \in \mathbb{Z}^{d_1}[\underline{D}]$ (see Exercise 8.8).

When $G^- = \{e\}$ and Σ is finite, the above theorem is a special case of Theorem 2.1 of [P 1986a]. As in the proof of Theorem 5.1, there is little additional difficulty in assuming that Σ may be infinite. We will use this in § 8.4 to deduce the general case of Theorem 8.1 from the case where $G^- = \{e\}$. The proof of this special case will follow essentially the same pattern as the proof of Theorem 5.1. However, the arguments will be less geometric. In order to handle the multiplicities, we shall need to concentrate not only on algebraic sets but also on the ideals defining them.

Note that Theorem 5.1 follows from Theorem 8.1 by taking $G^- = \{e\}$, $G^+ = G$, $S_0 = 0$ and $\mathcal{W} = \{0\}$, so that $\ell_0' = 0$.

8.1.1 An Example of Application

We give below a simple example of application of Theorem 8.1 with the group $G = \mathbb{G}_m^2$. Another example, with the group $G = \mathbb{G}_a \times \mathbb{G}_m$, is given in Exercise 8.1.

Example. Let $G = \mathbb{G}_m^2 = (K^\times)^2$ and let $(\alpha, \beta) \in G$. Assume that α and β are multiplicatively independent. Choose $b \in K$ with $b \notin \mathbb{Q}$ and let $\mathcal{W}$ be the subspace of $T_e(G) = K^2$ generated by the vector $w = (1, b)$. Fix two positive integers S_0 and S_1 and consider the subset Σ of G given by

$$\Sigma = \{(\alpha^s, \beta^s);\ s \in \mathbb{Z},\ |s| \le S_1\}.$$

Suppose that, for some positive integers D_1, D_2, there exists a nonzero polynomial $P \in K[G] = K[Y_1^{\pm 1}, Y_2^{\pm 1}]$ of bidegree $\le (D_1, D_2)$ which vanishes to order $> 2S_0$ with respect to $\mathcal{W}$ at each point of

$$\Sigma[2] = \{(\alpha^s, \beta^s);\ s \in \mathbb{Z},\ |s| \le 2S_1\}.$$

Then all the hypotheses of Theorem 8.1 are satisfied with $G^- = \{e\}$ and $G^+ = G$. So, there exists a connected algebraic subgroup G^* of G of dimension < 2 such that

$$\binom{S_0 + \ell_0'}{\ell_0'} \operatorname{Card}\left(\frac{\Sigma + G^*}{G^*}\right) \mathcal{H}(G^*;\ D_1, D_2) \le 8 D_1 D_2.$$

where ℓ_0' denotes the dimension over K of the quotient $(\mathcal{W} + T_e(G^*))/T_e(G^*)$. Since G^* has dimension at most one, it is of the form $G^* = T_\Phi$ for some subgroup Φ of $\mathbb{Z}^2$ of rank at least one. Let (k_1, k_2) be a nonzero element of Φ. Then $T_e(G^*)$ is contained in the kernel of the linear form $k_1 Y_1 + k_2 Y_2$. Since $b \notin \mathbb{Q}$, we have $k_1 + k_2 b \ne 0$, thus $w \notin T_e(G^*)$ and therefore $\ell_0' = 1$. On the other hand, since α, β are multiplicatively independent, we have $(\alpha^s)^{k_1}(\beta^s)^{k_2} \ne 1$ for any $s \in \mathbb{Z}$ with $s \ne 0$. Thus the elements of Σ are pairwise incongruent modulo G^* and therefore $(\Sigma + G^*)/G^*$ has cardinality $2S_1 + 1$. Since $\mathcal{H}(G^*;\ D_1, D_2)$ is a positive integer, we deduce that:

$$8 D_1 D_2 \ge (S_0 + 1)(2S_1 + 1).$$

Conversely, suppose that D_1, D_2 are positive integers with $D_1 D_2 \ge (S_0 + 1)(2S_1 + 1)$. Then the vector space of polynomials of $K[G]$ of bidegree $\le (D_1, D_2)$ has dimension $(2D_1 + 1)(2D_2 + 1) > (2S_0 + 1)(4S_1 + 1)$. Since the right hand side of this inequality is the number of linear conditions that a polynomial must satisfy in order to vanish with multiplicity $> 2S_0$ with respect to $\mathcal{W}$ on the set $\Sigma[2]$, there is a nonzero polynomial of $K[G]$ of bidegree $\le (D_1, D_2)$ which satisfies all these conditions. Thus the constraint given by the zero estimate is optimal up to the value of the multiplicative constant.

8.2 Some Commutative Algebra

The purpose of this section is to extend Philippon's upper bound for the function $\mathcal{H}$ in § 5.2.4 to a certain class of ideals of $K[\underline{X}] = K[X_1, \dots, X_n]$ called *complete intersections*. We start by recalling some facts and definitions from commutative algebra.

8.2.1 Primary Decomposition and Rank of an Ideal

Let R be a Noetherian ring. We say that an ideal I of R is *proper* if $I \neq R$. Fix such an ideal I. Then, I can be written as an intersection of primary ideals of R:

$$I = \mathfrak{q}_1 \cap \dots \cap \mathfrak{q}_s. \tag{8.2}$$

Moreover, it is possible to choose $\mathfrak{q}_1, \dots, \mathfrak{q}_s$ in such a way that none of these primary ideals contains the intersection of the others and that their respective radicals $\mathfrak{p}_1, \dots, \mathfrak{p}_s$ are distinct prime ideals of R. When this is the case, the decomposition is said to be *irredundant* and the prime ideals $\mathfrak{p}_1, \dots, \mathfrak{p}_s$ are uniquely determined by I. They are called the *associated prime ideals* of I. These prime ideals are characterized by the following property: given a homogeneous polynomial $P \in R$, the multiplication by P in the quotient R/I is an injective map if and only if P does not belong to any of $\mathfrak{p}_1, \dots, \mathfrak{p}_s$. However, the corresponding primary ideals $\mathfrak{q}_1, \dots, \mathfrak{q}_s$ in the decomposition (8.2), called *primary components* of I, may differ from one decomposition to another.

By construction, any associated prime ideal of I contains I. Moreover, any prime ideal $\mathfrak{p}$ of R containing I must contain an associated prime ideal of I. Therefore the set of all prime ideals of R containing I and the set of associated prime ideals of I have the same minimal elements with respect to inclusion. These elements are called *minimal prime ideals* of I. Recall that, for these prime ideals, the corresponding primary ideals are unique: they do not depend on the choice of a particular irredundant primary decomposition of I (see Theorem 8, § 5, Chap. IV of [ZSa 1958]).

The *rank* of a prime ideal $\mathfrak{p}$ of R is the largest integer r for which there exists a strictly increasing chain of $r + 1$ prime ideals of R ending with $\mathfrak{p}$:

$$\mathfrak{p}_0 \subset \mathfrak{p}_1 \subset \cdots \subset \mathfrak{p}_r = \mathfrak{p}.$$

In general, if I is a proper ideal of R, the rank of I is defined as the minimum of the ranks of the prime ideals of R containing I. It is denoted $\mathrm{rank}(I)$. Equivalently, $\mathrm{rank}(I)$ is the minimum of the ranks of the minimal prime ideals of I. A theorem of Krull (see Theorem 30, § 14, Chap. IV of [ZSa 1958]) shows that one has $\mathrm{rank}(I) \leq r$ if I is a proper ideal of R generated by r elements. In particular, the rank of a proper ideal of R is always finite. An ideal of R is said to be *unmixed* if all its associated prime ideals have the same rank. In particular, when an ideal I is unmixed, all its associated prime ideals are minimal prime ideals of I, and thus I admits a unique irredundant primary decomposition.

In this section and the next one, we work over the Noetherian ring $R = K[\underline{X}] = K[X_1, \ldots, X_n]$ for some positive integer n. A theorem of Hilbert (the "Nullstellensatz") shows that an ideal I of $K[\underline{X}]$ is proper if and only if it admits at least one zero in K^n. Moreover, when I is proper ideal of $K[\underline{X}]$, there is a bijection between the minimal prime ideals of I and the irreducible components of its zero set in K^n. Under this bijection, a minimal prime ideal $\mathfrak{p}$ of I is mapped to an irreducible component V of the zero set of I if and only if $\mathfrak{p} = I(V)$ is the set of polynomials vanishing identically on V or, equivalently, if and only if V is the zero set of $\mathfrak{p}$. If I has rank r, its zero set has dimension $n - r$. Finally, a theorem of Macaulay (see Theorem 26, § 8, Chap. VII of [ZSa 1958]) shows that, if I is an ideal of $K[\underline{X}]$ of rank r generated by r polynomials $P_1, \ldots, P_r$, then I is unmixed. An ideal of this type is said to be a *complete intersection*.

8.2.2 Multihomogeneous Hilbert-Samuel Polynomial

As in § 5.2.3, we decompose the set of variables $\underline{X} = (X_1, \ldots, X_n)$ into subsets

$$\underline{X}^{(1)} = (X_1^{(1)}, \ldots, X_{n_1}^{(1)}), \ \ldots, \ \underline{X}^{(k)} = (X_1^{(k)}, \ldots, X_{n_k}^{(k)})$$

with $n_1 + \ldots + n_k = n$. Recall that, for a given k-tuple of integers $\underline{D} = (D_1, \ldots, D_k) \in \mathbb{N}^k$, we denote by $K[\underline{X}]_{\leq \underline{D}}$ the vector space of elements of $K[\underline{X}] = K[\underline{X}^{(1)}, \ldots, \underline{X}^{(k)}]$ consisting of all polynomials of multidegree $\leq \underline{D}$, i.e. those polynomials having total degree $\leq D_i$ in the set of variables $\underline{X}^{(i)}$ for $i = 1, \ldots, k$. We define the *Hilbert function* of an ideal I of $K[\underline{X}]$ as the map $H(I; -)\colon \mathbb{N}^k \to \mathbb{N}$ given, for any $\underline{D} \in \mathbb{N}^k$, by

$$H(I; \underline{D}) = \dim_K \left((K[\underline{X}]_{\leq \underline{D}} + I)/I \right).$$

In § 5.2.3, we discussed the case where I is the ideal $I(V)$ of all polynomials vanishing identically on some nonempty algebraic subset V of K^n. In the more general situation that we consider here, it can be shown again that $H(I; \underline{D})$ is given by a polynomial in $D_1, \ldots, D_k$ for sufficiently large integral values of $D_1, \ldots, D_k$. This polynomial is also called the *multihomogeneous Hilbert-Samuel polynomial* of I associated with the above partition of $\underline{X}$, and the reason is similar. This polynomial is clearly 0 if $I = K[\underline{X}]$. Otherwise, one shows that its degree is the dimension $d \geq 0$ of the set of zeros of I in K^n. Its degree is also given by $d = n - r$ where r denotes the rank of I. A reference for this is [Vd 1928].

If I is a proper ideal of $K[\underline{X}]$ of rank r, we denote by $\mathcal{H}(I;\, \underline{D})$ the product by $(n - r)!$ of the homogeneous part of degree $n - r$ of its multihomogeneous Hilbert-Samuel polynomial. Otherwise, if $I = K[\underline{X}]$, we define $\mathcal{H}(I;\, \underline{D}) = 0$. In all cases, if we fix a point $\underline{C} \in \mathbb{N}^k$ with sufficiently large coordinates, then $H(I;\, \underline{C} + \underline{T})$ is given by a polynomial in $\underline{T}$ for $\underline{T} \in \mathbb{N}^k$ and we get

$$\mathcal{H}(I;\, \underline{D}) = (n - r)! \lim_{\substack{t \to \infty \\ t \in \mathbb{N}}} \frac{H(I;\, \underline{C} + t\underline{D})}{t^{n-r}}$$

for any $\underline{D} \in \mathbb{N}^k$. In accordance with § 5.2.3, when $I = I(V)$ for some algebraic subset $V \neq \emptyset$ of K^n, we also write $H(V; \underline{D})$ to denote $H(I; \underline{D})$ and $\mathcal{H}(V; \underline{D})$ to denote $\mathcal{H}(I; \underline{D})$.

In the sequel, we shall need the following important fact which generalizes Proposition 5.2:

Proposition 8.3. *Let I be a proper ideal of $K[\underline{X}]$ of rank r and let $I = \mathfrak{q}_1 \cap \cdots \cap \mathfrak{q}_s$ be an irredundant primary decomposition of I. Assume that $\mathfrak{q}_i$ has rank r for $i = 1, \ldots, t$ and rank $> r$ for $i = t+1, \ldots, s$. Then, we have*

$$\mathcal{H}(I; \underline{D}) = \sum_{i=1}^{t} \mathcal{H}(\mathfrak{q}_i; \underline{D}).$$

The easiest way to prove this result is to work over the larger ring

$$K[\underline{\widetilde{X}}^{(1)}, \ldots, \underline{\widetilde{X}}^{(k)}]$$

introduced in § 5.2.3 and to replace I by the corresponding multihomogeneous ideal $\widetilde{I}$ also defined in § 5.2.3. Then, one shows that $\widetilde{I} = \widetilde{\mathfrak{q}}_1 \cap \cdots \cap \widetilde{\mathfrak{q}}_s$ is a primary decomposition of $\widetilde{I}$ (see part 9 of Theorem 17, § 5, Chap. VII of [ZSa 1958] for the homogeneous case; the general case is similar). The conclusion follows by applying Theorem 8 of [Vd 1928] to this decomposition of $\widetilde{I}$. An alternative and more direct approach is suggested by the exercises 8.4 and 8.5.

8.2.3 Philippon's Upper Bound

The following result is again a special case of P. Philippon's general upper bound for the function $\mathcal{H}$ (Proposition 3.3 of [P 1986a]):

Theorem 8.4. *Let I be an ideal of $K[\underline{X}] = K[X_1, \ldots, X_n]$ which is a complete intersection and let J be the ideal of $K[\underline{X}]$ generated by I and by a family $\mathcal{F}$ of polynomials of multidegree $\leq \underline{D}$. Assume that J is a proper ideal of $K[\underline{X}]$. Then, we have*

$$\mathcal{H}(J; \underline{D}) \leq \mathcal{H}(I; \underline{D}).$$

For the proof, we will need the following lemma:

Lemma 8.5. *Let $k \geq 0$ be an integer, let $P_1, \ldots, P_k$ be elements of $K[\underline{X}]$ which generate an ideal I_k of rank k, and let $P_{k+1} \in K[\underline{X}]$ be a polynomial of multidegree $\leq \underline{D}$ which does not belong to any of the associated prime ideals of I_k. Assume that the ideal I_{k+1} generated by $P_1, \ldots, P_{k+1}$ is proper. Then, I_{k+1} is a complete intersection of rank $k+1$ and we have*

$$\mathcal{H}\left(I_{k+1}; \underline{D}\right) \leq \mathcal{H}(I_k; \underline{D}).$$

Proof. Since I_{k+1} is proper and generated by $k+1$ elements, Krull's theorem shows that its rank is at most $k+1$ (see § 8.2.1). On the other hand, since P_{k+1} does not belong to any minimal prime ideal of I_k, none of the minimal prime ideals of I_{k+1} is a minimal prime ideal of I_k. The rank of I_k being k, this implies that I_{k+1} has rank at least $k+1$ and thus its rank is $k+1$. Hence, I_{k+1} is a complete intersection and it is unmixed of rank $k+1$.

Since P_{k+1} does not belong to any of the associated prime ideals of I_k, the multiplication by P_{k+1} defines an injective endomorphism of the $K[\underline{X}]$-module $K[\underline{X}]/I_k$. Its image being I_{k+1}/I_k, we get an exact sequence of $K[\underline{X}]$-modules

$$0 \longrightarrow K[\underline{X}]/I_k \xrightarrow{\times P_{k+1}} K[\underline{X}]/I_k \xrightarrow{\nu} K[\underline{X}]/I_{k+1} \longrightarrow 0,$$

where ν denotes the canonical map sending a class $Q + I_k \in K[\underline{X}]/I_k$ to $Q + I_{k+1} \in K[\underline{X}]/I_{k+1}$. By restriction, ν induces, for each $\underline{T} \in \mathbb{N}^k$ a surjective K-linear map

$$\left(K[\underline{X}]_{\leq \underline{D}+\underline{T}} + I_k\right)/I_k \longrightarrow \left(K[\underline{X}]_{\leq \underline{D}+\underline{T}} + I_{k+1}\right)/I_{k+1}.$$

Moreover, since P_{k+1} has multidegree $\leq \underline{D}$, the multiplication by P_{k+1} induces an injective K-linear map

$$\left(K[\underline{X}]_{\leq \underline{T}} + I_k\right)/I_k \longrightarrow \left(K[\underline{X}]_{\leq \underline{D}+\underline{T}} + I_k\right)/I_k,$$

whose image is contained in the kernel of the previous map. Comparing dimensions, this implies

$$H(I_{k+1};\, \underline{D}+\underline{T}) \leq H(I_k;\, \underline{D}+\underline{T}) - H(I_k;\, \underline{T}).$$

From this, we conclude as in the last part of the proof of Lemma 5.4. □

Proof of Theorem 8.4. Let r and s be the respective ranks of I and J. Since I is a complete intersection, there exist polynomials $P_1, \ldots, P_r \in K[\underline{X}]$ such that $I = (P_1, \ldots, P_r)$. We claim that there also exist polynomials $P_{r+1}, \ldots, P_s$, all of multidegree $\leq \underline{D}$, such that for $k = r, \ldots, s$, the ideal $I_k = (P_1, \ldots, P_k)$ is a complete intersection of rank k with $I \subseteq I_k \subseteq J$ and $\mathcal{H}(I_k;\, \underline{D}) \leq \mathcal{H}(I;\, \underline{D})$.

We proceed by induction on k. For $k = r$, there is nothing to prove. Assume that $P_1, \ldots, P_k$ have been constructed for some integer k with $r \leq k < s$ and that the corresponding ideal I_k has the required properties. Let $\mathfrak{p}_1, \ldots, \mathfrak{p}_t$ be the associated prime ideals of I_k. Since they all have rank $k < s$, none of these prime ideals contains the set $\mathcal{F}$. So, for each $i = 1, \ldots, t$, there is a polynomial $Q_i \in \mathcal{F}$ which does not belong to $\mathfrak{p}_i$. Consider the sequence of polynomials $\left(\sum_{i=1}^{t} m^{i-1} Q_i\right)_{m \in \mathbb{N}^*}$. This sequence has the property that any subsequence of t elements span the same vector subspace of $K[\underline{X}]$ as $Q_1, \ldots, Q_t$. Thus, each of the prime ideals $\mathfrak{p}_1, \ldots, \mathfrak{p}_t$ contains at most t elements of the sequence and, consequently, all but finitely many polynomials of that sequence do not belong to any of these prime ideals. Let P_{k+1} be one such polynomial, and let $I_{k+1} = (I_k, P_{k+1})$. By construction, P_{k+1} has multidegree $\leq \underline{D}$. Moreover I_{k+1} is a proper ideal of $K[\underline{X}]$ since it is contained in J. By Lemma 8.5, this implies $\mathcal{H}(I_{k+1};\, \underline{D}) \leq \mathcal{H}(I_k\,;\, \underline{D})$, as required.

For $k = s$, the above construction provides an ideal I_s of $K[\underline{X}]$ of the same rank as J with $I_s \subseteq J$ and

$$\mathcal{H}(I_s;\,\underline{D}) \leq \mathcal{H}(I;\,\underline{D}).$$

The fact that I_s and J share the same rank and satisfy $I_s \subseteq J$ also implies

$$\mathcal{H}(J;\,\underline{D}) \leq \mathcal{H}(I_s;\,\underline{D})$$

(see Exercise 8.3). The conclusion follows by combining the above two inequalities. □

8.3 The Group G and its Invariant Derivations

In this section, we extend the definition of the function $\mathcal{H}$ to ideals of $K[G]$. We prove an upper bound for the function $\mathcal{H}$ when the ideal is generated by the ideal of an algebraic subgroup of G and by elements of $K[G]$ of bounded multidegree. We also prove a version of Wüstholz' lemma which gives a lower bound for $\mathcal{H}$ when the ideal consists of polynomials vanishing with multiplicity on certain components of its zero set, assuming that these components are translates of an algebraic subgroup of G.

8.3.1 Intersections on an Algebraic Subgroup

As in § 5.3, we introduce a new set of variables $\underline{Z} = (Z_1, \ldots, Z_{d_1})$ besides $\underline{X} = (X_1, \ldots, X_{d_0})$ and $\underline{Y} = (Y_1, \ldots, Y_{d_1})$. These three sets of variables generate over K a polynomial ring $K[\underline{X}, \underline{Y}, \underline{Z}]$ in $n := d_0 + 2d_1$ variables. Given integers $D_0 \in \mathbb{N}$ and $\underline{D} = (D_1, \ldots, D_{d_1}) \in \mathbb{N}^{d_1}$, we say that an element of this ring has multidegree $\leq (D_0, \underline{D})$ if it has total degree $\leq D_0$ in the set of variables $\underline{X}$ and total degree $\leq D_j$ in the variables (Y_j, Z_j) for $j = 1, \ldots, d_1$. We also denote by $K[\underline{X}, \underline{Y}, \underline{Z}]_{\leq(D_0,\underline{D})}$ the subspace of $K[\underline{X}, \underline{Y}, \underline{Z}]$ consisting of all elements of multidegree $\leq (D_0, \underline{D})$.

Consider the surjective map of K-algebras

$$\psi\colon K[\underline{X}, \underline{Y}, \underline{Z}] \longrightarrow K[\underline{X}, \underline{Y}^{\pm 1}] \simeq K[G]$$

which sends Z_j to Y_j^{-1} for $j = 1, \ldots, d_1$ and sends the remaining variables $\underline{X}$, $\underline{Y}$ to themselves. His kernel is the ideal

$$(Y_1 Z_1 - 1, \ldots, Y_{d_1} Z_{d_1} - 1)$$

and the zero set of this ideal in $K^n = K^{d_0} \times K^{d_1} \times K^{d_1}$ is the algebraic subgroup U of $K^{d_0} \times \left(K^\times\right)^{2d_1}$ defined in § 5.3. Recall also the group isomorphism $\pi\colon U \to G$ defined in § 5.3 by sending a point $(\underline{x}, \underline{y}, \underline{z}) \in U$ to the point $(\underline{x}, \underline{y}) \in G$. This map is related to ψ by the property that, for any polynomial $P \in K[\underline{X}, \underline{Y}, \underline{Z}]$, we have $\psi(P) = P \circ \pi^{-1}$. Moreover, if E is an algebraic subset of G, the ideal $I(E)$ of E in $K[G]$ is related to the ideal $I(\pi^{-1}(E))$ of $\pi^{-1}(E)$ in $K[\underline{X}, \underline{Y}, \underline{Y}]$ by

$$I\big(\pi^{-1}(E)\big) = \psi^{-1}\big(I(E)\big). \tag{8.6}$$

In terms of the map ψ, an element P of $K[G]$ has multidegree $\leq (D_0, \underline{D})$ if $P = \psi(Q)$ for some element Q of $K[\underline{X}, \underline{Y}, \underline{Z}]$ of multidegree $\leq (D_0, \underline{D})$ (see § 5.1 and § 5.3). Accordingly, we put

$$K[G]_{\leq(D_0,\underline{D})} = \psi\big(K[\underline{X}, \underline{Y}, \underline{Z}]_{\leq(D_0,\underline{D})}\big).$$

Finally, we define the *Hilbert function* of an ideal I of $K[G]$ as the map $H(I;\ -)\colon \mathbb{N}\times\mathbb{N}^{d_1} \to \mathbb{N}$ given by

$$\begin{aligned} H(I;\ D_0, \underline{D}) &= \dim_K\big((K[G]_{\leq(D_0,\underline{D})} + I)/I\big) \\ &= \dim_K\big((K[\underline{X}, \underline{Y}, \underline{Z}]_{\leq(D_0,\underline{D})} + \psi^{-1}(I))/\psi^{-1}(I)\big). \end{aligned}$$

For large values of $D_0, \dots, D_{d_1}$, this function coincides with a polynomial in $D_0, \dots, D_{d_1}$ whose degree, say m, is both the dimension of the zero set of $\psi^{-1}(I)$ in K^n and the dimension of the zero set of I in G. We denote by $\mathcal{H}(I;\ D_0, \underline{D})$ the product by $m!$ of the homogeneous part of degree m of this polynomial. When $I = I(E)$ is the ideal of an algebraic subset $E \neq \emptyset$ of G, the formula (8.6) shows that $H(I;\ D_0, \underline{D})$ coincides with the function $H(E;\ D_0, \underline{D})$ defined in § 5.3. Then, $\mathcal{H}(I;\ D_0, \underline{D})$ also coincides with $\mathcal{H}(E;\ D_0, \underline{D})$.

Lemma 8.7. *Let G^* be an algebraic subgroup of G and let $I = \psi^{-1}\big(I(G^*)\big)$. Then, the ideal I is a complete intersection in $K[\underline{X}, \underline{Y}, \underline{Z}]$.*

Proof. Write $G^* = \mathcal{V} \times T_\Phi$ where $\mathcal{V}$ is a subspace of K^{d_0} and Φ is a subgroup of $\mathbb{Z}^{d_1}$. Define $\mathcal{V}^\perp$ to be the subspace of $K[\underline{X}] = K[X_1, \dots, X_{d_0}]$ consisting of the linear forms $a_1X_1 + \cdots + a_{d_0}X_{d_0}$ which vanish identically on $\mathcal{V}$. Choose a basis $\{L_1, \dots, L_r\}$ of $\mathcal{V}^\perp$ and a basis $\{\underline{\varphi}_1, \dots, \underline{\varphi}_s\}$ of the group Φ. Proposition 5.6 shows that $I(G^*)$ is the ideal of $K[G]$ generated by the $r+s$ elements $L_1, \dots, L_r$ and $\underline{Y}^{\underline{\varphi}_1} - 1, \dots, \underline{Y}^{\underline{\varphi}_s} - 1$. Since the kernel of ψ is generated by d_1 elements, we conclude that I is generated by $r+s+d_1$ polynomials. On the other hand, Theorem 5.13 shows that G^* is equidimensional of dimension

$$\dim_K(\mathcal{V}) + (d_1 - \operatorname{rank}(\Phi)) = (d_0 - r) + (d_1 - s) = n - (r + s + d_1).$$

Since $\pi^{-1}(G^*)$ is an algebraic subset of K^n of the same dimension and since I is the ideal of all polynomials of $K[\underline{X}, \underline{Y}, \underline{Z}]$ vanishing on that set, this shows that I is a complete intersection. □

Theorem 8.8. *Let G^* be an algebraic subgroup of G and let $\mathfrak{A}$ be an ideal of $K[G]$ generated by the ideal $I(G^*)$ of G^* and by elements of $K[G]$ of multidegree $\leq (D_0, \underline{D})$. Assume that $\mathfrak{A}$ admits at least one zero in G. Then, we have*

$$\mathcal{H}(\mathfrak{A};\ D_0, \underline{D}) \leq \mathcal{H}(G^*;\ D_0, \underline{D}).$$

Proof. Put $I = \psi^{-1}(I(G^*))$ and $J = \psi^{-1}(\mathfrak{A})$. Then, J is an ideal of $K[\underline{X}, \underline{Y}, \underline{Z}]$ generated by I and by elements of $K[\underline{X}, \underline{Y}, \underline{Z}]$ of multidegree $\leq (D_0, \underline{D})$. Since, by Lemma 8.7, the ideal I is a complete intersection, Theorem 8.4 gives $\mathcal{H}(J;\, D_0, \underline{D}) \leq \mathcal{H}(I;\, D_0, \underline{D})$. The conclusion follows. □

8.3.2 Invariant Derivations

Recall that a *K-derivation* of a commutative K-algebra R is a K-linear map $\mathcal{D}: R \to R$ which satisfies

$$\mathcal{D}(PQ) = \mathcal{D}(P)Q + P\mathcal{D}(Q)$$

for any $P, Q \in R$ (see Chap. VIII, § 5 of [L 1993]). They form an R-module with the sum of two derivations $\mathcal{D}_1$ and $\mathcal{D}_2$ defined by $(\mathcal{D}_1 + \mathcal{D}_2)(P) = \mathcal{D}_1(P) + \mathcal{D}_2(P)$ for any $P \in R$, and the product of a derivation $\mathcal{D}$ by an element a of R defined by $(a\mathcal{D})(P) = a\mathcal{D}(P)$ for any $P \in R$.

We consider here the case where $R = K[G]$ is the K-algebra of polynomial functions $P: G \to K$ on the group G. If we identify $K[G]$ with $K[\underline{X}, \underline{Y}^{\pm 1}]$ in the usual way (see § 5.3), then one sees that the differential operators

$$\sum_{i=1}^{d_0} A_i \frac{\partial}{\partial X_i} + \sum_{j=1}^{d_1} B_j \frac{\partial}{\partial Y_j} \tag{8.9}$$

with $A_1, \ldots, A_{d_0}, B_1, \ldots, B_{d_1} \in K[G]$ are K-derivations of $K[G]$ and that any K-derivation $\mathcal{D}$ of $K[G]$ can be written in this way by taking $A_i = \mathcal{D}(X_i)$ for $i = 1, \ldots, d_0$ and $B_j = \mathcal{D}(Y_j)$ for $j = 1, \ldots, d_1$. In the sequel, we use the word derivation to mean a K-derivation.

We say that a derivation $\mathcal{D}$ of $K[G]$ is *invariant* if it commutes with the operator $\tau_g: G \to G$ of translation by g for any $g \in G$ (see the definition of τ_g in § 5.3.2). More precisely, a derivation $\mathcal{D}$ of $K[G]$ is said to be invariant if it satisfies

$$\mathcal{D}(P) \circ \tau_g = \mathcal{D}(P \circ \tau_g) \tag{8.10}$$

for any $P \in K[G]$ and any $g \in G$.

Lemma 8.11. *The invariant derivations of $K[G]$ are precisely the derivations*

$$\mathcal{D}_w = \sum_{i=1}^{d_0} \xi_i \frac{\partial}{\partial X_i} + \sum_{j=1}^{d_1} \eta_j Y_j \frac{\partial}{\partial Y_j}$$

with $w = (\xi_1, \ldots, \xi_{d_0}, \eta_1, \ldots, \eta_{d_1}) \in K^d$, introduced in § 8.1.

Proof. We first observe that a derivation $\mathcal{D}$ is invariant if it satisfies the condition (8.10) for any $g \in G$ and for each of the polynomials $X_1, \ldots, X_{d_0}, Y_1, \ldots, Y_{d_1}$. Write a derivation $\mathcal{D}$ in the form (8.9) and let $g = (x_1, \ldots, x_{d_0}, y_1, \ldots, y_{d_1})$ be an arbitrary element of G. The condition (8.10) applied respectively with $P = X_i$ and $P = Y_j$ reduces to

$$A_i \circ \tau_g = \mathcal{D}(X_i + x_i) = A_i \quad \text{and} \quad B_j \circ \tau_g = \mathcal{D}(y_j Y_j) = y_j B_j. \tag{8.12}$$

Evaluating these equalities at the neutral element e of G gives respectively

$$A_i(g) = A_i(e) \quad \text{and} \quad B_j(g) = B_j(e) y_j.$$

So, if $\mathcal{D}$ is an invariant derivation of $K[G]$, we must have $A_i = A_i(e)$ for $i = 1, \dots, d_0$ and $B_j = B_j(e) Y_j$ for $j = 1, \dots, d_1$, and therefore $\mathcal{D}$ has the form stated in the lemma. Conversely, if $\mathcal{D}$ is a derivation of this form, then $A_i = \xi_i$ and $B_j = \eta_j Y_j$ clearly satisfy the conditions (8.12), and thus $\mathcal{D}$ is invariant. □

In § 8.1, we defined the *tangent space at the identity* of an algebraic subgroup G^* of G to be the subspace $T_e(G^*)$ of K^d consisting of all points $w \in K^d$ for which the corresponding invariant derivation $\mathcal{D}_w$ maps the ideal $I(G^*)$ to itself. When $G^* = G$, we have $I(G^*) = 0$ and so $T_e(G) = K^d$. In general, we have the following description of $T_e(G^*)$:

Lemma 8.13. *Let $\mathcal{V}$ be a subspace of K^{d_0}, let Φ be a finitely generated subgroup of $\mathbb{Z}^{d_1}$, and let $G^* = \mathcal{V} \times T_\Phi$ be the corresponding algebraic subgroup of G. Then, $T_e(G^*) = \mathcal{V} \times \Phi^\perp$ where $\Phi^\perp$ denotes the subspace of K^{d_1} consisting of the common zeros of the linear forms $\varphi_1 Y_1 + \cdots + \varphi_{d_1} Y_{d_1}$ with $(\varphi_1, \dots, \varphi_{d_1}) \in \Phi$. Moreover, the group G^* and its neutral component G_0^* have the same tangent space at the identity.*

Proof. Let $\{L_1, \dots, L_r\}$ be a basis of the space of linear forms in $K[\underline{X}]$ which vanish identically on $\mathcal{V}$, and let $\{\underline{\varphi}_1, \dots, \underline{\varphi}_s\}$ be a basis of Φ. By Proposition 5.6, the ideal $I(G^*)$ is generated by $L_1, \dots, L_r$ and the polynomials $\underline{Y}^{\underline{\varphi}_1} - 1, \dots, \underline{Y}^{\underline{\varphi}_s} - 1$. Therefore, a point $w = (\underline{\xi}, \underline{\eta}) \in K^{d_0} \times K^{d_1}$ belongs to $T_e(G^*)$ if and only if it satisfies

$$\mathcal{D}_w(L_i) \in I(G^*) \text{ for } i = 1, \dots, r \quad \text{and} \quad \mathcal{D}_w\big(\underline{Y}^{\underline{\varphi}_j} - 1\big) \in I(G^*) \text{ for } j = 1, \dots, s.$$

Now, we have $\mathcal{D}_w(L) = L(\underline{\xi})$ for any linear form L in $K[\underline{X}]$. We also find, for any $\underline{\varphi} \in \Phi$,

$$\mathcal{D}_w\big(\underline{Y}^{\underline{\varphi}} - 1\big) = (\underline{\varphi}, \underline{\eta})\underline{Y}^{\underline{\varphi}} \equiv (\underline{\varphi}, \underline{\eta}) \bmod (I(G^*)),$$

where $(\underline{\varphi}, \underline{\eta})$ stands for $\varphi_1\eta_1 + \cdots + \varphi_{d_1}\eta_{d_1}$. Thus, the conditions on w amount to $L_i(\underline{\xi}) = 0$ for $i = 1, \dots, r$ and $(\underline{\varphi}_j, \underline{\eta}) = 0$ for $j = 1, \dots, s$. They are satisfied if and only if $w \in \mathcal{V} \times \Phi^\perp$. This proves the first assertion of the lemma.

For the second assertion, we use Theorem 5.13. It shows that $G_0^* = \mathcal{V} \times T_{\overline{\Phi}}$ where $\overline{\Phi}$ is the largest subgroup of $\mathbb{Z}^{d_1}$ containing Φ with the same rank as Φ. Since Φ is of finite index in $\overline{\Phi}$, we have $\Phi^\perp = \overline{\Phi}^\perp$ and, by the first part of the lemma, we deduce $T_e(G^*) = T_e(G_0^*)$. □

We conclude this section with three lemmas. The first one is essentially Lemma 4.6 of [P 1986a].

Lemma 8.14. *Let G^* be an algebraic subgroup of G, let $\mathcal{W}$ be a subspace of K^d, and put*

$$\ell = \dim_K \left(\frac{\mathcal{W} + T_e(G^*)}{T_e(G^*)} \right).$$

Then, there are elements $w_1, \ldots, w_\ell$ of $\mathcal{W}$ and polynomials $P_1, \ldots, P_\ell$ of $I(G^*)$ such that

$$\mathcal{D}_{w_i}(P_j) \equiv \delta_{i,j} \bmod I(G^*), \quad \text{for } i, j = 1, \ldots, \ell.$$

Note that, in the above statement, we do not assume that the group G^* is connected.

Proof. Write $G^* = \mathcal{V} \times T_\Phi$ where $\mathcal{V}$ is a subspace of K^{d_0} and Φ is a finitely generated subgroup of $\mathbb{Z}^{d_1}$. As in the proof of Lemma 8.13, choose a basis $\{L_1, \ldots, L_r\}$ of the space of linear forms in $K[\underline{X}]$ vanishing identically on $\mathcal{V}$, and a basis $\{\underline{\varphi}_1, \ldots, \underline{\varphi}_s\}$ of Φ. Consider the vector-subspace of $I(G^*)$

$$E = \langle L_1, \ldots, L_r, \underline{Y}^{\underline{\varphi}_1} - 1, \ldots, \underline{Y}^{\underline{\varphi}_s} - 1 \rangle \subseteq I(G^*),$$

spanned over K by $L_1, \ldots, L_r$ and the polynomials $\underline{Y}^{\underline{\varphi}_j} - 1$ with $j = 1, \ldots, s$. We observed in the proof of Lemma 8.13 that, for any point $w = (\underline{\xi}, \underline{\eta}) \in K^{d_0} \times K^{d_1}$, any $L \in \mathcal{V}^\perp$ and any $\underline{\varphi} \in \Phi$, we have

$$\mathcal{D}_w(L) = L(\underline{\xi}) \quad \text{and} \quad \mathcal{D}_w\left(\underline{Y}^{\underline{\varphi}} - 1\right) \equiv (\underline{\varphi}, \underline{\eta}) \bmod I(G^*),$$

where $(\underline{\varphi}, \underline{\eta}) = \varphi_1\eta_1 + \cdots + \varphi_{d_1}\eta_{d_1}$. Thus, for any $w \in K^d$ and any $P \in E$, there exists a constant $c \in K$ such that $\mathcal{D}_w(P) \equiv c \bmod I(G^*)$. Consider the bilinear map

$$b : K^d \times E \longrightarrow K$$

which maps a point $(w, P) \in K^d \times E$ to the unique element c of K satisfying the above condition. Since, $I(G^*)$ is generated by E as an ideal, it is clear that the left kernel of b is $E^\perp = T_e(G^*)$. Thus, if $w_1, \ldots, w_\ell$ are elements of $\mathcal{W}$ whose images in $K^d/T_e(G^*)$ are linearly independent over K, then there exist $P_1, \ldots, P_\ell \in E$ such that $b(w_i, P_j) = \delta_{i,j}$ for $i, j = 1, \ldots, \ell$. □

The next lemma extends the validity of Lemma 8.14 to a finite union of translates of an algebraic subgroup of G.

Lemma 8.15. *Let G^* be an algebraic subgroup of G, let Σ be a subset of G, and let $\mathcal{W}$ be a subspace of K^d. Assume that $\Sigma + G^*$ consists of a finite union of translates of G^*. Denote by J the ideal $I(\Sigma + G^*)$ and define ℓ as in Lemma 8.14. Then, there are elements $w_1, \ldots, w_\ell$ of $\mathcal{W}$ and polynomials $Q_1, \ldots, Q_\ell$ of J such that*

$$\mathcal{D}_{w_i}(Q_j) \equiv \delta_{i,j} \bmod J, \quad \text{for } i, j = 1, \ldots, \ell.$$

Proof. Proposition 5.6 shows that the ideal $I(G^*)$ of G^* is generated by elements of $K[G]$ which are constant on each translate of G^*. Thus, for each $\sigma \in G$ with $\sigma \notin G^*$, there is an element A_σ of $I(G^*)$ which induces the constant function 1 on $\sigma + G^*$.

Let $\{\sigma_1, \dots, \sigma_s\}$ be a maximal subset of Σ consisting of elements which are pairwise incongruent modulo G^*. For each $i = 1, \dots, s$, define

$$B_i = \prod_{j \neq i} A_{\sigma_i - \sigma_j} \circ \tau_{-\sigma_j}.$$

Then, B_i is identically 1 on $\sigma_i + G^*$ and identically 0 on each of the other translates $\sigma_j + G^*$ with $j \neq i$.

Let $w_1, \dots, w_\ell$ and $P_1, \dots, P_\ell$ be as in Lemma 8.14, and consider the polynomials $Q_1, \dots, Q_\ell$ given by

$$Q_j = \sum_{k=1}^{s} B_k^2 \big(P_j \circ \tau_{-\sigma_k}\big), \quad (1 \leq j \leq \ell).$$

Since $P_j \circ \tau_{-\sigma_k}$ is identically 0 on $\sigma_k + G^*$, the product $B_k^2\big(P_j \circ \tau_{-\sigma_k}\big)$ is identically 0 on $\Sigma + G^*$ and thus belongs to the ideal J for any choice of k and j. This implies that $Q_1, \dots, Q_\ell$ all belong to J. Moreover, the function

$$B_k^2 \mathcal{D}_{w_i}\big(P_j \circ \tau_{-\sigma_k}\big) = B_k^2\big(\mathcal{D}_{w_i}(P_j) \circ \tau_{-\sigma_k}\big)$$

is, by construction, constant equal to $\delta_{i,j}$ on $\sigma_k + G^*$. It is also identically 0 on the other translates of G^* contained in $\Sigma + G^*$. Therefore, we find that

$$\mathcal{D}_{w_i}(Q_j) = \sum_{k=1}^{s} \Big(2B_k \mathcal{D}_{w_i}(B_k)\big(P_j \circ \tau_{-\sigma_k}\big) + B_k^2 \mathcal{D}_{w_i}\big(P_j \circ \tau_{-\sigma_k}\big)\Big)$$

is constant equal to $\delta_{i,j}$ on $\Sigma + G^*$. This means simply $\mathcal{D}_{w_i}(Q_j) \equiv \delta_{i,j} \bmod J$, as required. □

Lemma 8.16. *Let J be an ideal of $K[G]$. Assume that there exist elements $w_1, \dots, w_\ell$ of K^d and polynomials $Q_1, \dots, Q_\ell$ in J such that $\mathcal{D}_{w_i}(Q_j) \equiv \delta_{i,j} \bmod J$ for $i, j = 1, \dots, \ell$. Then, for any $\underline{\alpha}, \underline{\beta} \in \mathbb{N}^\ell$ with $|\underline{\beta}| \leq |\underline{\alpha}|$, we have, modulo J,*

$$\big(\mathcal{D}_{w_1}^{\beta_1} \cdots \mathcal{D}_{w_\ell}^{\beta_\ell}\big)\big(P Q_1^{\alpha_1} \cdots Q_\ell^{\alpha_\ell}\big) \equiv \begin{cases} 0 & \text{if } \underline{\beta} \neq \underline{\alpha}, \\ \alpha_1! \cdots \alpha_\ell! P & \text{if } \underline{\beta} = \underline{\alpha}. \end{cases} \tag{8.17}$$

Proof. We first observe that, for any derivation $\mathcal{D}$ of $K[G]$ and any integer $t \geq 2$, we have

$$\mathcal{D}(J^t) \subseteq J^{t-1}.$$

By induction on s, this implies that, for any derivations $\mathcal{D}_1, \dots, \mathcal{D}_s$ of $K[G]$ and any integer $t > s$, we have

$$\big(\mathcal{D}_1 \cdots \mathcal{D}_s\big)(J^t) \subseteq J^{t-s}.$$

Note also that the invariant derivations of $K[G]$ commute (see Exercise 8.6).

We are now ready to prove the lemma by induction on $|\underline{\beta}|$. If $|\underline{\beta}| = 0$, the conclusion is clear. Assume that $|\underline{\beta}| = t$ for some integer $t > 0$ and that the lemma holds for differential operators of order $< t$. Choose an index i such that $\beta_i \neq 0$. Then, for any $\underline{\alpha} \in \mathbb{N}^\ell$ with $|\underline{\alpha}| \geq t$, we find, modulo J^t,

$$\mathcal{D}_{w_i}\left(PQ_1^{\alpha_1}\cdots Q_\ell^{\alpha_\ell}\right) \equiv \begin{cases} 0 & \text{if } \alpha_i = 0, \\ \alpha_i PQ_1^{\alpha_1}\cdots Q_i^{\alpha_i-1}\cdots Q_\ell^{\alpha_\ell} & \text{if } \alpha_i \geq 1. \end{cases}$$

Applying $\mathcal{D}_{w_1}^{\beta_1}\cdots\mathcal{D}_{w_i}^{\beta_i-1}\cdots\mathcal{D}_{w_\ell}^{\beta_\ell}$ on both sides of this congruence and using the previous observations, we deduce that the left hand side of (8.17) is congruent to 0 modulo J if $\alpha_i = 0$ and that it is congruent to

$$\alpha_i\left(\mathcal{D}_{w_1}^{\beta_1}\cdots\mathcal{D}_{w_i}^{\beta_i-1}\cdots\mathcal{D}_{w_\ell}^{\beta_\ell}\right)\left(PQ_1^{\alpha_1}\cdots Q_i^{\alpha_i-1}\cdots Q_\ell^{\alpha_\ell}\right)$$

modulo J if $\alpha_i \geq 1$. The conclusion then follows from the induction hypothesis. □

8.3.3 Wüstholz' Lemma

We now prove the following special case of Wüstholz' lemma (see [Wü 1989]):

Theorem 8.18. *Let $\mathfrak{A}$ be an ideal of $K[G]$ and let $\mathcal{W}$ be a subspace of K^d. Assume that there exist a subset Σ of G, an algebraic subgroup G^* of G, and an integer $S_0 \geq 0$ such that each element Q of $\mathfrak{A}$ vanishes to order $> S_0$ at each point of $\Sigma + G^*$ with respect to $\mathcal{W}$. Assume moreover that $\Sigma + G^*$ is a finite union of translates of G^* and that the set of zeros of $\mathfrak{A}$ in G has the same dimension as G^*. Then, for any $(D_0, \underline{D}) \in \mathbb{N} \times \mathbb{N}^{d_1}$, we have*

$$\mathcal{H}(\mathfrak{A};\, D_0, \underline{D}) \geq \binom{S_0 + \ell_0'}{\ell_0'} \mathcal{H}(\Sigma + G^*;\, D_0, \underline{D}) \tag{8.19}$$

where $\ell_0' = \dim_K((\mathcal{W} + T_e(G^*))/T_e(G^*))$.

Note that the condition that the zero set of $\mathfrak{A}$ have the same dimension as G^* is satisfied when this zero set is a union of translates of G^*. In general, the hypotheses of the theorem imply that $\Sigma + G^*$ is a union of irreducible components of this zero set.

Proof. For simplicity, put $\ell = \ell_0'$ and $J = I(\Sigma + G^*)$. By Lemma 8.15, there exist elements $w_1, \ldots, w_\ell$ of $\mathcal{W}$ and polynomials $Q_1, \ldots, Q_\ell \in J$ such that $\mathcal{D}_{w_i}(Q_j) \equiv \delta_{i,j} \bmod J$ for $i, j = 1, \ldots, \ell$. Choose an upper bound $(C_0, \underline{C}) \in \mathbb{N} \times \mathbb{N}^{d_1}$ for the multidegrees of $Q_1, \ldots, Q_\ell$, and put $(T_0, \underline{T}) = S_0(C_0, \underline{C})$. We will show that, for each $(D_0, \underline{D}) \in \mathbb{N} \times \mathbb{N}^{d_1}$, we have

$$H(\mathfrak{A};\, D_0 + T_0, \underline{D} + \underline{T}) \geq \binom{S_0 + \ell}{\ell} H(J;\, D_0, \underline{D}). \tag{8.20}$$

Since, for large values of $D_0, \dots, D_{d_1}$, the functions $H(\mathfrak{A}; D_0 + T_0, \underline{D} + \underline{T})$ and $H(\mathfrak{A}; D_0, \underline{D})$ coincide with polynomials in $D_0, \dots, D_{d_1}$ with the same homogeneous part of largest degree, this will imply (8.19) independently of the choice of $(D_0, \underline{D})$.

Fix $(D_0, \underline{D}) \in \mathbb{N} \times \mathbb{N}^{d_1}$ and choose a subspace E of $K[G]_{\le(D_0,\underline{D})}$ of maximal dimension such that $E \cap J = 0$. By definition, the dimension of E is $N = H(J; D_0, \underline{D})$. Let $\{P_1, \dots, P_N\}$ be a basis of E. We claim that the products

$$P_i Q_1^{\alpha_1} \cdots Q_\ell^{\alpha_\ell}$$

with $1 \le i \le N$, $\underline{\alpha} \in \mathbb{N}^\ell$ and $|\underline{\alpha}| \le S_0$ are linearly independent modulo $\mathfrak{A}$. Since these polynomials have multidegree $\le (D_0 + T_0, \underline{D} + \underline{T})$ and since their number is $\binom{S_0+\ell}{\ell}N$, this will imply (8.20).

Assume on the contrary that there exist elements $P_{\underline{\alpha}}$ of E, not all zero, such that

$$\sum_{|\underline{\alpha}| \le S_0} P_{\underline{\alpha}} Q_1^{\alpha_1} \cdots Q_\ell^{\alpha_\ell} \in \mathfrak{A}.$$

Among the ℓ-tuples $\underline{\alpha}$ with $P_{\underline{\alpha}} \ne 0$, choose one, say $\underline{\beta}$, of minimal length $|\underline{\beta}|$. Since $|\underline{\beta}| \le S_0$, the hypothesis of the theorem gives

$$\left(\mathcal{D}_{w_1}^{\beta_1} \cdots \mathcal{D}_{w_\ell}^{\beta_\ell}\right) \left(\sum_{|\underline{\alpha}| \le S_0} P_{\underline{\alpha}} Q_1^{\alpha_1} \cdots Q_\ell^{\alpha_\ell} \right) \in J.$$

By Lemma 8.16, this is impossible since the above polynomial is congruent to $\beta_1! \cdots \beta_\ell! P_{\underline{\beta}}$ modulo J. The proof is complete. □

8.4 Proof of the Main Result

8.4.1 Case where $G^- = \{e\}$

In the case where $G^- = \{e\}$, the hypotheses of Theorem 8.1 are that $P \notin I(G^+)$ and that P vanishes to order $> (d^+)S_0$ along $\mathcal{W}$ at each point of $\Sigma[d^+]$.

Let I_1 be the ideal of $K[G]$ generated by $I(G^+)$ and P. For each integer $r \ge 2$, define also I_r to be the ideal of $K[G]$ generated by $I(G^+)$ and the polynomials

$$\mathcal{D}_{w_1} \cdots \mathcal{D}_{w_t} \left(P \circ \tau_\gamma\right) = \left(\mathcal{D}_{w_1} \cdots \mathcal{D}_{w_t} P\right) \circ \tau_\gamma$$

where $\gamma \in \Sigma[r-1]$, $0 \le t \le (r-1)S_0$ and $w_1, \dots, w_t \in \mathcal{W}$. By construction, these ideals form an increasing sequence

$$I(G^+) \subseteq I_1 \subseteq I_2 \subseteq \cdots \subseteq I_{1+d^+} \subseteq \cdots$$

Let X_r be the set of zeros of I_r in G for $r = 1, 2, \dots$ These algebraic subsets of G form in turn a decreasing sequence

$$G^+ \supseteq X_1 \supseteq \cdots \supseteq X_{1+d^+} \supseteq \cdots$$

By hypothesis, X_1 is a proper algebraic subset of G^+ and we have $e \in X_{1+d^+}$. This means $\dim(X_{1+d^+}) \geq 0$ and $\dim(X_1) < d^+$. Therefore, there is an index r with $1 \leq r \leq d^+$ such that X_r and X_{r+1} have the same dimension. Let V be a common irreducible component of X_r and X_{r+1} of this dimension. As in the proof of Theorem 5.1, we consider the sets

$$E = \{g \in G\,;\, g + V \subseteq X_r\} \quad \text{and} \quad G^* = \{g \in G\,;\, g + V = V\}.$$

The same reasoning as in § 5.4 shows that G^* is an algebraic subgroup of G and that E is a finite union of translates of G^*. Moreover G^* is a proper algebraic subgroup of G^+ because $V \subseteq X_r \subset G^+$. Since

$$E = \bigcap_{v \in V} \big(-v + X_r \big),$$

we may also describe E as the set of zeros in G of the ideal $\mathfrak{A}$ generated by $I(G^+)$ and the polynomials $Q \circ \tau_v$ with $Q \in I_r$ and $v \in V$. Since $V \subset G^+$, the ideal $\mathfrak{A}$ is also generated by $I(G^+)$ and the family $\mathcal{F}$ consisting of all polynomials of the form

$$\big(\mathcal{D}_{w_1} \cdots \mathcal{D}_{w_t} P\big) \circ \tau_{\gamma+v}$$

with $\gamma \in \Sigma[r-1]$, $v \in V$, $0 \leq t \leq (r-1)S_0$ and $w_1, \ldots, w_t \in \mathcal{W}$. The polynomials of this family having multidegree $\leq (D_0, \underline{D})$, Theorem 8.8 gives

$$\mathcal{H}(\mathfrak{A};\, D_0, \underline{D}) \leq \mathcal{H}(G^+;\, D_0, \underline{D}). \tag{8.21}$$

We claim that each element Q of $\mathfrak{A}$ vanishes to order $> S_0$ at each point of $\Sigma + G^*$ with respect to $\mathcal{W}$.

It suffices to prove this claim for a set of generators of $\mathfrak{A}$, namely for $Q \in I(G^+)$ and $Q \in \mathcal{F}$. Since $\mathcal{W} \subseteq T_e(G^+)$, we have $\mathcal{D}_w(Q) \in I(G^+)$ for each $w \in \mathcal{W}$ and each $Q \in I(G^+)$. Thus, for any $w_1, \ldots, w_s \in \mathcal{W}$ and any $Q \in I(G^+)$, the polynomial $\mathcal{D}_{w_1} \cdots \mathcal{D}_{w_s} Q$ vanishes identically on G^+. Since $\Sigma + G^*$ is contained in G^+, this proves the claim when $Q \in I(G^+)$. For the elements of $\mathcal{F}$, the claim amounts to showing that P vanishes to order $> rS_0$ with respect to $\mathcal{W}$ at each point of $v + \gamma + \Sigma + G^*$, for any $v \in V$ and any $\gamma \in \Sigma[r-1]$. Fix such a choice of v and γ. Since V is a component of X_{r+1}, the definition of I_{r+1} shows that P vanishes to order $> rS_0$ with respect to $\mathcal{W}$ on each translate of V by an element of $\Sigma[r]$. Since $v + G^* \subseteq V$ and $\gamma + \Sigma \subseteq \Sigma[r]$, this implies in particular that P vanishes to order $> rS_0$ with respect to $\mathcal{W}$ on $\gamma + v + \Sigma + G^*$.

Recall that the set E of zeros of $\mathfrak{A}$ in G is a finite union of translates of G^*. By Theorem 8.18, this fact together with the above claim imply

$$\mathcal{H}(\mathfrak{A};\, D_0, \underline{D}) \geq \binom{S_0 + \ell_0'}{\ell_0'} \mathcal{H}(\Sigma + G^*;\, D_0, \underline{D}),$$

where

$$\ell_0' = \dim_K \left(\frac{\mathcal{W} + T_e(G^*)}{T_e(G^*)} \right).$$

By Lemma 5.10, we also have

$$\mathcal{H}(\Sigma + G^*;\ D_0, \underline{D}) = \operatorname{Card}\left(\frac{\Sigma + G^*}{G^*}\right) \mathcal{H}(G^*;\ D_0, \underline{D})$$

Combining (8.21) with the above two relations gives

$$\binom{S_0 + \ell_0'}{\ell_0'} \operatorname{Card}\left(\frac{\Sigma + G^*}{G^*}\right) \mathcal{H}(G^*;\ D_0, \underline{D}) \leq \mathcal{H}(G^+;\ D_0, \underline{D}).$$

The argument at the end of § 5.4 shows that this inequality stays valid if G^* is replaced by its neutral component G_0^*. Moreover, Lemma 8.13 shows that G^* and G_0^* have the same tangent space at the identity. So, in the formula for ℓ_0', we may as well replace G^* by G_0^*: this does not affect the value of this expression. The last assertion of the theorem is also verified since G_0^* is an irreducible component of E, since E is the set of zeros of $\mathcal{F}$ in G^+, and since the elements of $\mathcal{F}$ have multidegree $\leq (D_0, \underline{D})$. □

8.4.2 General Case

Define

$$\Sigma' = \bigcup_{\gamma \in \Sigma} \gamma + G^-.$$

By hypothesis, the polynomial P vanishes to order $> (d^+)S_0$ with respect to $\mathcal{W}$ at each point of $\Sigma'[d^+]$. By virtue of the special case of Theorem 8.1 established above, there exists a proper connected algebraic subgroup G^* of G^+ such that Σ' is contained in a finite union of translates of G^* and such that

$$\binom{S_0 + \ell_0'}{\ell_0'} \operatorname{Card}\left(\frac{\Sigma' + G^*}{G^*}\right) \mathcal{H}(G^*;\ D_0, \underline{D}) \leq \mathcal{H}(G^+;\ D_0, \underline{D}),$$

where $\ell_0' = \dim_K\big((\mathcal{W} + T_e(G^*))/T_e(G^*)\big)$. We may also assume that G^* is incompletely defined in G^+ by polynomials of multidegree $\leq (D_0, \underline{D})$. Since $e \in \Sigma$, we have $G^- \subseteq \Sigma'$ and therefore G^- is contained in a finite union of translates of G^*. Since these translates are disjoint algebraic subsets of G and since G^- is connected, this implies $G^- \subseteq G^*$. We deduce $(\Sigma' + G^*)/G^* = (\Sigma + G^*)/G^*$ and the proof is complete. □

Exercises

Exercise 8.1. Let $G = \mathbb{G}_a \times \mathbb{G}_m = K \times K^\times$, and let $(\beta, \alpha) \in G$. Assume that β is nonzero and that α is not a root of unity. Fix two positive integers S_0 and S_1. Denote by $\mathcal{W}$ the subspace of $T_e(G) = K^2$ generated by the vector $w = (1, 1)$, and consider the subset Σ of G given by

$$\Sigma = \big\{(s\beta, \alpha^s)\,;\ s \in \mathbb{Z}^2,\ |s| \leq S_1\big\}.$$

Suppose that, for some positive integers D_0, D_1, there exists a nonzero polynomial $P \in K[G] = K[X, Y, Y^{-1}]$ of bidegree $\leq (D_0, D_1)$ which vanishes to order $> 2S_0$ at each point of $\Sigma[2]$. Show that this implies $4D_0D_1 \geq (S_0 + 1)(2S_1 + 1)$. Conversely, find a condition on D_0 and D_1 which ensures the existence of such a polynomial.

Exercise 8.2. Produce alternative proofs for Theorem 4.1, either with an auxiliary function (like in Chapter 4), but without Schwarz lemma (Proposition 4.6), or else with an interpolation determinant.

Hint. *Use Theorem* 8.1 *for the subspace $\mathcal{W}$ of K^d generated by $w_1, \ldots, w_n$, with $d_0 = n$, where the coordinates in K^d of $w_1, \ldots, w_n$ are given by the rows of the $n \times d$ matrix*

$$\begin{pmatrix} I_n & x_1 & \cdots & x_{d_1} \end{pmatrix},$$

where I_n is the $n \times n$ identity matrix.

Exercise 8.3. Let $k[\underline{X}] = K[\underline{X}^{(1)}, \ldots, \underline{X}^{(k)}]$, as in § 8.2.2. Show that if I, J are proper ideals of $k[\underline{X}]$ of the same rank with $I \subseteq J$, then

$$\mathcal{H}(J; \underline{D}) \leq \mathcal{H}(I; \underline{D})$$

for any $\underline{D} \in \mathbb{N}^k$.

The next two exercises provide a proof of Proposition 8.3. Again we assume

$$K[\underline{X}] = K[\underline{X}^{(1)}, \ldots, \underline{X}^{(k)}],$$

as in § 8.2.2.

Exercise 8.4. Let I be a proper ideal of rank r of $K[\underline{X}]$ and let $I = \mathfrak{q}_1 \cap \cdots \cap \mathfrak{q}_s$ be an irredundant primary decomposition of I.

(a) Show that, for each $\underline{D} \in \mathbb{N}^k$, there is an injective K-linear map

$$\left(K[\underline{X}]_{\leq \underline{D}} + I\right)/I \longrightarrow \prod_{i=1}^{s} \left(\left(K[\underline{X}]_{\leq \underline{D}} + \mathfrak{q}_i\right)/\mathfrak{q}_i\right).$$

(b) Suppose that $\mathfrak{q}_i$ has rank r for $i = 1, \ldots, t$ and rank $> r$ for $i = t + 1, \ldots, s$. Deduce that we have

$$\mathcal{H}(I; \underline{D}) \leq \sum_{i=1}^{t} \mathcal{H}(\mathfrak{q}_i; \underline{D})$$

for any $\underline{D} \in \mathbb{N}^k$.

Exercise 8.5. Let the notation be as in Exercise 8.4.b. For each $i = 1, \ldots, t$, denote by $\mathfrak{p}_i$ the radical of $\mathfrak{q}_i$ and choose a polynomial P_i such that

$$P_i \in \bigcap_{j \neq i} \mathfrak{q}_j \quad \text{and} \quad P_i \notin \mathfrak{p}_i.$$

Choose also $\underline{C} \in \mathbb{N}^k$ such that $P_i \in K[\underline{X}]_{\leq \underline{C}}$ for $i = 1, \ldots, t$.

(a) Show that, for any $\underline{D} \in \mathbb{N}^k$, there is an injective K-linear map

$$\begin{array}{ccc} \prod_{i=1}^{t} \Big(\big(K[\underline{X}]_{\leq \underline{D}} + \mathfrak{q}_i\big) / \mathfrak{q}_i \Big) & \longrightarrow & \big(K[\underline{X}]_{\leq (\underline{D}+\underline{C})} + I\big) / I \\ \big(A_i + \mathfrak{q}_i\big)_{1 \leq i \leq t} & \longmapsto & \Big(\sum_{i=1}^{t} A_i P_i\Big) + I \end{array}$$

(b) Deduce that we have

$$\sum_{i=1}^{t} \mathcal{H}(\mathfrak{q}_i; \underline{D}) \leq \mathcal{H}(I; \underline{D})$$

for any $\underline{D} \in \mathbb{N}^k$.

Exercise 8.6. Let $\mathcal{D}_1$ and $\mathcal{D}_2$ be invariant derivations of $K[G]$. Show that, for any $P \in K[G]$, we have $\mathcal{D}_1\mathcal{D}_2 P = \mathcal{D}_2\mathcal{D}_1 P$. This property is expressed by saying that the invariant derivations of $K[G]$ *commute.*

Exercise 8.7. Show that the map $G^* \mapsto T_e(G^*)$ establishes a bijection between the connected algebraic subgroups of $G = \mathbb{G}_{\mathrm{a}}^{d_0} \times \mathbb{G}_{\mathrm{m}}^{d_1}$ and the subspaces of $T_e(G) = K^{d_0} \times K^{d_1}$ of the form $\mathcal{V} \times L$ where $\mathcal{V}$ is a subspace of K^{d_0} and L a subspace of K^{d_1} defined over $\mathbb{Q}$.

Exercise 8.8. Let G^+ be an algebraic subgroup of $G = \mathbb{G}_{\mathrm{a}}^{d_0} \times \mathbb{G}_{\mathrm{m}}^{d_1}$, let G^* be a connected algebraic subgroup of G^+, and let $(D_0, \underline{D}) \in \mathbb{N} \times \mathbb{N}^{d_1}$. Show that the following three conditions are equivalent:

(i) G^* is incompletely defined in G^+ by polynomials of multidegree $\leq (D_0, \underline{D})$;

(ii) if we write $G^+ = \mathcal{V}^+ \times \boldsymbol{T}_{\Phi^+}$, then there exist a subspace $\mathcal{V}$ of $\mathcal{V}^+$ and a subgroup Φ of $\mathbb{Z}^{d_1}$ generated by Φ^+ and by elements of $\mathbb{Z}^{d_1}[\underline{D}]$ such that G^* is of finite index in $\mathcal{V} \times \boldsymbol{T}_\Phi$;

(iii) if we write $T_e(G^+) = \mathcal{V}^+ \times L^+$, then $T_e(G^*)$ has the form $\mathcal{V}^* \times L^*$ where $\mathcal{V}^*$ is a subspace of $\mathcal{V}^+$ and where L^* is the intersection of L^+ with a subspace of K^{d_1} defined by linear forms $\varphi_1 Y_1 + \cdots + \varphi_{d_1} Y_{d_1}$ with $(\varphi_1, \ldots, \varphi_{d_1}) \in \mathbb{Z}^{d_1}[\underline{D}]$.

Hint. *To prove that (i) implies (ii), write $G^* = \mathcal{V} \times \boldsymbol{T}_{\Phi^*}$, and define Φ to be the subgroup of Φ^* generated by Φ^+ and $\Phi^* \cap \mathbb{Z}^{d_1}[\underline{D}]$. Then show that $\mathcal{V} \times \boldsymbol{T}_\Phi$ is contained both in G^+ and in the zero set of any polynomial $P \in K[G]_{\leq (D_0, \underline{D})}$ which vanishes identically on G^*.*

9. Refined Measures

The purpose of this chapter is twofold. On one hand we prove Baker's nonhomogeneous Theorem 1.6. This is the second proof (§ 9.1) of the transcendence result, after the proof given in Chap. 4. Another proof (akin to Baker's own argument) will be given in Chap. 10.

On the other hand we give a sharp measure for linear independence of logarithms, both in the homogeneous and in the general case. It is a remarkable fact that the same type of argument which enables one to deal with nonhomogeneous forms also yields refined estimates, even in the homogeneous situation with $\beta_0 = 0$.

Dealing with two logarithms, A. O. Gel'fond [G 1952] was using functions of a single variable, and he could not reach a dependence on the maximal height B of the coefficients β_i better than $\exp\{-C(\log B)^2\}$. In the present state of the theory, in order to achieve the best possible dependence in B, namely $B^{-C} = \exp\{-C\log B\}$, it is necessary to use functions of several variables (m variables when dealing with m logarithms), together with Fel'dman's Delta polynomials.

Here is the main result of this chapter.

Theorem 9.1. *For each $m \geq 1$ there exists a positive number $C(m)$ with the following property. Let $\lambda_1, \ldots, \lambda_m$ be $\mathbb{Q}$-linearly independent logarithms of algebraic numbers; define $\alpha_j = \exp(\lambda_j)$ $(1 \leq j \leq m)$. Let $\beta_0, \ldots, \beta_m$ be algebraic numbers, not all of which are zero. Denote by D the degree of the number field $\mathbb{Q}(\alpha_1, \ldots, \alpha_m, \beta_0, \ldots, \beta_m)$ over $\mathbb{Q}$. Further, let B, E, E^* be positive real numbers, each $\geq e$ and let $A_1, \ldots, A_m$ be positive real numbers. Assume*

$$\log A_j \geq \max\left\{\mathrm{h}(\alpha_j),\ \frac{E|\lambda_j|}{D},\ \frac{\log E}{D}\right\} \quad (1 \leq j \leq m),$$

$$\log E^* \geq \max\left\{\frac{1}{D}\log E,\ \log\left(\frac{D}{\log E}\right)\right\}$$

and $B \geq E^*$. Further, assume either

(i) (general case)

$$B \geq \max_{1\leq i\leq m} \frac{D\log A_i}{\log E} \quad \textit{and} \quad \log B \geq \max_{0\leq i\leq m} \mathrm{h}(\beta_i)$$

or

(ii) (homogeneous rational case)

$$\beta_0 = 0, \quad \beta_i = b_i \in \mathbb{Z} \quad (1 \le i \le m), \quad b_m \neq 0$$

and

$$B \ge \max_{1 \le j \le m-1} \left(\frac{|b_m|}{\log A_j} + \frac{|b_j|}{\log A_m} \right) \cdot \frac{\log E}{D}.$$

Then the number

$$\Lambda = \beta_0 + \beta_1 \lambda_1 + \cdots + \beta_m \lambda_m$$

is nonzero and has absolute value bounded from below by

$$|\Lambda| > \exp\{-C(m) D^{m+2} (\log B)(\log A_1) \cdots (\log A_m)(\log E^*)(\log E)^{-m-1}\}.$$

Apart from the exact value of $C(m)$, this estimate includes all known results on this topic (we postpone a discussion of this issue to § 10.4).

In § 9.2 we give a sketch of proof of Theorem 9.1, and establish several auxiliary results. This proof involves interpolation determinants with one derivative. Another proof of Theorem 9.1 will be given in Chap. 10, by means of Baker's method.

In § 9.3 we compute an admissible value for $C(m)$: the conclusion of Theorem 9.1 holds with

$$C(m) = 2^{26m} m^{3m}.$$

Corollaries and comments on Theorem 9.1 are given in § 9.4.

9.1 Second Proof of Baker's Nonhomogeneous Theorem

We gave a first proof of Baker's Theorem 1.6 in Chap. 4. The method was an extension of Gel'fond's solution to Hilbert's seventh problem. Here is an extension of Schneider's solution to the same problem.

9.1.1 Idea of the Proof

Let $\beta_0, \beta_1, \ldots, \beta_{m-1}$ be complex numbers with $\beta_0 \neq 0$. Assume that the m numbers $1, \beta_1, \ldots, \beta_{m-1}$ are $\mathbb{Q}$-linearly independent. Further let $\lambda_1, \ldots, \lambda_m$ be $\mathbb{Q}$-linearly independent complex numbers with

$$\beta_0 + \beta_1 \lambda_1 + \cdots + \beta_{m-1} \lambda_{m-1} - \lambda_m = 0.$$

Define $\alpha_j = \exp(\lambda_j)$ $(1 \le j \le m)$. By Lemma 1.7 (with $k = \mathbb{Q}$, $K = \overline{\mathbb{Q}}$, $\mathcal{E} = \mathbb{C}$, while $\mathcal{M}$ is the $\mathbb{Q}$-vector space spanned by 1 and $\mathcal{L}$), Baker's Theorem will be proved if we show that one at least of the numbers in the set

$$\{\theta_1, \ldots, \theta_{2m}\} = \{\alpha_1, \ldots, \alpha_m, \beta_0, \beta_1, \ldots, \beta_{m-1}\}$$

is transcendental.

We consider $m+1$ functions

$$z_0, z_1, \dots, z_{m-1}, \exp\{z_0 + \lambda_1 z_1 + \cdots + \lambda_{m-1} z_{m-1}\}$$

of m variables $z_0, \dots, z_{m-1}$. In the tangent space $T_e(G)$ of the algebraic group $G = \mathbb{G}_{\mathrm{a}}^m \times \mathbb{G}_{\mathrm{m}}$, these functions are the restrictions of

$$z_0, z_1, \dots, z_{m-1}, \; e^{z_m}$$

to the hyperplane

$$z_0 + \lambda_1 z_1 + \cdots + \lambda_{m-1} z_{m-1} = z_m.$$

We shall take the values of these functions (and of monomials in these functions) at the m points in $\mathbb{C}^m$:

$$(0,1,0,\dots,0), (0,0,1,\dots,0), \dots, (0,0,\dots,1) \quad \text{and} \quad (\beta_0, \beta_1, \dots, \beta_{m-1}).$$

We notice that the values of the function $\exp\{z_0 + \lambda_1 z_1 + \cdots + \lambda_{m-1} z_{m-1}\}$ at these m points are respectively

$$\alpha_1, \; \alpha_2, \; \dots, \; \alpha_{m-1}, \; \alpha_m.$$

We shall also introduce linear combinations of these points: for $\underline{s} = (s_1, \dots, s_m) \in \mathbb{Z}^m$ we denote by $\xi_{\underline{s}}$ the point in $\mathbb{C}^m$ of coordinates

$$(s_m\beta_0, s_1 + s_m\beta_1, \dots, s_{m-1} + s_m\beta_{m-1}).$$

It is necessary to use somewhere the fact that each of these functions satisfies a partial differential equation with respect to the differential operator $\partial/\partial z_0$ with coefficients in the ring $\mathbb{Z}[\theta_1, \dots, \theta_{2m}]$. If this information were not used, one could multiply the variable z_0 by a transcendental constant, and the assumption that β_0 is in the set $\{\theta_1, \dots, \theta_{2m}\}$ would not be used !

For $(\underline{\tau}, t) = (\tau_0, \dots, \tau_{m-1}, t) \in \mathbb{N}^m \times \mathbb{Z}$ and $\underline{z} = (z_0, \dots, z_{m-1}) \in \mathbb{C}^m$, define

$$f_{\underline{\tau}t}(\underline{z}) = \underline{z}^{\underline{\tau}} \exp\bigl\{t(z_0 + \lambda_1 z_1 + \cdots + \lambda_{m-1} z_{m-1})\bigr\}.$$

Then

$$\left(\frac{\partial}{\partial z_0}\right)^{\sigma} f_{\underline{\tau}t}(\underline{z}) =$$
$$\sum_{\kappa=0}^{\min\{\tau_0,\sigma\}} \frac{\sigma!}{\kappa!(\sigma-\kappa)!} \cdot \frac{\tau_0!}{(\tau_0-\kappa)!} t^{\sigma-\kappa} z_0^{\tau_0-\kappa} z_1^{\tau_1} \cdots z_{m-1}^{\tau_{m-1}} e^{t(z_0+\lambda_1 z_1+\cdots+\lambda_{m-1}z_{m-1})}.$$

Hence, for $\underline{\tau} \in \mathbb{N}^m$, $t \in \mathbb{Z}$, $\underline{s} \in \mathbb{Z}^m$ and $\sigma \in \mathbb{N}$, we have

$$\left(\frac{\partial}{\partial z_0}\right)^{\sigma} f_{\underline{\tau}t}(\xi_{\underline{s}}) =$$
$$\sum_{\kappa=0}^{\min\{\tau_0,\sigma\}} \frac{\sigma!}{\kappa!(\sigma-\kappa)!} \cdot \frac{\tau_0!}{(\tau_0-\kappa)!} t^{\sigma-\kappa} (s_m\beta_0)^{\tau_0-\kappa} \prod_{i=1}^{m-1} (s_i + s_m\beta_i)^{\tau_i} \cdot \prod_{j=1}^{m} \alpha_j^{ts_j}.$$

These numbers are all in the ring $\mathbb{Z}[\theta_1, \dots, \theta_{2m}]$. The sketch of proof is now clear: we consider a matrix $\boldsymbol{M}$ whose entries are among these numbers. Using a zero estimate, we show that $\boldsymbol{M}$ has maximal rank. We select a nonzero determinant of maximal size. We estimate from above the absolute value of this interpolation determinant, using analytic means (Schwarz' lemma). Finally we choose the parameters in such a way that the assumptions of Lemma 2.1 are satisfied.

We start with the analytic upper bound (§ 9.1.2), next we deal with the multiplicity estimate (§ 9.1.3) and then we complete the proof of Theorem 1.6 (§ 9.1.4).

9.1.2 Interpolation Determinants with Derivatives

We extend Lemma 6.4 by introducing multiplicities.

Lemma 9.2. *Let L be a positive integer. Let $f_1, \dots, f_L$ be entire functions in $\mathbb{C}^n$. For $1 \le \mu \le L$, let $\underline{\zeta}_\mu$ be an element of $\mathbb{C}^n$, σ_μ a nonnegative integer and $\mathcal{D}^{(\mu)}$ a derivative operator of order σ_μ. The function of one variable*

$$\Psi(z) = \det\left(\mathcal{D}^{(\mu)} f_\lambda(z\underline{\zeta}_\mu)\right)_{1\le\lambda,\mu\le L}$$

has a zero at the origin of multiplicity

$$\ge \Theta_n(L) - \sigma_1 - \cdots - \sigma_L.$$

Proof. By multilinearity we may assume $f_\lambda(\underline{z}) = \underline{z}^{\underline{\kappa}_\lambda}$ for some $\underline{\kappa}_\lambda = (\kappa_{\lambda 1}, \dots, \kappa_{\lambda n}) \in \mathbb{N}^n$ $(1 \le \lambda \le L)$.

By means of Leibniz formula for the derivative of a product, we deduce that for any $\mu = 1, \dots, L$, there exists a family $c_{\mu\underline{\iota}}$ of complex numbers such that, for any $\underline{\kappa} = (\kappa_1, \dots, \kappa_n) \in \mathbb{N}^n$,

$$\mathcal{D}^{(\mu)} \underline{z}^{\underline{\kappa}} = \sum_{\underline{\iota}} c_{\mu\underline{\iota}} \binom{\underline{\kappa}}{\underline{\iota}} \underline{z}^{\underline{\kappa}-\underline{\iota}},$$

where $\underline{\iota}$ ranges over the set of elements $(\iota_1, \dots, \iota_n) \in \mathbb{N}^n$ such that

$$\iota_1 + \cdots + \iota_n = \sigma_\mu,$$

and where we agree that

$$\binom{\underline{\kappa}}{\underline{\iota}} = \binom{\kappa_1}{\iota_1} \cdots \binom{\kappa_n}{\iota_n}$$

vanishes if there is an index i $(1 \le i \le n)$ such that $\iota_i > \kappa_i$. Accordingly we have

$$\Psi(z) = \sum_{\iota} c_{1\underline{\iota}_1} \cdots c_{L\underline{\iota}_L} \Psi_\iota(z)$$

where

$$\Psi_\iota(z) = \det\left(\binom{\underline{\kappa}_\lambda}{\underline{\iota}_\mu} (z\underline{\zeta}_\mu)^{\underline{\kappa}_\lambda - \underline{\iota}_\mu}\right)_{1\le\lambda,\mu\le L},$$

and where ι runs over the set of $(\underline{\iota}_1, \ldots, \underline{\iota}_L) \in (\mathbb{N}^n)^L$ whose components $\underline{\iota}_\mu = (\iota_{\mu 1}, \ldots, \iota_{\mu n}) \in \mathbb{N}^n$ satisfy

$$\iota_{\mu 1} + \cdots + \iota_{\mu n} = \sigma_\mu \quad (1 \le \mu \le L).$$

For each such ι we have

$$\Psi_\iota(z) z^{\sigma_1 + \cdots + \sigma_L} = z^{\|\underline{\kappa}_1\| + \cdots + \|\underline{\kappa}_L\|} \det\left(\binom{\underline{\kappa}_\lambda}{\underline{\iota}_\mu} \underline{\zeta}_\mu^{\underline{\kappa}_\lambda - \underline{\iota}_\mu} \right)_{1 \le \lambda, \mu \le L}.$$

Notice that the right hand side vanishes as soon as there are two indices $\lambda' \neq \lambda''$ with $\underline{\kappa}_{\lambda'} = \underline{\kappa}_{\lambda''}$. Moreover this formula shows that the multiplicity at the origin of Ψ_ι is at least

$$\|\underline{\kappa}_1\| + \cdots + \|\underline{\kappa}_L\| - (\sigma_1 + \cdots + \sigma_L).$$

Lemma 9.2 now follows from the definition of $\Theta_n(L)$ (see § 6.1.2). □

9.1.3 Multiplicity Estimate

For $P \in \mathbb{C}[X_0, \ldots, X_{m-1}, Y^{\pm 1}]$, define $F \in \mathbb{C}[z, X_0, \ldots, X_{m-1}, Y^{\pm 1}]$ by

$$F(z) = P\bigl(z + X_0, X_1, \ldots, X_{m-1}, e^z Y\bigr).$$

Then

$$\frac{\partial}{\partial z} F(z) = (\mathcal{D}P)\bigl(z + X_0, X_1, \ldots, X_{m-1}, e^z Y\bigr),$$

where $\mathcal{D}$ denotes the derivative operator

$$\frac{\partial}{\partial X_0} + Y \frac{\partial}{\partial Y}$$

on the ring of polynomials in the variables $X_0, \ldots, X_{m-1}, Y^{\pm 1}$ with coefficients in the field $\mathbb{C}(z)$.

We use the multiplicity estimate of Chap. 8 in a special case: here, there is a single derivative and also a single multiplicative factor.

Let K be an algebraically closed field of zero characteristic and $m \ge 1$ a positive integer. Again we denote by $\mathcal{D}$ the derivative operator $(\partial/\partial X_0) + Y(\partial/\partial Y)$ on the ring $K[X_0, \ldots, X_{m-1}, Y^{\pm 1}]$. Let $\alpha_1, \ldots, \alpha_m$ be nonzero elements of K and $\beta_0, \ldots, \beta_{m-1}$ be elements of K. For $\underline{s} \in \mathbb{Z}^m$, define

$$\underline{\xi}_{\underline{s}} = \bigl(s_m\beta_0, s_1 + s_m\beta_1, \ldots, s_{m-1} + s_m\beta_{m-1}\bigr) \in K^m.$$

Proposition 9.3. *Assume $\alpha_1, \ldots, \alpha_m$ generate a multiplicative subgroup of $K^\times$ of rank $\ge m - 1$. Assume further $1, \beta_1, \ldots, \beta_{m-1}$ are $\mathbb{Q}$-linearly independent and $\beta_0 \neq 0$. Let T_0, T_1, S_0 and S_1 be positive integers satisfying the following conditions:*

$$S_0 < 4T_0T_1, \quad 2S_1 < T_0$$

and

$$2(m+1)T_0^m T_1 < (S_0+1)(2S_1+1)^m. \tag{9.4}$$

For $\sigma \in \mathbb{N}$, $\underline{\tau} \in \mathbb{N}^m$, $t \in \mathbb{Z}$ and $\underline{s} \in \mathbb{Z}^m$, define $a_{\underline{\tau}t}^{(\sigma\underline{s})}$ as the value, at the point

$$\left(\underline{\xi}_{\underline{s}}, \underline{\alpha}^{\underline{s}}\right) \in K^m \times K^\times,$$

of the polynomial

$$\mathcal{D}^\sigma\left(X_0^{\tau_0} \cdots X_{m-1}^{\tau_{m-1}} Y^t\right) \in K[X_0, \ldots, X_{m-1}, Y^{\pm 1}].$$

Consider the following matrix:

$$\boldsymbol{M} = \left(a_{\underline{\tau}t}^{(\sigma\underline{s})}\right)_{\substack{(\underline{\tau},t)\\(\sigma,\underline{s})}}$$

where the index of rows $(\underline{\tau}, t)$ ranges over the elements $(\underline{\tau}, t)$ in $\mathbb{N}^m \times \mathbb{Z}$ with $\|\underline{\tau}\| \le T_0$ and $|t| \le T_1$, while the index of columns $(\sigma, \underline{s})$ runs over the elements of $\mathbb{N} \times \mathbb{Z}^m$ with $0 \le \sigma \le (m+1)S_0$ and $|s_j| \le (m+1)S_1$ $(1 \le j \le m)$. Then the matrix $\boldsymbol{M}$ has rank $\binom{T_0+m}{m}(2T_1+1)$.

Remark. In case $m = 1$, Proposition 2.14 yields a slightly sharper result.

Proof. Define

$$\Sigma = \left\{\left(\underline{\xi}_{\underline{s}}, \underline{\alpha}^{\underline{s}}\right); \underline{s} \in \mathbb{Z}^m[S_1]\right\} \subset K^m \times K^\times$$

and denote by

$$\mathcal{E} = \{\underline{\xi}_{\underline{s}}; \underline{s} \in \mathbb{Z}^m[S_1]\} \subset K^m$$

the projection of Σ onto K^m.

If the rank of $\boldsymbol{M}$ is not equal to the number of rows, then there is a nonzero polynomial $P \in K[X_0, \ldots, X_{m-1}, Y^{\pm 1}]$, of total degree at most T_0 in $X_0, \ldots, X_{m-1}$ and of degree at most T_1 in $Y^{\pm 1}$, which vanishes, together with its $(m+1)S_0 + 1$ first derivatives $\mathcal{D}^\sigma$ $(0 \le \sigma \le (m+1)S_0)$, at all points of the set

$$\Sigma(m+1) = \left\{\left(\underline{\xi}_{\underline{s}}, \underline{\alpha}^{\underline{s}}\right); \underline{s} \in \mathbb{Z}^m[(m+1)S_1]\right\}.$$

The assumptions of Theorem 8.1 are satisfied with $d_0 = m$, $d_1 = 1$, $G = G^+ = \mathbb{G}_a^m \times \mathbb{G}_m$, $G^- = \{e\}$, $d = m+1$, $D_0 = T_0$, $D_1 = T_1$ and $\mathcal{W} = K(1, 0, \ldots, 0, 1)$. Therefore there exists a connected algebraic subgroup G^* of G of dimension $d^* < d$ which satisfies

$$\binom{S_0+\ell_0'}{\ell_0'} \operatorname{Card}\left(\frac{\Sigma + G^*}{G^*}\right) \mathcal{H}(G^*; T_0, T_1) \le \mathcal{H}(G; T_0, T_1),$$

where

$$\ell_0' = \dim_K\left(\frac{T_e(G^*) + \mathcal{W}}{T_e(G^*)}\right) = \begin{cases} 0 & \text{if } (1,0,\ldots,0,1) \in T_e(G^*) \\ 1 & \text{if } (1,0,\ldots,0,1) \notin T_e(G^*). \end{cases}$$

By Proposition 5.6, the algebraic subgroup G^* of G can be written $G_0^* \times G_1^*$, where G_0^* is a vector subspace of K^m of dimension say d_0^* while G_1^* is a connected algebraic subgroup of $\mathbb{G}_\mathrm{m}$ of dimension $d_1^* \in \{0, 1\}$. The dimension of G^* is $d^* = d_0^* + d_1^*$.

If $d_1^* = 0$ we have

$$G_1^* = \{1\}, \quad d^* = d_0^*, \quad G^* = G_0^* \times \{1\} \quad \text{and} \quad 0 \le d_0^* \le m.$$

If $d_1^* = 1$, then

$$G_1^* = \mathbb{G}_\mathrm{m}, \quad d^* = d_0^* + 1, \quad G^* = G_0^* \times \mathbb{G}_\mathrm{m} \quad \text{and} \quad 0 \le d_0^* \le m - 1$$

(recall that $d^* < d$).

Notice also that in case $\ell_0' = 0$, since $(1, 0, \ldots, 0, 1) \in \mathcal{W} \subset T_e(G^*)$, we have $(1, 0, \ldots, 0) \in G_0^*$ and $d_1^* = 1$.

Recall (§ 5.1.1) that

$$\mathcal{H}(G; T_0, T_1) = 2(m+1)T_0^m T_1$$

and

$$\mathcal{H}(G^*; T_0, T_1) = \begin{cases} T_0^{d_0^*} & \text{if } d_1^* = 0 \\ 2(d_0^*+1)T_0^{d_0^*} T_1 & \text{if } d_1^* = 1. \end{cases}$$

We distinguish three cases.

First case: $d_1^* = 0$ and $\ell_0' = 1$

In this case $G^* = G_0^* \times \{1\}$, $0 \le d_0^* \le m$ and the conclusion of Theorem 8.1 yields

$$(S_0+1)\mathrm{Card}\left(\frac{\Sigma + (G_0^* \times \{1\})}{G_0^* \times \{1\}}\right) \le 2(m+1)T_0^{m-d_0^*} T_1.$$

If $d_0^* = 0$ (which means that $G^* = \{e\}$ is the trivial subgroup) we deduce from (9.4) that Σ has less than $(2S_1+1)^m$ elements. Therefore $\beta_0 = 0$ and each of the numbers $\beta_1, \ldots, \beta_{m-1}$ is rational, which is a contradiction.

If $d_0^* \ge 1$ we use the assumption that $\alpha_1, \ldots, \alpha_m$ generate a multiplicative group of rank $\ge m-1$: denote by π the projection from $K^\times$ onto $K^\times/K^\times_{\mathrm{tors}}$. Then (see Lemma 7.8 and Exercise 7.5)

$$\mathrm{Card}\left(\frac{\Sigma + (G_0^* \times \{1\})}{G_0^* \times \{1\}}\right) \ge \mathrm{Card}\left\{\pi\left(\underline{\alpha}^{\underline{s}}\right);\ \underline{s} \in \mathbb{Z}^m[S_1]\right\} \ge (2S_1+1)^{m-1}.$$

On the other hand, using (9.4) together with the condition $T_0 \ge 2S_1 + 1$, we deduce

$$2(m+1)T_0^{m-1}T_1 < (S_0+1)(2S_1+1)^{m-1},$$

which shows that the conditions $d_0^* \ge 1$, $d_1^* = 0$ and $\ell_0' = 1$ are not compatible.

Second case: $d_1^* = 1$, $\ell_0' = 1$

Here $G^* = G_0^* \times K^\times$, $G_0^* \not\ni (1, 0, \ldots, 0)$, $0 \le d_0^* \le m-1$ and

$$(S_0+1)\mathrm{Card}\big(\pi_{G_0^*}(\mathcal{E})\big) \le \frac{m+1}{d_0^*+1}T_0^{m-d_0^*},$$

where $\pi_{G_0^*}$ is the canonical map $K^d \to K^d/G_0^*$.

From (9.4), using the lower bound $T_1 \ge 1$ we deduce

$$(m+1)T_0^m < (S_0+1)(2S_1+1)^m.$$

From the condition $2S_1+1 \le T_0$ we get

$$\frac{m+1}{d_0^*+1}T_0^{m-d_0^*} < (S_0+1)(2S_1+1)^{m-d_0^*}.$$

Therefore

$$\mathrm{Card}\big(\pi_{G_0^*}(\mathcal{E})\big) < (2S_1+1)^{m-d_0^*}.$$

This is impossible for $\beta_0 \neq 0$ (see Exercise 9.1.a).

Third case: $d_1^* = 1$, $\ell_0' = 0$

Now we have $G^* = G_0^* \times K^\times$, $G_0^* \ni (1,0,\dots,0)$, $1 \le d_0^* \le m-1$ and

$$\mathrm{Card}\big(\pi_{G_0^*}(\mathcal{E})\big) \le \frac{m+1}{d_0^*+1}T_0^{m-d_0^*}.$$

Since $S_0+1 \le 4T_0T_1$, (9.4) implies

$$(m+1)T_0^{m-1} < 2(2S_1+1)^m.$$

From $d_0^* \ge 1$, using $2S_1+1 \le T_0$, we deduce

$$\frac{m+1}{d_0^*+1}T_0^{m-d_0^*} < (2S_1+1)^{m+1-d_0^*}$$

and

$$\mathrm{Card}\big(\pi_{G_0^*}(\mathcal{E})\big) < (2S_1+1)^{m+1-d_0^*}.$$

Since $G_0^* \ni (1,0,\dots,0)$, Lemma 6.2 shows that the numbers $1,\beta_1,\dots,\beta_{m-1}$ are linearly dependent over $\mathbb{Q}$ (see Exercise 9.1.b). □

9.1.4 Completion of the Proof of Baker's Nonhomogeneous Theorem

Combining the multiplicity estimate with the analytic upper bound, we shall deduce the following result which generalizes Proposition 2.11.

Proposition 9.5. *Let $\beta_0,\beta_1,\dots,\beta_{m-1},\lambda_1,\dots,\lambda_m$ be complex numbers satisfying*

$$\beta_0+\beta_1\lambda_1+\cdots+\beta_{m-1}\lambda_{m-1}-\lambda_m = 0.$$

Assume firstly that $\beta_0 \neq 0$, secondly that the m numbers $1,\beta_1,\dots,\beta_{m-1}$ are $\mathbb{Q}$-linearly dependent and thirdly that the numbers $\lambda_1,\dots,\lambda_m$ are $\mathbb{Q}$-linearly *independent. Define* $\alpha_j = \exp(\lambda_j)$ $(1 \le j \le m)$. *Let* T_0, T_1, S_0, S_1, L *be positive rational integers. Further let* E *be a real number with* $E \ge e$. *Assume*

$$T_0 \ge 16m^3, \quad 4T_0T_1 > S_0, \quad T_0 > 2S_1$$
$$S_0S_1^m \ge 2T_0^m T_1$$

and

$$L = \binom{T_0+m}{m}(2T_1+1).$$

Then there exists a polynomial $f \in \mathbb{Z}[\mathrm{X}_1^{\pm 1}, \ldots, \mathrm{X}_m^{\pm 1}, \mathrm{Y}_0, \mathrm{Y}_1, \ldots, \mathrm{Y}_{m-1}]$ *of degree and length bounded by*

$$\deg_{\mathrm{X}_i^{\pm 1}} f \le \frac{1}{2}(m+1)L(T_1+1)S_1, \quad \deg_{\mathrm{Y}_j} f \le LT_0,$$

$$\mathrm{L}(f) \le L!\bigl(2(m+1)S_1\bigr)^{LT_0}(T_1+T_0)^{(m+1)LS_0},$$

such that

$$0 < |f(\alpha_1, \ldots, \alpha_m, \beta_0, \beta_1, \ldots, \beta_{m-1})| \le$$
$$E^{-\frac{1}{3}L^{1+(1/m)}} \cdot E^{L\left((m+1)S_0+T_0\right)}\bigl(2(m+1)S_1\bigr)^{LT_0}(T_0+T_1)^{(m+1)LS_0}e^{c_0L(T_0+T_1S_1E)}$$

with

$$c_0 = 1 + \max\Bigl\{(m+1)\bigl(|\lambda_1| + \cdots + |\lambda_m|\bigr), \ \log\bigl(1 + \max_{0 \le i \le m-1} |\beta_i|\bigr)\Bigr\}.$$

Proof. From the assumptions of Proposition 9.5 we deduce that the matrix $\boldsymbol{M}$ of Proposition 9.3 (with $K = \mathbb{C}$) has maximal rank. Let $(\sigma_\mu, \underline{s}_\mu)$ be elements in $\mathbb{N} \times \mathbb{Z}^m$ with

$$0 \le \sigma_\mu \le (m+1)S_0 \quad \text{and} \quad \max_{1 \le j \le m} |s_{\mu j}| \le (m+1)S_1$$

for $1 \le \mu \le L$, such that the determinant Δ of the matrix

$$\left(\left(\frac{\partial}{\partial z_0}\right)^{\sigma_\mu} f_{\underline{\tau}t}(\underline{\xi}_{\underline{s}_\mu})\right)_{\substack{(\underline{\tau},t)\\ 1 \le \mu \le L}} = \left(a_{\underline{\tau}t}^{(\sigma_\mu, \underline{s}_\mu)}\right)_{\substack{(\underline{\tau},t)\\ 1 \le \mu \le L}}$$

is not zero.

Each entry of this matrix is the value of a polynomial in $3m$ variables $\mathrm{X}_1^{\pm 1}, \ldots, \mathrm{X}_m^{\pm 1}, \mathrm{Y}_0, \mathrm{Y}_1, \ldots, \mathrm{Y}_{m-1}$ at the point

$$\bigl(\alpha_1, \ldots, \alpha_m, \beta_0, \beta_1, \ldots, \beta_{m-1}\bigr).$$

This polynomial has degree at most $(m+1)|t|S_1$ in each of the first $2m$ variables and total degree at most T_0 in the last m ones, its coefficients are rational integers and from Lemma 4.9 it follows that the length is at most

$$\bigl(2(m+1)S_1\bigr)^{T_0}(T_1+T_0)^{(m+1)S_0}.$$

From Lemma 3.15 we deduce

$$\Delta = f(\alpha_1, \dots, \alpha_m, \beta_0, \beta_1, \dots, \beta_{m-1})$$

where $f \in \mathbb{Z}[X_1^{\pm 1}, \dots, X_m^{\pm 1}, Y_0, Y_1, \dots, Y_{m-1}]$ has degree and length bounded as announced.

Since $T_0 \geq 16m^3$ we have $L \geq (4m)^{2m}$. By Lemma 9.2, the function of a single variable z

$$\Psi(z) = \det\left(\left(\frac{\partial}{\partial z_0}\right)^{\sigma_\mu} f_{\underline{\tau}t}(z\underline{\xi}_{\underline{s}_\mu})\right)_{\substack{(\underline{\tau},t)\\ 1\leq\mu\leq L}}$$

has a zero at the origin of multiplicity at least

$$\Theta_m(L) - (m+1)LS_0 \geq L\left(\frac{1}{3}L^{1/m} - (m+1)S_0\right)$$

(the coefficient $1/3$ is a lower bound for m/e – see Lemma 6.5). We use Schwarz' lemma as in the proof of Lemma 6.1:

$$\frac{1}{L}\log|\Delta| = \frac{1}{L}\log|\Psi(1)| \leq -\frac{1}{3}L^{1/m}\log E + (m+1)S_0\log E + \frac{1}{L}\log|\Psi|_E.$$

From the relation

$$\left(\frac{\partial}{\partial z_0}\right)^{\sigma} f_{\underline{\tau}t}(z\underline{\xi}_{\underline{s}}) =$$

$$\sum_{\kappa=0}^{\min\{\sigma,\tau_0\}} \frac{\sigma!}{\kappa!(\sigma-\kappa)!}\cdot\frac{\tau_0!}{(\tau_0-\kappa)!}t^{\sigma-\kappa}(s_m\beta_0)^{\tau_0-\kappa}z^{\|\underline{\tau}\|-\kappa}\prod_{i=1}^m(s_i+s_m\beta_i)^{\tau_i}\cdot\prod_{j=1}^m e^{t\lambda_j s_j z}$$

using Lemma 4.9 we deduce, for $|z| \leq E$ and all $(\underline{\tau}, t)$, μ,

$$\left|\left(\frac{\partial}{\partial z_0}\right)^{\sigma_\mu} f_{\underline{\tau}t}(z\underline{\xi}_{\underline{s}_\mu})\right| \leq (T_0+T_1)^{(m+1)S_0}\big(2(m+1)S_1E\big)^{T_0}e^{c_1T_0+c_2T_1S_1E}$$

with

$$c_1 = \log\max\{1, |\beta_0|, \dots, |\beta_{m-1}|\} \quad\text{and}\quad c_2 = (m+1)\big(|\lambda_1| + \cdots + |\lambda_m|\big).$$

We conclude

$$\frac{1}{L}\log|\Psi|_E \leq$$
$$\log L + (m+1)S_0\log(T_0+T_1) + T_0\log\big((2(m+1)S_1E)\big) + c_1T_0 + c_2T_1S_1E.$$

Finally we bound $\log L$ by $T_0 + T_1S_1E$. □

Proof of Theorem 1.6. We apply Proposition 9.5 as follows. We want to use Lemma 2.1 with

$$\{\theta_1, \dots, \theta_{3m}\} = \{\alpha_1^{\pm 1}, \dots, \alpha_m^{\pm 1}, \beta_0, \dots, \beta_{m-1}\}.$$

Let $\kappa > 0$. We need to choose the parameters so that $L := \binom{T_0+m}{m}(2T_1+1)$ satisfies

$$2T_0^m T_1 \le S_0 S_1^m$$

and

$$L^{1/m} \log E > \kappa\big(S_0 \log(ET_0T_1) + T_0 \log(S_1 E) + T_1 S_1 E\big).$$

We take $E = e$ and we replace the last inequality by the sharper requirements:

$$\kappa' T_0 \log S_1 \le L^{1/m}, \quad \kappa' S_0 \log(T_0T_1) \le L^{1/m}, \quad \kappa' T_1 S_1 \le L^{1/m}$$

with $\kappa' = 5\kappa$,.and T_0, T_1, S_0, S_1 are sufficiently large integers.

The condition $\kappa' S_0 \log(T_0T_1) \le L^{1/m}$ implies $S_0 < 4T_0T_1$, and the condition $\kappa' T_1 S_1 \le L^{1/m}$ implies $2S_1 < T_0$.

As a first example of a solution to this system of conditions, we choose a large integer N and we look for parameters which are powers of N. We replace the unknowns T_0, T_1, S_0, S_1 by

$$T_0 = N^{t_0}, \quad T_1 = N^{t_1}, \quad S_0 = N^{s_0}, \quad S_1 = N^{s_1}.$$

where t_0, t_1, s_0, s_1 are (free) positive integers. The previous requirements can be summarized as follows:

$$\max\{t_0, s_0, t_1 + s_1\} < t_0 + \frac{t_1}{m} < \frac{s_0}{m} + s_1.$$

The conditions on s_1 amount to say that s_1 lies in the interval

$$t_0 + \frac{t_1}{m} - \frac{s_0}{m} < s_1 < t_0 - \frac{(m-1)t_1}{m}.$$

We seek for natural integers in order to avoid integral parts (in fact by homogeneity we could replace N by a power of N). Hence we shall require that the above interval has length 2, which gives $s_0 = m(t_1+2)$. Now the remaining conditions just become

$$t_0 > t_1\left(m - \frac{1}{m}\right) + 2m.$$

A simple choice is $t_1 = m$, $t_0 = s_0 = m(m+2)$, $s_1 = m(m+1)$, which gives the solution

$$T_0 = N^{m(m+2)}, \quad T_1 = N^m, \quad S_0 = T_0, \quad S_1 = N^{m(m+1)}.$$

Here is another solution: we look for T_0 and T_1 of the following shape:

$$T_0 = \big[S_1(\log S_1)^{t_0}\big], \quad T_1 = \big[(\log S_1)^{t_1}\big]$$

where S_1 is a sufficiently large integer, and t_0, t_1 are positive fixed integers which we have to choose. The conditions on t_0 and t_1 are

$$t_1 > m, \quad t_0 > (m-1)\frac{t_1}{m}\cdot$$

For instance $t_1 = m + 1$ and $t_0 = m$ will do. The condition on S_0 is that it belongs to an interval

$$\kappa(\log S_1)^{mt_0+t_1} < S_0 < \frac{1}{3\kappa} S_1(\log S_1)^{t_0-1+(t_1/m)}.$$

Therefore another admissible choice is

$$T_0 = \left[S_1(\log S_1)^m\right], \quad T_1 = \left[(\log S_1)^{m+1}\right], \quad S_0 = S_1,$$

with S_1 a sufficiently large positive integer. With the same values for T_0 and T_1, but with $S_0 = \left[(\log S_1)^{m^2+m+2}\right]$, a weaker multiplicity estimate suffices. □

9.2 Proof of Theorem 9.1

9.2.1 Sketch of Proof of Theorem 9.1

Consider first the general case of Theorem 9.1. Assume $\beta_m = -1$:

$$\Lambda = \beta_0 + \beta_1\lambda_1 + \cdots + \beta_{m-1}\lambda_{m-1} - \lambda_m$$

where $\beta_0, \ldots, \beta_{m-1}$ are algebraic numbers and $\lambda_1, \ldots, \lambda_m$ are logarithms of algebraic numbers. Define $\alpha_i = e^{\lambda_i}$ $(1 \le i \le m)$.

a) Exponential Polynomials

For $\underline{\tau} = (\tau_0, \ldots, \tau_{m-1}) \in \mathbb{N}^m$, $\underline{z} = (z_0, \ldots, z_{m-1}) \in \mathbb{C}^m$ and $t \in \mathbb{Z}$ consider the exponential monomial in $m+1$ variables $\underline{z}^{\underline{\tau}} e^{tz_m}$ where

$$\underline{z}^{\underline{\tau}} = z_0^{\tau_0} z_1^{\tau_1} \cdots z_{m-1}^{\tau_{m-1}}.$$

The restriction of this function to the hyperplane of $\mathbb{C}^{m+1}$ of equation

$$z_0 + \lambda_1 z_1 + \cdots + \lambda_{m-1} z_{m-1} = z_m$$

gives rise to a complex function of m variables

$$f_{\underline{\tau}t}(\underline{z}) = \underline{z}^{\underline{\tau}} e^{t(z_0+\lambda_1 z_1+\cdots+\lambda_{m-1}z_{m-1})}.$$

The difference with Chap. 7 is only the appearance of the variable z_0 which will be useful even in the case $\beta_0 = 0$. The main point is that we can take derivatives with respect to z_0 without introducing transcendental numbers: for $\sigma \in \mathbb{N}$ we have

$$\left(\frac{\partial}{\partial z_0}\right)^{\sigma} f_{\underline{\tau}t}(\underline{z}) = \left(\frac{\partial}{\partial z_0}\right)^{\sigma} \left(z_0^{\tau_0} e^{tz_0}\right) \cdot z_1^{\tau_1} \cdots z_{m-1}^{\tau_{m-1}} e^{t(\lambda_1 z_1+\cdots+\lambda_{m-1}z_{m-1})}$$

with

$$\left(\frac{\partial}{\partial z_0}\right)^{\sigma} \left(z_0^{\tau_0} e^{tz_0}\right) = \sum_{\kappa=0}^{\min\{\sigma,\tau_0\}} \frac{\sigma!}{\kappa!(\sigma-\kappa)!} \cdot \frac{\tau_0!}{(\tau_0-\kappa)!} z_0^{\tau_0-\kappa} t^{\sigma-\kappa} e^{tz_0}.$$

Define

$$\underline{y}_j = (0, \underline{e}_j) \quad (1 \le j \le m-1) \quad \text{and} \quad \underline{y}_m = (\beta_0, \beta_1, \dots, \beta_{m-1})$$

where $\underline{e}_1, \dots, \underline{e}_{m-1}$ is the canonical basis of $\mathbb{C}^{m-1}$. For $\underline{s} \in \mathbb{Z}^m$ define $\underline{s}\,y \in \mathbb{C}^m$ by

$$\underline{s}\,y = s_1\underline{y}_1 + \cdots + s_m\underline{y}_m = (s_m\beta_0, s_1 + s_m\beta_1, \dots, s_{m-1} + s_m\beta_{m-1}).$$

Then

$$\left(\frac{\partial}{\partial z_0}\right)^{\sigma} f_{\underline{\tau}t}(\underline{s}\,y) = \gamma_{\underline{\tau}t}^{(\sigma\underline{s})} \cdot e^{ts_m\Lambda}$$

where $\gamma_{\underline{\tau}t}^{(\sigma\underline{s})}$ is the algebraic number

$$\sum_{\kappa=0}^{\min\{\sigma,\tau_0\}} \frac{\sigma!}{\kappa!(\sigma-\kappa)!} \cdot \frac{\tau_0!}{(\tau_0-\kappa)!} (s_m\beta_0)^{\tau_0-\kappa} t^{\sigma-\kappa} \prod_{i=1}^{m-1} (s_i + s_m\beta_i)^{\tau_i} \cdot \prod_{j=1}^{m} \alpha_j^{ts_j}.$$

There is another expression for $\gamma_{\underline{\tau}t}^{(\sigma\underline{s})}$. Introduce the derivative operator $\mathcal{D}$ on the ring of entire functions in $m+1$ variables $z_0, \dots, z_m$ by

$$\mathcal{D} = \frac{\partial}{\partial z_0} + \frac{\partial}{\partial z_m}.$$

For $1 \le j \le m$, define

$$\underline{\eta}_j = (\underline{y}_j, \lambda_j) \in \mathbb{C}^{m+1}$$

and for $\underline{s} \in \mathbb{Z}^m$, write $\underline{s}\eta \in \mathbb{C}^m$ in place of

$$s_1\underline{\eta}_1 + \cdots + s_m\underline{\eta}_m = (\underline{s}\,y,\ s_1\lambda_1 + \cdots + s_m\lambda_m).$$

Then

$$\gamma_{\underline{\tau}t}^{(\sigma\underline{s})} = \mathcal{D}^{\sigma}\left(\underline{z}^{\underline{\tau}} e^{tz_m}\right)(\underline{s}\eta).$$

The starting idea is to construct a matrix $\boldsymbol{M}$ having these numbers as entries in order to prove that the rank of $\boldsymbol{M}$ is less than the number of rows. This will enable us to use the zero estimate.

The points

$$\underline{\eta}'_j = \underline{\eta}_j \quad (1 \le j \le m-1) \qquad \text{and} \quad \underline{\eta}'_m = (\beta_0, \beta_1, \dots, \beta_{m-1}, \lambda_m + \Lambda_m)$$

belong to the hyperplane of $\mathbb{C}^{m+1}$ of equation

$$z_0 + \lambda_1 z_1 + \cdots + \lambda_{m-1} z_{m-1} = z_m.$$

If we set

$$\eta' = s_1\underline{\eta}'_1 + \cdots + s_m\underline{\eta}'_m,$$

we have

$$\mathcal{D}^{\sigma}\left(\underline{z}^{\underline{\tau}} e^{tz_m}\right)(\underline{s}\eta') = \gamma_{\underline{\tau}t}^{(\sigma\underline{s})} \cdot e^{ts_m\Lambda}.$$

This fact will be germane to the analytic upper bound for the absolute value of an interpolation determinant Δ. The expression of the number $\gamma_{\underline{\tau}t}^{(\sigma\underline{s})}$ as a derivative of an exponential polynomial in $m+1$ variables will be pertinent to the zero estimate. Of course the mere fact that $\gamma_{\underline{\tau}t}^{(\sigma\underline{s})}$ is an algebraic number will be relevant to the arithmetic lower bound for $|\Delta|$.

b) Basic Estimates and Choice of Parameters

Introduce parameters $T_0, T_1, S_0, S_1, \ldots, S_m$ (which are positive integers) and consider the sets of $\underline{\tau} \in \mathbb{N}^m$, $t \in \mathbb{Z}$, $\sigma \in \mathbb{N}$ and $\underline{s} \in \mathbb{Z}^m$ which satisfy

$$\|\underline{\tau}\| \le T_0, \quad |t| \le T_1, \qquad 0 \le \sigma \le S_0, \quad |s_j| \le S_j \quad (1 \le j \le m).$$

Define $L = \binom{T_0+m}{m}(2T_1+1)$ and denote by Δ the determinant of some $L \times L$ submatrix of

$$M = \left(\gamma_{\underline{\tau}t}^{(\sigma\underline{s})}\right)_{\substack{(\underline{\tau},t)\\(\sigma,\underline{s})}}.$$

Under the assumptions of Theorem 9.1 (in the general case, and with $\beta_m = -1$), we deduce from Liouville's estimate a lower bound for $|\Delta|$ assuming it is not zero. The main terms in this estimate arise from

$$(s_i + s_m\beta_i)^{\tau_i}, \qquad t^\sigma, \qquad \prod_{j=1}^m \alpha_j^{ts_j}$$

which introduce, in the lower bound for $\log|\Delta|$,

$$T_0 \log S^* + T_0 \max_{0\le i\le m} \mathrm{h}(\beta_i), \qquad S_0 \log T_1, \qquad T_1 \sum_{j=1}^m S_j \mathrm{h}(\alpha_j)$$

respectively, where $S^* = S_1 + \cdots + S_m$. Using the definition[14] of $B, A_1, \ldots, A_m$ with

$$\log B \ge \max_{0\le i\le m} \mathrm{h}(\beta_i), \qquad \log A_j \ge \mathrm{h}(\alpha_j) \quad (1 \le j \le m),$$

we deduce that either $\Delta = 0$ or else

$$\frac{1}{L}\log|\Delta| \ge -U_1$$

with

$$U_1 = cD\Big(T_0 \log(BS^*) + S_0 \log T_1 + T_1 \sum_{j=1}^m S_j \log A_j\Big).$$

Here (and below), c is a suitable (sufficiently large) positive absolute constant[15].

An analytic argument will enable us to deduce an upper bound for $|\Delta|$, under the assumption

[14] Say, for the general case.

[15] We use the same letter c to denote such absolute constants throughout this section. Explicit computations are carried out in § 9.3 only.

$$|\Lambda| < e^{-V} \quad \text{where} \quad V = \frac{1}{2m}(T_0 + m)(2T_1 + 1)\log E.$$

This value of V already occurred (with m replaced by n) in Proposition 7.6. From Lemma 9.2 we see that we shall lose $S_0 \log E$ because of the derivatives. The main terms in the upper bound will arise from the maximum of an exponential polynomial on a disc of radius $\geq E(S_j + S_m|\beta_j|)$ involving

$$|z|^\tau, \qquad |t|^{\sigma-\kappa}, \qquad e^{tz}.$$

The upper bound for $\log|\Delta|$ will involve

$$S_0 \log(ET_1), \qquad T_0 \log(ES^*), \qquad T_1 \sum_{j=1}^m S_j E|\lambda_j|$$

respectively. Using the assumption

$$E|\lambda_j| \leq D \log A_j,$$

we deduce

$$\log|\Delta| < -\frac{1}{2}LV + LU_2$$

with

$$U_2 = c\Big(S_0 \log(ET_1) + T_0 \log(ES^*) + T_1 \sum_{j=1}^m S_j \log A_j\Big).$$

With a suitable choice of the parameters for which $V \geq 2(U_1 + U_2)$, we conclude that the arithmetic and the analytic estimates are not compatible, hence Δ is zero and therefore M has rank $< L$.

Let us introduce B_1 and B_2 satisfying

$$B_1 \geq \max\{B, S^*, E^{1/D}\} \quad \text{and} \quad B_2 \geq \max\{T_1, E^{1/D}\}$$

so that each occurrence of T_0 involves at most $DT_0 \log B_1$ and similarly S_0 occurs only with a factor bounded by $D \log B_2$.

A natural and simple choice for the parameters is therefore as follows. Introduce one more parameter U, a suitable value of which will occur as a consequence of the zero estimate. Then take

$$T_0 = \frac{U}{D\log B_1}, \quad S_0 = \frac{U}{D\log B_2}, \quad S_j = \frac{U}{mDT_1 \log A_j} \quad (1 \leq j \leq m).$$

This is very close to our final choice (however we shall select in § 9.3 integers by taking integral parts).

The parameter T_1 is not yet fixed. We shall need the upper bound $cmU \leq V$. The coefficient m arises for the need, in the zero estimate, to work with $\Sigma[m+1]$ ($m+1$ will be the dimension of the underlying algebraic group G), and not only with Σ – see Theorem 8.1. Hence a natural choice for T_1 is

$$T_1 = \frac{cm^2 D \log B_1}{\log E}.$$

With this sketch of proof one deduces the result with B replaced by B_1 and E^* by B_2. In the general case, the conditions $B_1 \geq S^*$ and $B_1 \geq E^{1/D}$ are not too strong: one can replace B_1 by B^{cm}. In the homogeneous rational case these conditions can be relaxed by means of Fel'dman's polynomials, almost in the same way as in Chap. 7.

The condition $B_2 \geq T_1$ is more serious. If we wish to bound B_2 by a power of E^*, we need to require a further condition on E^*, namely

$$E^* \geq \log B. \tag{9.6}$$

We shall see later (part d of this section) how to avoid this condition (9.6), but right now we pursue the sketch of proof under this extra assumption.

The matrix M above has L rows and $(S_0+1)(2S_1+1)\cdots(2S_m+1)$ columns. We cannot get any interesting conclusion unless the number of columns is at least L. In fact the zero estimate requires slightly more: in order to avoid the trivial subgroup G^* when we apply Theorem 8.1, we need to assume

$$(S_0+1)(2S_1+1)\cdots(2S_m+1) > 2(m+1)T_0^m T_1.$$

If we replace the parameters T_0, S_0 and S_j by their values above, we find that U should satisfy a condition

$$U > c\cdot 2^{-m} m^{m+1} D(\log A_1)\cdots(\log A_m)(\log B_2)\cdot T_1^{m+1}(\log B_1)^{-m}.$$

We now replace T_1 by $cm^2 D \log B_1/\log E$ and deduce that

$$U = c^m m^{3m} D^{m+2}(\log B_1)(\log A_1)\cdots(\log A_m)(\log B_2)(\log E)^{-m-1}$$

is an admissible value.

c) Consequence of the Zero Estimate

Consider the algebraic group $G = G_0 \times G_1$ with $G_0 = \mathbb{G}_{\mathrm{a}}^m$ and $G_1 = \mathbb{G}_{\mathrm{m}}$. Once we know that M has rank $< L$, the zero estimate produces an algebraic subgroup $G^* = G_0^* \times G_1^*$ of G, such that G_0^* contains points of the form

$$\underline{s}y = (s_m\beta_0, s_1 + s_m\beta_1, \ldots, s_{m-1} + s_m\beta_{m-1}) \in \mathbb{C}^m$$

for *many* $\underline{s} \in \mathbb{Z}^m[\underline{S}]$.

As we have seen in § 9.1.3, this can happen only when there are linear dependence relations between $\beta_1, \ldots, \beta_{m-1}$ and 1. We control these relations by considering the set of $\underline{s}$ rather than the set of $\underline{s}y$.

The strategy now is to select such an algebraic subgroup G_0^* of $\mathbb{G}_{\mathrm{a}}^m$ (call it G_0^+) of minimal dimension which contains *many* such points $\underline{s}y$. Next we repeat the above construction with G_0 replaced by G_0^+. If we prove that the rank of the new matrix is not maximal, then we shall be able to use the zero estimate again, and to produce an algebraic subgroup G^* of G^+ which has a similar property as G^+ but has smaller dimension. This will give a contradiction: hence the assumption $|\Lambda| < e^{-cV}$ (which occurred in the analytic estimate) is not fulfilled and our final goal will be achieved.

Here is the construction of the new matrix. Consider G_0^+ as a subspace of $\mathbb{C}^m$ and define another vector subspace $\mathcal{V}$ of $\mathbb{C}^m$ by

$$\mathcal{V} = \left\{\underline{z} \in \mathbb{C}^m\,;\, \exists z_0 \in \mathbb{C},\ (z_0, z_1 + z_m\beta_1, \ldots, z_{m-1} + z_m\beta_{m-1}) \in G_0^+\right\}.$$

Hence $\mathcal{V}$ contains the point $(\beta_1, \ldots, \beta_{m-1}, -1)$. Denote by d the dimension of $\mathcal{V}$. Since $\mathcal{V}$ is not contained into $\mathbb{C}^{m-1} \times \{0\}$, the restriction to $\mathbb{C}^{m-1} \times \{0\}$ of the the canonical map $\pi_{\mathcal{V}}$ from $\mathbb{C}^m$ onto $\mathbb{C}^m/\mathcal{V}$ is surjective. Hence there exists a subset $\{\underline{e}_1, \ldots, \underline{e}_{m-d}\}$ of the canonical basis of $\mathbb{C}^{m-1} \times \{0\}$ such that $\pi_{\mathcal{V}}(\underline{e}_1), \ldots, \pi_{\mathcal{V}}(\underline{e}_{m-d})$ is a basis of $\mathbb{C}^m/\mathcal{V}$.

Complete into a basis $\{\underline{e}_1, \ldots, \underline{e}_m\}$ of $\mathbb{C}^m$ with the other elements of the canonical basis of $\mathbb{C}^m$ including $\underline{e}_m = (0, \ldots, 0, 1)$. If $\underline{z} = z_1\underline{e}_1 + \cdots + z_m\underline{e}_m \in \mathcal{V}$ satisfies $z_{m-d+1} = \cdots = z_m = 0$, then $\underline{z} = 0$. Hence the linear mapping

$$\begin{array}{ccc} \mathcal{V} & \longrightarrow & \mathbb{C}^{d-1} \\ z_1\underline{e}_1 + \cdots + z_m\underline{e}_m & \longmapsto & \left(z_{m-d+1} + z_m\beta_{m-d+1}, \ldots, z_{m-1} + z_m\beta_{m-1}\right) \end{array}$$

is surjective of kernel $\mathbb{C}(\beta_1, \ldots, \beta_{m-1}, -1)$.

Define $L_d = \binom{T_0+d+1}{d+1}(2T_1 + 1)$. One repeats the preceding construction, but with $G = \mathbb{G}_{\mathrm{a}}^m \times \mathbb{G}_{\mathrm{m}}$ replaced by $\mathbb{G}_{\mathrm{a}}^d \times \mathbb{G}_{\mathrm{m}}$. Now Δ is the determinant of a $L_d \times L_d$ submatrix of M involving only

$$\underline{\tau} = (\tau_0, \tau_{m-d+1}, \ldots, \tau_m) \in \mathbb{N}^d$$

for the index of rows and $\underline{s} \in \mathcal{V}$ for the column index.

To tell the truth we shall not exactly follow this pattern: instead of using twice the zero estimate (once for $\mathbb{G}_{\mathrm{a}}^m \times \mathbb{G}_{\mathrm{m}}$ and a second time for $\mathbb{G}_{\mathrm{a}}^d \times \mathbb{G}_{\mathrm{m}}$), we shall use it only once, together with the choice of G^+, in order to construct directly a nonzero determinant Δ (see Proposition 9.16).

d) Fel'dman's Polynomials

Fel'dman's polynomials will occur in two ways. Firstly, they are needed (in both the general and the homogeneous rational case) for removing the extra condition (9.6) on E^*. Secondly, further Δ polynomials will be required for the refinement related to the homogeneous rational case.

In the above estimates for $\gamma_{\underline{\tau}t}^{(\sigma\underline{s})}$, the term $S_0 \log T_1$ arises from $t^{\sigma-\kappa}$. Since T_1 is a multiple of $\log B_1$, the previous sketch of proof requires $E^* \geq \log B_1$, and the final result following this pattern would include an extra $\log\log B$ arising from $\log E^*$. The way to remove this factor is to introduce Fel'dman's Delta polynomials. However the situation here is different from Chap. 7, where we just replaced (in the homogeneous rational case) z^τ by $\Delta(z;\tau)$. Here, $t^{\sigma-\kappa}$ comes from the derivative of e^{tz}. More precisely, we want to replace $t^{\sigma-\kappa}$ by the value of a Delta polynomial in the formula (valid for $t \in \mathbb{C}$, $\sigma \in \mathbb{N}$ and $\tau \in \mathbb{N}$)

$$e^{-tz}\left(\frac{d}{dz}\right)^{\sigma}\left(z^\tau e^{tz}\right) = \sum_{\kappa=0}^{\min\{\sigma,\tau\}} \frac{\sigma!\tau!}{\kappa!(\sigma-\kappa)!(\tau-\kappa)!} z^{\tau-\kappa}t^{\sigma-\kappa}.$$

Both hand sides define a function whose value at a point $z_0 \in \mathbb{C}$ can be written

$$\sum_{\kappa=0}^{\min\{\sigma,\tau\}} \frac{1}{\kappa!}\left(\left(\frac{d}{dz}\right)^{\kappa} z^{\tau}\right)_{z=z_0}\left(\left(\frac{d}{dz}\right)^{\kappa} z^{\sigma}\right)_{z=t}.$$

This provides the clue: to each polynomial $\delta(z) \in \mathbb{C}[z]$ and each $t \in \mathbb{C}$ is associated a derivative operator

$$\mathcal{D}_{\delta t} = \sum_{\kappa\geq 0} \frac{1}{\kappa!}\delta^{(\kappa)}(t)\left(\frac{d}{dz}\right)^{\kappa},$$

where

$$\delta^{(\kappa)} = \left(\frac{d}{dz}\right)^{\kappa} \delta.$$

Notice that for $\delta(z) = z^{\sigma}$, we have by construction

$$\left(\mathcal{D}_{\delta t}\varphi\right)(z) = e^{-t}\left(\frac{d}{dz}\right)^{\sigma}\left(\varphi(z)e^{tz}\right).$$

Lemma 9.7. *Let S_0 be a nonnegative integers, t a complex number, $p \in \mathbb{C}[z]$ a polynomial, $\{\delta(z;\sigma)\,;\, 0 \leq \sigma \leq S_0\}$ a basis of the space of polynomials in $\mathbb{C}[z]$ of degree $\leq S_0$ and $Q \in \mathrm{GL}_{S_0+1}(\mathbb{C})$ the transition matrix from the basis $\{1, z, \ldots, z^{S_0}\}$:*

$$\begin{pmatrix}1 & z & \cdots & z^{S_0}\end{pmatrix} Q = \begin{pmatrix}\delta(z;0) & \delta(z;1) & \cdots & \delta(z;S_0)\end{pmatrix}.$$

Define

$$\Psi(z) = p(z)e^{tz} \quad \text{and} \quad \Phi_{\sigma}(z) = \sum_{\kappa=0}^{\min\{\sigma,\tau\}} \frac{1}{\kappa!}\delta(t;\sigma,\kappa)p^{(\kappa)}(z)e^{tz}$$

for $\leq \sigma \leq S_0$, where

$$p^{(\kappa)}(z) = \left(\frac{d}{dz}\right)^{\kappa} p(z) \quad \text{and} \quad \delta(z;\sigma,\kappa) = \left(\frac{d}{dz}\right)^{\kappa}\delta(z;\sigma).$$

Then

$$\left(\Psi(z) \quad \frac{d}{dz}\Psi(z) \quad \cdots \quad \left(\frac{d}{dz}\right)^{S_0}\Psi(z)\right) Q = \begin{pmatrix}\Phi_0(z) & \Phi_1(z) & \cdots & \Phi_{S_0}(z)\end{pmatrix}.$$

Proof. For $z_0 \in \mathbb{C}$ and $0 \leq \sigma \leq S_0$, we have

$$\left(\frac{d}{dz}\right)^{\sigma}\Psi(z_0) = \sum_{\kappa=0}^{\min\{\sigma,\tau\}} \frac{1}{\kappa!}\left(\left(\frac{d}{dz}\right)^{\kappa} z^{\sigma}\right)_{z=t} p^{(\kappa)}(z_0)e^{tz_0}.$$

This proves Lemma 9.7 in the special case $\delta(z;\sigma) = z^{\sigma}$ (where $Q = \mathrm{I}_{S_0+1}$). The general case immediately follows. □

We shall use this lemma with $p(z) = z^\tau$. One would be tempted to take for $\delta(z;\sigma)$ the elements of the basis $\{\Delta(z;\sigma);\ \sigma \geq 0\}$ given by Delta polynomials. However, as we shall see, the value $\Delta(t;\sigma,\kappa)$ of a derivative of $\Delta(z;\sigma)$ at an integer $t \in \mathbb{Z}$ is a rational number, and the estimate for the denominator would not be sharp enough (see Lemma 9.8 and Exercise 9.3.b).

There are several possibilities here. The first one is to split the parameter S_0 into two parts $S_0 = S_0' S_0''$ and to consider $\Delta(z;\sigma')^{\sigma''}$ where $0 \leq \sigma' < S_0'$ and $0 \leq \sigma'' \leq S_0''$. The second possibility is a variant of the first one: select the following basis (see Exercise 9.2):

$$\Delta(z+\sigma';S_0')^{\sigma''},\ (0 \leq \sigma' < S_0',\ 1 \leq \sigma'' \leq S_0'') \quad \text{and} \quad (\sigma',\sigma'') = (0,0).$$

Working with interpolation determinants, these two solutions do not make too much difference. But the first one is not suitable for the classical method involving an auxiliary function (see § 12.3).

Here, following E. M. Matveev [Mat 1993a], we shall use a third solution.

Definition. Let $a \geq 0$ and $b > 0$ be two integers. Define a polynomial $\delta_b(z;a) \in \mathbb{Q}[z]$ of degree a by

$$\begin{aligned}\delta_b(z;a) &= \big(\Delta(z-1;b)\big)^q\, \Delta(z-1;r)\\ &= \left(\frac{z(z+1)\cdots(z+b-1)}{b!}\right)^q \cdot \left(\frac{z(z+1)\cdots(z+r-1)}{r!}\right),\end{aligned}$$

where q and r are the quotient and remainder of the division of a by b:

$$a = bq + r, \quad 0 \leq r < b.$$

For $c \geq 0$, define

$$\delta_b(z;a,c) = \left(\frac{d}{dz}\right)^c \delta_b(z;a).$$

From this definition we deduce at once

$$\delta_b(z;0) = 1 \quad \text{for any} \quad b \geq 1$$

and

$$\delta_1(z;a) = z^a \quad \text{for any} \quad a \geq 0.$$

For $b > a$ we have

$$\delta_b(z;a) = \Delta(z-1;a).$$

Since $\delta_b(z;a)$ has degree a, for fixed b and A the polynomials $\{\delta_b(z;a);\ a = 0,1,\ldots,A\}$ constitute a basis of the space of polynomials of degree $\leq A$.

The main interest of delta polynomials is that they are integer valued:

$$\delta_b(m;a) \in \mathbb{Z} \quad \text{for any} \quad m \in \mathbb{Z};$$

indeed, for $m \geq 1$, we have

$$\delta_b(m;a) = \binom{m+b-1}{b}^q \binom{m+r-1}{r}.$$

We now estimate the denominator of the values of $\delta_b(z;a,c)$ at rational integers. The following lemma is due to E. M. Matveev [Mat 1993a]. We reproduce his proof (see also [Mat 1998], Lemma 7.1 and [NeW 1996], Lemma 4).

Lemma 9.8 involves the following arithmetic function: for any positive integer n, denote by $\nu(n)$ the least common multiple of $1, 2, \ldots, n$.

Lemma 9.8. *Let $a \geq 0$, $b > 0$, $C \geq 0$ be nonnegative integers. For any integer c in the interval $0 \leq c \leq C$ and any rational integer $m \in \mathbb{Z}$, the number*

$$\nu(b)^C \cdot \frac{1}{c!}\delta_b(m;a,c)$$

is a rational integer. Moreover, for any complex number z, we have

$$\sum_{c=0}^{C} \binom{C}{c} |\delta_b(z;a,c)| \leq C! e^{a+b} \left(\frac{|z|}{b}+1\right)^a.$$

Proof. Let us check that $\nu(b)^C$ is a common denominator to $(1/c!)\delta_b(m;a,c)$ by looking at the p-adic valuation of both numbers (recall the notation v_p from § 3.1.a). So we fix a prime number p.

We start with a well known estimate (see [HaWr 1938], Th. 416):

$$v_p(n!) = \sum_{\ell=1}^{\infty} \left[\frac{n}{p^\ell}\right].$$

This relation is proved as follows:

$$v_p(n!) = \sum_{m=1}^{n} v_p(m) = \sum_{m=1}^{n}\sum_{\ell=1}^{v_p(m)} 1 = \sum_{\ell=1}^{\infty} \sum_{\substack{m\leq n\\ v_p(m)\geq \ell}} 1$$

and

$$\sum_{\substack{m\leq n\\ v_p(m)\geq \ell}} 1 = \left[\frac{n}{p^\ell}\right].$$

Therefore, for $b \geq 1$ and $0 \leq r < b$,

$$v_p\left(b!^q r!\right) = \sum_{\ell=1}^{\infty}\left(q\left[\frac{b}{p^\ell}\right] + \left[\frac{r}{p^\ell}\right]\right).$$

On the right hand side one can restrict the sum over ℓ to $p^\ell \leq b$ since the other terms vanish.

It is well known and easy to check that the p-adic valuation of $\nu(n)$ is

$$v_p\big(\nu(n)\big) = \left[\frac{\log n}{\log p}\right] = \sum_{p^\ell \le n} 1.$$

Consider a product $P = b_1 \cdots b_a$ of a rational integers. For any positive integer ℓ denote by ϱ_ℓ the number of b_i's which are multiple of p^ℓ. Then

$$v_p(P) = \sum_{\ell \ge 1} \varrho_\ell.$$

If we delete any c numbers from $b_1, \dots, b_a$ and if P' denote the product of the remaining $a - c$ numbers, we derive

$$v_p(P') \ge \sum_{\ell \ge 1} \max(\varrho_\ell - c, 0).$$

The derivative $\delta_b(z; a, c)$ of $\delta_b(z; a)$ is given by the formula

$$\delta_b(z; a, c) = c! \cdot \delta_b(z; a) \cdot \sum (z + b_1)^{-1} \cdots (z + b_c)^{-1},$$

where $(b_1, \dots, b_c)$ runs over the tuples of c elements in $\{0, \dots, b-1\}$ such that the polynomial $(z + b_1) \cdots (z + b_c)$ divides $\delta_b(z; a)$.

Denote by $b_1, \dots, b_a$ the a factors in the product

$$P = \big(m(m-1)\cdots(m-b+1)\big)^q m(m-1)\cdots(m-r+1);$$

the value of this product P is nothing else than $b!^q r! \delta_b(m; a)$. Since $\delta_b(m; a) \in \mathbb{Z}$, for any positive integer ℓ the number ϱ_ℓ of b_i's which are multiple of p^ℓ satisfies

$$\varrho_\ell \ge q\left[\frac{b}{p^\ell}\right] + \left[\frac{r}{p^\ell}\right] \quad \text{for any} \quad \ell \ge 1.$$

Therefore

$$v_p\left(\frac{1}{c!} b!^q r! \delta_b(m; a, c)\right) \ge \sum_{p^\ell \le b} \max\left\{q\left[\frac{b}{p^\ell}\right] + \left[\frac{r}{p^\ell}\right] - c\,;\, 0\right\}.$$

Define $d = \nu(b)^C$. Then

$$v_p\left(d \frac{1}{c!} \delta_b(m; a, c)\right) \ge \sum_{p^\ell \le b} \max\left(C - c\,;\, C - q\left[\frac{b}{p^\ell}\right] - \left[\frac{r}{p^\ell}\right]\right) \ge 0.$$

This proves the assertion $(d/c!)\delta_b(m; a, c) \in \mathbb{Z}$ for $0 \le c \le C$ and $m \in \mathbb{Z}$.

Using the estimate

$$\frac{1}{b!^q r!} \le \frac{1}{b^a} e^{a+b},$$

we obtain, for any $z \in \mathbb{C}$,

$$
\begin{aligned}
\left|\delta_b(z;a,c)\right| &\le c!\binom{a}{c}\bigl(|z|+b-1\bigr)^{a-c}\frac{1}{b!^q r!}\\
&\le \frac{a!}{(a-c)!}\cdot\frac{\bigl(|z|+b-1\bigr)^{a-c}}{b^a}\cdot e^{a+b}.
\end{aligned}
$$

This completes the proof of Lemma 9.8. □

We need an upper bound for $\nu(n)$. Since $v_p(\nu(n)) \le (\log n)/(\log p)$, we have for any $n \ge 1$

$$
\nu(n) = \prod_{p\le n} p^{v_p(\nu(n))} \le n^{\pi(n)},
$$

where $\pi(n)$ is the counting function of prime numbers:

$$
\pi(n) = \sum_{p\le n} 1.
$$

From the prime number Theorem ([HaWr 1938], [GLin 1962]), we deduce that for any $\epsilon > 0$ we have

$$
\nu(n) \le e^{(1+\epsilon)n}
$$

for $n \ge n_0(\epsilon)$. We need an estimate valid for any $n \ge 1$. We shall use $\nu(n) \le 3^n$, which is elementary (see for instance Th. 7.5 of [Duv 1998]), but in fact the sharper inequality

$$
\nu(n) \le e^{107n/103}
$$

holds uniformly for $n \ge 1$ (see [Y 1989], Lemma 2.3 p. 127).

We shall use Lemma 9.8 with

$$
m=t,\quad a=\sigma\quad (0\le\sigma\le S_0),\quad b=S_0^\sharp,\quad C=\tau_0,\quad c=\kappa.
$$

For any $z \in \mathbb{C}$ we have

$$
\sum_{\kappa=0}^{\tau_0}\binom{\tau_0}{\kappa}\left|\delta_{S_0^\sharp}(z;\sigma,\kappa)\right| \le \tau_0!\left(\frac{|z|}{S_0^\sharp}+1\right)^{\sigma} e^{\sigma+S_0^\sharp}.
$$

The estimate for the common denominator of the rational numbers $\delta_{S_0^\sharp}(t;\sigma,\kappa)$ involves

$$
\nu(S_0^\sharp)^{\tau_0} \le 3^{\tau_0 S_0^\sharp} \le 3^{S_0^\sharp T_0}.
$$

For this reason, one cannot take for $S_0^\sharp$ a too large integer, and this is why the polynomials $\delta_{S_0^\sharp}(z;\sigma)$ are sometimes better than $\Delta(z;\sigma)$ (which corresponds to $S_0^\sharp = S_0$).

Definition 9.9. In the general case, for

$$
\underline{\tau}\in\mathbb{N}^d,\quad t\in\mathbb{Z},\quad \sigma\in\mathbb{N}\quad\text{and}\quad \underline{s}\in\mathbb{Z}^m,
$$

we introduce the numbers

$$\widetilde{\gamma}_{\underline{\tau}t}^{(\sigma\underline{s})} = \sum_{\kappa=0}^{\tau_0} \binom{\tau_0}{\kappa} \delta_{S_0^\sharp}(t;\sigma,\kappa)(s_m\beta_0)^{\tau_0-\kappa} \prod_{i=m-d+1}^{m-1} (s_i + s_m\beta_i)^{\tau_i} \cdot \prod_{j=1}^{m} \alpha_j^{ts_j}.$$

In the homogeneous case $\beta_0 = 0$, the only non-vanishing term in the sum occurs for $\kappa = \tau_0$ and the formula is simpler:

$$\widetilde{\gamma}_{\underline{\tau}t}^{(\sigma\underline{s})} = \delta_{S_0^\sharp}(t;\sigma,\tau_0) \prod_{i=m-d+1}^{m-1} (s_i + s_m\beta_i)^{\tau_i} \cdot \prod_{j=1}^{m} \alpha_j^{ts_j}.$$

If $\beta_1, \dots, \beta_{m-1}$ are all rational numbers, we define b_m as their least (positive) common denominator, so that the rational integers $b_i = -b_m\beta_i \in \mathbb{Z}$ $(1 \le i \le m-1)$ satisfy $\gcd(b_1, \dots, b_m) = 1$.

As already mentioned, further Δ polynomials are needed in the homogeneous rational case

Definition 9.10. For the homogeneous rational case, we define

$$\widetilde{\gamma}_{\underline{\tau}t}^{(\sigma\underline{s})} = \frac{1}{\tau_0!}\delta_{S_0^\sharp}(t;\sigma,\tau_0) \prod_{i=m-d+1}^{m-1} \Delta(s_ib_m - s_mb_i;\tau_i) \cdot \prod_{j=1}^{m} \alpha_j^{ts_j}.$$

In the transcendence proof, analytic as well as arithmetic estimates related to these numbers $\widetilde{\gamma}_{\underline{\tau}t}^{(\sigma\underline{s})}$ will rest on the following result.

Lemma 9.11.
a) (General case)

$$\sum_{\kappa=0}^{\tau_0} \binom{\tau_0}{\kappa} \delta_{S_0^\sharp}(t;\sigma,\kappa)(s_mY_0)^{\tau_0-\kappa} \prod_{i=m-d+1}^{m-1} (s_i + s_mY_i)^{\tau_i}$$

is a polynomial in $Y_0, Y_{m-d+1}, \dots, Y_{m-1}$ with rational coefficients. The total degree of this polynomial is $\le T_0$. The length is bounded by

$$T_0!e^{S_0+S_0^\sharp}\left(\frac{T_1}{S_0^\sharp}+1\right)^{S_0}(S^*)^{T_0},$$

where $S^* = S_1 + \cdots + S_m$. Moreover the product of this polynomial by $\nu(S_0^\sharp)^{\tau_0}$ has integer coefficients.
b) (Homogeneous rational case).

$$\frac{1}{\tau_0!}\delta_{S_0^\sharp}(t;\sigma,\tau_0) \prod_{i=m-d+1}^{m-1} \Delta(s_ib_m - s_mb_i;\tau_i)$$

is a rational number of absolute value bounded by

$$e^{S_0+S_0^\sharp+T_0}\left(\frac{T_1}{S_0^\sharp}+1\right)^{S_0}\max_{1\le i\le m-1}\left(\frac{S_i|b_m|+S_m|b_i|}{T_0}+1\right)^{T_0}.$$

A denominator is

$$\nu(S_0^\sharp)^{\tau_0}\le 3^{T_0S_0^\sharp}.$$

Proof. This follows easily from Lemma 9.8. In the general case, since

$$\delta_{S_0^\sharp}(t;\sigma,\kappa)\le\kappa!e^{\sigma+S_0^\sharp}\left(\frac{|t|}{S_0^\sharp}+1\right)^{\sigma}$$

the length of the polynomial is bounded by

$$\sum_{\kappa=0}^{\tau_0}\binom{\tau_0}{\kappa}\kappa!e^{\sigma+S_0^\sharp}\left(\frac{|t|}{S_0^\sharp}+1\right)^{\sigma}S_m^{\tau_0-\kappa}\prod_{i=m-d+1}^{m-1}(S_i+S_m)^{\tau_i}$$

and we have

$$\max\left\{1+S_m,\ S_{m-d+1}+S_m,\dots,S_{m-1}+S_m\right\}\le S^*.$$

In the homogeneous rational case, we use the estimates

$$\frac{1}{\tau_0!}\left|\delta_{S_0^\sharp}(t;\sigma,\tau_0)\right|\le e^{\sigma+S_0^\sharp}\left(\frac{T_1}{S_0^\sharp}+1\right)^{S_0}$$

and

$$\prod_{i=m-d+1}^{m-1}\left|\Delta(s_ib_m-s_mb_i;\tau_i)\right|\le\max_{1\le i\le m-1}\left(\frac{S_i|b_m|+S_m|b_i|}{T_0}+1\right)^{T_0}e^{T_0}.$$

□

Remark. The first idea of eliminating the factorials from the derivatives of auxiliary functions with the help of such polynomials is due to Feldman [F 1960a], [F 1960b]. In order to improve the transcendence measure of π and of logarithms of algebraic numbers, he introduced the polynomials $F_a(z)=z(z-1)\cdots(z-a+1)$ and proved a lower bound for the greatest common divisor of the values of its derivatives $(d/dz)^cF_a$ at integers m. This estimate was uniform in m, polynomial in c, but only for fixed a (see Lemma 10.7 of [F 1982]). This was not a common factor for $a=0,1,\dots A$. Later, to avoid this difficulty, he used $(A-a)!F_a$ for $a=0,1,\dots A$. Other authors used $\Delta(z;a)=F_a(z)/a!$ with the corresponding upper bound for a denominator. Since this estimate is valid only for fixed a, it was necessary to introduce $\Delta(z+a;A)$ for $0\le a\le A$ where the polynomials have the same degree A.

These Delta polynomials were one of the key tools of Fel'dman [F 1968] when he achieved a best possible dependence for the estimate in terms of the heights of the coefficients β_i in lower bounds for linear combinations $\beta_0+\beta_1\log\alpha_1+\cdots+\beta_n\log\alpha_n$.

In turn, such an optimal estimate has dramatic consequences: in particular it yields the first effective improvement to Liouville's inequality. The introduction of polynomials $\Delta(z;a)^b$ into transcendence theory is due to A. Baker [B 1972], who improved in this way the dependence of lower bounds for linear forms in logarithms in terms of the heights of the α_i.

The polynomials

$$\left(\frac{z(z+1)\cdots(z+b-1)}{b!}\right)^q \cdot \left(\frac{z(z+1)\cdots(z+r-1)}{r!}\right)$$

for $a = bq+r$ and $0 \le r \le b$ were introduced in [Mat 1993a] (for $a = (b+1)q = bq+q$ one can choose either $r = 0$ or $r = b$, the result is the same; so there is no jump in the sequence).

e) Sketch of Proof (Conclusion)

The sketch of proof of Theorem 9.1 is now the usual one: with the numbers $\widetilde{\gamma}_{\underline{\tau}t}^{(\sigma\underline{s})}$ we build a matrix $\boldsymbol{M}$, we use a zero estimate to show that it has maximal rank, we take a maximal square submatrix with non-vanishing determinant Δ, we produce a lower bound for $|\Delta|$ by means of Liouville's estimate and an upper bound by means of Schwarz' Lemma, and the conclusion will follow.

In the next section we investigate the consequences of our change of basis (introducing delta polynomials) in the main analytic upper bound for the absolute value of the interpolation determinant.

9.2.2 Analytic Estimates

Recall the derivative operator $\mathcal{D}$ on the ring of entire functions in $m+1$ variables $z_0, \ldots, z_m$:

$$\mathcal{D} = \frac{\partial}{\partial z_0} + \frac{\partial}{\partial z_m}.$$

We shall work here with only $d+1$ variables $\underline{z} = (z_0, z_{m-d+1}, \ldots, z_m)$. Let $\delta^{(1)}(\underline{z};\tau)$ $1 \le \tau \le \binom{T_0+d}{d}$ be a basis of the space of polynomials in d variables $z_0, z_{m-d+1}, \ldots, z_{m-1}$ of total degree $\le T_0$ (and degree 0 in z_m). In our applications later this basis will be either

$$z_0^{\tau_0} z_{m-d+1}^{\tau_{m-d+1}} \cdots z_{m-1}^{\tau_{m-1}}$$

or else

$$z_0^{\tau_0}\, \Delta\left(z_{m-d+1}; \tau_{m-d+1}\right) \cdots \Delta\left(z_{m-1}; \tau_{m-1}\right)$$

with $\underline{\tau} \in \mathbb{N}^d$, $\|\underline{\tau}\| \le T_0$.

For $\kappa \in \mathbb{N}$ define

$$\delta^{(1)}(\underline{z};\tau,\kappa) = \left(\frac{\partial}{\partial z_0}\right)^{\kappa} \delta^{(1)}(\underline{z};\tau).$$

Denote by $\boldsymbol{M}$ the matrix with $L_d = \binom{T_0+d}{d}(2T_1+1)$ rows and

$$(S_0+1)(2S_1+1)\cdots(2S_m+1)$$

columns:

$$\boldsymbol{M} = \left(\mathcal{D}^{\sigma}\left(\delta^{(1)}(\underline{z};\tau)e^{tz_m}\right)(\underline{s}\eta)\right)_{\substack{(\tau,t)\\(\sigma,\underline{s})}},$$

where $\underline{s}\eta$ stands now for

$$(s_m\beta_0, s_{m-d+1}+s_m\beta_{m-d+1},\ldots,s_{m-1}+s_m\beta_{m-1}, s_1\lambda_1+\cdots+s_m\lambda_m)\in\mathbb{C}^{d+1}.$$

The index of rows (τ, t) ranges over the set of pairs in $\mathbb{N}\times\mathbb{Z}$ with $1\le\tau\le\binom{T_0+d}{d}$ and $|t|\le T_1$, while the index of columns $(\sigma,\underline{s})$ runs over the set of tuples in $\mathbb{N}\times\mathbb{Z}^m$ with $0\le\sigma\le S_0$ and $|s_j|\le S_j$. We select an ordering for the tuples $(\sigma,\underline{s})$ and we order accordingly the columns of $\boldsymbol{M}$ so that the tuples with the same $\underline{s}$ are consecutive in order

$$(0,\underline{s}),\quad(1,\underline{s}),\quad\ldots,\quad(S_0,\underline{s}).$$

This yields $(2S_1+1)\cdots(2S_m+1)$ blocs of $L_d\times(S_0+1)$ matrices.

Consider now a basis $\left\{\delta^{(2)}(z;\sigma)\,;\,0\le\sigma\le S_0\right\}$ of the space of polynomials in a single variable of degree $\le S_0$. Later we shall select the basis

$$\delta_{S_0^\sharp}(z;\sigma)\quad(0\le\sigma\le S_0).$$

Denote by $\boldsymbol{Q}$ the regular square $(S_0+1)\times(S_0+1)$ matrix occurring in Lemma 9.7 and by $\widetilde{\boldsymbol{Q}}$ the regular square matrix which is a diagonal bloc consisting of $(2S_1+1)\cdots(2S_m+1)$ blocs $\boldsymbol{Q}$:

$$\widetilde{\boldsymbol{Q}} = \operatorname{diag}\left(\boldsymbol{Q}\ \cdots\ \boldsymbol{Q}\right).$$

Define

$$\Phi_{\tau t}^{(\sigma)}(\underline{z}) = \sum_{\kappa=0}^{\tau_0}\frac{1}{\kappa!}\delta^{(2)}(t;\sigma,\kappa)\delta^{(1)}(\underline{z};\tau,\kappa)e^{tz_m}.$$

Let $\widetilde{\boldsymbol{M}}$ be the matrix

$$\widetilde{\boldsymbol{M}} = \left(\Phi_{\tau t}^{(\sigma)}(\underline{s}\eta)\right)_{\substack{(\tau,t)\\(\sigma,\underline{s})}}$$

with the same size as $\boldsymbol{M}$. Then we deduce from Lemma 9.7

$$\boldsymbol{M}\widetilde{\boldsymbol{Q}} = \widetilde{\boldsymbol{M}}. \tag{9.12}$$

Let

$$\Delta = \det\left(\Phi_{\tau t}^{(\sigma_\mu)}(\underline{s}_\mu\eta)\right)_{\substack{(\tau,t)\\1\le\mu\le L_d}}$$

be a $L_d\times L_d$ minor of $\widetilde{\boldsymbol{M}}$. We want an upper bound for $|\Delta|$. Write

$$\underline{s}\eta' = \left(s_m\beta_0, s_1+s_m\beta_1,\ldots,s_{m-1}+s_m\beta_{m-1}, s_1\lambda_1+\cdots+s_m\lambda_m+s_m\Lambda\right)$$

and $\underline{\zeta}_\mu = \underline{s}_\mu\eta'$. Let ϵ and $b_{\tau t\mu}$ be complex numbers satisfying

$$\Phi_{\tau t}^{(\sigma_\mu)}(\underline{s}_\mu\eta) = \Phi_{\tau t}^{(\sigma_\mu)}(\underline{s}_\mu\eta') + \epsilon b_{\tau t\mu}$$

so that

$$\Delta = \det\Big(\Phi_{\tau t}^{(\sigma_\mu)}(\underline{\zeta}_\mu) + \epsilon b_{\tau t \mu}\Big)_{\substack{(\tau,t)\\ 1\le\mu\le L_d}}.$$

Proposition 9.13. *Define*

$$V_d = \frac{1}{2d}(T_0 + d)(2T_1 + 1)\log E$$

and assume $|\epsilon| \le e^{-V_d}$. For each (τ, t), let $M_{\tau t}$ be a positive real number such that

$$M_{\tau t} \ge \log \sup_{|z|=E} \max_{1\le\mu\le L_d} |\Phi_{\tau t}^{(\sigma_\mu)}(z\underline{\zeta}_\mu)| \quad \text{and} \quad M_{\tau t} \ge \log \max_{1\le\mu\le L_d} |b_{\tau t\mu}|.$$

Then

$$\log|\Delta| \le -\frac{1}{2}L_d V_d + L_d S_0 \log E + L_d \log(2L_d) + \sum_{\tau=1}^{\binom{T_0+d}{d}} \sum_{t=-T_1}^{T_1} M_{\tau t}.$$

Proposition 9.13 rests on a lower bound for the order of vanishing at the origin of some interpolation determinant.

Let I be a subset of $\{(\tau, t)\,;\, 1 \le \tau \le \binom{T_0+d}{d},\ |t| \le T_1\}$. For $z \in \mathbb{C}$ define

$$d_{\tau t\mu}(z) = \begin{cases} \Phi_{\tau t}^{(\sigma_\mu)}(z\underline{\zeta}_\mu) & \text{for } (\tau,t) \in I,\\ b_{\tau t\mu} & \text{for } (\tau,t) \notin I \end{cases}$$

and

$$D_I(z) = \det\Big(d_{\tau t\mu}(z)\Big)_{\substack{(\tau,t)\\ 1\le\mu\le L_d}}.$$

Lemma 9.14. *The function D_I has a zero of multiplicity*

$$\ge \Theta(d; T_0, |I|) - L_d S_0$$

at the origin.

Lemma 9.14 is a special case of the following more general result, which extends at the same time Lemmas 6.2, 7.2 and 9.2.

Lemma 9.15. *Let n, T_0, S_0, L and L' be positive integers, $\underline{\xi}_1, \dots, \underline{\xi}_{L'}$ elements of $\mathbb{C}^n$, $\varphi_1, \dots, \varphi_L$ analytic functions in $\mathbb{C}$, $\theta_1, \dots, \theta_n$ complex numbers, $p_1, \dots, p_L$ polynomials in $\mathbb{C}[z_1, \dots, z_n]$ of total degree $\le T_0$ and $\mathfrak{D}^{(1)}, \dots, \mathfrak{D}^{(L')}$ derivative operators (acting on functions of n variables) of order $\le S_0$. For $1 \le \lambda \le L$, define,*

$$f_\lambda(z_1, \dots, z_n) = p_\lambda(z_1, \dots, z_n)\varphi_\lambda(\theta_1 z_1 + \cdots + \theta_n z_n).$$

Let I be a subset of $\{1, \dots, L\}$, and let $\delta_{\lambda\mu}$ ($1 \le \lambda \le L$ with $\lambda \notin I$, and $1 \le \mu \le L'$) be complex numbers. Define, for $1 \le \lambda \le L$, $1 \le \mu \le L'$ and $z \in \mathbb{C}$,

$$d_{\lambda\mu}(z) = \begin{cases} \left(\mathcal{D}^{(\mu)} f_\lambda\right)(z\underline{\zeta}_\mu) & \textit{for } \lambda \in I, \\ \delta_{\lambda\mu} & \textit{for } \lambda \notin I. \end{cases}$$

Let Q be a $L' \times L$ matrix with constant coefficients in $\mathbb{C}$ and $D(z)$ the determinant of the product

$$\Big(d_{\lambda\mu}(z)\Big)_{\substack{1\le\lambda\le L\\ 1\le\mu\le L'}} \cdot Q.$$

Then D has a zero at the origin of multiplicity $\geq \Theta(n; T_0, |I|) - LS_0$.

Proof of Lemma 9.15. After a change of coordinates we may assume

$$\theta_1 = 1, \quad \theta_2 = \cdots = \theta_n = 0.$$

Also, expanding each φ_λ in power series at the origin, without loss of generality we may assume

$$f_\lambda(\underline{z}) = \underline{z}^{\underline{\kappa}_\lambda} \quad \text{where} \quad \kappa_{\lambda 2} + \cdots + \kappa_{\lambda n} \leq T_0.$$

Finally we may also assume $I = \{1, \ldots, |I|\}$.

Like in the proof of Lemma 9.2, we can write

$$\mathcal{D}^{(\mu)} \underline{z}^{\underline{\kappa}} = \sum_{\|\underline{\iota}\| \leq S_0} c_{\mu\underline{\iota}} \binom{\underline{\kappa}}{\underline{\iota}} \underline{z}^{\underline{\kappa}-\underline{\iota}},$$

hence

$$\Big(d_{\lambda\mu}(z)\Big)_{\substack{1\le\lambda\le L\\ 1\le\mu\le L'}} = P(z)\widetilde{M}(z),$$

where $P(z)$ is the diagonal $L \times L$ matrix

$$P(z) = \operatorname{diag}\Big(z^{\|\underline{\kappa}_1\| - S_0} \quad \cdots \quad z^{\|\underline{\kappa}_{|I|}\| - S_0} \quad 1 \quad \cdots \quad 1\Big)$$

and where the first $|I|$ rows of the $L \times L'$ matrix $\widetilde{M}(z)$ are

$$\left(\sum_{\|\underline{\iota}\| \leq S_0} c_{\mu\underline{\iota}} \binom{\underline{\kappa}_\lambda}{\underline{\iota}} z^{S_0 - \|\underline{\iota}\|} \underline{\zeta}_\mu^{\underline{\kappa}_\lambda - \underline{\iota}} \right)_{1 \leq \mu \leq L'} \qquad (1 \leq \lambda \leq |I|)$$

while the last $L - |I|$ rows of $\widetilde{M}(z)$ are

$$\Big(\delta_{\lambda 1} \quad \cdots \quad \delta_{\lambda L'}\Big) \qquad (|I| < \lambda \leq L).$$

The $L \times L$ matrix $\widetilde{M}(z) \cdot Q$ has entries in $\mathbb{C}[z]$; it has rank $< L$ as soon as there exists $\lambda \neq \lambda'$ with $\underline{\kappa}_\lambda = \underline{\kappa}_{\lambda'}$. The conclusion of Lemma 9.15 plainly follows. □

Proof of Proposition 9.13. Let I be a subset of

$$\left\{(\tau, t) \in \mathbb{N} \times \mathbb{Z}, 1 \leq \tau \leq \binom{T_0 + d}{d}, \ |t| \leq T_1\right\}.$$

From Lemma 9.14, using Schwarz Lemma 2.4, we deduce

$$\log|D_I(1)| \le -\Theta(d; T_0, |I|)\log E + L_d S_0 \log E + \log \sup_{|z|=E} |D(z)|.$$

For $|z| = E$, we have

$$\log|D_I(z)| \le \log(L_d!) + \sum_{\tau=1}^{\binom{T_0+d}{d}} \sum_{t=-T_1}^{T_1} M_{\tau t}.$$

Define

$$\Delta_I = \det\bigl(c_{\tau t\mu}^{(I)}\bigr)_{\substack{(\tau,t)\\ 1\le\mu\le L_d}}$$

where

$$c_{\tau t\mu}^{(I)} = \begin{cases} \Phi_{\tau t}^{(\sigma_\mu)}(\underline{\zeta}_\mu) & \text{for } (\tau,t)\in I,\\ b_{\tau t\mu} & \text{for } (\tau,t)\notin I.\end{cases}$$

From Lemmas 9.15 and 7.3, we see that the hypotheses of Lemma 7.4 are satisfied with $r = 1$, $V = V_d$ and

$$\chi_0 = \frac{1}{2}(\log E)\binom{T_0+d-1}{d-1}^{-1},$$

$$\chi_1 = V_d - \chi_0 + \frac{1}{2d}(T_0+d)\log E \le \frac{4}{3}V_d$$

and

$$\chi_2 = L_d S_0 \log E + \log(L_d!) + \sum_{\tau=1}^{\binom{T_0+d}{d}} \sum_{t=-T_1}^{T_1} M_{\tau t}.$$

Proposition 9.13 follows with the same estimates as in the proof of Proposition 7.6. □

9.2.3 The Zero Estimate

Here is the zero estimate needed for the proof of Theorem 9.1. Let K be an algebraically closed field of zero characteristic, m and d positive integers with $1 \le d \le m$. We denote by $(\underline{e}_1, \ldots, \underline{e}_m)$ the canonical basis of K^m.

Let $\alpha_1, \ldots, \alpha_m$ be elements of $K^\times$, $\beta_0, \beta_1, \ldots, \beta_{m-1}$ elements of K, T_0, T_1, S_0 positive integers and $\mathfrak{S}$ a subset of $\mathbb{Z}^m$. Define

$$L_d = \binom{T_0+d}{d}(2T_1+1).$$

Let

$$\delta^{(1)}(\underline{z};\tau) \qquad 1 \le \tau \le \binom{T_0+d}{d}$$

be a basis of the space of polynomials in d variables $z_0, z_{m-d+1}, \ldots, z_{m-1}$ of total degree $\le T_0$. For $\kappa \in \mathbb{N}$ define

$$\delta^{(1)}(\underline{z};\tau,\kappa) = \left(\frac{\partial}{\partial z_0}\right)^{\kappa} \delta^{(1)}(\underline{z};\tau).$$

Further, let

$$\delta^{(2)}(z;\sigma) \qquad 0 \le \sigma \le (m+1)S_0$$

be a basis of the space of polynomials in a single variable of degree $\le (m+1)S_0$. Again, for $\kappa \in \mathbb{N}$, we define

$$\delta^{(2)}(z;\sigma,\kappa) = \left(\frac{d}{dz}\right)^{\kappa} \delta^{(2)}(z;\sigma).$$

Denote by $\underline{y}_{m-d+1}, \ldots, \underline{y}_m$ the column vectors of the $d \times d$ matrix

$$\begin{pmatrix} 0 & \cdots & \beta_0 \\ & & \beta_{m-d+1} \\ & \mathsf{I}_{d-1} & \vdots \\ & & \beta_{m-1} \end{pmatrix}.$$

For $\underline{s} \in \mathbb{Z}^m$, define $\underline{s}\,\underline{y} \in K^d$ by

$$\underline{s}\,\underline{y} = \big(s_m\beta_0, s_{m-d+1} + s_m\beta_{m-d+1}, \ldots, s_{m-1} + s_m\beta_{m-1}\big).$$

For $\tau \in \mathbb{N}$, $t \in \mathbb{Z}$, $\sigma \in \mathbb{N}$ and $\underline{s} \in \mathbb{Z}^m$, define

$$\widetilde{\gamma}_{\tau t}^{(\sigma \underline{s})} = \sum_{\kappa=0}^{\tau_0} \frac{1}{\kappa!} \delta^{(2)}(t;\sigma,\kappa)\delta^{(1)}(\underline{s}\,\underline{y};\tau,\kappa) \prod_{j=1}^{m} \alpha_j^{ts_j}.$$

Let $\mathcal{V}$ a subspace of K^m of dimension d containing the point

$$(\beta_1, \ldots, \beta_{m-1}, -1)$$

and such that $\pi_{\mathcal{V}}(\underline{e}_1), \ldots, \pi_{\mathcal{V}}(\underline{e}_{m-d})$ is a basis of $K^m/\mathcal{V}$, where $\pi_{\mathcal{V}}$ denotes the linear canonical map $K^m \to K^m/\mathcal{V}$. Define $\widetilde{\mathfrak{S}} = \{\underline{s}' - \underline{s}''\ ;\ \underline{s}' \in \mathfrak{S},\ \underline{s}'' \in \mathfrak{S}\}$ and $\widetilde{\mathfrak{S}}_{\mathcal{V}} = \widetilde{\mathfrak{S}} \cap \mathcal{V}$. For any $n \ge 1$, denote

$$\widetilde{\mathfrak{S}}[n] = \{\underline{s}_1 + \cdots + \underline{s}_n\ ;\ \underline{s}_i \in \widetilde{\mathfrak{S}}\ (1 \le i \le n)\}.$$

Let $\boldsymbol{M}$ be the matrix

$$\boldsymbol{M} = \left(\widetilde{\gamma}_{\tau t}^{(\sigma \underline{s})}\right)_{\substack{(\tau,t)\\(\sigma,\underline{s})}},$$

with L_d rows indexed by $(\tau, t) \in \mathbb{N} \times \mathbb{Z}$,

$$1 \le \tau \le \binom{T_0+d}{d} \quad \text{and} \quad |t| \le T_1,$$

while the index of columns is $(\sigma, \underline{s}) \in \mathbb{N} \times \mathbb{Z}^m$, running in the range

$$0 \le \sigma \le (m+1)S_0, \qquad \underline{s} \in \widetilde{\mathfrak{S}}_{\mathcal{V}}[m+1].$$

For Proposition 9.16 as well as for Lemma 9.17, we assume that $\alpha_1, \ldots, \alpha_m$ generate a multiplicative subgroup of $K^\times$ of rank $\ge m-1$. We assume further $S_0 + 1 \ge 2T_0$, and $|\underline{s}| < T_0/4$ for any $\underline{s} \in \mathfrak{S}$. Furthermore, assume

$$\text{either } \beta_0 \neq 0 \text{ or else } \widetilde{\mathfrak{S}}[2] \cap K(\beta_1, \ldots, \beta_{m-1}, -1) = \{0\}.$$

Proposition 9.16. *Assume also*

$$(S_0+1)\mathrm{Card}(\mathfrak{S}) > 2(m+1)T_0^m T_1,$$

$$\mathrm{Card}\big(\pi_{\mathcal{V}}(\mathfrak{S})\big) \le \frac{m+1}{d+1} T_0^{m-d},$$

and that there is no subspace $\mathcal{V}'$ of $\mathcal{V}$, other than $\mathcal{V}$ itself, containing

$$(\beta_1, \ldots, \beta_{m-1}, -1),$$

which satisfies this inequality with d replaced by $d' = \dim_K(\mathcal{V}')$.

Then $\boldsymbol{M}$ has rank L_d.

The proof of Proposition 9.16 combines the arguments of the proofs of Propositions 7.7 and 9.3. The main tool is the following auxiliary result (see [W 1993], Corollaire 5.4 for the case $\beta_0 = 0$).

Lemma 9.17. *Assume $\boldsymbol{M}$ has rank $< L_d$. Assume also*

$$(S_0+1)\mathrm{Card}\big(\widetilde{\mathfrak{S}}_{\mathcal{V}}\big) > 2(d+1)T_0^d T_1.$$

Then there exists a vector subspace $\mathcal{V}'$ of $\mathcal{V}$, containing $(\beta_1, \ldots, \beta_{m-1}, -1)$, of dimension d' with $1 \le d' \le d-1$, such that

$$\mathrm{Card}\Big(\pi_{\mathcal{V}'}\big(\widetilde{\mathfrak{S}}_{\mathcal{V}}\big)\Big) \le \frac{d+1}{d'+1} \cdot T_0^{d-d'}.$$

Proof of Lemma 9.17. Define

$$\mathcal{E} = \Big\{\underline{s}\underline{y}\,;\, \underline{s} \in \widetilde{\mathfrak{S}}_{\mathcal{V}}\Big\} \subset K^d.$$

Step 1.

We first check that the elements $\underline{s}\underline{y}$, for $\underline{s} \in \widetilde{\mathfrak{S}}_{\mathcal{V}}$, are pairwise distinct.

For $\underline{s}'$ and $\underline{s}''$ in $\widetilde{\mathfrak{S}}_{\mathcal{V}}$ the difference $\underline{s}' - \underline{s}''$ is in $\widetilde{\mathfrak{S}}_{\mathcal{V}}[2]$. So it is sufficient to prove that if $\underline{s} \in \widetilde{\mathfrak{S}}_{\mathcal{V}}[2]$ satisfies $\underline{s}\underline{y} = 0$:

$$s_m\beta_0 = s_{m-d+1} + s_m\beta_{m-d+1} = \cdots = s_{m-1} + s_m\beta_{m-1} = 0,$$

then $\underline{s} \neq 0$.

If $\beta_0 \neq 0$, then we have $s_m = 0$, $s_{m-d+1} = \cdots = s_{m-1} = 0$. Since $\underline{s} = (s_1, \dots, s_m)$ is in $\mathcal{V}$, and since $\pi_{\mathcal{V}}(\underline{e}_1), \dots, \pi_{\mathcal{V}}(\underline{e}_{m-d})$ is a basis of $K^m/\mathcal{V}$, this implies $\underline{s} = 0$.

If $\beta_0 = 0$ we deduce from the conditions $\underline{s} \in \mathcal{V}$ and $(\beta_1, \dots, \beta_{m-1}, -1) \in \mathcal{V}$

$$(s_1 + s_m\beta_1, \dots, s_{m-1} + s_m\beta_{m-1}, 0) \in \mathcal{V}.$$

Since the last d components are zero, we also have $s_j + s_m\beta_j = 0$ for $1 \leq j \leq m$, hence $s_m(\beta_1, \dots, \beta_{m-1}, -1) \in \widetilde{\mathfrak{S}}[2]$. By assumption this is possible only for $s_m = 0$, and therefore $s_j = 0$ for $1 \leq j \leq m$.

Step 2.

If $\alpha_1, \dots, \alpha_m$ are multiplicatively dependent, they generate a multiplicative group of rank $m-1$. In this case, among the tuples $(a_1, \dots, a_m) \in \mathbb{Z}^m \setminus \{0\}$ for which

$$\alpha_1^{a_1} \cdots \alpha_m^{a_m} = 1,$$

there is one (which is unique, up to sign) for which $\max\{|a_1|, \dots, |a_m|\}$ is minimal. Recall that for any $\underline{s} \in \widetilde{\mathfrak{S}}$ we have $|\underline{s}| < T_0/2$. It follows that for each $\gamma \in K^\times$ the number of elements $\underline{s} \in \widetilde{\mathfrak{S}}_{\mathcal{V}}$ such that $\underline{\alpha}^{\underline{s}} = \gamma$ is $\leq T_0$. Of course this is true also if $\alpha_1, \dots, \alpha_m$ are multiplicatively independent (!). Using Lemma 7.8 for the mapping

$$\begin{array}{ccc} \widetilde{\mathfrak{S}}_{\mathcal{V}} & \longrightarrow & K^\times \\ \underline{s} & \longmapsto & \underline{\alpha}^{\underline{s}} \end{array}$$

we derive

$$\mathrm{Card}\left\{\underline{\alpha}^{\underline{s}};\ \underline{s} \in \widetilde{\mathfrak{S}}_{\mathcal{V}}\right\} \geq \frac{1}{T_0}\mathrm{Card}\big(\widetilde{\mathfrak{S}}_{\mathcal{V}}\big).$$

Step 3.

We already introduced the derivative operator

$$\mathcal{D} = \frac{\partial}{\partial X_0} + Y\frac{\partial}{\partial Y}\cdot$$

Using Lemma 9.7 (compare with (9.12)), we deduce from the assumption $\mathrm{rank}(\boldsymbol{M}) < L_d$ that there exists a nonzero polynomial P in the ring

$$K[X_0, X_{m-d+1}, \dots, X_{m-1}, Y^{\pm 1}],$$

of total degree $\leq T_0$ with respect to $X_0, X_{m-d+1}, \dots, X_{m-1}$, of degree $\leq T_1$ with respect to $Y^{\pm 1}$, which satisfies

$$\mathcal{D}^\sigma P\big(\underline{s}\,\underline{y}, \underline{\alpha}^{\underline{s}}\big) = 0$$

for $0 \leq \sigma \leq (m+1)S_0$ and $\underline{s} \in \widetilde{\mathfrak{S}}_{\mathcal{V}}[m+1]$.

The assumptions of Theorem 8.1 are satisfied with d replaced by $d+1$,

$$G_0 = \mathbb{G}_{\mathrm{a}}^d, \quad G_1 = \mathbb{G}_{\mathrm{m}}, \quad G^+ = G = G_0 \times G_1, \quad G^- = \{e\},$$

$$\mathcal{W} = K(1, 0, \dots, 0) \subset K^{d+1}$$

and

$$\Sigma = \left\{ \left(\underline{s}y, \underline{\alpha}^{\underline{s}} \right) ; \underline{s} \in \widetilde{\mathfrak{S}}_V \right\} \subset G(K).$$

We deduce that there exists a connected algebraic subgroup $G^* = G_0^* \times G_1^*$ of G (where G_i^* is an algebraic subgroup of G_i for $i = 0, 1$) such that the conclusion of Theorem 8.1 holds:

$$\binom{S_0 + \ell_0'}{\ell_0'} \mathrm{Card}\left(\frac{\Sigma + G^*}{G^*} \right) \mathcal{H}(G^*; \underline{T}) \leq \mathcal{H}(G; \underline{T})$$

where $\underline{T} = (T_0; T_1)$ and

$$\ell_0' = \begin{cases} 0 & \text{if } (1, 0, \ldots, 0, 1) \in T_e(G^*) \\ 1 & \text{if } (1, 0, \ldots, 0, 1) \notin T_e(G^*). \end{cases}$$

Denote by d_i^* the dimension of G_i^*. We have (see § 5.1.1)

$$\mathcal{H}(G; \underline{T}) = 2(d+1) T_0^d T_1$$

and

$$\mathcal{H}(G^*; \underline{T}) = \begin{cases} T_0^{d_0^*} & \text{if } d_1^* = 0 \\ \\ 2(d_0^* + 1) T_0^{d_0^*} T_1 & \text{if } d_1^* = 1. \end{cases}$$

Step 4

We claim $d_0^* \geq 1$. Indeed for $d_0^* = 0$ we have $(1, 0, \ldots, 0, 1) \notin T_e(G^*)$ hence $\ell_0' = 1$ and

$$\mathrm{Card}\left(\frac{\Sigma + G^*}{G^*} \right) \geq \mathrm{Card}(\mathcal{E}).$$

From step 1 we deduce $\mathrm{Card}(\mathcal{E}) = \mathrm{Card}\left(\widetilde{\mathfrak{S}}_V\right)$, and therefore the assumption of Lemma 9.17 implies

$$(S_0 + 1) \mathrm{Card}(\mathcal{E}) > 2(d+1) T_0^d T_1.$$

Step 5.

We claim $G_1^* = G_1$. Indeed, for $G_1^* = \{1\}$ we have $d_1^* = 0$ and $\ell_0' = 1$. On the other hand we deduce from step 2:

$$\mathrm{Card}\left\{ \underline{\alpha}^{\underline{s}} ; \underline{s} \in \widetilde{\mathfrak{S}}_V \right\} \geq \frac{1}{T_0} \mathrm{Card}\left(\widetilde{\mathfrak{S}}_V\right) > \frac{2(d+1) T_0^{d-1} T_1}{S_0 + 1},$$

and therefore, since $d_0^* \geq 1$ (cf. step 4),

$$\begin{aligned} (S_0 + 1) \mathrm{Card}\left(\frac{\Sigma + (G_0^* \times \{1\})}{G_0^* \times \{1\}} \right) &> 2(d+1) T_0^{d-1} T_1 \\ &\geq 2(d+1) T_0^{d-d_0^*} T_1. \end{aligned}$$

Step 6.

From steps 3 and 5 we deduce $d_1^* = 1$ and

$$(S_0+1)^{\ell'_0}\mathrm{Card}\bigl(\pi_{G_0^*}(\mathcal{E})\bigr) \le \frac{d+1}{d_0^*+1}T_0^{d-d_0^*},$$

where $\pi_{G_0^*}$ is the canonical map $K^d \to K^d/G_0^*$. Since $G^* \neq G$ we also have $d_0^* \le d-1$.

Define a vector subspace $\mathcal{V}'$ of $\mathcal{V}$ by

$$\mathcal{V}' = \bigl\{\underline{z} \in \mathcal{V}\,;\ \exists z_0 \in K,\ (z_0, z_{m-d+1}+z_m\beta_{m-d+1}, \dots, z_{m-1}+z_m\beta_{m-1}) \in G_0^*\bigr\}.$$

Hence $\mathcal{V}'$ contains the point $(\beta_1, \dots, \beta_{m-1}, -1)$. We denote by d' the dimension of $\mathcal{V}'$.

The linear map

$$\begin{array}{rccc} \psi\colon & \mathcal{V} & \longrightarrow & K^{d-1} \\ & \underline{z} & \longmapsto & \bigl(z_{m-d+1}+z_m\beta_{m-d+1}, \dots, z_{m-1}+z_m\beta_{m-1}\bigr) \end{array}$$

is surjective with kernel $K(\beta_1, \dots, \beta_{m-1}, -1)$.

Let $\pi\colon K^d \to K^{d-1}$ denote the projection of kernel $K(1,0,\dots,0)$ and $\mathcal{U}$ be the image of G_0^* under π. Plainly we have

$$\mathcal{V}' = \psi^{-1}(\mathcal{U}),$$

hence $\dim(\mathcal{U}) = d'-1$. Since

$$\pi(\mathcal{E}) = \psi\bigl(\widetilde{\mathfrak{S}}_{\mathcal{V}}\bigr),$$

we deduce from the diagram

$$\begin{array}{ccccc}
\mathcal{E} & & \pi(\mathcal{E}) & & \widetilde{\mathfrak{S}}_{\mathcal{V}} \\
\cap & & \cap & & \cap \\
K^d & \xrightarrow{\ \pi\ } & K^{d-1} & \xleftarrow{\ \psi\ } & \mathcal{V} \\
{\scriptstyle \pi_{G_0^*}}\downarrow & & {\scriptstyle \pi_{\mathcal{U}}}\downarrow & & {\scriptstyle \pi_{\mathcal{V}'}}\downarrow \\
\dfrac{K^d}{G_0^*} & \longrightarrow & \dfrac{K^{d-1}}{\mathcal{U}} & \xleftarrow{\ \sim\ } & \dfrac{\mathcal{V}}{\mathcal{V}'} \\
\cup & & \cup & & \cup \\
\pi_{G_0^*}(\mathcal{E}) & & \pi_{\mathcal{U}}\bigl(\pi(\mathcal{E})\bigr) & & \pi_{\mathcal{V}'}\bigl(\widetilde{\mathfrak{S}}_{\mathcal{V}}\bigr)
\end{array}$$

the inequality

$$\mathrm{Card}\Big(\pi_{\mathcal{V}'}(\widetilde{\mathfrak{S}}_{\mathcal{V}})\Big) \le \mathrm{Card}\big(\pi_{G_0^*}(\mathcal{E})\big).$$

Notice that the map $K^d/G_0^* \longrightarrow K^{d-1}/\mathcal{U}$ is surjective with kernel $\pi_{G_0^*}(K \times \mathcal{U})$, while $\mathcal{V}/\mathcal{V}' \longrightarrow K^{d-1}/\mathcal{U}$ is an isomorphism.

The surjective linear map $G_0^* \rightarrow \mathcal{U}$ which maps x onto $\pi(x)$ has kernel $G_0^* \cap K(1, 0, \dots, 0)$. We distinguish two cases.

Case a): $\ell_0' = 1$

In this case $G_0^* \cap K(1, 0, \dots, 0) = \{0\}$, hence $\mathcal{U}$ has dimension d_0^* and $d' = d_0^* + 1$. From $1 \le d_0^* \le d - 1$ we deduce $2 \le d' \le d$. From the condition $S_0 + 1 \ge 2T_0$ we deduce

$$\mathrm{Card}\Big(\pi_{\mathcal{V}'}(\widetilde{\mathfrak{S}}_{\mathcal{V}})\Big) \le \mathrm{Card}\big(\pi_{G_0^*}(\mathcal{E})\big) \le \frac{d+1}{2d'} T_0^{d-d'}.$$

Moreover the inequality $\mathrm{Card}\big(\pi_{G_0^*}(\mathcal{E})\big) \ge 1$ implies $d' < d$, hence we have $2 \le d' \le d - 1$.

Case b): $\ell_0' = 0$

Now we have $G_0^* \ni (1, 0, \dots, 0)$, hence $G_0^* = K \times \mathcal{U}$ and $\mathcal{U}$ has dimension $d_0^* - 1$. Therefore $d' = d_0^*$ with $1 \le d' \le d - 1$, and

$$\mathrm{Card}\Big(\pi_{\mathcal{V}'}(\widetilde{\mathfrak{S}}_{\mathcal{V}})\Big) = \mathrm{Card}\big(\pi_{G_0^*}(\mathcal{E})\big) \le \frac{d+1}{d'+1} T_0^{d-d'}.$$

□

Proof of Proposition 9.16. We shall use Lemma 7.8 twice. First we consider the restriction of $\pi_{\mathcal{V}}$ to $\mathfrak{S}$ and get

$$\mathrm{Card}\big(\widetilde{\mathfrak{S}}_{\mathcal{V}}\big) \cdot \mathrm{Card}\Big(\pi_{\mathcal{V}}(\mathfrak{S})\Big) \ge \mathrm{Card}(\mathfrak{S}).$$

Hence from the assumption of Proposition 9.16 we obtain

$$\begin{aligned}(S_0+1)\mathrm{Card}\big(\widetilde{\mathfrak{S}}_{\mathcal{V}}\big) &\ge (S_0+1)\mathrm{Card}(\mathfrak{S}) \cdot \frac{d+1}{m+1} \cdot T_0^{d-m} \\ &> 2(d+1)T_0^d T_1.\end{aligned}$$

Assume the conclusion of Proposition 9.16 does not hold: $\mathrm{rank}(M) < L_d$. We use Lemma 9.17: there exists a vector subspace $\mathcal{V}'$ of $\mathcal{V}$, containing $(\beta_1, \dots, \beta_{m-1}, -1)$, of dimension d' with $1 \le d' \le d - 1$, such that

$$\mathrm{Card}\Big(\pi_{\mathcal{V}'}(\widetilde{\mathfrak{S}}_{\mathcal{V}})\Big) \le \frac{d+1}{d'+1} \cdot T_0^{d-d'}.$$

We use Lemma 7.8 again for the canonical mapping

$$\psi\colon \frac{K^m}{\mathcal{V}'} \longrightarrow \frac{K^m}{\mathcal{V}}$$

with (using the notation after (7.9))

$$\mathfrak{C} = \pi_{\mathcal{V}'}(\mathfrak{S}), \quad \psi(\mathfrak{C}) = \pi_{\mathcal{V}}(\mathfrak{S}), \quad \widetilde{\mathfrak{C}} \cap \ker \psi = \pi_{\mathcal{V}'}\big(\widetilde{\mathfrak{S}}_{\mathcal{V}}\big),$$

$$\operatorname{Card}\Big(\widetilde{\mathfrak{C}} \cap \ker \psi\Big) \leq \frac{d+1}{d'+1} \cdot T_0^{d-d'},$$

$$\operatorname{Card}\big(\psi(\mathfrak{C})\big) \leq \frac{m+1}{d+1} \cdot T_0^{m-d}$$

and we conclude

$$\operatorname{Card}\big(\pi_{\mathcal{V}'}(\mathfrak{S})\big) \leq \frac{m+1}{d'+1} \cdot T_0^{m-d'}.$$

This contradicts the minimality of $\mathcal{V}$. □

Remark. Under the assumptions of Proposition 9.16, assume $4T_0T_1 \geq S_0 + 1$ and $d = 1$. In this case $\mathcal{V} = K(\beta_1, \ldots, \beta_{m-1}, -1)$. Using Lemma 7.8 we deduce

$$\operatorname{Card}\big(\widetilde{\mathfrak{S}}_{\mathcal{V}}\big)\operatorname{Card}\big(\pi_{\mathcal{V}}(\mathfrak{S})\big) \geq \operatorname{Card}(\mathfrak{S}) > \frac{2(m+1)T_0^m T_1}{S_0+1} \geq \frac{m+1}{2}T_0^{m-1}.$$

Therefore $\operatorname{Card}\big(\widetilde{\mathfrak{S}}_{\mathcal{V}}\big) > 1$ and

$$\widetilde{\mathfrak{S}} \cap K(\beta_1, \ldots, \beta_{m-1}, -1) \neq \{0\}.$$

Notice that this has been excluded when $\beta_0 \neq 0$.

9.3 Value of $C(m)$

We prove Theorem 9.1 with an explicit value for $C(m)$. Our goal is only to show that everything can be explicitly computed, not to seek for a sharp numerical value.

Proposition 9.18. *The conclusion of Theorem 9.1 holds with*

$$C(m) = 2^{m+25}m^{3m+9}.$$

Moreover if $E = e$, $m \geq 10$ and $\log A_j \geq m/(2D)$ for $1 \leq j \leq m$, then the conclusion holds with

$$C(m) = 2^{5m+21}m^{2m+8}.$$

We shall prove both estimates at the same time. For this we introduce a real number $M \geq 1$ and define

$$M^* = \begin{cases} 1 & \text{if } M = 1, \\ 2^{m+3}M^2 & \text{if } M > 1. \end{cases}$$

We shall prove the conclusion of Theorem 9.1 with $C(m)(\log E)^{-m-1}$ replaced by

$$2^{m+25}m^{3m+9}M^*\big(\log(ME)\big)^{-m-1}.$$

Choosing $M = M^* = 1$ we get the first part of Proposition 9.18. For the second part where $E = e$, we choose $M = e^{(m/2)-1}$ so that

$$2^{m+3} M^2 \big(\log(ME)\big)^{-m-1} = 2^{2m+4} e^{m-2} m^{-m-1},$$

which is < 1 exactly when $m \geq 10$.

This parameter M arises as follows. In § 9.2, from

$$E|\lambda_j| \leq D \log A_j \quad \text{and} \quad S_j \leq \frac{U}{mDT_1 \log A_j}$$

we have derived the crude estimate

$$E|s_1\lambda_1 + \cdots + s_m\lambda_m| \leq E \sum_{j=1}^{m} S_j|\lambda_j| \leq \sum_{j=1}^{m} DS_j \log A_j \leq \frac{U}{T_1}$$

for any $\underline{s} \in \mathbb{Z}^m[\underline{S}]$. A preliminary remark is that we can replace the assumption

$$E|\lambda_j| \leq D \log A_j \quad (1 \leq j \leq m)$$

by the weaker requirement

$$\sum_{j=1}^{m} \frac{|\lambda_j|}{\log A_j} \leq \frac{D}{E}.$$

This is relevant only as far as we are interested in the dependence of $C(m)$ with respect to m. Next the main point is the following trick which we borrow from E. M. Matveev [Mat 1998]: if we restrict the range of $\underline{s}$ to the subset $\mathfrak{S}$ of

$$\mathbb{Z}^m[\underline{S}] = \big\{\underline{s} \in \mathbb{Z}^m;\ |s_j| \leq S_j\ (1 \leq j \leq m)\big\}$$

for which

$$|s_1\lambda_1 + \cdots + s_m\lambda_m| \leq \frac{U}{MET_1},$$

then we are able to use a larger radius for the analytic estimate, namely with E replaced by ME. This is why in the conclusion we can replace $\log E$ by $\log(ME)$. There will be a cost in counting $\mathrm{Card}(\mathfrak{S})$, which is needed to apply the zero estimate. Matveev [Mat 1998] gives a sharp lower bound for this number (see Exercise 9.5), but we shall simply use Dirichlet's box principle (step 2 of § 9.3.3; compare with [Y 1998], II).

9.3.1 Main Estimate

For the results of § 9.3.1 and § 9.3.3 we introduce some data and assume the following conditions are satisfied.

Let $\lambda_1, \ldots, \lambda_m$ be logarithms of nonzero algebraic numbers. Define $\alpha_i = \exp(\lambda_i)$ for $1 \leq i \leq m$ and assume that $\alpha_1, \ldots, \alpha_m$ span a multiplicative group of rank $\geq m - 1$. Let $\beta_0, \ldots, \beta_{m-1}$ be algebraic numbers. Set

$$D = [\mathbb{Q}(\alpha_1, \ldots, \alpha_m, \beta_0, \ldots, \beta_{m-1}) : \mathbb{Q}]$$

and

$$\Lambda = \beta_0 + \beta_1\lambda_1 + \cdots + \beta_{m-1}\lambda_{m-1} - \lambda_m.$$

Assume $\Lambda \neq 0$. Let $A_1, \ldots, A_m$, B_1, B_2, E and M be positive real numbers with $B_1 \geq e$, $B_2 \geq e$, $E \geq e$, $M \geq 1$, which satisfy

$$\log A_i \geq \mathrm{h}(\alpha_i), \quad D\log A_i \geq 1 \quad (1 \leq i \leq m),$$

$$\sum_{i=1}^{m} \frac{|\lambda_i|}{\log A_i} \leq \frac{D}{E},$$

$$B_1^D \geq ME \quad \text{and} \quad B_2^D \geq ME.$$

Our main estimate is the following.

Theorem 9.19. *Let T_0, T_1, $S_0^\sharp$, S_0, $S_1, \ldots, S_m$ be positive rational integers, U a positive real number and $\mathfrak{S}$ a subset of $\mathbb{Z}^m[\underline{S}]$ satisfying the following conditions:*

$$S_0 + 1 \geq 2T_0, \quad T_0 > 4S_i \qquad (1 \leq i \leq m),$$

$$\left|s_1\lambda_1 + \cdot + s_m\lambda_m\right| \leq \frac{U}{ME(T_1+1)} \quad \textit{for any} \quad \underline{s} \in \mathfrak{S},$$

and

$$(S_0+1)\mathrm{Card}(\mathfrak{S}) > 2(m+1)T_0^m T_1.$$

Assume also

$$B_2 \geq e^2\left(1 + \frac{T_1}{S_0^\sharp}\right).$$

Define

$$V = \frac{1}{2m}(T_0+1)(2T_1+1)\log(ME),$$

$$L = \binom{T_0+m}{m}(2T_1+1)$$

and assume

$$\frac{V}{2} \geq 4DT_0\log B_1 + 2(m+1)DS_0\log B_2 + DS_0^\sharp$$
$$+D\log(2L) + (m+1)U + 2(m+1)D(T_1+1)\sum_{j=1}^{m} S_j\log A_j.$$

Finally, assume

- *Either (general case)*

$$\log B_1 \geq \mathrm{h}(1 : \beta_0 : \cdots : \beta_{m-1}) \quad \textit{and} \quad B_1 \geq 2(m+1)T_0 3^{S_0^\sharp}(S_1 + \cdots + S_m).$$

- *Or else (homogeneous rational case)*

$$\beta_0 = 0, \quad \beta_i = -\frac{b_i}{b_m} \quad (1 \le i \le m) \quad \text{with} \quad (b_1, \dots, b_m) \in \mathbb{Z}^m,$$

$$B_1 \ge e + \frac{2e(m+1)}{T_0} \max_{1 \le j < m} \big(|b_m| S_j + |b_j| S_m\big) \quad \text{and} \quad B_1 \ge 3^{S_0^\sharp}.$$

Then

$$|\Lambda| > e^{-mV}.$$

For practical applications (for instance for solving diophantine equations) it is much more efficient to use Theorem 9.19 than Proposition 9.18.

9.3.2 Proof of the Main Estimate

Proof of Theorem 9.19.

We assume that the hypotheses of Theorem 9.19 are satisfied. We split the proof into several steps.

Step 1. Liouville's lower bound for $|\Lambda|$

If $\beta_0 = 0$ and at the same time $s(\beta_1, \dots, \beta_{m-1}, -1) \in \mathbb{Z}^m[4\underline{S}]$ for some $s \neq 0$, then Liouville's estimate readily provides the conclusion of Theorem 9.1 (the argument is the same as in the first step of the proof of Theorem 7.10; notice also that this is the only place in the proof where we need the assumption $\Lambda \neq 0$ – see Exercice 7.1).

Therefore we may assume, without loss of generality, that either $\beta_0 \neq 0$ or else $s(\beta_1, \dots, \beta_{m-1}, -1) \notin \mathbb{Z}^m[4\underline{S}]$ for any $s \neq 0$. This condition will be needed to apply Proposition 9.16 in the next step.

Step 2. The Matrix M and the Determinant Δ_{ar}

Let $\mathcal{V}$ a subspace of $\mathbb{C}^m$ containing the point

$$(\beta_1, \dots, \beta_{m-1}, -1)$$

and such that

$$\operatorname{Card}\big(\pi_{\mathcal{V}}(\mathfrak{S})\big) \le \frac{m+1}{d+1} T_0^{m-d}$$

where d is the dimension of $\mathcal{V}$ and $\pi_{\mathcal{V}}: \mathbb{C}^m \to \mathbb{C}^m/\mathcal{V}$ the canonical surjection. Such subspaces $\mathcal{V}$ exist: an example is $\mathbb{C}^m$ itself. We select $\mathcal{V}$ of minimal dimension for this property and we define $\mathfrak{S}_{\mathcal{V}} = \mathfrak{S} \cap \mathcal{V}$, $\widetilde{\mathfrak{S}}_{\mathcal{V}} = \mathfrak{S}[2] \cap \mathcal{V}$.

Since we shall work with functions of d variables, it suffices to know $d \ge 1$; hence there is no need here to prove $\mathcal{V} \neq \mathbb{C}(\beta_1, \dots, \beta_{m-1}, -1)$ (compare with step 1 of § 7.5 and with the remark at the end of § 9.2).

We permute, if necessary, the elements of the canonical basis of $\mathbb{C}^m$, so that $\pi_{\mathcal{V}}(\underline{e}_1), \dots, \pi_{\mathcal{V}}(\underline{e}_{m-d})$ is a basis of $\mathbb{C}^m/\mathcal{V}$.

For $\underline{\tau} \in \mathbb{N}^d$, $t \in \mathbb{Z}$, $\sigma \in \mathbb{N}$ and $\underline{s} \in \mathbb{Z}^m$, define $\widetilde{\gamma}_{\underline{\tau}t}^{(\sigma\underline{s})}$ as in (9.9) and (9.10). Let $\boldsymbol{M}$ be the matrix

$$\boldsymbol{M} = \left(\widetilde{\gamma}_{\underline{\tau}t}^{(\sigma\underline{s})}\right)_{\substack{(\underline{\tau},t)\\(\sigma,\underline{s})}}$$

where the index of rows $(\underline{\tau}, t)$ ranges over the elements in $\mathbb{N}^d \times \mathbb{Z}$ with $\|\underline{\tau}\| \leq T_0$ and $|t| \leq T_1$, while the index of columns $(\sigma, \underline{s})$ runs over the elements of $\mathbb{N} \times \mathbb{Z}^m$ with

$$0 \leq \sigma \leq (m+1)S_0 \quad \text{and} \quad \underline{s} \in \widetilde{\mathfrak{S}}_V[m+1].$$

The number of rows is

$$L_d = \binom{T_0 + d}{d}(2T_1 + 1).$$

Let us check the hypotheses of Proposition 9.16.

The first step as well as the assumptions

$$S_0 + 1 \geq 2T_0, \quad T_0 > 4 \max_{1 \leq i \leq m} S_i$$

and

$$(S_0 + 1)\mathrm{Card}(\mathfrak{S}) > 2(m+1)T_0^m T_1$$

are needed here.

For $0 \leq \sigma \leq S_0$ we set

$$\delta^{(2)}(z;\sigma) = \delta_{S_0^\natural}(z;\sigma) \in \mathbb{C}[z].$$

For $\underline{\tau} \in \mathbb{N}^d$ with $\|\underline{\tau}\| \leq T_0$ we define a polynomial $\delta^{(1)}(\underline{z}, \underline{\tau}) \in \mathbb{C}[\underline{z}]$ in d variables as

$$\begin{cases} z_0^{\tau_0} z_{m-d+1}^{\tau_{m-d+1}} \cdots z_{m-1}^{\tau_{m-1}}, \\ \dfrac{1}{\tau_0!} b_m^{\|\underline{\tau}\|-\tau_0} z_0^{\tau_0} \,\Delta\left(z_{m-d+1}; \tau_{m-d+1}\right) \cdots \Delta\left(z_{m-1}; \tau_{m-1}\right) \end{cases}$$

in the general case and in the homogeneous rational case respectively.

By Proposition 9.16 the matrix $\boldsymbol{M}$ has rank L_d. Let Δ_{ar} be a nonzero determinant of a square $L_d \times L_d$ submatrix of $\boldsymbol{M}$:

$$\Delta_{\mathrm{ar}} = \left(\widetilde{\gamma}_{\underline{\tau}t}^{(\mu)}\right)_{\substack{(\underline{\tau},t)\\1 \leq \mu \leq L_d}}$$

where $\widetilde{\gamma}_{\underline{\tau}t}^{(\mu)}$ stands for $\widetilde{\gamma}_{\underline{\tau}t}^{(\sigma_\mu \underline{s}_\mu)}$.

Step 3. Arithmetic Lower Bound

Recall the crucial fact that $\Delta_{\mathrm{ar}} \neq 0$, which follows from our construction together with the zero estimate. Our goal is to show

$$\log|\Delta_{\mathrm{ar}}| \geq -L_d U_1$$

where

$$U_1 = (2D-1)T_0 \log B_1 + (m+1)(D-1)S_0 \log B_2 + (D-1)S_0^\sharp + (D-1)\log L_d + 2(m+1)D(T_1+1)\sum_{j=1}^m S_j \log A_j.$$

We check this inequality as follows.

Consider first the general case. Each entry $\widetilde{\gamma}_{\underline{\tau} t}^{(\mu)}$ of the matrix whose determinant in Δ_{ar} is the value at the point $(\alpha_1, \cdots, \alpha_m, \beta_0, \ldots, \beta_{m-1})$ of a polynomial with rational coefficients in the $3m$ variables $\mathrm{X}_1^{\pm 1}, \ldots, \mathrm{X}_m^{\pm 1}, \mathrm{Y}_0, \ldots, \mathrm{Y}_{m-1}$. We multiply this polynomial by $\nu(S_0^\sharp)^{T_0}$ and get a polynomial $p_{\underline{\tau} t}^{(\mu)}$ with rational integer coefficients. Lemma 9.11, with S_0 replaced by $(m+1)S_0$ and S_j by $2(m+1)S_j$ for $1 \le j \le m$, shows that the degrees and length of this polynomial are bounded as follows:

$$\deg_{\mathrm{X}_j^{\pm 1}}\left(p_{\underline{\tau} t}^{(\mu)}\right) \le 2(m+1)|t|S_j \qquad (1 \le j \le m),$$
$$\deg_{\underline{\mathrm{Y}}}\left(p_{\underline{\tau} t}^{(\mu)}\right) \le T_0$$

and

$$\mathrm{L}\left(p_{\underline{\tau} t}^{(\mu)}\right) \le \left(\nu(S_0^\sharp)\overline{S}\right)^{T_0} T_0! e^{(m+1)S_0+S_0^\sharp} \left(\frac{T_1}{S_0^\sharp}+1\right)^{(m+1)S_0}$$

with

$$\overline{S} = 2(m+1)(S_1 + \cdots + S_m).$$

Hence

$$\mathrm{L}\left(p_{\underline{\tau} t}^{(\mu)}\right) \le e^{S_0^\sharp} B_1^{T_0} B_2^{(m+1)S_0}.$$

We apply Lemma 3.15 and find that $\nu(S_0^\sharp)^{L_d T_0}\Delta_{\mathrm{ar}}$ is the value, at the point

$$\left(\alpha_1, \cdots, \alpha_m, \ \beta_0, \ldots, \beta_{m-1}\right),$$

of a polynomial $P \in \mathbb{Z}[\mathrm{X}_1^{\pm 1}, \ldots, \mathrm{X}_m^{\pm 1}, \mathrm{Y}_0, \ldots, \mathrm{Y}_{m-1}]$, with

$$\deg_{\mathrm{X}_j^{\pm 1}} P \le (m+1)(T_1+1)L_d S_j \qquad (1 \le j \le m),$$
$$\deg_{\underline{\mathrm{Y}}} P \le L_d T_0,$$

and

$$\mathrm{L}(P) \le L_d! e^{L_d S_0^\sharp} B_1^{L_d T_0} B_2^{(m+1)L_d S_0}.$$

We conclude by means of Proposition 3.14 with

$$\ell = 2m+1, \quad \nu_1 = \cdots = \nu_{2m} = 1, \quad \nu_{2m+1} = m$$

and the $3m$ variables $\mathrm{X}_1, \ldots, \mathrm{X}_m, \mathrm{X}_1^{-1}, \ldots, \mathrm{X}_m^{-1}, \mathrm{Y}_0, \ldots, \mathrm{Y}_{m-1}$. We bound from above

$$\mathrm{h}(1 : \beta_0 : \cdots : \beta_{m-1}) \quad \text{by} \quad \log B_1.$$

In the homogeneous rational case, each $\nu(S_0^\sharp)^{T_0}\widetilde{\gamma}_{\underline{\tau} t}^{(\mu)}$ can be written as the value, at the point $(\alpha_1, \cdots, \alpha_m)$, of a polynomial $p_{\underline{\tau} t}^{(\mu)}$ in $\mathbb{Z}[\mathrm{X}_1^{\pm 1}, \ldots, \mathrm{X}_m^{\pm 1}]$, whose degree with respect to $\mathrm{X}_j^{\pm 1}$ is

$$\deg_{X_j^{\pm 1}}\left(p_{\underline{\tau}t}^{(\mu)}\right) \le 2(m+1)|t|S_j$$

and whose length is bounded by

$$\begin{aligned} \mathrm{L}\left(p_{\underline{\tau}t}^{(\mu)}\right) &\le \nu(S_0^\sharp)^{T_0} e^{(m+1)S_0+S_0^\sharp+T_0}\left(\frac{T_1}{S_0^\sharp}+1\right)^{(m+1)S_0} \cdot \\ &\qquad \max_{1\le i\le m-1}\left(2(m+1)\cdot\frac{S_i|b_m|+S_m|b_i|}{T_0}+1\right)^{T_0} \\ &\le B_1^{2T_0} B_2^{(m+1)S_0} e^{S_0^\sharp}. \end{aligned}$$

Hence $\nu(S_0^\sharp)^{L_dT_0}\Delta_{\mathrm{ar}}$ itself can be written as the value of a polynomial P in $\mathbb{Z}[X_1^{\pm 1},\ldots,X_m^{\pm 1}]$ at the point $(\alpha_1,\ldots,\alpha_m)$, where

$$\deg_{X_j^{\pm 1}} P \le (m+1)(T_1+1)L_dS_j \qquad (1\le j\le m)$$

and

$$\mathrm{L}(P) \le L_d!B_1^{2L_dT_0}B_2^{(m+1)L_dS_0}e^{L_dS_0^\sharp}.$$

Step 4. Analytic Upper Bound

Assume $|\Lambda| \le e^{-V_d}$ with

$$V_d = \frac{1}{2d}(T_0+d)(2T_1+1)\log(ME).$$

Our goal is to deduce

$$\log|\Delta_{\mathrm{ar}}| < -\frac{1}{2}L_dV_d + L_dU_2$$

with

$$U_2 = T_0\log(B_1^{D+1}ME) + (m+1)S_0\log(B_2ME) + S_0^\sharp + \log(2L_d) + (m+1)U.$$

We use Proposition 9.13 with S_0 replaced by $(m+1)S_0$, with S_j replaced by $2(m+1)S_j$ $(1\le j\le m)$ and with E replaced by ME. We set $\epsilon = \Lambda$,

$$M_{\underline{\tau}t} = T_0\log(B_1^{D+1}ME) + (m+1)S_0\log B_2 + S_0^\sharp + \frac{2(m+1)U|t|}{T_1+1}$$

and we take the same polynomials $\delta^{(1)}(\underline{z};\underline{\tau})$ and $\delta^{(2)}(z;\sigma)$ as in step 2.

We need to bound

$$\sup_{|z|=ME}\left|\Phi_{\underline{\tau}t}^{(\sigma)}(z\underline{s}\eta')\right|$$

which is related to Δ_{ar} by

$$\widetilde{\gamma}_{\underline{\tau}t}^{(\mu)} = \Phi_{\underline{\tau}t}^{(\sigma_\mu)}(\underline{s}_\mu\eta) = \Phi_{\underline{\tau}t}^{(\sigma_\mu)}(\underline{s}_\mu\eta')e^{ts_m^{(\mu)}\Lambda}.$$

In the general case we have

$$\Phi^{(\sigma)}_{\underline{\tau}t}(z\underline{s}\eta') =$$
$$\sum_{\kappa=0}^{\tau_0}\binom{\tau_0}{\kappa}\delta_{S_0^\sharp}(t;\sigma,\kappa)\cdot(s_m\beta_0)^{\tau_0-\kappa}\left(\prod_{i=m-d+1}^{m-1}(s_i+s_m\beta_i)^{\tau_i}\right)z^{\|\underline{\tau}\|-\kappa}e^{tz(s_1\lambda_1+\cdots+s_m\lambda_m+s_m\Lambda)}.$$

Recall

$$\max\{1,|\beta_0|,\ldots,|\beta_{m-1}|\}\le B_1^D.$$

By Lemma 9.11,

$$\sum_{\kappa=0}^{\tau_0}\binom{\tau_0}{\kappa}|\delta_{S_0^\sharp}(t;\sigma,\kappa)|\cdot|s_m\beta_0|^{\tau_0-\kappa}\prod_{i=m-d+1}^{m-1}|s_i+s_m\beta_i|^{\tau_i}$$
$$\le T_0!(\overline{S}B_1^D)^{T_0}e^{(m+1)S_0+S_0^\sharp}\left(\frac{T_1}{S_0^\sharp}+1\right)^{(m+1)S_0}$$
$$\le B_1^{(D+1)T_0}B_2^{(m+1)S_0}e^{S_0^\sharp}.$$

Hence $\sup_{|z|=ME}\left|\Phi^{(\sigma)}_{\underline{\tau}t}(z\underline{s}\eta')\right|$ is bounded by

$$(B_1^{D+1}ME)^{T_0}B_2^{(m+1)S_0}e^{S_0^\sharp}\exp\{ME\left|t(s_1\lambda_1+\cdots+s_m\lambda_m+s_m\Lambda)\right|\}.$$

By our condition on $\mathfrak{S}$ we have, for $|t|\le T_1$ and $\underline{s}\in\widetilde{\mathfrak{S}}[m+1]$,

$$\left|t(s_1\lambda_1+\cdots+s_m\lambda_m)\right|\le\frac{2(m+1)U|t|}{ME(T_1+1)}.$$

Hence

$$\log\sup_{|z|=ME}\left|\Phi^{(\sigma)}_{\underline{\tau}t}(z\underline{s}\eta')\right|\le M_{\underline{\tau}t}.$$

The numbers $b^{(\mu)}_{\underline{\tau}t}$ which occur in Proposition 9.13 satisfy

$$\widetilde{\gamma}^{(\mu)}_{\underline{\tau}t}=\widetilde{\gamma}^{(\mu)}_{\underline{\tau}t}e^{ts_m^{(\mu)}\Lambda}+\epsilon b^{(\mu)}_{\underline{\tau}t}.$$

Here $\epsilon=\Lambda$. Hence (recall Exercise 1.1.a)

$$|b^{(\mu)}_{\underline{\tau}t}|\le 2\left|ts_m^{(\mu)}\widetilde{\gamma}^{(\mu)}_{\underline{\tau}t}\right|\le 2\left|ts_m^{(\mu)}\Phi^{(\sigma_\mu)}_{\underline{\tau}t}(\underline{s}_\mu\eta)\right|.$$

Here we use very crude estimates. From the assumptions $4S_j<T_0$ and

$$2(m+1)T_0^mT_1<(S_0+1)(2S_1+1)\cdots(2S_m+1)$$

we deduce $2S_j+1<T_0$, $T_1<S_0+1$ and

$$2T_1S_m<T_0(S_0+1)<E^{T_0+(m+1)S_0}.$$

The estimate

$$|b^{(\mu)}_{\underline{\tau}t}|\le e^{M_{\underline{\tau}t}}$$

follows.

In the homogeneous rational case we have

$$\Phi^{(\sigma)}_{\underline{\tau} t}(z\underline{s}\eta) = \frac{1}{\tau_0!}\delta_{S_0^\sharp}(t;\sigma,\tau_0)\left(\prod_{i=m-d+1}^{m-1}\Delta(s_i b_m - s_m b_i;\tau_i)\right) z^{\|\underline{\tau}\|} e^{tz(s_1\lambda_1+\cdots+s_m\lambda_m)}$$

and we use Lemma 9.11:

$$\frac{1}{\tau_0!}|\delta_{S_0^\sharp}(t;\sigma,\tau_0)| \prod_{i=m-d+1}^{m-1} |\Delta(s_i b_m - s_m b_i;\tau_i)| \le B_2^{(m+1)S_0} e^{S_0^\sharp} B_1^{T_0},$$

while

$$\sup_{|z|=ME} \left|z^{\|\underline{\tau}\|}\right| \le (ME)^{T_0}.$$

The rest of the proof is just the same as in the general case.

Step 5. Conclusion: Lower Bound for $|\Lambda|$

Since

$$U_1 + U_2 \le \frac{V}{2} \le \frac{V_d}{2},$$

the conclusions of steps 3 and 4 are not compatible, and the assumption $|\Lambda| \le e^{-V_d}$ in step 4 is not satisfied. Therefore

$$|\Lambda| \ge e^{-V_d} \ge e^{-mV}.$$

□

9.3.3 Consequence of the Main Estimate

Recall the assumptions in § 9.3.1 before the statement of Theorem 9.19.

Corollary 9.20. *Let* $N_0 = 8(m+1)^3$,

$$C_0 = (m+1)2^{-2m+1}(N_0+1)^{m+1}, \quad C_1 = \frac{1}{4}(2N_0+1)(C_0+2)$$

and

$$M^* = \begin{cases} 1 & \text{if } M = 1, \\ 2^{m+3}M^2 & \text{if } M > 1. \end{cases}$$

Assume $B_1 \ge B_2$,

$$\log B_2 \ge \frac{1}{D}\log(ME), \qquad \log A_j \ge \frac{1}{D}\log(ME) \qquad (1 \le j \le m)$$

and

$$B_2 \ge \frac{71N_0}{m+1}\left(1 + \frac{D}{\log(ME)}\right).$$

Assume further

- *Either (general case)*

$$\log B_1 \geq \mathrm{h}(1\colon\beta_0\colon\cdots\colon\beta_{m-1})$$

and

$$B_1 \geq C_0^3 M^5 \left(\frac{D\log A}{\log(ME)}\right)^{2m+2} (\log B_2)^3.$$

- *Or else (homogeneous rational case)*

$$\beta_0 = 0, \quad \beta_i = -\frac{b_i}{b_m} \quad (1 \leq i \leq m) \quad \textit{with} \quad (b_1, \ldots, b_m) \in \mathbb{Z}^m,$$

and

$$B_1 \geq e + \frac{4e(m+1)\log(ME)}{N_0 D} \max_{1\leq j\leq m-1} \left\{1 + \frac{|b_m|}{\log A_j} + \frac{|b_j|}{\log A_m}\right\}.$$

Then

$$|\Lambda| > \exp\left\{-C_1 D^{m+2}(\log B_1)(\log B_2)(\log A_1)\cdots(\log A_m)M^*\bigl(\log(ME)\bigr)^{-m-1}\right\}.$$

Proof of Corollary 9.20. We deduce Corollary 9.20 from Theorem 9.19.

Step 1. Choice of Parameters

Define U, T_0, T_1, $S_0^\sharp$, S_0, $S_1, \ldots, S_m$ as follows:

$$U = C_0 D^{m+2}(\log B_1)(\log A_1)\cdots(\log A_m)(\log B_2)M^*\bigl(\log(ME)\bigr)^{-m-1},$$

$$T_0 = \left[\frac{U}{2D\log B_1}\right], \quad T_1 = \left[\frac{N_0 D\log B_1}{\log(ME)}\right],$$

$$S_0 = \left[\frac{U}{D\log B_2}\right],$$

$$S_0^\sharp = \begin{cases} \left[\dfrac{\log B_1}{6m}\right] & \text{in the general case,} \\ \left[\dfrac{\log B_1}{\log 3}\right] & \text{in the homogeneous rational case,} \end{cases}$$

$$S_j = \left[\frac{U}{D(T_1+1)\log A_j}\right] \quad (1 \leq j \leq m).$$

Other options may lead to better numerical values (the coefficients 2 in T_0, $6m$ and $\log 3$ in $S_0^\sharp$,... and 1 at many places!), but we just want to give an explicit result without trying to optimize the final estimate.

It will be useful to know that these parameters are somewhat large. Notice for instance that our assumptions imply $S_0^\sharp \geq 2$.

Using the hypotheses

$$\log B_1 \geq \log B_2 \geq \frac{1}{D}\log(ME) \quad \text{and} \quad \log A_j \geq \frac{1}{D}\log(ME)$$

we deduce

$$T_0 > \frac{C_0}{2} - 1, \quad T_1 \geq N_0, \quad S_0 > C_0 - 1, \quad S_j > \frac{C_0}{N_0} - 1.$$

We shall use repeatedly estimates like

$$T_1 + 1 \leq \left(1 + \frac{1}{N_0}\right) T_1 \leq \frac{(N_0+1)D\log B_1}{\log(ME)}.$$

The following inequalities, occurring in the hypotheses of Theorem 9.19, are plain:

$$S_0 + 1 \geq 2T_0, \quad T_0 > 4S_i \qquad (1 \leq i \leq m).$$

Step 2. Choice of $\mathfrak{S}$

We show that there exists a subset $\mathfrak{S}$ of $\mathbb{Z}^m[\underline{S}]$ satisfying

$$(S_0+1)\mathrm{Card}(\mathfrak{S}) > 2(m+1)T_0^m T_1$$

such that, for any $\underline{s} \in \mathfrak{S}$,

$$|s_1\lambda_1 + \cdots + s_m\lambda_m| \leq \frac{U}{ME(T_1+1)}.$$

a) Consider first the case $M = M^* = 1$. We take $\mathfrak{S} = \mathbb{Z}^m[\underline{S}]$. For $\underline{s} \in \mathbb{Z}^m[\underline{S}]$ it suffices to use the trivial estimate

$$\left|\sum_{j=1}^m s_j\lambda_j\right| \leq \sum_{j=1}^m |s_j\lambda_j| \leq \sum_{j=1}^m S_j|\lambda_j| \leq \frac{U}{D(T_1+1)}\sum_{j=1}^m \frac{|\lambda_j|}{\log A_j} \leq \frac{U}{E(T_1+1)}.$$

We also need to check

$$(S_0+1)(2S_1+1)\cdots(2S_m+1) > 2(m+1)T_0^m T_1.$$

Indeed we have

$$S_0 + 1 > \frac{U}{D\log B_2},$$

$$2S_j + 1 > \left(2 - \frac{N_0}{C_0}\right)(S_j+1) > \left(1 - \frac{N_0}{2C_0}\right)\frac{2U}{D(T_1+1)\log A_j} \qquad (1 \leq j \leq m),$$

$$T_0 \leq \frac{U}{2D\log B_1} \quad \text{and} \quad T_1 + 1 \leq \left(1 + \frac{1}{N_0}\right)T_1.$$

Moreover

$$\left(1 - \frac{N_0}{2C_0}\right)^m \left(1 + \frac{1}{N_0}\right) > 1,$$

so the inequality we want to check is a consequence of

$$2^{2m-1}U \geq (m+1)\left(1+\frac{1}{N_0}\right)^{m+1} D(\log B_2)(\log A_1)\cdots(\log A_m)\cdot T_1\left(\frac{T_1}{\log B_1}\right)^m.$$

Given the definitions of U and T_1, this explains the choice of C_0 (at least in case $M = M^* = 1$).

b) Assume now $M > 1$ and $M^* = 2^{m+4}M^2$. Define $S'_j = [S_j/2]$ $(1 \leq j \leq m)$. We use Dirichlet's box principle: the image of the map

$$\begin{array}{rccc} f\colon & \mathbb{Z}^m[\underline{S}'] & \longrightarrow & \mathbb{C} \\ & \underline{s} & \longmapsto & s_1\lambda_1+\cdots+s_m\lambda_m \end{array}$$

is contained in a disc of radius $\leq U/\big(2E(T_1+1)\big)$. Since $M \geq 1$ and $\sqrt{6} > 1+\sqrt{2}$, there is an integer ℓ in the range $\sqrt{2}M \leq \ell \leq \sqrt{6}M$. We decompose the square

$$\left\{x+iy\ ;\ |x| \leq \frac{U}{2E(T_1+1)},\ |y| \leq \frac{U}{2E(T_1+1)}\right\}$$

into ℓ^2 small squares of sides $\leq U/\big(E(T_1+1)\big)\ell$. One at least of the small squares contains elements $f(\underline{s})$ for $\underline{s}$ in a subset of $\mathbb{Z}^m[\underline{S}']$ of cardinal

$$\geq \frac{1}{\ell^2}\mathrm{Card}\big(\mathbb{Z}^m[\underline{S}']\big) \geq \frac{1}{\ell^2}S_1\cdots S_m$$

(notice the inequalities $2S'_j+1 \geq S_j$).

We fix one such small square, and one $\underline{s}' \in \mathbb{Z}^m[\underline{S}']$ for which $f(\underline{s}')$ falls in this small box. Then for any $\underline{s}'' \in \mathbb{Z}^m[\underline{S}']$ for which $f(\underline{s}'')$ we have

$$|f(\underline{s}')-f(\underline{s}'') = |f(\underline{s}'-\underline{s}'')| \leq \frac{\sqrt{2}U}{E(T_1+1)\ell} \leq \frac{U}{ME(T_1+1)}.$$

We let $\mathfrak{S}$ be the set of these $\underline{s}'-\underline{s}''$. Then by construction $\mathfrak{S} \subset \mathbb{Z}^m[\underline{S}]$ and

$$|s_1\lambda_1+\cdots+s_m\lambda_m| \leq \frac{U}{ME(T_1+1)}$$

for any $\underline{s} \in \mathfrak{S}$.

Let us check

$$\frac{1}{\ell^2}(S_0+1)S_1\cdots S_m > 2(m+1)T_0^mT_1.$$

From

$$S_j > \left(1-\frac{N_0}{C_0}\right)(S_j+1), \quad \left(1-\frac{N_0}{C_0}\right)^m > \frac{3}{4} \quad \text{and} \quad \ell^2 < 6M^2$$

one deduces

$$\frac{1}{\ell^2}(S_0+1)S_1\cdots S_m >$$

$$\frac{3}{4}\cdot\frac{U}{D\log B_2}\cdot\frac{1}{6M^2}\cdot\left(\frac{U}{D(T_1+1)}\right)^m\cdot\frac{1}{(\log A_1)\cdots(\log A_m)}.$$

On the other hand we have

$$2(m+1)T_0T_1^m \le 2(m+1)\left(\frac{U}{2D\log B_1}\right)^m T_1 \quad \text{and} \quad T_1+1 \le \left(1+\frac{1}{N_0}\right)T_1,$$

so it suffices to check

$$2^{m-4}U \ge (m+1)\left(1+\frac{1}{N_0}\right)^m M^2D(\log B_2)(\log A_1)\cdots(\log A_m)\cdot T_1\left(\frac{T_1}{\log B_1}\right)^m.$$

This explains the choice of M^* in case $M>1$.

Step 3. Estimate Involving B_1

In the homogeneous rational case the inequality $B_1 \ge 3^{S_0^\sharp}$ is plain.

In the general case we need to check

$$B_1 \ge 2(m+1)T_0S^*3^{S_0^\sharp}.$$

Indeed from

$$T_0 \le \frac{U}{2D\log B_1} \quad \text{and} \quad S_j < \frac{U\log(ME)}{N_0D^2(\log A_j)(\log B_1)},$$

we deduce

$$T_0S^* \le \frac{mC_0^2}{2N_0}(M^*)^2D^{2m+1}(\log A)^{2m-1}(\log B_2)^2\bigl(\log(ME)\bigr)^{-2m-1}.$$

Further we have

$$3^{S_0^\sharp} \le B_1^{1/(5m)}, \quad M^4 \le (M^5)^{1-(1/5m)}, \quad 2m+1 < (2m+2)\left(1-\frac{1}{5m}\right)$$

and

$$\frac{m(m+1)C_0^2}{N_0}\cdot\left(\frac{M^*}{M}\right)^2 < (C_0^3)^{1-(1/5m)}.$$

Step 4. Estimate Involving B_2

We check

$$B_2 \ge e^2\left(1+\frac{T_1}{S_0^\sharp}\right).$$

Since

$$S_0^\sharp \ge \frac{\log B_1}{9.6(m+1)}$$

we have

$$\frac{e^2T_1}{S_0^\sharp} \le \frac{71N_0D}{(m+1)\log(ME)}.$$

Step 5. End of the Proof of Corollary 9.20

We first check the (weak) bounds

$$DS_0^\sharp < \frac{1}{2}U \quad \text{and} \quad D\log(2L) \le \frac{1}{2}U.$$

The first one is plain. For the second one, we start with

$$2L \le 2(T_0+m)^m(2T_1+1) \le 4\left(1+\frac{3m}{C_0}\right)\left(1+\frac{1}{N_0}\right)T_0^mT_1 < 5T_0^mT_1.$$

We deduce

$$2L \le 5N_0C_0^m\left(\frac{D\log B_1}{\log E}\right)\left(\frac{D\log B_2}{\log E}\right)^m\left(\frac{D\log A}{\log E}\right)^{m^2}$$

with $A = \max_{1\le i\le m} A_i$. On the other hand the inequalities

$$\frac{U}{C_0D} \ge \frac{D\log B_1}{\log E}, \quad \frac{U}{C_0D} \ge \frac{D\log B_2}{\log E} \quad \text{and} \quad \frac{U}{C_0D} \ge \frac{D\log A}{\log E}$$

give

$$D\log(2L) \le (m^2+m+1)\frac{U}{C_0} + D\log(5N_0C_0^m).$$

Our claim $D\log(2L) \le \frac{1}{2}U$ then follows from $U \ge C_0D$.

One deduces that the number

$$4DT_0\log B_1 + 2(m+1)DS_0\log B_2 + DS_0^\sharp$$
$$+D\log(2L) + (m+1)U + 2(m+1)D(T_1+1)\sum_{j=1}^m S_j\log A_j$$

is at most

$$\bigl(2+2(m+1)+1+(m+1)+2m(m+1)\bigr)U = (2m^2+5m+6)U.$$

On the other hand since

$$T_0+1 > \frac{U}{D\log B_1} \quad \text{and} \quad 2T_1+1 > \frac{(2N_0-1)D\log B_1}{\log(ME)}$$

the number

$$V = \frac{1}{2m}(T_0+1)(2T_1+1)\log(ME)$$

satisfies

$$V > \frac{(2N_0-1)U}{4m}.$$

Hence our choice of N_0 yields

$$\frac{V}{2} > 2(m^2+3m+3)U.$$

Finally the choice of C_1 arises from the estimates

$$T_0 + 1 \le \left(1 + \frac{2}{C_0}\right) \frac{U}{2D \log B_1}, \qquad 2T_1 + 1 \le \frac{(2N_0+1)D \log B_1}{\log(ME)}$$

and

$$mV \le \frac{1}{4}\left(1 + \frac{2}{C_0}\right)(2N_0+1)U \le \frac{C_1}{C_0}U.$$

Corollary 9.20 follows. □

9.3.4 Proof of Proposition 9.18

Here, we deduce Proposition 9.18 from Corollary 9.20.

Proof of Proposition 9.18. The condition on linear independence of $\lambda_1, \ldots, \lambda_m$ ensures that the rank of the multiplicative subgroup of $\mathbb{C}^\times$ generated by $\alpha_1, \ldots, \alpha_m$ is at least $m-1$. We apply Corollary 9.20 with

$$\log B_1 = 37m^{3/2} \log B, \qquad \text{and} \qquad \log B_2 = 10\sqrt{m} \log E^*.$$

Since $M \le e^{(m/2)-1}$ and $E \ge e$ we have

$$\log(ME) \le \frac{m}{2} \log E.$$

In case $M = e^{(m/2)-1}$ and $E = e$, we check

$$\log A_j \ge \frac{1}{D} \log(ME) \qquad (1 \le j \le m)$$

thanks to the extra assumption $\log A_j \ge m/(2D)$ in Proposition 9.18.

We wish to check that the assumptions on B_1 and B_2 in Corollary 9.20 follow from the hypotheses of Proposition 9.18. We start with B_2.

From $E^* \ge D/\log E$, $E^* \ge e$ and $M \ge 1$ we deduce

$$E^* \ge \frac{e}{e+1}\left(1 + \frac{D}{\log E}\right) \ge \frac{e}{e+1}\left(1 + \frac{D}{\log(ME)}\right).$$

From the estimate

$$(m+1)e^{10\sqrt{m}} \ge 71(e+1)N_0$$

we deduce

$$B_2 \ge e^{10\sqrt{m}-1} \frac{e}{e+1}\left(1 + \frac{D}{\log(ME)}\right) \ge \frac{71N_0}{m+1}\left(1 + \frac{D}{\log(ME)}\right).$$

Now we deal with B_1. To start with, we consider the general case. From $B \ge D(\log A)(\log E)^{-1}$, $B \ge E^* \ge e$ and $M \ge 1$ we deduce $B \ge (e^3/27)(\log E^*)^3$ and

$$B_1 \geq \frac{1}{27} e^{37m^{3/2}-2m} \left(\frac{D \log A}{\log(ME)} \right)^{2m+2} (\log E^*)^3.$$

The inequality

$$B_1 \geq C_0^3 M^5 \left(\frac{D \log A}{\log(ME)} \right)^{2m+2} (\log B_2)^3$$

follows from the estimate

$$30^3 m^{3/2} C_0^3 M^5 < e^{37m^{3/2}-2m}.$$

In the homogeneous rational case the inequality

$$B_1 \geq e + \frac{4e(m+1)\log(ME)}{N_0 D} \max_{1 \leq j \leq m-1} \left\{ 1 + \frac{|b_m|}{\log A_j} + \frac{|b_j|}{\log A_m} \right\}$$

is a consequence of

$$B^{m+1} \geq \max_{1 \leq j \leq m-1} \left(1 + \frac{|b_m|}{\log A_j} + \frac{|b_j|}{\log A_m} \right) \cdot \frac{\log(ME)}{D}.$$

and

$$B^3 \geq \frac{4e(m+1)}{N_0} + \frac{eD}{\log(ME)}.$$

We deduce that Proposition 9.18 holds (under the extra assumption $\beta_m = -1$ in the general case), with the constant $C(m)$ replaced by

$$\begin{cases} 370m^2 C_1 & \text{if } M = 1, \\ 370m^2 C_1 2^{2m+4} e^{m-2} m^{-m-1} & \text{if } M \neq 1. \end{cases}$$

In any case this constant is $< C(m) - 1$, where $C(m)$ is the constant given in Proposition 9.18. It should be noticed here that in the general case with $\beta_m = -1$, since $B_1 > B^{m+1}$, we have used only the weaker assumption

$$\mathrm{h}(1 : \beta_0 : \cdots : \beta_{m-1}) \leq (m+1) \log B$$

in place of $\max_{0 \leq i \leq m-1} \mathrm{h}(\beta_i) \leq \log B$.

In the general case we remove the restriction $\beta_m = -1$ by means of Liouville's inequality like in step 5 of § 7.6: assume (as we may without loss of generality) $\beta_m \neq 0$, so that

$$|\beta_m| \geq B^{-D}.$$

Define $\beta'_j = -\beta_j / \beta_m$ $(0 \leq j \leq m)$ and

$$\Lambda' = \beta'_0 + \beta'_1 \lambda_1 + \cdots + \beta'_{m-1} \lambda_{m-1} - \lambda_m.$$

Hence $\Lambda = -\beta_m \Lambda'$. From the assumption $\max_{0 \leq i \leq m} \mathrm{h}(\beta_j) \leq \log B$ we deduce

$$\mathrm{h}(1 : \beta'_0 : \cdots : \beta'_{m-1}) = \mathrm{h}(\beta_0 : \cdots : \beta_m) \leq (m+1) \log B.$$

Since

$$|\Lambda'| \geq \exp\{-\big(C(m) - 1\big) D^{m+2} (\log B)(\log A_1) \cdots (\log A_m)(\log E^*)(\log E)^{-m-1}\}$$

and

$$D \log B \leq D^{m+2} (\log B)(\log A_1) \cdots (\log A_m)(\log E^*)(\log E)^{-m-1},$$

Theorem 9.1 follows. □

9.4 Corollaries

To conclude this chapter, we give a few comments on Theorem 9.1, we remove the condition of linear independence on the λ's, and we state and prove some consequences.

9.4.1 Comments on Theorem 9.1

Remark 1. Let us compare Theorem 7.1 with Theorem 9.1. The two main differences are the following:

1. We have replaced the factor $(\log B)^2$ which occurred in Chap. 7 by the product $(\log B)(\log E^*)$.
2. The condition on B in the homogeneous rational case is weaker in Theorem 9.1.

The second refinement could have easily been included in Chap. 7: Theorem 7.16 suffices. But the first refinement requires the introduction simultaneously of the extra factor $\mathbb{G}_a$ and of one derivative.

In § 9.3 we had two parameters B_1 and B_2. The assumptions of Theorem 9.19 involve $DT_0 \log B_1$ and $DS_0 \log B_2$. In § 9.3.4 we replaced $\log B_1$ by a multiple of $\log B$ and $\log B_2$ by a multiple of $\log E^*$. In the next chapter (§ 10.2) again the quantities $DT_0 \log B_1$ and $DS_0 \log B_2$ will occur, but, in § 10.2.6, B_1 and B_2 will be related to E^* and B respectively. In § 14.4 we shall explain what is going on.

Let us come back to the hypotheses of Theorem 9.1. The assumptions

$$\log A_j \geq \mathrm{h}(\alpha_j) \quad \text{and} \quad \log A_j \geq \frac{E|\lambda_j|}{D}$$

arise in a natural way from the arithmetic and analytic estimates respectively, as we saw in § 9.2.1. At the same place we explained how the condition $B \geq D(\log A_i)/\log E$ in the general case occurs from the estimates for

$$(s_j + s_m\beta_j)^{\tau_j},$$

which involve $(S^*)^{T_0}$: in order to bound $DT_0 \log S^*$ by U we need S^* to be less than B_1.

The conditions $E^* \geq E^{1/D}$ and $B \geq E^{1/D}$ arise from the analytic estimates: in step 4 of the proof of Theorem 9.19 in § 9.3.2, the definition of U_2 involves $T_0 \log(ME)$ and $S_0 \log(ME)$, which we want to bound from above by U.

The origin of the requirement $E^* \geq D/\log E$ is the following: using Lemma 9.8 to estimate $\delta_{S_0^\sharp}(t;\sigma,\kappa)$, we get a factor

$$\left(\frac{|t|}{S_0^\sharp}+1\right)^\sigma,$$

which is responsible for the factor

$$\left(\frac{T_1}{S_0^\sharp}+1\right)^{S_0}$$

occurring in Lemma 9.11. This is why we need to impose $B_2 \geq T_1/S_0^\sharp$.

The condition $B \geq D/\log E$ is easy to explain: the hypothesis $S_0 + 1 \geq 2T_0$ in Proposition 9.16 leads us to require $B \geq E^*$.

Finally the condition $D \log A_j \geq \log E$ does not occur in the hypotheses of Theorem 9.19 where we only required $D \log A_j \geq 1$, but it has been used several times in the proof of Corollary 9.20.

Remark 2. In the special case $m = 1$, apart from the explicit value of the absolute constant $C(1)$, Theorem 9.1 is slightly stronger than Theorem 5 of [NeW 1996], where the extra hypotheses

$$B \geq E, \quad B \geq \log A_1 \quad \text{and} \quad E^* \geq D$$

are required. A close look at the proof of [NeW 1996] shows that these conditions may be dispensed with: our hypotheses are sufficient to imply

$$E \max\{1, |\beta|\} \leq B^{2D} \quad \text{and} \quad D^2(\log A_1)(\log E^*)(\log E)^{-2} \leq B^2 \log B.$$

Remark 3. From the assumptions $B \geq E^*$ and

$$\log E^* \geq \max\left\{\frac{1}{D}\log E,\ \log\left(\frac{D}{\log E}\right)\right\}$$

we deduce

$$B \geq E^{1/D} \geq \frac{\log E}{D}.$$

In the homogeneous rational case, if B' is a positive real number satisfying $B' \geq E^*$ and

$$B' \geq \max_{1\leq j\leq m-1}\left(\frac{|b_m|}{\log A_j}+\frac{|b_j|}{\log A_m}\right),$$

then

$$(B')^2 \geq \max_{1\leq j\leq m-1}\left(\frac{|b_m|}{\log A_j}+\frac{|b_j|}{\log A_m}\right)\cdot\frac{\log E}{D}$$

and we may apply Theorem 9.1 with $B = (B')^2$ (which means that we get the conclusion with the factor $\log B$ replaced by $\log B'$ and also $C(m)$ by $2C(m)$).

Similarly, if B'' satisfies $B'' \geq D/\log E$, $B'' \geq \log E$ and

$$B'' \geq \max_{1 \leq j \leq m-1} \left(\frac{|b_m|}{D \log A_j} + \frac{|b_j|}{D \log A_m} \right),$$

again we may apply Theorem 9.1 with $B = (B'')^2$.

Remark 4. In the homogeneous rational case, one deduces from the hypotheses

$$B \geq e, \quad \log A_i \geq D \log E \quad \text{and} \quad \log E^* \geq D \log E$$

that the conclusion of Theorem 9.1 is stronger than Liouville's estimate only when

$$\max_{1 \leq i \leq m} |b_i| \geq \frac{C(m)}{2m} \cdot \frac{D^2 \log A}{(\log E)^2}$$

with $A = \max\{A_1, \ldots, A_m\}$ (see for instance Remark 2 in § 7.1.1). Therefore the homogeneous rational case of Theorem 9.1 is interesting only if

$$\max_{1 \leq j \leq m-1} \left(\frac{|b_m|}{\log A_j} + \frac{|b_j|}{\log A_m} \right) \cdot \frac{\log E}{D} \geq \frac{C(m)}{2m} \cdot \frac{D}{\log E}.$$

Hence the condition $B \geq D/\log E$ (which follows from our hypothesis $B \geq E^*$) could be omitted in the homogeneous rational case (it does not occur explicitly in Corollary 3 of [LauMN 1995]).

9.4.2 Relaxing the Hypothesis of Linear Independence of Logarithms

Let us first replace the assumption that the numbers $\lambda_1, \ldots, \lambda_m$ in Theorem 9.1 are $\mathbb{Q}$-linearly independent by a further (mild) condition on the parameters.

Proposition 9.21. *Theorem 9.1 still holds if we replace the hypothesis that the numbers $\lambda_1, \ldots, \lambda_m$ are linearly independent over $\mathbb{Q}$ by the extra assumptions $\Lambda \neq 0$ and*

$$D^3 (\log B)(\log A_i)(\log E^*) \geq (\log D)(\log E)^2$$

for $1 \leq i \leq m$.

Proof.

We start with the general case. For $\emptyset \neq I \subset \{1, \ldots, m\}$, define

$$\Phi_I(B_0) = C(n) D^{n+2} (\log B_I) \left(\prod_{i \in I} \log A_i \right) (\log E^*)(\log E)^{-n-1},$$

where $n = |I|$,

$$B_I = \max\left\{B_0,\ E^*,\ \frac{D\log A_I}{\log E}\right\}$$

and $A_I = \max\{e,\ \max_{i\in I} A_i\}$. Define also

$$B'_I = \max\left\{2N_I B_0^2,\ E^*,\ \frac{D\log A_I}{\log E}\right\}$$

with

$$N_I = \left[(11nD^3\log A_I)^{n-1}\right].$$

From the assumption

$$D(\log B)(\log A_i) \geq (\log D)(\log E) \quad (1 \leq i \leq m)$$

we deduce

$$C(n')\log B_{I'} \leq C(n)(\log B_I)\prod_{i\in I\setminus I'}\frac{D\log A_i}{\log E}$$

for $I' \subset I$,

$$\log N_I \leq \frac{1}{2}\Phi_I(B)$$

and

$$C(n-1)\log B'_I \leq \frac{1}{2}C(n)(\log B_I)\cdot\frac{D\log A_I}{\log E}.$$

Hence Lemma 7.20 provides the conclusion in the general case.

Finally we consider the homogeneous rational case. We follow [W 1993], § 10, proof of Corollaire 10.1.

By induction on m we may assume, without loss of generality, that the $\mathbb{Q}$-vector space spanned by $\lambda_1, \ldots, \lambda_m$ has dimension $m-1$. There is a unique (up to a multiplicative coefficient ± 1) linear dependence relation

$$a_1\lambda_1 + \cdots + a_m\lambda_m = 0$$

with relatively prime rational integers $a_1, \ldots, a_m$ satisfying, by Lemma 7.19,

$$0 < \max\{|a_1|, \ldots, |a_m|\} \leq N,$$

where

$$N = \left[(11mD^3\max_i \log A_i)^{m-1}\right]$$

and i runs over the set of indices in $\{1, \ldots, m\}$ for which $a_i \neq 0$.

We distinguish three cases.

Case 1. Assume $a_m \neq 0$ and $N \leq (11mD^3\log A_m)^{m-1}$.

We eliminate b_m. Set

$$b'_i = a_m b_i - a_i b_m \quad (1 \leq i \leq m-1),$$

so that

$$a_m\Lambda = b'_1\lambda_1 + \cdots + b'_{m-1}\lambda_{m-1}.$$

Define

$$B'' = \max_{1\le i\le m-2}\left(\frac{|b'_{m-1}|}{\log A_i} + \frac{|b'_i|}{\log A_{m-1}}\right)\cdot\frac{\log E}{D}$$

and $B' = \max\{E^*\ ,\ B''\}$. We first estimate B'. Since

$$\left(\frac{|b_m|}{\log A_i} + \frac{|b_i|}{\log A_m}\right)\cdot\frac{\log E}{D} \le B$$

for $1 \le i \le m-1$, we deduce

$$\frac{|b_i|}{\log A_i}\cdot\frac{\log E}{D} \le |b_i| \le B\cdot\frac{D\log A_m}{\log E} \qquad (2\le i\le m-1),$$
$$\frac{|b_m|}{\log A_i}\cdot\frac{\log E}{D} \le B,$$

and similarly

$$\frac{|b_i|}{\log A_{m-1}}\cdot\frac{\log E}{D} \le |b_i| \le B\cdot\frac{D\log A_m}{\log E} \qquad (2\le i\le m-1),$$
$$\frac{|b_m|}{\log A_{m-1}}\cdot\frac{\log E}{D} \le B.$$

Therefore

$$B' \le 2NB\left(1 + \frac{D\log A_m}{\log E}\right) \le 4NBD\log A_m \le (11mD^3\log A_m)^m B.$$

Since $a_m \ne 0$, the numbers $\lambda_1, \ldots, \lambda_{m-1}$ are $\mathbb{Q}$-linearly independent. From the inductive hypothesis we deduce

$$|a_m\Lambda| \ge e^{-\Phi_{m-1}},$$

where

$$\Phi_{m-1} = C(m-1)D^{m+1}(\log B')(\log A_1)\cdots(\log A_{m-1})(\log E^*)(\log E)^{-m}.$$

Notice that the number

$$\Phi_m = C(m)D^{m+2}(\log B)(\log A_1)\cdots(\log A_m)(\log E^*)(\log E)^{-m-1}$$

satisfies

$$\begin{aligned}\Phi_m &\ge C(m)\cdot\max_{1\le i\le m} D^3(\log B)(\log A_i)(\log E^*)(\log E)^{-2}\\ &\ge C(m)\max\bigl\{D\log B,\ D\log A_m,\ \log D\bigr\},\end{aligned}$$

thanks to the extra assumption of Proposition 9.21

$$D^3(\log B)(\log A_i)(\log E^*) \ge (\log D)(\log E)^2.$$

We deduce

$$\log |a_m| \le \log N \le \frac{1}{2}\Phi_m$$

and

$$C(m-1)(\log B') \le \frac{1}{2}C(m)(\log B) \cdot \frac{D \log A_m}{\log E}.$$

We conclude

$$|\Lambda| \ge e^{-\Phi_m},$$

which is what we wanted.

Case 2. After renumbering the λ's if necessary, we may assume $a_1 \neq 0$, and we may also assume that A_1 is the maximum of the numbers A_j where j ranges over the integers in $\{1, \dots, m\}$ for which $a_j \neq 0$. In this second case, we have either $a_m = 0$ or else $A_m \le A_1$; hence

$$N \le (11mD^3 \log A_1)^{m-1}.$$

Let us eliminate b_1. Define

$$b'_j = a_1 b_j - a_j b_1 \quad (2 \le j \le m),$$

so that

$$a_1 \Lambda = b'_2 \lambda_2 + \cdots + b'_m \lambda_m.$$

From the inductive hypothesis we deduce

$$|a_1 \Lambda| \ge e^{-\Phi_{m-1}},$$

where

$$\Phi_{m-1} = C(m-1)D^{m+1}(\log B')(\log A_2) \cdots (\log A_m)(\log E^*)(\log E)^{-m}$$

and $B' = \max\{E^*\ ,\ B''\}$. We need to define B'' and estimate it from above. Again we distinguish two cases

a) If $b'_m \neq 0$, we set

$$B'' = \max_{2 \le i \le m-1} \left(\frac{|b'_m|}{\log A_i} + \frac{|b'_i|}{\log A_m} \right) \cdot \frac{\log E}{D}$$

and we have

$$\begin{aligned} B'' &\le N \max_{2 \le i \le m-1} \left(\frac{|b_m|}{\log A_i} + \frac{|b_1|}{\log A_i} + \frac{|b_i|}{\log A_m} + \frac{|b_1|}{\log A_m} \right) \cdot \frac{\log E}{D} \\ &\le NB\left(2 + \frac{D \log A_1}{\log E}\right), \end{aligned}$$

because

$$\frac{|b_1|}{\log A_i} \cdot \frac{\log E}{D} \le |b_1| \le B \cdot \frac{D \log A_1}{\log E} \quad (2 \le i \le m-1).$$

b) If $b'_m = 0$, we select an integer n in the range $2 \leq n \leq m-1$ for which $b'_n \neq 0$ and we set

$$B'' = \max_{\substack{2\leq i\leq m \\ i\neq n}} \left(\frac{|b'_n|}{\log A_i} + \frac{|b'_i|}{\log A_n} \right) \cdot \frac{\log E}{D}.$$

Hence

$$B'' \leq 4N \max\{|b_1|, \dots, |b_{m-1}|\}.$$

For $1 \leq i \leq m-1$ we have

$$|b_i| \leq B \cdot \frac{D \log A_m}{\log E} \leq B \cdot \frac{D \log A_1}{\log E},$$

because the condition $b'_m = 0$ implies $a_m \neq 0$.

We deduce that in each of the two cases a) and b), we have

$$B' \leq (11 m D^3 \log A_1)^m B.$$

We conclude, like in case 1,

$$C(m-1) \log B' \leq \frac{1}{2} C(m)(\log B) \cdot \frac{D \log A_1}{\log E}.$$

This completes the proof of Proposition 9.21. □

9.4.3 Statements of Corollaries

For the next four corollaries we use the following notation: $\alpha_1, \dots, \alpha_m$ are nonzero algebraic numbers in a number field of degree $\leq D$ over $\mathbb{Q}$ and $b_1, \dots, b_m$ are rational integers such that

$$\alpha_1^{b_1} \cdots \alpha_m^{b_m} \neq 1.$$

We define $A_1, \dots, A_m, B_0$ by

$$B_0 = \max\{e, |b_1|, \dots, |b_m|\}$$

and

$$\log A_j = \max\left\{ \frac{1}{D}, \mathrm{h}(\alpha_j) \right\} \qquad (1 \leq j \leq m).$$

We denote by $C(m)$ the constant occurring in Theorem 9.1 (and Proposition 9.21). We do not need to know the exact value of $C(m)$ given in § 9.3. All we shall need is

$$C(m) \geq e^{11} \quad \text{and} \quad C(m) \geq m^2 + \log 2.$$

Here is a lower bound for $|\alpha_1^{b_1} \cdots \alpha_m^{b_m} - 1|$. The importance of such estimates has been stressed in § 1.2.

Corollary 9.22. *We have*

$$|\alpha_1^{b_1} \cdots \alpha_m^{b_m} - 1| \geq \exp\{-C_1(m)(\log B_0)(\log A_1) \cdots (\log A_m) D^{m+2} \max\{1, \log D\}\},$$

where

$$C_1(m) = 2 \cdot 3^{2m+2} C(m+1).$$

This estimate is best possible with respect to the parameter B_0, namely it is of the shape B_0^{-C} where C does not depend on $b_1, \ldots, b_m$, but only on $\alpha_1, \ldots, \alpha_m$. This result, as pointed out in S 1.2, is due to N. I. Feldman [F 1968] and has important consequences.

In Corollary 9.22 the value of $C(m)$ requires only $E = e$, hence the sharper estimate provided by the second part of Proposition 9.18 applies.

We state the next corollaries in terms of lower bounds for linear combinations of logarithms, but similar statements hold in terms of lower bounds for the distance between two products of algebraic numbers, like in Corollary 9.22 (see Exercise 9.8 for the simpler case of real positive algebraic numbers).

For $1 \le j \le m$, let $\lambda_j \in \mathcal{L}$ satisfy $e^{\lambda_j} = \alpha_j$. Assume $\log A_j \ge (e/D)|\lambda_j|$ $(1 \le j \le m)$. Further, let $E \ge e$ satisfy

$$\sum_{j=1}^{m} \frac{|\lambda_j|}{\log A_j} \le \frac{D}{E} \quad \text{and} \quad \log E \le D \log A_j \quad (1 \le j \le m).$$

Assume also $b_m > 0$. Let $B \ge e$ be a real number satisfying

$$B \ge \max_{1 \le i \le m-1} \left\{ \frac{b_m}{\log A_i} + \frac{|b_i|}{\log A_m} \right\}.$$

Let $E_1 \ge e$ and $E_2 \ge e$ satisfy

$$E_1 \ge \max\left\{ \frac{E^{1/D}}{\log E}, D \right\} \quad \text{and} \quad E_2 \ge \max\{D, E_1\}.$$

Define

$$X = C(m) D^{m+2} (\log A_1) \cdots (\log A_m)(\log E_1)(\log E_2)(\log E)^{-m-1}.$$

Corollary 9.23. *Then*

$$\left|b_1\lambda_1 + \cdots + b_m\lambda_m\right| \ge B^{-X}.$$

Finally define

$$B' = \max\left\{2, |b_1|, \ldots, |b_{m-1}|\right\}$$

and

$$Y = 2X \log\left(\frac{X}{\log A_m}\right).$$

Corollary 9.24. *For any δ in the range $0 < \delta \le 1/2$,*

$$\left|b_1\lambda_1 + \cdots + b_m\lambda_m\right| \ge e^{-\delta B'} \left(\frac{\delta}{b_m}\right)^{Y}.$$

An easy consequence of Corollary 9.24 reads as follows (compare with [B 1972], II):

Corollary 9.25. *Let* ϵ *satisfy* $0 < \epsilon < 1$ *and*

$$\left|b_1\lambda_1 + \cdots + b_m\lambda_m\right| < e^{-\epsilon B'}.$$

Then

$$B' \leq \frac{2Y}{\epsilon} \log\left(\frac{2b_m}{\epsilon}\right).$$

A further discussion of this topic, in particular of the relevance of the last three corollaries, will take place in § 10.4.

9.4.4 Proofs

Proof of Corollary 9.22. The special case where $\alpha_1, \ldots, \alpha_m$ are positive real (algebraic) numbers is easier and yields a smaller value for $C_1(m)$ (see Exercise 9.7). We consider here the general case.

Without loss of generality we may assume $b_i \neq 0$ for $1 \leq i \leq m$. Further, since the statement of Corollary 9.22 is symmetric (this is not the case with the next ones), we may also assume $A_1 \leq \cdots \leq A_m$.

We distinguish two cases:

a) Assume $B_0 \leq mD$. From Liouville's inequality (e.g. Exercise 3.7.b) we deduce

$$\left|\alpha_1^{b_1} \cdots \alpha_m^{b_m} - 1\right| \geq 2^{-D} A^{-mDB_0} \quad \text{where} \quad A = \max_{1 \leq i \leq m} A_i.$$

Since $C(m) \geq m^2 + \log 2$ we have

$$D \log 2 + m^2 D^2 \log A \leq C(m) D^{m+2} (\log B_0)(\log A_1) \cdots (\log A_m) \max\{1, \log D\},$$

and the conclusion of Corollary 9.22 is plain.

b) From now on we assume $B_0 \geq mD$. For $1 \leq i \leq m$, define $\lambda_i \in \mathbb{C}$ by the conditions

$$e^{\lambda_i} = \alpha_i, \quad -\pi \leq \mathrm{Im}\lambda_i < \pi.$$

Since

$$|\lambda_i|^2 \leq \pi^2 + (\log|\alpha_i|)^2 \quad \text{and} \quad |\alpha_i| \leq e^{D\mathrm{h}(\alpha_i)} \leq A_i^D,$$

we have

$$|\lambda_i| \leq \sqrt{\pi^2 + 1} D \log A_i.$$

We use Exercise 1.1.b with $\theta = 1/2$ and

$$z = \alpha_1^{b_1} \cdots \alpha_m^{b_m}.$$

Clearly without loss of generality we may assume $|z - 1| < 1/2$. We deduce that there exists an even integer $b_0 \in 2\mathbb{Z}$ such that the number

$$\Lambda = b_0 i\pi + b_1\lambda_1 + \cdots + b_m\lambda_m$$

satisfies

$$|\Lambda| \le 2\left|\alpha_1^{b_1}\cdots\alpha_m^{b_m}-1\right|.$$

Since $\alpha_1^{b_1}\cdots\alpha_m^{b_m}\ne 1$ we have $\Lambda \ne 0$. An estimate for $|b_0|$ is the following:

$$\begin{aligned}\pi|b_0| &\le 1+\left|b_1\lambda_1+\cdots+b_m\lambda_m\right|\\ &\le 1+m\sqrt{\pi^2+1}DB_0\log A_m\\ &\le 2\pi mDB_0\log A_m.\end{aligned}$$

We shall use Proposition 9.21 with

$$E=e,\quad E^*=\max\{e,D\},\quad B=B_0^2,$$

m replaced by $m+1$ and A_i replaced by A_i^*, where

$$\log A_0^*=\frac{e\pi}{D},\quad\text{and}\quad \log A_i^*=e\sqrt{\pi^2+1}\log A_i\quad\text{for}\quad 1\le i\le m,$$

so that if we set $\lambda_0=i\pi$ the inequalities

$$e|\lambda_i|\le D\log A_i^*\quad\text{are satisfied for}\quad 0\le i\le m.$$

Notice that the extra assumption

$$D^3(\log B)(\log A_i)(\log E^*)\ge(\log D)(\log E)^2$$

of Proposition 9.21 is satisfied, and that the condition $B_0\ge mD$ yields $B\ge mDB_0$, hence the inequalities

$$B\ge\frac{|b_i|}{\log A_m^*}+\frac{|b_m|}{\log A_i^*}$$

are satisfied for $0\le i\le m$. The product

$$C(m+1)D^{m+3}(\log B)(\log A_0^*)\cdots(\log A_m^*)\log E^*$$

arising from the conclusion of Theorem 9.1 is bounded by

$$2e^{m+1}\pi(\pi^2+1)^{m/2}C(m+1)D^{m+2}(\log B_0)(\log A_1)\cdots(\log A_m)\max\{1,\log D\}.$$

Finally we use the estimates

$$D^{m+2}(\log B_0)(\log A_1)\cdots(\log A_m)\max\{1,\log D\}\ge 1$$

and

$$2e^{m+1}\pi(\pi^2+1)^{m/2}C(m+1)+\log 2\le 18\cdot 9^mC(m+1).$$

□

Proof of Corollary 9.23. We shall use Proposition 9.21 with B replaced by B' which is defined by

$$\log B'=(\log B)(\log E_2).$$

Since $B\ge e$ and $E_2\ge e$, we have

$$\max\{B, E_2\} \le B'$$

and the conclusion of Corollary 9.23 follows essentially from the homogeneous rational case (part b) in Theorem 9.1 with $E^* = E_1$. More precisely we have replaced here the assumption

$$\log A_j \ge \frac{E|\lambda_j|}{D}, \quad (1 \le j \le m)$$

(which would be required for applying Theorem 9.1) by the weaker condition

$$\sum_{j=1}^{m} \frac{|\lambda_j|}{\log A_j} \le \frac{D}{E}.$$

This is allowed thanks to a remark at the beginning of § 9.3. Since

$$\begin{aligned} D^3(\log B')(\log A_i)(\log E^*) &= D^3(\log B)(\log E_2)(\log A_i)(\log E^*) \\ &\ge (\log D)(\log E)^2, \end{aligned}$$

we may apply Proposition 9.21. □

The proof of Corollary 9.24 will use the following elementary result.

Lemma 9.26. *Let X, Y, ℓ, b, B, B' and δ be positive real numbers satisfying*

$$Y \ge X \ge \ell, \quad b \ge 1, \quad 0 < \delta \le \frac{1}{2},$$

$$Y \log 2 \ge X \log \frac{3Y}{\ell} \quad \text{and} \quad B = \max\left\{e, b + \frac{B'}{\ell}\right\}.$$

Then

$$X \log B \le \delta B' + Y \log \frac{b}{\delta}.$$

Proof. The real function $x \mapsto xB' - Y\log x$ reaches its minimum at $x = Y/B'$. Accordingly we distinguish two cases:
a) Assume $B' \le 2Y$. The conclusion is

$$X \log B \le \frac{1}{2} B' + Y \log(2b).$$

Since

$$b \le \frac{bY}{\ell}, \quad \frac{B'}{\ell} \le \frac{2bY}{\ell} \quad \text{and} \quad \frac{3bY}{\ell} \ge e,$$

we have $B \le 3bY/\ell$. Using the assumptions

$$Y \ge X \quad \text{and} \quad Y \log 2 \ge X \log \frac{3Y}{\ell},$$

we get

$$X \log B \leq X \log b + X \log\left(\frac{3Y}{\ell}\right) \leq Y \log b + Y \log 2 = Y \log(2b).$$

b) Assume $B' \geq 2Y$. The conclusion is now

$$X \log B \leq Y + Y \log \frac{bB'}{Y}.$$

From $\ell \leq Y \leq B'/2$ and $b \geq 1$ we deduce

$$b \leq \frac{bB'}{2\ell}, \quad \frac{B'}{\ell} \leq \frac{bB'}{\ell} \quad \text{and} \quad \frac{3bB'}{2\ell} \geq 3b \geq e,$$

hence $B \leq 3bB'/2\ell$.

Since $X \leq Y$ and $bB' \geq B' \geq 2Y$, we have

$$X \log \frac{bB'}{2Y} \leq Y \log \frac{bB'}{2Y}.$$

Using once more the assumption $Y \log 2 \geq X \log(3Y/\ell)$, we conclude

$$X \log B \leq X \log \frac{bB'}{2Y} + X \log \frac{3Y}{\ell} \leq Y \log \frac{bB'}{2Y} + Y \log 2 \leq Y \log \frac{bB'}{Y}.$$

□

Proof of Corollary 9.24. We apply Lemma 9.26 with $\ell = \log A_m$ and $b = b_m$. Hence, by Corollary 9.23, it suffices to check

$$Y \log 2 \geq X \log \frac{3Y}{\ell}.$$

Define $t = \log(X/\log A_m)$ so that $Y = 2tX$. Since $t \geq 11$, we have

$$(2 \log 2 - 1)t \geq \log(6t).$$

Therefore

$$\frac{Y}{X} \log 2 = 2t \log 2 \geq t + \log(6t) = \log \frac{6tX}{\log A_m} = \log \frac{3Y}{\log A_m}.$$

□

Proof of Corollary 9.25. Corollary 9.25 follows from Corollary 9.24 with $\delta = \epsilon/2$.

□

Exercises

Exercise 9.1. Let K be a field of characteristic zero and let $\beta_0, \ldots, \beta_{m-1}$ be elements of K. Define

$$\begin{aligned} Y &= \{0\} \times \mathbb{Z}^{m-1} + \mathbb{Z}(\beta_0, \beta_1, \ldots, \beta_{m-1}) \subset K^m \\ &= \left\{(s_m\beta_0, s_1 + s_m\beta_1, \ldots, s_{m-1} + s_m\beta_{m-1}) \,;\, \underline{s} \in \mathbb{Z}^m\right\} \end{aligned}$$

and, for $S \in \mathbb{Z}$, $S \geq 1$, consider the subset

$$Y[S] = \left\{(s_m\beta_0, s_1 + s_m\beta_1, \ldots, s_{m-1} + s_m\beta_{m-1}) \,;\, \underline{s} \in \mathbb{Z}^m[S]\right\}.$$

Let $\mathcal{V}$ be a vector subspace of K^m of codimension $r \geq 1$.
a) Assume $\beta_0 \neq 0$. Check

$$\operatorname{Card}\left(\frac{Y[S] + \mathcal{V}}{\mathcal{V}}\right) \geq (2S+1)^r.$$

Hint. *Notice that Y is spanned over $\mathbb{Z}$ by a basis of K^m over K. Hence r at least of the generators of Y are linearly independent modulo $\mathcal{V}$.*

b) Assume $1, \beta_1, \ldots, \beta_{m-1}$ are linearly independent over $\mathbb{Q}$ and $(1, 0, \ldots, 0) \in \mathcal{V}$. Check

$$\operatorname{Card}\left(\frac{Y[S] + \mathcal{V}}{\mathcal{V}}\right) \geq (2S+1)^{r+1}.$$

Hint. *Consider the projection of K^m onto K^{m-1} with kernel $K(1, 0, \ldots, 0)$ and use Lemma 6.2.*

Exercise 9.2. Let A and B be positive integers. Then the $AB + 1$ polynomials

$$1 \quad \text{and} \quad \Delta(z + a; A)^b, \quad (0 \leq a < A, \quad 1 \leq b \leq B)$$

give a basis of the space of polynomials of degree $\leq AB$.

Exercise 9.3. Introduce the following notation: for a, b and c nonnegative integers, define

$$\Delta(z; a, b, c) = \left(\frac{d}{dz}\right)^c \big(\Delta(z; a)\big)^b.$$

a) Let a be a positive integer and b, c nonnegative integers . For any complex number z, check

$$|\Delta(z; a, b, c)| \leq \left(\frac{|z|}{a} + 1\right)^{ab} (2e)^{ab}.$$

b) For $m \in \mathbb{Z}$, show that the number $\nu(a)^c\, \Delta(m; a, b, c)$ is a rational integer.
c) Let a, b, τ be nonnegative rational integers and t a complex number. Consider the entire function of one variable

$$\Psi(z) = z^\tau e^{tz}.$$

Check the relation

$$\Delta(z + t; a, b, \tau) = \sum_{c=0}^{ab} \Delta(z; a, b, c) \left(\frac{d}{dz}\right)^c \Psi(0)$$

Hint. *(See [W 1993], Lemme 3.4). Use the relation*

$$\Delta(z_1+z_2;a)^b = \sum_{c=0}^{ab} \Delta(z_1;a,b,c) z_2^c$$

for $z_1 \in \mathbb{C}$ and $z_2 \in \mathbb{C}$ together with

$$\left(\frac{d}{dz}\right)^{\tau} z^c \Big|_{z=t} = \left(\frac{d}{dz}\right)^{c} \Psi(0).$$

d) Let S' and S'' be positive integers. Set $S = S'S''+1$. Denote by $\mathcal{S}$ the set of $\underline{\sigma} = (\sigma', \sigma'') \in \mathbb{N}^2$ satisfying either $0 \le \sigma' < S'$ and $1 \le \sigma'' \le S''$ or else $\sigma' = \sigma'' = 0$. Check that the $S \times S$ matrix

$$\Big(\Delta(\sigma'; S', \sigma'', \sigma)\Big)_{\substack{\underline{\sigma}\in\mathcal{S} \\ 0 \le \sigma \le S}}$$

is regular.

Exercise 9.4 (Nonhomogeneous rational case).
Assume $b_0, b_1, \ldots, b_m$ are all in $\mathbb{Z}$. Let $T_0^\sharp$ and $S_0^\sharp$ be positive integers. Check that the rational number

$$\sum_{\kappa=0}^{\tau_0} \frac{1}{\kappa!} \delta_{S_0^\sharp}(t;\sigma,\kappa)\delta_{T_0^\sharp}(s_m b_0;\tau_0,\kappa) \prod_{i=1}^{m-1} \Delta(s_i b_m - s_m b_i;\tau_i),$$

for

$$\underline{\tau} \in \mathbb{N}^m, \quad t \in \mathbb{Z}, \quad \sigma \in \mathbb{Z}, \quad \underline{s} \in \mathbb{Z}^m$$

has absolute value bounded from above by

$$\left(\frac{|t|}{S_0^\sharp}+1\right)^{\sigma} e^{2\sigma+2\|\underline{\tau}\|} \max\left\{\frac{\|\underline{\tau}\|}{T_0^\sharp} ; \frac{|s_m b_0|}{T_0^\sharp}+1\right\}^{\|\underline{\tau}\|}$$

and has a denominator which divides

$$\big(\nu(S_0^\sharp)\nu(T_0^\sharp)\big)^{\tau_0}.$$

Hint. *The sum*

$$\sum_{\kappa=0}^{\tau_0} \frac{\sigma!\tau_0!}{\kappa!(\sigma-\kappa)!(\tau_0-\kappa)!}\big(|t|+S_0^\sharp-1\big)^{\sigma-\kappa}\big(|s_m b_0|+T_0^\sharp-1\big)^{\tau_0-\kappa}$$

is bounded by

$$(|t|+S_0^\sharp)^{\sigma} \max\left\{\tau_0 ; |s_m b_0| + T_0^\sharp - 1\right\}^{\tau_0}.$$

Exercise 9.5. Refine the lower bound for Card($\mathfrak{S}$) in step 2 of § 9.3.3: replace Dirichlet's box principle by geometry of numbers (see [Mat 1998]).

Exercise 9.6. Extend Lemma 7.20 to the nonhomogeneous case:

$$\Lambda = \beta_0 + \beta_1\lambda_1 + \cdots + \beta_m\lambda_m.$$

Deduce that for the proof of Theorem 9.1, there is no loss of generality to assume that the numbers $\lambda_1, \ldots, \lambda_m$ are linearly independent over $\mathbb{Q}$.

Hint. *Define*

$$\Phi_n(B) = C(n)D^{n+2}(\log B)(\log E^*)(\log E)^{-n-1},$$
$$N_{kn} = \left[11nD^3 \log_+ A_k\right] \quad \textit{and} \quad \widetilde{B} = 2N_{kn}B^2.$$

Check

$$C(n-1)\log \widetilde{B} \le \frac{1}{2}C(n)D(\log B)(\log E^*)(\log_+ A_k)(\log E)^{-1}$$

and

$$\log N_{kn} \le \frac{1}{2}C(n)D^3(\log B)(\log E^*)(\log_+ A_k)(\log E)^{-2}.$$

Exercise 9.7. Under the assumptions of Corollary 9.22, assume $\alpha_1, \ldots, \alpha_m$ are positive real numbers. Check that the conclusion holds with $C_1(m)$ replaced by

$$(1 + \log 2)C(m) + \log 2.$$

Hint. *Define $B = 2B_0$. In case $B \le D$, use Liouville's inequality. Otherwise, use Theorem 9.1 with*

$$E = e, \quad E^* = \max\{e, D\}, \quad \log B \le (1 + \log 2)\log B_0$$

(see the proof of Corollary 9.22 as well as Corollaire 10.4 of [W 1993]*).*

Exercise 9.8. Assuming $\alpha_1, \ldots, \alpha_m$ are positive real numbers, show that in the conclusions of Corollaries 9.23, 9.24 and 9.25, one can replace

$$|b_1\lambda_1 + \cdots + b_m\lambda_m| \quad \text{by} \quad \left|\alpha_1^{b_1}\cdots\alpha_m^{b_m} - 1\right|,$$

provided that, at the same time, one replaces $C(m)$ in the definition of X by $C(m) + \log 2$.

Exercise 9.9. Let α be a complex algebraic number which is not a root of unity.
a) Show that there exists a constant $c = c(\alpha) > 0$ such that, for any integer $q \ge 2$ and any root of unity ζ of order q,

$$|\alpha - \zeta| > q^{-c}.$$

b) Show that there exists a constant $c' = c'(\alpha) > 0$ such that, for any integer $q \ge 2$,

$$|\alpha^q - 1| > q^{-c'}.$$

Remark. Nontrivial (but also noneffective) lower bounds for non-vanishing sums

$$\left|a_1\alpha_1^{m_1} + \cdots + a_k\alpha_k^{m_k}\right|$$

can be deduced from Schmidt's subspace Theorem (see [Sc 1991], Chap. V § 1).

10. On Baker's Method

In Chap. 4 we deduced Baker's Theorems 1.5 and 1.6 from Schneider-Lang's Criterion. The proof used an extension of Gel'fond's method in several variables. In Chapters 6 and 7, we extended Schneider's method in several variables in order to prove the homogeneous transcendence result (Theorem 1.5) as well as quantitative refinements. The proofs did not involve any derivative at all. In Chap. 9, a single derivative was introduced, so that a second proof of Theorem 1.6 could be achieved, and at the same time measures for nonhomogeneous linear independence of logarithms could be derived. As we saw, it turned out that this approach was useful also for getting sharper estimates for homogeneous measures of linear independence.

Now we consider Baker's original method of proving not only his transcendence results, but also quantitative refinements. This method is now more than 30 years old now, and many improvements have been incorporated in order to refine the initial estimate. These sharpening have not always contributed to simplify the proof, and we plan to explain some of the main features of successive refinements. However our approach does not follow an historical path: for instance we use Laurent's interpolation determinants.

We first explain in § 10.1 the proof of the transcendence result, next (in § 10.2) we give an estimate using interpolation determinants. A brief outline of the more classical proof involving an auxiliary function is given in § 10.3. Finally (§ 10.4) we give further comments on earlier developments of the subject.

10.1 Linear Independence of Logarithms of Algebraic Numbers

This section is devoted to a proof of Baker's transcendence results which is close to Baker's original arguments ([B 1975], Chap. 2), apart from the fact that we use an interpolation determinant in place of an auxiliary function.

In § 10.1.1 we give a sketch of proof of Baker's homogeneous Theorem 1.5. Then we shall explain how to deal with the general situation of Theorem 1.6. In §§ 10.1.3 and 10.1.4 we shall provide the details directly for the general case.

In table 10.1, we provide a summary of the main features of these methods (we refer also to §§ 11.4 and 14.4.6 where we shall give a detailed comparison between the different methods considered in Chap. 6 and 7, Chap. 9, § 10.1.1 and § 10.1.2).

Given algebraic numbers $\beta_0, \ldots, \beta_{m-1}$, and logarithms of algebraic numbers $\lambda_1, \ldots, \lambda_m$, define

$$\Lambda = \beta_0 + \beta_1\lambda_1 + \cdots + \beta_{m-1}\lambda_{m-1} - \lambda_m.$$

The first and third columns are related with the homogeneous case (Theorem 1.5) where $\beta_0 = 0$, while the second and fourth ones deal with the general case (Theorem 1.6). The symbols $\boxed{1'}$, $\boxed{2'}$, $\boxed{1}$, $\boxed{2}$ refer to the numbering of the different methods in § 11.4 and § 14.4. The first two rows refer to the algebraic group $G = \mathbb{G}_{\mathrm{a}}^{d_0} \times \mathbb{G}_{\mathrm{m}}^{d_1}$, the third to the dimension ℓ_0 of the space of derivation $\mathcal{W}$ in $T_e(G)$, and the fourth row to the number ℓ_1 of $\mathbb{Q}$-linearly independent points $\eta_1, \ldots, \eta_{\ell_1}$ in $\exp_G^{-1}\big(G(\overline{\mathbb{Q}})\big)$.

Table 10.1.

	Chap. 6 and 7 $\boxed{1'}$	Chap. 9 $\boxed{2'}$	§ 10.1.1 $\boxed{1}$	§ 10.1.2 $\boxed{2}$
d_0	$m-1$	m	0	1
d_1	1	1	m	m
ℓ_0	0	1	$m-1$	m
ℓ_1	m	m	1	1

In the transcendence method of Chapters 6 and 7, which extends Schneider's solution of Hilbert's seventh problem, the analytic functions are

$$z_1, \ldots, z_{m-1}, \; e^{z_m},$$

the underlying algebraic group (occurring in the zero estimate) is $\mathbb{G}_{\mathrm{a}}^{m-1} \times \mathbb{G}_{\mathrm{m}}$, there is no derivative, and the main observation is that the hyperplane of equation

$$\lambda_1 z_1 + \cdots + \lambda_{m-1} z_{m-1} = z_m,$$

in $\mathbb{C}^m$, which contains the $m-1$ points

$$\eta_j = (\delta_{j1}, \ldots, \delta_{j,m-1}, \lambda_j) \quad (1 \le j \le m-1),$$

also contains

$$\eta_m = (\beta_1, \ldots, \beta_{m-1}, \lambda_m)$$

if and only if $\Lambda = 0$.

In Chap. 9, there are $m+1$ functions

$$z_0, \ldots, z_{m-1}, \; e^{z_m},$$

the algebraic group is $\mathbb{G}_{\mathrm{a}}^m \times \mathbb{G}_{\mathrm{m}}$, the space of derivations is the complex line

$$\mathbb{C}(1, 0, \ldots, 0, 1),$$

and the m points are

$$\eta_j = (0, \delta_{j1}, \ldots, \delta_{j,m-1}, \lambda_j)\ (1 \leq j \leq m-1),$$

$$\eta_m = (\beta_0, \beta_1, \ldots, \beta_{m-1}, \lambda_m)$$

while the related hyperplane has equation

$$z_0 + \lambda_1 z_1 + \cdots + \lambda_{m-1} z_{m-1} = z_m.$$

Baker's method, which is the subject of the present chapter, is quite different. For the homogeneous case (§ 10.1.1) we shall work with the exponential functions

$$e^{z_1}, \ldots, e^{z_m}$$

and the algebraic group $\mathbb{G}_{\mathrm{m}}^m$, the space $\mathcal{W}$ of derivations is the hyperplane of equation

$$\beta_1 z_1 + \cdots + \beta_{m-1} z_{m-1} = z_m$$

and there is a single point, namely $(\lambda_1, \ldots, \lambda_m)$.

From § 10.1.2 on, we shall consider the general case using the functions

$$z_0,\ e^{z_1}, \ldots, e^{z_m},$$

the algebraic group $\mathbb{G}_{\mathrm{a}} \times \mathbb{G}_{\mathrm{m}}^m$, the hyperplane of equation

$$\beta_0 z_0 + \beta_1 z_1 + \cdots + \beta_{m-1} z_{m-1} = z_m$$

and the point $(1, \lambda_1, \ldots, \lambda_m)$.

The permutation $(1, 2, 3, 4) \mapsto (3, 4, 1, 2)$ on the columns as well as on the rows reveals a symmetry in Table 10.1, which can be written as

A	B
B	A

with $A =$

$m-1$	m
1	1

and $B =$

0	1
m	m

.

We shall discuss this *duality* in § 13.7.

Notice finally that in the simplest case $m = 1$, the second and fourth column become identical with $d_0 = d_1 = \ell_0 = \ell_1 = 1$, and they both correspond to Hermite-Lindemann's Theorem 1.2. In case $m = 2$ and $\beta_0 = 0$, the first column is related to Schneider's solution of Hilbert's seventh problem (§ 2.3) and the third column to Gel'fond's solution (§ 2.4).

10.1.1 Sketch of Proof of Baker's Theorem 1.5

We start with a sketch of proof of Baker's homogeneous Theorem 1.5.

Assume that $\beta_1, \ldots, \beta_{m-1}$ and $\lambda_1, \ldots, \lambda_m$ are complex numbers which satisfy

$$\beta_1\lambda_1 + \cdots + \beta_{m-1}\lambda_{m-1} = \lambda_m.$$

Consider the hyperplane $\mathcal{W}$ of equation

$$\beta_1 z_1 + \cdots + \beta_{m-1} z_{m-1} = z_m$$

in $\mathbb{C}^m$. This hyperplane contains the points $(s\lambda_1, \ldots, s\lambda_m)$ $(s \in \mathbb{Z})$. A basis of $\mathcal{W}$ is

$$(\delta_{i1}, \ldots, \delta_{i,m-1}, \beta_i) \quad (1 \le i \le m-1),$$

the map

$$\begin{array}{ccc} \mathbb{C}^{m-1} & \longrightarrow & \mathcal{W} \\ \underline{z} & \longmapsto & \bigl(z_1, \ldots, z_{m-1}, \beta_1 z_1 + \cdots + \beta_{m-1} z_{m-1}\bigr) \end{array}$$

is an isomorphism, and the restrictions to $\mathcal{W}$ of the functions $e^{z_1}, \ldots, e^{z_m}$ produce m functions of $m-1$ variables:

$$e^{z_1}, \ldots, e^{z_{m-1}}, e^{\beta_1 z_1 + \cdots + \beta_{m-1} z_{m-1}}.$$

A monomial in these functions (and their inverse) can be written

$$f_{\underline{t}} = \exp\bigl\{t_1 z_1 + \cdots + t_{m-1} z_{m-1} + t_m(\beta_1 z_1 + \cdots + \beta_{m-1} z_{m-1})\bigr\}$$

for some $\underline{t} = (t_1, \ldots, t_m) \in \mathbb{Z}^m$. Notice that for $s \in \mathbb{Z}$ we have

$$f_{\underline{t}}(s\lambda_1, \ldots, s\lambda_{m-1}) = \alpha_1^{t_1 s} \cdots \alpha_m^{t_m s}$$

where $\alpha_j = e^{\lambda_j}$. Take a derivative of this exponential monomial:

$$\left(\frac{\partial}{\partial z_1}\right)^{\sigma_1} \cdots \left(\frac{\partial}{\partial z_{m-1}}\right)^{\sigma_{m-1}} f_{\underline{t}} = (t_1 + t_m\beta_1)^{\sigma_1} \cdots (t_{m-1} + t_m\beta_{m-1})^{\sigma_{m-1}} f_{\underline{t}}$$

for $\underline{\sigma} = (\sigma_1, \ldots, \sigma_{m-1}) \in \mathbb{N}^{m-1}$. The value of this derivative at the point $(s\lambda_1, \ldots, s\lambda_{m-1})$ for $s \in \mathbb{Z}$ is:

$$\gamma_{\underline{t}}^{(\underline{\sigma} s)} = (t_1 + t_m\beta_1)^{\sigma_1} \cdots (t_{m-1} + t_m\beta_{m-1})^{\sigma_{m-1}} \alpha_1^{t_1 s} \cdots \alpha_m^{t_m s}.$$

This number belongs to the field

$$\mathbb{Q}(\alpha_1, \ldots, \alpha_m, \beta_1, \ldots, \beta_{m-1}).$$

More precisely $\gamma_{\underline{t}}^{(\underline{\sigma} s)}$ is the value, at the point $(\alpha_1, \ldots, \alpha_m, \beta_1, \ldots, \beta_{m-1})$, of a polynomial in the ring

$$\mathbb{Z}[\mathrm{X}_1^{\pm 1}, \ldots, \mathrm{X}_m^{\pm 1}, \mathrm{Y}_1, \ldots, \mathrm{Y}_{m-1}].$$

We put these numbers into a matrix. We first restrict $\underline{t}$, $\underline{\sigma}$, s to finite subsets of $\mathbb{Z}^{m-1}$, $\mathbb{N}^{m-1}$ and $\mathbb{Z}$ respectively, say

$$|\underline{t}| \le T_1, \quad \|\underline{\sigma}\| \le S_0, \quad |s| \le S_1,$$

and we introduce the corresponding matrix (depending on choices for the orderings of the rows and columns)

$$M = \left(\gamma_{\underline{t}}^{(\underline{\sigma} s)}\right)_{\substack{\underline{t} \\ (\underline{\sigma}, s)}}.$$

Assume now that $1, \beta_1, \ldots, \beta_{m-1}$ are $\mathbb{Q}$-linearly independent and $\lambda_1, \ldots, \lambda_m$ are $\mathbb{Q}$-linearly independent. From the multiplicity estimate (Theorem 8.1) one deduces that M has maximal rank L, where $L = (2T_1 + 1)^{m-1}$. Let Δ be the determinant of a regular square $L \times L$ matrix extracted from M. This number Δ is the value, at the point $(\alpha_1, \ldots, \alpha_m, \beta_1, \ldots, \beta_{m-1})$, of a polynomial f in the ring

$$\mathbb{Z}[\mathrm{X}_1^{\pm 1}, \ldots, \mathrm{X}_m^{\pm 1}, \mathrm{Y}_1, \ldots, \mathrm{Y}_{m-1}].$$

Using Schwarz' Lemma one obtains a sharp upper bound for $|\Delta|$. Liouville's estimate (Lemma 2.1) implies that one at least of the numbers $\alpha_1, \ldots, \alpha_m, \beta_1, \ldots, \beta_{m-1}$ is transcendental. Finally, using Lemma 1.7, one gets the conclusion of Theorem 1.5 (for more information, see Exercise 10.1).

10.1.2 Sketch of Proof of Baker's Theorem 1.6

In § 10.1.1, the multiplicity estimate involves the algebraic group $\mathbb{G}_{\mathrm{m}}^m$. In order to deal with the coefficient β_0 in the general case, one works with the algebraic group $\mathbb{G}_{\mathrm{a}} \times \mathbb{G}_{\mathrm{m}}^m$.

Let $\beta_0, \ldots, \beta_{m-1}$ and $\lambda_1, \ldots, \lambda_m$ be complex numbers which satisfy

$$\beta_0 + \beta_1\lambda_1 + \cdots + \beta_{m-1}\lambda_{m-1} = \lambda_m.$$

The hyperplane $\mathcal{W}$ of equation

$$\beta_0 z_0 + \beta_1 z_1 + \cdots + \beta_{m-1} z_{m-1} = z_m$$

in $\mathbb{C}^{m+1}$ contains the points $(s, s\lambda_1, \ldots, s\lambda_m)$ $(s \in \mathbb{Z})$. A basis of $\mathcal{W}$ is

$$(1, 0, \ldots, 0, \beta_0), \qquad (0, \delta_{i1}, \ldots, \delta_{i,m-1}, \beta_i) \quad (1 \le i \le m-1).$$

The restrictions to $\mathcal{W}$ of the functions $z_0, e^{z_1}, \ldots, e^{z_m}$ produce the $m+1$ functions of m variables

$$z_0, e^{z_1}, \dots, e^{z_{m-1}}, e^{\beta_0 z_0 + \beta_1 z_1 + \cdots + \beta_{m-1} z_{m-1}}.$$

For $\tau \in \mathbb{N}$ and $\underline{t} = (t_1, \dots, t_m) \in \mathbb{Z}^m$, define

$$f_{\tau\underline{t}} = z_0^\tau \exp\bigl\{t_1 z_1 + \cdots + t_{m-1} z_{m-1} + t_m(\beta_0 z_0 + \beta_1 z_1 + \cdots + \beta_{m-1} z_{m-1})\bigr\}.$$

For $s \in \mathbb{Z}$ we have

$$f_{\tau\underline{t}}(s, s\lambda_1, \dots, s\lambda_{m-1}) = s^\tau \alpha_1^{t_1 s} \cdots \alpha_m^{t_m s}.$$

Using the relation (which is a simple special case of Lemma 4.9)

$$\left(\frac{d}{dz}\right)^\sigma \left(z^\tau e^{tz}\right) = \sum_{\kappa=0}^{\min\{\sigma,\tau\}} \frac{\sigma!\tau!}{\kappa!(\sigma-\kappa)!(\tau-\kappa)!} t^{\sigma-\kappa} z^{\tau-\kappa} e^{tz},$$

we deduce that for $\underline{\sigma} = (\sigma_0, \dots, \sigma_{m-1}) \in \mathbb{N}^m$ and $s \in \mathbb{Z}$, we have

$$\left(\frac{\partial}{\partial z_0}\right)^{\sigma_0} \cdots \left(\frac{\partial}{\partial z_{m-1}}\right)^{\sigma_{m-1}} f_{\tau\underline{t}} =$$

$$\sum_{\kappa=0}^{\min\{\sigma_0,\tau\}} \frac{\sigma_0!\tau!}{\kappa!(\sigma_0-\kappa)!(\tau-\kappa)!} (t_m\beta_0)^{\sigma_0-\kappa} (t_1 + t_m\beta_1)^{\sigma_1} \cdots (t_{m-1} + t_m\beta_{m-1})^{\sigma_{m-1}} f_{\tau-\kappa,\underline{t}}.$$

Define

$$\gamma_{\tau\underline{t}}^{(\underline{\sigma}s)} = \left(\frac{\partial}{\partial z_0}\right)^{\sigma_0} \cdots \left(\frac{\partial}{\partial z_{m-1}}\right)^{\sigma_{m-1}} f_{\tau\underline{t}}(s, s\lambda_1, \dots, s\lambda_{m-1}).$$

This number $\gamma_{\tau\underline{t}}^{(\underline{\sigma}s)}$ is the value, at the point $(\alpha_1, \dots, \alpha_m, \beta_0, \dots, \beta_{m-1})$, of a polynomial in the ring

$$\mathbb{Z}[X_1^{\pm 1}, \dots, X_m^{\pm 1}, Y_0, \dots, Y_{m-1}].$$

The rest of the proof is the usual one: we put these numbers into a matrix; assuming $\lambda_1, \dots, \lambda_m$ are $\mathbb{Q}$-linearly independent and either $\beta_0 \neq 0$ or else $1, \beta_1, \dots, \beta_{m-1}$ are $\mathbb{Q}$-linearly independent, we deduce from the multiplicity estimate that this matrix has maximal rank. Schwarz' Lemma provides an upper bound for $|\Delta|$, and Liouville's estimate yields the conclusion.

An interesting point is that this proof yields the general case of Baker's Theorem 1.6; it works for $\beta_0 \neq 0$, but also for $\beta_0 = 0$. It turns out that for producing measures of linear independence, this method is more efficient than the method of § 10.1.1. More precisely the method of § 10.1.1 yields the quantitative estimate which has been proved in Chap. 7, while the method of § 10.1.2 will enable us to prove Theorem 9.1.

10.1.3 A Consequence of the Multiplicity Estimate

We explain here how the multiplicity estimate (Theorem 8.1) will be used in § 10.1.4.

Let K be an algebraically closed field of zero characteristic and let $\beta_0, \ldots, \beta_{m-1}$ be elements of K. On the ring $K[X_0, X_1^{\pm 1}, \ldots, X_m^{\pm 1}]$ we introduce derivative operators:

$$\mathcal{D}_0 = \frac{\partial}{\partial X_0} + \beta_0 X_m \frac{\partial}{\partial X_m}, \qquad \mathcal{D}_i = X_i \frac{\partial}{\partial X_i} + \beta_i X_m \frac{\partial}{\partial X_m} \quad (1 \le i \le m-1)$$

and, for $\underline{\sigma} \in \mathbb{N}^m$,

$$\mathcal{D}^{\underline{\sigma}} = \mathcal{D}_0^{\sigma_0} \cdots \mathcal{D}_{m-1}^{\sigma_{m-1}}.$$

In case $K = \mathbb{C}$, if one substitutes

$$X_0 = z_0, \quad X_i = e^{z_i}, \quad (1 \le i \le m-1)$$

and

$$X_m = e^{\beta_0 z_0 + \beta_1 z_1 + \cdots + \beta_{m-1} z_{m-1}}$$

in the polynomial $\mathcal{D}^{\underline{\sigma}}\big(X_0^\tau X_1^{t_1} \cdots X_m^{t_m}\big)$, one gets

$$\left(\frac{\partial}{\partial z_0}\right)^{\sigma_0} \cdots \left(\frac{\partial}{\partial z_{m-1}}\right)^{\sigma_{m-1}} f_{\tau \underline{t}}(z_0, \ldots, z_{m-1}).$$

Proposition 10.2. *Let $\alpha_1, \ldots, \alpha_m$ be nonzero elements of K which generate a multiplicative subgroup of $K^\times$ of rank $\ge m-1$ and let $\beta_0, \ldots, \beta_{m-1}$ be elements of K with $\beta_0 \ne 0$. Assume also that $1, \beta_1, \ldots, \beta_{m-1}$ are linearly independent over $\mathbb{Q}$. Let T_0, T_1, S_0, S_1 be positive integers satisfying the following conditions:*

$$S_0 \ge 2(m+1)T_1, \quad (S_0+1)(2S_1+1) > m!(m+1)! \max\left\{\frac{T_0}{2}, 2T_1\right\}$$

and

$$(S_0+1)^m(2S_1+1) > m!(m+1)! 2^m T_0 T_1^m.$$

For $\tau \in \mathbb{N}$, $\underline{t} \in \mathbb{Z}^m$, $\underline{\sigma} \in \mathbb{N}^m$ and $s \in \mathbb{Z}$, define $\gamma_{\tau \underline{t}}^{(\underline{\sigma} s)} \in K$ as the value, at the point

$$\big(s\beta_0, \alpha_1^s, \ldots, \alpha_m^s\big) \in K \times (K^\times)^m,$$

of the polynomial

$$\mathcal{D}^{\underline{\sigma}}\big(X_0^\tau X_1^{t_1} \cdots X_m^{t_m}\big) \in K[X_0, X_1^{\pm 1}, \ldots, X_m^{\pm 1}].$$

Consider the following matrix:

$$M = \Big(\gamma_{\tau \underline{t}}^{(\underline{\sigma} s)}\Big)_{\substack{(\tau, \underline{t}) \\ (\underline{\sigma}, s)}}$$

where the index of rows $(\tau, \underline{t})$ runs over the elements in $\mathbb{N} \times \mathbb{Z}^m$ with $0 \le \tau \le T_0$, $|t_i| \le T_1$ $(1 \le i \le m)$, while the index of columns $(\underline{\sigma}, s)$ runs over the elements in

$\mathbb{N}^{m-1} \times \mathbb{Z}$ *with* $\|\underline{\sigma}\| \leq (m+1)S_0$ *and* $|s| \leq (m+1)S_1$. *Then the matrix* $\boldsymbol{M}$ *has rank* $(T_0+1)(2T_1+1)^m$.

Proof. We apply Theorem 8.1 to the algebraic groups $G = G^+ = \mathbb{G}_a \times \mathbb{G}_m^m$, $G^- = 0$, with $d_0 = 1$, $d_1 = m$, with the hyperplane $\mathcal{W}$ in K^{m+1} of equation

$$\beta_0 z_0 + \beta_1 z_1 + \cdots + \beta_{m-1} z_{m-1} = z_m$$

and with the set

$$\Sigma = \left\{(s\beta_0, \alpha_1^s, \ldots, \alpha_m^s);\ s \in \mathbb{Z},\ |s| \leq S_1\right\} \subset G(K) = K \times (K^\times)^m.$$

If the rank of the matrix $\boldsymbol{M}$ is less than $(T_0+1)(2T_1+1)^m$, then there exists a nonzero polynomial $P \in K[G] = K[X_0, X_1^{\pm 1}, \ldots, X_m^{\pm 1}]$ which satisfies the hypotheses of Theorem 8.1 with $D_0 = T_0$, $D_i = T_1$ $(1 \leq i \leq m)$. Hence there exists an algebraic subgroup G^* of G of dimension $d^* < d$ such that

$$\binom{S_0 + \ell_0'}{\ell_0'} \mathrm{Card}\left(\frac{\Sigma + G^*}{G^*}\right) \mathcal{H}(G^*; T_0; T_1) \leq \mathcal{H}(G; T_0; T_1),$$

where

$$\ell_0' = \dim_K \left(\frac{\mathcal{W} + T_e(G^*)}{T_e(G^*)}\right).$$

We first check that this inequality is not satisfied with $G^* = \{e\}$: indeed in this case we have

$$d^* = 0, \quad \ell_0' = m, \quad \mathrm{Card}\left(\frac{\Sigma + G^*}{G^*}\right) = 2S_1 + 1, \quad \mathcal{H}(G^*; T_0; T_1) = 1,$$

so that

$$\begin{aligned} \binom{S_0 + \ell_0'}{\ell_0'} \mathrm{Card}\left(\frac{\Sigma + G^*}{G^*}\right) \mathcal{H}(G^*; T_0; T_1) &= \binom{S_0 + m}{m}(2S_1 + 1) \\ &\geq \frac{(S_0+1)^m}{m!}(2S_1 + 1) \\ &> (m+1)!2^m T_0 T_1^m, \end{aligned}$$

while, by (5.8),

$$\mathcal{H}(G; T_0; T_1) = (m+1)!2^m T_0 T_1^m.$$

Therefore $d^* \geq 1$.

Let us write $G = G_0 \times G_1$, $G^* = G_0^* \times G_1^*$, where $G_0 = \mathbb{G}_a$ and $G_1 = \mathbb{G}_m^m$, while G_0^* is an algebraic subgroup of G_0 and G_1^* an algebraic subgroup of G_1. Denote by d^*, d_0^*, d_1^* the dimensions of G^*, G_0^* and G_1^* respectively, and by d', d_0', d_1' their codimensions:

$$d^* + d' = d = m + 1, \quad d_0^* + d_0' = d_0 = 1, \quad d_1^* + d_1' = d_1 = m.$$

From the definition of ℓ_0' we derive

$$\ell_0' = \begin{cases} d'-1 & \text{if } T_e(G^*) \subset \mathcal{W}, \\ d' & \text{otherwise.} \end{cases}$$

However $1, \beta_1, \ldots, \beta_{m-1}$ are linearly independent over $\mathbb{Q}$, which means that the hyperplane of K^m of equation

$$\beta_1 z_1 + \cdots + \beta_{m-1} z_{m-1} = z_m$$

does not contain any nonzero element of $\mathbb{Q}^m$. Since $\beta_0 \neq 0$ and $T_e(G^*) \neq 0$, we deduce $\ell_0' = d'$.

We first consider the case $G_0^* = \{0\}$. We have $d_0^* = 0, d^* = d_1^*$ and $d' = m+1-d^* = m+1-d_1^*$. Further, by Proposition 5.7,

$$\mathcal{H}(G^*; T_0; T_1) \geq (d_1^* + 1)!(2T_1)^{d_1^*}.$$

Furthermore, since $\beta_0 \neq 0$,

$$\operatorname{Card}\left(\frac{\Sigma + G^*}{G^*}\right) = 2S_1 + 1.$$

Therefore the conclusion of the multiplicity estimate implies

$$\binom{S_0 + d'}{d'}(2S_1 + 1) \leq \frac{(m+1)!}{(m+2-d')!} T_0 (2T_1)^{d'-1}.$$

Since $d' \leq m$ this estimate yields

$$(S_0 + 1)^{d'}(2S_1 + 1) \leq \frac{(m+1)!d'!}{(m+2-d')!} T_0 (2T_1)^{d'-1}.$$

However we have $S_0 + 1 \geq 2T_1$, $d' \geq 1$ and

$$\frac{d'!}{(m+2-d')!} \leq \frac{1}{2} m!,$$

hence we get a contradiction with the inequality

$$(S_0 + 1)(2S_1 + 1) > \frac{1}{2} m!(m+1)! T_0.$$

So we may assume $d_0^* = 1$, so that $d^* = d_1^* + 1$ and $d' = m + 1 - d^* = m - d_1^*$. Further (Proposition 5.7 again)

$$\mathcal{H}(G^*; T_0; T_1) \geq (d_1^* + 1)! T_0 (2T_1)^{d_1^*}.$$

The conclusion of the multiplicity estimate gives now

$$\binom{S_0 + d'}{d'} \operatorname{Card}\left(\frac{\Sigma + G^*}{G^*}\right) \leq \frac{(m+1)!}{(m+1-d')!} (2T_1)^{d'}$$

from which we deduce

$$(S_0 + 1)^{d'} \mathrm{Card}\left(\frac{\Sigma + G^*}{G^*}\right) \leq \frac{(m+1)!d'!}{(m+1-d')!}(2T_1)^{d'}.$$

From the estimates

$$S_0 + 1 \geq 2T_1, \quad d' \geq 1, \quad \frac{d'!}{(m+1-d')!} \leq m!$$

and

$$(S_0 + 1)(2S_1 + 1) > m!(m+1)!2T_1$$

we obtain

$$\mathrm{Card}\left(\frac{\Sigma + G^*}{G^*}\right) < 2S_1 + 1,$$

which means that $\Sigma \cap G^* \neq \{e\}$. The assumption on the rank of the subgroup of K^* generated by $\alpha_1, \ldots, \alpha_m$ then implies $d_1^* = m - 1$, $d' = 1$ and we get the estimate

$$S_0 + 1 \leq 2(m+1)T_1$$

which is not compatible with our assumptions. □

10.1.4 The Transcendence Argument

Here is a extension of Proposition 2.17.

Proposition 10.3. *Let $\lambda_1, \ldots, \lambda_m, \beta_0, \ldots, \beta_{m-1}$ be complex numbers satisfying*

$$\lambda_m = \beta_0 + \beta_1\lambda_1 + \cdots + \beta_{m-1}\lambda_{m-1}.$$

Define $\alpha_i = e^{\lambda_i}$ $(1 \leq i \leq m)$. Assume $\alpha_1, \ldots, \alpha_m$ generate a multiplicative subgroup of $\mathbb{C}^\times$ of rank $\geq m - 1$. Assume also either $\beta_0 \neq 0$ and $1, \beta_1, \ldots, \beta_{m-1}$ linearly independent over $\mathbb{Q}$. Let $E \geq e$ be a real number and T_0, T_1, S_0, S_1 L be five integers, all greater than one, satisfying

$$L = (T_0 + 1)(2T_1 + 1)^m,$$

$$T_1 \geq 8m^2, \quad S_0 \geq 4mT_1, \quad S_0S_1 > 2m^{2m}\max\{T_0, T_1\}$$

and

$$S_0^m S_1 > (2m)^{2m} T_0 T_1^m.$$

Then there exists a polynomial $f \in \mathbb{Z}[\mathrm{X}_1^{\pm 1}, \ldots, \mathrm{X}_m^{\pm 1}, \mathrm{Y}_0, \mathrm{Y}_1, \ldots, \mathrm{Y}_{m-1}]$ satisfying

$$\deg f \leq L\big((m+1)S_0 + T_0 + m(m+1)(T_1+1)S_1\big),$$

$$\mathrm{L}(f) \leq L!(2T_1)^{(m+1)LS_0}\big((m+1)S_1\big)^{LT_0},$$

and

$$0 < |f(\alpha_1, \ldots, \alpha_m, \beta_0, \ldots, \beta_{m-1})| \leq$$

$$\exp\left\{-\frac{1}{3}L^{1+(1/m)}\log E + L\big(c_0S_0\log(ET_0T_1) + T_0\log(c_0ES_1) + c_0T_1S_1E\big)\right\}$$

with

$$c_0 = \max\Bigl\{e(m+1)\max_{0\le i\le m-1}(1+|\beta_i|)^{m+1},\ 1+(m+1)(|\lambda_1|+\cdots+|\lambda_m|)\Bigr\}.$$

Proof of Proposition 10.3. For $\tau\in\mathbb{N}$, $\underline{t}\in\mathbb{Z}^m$, $\underline{\sigma}\in\mathbb{N}^m$ and $s\in\mathbb{Z}$, define the polynomial $P_{\tau\underline{t}}^{(\underline{\sigma}s)}$ in the ring $\mathbb{Z}[\mathrm{X}_1^{\pm1},\dots,\mathrm{X}_m^{\pm1},\mathrm{Y}_0,\dots,\mathrm{Y}_{m-1}]$ by

$$P_{\tau\underline{t}}^{(\underline{\sigma}s)} = \sum_{\kappa=0}^{\min\{\sigma_0,\tau\}}\frac{\sigma_0!\tau!}{\kappa!(\sigma_0-\kappa)!(\tau-\kappa)!}(t_m\mathrm{Y}_0)^{\sigma_0-\kappa}s^{\tau-\kappa}\cdot$$
$$(t_1+t_m\mathrm{Y}_1)^{\sigma_1}\cdots(t_{m-1}+t_m\mathrm{Y}_{m-1})^{\sigma_{m-1}}\mathrm{X}_1^{t_1s}\cdots\mathrm{X}_m^{t_ms},$$

so that the number $\gamma_{\tau\underline{t}}^{(\underline{\sigma}s)}$, which have been introduced in § 10.1.2, satisfies

$$\gamma_{\tau\underline{t}}^{(\underline{\sigma}s)} = P_{\tau\underline{t}}^{(\underline{\sigma}s)}(\alpha_1,\dots,\alpha_m,\beta_0,\dots,\beta_{m-1}).$$

By Lemma 4.9, for

$$0\le\tau\le T_0,\quad |\underline{t}|\le T_1,\quad \|\underline{\sigma}\|\le(m+1)S_0\quad\text{and}\quad |s|\le(m+1)S_1,$$

this polynomial $P_{\tau\underline{t}}^{(\underline{\sigma}s)}$ has degree at most $(m+1)|t_i|S_1$ in each of the two variables $\mathrm{X}_i^{\pm1}$ $(1\le i\le m)$ and degree at most T_0 in Y_0, total degree at most $(m+1)S_0$ in $\mathrm{Y}_1,\dots,\mathrm{Y}_{m-1}$ and length

$$\mathrm{L}(P_{\tau\underline{t}}^{(\underline{\sigma}s)})\le(T_1+T_0)^{(m+1)S_0}\bigl((m+1)S_1\bigr)^{T_0}.$$

Consider the matrix

$$M=\Bigl(\gamma_{\tau\underline{t}}^{(\underline{\sigma}s)}\Bigr)_{\substack{(\tau,\underline{t})\\(\underline{\sigma},s)}}.$$

We deduce from Proposition 10.2 that M has maximal rank L. Let Δ be the determinant of a regular square $L\times L$ matrix extracted from M, say

$$\Delta=\det\Bigl(\gamma_{\tau\underline{t}}^{(\underline{\sigma}_\mu s_\mu)}\Bigr)_{\substack{(\tau,\underline{t})\\1\le\mu\le L}}.$$

This number Δ is the value, at the point $(\alpha_1,\dots,\alpha_m,\beta_0,\dots,\beta_{m-1})$, of the polynomial

$$f=\det\Bigl(P_{\tau\underline{t}}^{(\underline{\sigma}_\mu s_\mu)}\Bigr)_{\substack{(\tau,\underline{t})\\1\le\mu\le L}}$$

in the ring $\mathbb{Z}[\mathrm{X}_1^{\pm1},\dots,\mathrm{X}_m^{\pm1},\mathrm{Y}_0,\dots,\mathrm{Y}_{m-1}]$. The degree of f is at most

$$(m+1)S_1(2T+1)^{m-1}\sum_{t_i=-T}^{T}|t_i|\le(m+1)S_1T(T+1)(2T+1)^{m-1}$$
$$\le\frac{1}{2}(m+1)L(T+1)S_1$$

in each of the $2m$ variables $X_1^{\pm 1}, \dots, X_m^{\pm 1}$, at most LT_0 in Y_0 and the total degree is at most $(m+1)LS_0$ in $Y_1, \dots, Y_{m-1}$. Hence the total degree of f is at most $L(T_0 + (m+1)S_0 + m(m+1)(T_1+1)S_1)$. Moreover the length of f is bounded by

$$L!(T_1 + T_0)^{(m+1)LS_0}\big((m+1)S_1\big)^{LT_0}.$$

We bound $|\Delta|$ from above. For each $\mu = 1, \dots, L$, define

$$\mathcal{D}^{(\mu)} = \left(\frac{\partial}{\partial z_0}\right)^{\sigma_{\mu 0}} \cdots \left(\frac{\partial}{\partial z_{m-1}}\right)^{\sigma_{\mu,m-1}}$$

and

$$\underline{\zeta}_\mu = s_\mu(\beta_0, \lambda_1, \dots, \lambda_{m-1}),$$

so that

$$\gamma_{\tau \underline{t}}^{(\underline{\sigma}_\mu s_\mu)} = \mathcal{D}^{(\mu)} f_{\tau, \underline{t}}(\underline{\zeta}_\mu).$$

Lemma 9.2 shows that the entire function of one variable z:

$$\Psi(z) = \det\Big(\mathcal{D}^{(\mu)} f_{\tau \underline{t}}(z\underline{\zeta}_\mu)\Big)_{\substack{(\tau, \underline{t}) \\ 1 \le \mu \le L}}.$$

has a zero at the origin of multiplicity at least $\Theta_m(L) - (m+1)LS_0$. From Schwarz's Lemma (see the proof of Lemma 6.1) we deduce

$$|\Delta| = |\Psi(1)| \le E^{-\Theta_m(L)} E^{(m+1)LS_0} L! \prod_{\lambda=1}^{L} \sup_{|z|=E} \left|\mathcal{D}^{(\mu)} f_{\tau \underline{t}}(z\underline{\zeta}_\mu)\right|.$$

From the explicit formula

$$\mathcal{D}^{(\mu)} f_{\tau \underline{t}}(z\underline{\zeta}_\mu) = \sum_{\kappa=0}^{\min\{\sigma_{\mu 0}, \tau\}} \frac{\sigma_{\mu 0}! \tau!}{\kappa!(\sigma_{\mu 0} - \kappa)!(\tau - \kappa)!} (t_m \beta_0)^{\sigma_{\mu 0} - \kappa} z_0^{\tau - \kappa}.$$
$$(t_1 + t_m \beta_1)^{\sigma_{\mu 1}} \cdots (t_{m-1} + t_m \beta_{m-1})^{\sigma_{\mu, m-1}} e^{(t_1 \lambda_1 + \cdots + t_m \lambda_m) s_\mu z}$$

we deduce the estimate

$$\sup_{|z|=E} \left|\mathcal{D}^{(\mu)} f_{\tau \underline{t}}(z\underline{\zeta}_\mu)\right| \le (T_1 + T_0)^{(m+1)S_0} \big((m+1)S_1\big)^{T_0} c_1^{(m+1)S_0 + T_0} E^{T_0} e^{c_2 T_1 S_1 E}.$$

with

$$c_1 = 1 + \max_{0 \le i \le m-1} |\beta_i| \quad \text{and} \quad c_2 = (m+1)(|\lambda_1| + \cdots + |\lambda_m|).$$

Therefore

$$\log|\Delta| \le -\Theta_m(L)\log E + L\Big\{\log L + (m+1)S_0 \log\big(E(T_0 + T_1)\big)$$
$$+ T_0 \log\big((m+1)ES_1\big) + c_3(T_0 + S_0) + c_2 T_1 S_1 E\Big\}$$

with $c_3 = (m+1)\log c_1$. From the assumption $T_1 \ge 8m^2$ we deduce $L \ge (4m)^{2m}$ hence from Lemma 6.5 we derive

$$\Theta_m(L) \geq \frac{1}{3} L^{1+(1/m)}.$$

We use also the trivial bounds

$$\log L \leq T_0 + S_0 + T_1 S_1 E \quad \text{and} \quad e(m+1)c_1^{m+1} \geq 2 + m + c_1^{m+1}.$$

The conclusion of Proposition 10.3 follows with

$$c_0 = \max\{e(m+1)c_1^{m+1}, 1 + c_2\}.$$

□

10.1.5 Proof of Baker's Theorem 1.6

We are now ready to complete the proof of Baker's Theorem.

Proof of Theorem 1.6. Let S_1 be a fixed sufficiently large integer and T_0 an integer which tends to infinity. Define

$$T_1 = T_0, \quad S_0 = \left[\frac{1}{S_1} T_0^{1+(1/m)}\right], \quad E = T_0^{1/(2m)}.$$

Using Lemma 2.1, we deduce that under the assumptions of Proposition 10.3, one at least of the numbers $\beta_0, \ldots, \beta_{m-1}, \alpha_1, \ldots, \alpha_m$ is transcendental. This completes the proof of Theorem 1.6. □

10.2 Baker's Method with Interpolation Determinants

Our goal is to give a new proof of Theorem 9.1, combining Baker's method with Laurent's interpolation determinants. In the next section (§ 10.3), we shall describe very briefly the classical approach which involves an auxiliary function and an extrapolation argument.

The method involving an interpolation determinant we are going to use is simpler than the classical one involving an auxiliary function. As we shall see in § 10.3, the only disadvantage would be for an explicit computation of $C(m)$, but we shall not address this issue here.

One could introduce further simplifications in our method of proof, which then would lead a slightly weaker result.

- Firstly, we could avoid Fel'dman's polynomials, but then we should assume $E^* \geq \log B$. Because of the term $\log E^*$, the final estimate would involve $(\log B)(\log\log B)$ in place of $\log B$.
- Secondly, we could work with a torus $\mathbb{G}_{\mathrm{m}}^m$ in place of $\mathbb{G}_{\mathrm{a}} \times \mathbb{G}_{\mathrm{m}}^m$. In this case the stronger condition $E^* \geq B$ would be required, and in place of $\log B$ one would have a factor $(\log B)^2$ in the final measure.

From this point of view, the situation is therefore quite similar with what we observed in Chapters 7 and 9: we shall come back to this point in § 14.4.

10.2.1 Sketch of Proof of Theorem 9.1

Consider a nonzero number

$$\Lambda = \beta_0 + \beta_1\lambda_1 + \cdots + \beta_m\lambda_m$$

where $\beta_0, \dots, \beta_m$ are algebraic numbers and $\lambda_1, \dots, \lambda_m$ are in $\mathcal{L}$. Assume $\beta_m = -1$, so that Λ is the value at the point $\underline{\eta} = (1, \lambda_1, \dots, \lambda_m)$ of the linear form $\beta_0 z_0 + \beta_1 z_1 + \cdots + \beta_{m-1}z_{m-1} - z_m$. In the rational homogeneous case we have $\beta_0 = 0$ and $\beta_i = -b_i/b_m$ $(1 \le i \le m)$.

For $i = 1, \dots, m$ define $\alpha_i = e^{\lambda_i}$. Let K be the number field generated by $\beta_0, \dots, \beta_{m-1}, \alpha_1, \dots, \alpha_m$ and let G be the algebraic group $\mathbb{G}_{\mathrm{a}} \times \mathbb{G}_{\mathrm{m}}^m$ over K. The point $(1, \alpha_1, \dots, \alpha_m)$ lies in $G(K)$. The exponential map of $G(\mathbb{C})$ involves the functions

$$z_0,\ e^{z_1}, \dots, e^{z_m}.$$

In the tangent space $T_e(G) = \mathbb{C}^{m+1}$ of G, the hyperplane $\mathcal{W}$ of equation

$$z_m = \beta_0 z_0 + \beta_1 z_1 + \cdots + \beta_{m-1}z_{m-1}$$

is rational over K. A basis of the hyperplane $\mathcal{W}$ is given by the m column vectors $\underline{w}_k$ $(0 \le k \le m-1)$ of the matrix

$$\begin{pmatrix} & \mathrm{I}_m & \\ \beta_0 & \cdots & \beta_{m-1} \end{pmatrix}.$$

The restriction to $\mathcal{W}$ of the exponential map of G leads us to consider the $m+1$ functions of m variables

$$z_0,\ e^{z_1}, \dots, e^{z_{m-1}},\ e^{\beta_0 z_0 + \beta_1 z_1 + \cdots + \beta_{m-1}z_{m-1}}.$$

They satisfy differential equations with coefficients in K. We introduce monomials in these functions, we take their derivatives, and we consider the values of these derivatives at the points $s\underline{\eta}$, $s \in \mathbb{Z}$. We put these numbers into a matrix $\boldsymbol{M}$ and we investigate the rank of $\boldsymbol{M}$.

We shall choose later suitable parameters $T_0, T_1, \dots, T_m, S_0, S_1$ (positive integers) which will enable us to perform the construction of $\boldsymbol{M}$ as follows. Define $L = (T_0+1)(2T_1+1)\cdots(2T_m+1)$ (this will be the number of rows of $\boldsymbol{M}$). Denote by $\delta(X;\tau)$ $(0 \le \tau \le T_0)$ any basis of the space of polynomials in $\mathbb{Q}[X]$ of degree $\le T_0$. For $(\tau, \underline{t}) \in \mathbb{N} \times \mathbb{Z}^m$ satisfying $0 \le \tau \le T_0$ and $|t_i| \le T_i$ $(1 \le i \le m)$, define

$$F_{\tau\underline{t}}(z_0, \dots, z_m) = \delta(z_0;\tau)e^{t_1 z_1 + \cdots + t_m z_m}$$

and

$$f_{\tau\underline{t}}(z_0, \ldots, z_{m-1}) = F_{\tau\underline{t}}(z_0, \ldots, z_{m-1}, \beta_0 z_0 + \beta_1 z_1 + \cdots + \beta_{m-1} z_{m-1})$$
$$= \delta(z_0; \tau) e^{t_m\beta_0 z_0 + (t_1 + t_m\beta_1) z_1 + \cdots + (t_{m-1} + t_m\beta_{m-1}) z_{m-1}}.$$

On the space $\mathbb{C}[X, Y_1^{\pm 1}, \ldots, Y_m^{\pm 1}]$, the derivative operators

$$\mathcal{D}_0 = \frac{\partial}{\partial X} + \beta_0 Y_m \frac{\partial}{\partial Y_m}, \qquad \mathcal{D}_i = Y_i \frac{\partial}{\partial Y_i} + \beta_i Y_m \frac{\partial}{\partial Y_m} \quad (1 \le i \le m-1)$$

act as follows. Define, for $\underline{\sigma} \in \mathbb{N}^m$,

$$\mathcal{D}^{\underline{\sigma}} = \mathcal{D}_0^{\sigma_0} \cdots \mathcal{D}_{m-1}^{\sigma_{m-1}},$$

and for $\kappa \in \mathbb{N}$,

$$\delta(X; \tau, \kappa) = \left(\frac{d}{dX}\right)^{\kappa} \delta(X; \tau).$$

Then

$$\mathcal{D}^{\underline{\sigma}}\big(\delta(X;\tau)\underline{Y}^{\underline{t}}\big) = \sum_{\kappa=0}^{\sigma_0} \binom{\sigma_0}{\kappa} (t_m\beta_0)^{\sigma_0-\kappa} \left(\prod_{i=1}^{m-1} (t_i + t_m\beta_i)^{\sigma_i}\right) \delta(X; \tau, \kappa)\underline{Y}^{\underline{t}}.$$

For $s \in \mathbb{Z}$ define

$$\gamma_{\tau\underline{t}}^{(\underline{\sigma}s)} = \mathcal{D}^{\underline{\sigma}}\big(\delta(X;\tau)\underline{Y}^{\underline{t}}\big)(s, \alpha_1^s, \ldots, \alpha_m^s).$$

If we replace X by z_0 and Y_i by e^{z_i} for $1 \le i \le m$, setting

$$\mathcal{D}_{\underline{w}_i} = \frac{\partial}{\partial z_i} + \beta_i \frac{\partial}{\partial z_m} \qquad (0 \le i \le m-1)$$

and $\mathcal{D}_{\underline{w}}^{\underline{\sigma}}$ in place of $\mathcal{D}_{\underline{w}_0}^{\sigma_0} \cdots \mathcal{D}_{\underline{w}_{m-1}}^{\sigma_{m-1}}$, we obtain

$$\gamma_{\tau\underline{t}}^{(\underline{\sigma}s)} = \mathcal{D}_{\underline{w}}^{\underline{\sigma}} F_{\tau\underline{t}}(s\underline{\eta})$$
$$= \sum_{\kappa=0}^{\sigma_0} \binom{\sigma_0}{\kappa} \delta(s; \tau, \kappa)(t_m\beta_0)^{\sigma_0-\kappa} \left(\prod_{i=1}^{m-1} (t_i + t_m\beta_i)^{\sigma_i}\right) \left(\alpha_1^{t_1} \cdots \alpha_m^{t_m}\right)^s.$$

For each $\tau \in \mathbb{N}$, $\underline{t} \in \mathbb{Z}^m$, $\underline{\sigma} \in \mathbb{N}^m$, $s \in \mathbb{Z}$, this number $\gamma_{\tau\underline{t}}^{(\underline{\sigma}s)}$ is in K. The connection with the function $f_{\tau\underline{t}}$ (which depends only on m variables) is the following: if we introduce the projection $\underline{\eta}' = (1, \lambda_1, \ldots, \lambda_{m-1}) \in \mathbb{C}^m$ of $\underline{\eta}$ on the space $\mathbb{C}^m \times \{0\}$, then

$$\left(\frac{\partial}{\partial z_0}\right)^{\sigma_0} \cdots \left(\frac{\partial}{\partial z_{m-1}}\right)^{\sigma_{m-1}} f_{\tau\underline{t}}(s\underline{\eta}') = \gamma_{\tau\underline{t}}^{(\underline{\sigma}s)} e^{t_m s\Lambda}.$$

When $|\Lambda|$ is small then $e^{t_m s\Lambda}$ is close to 1.

We introduce a further parameter $T_0^\sharp$ (again a positive integer), and we select for $\delta(z;\tau)$ the polynomials $\delta_{T_0^\sharp}(z;\tau)$ which have been introduced in § 9.2.1. We also recall the notation (§ 9.2.1):

$$\delta_{T_0^\sharp}(z; \tau, \kappa) = \left(\frac{d}{dz}\right)^{\kappa} \delta_{T_0^\sharp}(z; \tau).$$

In the general case, define

$$\widetilde{\gamma}_{\tau\underline{t}}^{(\underline{\sigma}s)} = \sum_{\kappa=0}^{\sigma_0} \binom{\sigma_0}{\kappa} \delta_{T_0^\sharp}(s;\tau,\kappa)(t_m\beta_0)^{\sigma_0-\kappa} \left(\prod_{i=1}^{m-1}(t_i+t_m\beta_i)^{\sigma_i}\right) \alpha_1^{t_1 s}\cdots\alpha_m^{t_m s}.$$

In the homogeneous rational case, define,

$$\widetilde{\gamma}_{\tau\underline{t}}^{(\underline{\sigma}s)} = \frac{1}{\sigma_0!}\delta_{T_0^\sharp}(s;\tau,\sigma_0)\left(\prod_{i=1}^{m-1}\Delta(t_i b_m - t_m b_i;\sigma_i)\right)\alpha_1^{t_1 s}\cdots\alpha_m^{t_m s}.$$

Our fundamental matrix M will be

$$M = \left(\widetilde{\gamma}_{\tau\underline{t}}^{(\underline{\sigma}s)}\right)_{\substack{(\tau,\underline{t})\\(\underline{\sigma},s)}}$$

where $\tau \in \mathbb{N}$ and $\underline{t} \in \mathbb{Z}^m$ satisfy $0 \le \tau \le T_0$ and $|t_i| \le T_i \quad (1 \le i \le m)$, while $\underline{\sigma} \in \mathbb{N}^m$ and $s \in \mathbb{Z}$ range over the sets of elements such that $\|\underline{\sigma}\| \le (m+1)S_0$ and $|s| \le (m+1)S_1$ respectively.

This matrix M has L rows and $\binom{(m+1)S_0+m}{m}\bigl(2(m+1)S_1+1\bigr)$ columns. Assume that the number of columns is (slightly) larger than the number of rows. Our first goal is then to prove that M has rank $< L$.

For this purpose, consider a $L \times L$ submatrix of M and denote by Δ its determinant. Using Schwarz' Lemma we shall get an upper bound for $|\Delta|$. Assuming $|\Lambda|$ is sufficiently small, this upper bound will be sharp.

To make things simple, assume the conclusion of Theorem 9.1 does not hold, which means that $|\Lambda|$ is very small. Let us write the upper bound for $|\Delta|$ we derive as:

$$|\Delta| \le e^{-LV}.$$

Our main conditions on V occurs in Proposition 10.5 (with $n = m$ so far) and the optimal choice will be to take V close to $(L\log E)/S_0^{m-1}$, which will be close to $S_0S_1\log E$. Next, we use Liouville's inequality and deduce $\Delta = 0$. This implies that M has rank $< L$.

Now we invoke the multiplicity estimate: Theorem 8.1 produces a connected algebraic subgroup G^* of G of positive codimension (which means $G^* \ne G$) such that

$$\binom{S_0+n'}{n'}\operatorname{Card}\left(\frac{\Sigma+G^*}{G^*}\right)\mathcal{H}(G^*;\underline{T}) \le \mathcal{H}(G;\underline{T}), \tag{10.4}$$

where

$$\Sigma = \left\{(s,\alpha_1^s,\ldots,\alpha_m^s);\ s\in\mathbb{Z},\ |s|\le S_1\right\} \subset G(K) = K\times(K^\times)^m,$$

$$\underline{T} = (T_0;T_1,\ldots,T_m) \quad \text{and} \quad n' = \dim_{\mathbb{C}}\left(\frac{\mathcal{W}+T_e(G^*)}{T_e(G^*)}\right).$$

It will turn out (see Lemma 10.8) that it suffices to consider the subgroups G^* which satisfy not only this condition (10.4), but also two more properties:

- $T_e(G^*)$ is contained in the hyperplane $\mathcal{W}$.
- $\Sigma[2] \cap G^*(K) = \{e\}$.

We shall call such a G^* *an obstruction subgroup*. Since $T_e(G^*) \subset \mathcal{W}$, the codimension of G^* in G is $n' + 1$. Applying Lemma 7.8, one deduces from $\Sigma[2] \cap G^*(K) = \{e\}$ the relation

$$\mathrm{Card}\left(\frac{\Sigma + G^*}{G^*}\right) = 2S_1 + 1.$$

In the proofs of the qualitative result (Proposition 10.3), the mere existence of G^* was sufficient to get the conclusion. Here, we need to work more.

The optimal situation, from the point of view of the multiplicity estimate, is when the algebraic subgroup G^* produced by conclusion of Theorem 8.1 is the trivial subgroup $\{e\}$ of dimension 0. When this is the case, the proof can be greatly simplified.

This simplification takes place for instance when there is no *small* linear dependence relation between the coefficients β_j. Indeed, the tangent space of an obstruction subgroup $G^* = G_0^* \times G_1^*$ is contained in $\mathcal{W}$ (by definition), and $T_e(G_1^*)$ is a vector subspace of $\mathbb{C}^m$ rational over $\mathbb{Q}$. Any element $\underline{k} \in \mathbb{Q}^m \cap T_e(G_1^*)$ gives rise to a linear dependence relation

$$k_1\beta_1 + \cdots + k_{m-1}\beta_{m-1} = k_m.$$

We don't want to loose any generality and we must take into account the possibility that such relations take place. We start the proof by looking at a maximal G^* which might occur in the conclusion of the multiplicity estimate. We use this G^* to construct our auxiliary function, so that the multiplicity estimate at the end of the proof will provide an algebraic subgroup which is already under control.

Among the obstruction subgroups G^*, we select a maximal one, say $G^- = G_0^- \times G_1^-$. Here G_0^- is an algebraic subgroup of $G_0 = \mathbb{G}_a$, hence G_0^- is either $\{0\}$ or else $\mathbb{G}_a$, while G_1^- is a connected algebraic subgroup of $G_1 = \mathbb{G}_m^m$.

The idea (arising in [PW 1988a]) is to replace G by its quotient G/G^-. A monomial $X^\tau \underline{Y}^{\underline{t}} \in \mathbb{C}[X, Y_1^{\pm 1}, \ldots, Y_m^{\pm 1}]$ is constant on the classes modulo G^- if and only if $(\tau, \underline{t})$ satisfies

- If $G_0^- = \mathbb{G}_a$ then $\tau = 0$.
- For any $\underline{y} \in G_1^-$, $y_1^{t_1} \cdots y_m^{t_m} = 1$.

Hence we shall restrict the exponents $(\tau, \underline{t})$ to a subset $\mathcal{T}$ of tuples in $\mathbb{N} \times \mathbb{Z}^m$ with $0 \le \tau \le T_0$, $|t_i| \le T_i$, for which these conditions are satisfied. As we shall see (Lemma 10.10), the number L' of such tuples $(\tau, \underline{t})$ is essentially

$$\frac{\mathcal{H}(G; \underline{T})}{\mathcal{H}(G^-; \underline{T})}.$$

We repeat the above construction with the algebraic group G of dimension $m+1$ replaced by G/G^- of dimension say $n+1$. More precisely we replace M by a submatrix M' where, in the definition, we restrict the set of $(\tau, \underline{t})$ to $\mathcal{T}$. The new determinant Δ' has an absolute value bounded from above by $e^{-L'V'}$, where V' is essentially $(L' \log E)/S_0^{n-1}$, hence V' is not much smaller than $S_0 S_1 \log E$. Therefore the previous arithmetic estimates will be sufficient to deduce $\Delta' = 0$ and to conclude that M' has rank $< L'$. We are back to the multiplicity estimate, but now we have the extra information that the new obstruction subgroup G^* contains G^-. Since G^- was maximal among the obstruction subgroups, we deduce that G^* is G^- itself. We need of course to eliminate also G^- in order to get a contradiction. So we introduce a further modification in our construction: we replace, if necessary, $\mathcal{T}$ by a smaller subset $\mathcal{T}'$ so that L' is replaced by a number which is not too large compared to $S_0^n S_1 \log E$. This will enable us to exclude G^- as well.

10.2.2 Analytic Upper Bound for the Interpolation Determinant

We are looking for an upper bound for the absolute value of an interpolation determinant. Lemma 9.2 is sufficient not only for the proof of the transcendence result (§ 10.1), but also to achieve a nontrivial explicit estimate. However the measure of linear independence we would obtain this way would not be extremely good (it is comparable with the rough estimate of Chap. 7 in [W 1992] – see § 14.4.1). It will be more efficient to produce a new upper bound for the absolute value of the interpolation determinant, taking into account the fact that the points $s\underline{\eta}$ lie on a complex line, even if the derivatives involve several variables. This is the analog, for interpolation determinants, to the fact that Baker's original method (see § 10.3) can be explained without mentioning functions of several variables (see [W 1974], Chap. 8).

Another refinement is included in our analytic estimate: we introduce a subspace $\mathcal{U}$ of the ambient space $\mathbb{C}^d$ (here $d = m+1$) for which our functions satisfy $\varphi(\underline{z} + \underline{u}) = \varphi(\underline{z})$ for any $\underline{u} \in \mathcal{U}$. This means that the functions φ are in fact defined on $\mathbb{C}^d/\mathcal{U}$, and indeed this is what shall be used in the proof. But introducing $\mathcal{U}$ will be more convenient for our application, where $\mathcal{U}$ will be $T_e(G^-)$.

Our aim is to prove only the analytic result needed for this chapter, we are not looking for the most general statement (see Proposition 13.2): the next analytic estimate will not include Proposition 9.13, let alone Proposition 7.6.

Let m and n be rational integers with $0 \le n \le m$ and $m \ge 1$. Let $\mathcal{X}$, $\mathcal{U}$, $\mathcal{W}$ be vector subspaces of $\mathbb{C}^{m+1}$ with

$$\mathcal{U} \subset \mathcal{W}, \quad \mathcal{X} \subset \mathcal{W},$$

$$\dim_{\mathbb{C}}(\mathcal{X}) = 1, \quad \dim_{\mathbb{C}}(\mathcal{U}) = m - n \quad \text{and} \quad \dim_{\mathbb{C}}(\mathcal{W}) = m.$$

Assume the hyperplane $\mathcal{W}$ has an equation

$$\beta_0 z_0 + \cdots + \beta_{m-1} z_{m-1} = z_m.$$

Let L be a positive integer and $\varphi_1, \ldots, \varphi_L$ be entire functions in $\mathbb{C}^{m+1}$ satisfying

$$\varphi_\lambda(\underline{z} + \underline{u}) = \varphi_\lambda(\underline{z})$$

for any $\underline{u} \in \mathcal{U}$, any $\underline{z} \in \mathbb{C}^{m+1}$ and any $\lambda = 1, \ldots, L$.

Let $\underline{\zeta}_1, \ldots, \underline{\zeta}_L$ be elements of $\mathcal{X}$ and S_0 a positive integer. Further, for $1 \leq \lambda \leq L$ and $1 \leq \mu \leq L$, let $\delta_{\lambda\mu}$ be a complex number. Recall that for $\underline{\sigma} \in \mathbb{N}^m$, $\mathcal{D}_{\underline{w}}^{\underline{\sigma}}$ has been defined in § 10.2.1:

$$\mathcal{D}_{\underline{w}}^{\underline{\sigma}} = \left(\frac{\partial}{\partial z_0} + \beta_0 \frac{\partial}{\partial z_m}\right)^{\sigma_0} \cdots \left(\frac{\partial}{\partial z_{m-1}} + \beta_{m-1} \frac{\partial}{\partial z_m}\right)^{\sigma_{m-1}}.$$

Dealing with the general case of Theorem 9.1, we shall consider numbers $\mathcal{D}_{\underline{w}}^{\underline{\sigma}_\mu} \varphi_\lambda(\underline{\zeta}_\mu)$, where $\underline{\sigma}_1, \ldots, \underline{\sigma}_L$ are L elements of $\mathbb{N}^m$ satisfying $\|\underline{\sigma}_\mu\| \leq S_0$ $(1 \leq \mu \leq L)$. But in view of the application of our analytic estimate to the homogeneous rational case, we consider a slightly more general situation. For $1 \leq \mu \leq L$ and $\underline{\sigma} \in \mathbb{N}^m$ with $\|\underline{\sigma}\| \leq S_0$, let $q_{\mu\underline{\sigma}}$ be a complex number. Define

$$\varphi_{\lambda\mu} = \sum_{\|\underline{\sigma}\| \leq S_0} q_{\mu\underline{\sigma}} \mathcal{D}_{\underline{w}}^{\underline{\sigma}} \varphi_\lambda \qquad (1 \leq \mu \leq L).$$

Let $E, V, M_1, \ldots, M_L$ be positive real numbers with $E > 1$ and let ϵ be a complex number satisfying

$$|\epsilon| \leq e^{-V}.$$

Assume, for $1 \leq \lambda \leq L$,

$$M_\lambda \geq \log \sup_{|z|=E} \max_{1 \leq \mu \leq L} |\varphi_{\lambda\mu}(z\underline{\zeta}_\mu)| \quad \text{and} \quad M_\lambda \geq \log \max_{1 \leq \mu \leq L} |\delta_{\lambda\mu}|.$$

Proposition 10.5. *Assume*

$$2(S_0 + 1) \log E + \log(2L) + M_\lambda \leq \frac{V}{4} \quad (1 \leq \lambda \leq L)$$

and

$$L \geq 2\binom{S_0 + n - 1}{n - 1} \cdot \frac{V}{\log E}.$$

Then the determinant

$$\Delta = \det\left(\varphi_{\lambda\mu}(\underline{\zeta}_\mu) + \epsilon\delta_{\lambda\mu}\right)_{1 \leq \lambda, \mu \leq L}$$

has absolute value bounded by

$$|\Delta| \leq e^{-LV/4}.$$

The proof of Proposition 10.5 requires two preliminary lemmas.

Let I a subset of $\{1, \ldots, L\}$. For $1 \leq \lambda, \mu \leq L$, let $d_{\lambda\mu}^{(I)}$ be the entire function of one variable which is defined by

$$d_{\lambda\mu}^{(I)}(z) = \begin{cases} \varphi_{\lambda\mu}(z\underline{\zeta}_\mu) & \text{for } \lambda \in I, \\ \delta_{\lambda\mu} & \text{for } \lambda \notin I. \end{cases}$$

Set

$$D_I(z) = \det\Big(d_{\lambda\mu}^{(I)}(z)\Big)_{1\le\lambda,\mu\le L}.$$

Lemma 10.6. *The function $D_I(z)$ has a zero at $z = 0$ of multiplicity at least $\Theta(n, S_0, |I|) - |I|S_0$.*

Proof. Choose a basis of $\mathbb{C}^{m+1}/\mathcal{U}$ giving an isomorphism $\iota: \mathbb{C}^{m+1}/\mathcal{U} \to \mathbb{C}^{n+1}$ and denote by $\pi: \mathbb{C}^{m+1} \to \mathbb{C}^{n+1}$ the composition of ι with the canonical surjection $\mathbb{C}^{m+1} \to \mathbb{C}^{m+1}/\mathcal{U}$. The relations $\varphi_\lambda(\underline{z} + \underline{u}) = \varphi_\lambda(\underline{z})$ mean that there exist entire functions $\widetilde{\varphi}_\lambda: \mathbb{C}^{n+1} \to \mathbb{C}$ such that $\widetilde{\varphi}_\lambda \circ \pi = \varphi_\lambda$ $(1 \le \lambda \le L)$. Define $\underline{\widetilde{w}}_k = \pi(\underline{w}_k)$ $(1 \le k \le m)$, $\widetilde{\underline{\zeta}}_\mu = \pi(\underline{\zeta}_\mu)$ $(1 \le \mu \le L)$,

$$\widetilde{\varphi}_{\lambda\mu} = \sum_{\|\underline{\sigma}\|\le S_0} q_{\mu\underline{\sigma}} \mathcal{D}_{\underline{\widetilde{w}}}^{\underline{\sigma}} \widetilde{\varphi}_\lambda \qquad (1 \le \lambda, \mu \le L).$$

and

$$\widetilde{d}_{\lambda\mu}^{(I)}(z) = \begin{cases} \widetilde{\varphi}_{\lambda\mu}(\widetilde{\underline{\zeta}}_\mu z) & \text{for } \lambda \in I, \\ \delta_{\lambda\mu} & \text{for } \lambda \notin I. \end{cases}$$

Then

$$D_I(z) = \det\Big(\widetilde{d}_{\lambda\mu}^{(I)}(z)\Big)_{1\le\lambda,\mu\le L}.$$

Therefore we may assume, without loss of generality, $\mathcal{U} = 0$ and $m = n$.

For $m = 1$, the proof is the same as for Lemma 2.8. For $m \ge 2$ we repeat the proof of Lemmas 7.2 and 9.2 with a tiny modification.

After a change of variable (which does not affect the multiplicity) we may assume $\mathcal{X} = \mathbb{C} \times \{0\}^{m-1}$. Using the Taylor expansion at the origin of each φ_λ, we are reduced to the case $\varphi_\lambda(\underline{z}) = \underline{z}^{\underline{\kappa}_\lambda}$. The point is that each $\mathcal{D}_{\underline{w}}^{\underline{\sigma}}\varphi_\lambda$ vanishes on $\mathcal{X}$, unless $\sigma_i \ge \kappa_{\lambda i}$ for $2 \le i \le m$, in which case $\kappa_{\lambda 2} + \cdots + \kappa_{\lambda m} \le S_0$. □

Repeating the proof of Lemma 7.5, we deduce:

Lemma 10.7. *We have*

$$\log|\widetilde{D}_I(1)| \le -\Theta(n, S_0, |I|)\log E + \log(L!) + |I| \cdot S_0 \log E + M_1 + \cdots + M_L.$$

Proof of Proposition 10.5. (Compare with the proof of Proposition 7.6). By Lemma 10.7, we can use Lemma 7.4 with $r = 1$ and

$$\chi_0 = \frac{1}{2}(\log E)\binom{S_0 + n - 1}{n - 1}^{-1},$$

$$\chi_1 = V - \chi_0 + S_0 \log E + \frac{1}{2n}(S_0 + n) \leq V + 2S_0 \log E$$

and

$$\chi_2 = \log(L!) + M_1 + \cdots + M_L.$$

The assumption $2(S_0+1)\log E + \log(2L) + M_\lambda \leq V/4$ implies $\chi_2 + L\log 2 \leq LV/4$.

□

10.2.3 Obstruction Subgroups

We explain here how to apply Theorem 8.1 in order to construct a matrix with maximal rank.

Let K be an algebraically closed field of zero characteristic, m a positive integer, $\alpha_1, \ldots, \alpha_m$ elements of $K^\times$ and $\beta_0, \ldots, \beta_{m-1}$ elements of K. Further let $T_0, T_1, \ldots, T_m, T_1', \ldots, T_m'$, S_0 and S_1 be nonnegative integers with $1 \leq T_i' \leq T_i$ for $1 \leq i \leq m$.

For $\underline{t} \in \mathbb{Z}^m$ define

$$\underline{\xi}_{\underline{t}} = \bigl(t_m\beta_0, t_1 + t_m\beta_1, \ldots, t_{m-1} + t_m\beta_{m-1}\bigr) \in K^m.$$

Let $\delta^{(1)}(z;\tau)$ $(0 \leq \tau \leq T_0)$ denote a basis of the space of polynomials in $K[z]$ of degree $\leq T_0$. For $\kappa \in \mathbb{N}$, define

$$\delta^{(1)}(z;\tau,\kappa) = \left(\frac{d}{dz}\right)^{\kappa} \delta^{(1)}(z;\tau).$$

Next, let $\delta^{(2)}(\underline{z};\sigma)$ $(1 \leq \sigma \leq \binom{(m+1)S_0+m}{m})$ denote a basis of the space of polynomials in $K[z_0, \ldots, z_{m-1}]$ of total degree $\leq (m+1)S_0$. For $\kappa \in \mathbb{N}$, define

$$\delta^{(2)}(\underline{z};\sigma,\kappa) = \left(\frac{\partial}{\partial z_0}\right)^{\kappa} \delta^{(2)}(\underline{z};\sigma).$$

For instance when $\delta^{(2)}(\underline{z};\sigma)$ is the element $z_0^{\sigma_0} \cdots z_{m-1}^{\sigma_{m-1}}$ of the standard basis, then

$$\frac{1}{\kappa!}\delta^{(2)}(\underline{\xi}_{\underline{t}};\sigma,\kappa)$$

is nothing else than

$$\binom{\sigma_0}{\kappa}(t_m\beta_0)^{\sigma_0-\kappa}(t_1+t_m\beta_1)^{\sigma_1}\cdots(t_{m-1}+t_m\beta_{m-1})^{\sigma_{m-1}}.$$

Denote by G the algebraic group $G_0 \times G_1$ with $G_0 = \mathbb{G}_{\mathrm{a}}$ and $G_1 = \mathbb{G}_{\mathrm{m}}^m$. Let $G^- = G_0^- \times G_1^-$ be a connected algebraic subgroup of G, where G_0^- is either $\mathbb{G}_{\mathrm{a}}$ or $\{0\}$, and where G_1^- is a connected algebraic subgroup of G_1. We shall use Hilbert-Samuel's polynomial

$$\mathcal{H}(G; D_0, D_1, \ldots, D_{d_1})$$

with the parameters $D_0, D_1, \dots, D_{d_1}$ replaced either by $\underline{T} = (T_0, T_1, \dots, T_m)$ or else by $\underline{T}' = (T_0, T_1', \dots, T_m')$.

Define

$$\Sigma = \left\{(s, \alpha_1^s, \dots, \alpha_m^s)\,;\ s \in \mathbb{Z},\ |s| \le S_1\right\} \subset G(K).$$

In K^{m+1} (which we identify with $T_e(G)$), let $\mathcal{W}$ be the hyperplane defined by the equation

$$\beta_0 z_0 + \beta_1 z_1 + \cdots + \beta_{m-1} z_{m-1} = z_m.$$

A basis of $\mathcal{W}$ is given by the m column vectors of the matrix

$$\begin{pmatrix} & \mathrm{I}_m & \\ \beta_0 & \cdots & \beta_{m-1} \end{pmatrix}$$

which are denoted by $\underline{w}_0, \dots, \underline{w}_{m-1}$. We assume

$$T_e(G^-) \subset \mathcal{W} \quad \text{and} \quad \Sigma[2] \cap G^-(K) = \{e\}.$$

We denote by $n+1$ the codimension of G^- in G. Let $\mathcal{T}'$ be a subset of $\mathbb{N} \times \mathbb{Z}^m$. We assume that for any $(\tau, \underline{t}) \in \mathcal{T}'$, we have

$$0 \le \tau \le T_0, \quad |t_i| \le T_i' \quad (1 \le i \le m).$$

We assume also

$$\begin{cases} \tau = 0 & \text{if } G_0^- = \mathbb{G}_{\mathrm{a}}, \\ y_1^{t_1} \cdots y_m^{t_m} = 1 & \text{for any } \underline{y} \in G_1^-. \end{cases}$$

For $(\tau, \underline{t}) \in \mathcal{T}'$ and $(\sigma, s) \in \mathbb{N} \times \mathbb{Z}$ with $1 \le \sigma \le \binom{(m+1)S_0+m}{m}$ and $|s| \le (m+1)S_1$, define

$$\widetilde{\gamma}_{\tau\underline{t}}^{(\sigma s)} = \sum_{\kappa \ge 0} \frac{1}{\kappa!} \delta^{(1)}(s; \tau, \kappa) \delta^{(2)}(\underline{\xi}_{\underline{t}}; \sigma, \kappa) \cdot \alpha_1^{t_1 s} \cdots \alpha_m^{t_m s}.$$

Notice that the only non-vanishing terms in the sum over κ occur for

$$\kappa \le \min\{T_0, (m+1)S_0\}.$$

We build a matrix with these numbers:

$$M' = \left(\widetilde{\gamma}_{\tau\underline{t}}^{(\sigma s)}\right)_{\substack{(\tau,\underline{t}) \\ (\sigma,s)}}.$$

This is compatible with the notation of § 10.2.1. The number of rows of M' is $\mathrm{Card}(\mathcal{T}')$, and the number of columns is

$$\binom{(m+1)S_0+m}{m}\bigl(2(m+1)S_1+1\bigr).$$

The following assumptions will be take place until the end of § 10.2.3:

$$(S_0+1)(2S_1+1) > (m+1)!2^m T_0, \qquad S_0 \ge 2(m+1) \max_{1 \le i \le m} T_i,$$

and $\alpha_1, \dots, \alpha_m$ generate a multiplicative subgroup of $K^\times$ of rank $\ge m-1$.

Further, in the special case where $\beta_0 = 0$ and $\alpha_1, \ldots, \alpha_m$ multiplicatively dependent, we assume that $\beta_1 z_1 + \cdots + \beta_{m-1} z_{m-1} = z_m$ is not the equation of the tangent space of an algebraic subgroup of $\mathbb{G}_{\mathrm{m}}^m$ containing $(\alpha_1^s, \ldots, \alpha_m^s)$ for some $s \in \mathbb{Z}$, $s \neq 0$.

We deduce from Theorem 8.1 the following result:

Lemma 10.8. *If M' has rank $<$ Card$(\mathcal{T}')$, then there exists a connected algebraic subgroup G^* of G, distinct from G, which contains G^-, such that*

$$T_e(G^*) \subset \mathcal{W}, \qquad \Sigma[2] \cap G^*(K) = \{e\}$$

and

$$\binom{S_0 + n'}{n'}(2S_1 + 1)\mathcal{H}(G^*; \underline{T}') \leq \mathcal{H}(G, \underline{T}'),$$

where $n' + 1$ is the codimension of G^* in G.

Proof. For $1 \leq \sigma \leq \binom{(m+1)S_0+m}{m}$, write

$$\delta^{(2)}(\underline{z}, \sigma) = \sum_{\|\underline{\sigma}'\| \leq (m+1)S_0} q_{\underline{\sigma}'\sigma} \underline{z}^{\underline{\sigma}'}.$$

This defines a regular square matrix of size $\binom{(m+1)S_0+m}{m}$:

$$Q = \Big(q_{\underline{\sigma}'\sigma}\Big)_{\substack{\|\underline{\sigma}'\| \leq (m+1)S_0 \\ 1 \leq \sigma \leq \binom{(m+1)S_0+m}{m}}}.$$

For $\kappa \in \mathbb{N}$ a simple computation yields

$$\delta^{(2)}(\underline{\xi}_{\underline{t}}, \sigma, \kappa) =$$

$$\sum_{\|\underline{\sigma}'\| \leq (m+1)S_0} q_{\underline{\sigma}'\sigma} \frac{\sigma_0'!}{(\sigma_0' - \kappa)!}(t_m\beta_0)^{\sigma_0' - \kappa}(t_1 + t_m\beta_1)^{\sigma_1'} \cdots (t_{m-1} + t_m\beta_{m-1})^{\sigma_{m-1}'}.$$

On the other hand, for $\underline{\sigma}' \in \mathbb{N}^m$, using the derivations $\mathcal{D}_i$ defined in § 10.2.1, we have

$$\mathcal{D}^{\underline{\sigma}'}\big(\delta^{(1)}(X;\tau)\underline{Y}^{\underline{t}}\big) = \sum_{\kappa=0}^{\sigma_0'} \binom{\sigma_0'}{\kappa}(t_m\beta_0)^{\sigma_0'-\kappa}\left(\prod_{i=1}^{m-1}(t_i + t_m\beta_i)^{\sigma_i'}\right)\delta^{(1)}(X;\tau,\kappa)\underline{Y}^{\underline{t}}.$$

We deduce

$$\widetilde{\gamma}_{\tau\underline{t}}^{(\sigma s)} = \sum_{\|\underline{\sigma}'\| \leq (m+1)S_0} q_{\underline{\sigma}'\sigma}\mathcal{D}^{\underline{\sigma}'}\big(\delta^{(1)}(X;\tau)\underline{Y}^{\underline{t}}\big)(s, \alpha_1^s, \ldots, \alpha_m^s).$$

Since M' has rank $<$ Card$(\mathcal{T}')$, there exist elements $c_{\tau\underline{t}}$ in K, not all of which are zero, such that

$$\sum_{(\tau,\underline{t}) \in \mathcal{T}'} c_{\tau\underline{t}}\widetilde{\gamma}_{\tau\underline{t}}^{(\sigma s)} = 0$$

for any $(\sigma, s) \in \mathbb{N} \times \mathbb{Z}$ with

$$1 \leq \sigma \leq \binom{(m+1)S_0 + m}{m} \quad \text{and} \quad |s| \leq (m+1)S_1.$$

These relations show that the polynomial

$$\sum_{\|\underline{\sigma}'\| \leq (m+1)S_0} q_{\underline{\sigma}'\sigma} \mathcal{D}^{\underline{\sigma}'} \Big(\sum_{(\tau,\underline{t}) \in \mathcal{T}'} c_{\tau\underline{t}} \big(\delta^{(1)}(X;\tau) \underline{Y}^{\underline{t}} \big) \Big)$$

vanishes at $(s, \alpha_1^s, \ldots, \alpha_m^s)$. Since Q is a regular matrix, we deduce

$$\mathcal{D}^{\underline{\sigma}'} \Big(\sum_{(\tau,\underline{t}) \in \mathcal{T}'} c_{\tau\underline{t}} \big(\delta^{(1)}(X;\tau) \underline{Y}^{\underline{t}} \big) \Big) (s, \alpha_1^s, \ldots, \alpha_m^s) = 0$$

for all $\underline{\sigma}'$ with $\|\underline{\sigma}'\| \leq (m+1)S_0$, which means that the polynomial

$$\sum_{(\tau,\underline{t}) \in \mathcal{T}'} c_{\tau\underline{t}} \delta^{(1)}(X;\tau) \underline{Y}^{\underline{t}}$$

in $K[X, Y_1^{\pm 1}, \ldots, Y_m^{\pm 1}]$ vanishes to order $\geq (m+1)S_0$ along $\mathcal{W}$ at each point of $\Sigma[m+1]$.

The hypotheses of Theorem 8.1 are therefore satisfied with $G^+ = G$. It follows that there exists a connected algebraic subgroup G^* of G of dimension $\leq m$, containing G^-, such that, if we set

$$\ell_0' = \dim_K \left(\frac{\mathcal{W} + T_e(G^*)}{T_e(G^*)} \right),$$

then

$$\binom{S_0 + \ell_0'}{\ell_0'} \operatorname{Card}\left(\frac{\Sigma + G^*}{G^*} \right) \mathcal{H}(G^*; \underline{T}') \leq \mathcal{H}(G; \underline{T}').$$

By Lemma 5.8,

$$\mathcal{H}(G; \underline{T}') = (m+1)! 2^m T_0 T_1' \cdots T_m'.$$

Denote by $m - n'$ the dimension of G^*, with $0 \leq n' \leq m$.

Let us check $\Sigma[2] \cap G^*(K) = \{e\}$. Indeed, if

$$(s, \alpha_1^s, \ldots, \alpha_m^s) \in G^*(K) \quad \text{for some } s \in \mathbb{Z} \text{ with } 0 < |s| \leq 2S_1,$$

then $G^* = \mathbb{G}_a \times G_1^*$ where G_1^* is a connected algebraic subgroup of G_1 of codimension 1 (recall that $\alpha_1, \ldots, \alpha_m$ generate a multiplicative subgroup of $K^\times$ of rank $\geq m-1$). Since $(\alpha_1^s, \ldots, \alpha_m^s) \in G_1^*(K)$, we have by assumption $T_e(G^*) \not\supset \mathcal{W}$, hence $T_e(G^*) + \mathcal{W} = T_e(G)$ and $\ell_0' = 1$. Since G_1^* has codimension 1, we have, by Proposition 5.7,

$$\mathcal{H}(G^*; \underline{T}') \geq m! 2^{m-1} \cdot \frac{T_0 T_1' \cdots T_m'}{\max_{1 \leq i \leq m} T_i'},$$

and we get a contradiction with our assumptions $T_i' \le T_i$ and $S_0 \ge 2(m+1)T_i$ for $1 \le i \le m$.

As already mentioned, from $\Sigma[2] \cap G^*(K) = \{e\}$ we deduce, by means of Lemma 7.8,

$$\mathrm{Card}\left(\frac{\Sigma + G^*}{G^*}\right) = 2S_1 + 1.$$

Next we check $T_e(G^*) \subset \mathcal{W}$. Otherwise $T_e(G^*) + \mathcal{W} = T_e(G)$ and $\ell_0' = n' + 1$ (recall that G^* has dimension $m - n'$). From Proposition 5.14 we derive

$$\mathcal{H}(G^*; \underline{T}') \ge (m+1)!2^m \min_{\underline{i}} T_{i_1}' \cdots T_{i_{m-n'}}',$$

where $\underline{i}$ runs over the subsets $\{i_1, \ldots, i_{m-n'}\}$ of $\{0, 1, \ldots, m\}$ with $m - n'$ elements. In this case the conclusion of the multiplicity estimate implies

$$\binom{S_0 + k}{k}(2S_1 + 1) \le (m+1)!2^m \max_{0 \le i_1 < \cdots < i_k \le m} T_{i_1} \cdots T_{i_k}.$$

with $k = n' + 1$, and this is not compatible with the assumptions

$$(S_0 + 1)(2S_1 + 1) > (m+1)!2^m T_0 \quad \text{and} \quad S_0 \ge 2(m+1) \max_{1 \le i \le m} T_i.$$

Hence we have $T_e(G^*) \subset \mathcal{W}$, and therefore $\ell_0' = n'$. The conclusion of the multiplicity estimate now reads

$$\binom{S_0 + n'}{n'}(2S_1 + 1)\mathcal{H}(G^*; \underline{T}') \le \mathcal{H}(G; \underline{T}').$$

This completes the proof of Lemma 10.8. □

Proposition 10.9. *Assume further that for any connected algebraic subgroup G^* of G, of codimension $n' + 1$ with $0 \le n' < n$, containing G^-, and satisfying*

$$T_e(G^*) \subset \mathcal{W} \quad \textit{and} \quad \Sigma[2] \cap G^*(K) = \{e\},$$

the estimate

$$\binom{S_0 + n'}{n'}(2S_1 + 1)\mathcal{H}(G^*; \underline{T}) > \mathcal{H}(G, \underline{T})$$

holds. Assume furthermore

$$\binom{S_0 + n}{n}(2S_1 + 1)\mathcal{H}(G^-; \underline{T}') > \mathcal{H}(G, \underline{T}').$$

Then the matrix M' has rank $\mathrm{Card}(\mathcal{T}')$.

Proof. By contradiction, assume that M' has not maximal rank. We use Lemma 10.8: there exists a connected algebraic subgroup G^* of G, of codimension $n' + 1$ in G, which contains G^-, such that

$$T_e(G^*) \subset \mathcal{W}, \qquad \Sigma[2] \cap G^*(K) = \{e\}$$

and

$$\binom{S_0 + n'}{n'}(2S_1 + 1)\mathcal{H}(G^*; \underline{T}') \le \mathcal{H}(G, \underline{T}').$$

Proposition 5.14 shows that each of the functions

$$T_i \longmapsto \frac{\mathcal{H}(G^*;\ T_0, T_1, \dots, T_{d_1})}{\mathcal{H}(G;\ T_0, T_1, \dots, T_{d_1})}$$

is a fractional linear transformation, hence is non-increasing. Therefore we also have

$$\binom{S_0 + n'}{n'}(2S_1 + 1)\mathcal{H}(G^*; \underline{T}) \le \mathcal{H}(G, \underline{T}).$$

By assumption this implies $G^* = G^-$. But now the conclusion of Lemma 10.8 becomes

$$\binom{S_0 + n}{n}(2S_1 + 1)\mathcal{H}(G^-; \underline{T}') \le \mathcal{H}(G, \underline{T}'),$$

which is not compatible with the last assumption of Proposition 10.9. □

10.2.4 Estimating a Hilbert Function

In order to apply Proposition 10.9 (with $K = \mathbb{C}$), we need to introduce G^-. Consider the so-called obstruction subgroups, which are the connected algebraic subgroups G^* of G of positive codimension $n' + 1$ for which

$$T_e(G^*) \subset \mathcal{W}, \quad \Sigma[2] \cap G^*(K) = \{e\}$$

and

$$\binom{S_0 + n'}{n'}(2S_1 + 1)\mathcal{H}(G^*;\ \underline{T}) \le \mathcal{H}(G;\ \underline{T}).$$

If there is no such G^*, we set $G^- = \{e\}$. Otherwise we define G^- as a maximal obstruction subgroup[16]. Let $n + 1$ be the codimension of G^- in G, so that either $n = m$ or else

$$\binom{S_0 + n}{n}(2S_1 + 1)\mathcal{H}(G^-;\ \underline{T}) \le \mathcal{H}(G;\ \underline{T}).$$

This algebraic group G^- can be written $G_0^- \times G_1^-$, where G_0^- is either $\{0\}$ or $\mathbb{G}_a$, and where G_1^- is a connected algebraic subgroup of $G_1 = \mathbb{G}_m^m$. By § 5.3, there is a subgroup Φ of $\mathbb{Z}^m$ such that $G_1^- = T_\Phi$, and the rank δ of Φ is

$$\delta = \operatorname{codim}_{G_1} G_1^- = \begin{cases} n + 1 & \text{if } G_0^- = \mathbb{G}_a, \\ n & \text{if } G_0^- = \{0\}. \end{cases}$$

For any $\underline{\alpha} \in \Phi$ and any $\underline{y} \in G_1^-$ we have

[16] The reader may first restrict to the case where $G^- = \{e\}$, which is much easier.

$$y_1^{\alpha_1} \cdots y_m^{\alpha_m} = 1$$

Denote by $\mathcal{T}_1$ the set of $\underline{\alpha} \in \Phi$ such that $|\alpha_i| \leq T_i$ for $i = 1, \ldots, m$. We need a lower bound for the number of elements in the finite set $\mathcal{T}_1$, which is the intersection of Φ with the polydisc

$$\left\{\underline{x} \in \mathbb{R}^m \; ; \; |x_i| \leq T_i \; (1 \leq i \leq m)\right\}$$

of $\mathbb{R}^m$, and Φ is a lattice in a vector subspace of $\mathbb{R}^m$. Recall the notation $\chi(\underline{T})$ of Proposition 5.7 for the number of cosets of Φ of the form $\underline{\alpha} + \Phi$ with $\underline{\alpha} \in \mathbb{Z}^m[\underline{T}]$.

Using Lemma 7.8 for the mapping

$$\begin{array}{ccc} \prod_{i=1}^{m}[0, T_i] & \longrightarrow & \dfrac{\mathbb{Z}^m}{\Phi} \\ \underline{\alpha} & \longmapsto & \underline{\alpha} + \Phi \end{array}$$

yields

$$\chi(\underline{T})\mathrm{Card}(\mathcal{T}_1) \geq (T_1 + 1) \cdots (T_m + 1).$$

Hence it suffices now to get an upper bound for $\chi(\underline{T})$. The number $\chi(\underline{T})$ is nothing else than Hilbert's function $H(G_1^-; \underline{T})$, and we need to compare it with Hilbert-Samuel's polynomial $\mathcal{H}(G_1^-; \underline{T})$. Here is Lemma 3.2 of [PW 1988a].

Lemma 10.10. *There exists a constant c which depends only on m such that*

$$\chi(\underline{T}) \leq c\mathcal{H}(G_1^-; \underline{T}).$$

The proof of Lemma 10.10 requires some preparation.

Let $\mathcal{V}$ be an Euclidean vector space of dimension m over $\mathbb{R}$. Denote by $(\epsilon_1, \ldots, \epsilon_m)$ an orthonormal basis. Let L be a discrete subgroup of $\mathcal{V}$ of rank ℓ. Denote by $(\lambda_1, \ldots, \lambda_\ell)$ a basis of L over $\mathbb{Z}$ and by $\mathcal{W}$ the vector space they span.

Let $(e_1, \ldots, e_\ell)$ be an orthonormal basis of $\mathcal{W}$. In $\bigwedge^\ell \mathcal{W}$, write

$$\lambda_1 \wedge \cdots \wedge \lambda_\ell = \mu \cdot e_1 \wedge \cdots \wedge e_\ell$$

with $\mu \in \mathbb{R}$. The number $|\mu|$ does not depend on the choice of the bases $(\lambda_1, \ldots, \lambda_\ell)$ and $(e_1, \ldots, e_\ell)$. Define

$$\mathrm{vol}(L) = |\mu|.$$

A *fundamental domain* for $\mathcal{W}/L$ is

$$\mathcal{P} = \left\{t_1\lambda_1 + \cdots + t_\ell\lambda_\ell \; ; \; \underline{t} \in \mathbb{R}^\ell, \; 0 \leq t_j < 1 \; (1 \leq j \leq \ell)\right\},$$

which means that $\mathcal{W}$ is the disjoint union of the sets $x + \mathcal{P}$, $x \in L$. The Euclidean volume of $\mathcal{P}$ in $\mathcal{W}$ is nothing else than $\mathrm{vol}(L)$. This is the key for estimating from above the number of elements in some finite subset E of L: it suffices to know an upper bound for the volume of the (disjoint) union of $x + \mathcal{P}$ where x ranges over E, and also a lower bound for $\mathrm{vol}(L)$.

Let $\mathsf{L} \in \mathrm{Mat}_{m\times\ell}(\mathbb{R})$ be the matrix whose column vectors are the components of $\lambda_1, \dots, \lambda_\ell$ in the basis $(\epsilon_1, \dots, \epsilon_m)$:

$$\lambda_j = \sum_{i=1}^{m} \lambda_{ij}\epsilon_i, \qquad \mathsf{L} = \bigl(\lambda_{ij}\bigr)_{\substack{1\le i\le m\\ 1\le j\le \ell}}.$$

Let us check

$$\mathrm{vol}(L)^2 = \det\bigl({}^{t}\mathsf{L}\,\mathsf{L}\bigr).$$

Denote by $\langle\cdot,\cdot\rangle$ the scalar product on $\mathcal{V}$. A scalar product on $\bigwedge^\ell \mathcal{W}$ is defined by

$$\langle x_1 \wedge \cdots \wedge x_\ell, y_1 \wedge \cdots \wedge y_\ell\rangle = \det\Bigl(\langle x_j, y_k\rangle\Bigr)_{1\le j,k\le \ell}.$$

From

$$\langle\lambda_j, \lambda_k\rangle = \sum_{i=1}^{m} \lambda_{ij}\lambda_{ik}$$

we deduce

$$\mu^2 = \langle\lambda_1 \wedge \cdots \wedge \lambda_\ell, \lambda_1 \wedge \cdots \wedge \lambda_\ell\rangle = \det\bigl({}^{t}\mathsf{L}\,\mathsf{L}\bigr).$$

This proves our claim.

Following [Sc 1980], Chap. IV, § 6, denote by $C(m,\ell)$ the subset of $[1,m]^\ell$ which consists of sequences $\underline{i} = (i_1, \dots, i_\ell)$ satisfying $1 \le i_1 < \cdots < i_\ell \le m$. For each $\underline{i} \in C(m,\ell)$, define $\mu_{\underline{i}} \in \mathbb{R}$ by

$$\lambda_1 \wedge \cdots \wedge \lambda_\ell = \mu_{\underline{i}} \cdot \epsilon_{i_1} \wedge \cdots \wedge \epsilon_{i_\ell}.$$

Lemma 10.11. *For each $\underline{i} \in C(m,\ell)$, $\mu_{\underline{i}}$ is the determinant of the square $\ell \times \ell$ matrix*

$$\bigl(\lambda_{i_k j}\bigr)_{1\le k,j\le \ell} = \begin{pmatrix} \lambda_{i_1 1} & \cdots & \lambda_{i_1 \ell}\\ \vdots & \ddots & \vdots\\ \lambda_{i_\ell 1} & \cdots & \lambda_{i_\ell \ell}\end{pmatrix}$$

and

$$\mathrm{vol}(L)^2 = \sum_{\underline{i}\in C(m,\ell)} |\mu_{\underline{i}}|^2.$$

Proof of Lemma 10.11. This follows from the Cauchy-Binet formula (see [Bou 1985], Algèbre, Chap. III, or J. Fresnel, Algèbre des Matrices, Hermann, Actualités Scientifiques et Industrielles **1439**, 1997): for any commutative ring A and any matrices $X \in \mathrm{Mat}_{n\times p}(A)$, $Y \in \mathrm{Mat}_{m\times n}(A)$, for any q in the range $1 \le q \le \min\{m,n,p\}$ and any $\underline{k} \in C(n,q)$, $\underline{\ell} \in C(q,n)$, the matrix $Z = XY$ satisfies

$$Z_{\underline{k}\underline{\ell}} = \sum_{\underline{h}\in C(p,q)} X_{\underline{k}\underline{h}} Y_{\underline{h}\underline{\ell}},$$

where the notation $X_{\underline{k}\underline{h}}$ stands for the determinant of the $q \times q$ matrix

$$\left(x_{ij}\right)_{\substack{i\in\underline{k}\\ j\in\underline{h}}}.$$

In particular for X and Y in $\mathrm{Mat}_{n\times m}(A)$

$$\det({}^{\mathrm{t}}XY) = \sum_{\underline{h}\in C(n,m)} X_{\underline{h}} Y_{\underline{h}}$$

where $X_{\underline{h}}$ stands for $X_{\underline{h}\underline{\ell}}$ with $\underline{\ell} = \{1,\ldots,m\}$. □

We wish to compare the volume of two discrete subgroups $L_0 \subset M$ of $\mathcal{V}$.

First, let $L_0 \subset L$ be two discrete subgroups of $\mathcal{V}$ of the same rank. Let us check

$$\mathrm{vol}(L_0) = [L : L_0]\mathrm{vol}(L).$$

Indeed, since L is a free $\mathbb{Z}$-module, the elementary divisors Theorem (already referred to in the proof of Theorem 5.13) shows that there exists a basis $(\lambda_1,\ldots,\lambda_\ell)$ of L and positive rational integers $a_1,\ldots,a_\ell$ such that $(a_1\lambda_1,\ldots,a_\ell\lambda_\ell)$ is a basis of L_0. Clearly

$$[L : L_0] = a_1\cdots a_\ell.$$

On the other hand if L (resp. L_0) is the matrix in $\mathrm{Mat}_{m\times\ell}(\mathbb{R})$ whose column vectors are the components of this basis of L (resp. L_0) in the basis $(\epsilon_1,\ldots,\epsilon_m)$, then

$$\mathrm{vol}(L_0)^2 = \det\bigl({}^{\mathrm{t}}\mathsf{L}_0\,\mathsf{L}_0\bigr) = (a_1\cdots a_\ell)^2\det\bigl({}^{\mathrm{t}}\mathsf{L}\,\mathsf{L}\bigr) = (a_1\cdots a_\ell)^2\mathrm{vol}(L)^2.$$

Our claim follows.

Next let $L \subset M$ be two discrete subgroups of $\mathcal{V}$ such that the quotient M/L has no torsion and M has rank m. Denote by $\mathcal{W}$ the subspace of $\mathcal{V}$ spanned by L. Let $(\lambda_1,\ldots,\lambda_\ell)$ be a basis of L over $\mathbb{Z}$ and $\lambda_{\ell+1},\ldots,\lambda_m$ be elements in M whose images modulo L give a basis of the free $\mathbb{Z}$-module M/L. Then $(\lambda_1,\ldots,\lambda_\ell)$ is a basis of $\mathcal{W}$ over $\mathbb{R}$ and $(\lambda_1,\ldots,\lambda_m)$ is a basis of M over $\mathbb{Z}$. As a consequence we have $L = M\cap\mathcal{W}$.

Let $p\colon\mathcal{V}\to\mathcal{V}$ denote the orthogonal projection on $\mathcal{W}^\perp$. For $\ell < i \le m$ define $\lambda'_i = p(\lambda_i)$. Since $\lambda'_{\ell+1},\ldots,\lambda'_m$ are $\mathbb{R}$-linearly independent, $p(M)$ is a discrete subgroup of $\mathcal{V}$. Further, let $(e_1,\ldots,e_\ell)$ an orthonormal basis of $\mathcal{W}$ and $(e_{\ell+1},\ldots,e_m)$ an orthonormal basis of $\mathcal{W}^\perp$. From the relations

$$\begin{aligned}
\lambda_1\wedge\cdots\wedge\lambda_m &= \lambda_1\wedge\cdots\wedge\lambda_\ell\wedge\lambda'_{\ell+1}\wedge\cdots\wedge\lambda'_m,\\
\lambda_1\wedge\cdots\wedge\lambda_m &= \pm\mathrm{vol}(M)\cdot e_1\wedge\cdots\wedge e_m\\
\lambda_1\wedge\cdots\wedge\lambda_\ell &= \pm\mathrm{vol}(L)\cdot e_1\wedge\cdots\wedge e_\ell
\end{aligned}$$

and

$$\lambda'_{\ell+1}\wedge\cdots\wedge\lambda'_m = \pm\mathrm{vol}\bigl(p(M)\bigr)\cdot e_{\ell+1}\wedge\cdots\wedge e_m,$$

one deduces

$$\mathrm{vol}(M) = \mathrm{vol}(L)\mathrm{vol}\bigl(p(M)\bigr)$$

(this is Lemma 3 of [BertP 1988]).

Proof of Lemma 10.10. Consider the Euclidean vector space $\mathcal{V} = \mathbb{R}^m$. A scaling will reduce the problem to counting points in the unit polydisc

$$\mathcal{C} = \left\{ \underline{x} \in \mathbb{R}^m \ ; \ |\underline{x}| \le 1 \right\}$$

in $\mathbb{R}^m$. Define

$$\begin{aligned} M &= \left\{ \left(\frac{\alpha_1}{T_1}, \dots, \frac{\alpha_m}{T_m} \right) \ ; \ \underline{\alpha} \in \mathbb{Z}^m \right\} \\ &= \mathbb{Z}\left(\frac{1}{T_1}, 0, \dots, 0 \right) + \cdots + \mathbb{Z}\left(0, \dots, 0, \frac{1}{T_m} \right) \end{aligned}$$

and

$$L_0 = \left\{ \left(\frac{\varphi_1}{T_1}, \dots, \frac{\varphi_m}{T_m} \right) \ ; \ \underline{\varphi} \in \Phi \right\}.$$

Hence $\chi(\underline{T})$ is the number of cosets of L_0 of the form $\underline{x} + L_0$ with $\underline{x} \in M \cap \mathcal{C}$.

Denote by $\mathcal{W}$ the $\mathbb{R}$-vector space spanned by L_0, by p the projection of $\mathbb{R}^m$ on $\mathcal{W}^\perp$ and define $L = \mathcal{W} \cap M$. Then L is a discrete subgroup of $\mathcal{W}$ containing L_0, hence L_0 has finite index in L. Moreover M/L has no torsion.

We use Lemma 7.8 for the mapping

$$\left\{ \underline{x} + L_0 \ ; \ \underline{x} \in M \cap \mathcal{C} \right\} \longrightarrow p(M \cap \mathcal{C})$$

deduced from the restriction of p to $M \cap \mathcal{C}$. For $\underline{x}$ and $\underline{x}'$ in $M \cap \mathcal{C}$ with $p(\underline{x}) = p(\underline{x}')$ we have $\underline{x} - \underline{x}' \in W \cap M = L$, hence

$$\chi(\underline{T}) \le [L : L_0] \mathrm{Card}\big(p(M \cap \mathcal{C})\big).$$

Any $\underline{z} \in \mathcal{W}^\perp$ can be written $\underline{z} = p(\underline{x}) + p(\underline{t})$ with $\underline{x} \in M$ and $\underline{t} \in \mathcal{C}'$ where

$$\mathcal{C}' = \left\{ \underline{x} \in \mathbb{R}^m \ ; \ 0 \le x_i < \frac{1}{T_i}, \ (1 \le i \le m) \right\}.$$

Hence there is a fundamental domain $\mathcal{P}$ for $p(M)$ in $\mathcal{W}^\perp$ which is contained in $p(\mathcal{C}')$. The sets $\underline{y} + \mathcal{P}$, where $\underline{y}$ ranges over $p(M \cap \mathcal{C})$, are pairwise disjoint; their Euclidean volume in $\mathcal{W}^\perp$ is $\mathrm{vol}\big(p(M)\big)$. Denote by N the square of the Euclidean norm on $\mathbb{R}^m$. For $\underline{x} \in \mathcal{C}$ and $\underline{t} \in \mathcal{C}'$ we have

$$N(\underline{x} + \underline{t}) < m + \frac{1}{T_1} + \cdots + \frac{1}{T_m} \le 2m.$$

Since p is an orthogonal projection, we deduce

$$N\big(p(\underline{x}) + p(\underline{t})\big) \le 2m.$$

Since $\mathcal{W}^\perp$ has dimension $\ell := m - \delta$, the volume of the Euclidean ball

$$\left\{ \underline{z} \in \mathcal{W}^\perp \ ; \ N(\underline{z}) \le 2m \right\}$$

is $c_\ell (2m)^{\ell/2}$, with

$$c_\ell = \frac{2\pi^{\ell/2}}{\ell\Gamma(\ell/2)}.$$

Therefore

$$\mathrm{Card}\big(p(M \cap \mathcal{C})\big)\mathrm{vol}\big(p(M)\big) \le c_\ell (2m)^{\ell/2}.$$

However we have

$$\frac{(L : L_0)}{\mathrm{vol}\big(p(M)\big)} = \frac{\mathrm{vol}(L_0)}{\mathrm{vol}(L)\mathrm{vol}\big(p(M)\big)} = \frac{\mathrm{vol}(L_0)}{\mathrm{vol}(M)}$$

and

$$\mathrm{vol}(M) = \frac{1}{T_1 \cdots T_m},$$

hence

$$\chi(\underline{T}) \le c_\ell (2m)^{\ell/2} T_1 \cdots T_m \mathrm{vol}(L_0).$$

We now want to relate $\mathrm{vol}(L_0)$ with $\mathcal{H}(G_1^-; \underline{T})$. Let A be a $\delta \times m$ matrix with integer coefficients whose columns vectors constitute a basis of Φ over $\mathbb{Z}$. From Proposition 5.14 we deduce

$$\mathcal{H}(G_1^-; \underline{T}) = \ell! 2^\ell T_1 \cdots T_m \sum_{\underline{i} \in C(m,\delta)} \mu_{\underline{i}}$$

with

$$\mu_{\underline{i}} = \frac{\det(\mathsf{A}_{\underline{i}})}{T_{i_1} \cdots T_{i_\delta}}.$$

Using Lemma 10.11 we obtain

$$\mathrm{vol}(L_0)^2 = \sum_{\underline{i} \in C(m,\delta)} \mu_{\underline{i}}^2 \le \left(\sum_{\underline{i} \in C(m,\delta)} \mu_{\underline{i}} \right)^2,$$

hence

$$T_1 \cdots T_m \mathrm{vol}(L_0) \le \frac{1}{\ell! 2^\ell} \mathcal{H}(G_1^-; \underline{T}).$$

Finally

$$\chi(\underline{T}) \le c_\ell \frac{m^{\ell/2}}{2^{\ell/2} \ell!} \mathcal{H}(G_1^-; \underline{T}).$$

This proves Lemma 10.10 with

$$c = \max_{1 \le \ell \le m} c_\ell \frac{m^{\ell/2}}{2^{\ell/2} \ell!}.$$

□

Remark. Sharper estimates follow from D. Bertrand's Chap. 9 in [NeP 2000]. Firstly, the simple upper bound

$$\chi(\underline{T}) \le \mathcal{H}(G_1^-; \underline{T}) + m - \delta$$

where $\ell = m - \delta = \dim(G_1^-)$ is proved as follows.

Let H be a hyperplane of $\mathbb{P}_2$. Consider the divisors

$$D_i = \mathbb{P}_2^{i-1} \times H \times \mathbb{P}_2^{m-1-i} \quad (1 \le i \le m)$$

on $\mathbb{P}_2^m$. The projective embedding ι of $\mathbb{G}_\mathrm{m}^m$ into $\mathbb{P}_2^m$ associated to the divisor $T_1 D_1 + \cdots + T_m D_m$ is given by the monomials $X_1^{t_1} \cdots X_m^{t_m}$ with $\underline{t} \in \mathbb{Z}^m$, $|t_i| \le T_i$ $(1 \le i \le m)$. By the very definition of the Hilbert function $\chi(\underline{T}) = H(G_1^-;\underline{T})$, $\mathbb{P}_2^{\chi(\underline{T})}$ is the image of G_1^- under ι. The *Italian* Lemma 1.1 of [NeP 2000], Chap. 9 gives

$$\chi(\underline{T}) \le \deg_{\underline{T}}(G_1^-) + \dim G_1^-,$$

where $\deg_{\underline{T}}(G_1^-)$ is the degree of G_1^- in the embedding ι:

$$\deg_{\underline{T}}(G_1^-) = \bigl(G_1^- \cdot (T_1 D_1 + \cdots + T_m D_m)^\ell\bigr).$$

Expanding by the multinomial formula provides the conclusion.

Since Hilbert's polynomial $\mathcal{H}(G_1^-;\underline{T})$ is the product by $\ell!$ of the homogeneous part of degree ℓ of the polynomial which coincides with $H(G_1^-;\underline{T})$ for sufficiently large $T_1, \ldots, T_m$, one should expect that asymptotically, the constant c in Lemma 10.10 should be replaced by $1/\ell!$. Such an estimate follows indeed from the second kind of estimates in D. Bertrand's Chap. 9 in [NeP 2000].

10.2.5 Main Estimate

Here we complete the transcendence argument and prove the following statement:

Theorem 10.12. *Let $\lambda_1, \ldots, \lambda_m$ be logarithms of nonzero algebraic numbers and $\beta_0, \ldots, \beta_{m-1}$ be algebraic numbers. Define $\alpha_i = \exp(\lambda_i)$ for $1 \le i \le m$,*

$$D = [\mathbb{Q}(\alpha_1, \ldots, \alpha_m, \beta_0, \ldots, \beta_{m-1}) : \mathbb{Q}]$$

and

$$\Lambda = \beta_0 + \beta_1\lambda_1 + \cdots + \beta_{m-1}\lambda_{m-1} - \lambda_m.$$

Assume $\alpha_1, \ldots, \alpha_m$ span a multiplicative group of rank $\ge m-1$ and $\Lambda \ne 0$. Let $A_1, \ldots, A_m$, B_1, B_2 and E be positive real numbers which satisfy

$$\log A_i \ge \mathrm{h}(\alpha_i), \quad D \log A_i \ge 1, \quad E|\lambda_i| \le D \log A_i, \quad (1 \le i \le m),$$

$$B_1 \ge e, \quad B_2 \ge e, \quad E \ge e, \quad B_1^D \ge E \quad \text{and} \quad B_2^D \ge E.$$

Then there exists two positive real numbers c_1 and c_2, which depend only on m, with the following property. Let T_0, $T_0^\sharp$, $T_1, \ldots, T_m$, S_0, S_1 and L be positive rational integers, U and V positive real numbers satisfying the following conditions:

$$L = (T_0+1)(2T_1+1)\cdots(2T_m+1), \quad L \ge 2\binom{(m+1)S_0 + m - 1}{m-1}\frac{V}{\log E},$$

$$S_1 \geq (m+1)!2^{m-1}, \quad S_0 \geq 2(m+1) \max_{1\leq i\leq m} T_i,$$

$$(S_0+1)(2S_1+1) > (m+1)!2^m T_0,$$
$$c_1 U \leq V \leq c_2 S_0 S_1 \log E,$$

$$DT_0 \log B_1 \leq U, \quad DS_0 \log B_2 \leq U, \quad DS_1 \max_{1\leq i\leq m} T_i \log A_i \leq U$$

and

$$B_1 \geq \frac{S_1}{T_0^\sharp}, \quad B_1 \geq E^{1/D} \quad \text{and} \quad B_2 \geq e^{T_0^\sharp}.$$

Finally, assume

- *Either (general case)*

$$\log B_2 \geq \max_{0\leq j\leq m} \mathrm{h}(\beta_j) \quad \text{and} \quad B_2 \geq S_0 \max_{1\leq i\leq m} T_i.$$

- *Or else (homogeneous rational case)*

$$\beta_0 = 0, \quad \beta_i = -\frac{b_i}{b_m} \quad (1 \leq i \leq m) \quad \text{with} \quad (b_1, \ldots, b_m) \in \mathbb{Z}^m,$$

$$B_2 \geq \frac{1}{S_0} \max_{1\leq j<m} (|b_m|T_j + |b_j|T_m).$$

Then

$$|\Lambda| > e^{-V}.$$

Proof. Given the parameters $T_0, \ldots, T_m$, S_0, S_1, we have defined in § 10.2.4 a connected algebraic subgroup G^- of codimension $n+1$ in $G = \mathbb{G}_a \times \mathbb{G}_m^m$, which is either $\{e\}$, or else a maximal obstruction subgroup.

Denote by $\mathcal{E}$ the set of tuples $\underline{T}' = (T_0, T_1', \ldots, T_m') \in \mathbb{N}^{m+1}$ satisfying $1 \leq T_i' \leq T_i$ for $1 \leq i \leq m$. Choose a lexicographic ordering on $\mathcal{E}$; the smallest element in $\mathcal{E}$ is $(T_0, 1, \ldots, 1)$ and the largest is $(T_0, T_1, \ldots, T_m)$. Consider the real valued mapping

$$\begin{array}{ccc} \mathcal{E} & \longrightarrow & \mathbb{R}_{>0} \\ \underline{T}' & \longmapsto & \dfrac{\mathcal{H}(G; \underline{T}')}{\mathcal{H}(G^-; \underline{T}')} \end{array}.$$

Let us check that for $\underline{T}' = (T_0, 1, \ldots, 1)$ we have

$$\frac{\mathcal{H}(G; (T_0, 1, \ldots, 1))}{\mathcal{H}(G^-; (T_0, 1, \ldots, 1))} < \binom{S_0+n}{n}(2S_1+1).$$

If $n = 0$, then $T_e(G^-) = \mathcal{W}$, hence $\beta_0 = 0$, $G_0^- = \mathbb{G}_a$ and in this case

$$\frac{\mathcal{H}(G; (T_0, 1, \ldots, 1))}{\mathcal{H}(G^-; (T_0, 1, \ldots, 1))} \leq (m+1)!2^m < 2S_1 + 1.$$

On the other hand, if $n \geq 1$, then

$$\frac{\mathcal{H}(G;\,(T_0,1,\dots,1))}{\mathcal{H}(G^-;\,(T_0,1,\dots,1))}\le(m+1)!2^mT_0<(S_0+1)(2S_1+1).$$

From now on we denote by $\underline{T}'$ the maximal element in $\mathcal{E}$ for which

$$\frac{\mathcal{H}(G;\,\underline{T}')}{\mathcal{H}(G^-;\,\underline{T}')}<\binom{S_0+n}{n}(2S_1+1).$$

In the case $G^-=\{e\}$, we have $\underline{T}'=\underline{T}$ because

$$\frac{\mathcal{H}(G;\,\underline{T})}{\mathcal{H}(\{e\};\,\underline{T})}=(m+1)!2^mT_0T_1\cdots T_m<\binom{S_0+m}{m}(2S_1+1).$$

Otherwise, G^- is an obstruction subgroup, and

$$\frac{\mathcal{H}(G;\,\underline{T}')}{\mathcal{H}(G^-;\,\underline{T}')}<\binom{S_0+n}{n}(2S_1+1)\le\frac{\mathcal{H}(G;\,\underline{T})}{\mathcal{H}(G^-;\,\underline{T})},$$

which implies $\underline{T}'\neq\underline{T}$. In this case there is at least one index $i\in\{1,\dots,m\}$ such that, if we replace T_i' by $T_i'+1$, we still get an element in $\mathcal{E}$ (which is larger than $\underline{T}'$). Using Proposition 5.14, we deduce

$$\frac{\mathcal{H}(G;\,\underline{T}')}{\mathcal{H}(G^-;\,\underline{T}')}\ge\frac{1}{2}\binom{S_0+n}{n}(2S_1+1).$$

Denote by $\mathcal{T}'$ the set of tuples $(\tau;t_1,\dots,t_m)$ in $\mathbb{N}\times\mathbb{Z}^m$ satisfying the following properties:

- Firstly we have

$$0\le\tau\le T_0\quad\text{and}\quad |t_i|\le T_i'\quad\text{for}\quad 1\le i\le m.$$

- Further if $G_0^-=\mathbb{G}_a$ then $\tau=0$.
- Furthermore

$$y_1^{t_1}\cdots y_m^{t_m}=1$$

for any $\underline{y}\in G_1^-$.

The number of elements of $\mathcal{T}'$ will be denoted by L'. From Lemma 10.10 we deduce

$$L'\ge\frac{1}{c}\cdot\frac{(T_0+1)(T_1'+1)\cdots(T_m'+1)}{\mathcal{H}(G^-;\underline{T}')},$$

where c is the constant of Lemma 10.10. Hence if $G^-\neq\{e\}$ we obtain

$$L'\ge mc_2\binom{(m+1)S_0+n}{n}(2S_1+1)\quad\text{with}\quad c_2=\frac{1}{(m+1)!2^{m+1}m(m+1)c}.$$

We deduce

$$L'\ge2\binom{(m+1)S_0+n-1}{n-1}\frac{V}{\log E}.$$

This last inequality is also true in case $G^- = \{e\}$, as shown by the assumption

$$L \geq 2\binom{(m+1)S_0 + m - 1}{m-1}\frac{V}{\log E}.$$

Consider the matrix with L' rows :

$$M' = \left(\widetilde{\gamma}_{\tau\underline{t}}^{(\underline{\sigma}s)}\right)_{\substack{(\tau,\underline{t})\\(\underline{\sigma},s)}},$$

where $(\tau, \underline{t})$ runs over the set $\mathcal{T}'$, while $(\underline{\sigma}, s)$ runs over the set of tuples in $\mathbb{N}^m \times \mathbb{Z}$ satisfying $\|\underline{\sigma}\| \leq (m+1)S_0$ and $|s| \leq (m+1)S_1$ (recall the definition of $\widetilde{\gamma}_{\tau\underline{t}}^{(\underline{\sigma}s)}$ in § 10.2.1).

We are going to use Proposition 10.9 with

$$\delta^{(1)}(z;\tau) = \delta_{T_0^\sharp}(z;\tau) \quad (0 \leq \tau \leq T_0)$$

and

$$\left\{\delta^{(2)}(\underline{z};\sigma)\,;\ 1 \leq \sigma \leq \binom{(m+1)S_0 + m}{m}\right\} =$$

$$\left\{\frac{1}{\sigma_0!}z_0^{\sigma_0}\cdots z_{m-1}^{\sigma_{m-1}}\,;\ \|\sigma\| \leq (m+1)S_0\right\},$$

Notice that in the case where $\beta_0 = 0$, if $\mathcal{W}$ is the tangent space of an algebraic subgroup of $\mathbb{G}_{\mathrm{a}} \times \mathbb{G}_{\mathrm{m}}^m$ containing $(1, \alpha_1^s, \ldots, \alpha_m^s)$ for some $s \in \mathbb{Z}$, $s \neq 0$, then $\Lambda \in 2\pi i\mathbb{Q}$ and then the conclusion of Theorem 10.12 is plain.

All hypotheses of Proposition 10.9 are satisfied; we deduce that M' has rank L'. Let Δ be the determinant of a regular square $L' \times L'$ submatrix

$$\left(\widetilde{\gamma}_{\tau\underline{t}}^{(\underline{\sigma}_\mu s_\mu)}\right)_{\substack{(\tau,\underline{t})\in\mathcal{T}'\\1\leq\mu\leq L'}}$$

of M'. We are going to use Proposition 10.5 with L, S_0 and S_1 replaced respectively by L', $(m+1)S_0$ and $(m+1)S_1$ for the following functions $\varphi_1, \ldots, \varphi_{L'}$:

$$\varphi_{\tau\underline{t}} = \delta_{T_0^\sharp}(z_0;\tau)e^{\underline{t}\underline{z}} \qquad (\tau, \underline{t}) \in \mathcal{T}'.$$

Define

$$\underline{\eta}' = (1, \lambda_1, \ldots, \lambda_{m-1}, \lambda_m + \Lambda) \in \mathcal{W}, \quad \underline{\zeta}'_\mu = s_\mu\underline{\eta}' \quad (1 \leq \mu \leq L'),$$

$$\mathcal{X} = \mathbb{C}\underline{\eta}', \quad \mathcal{U} = T_e(G^-),$$

$$\delta_{\tau\underline{t}\mu} = \widetilde{\gamma}_{\tau\underline{t}}^{(\underline{\sigma}_\mu s_\mu)} \cdot \frac{1 - e^{s_\mu t_m \Lambda}}{\Lambda},$$

$$\epsilon = \Lambda, \quad \text{and} \quad M_{\tau\underline{t}} = c_3 U \quad \text{for any} \quad (\tau, \underline{t}) \in \mathcal{T}'.$$

Let us check the hypotheses of Proposition 10.5. We have

$$2\big((m+1)S_0+1\big)\log E + \log(2L') + \max_{(\tau,\underline{t})\in\mathcal{T}'} M_{\tau\underline{t}} \le \frac{c_1}{4}\cdot U \le \frac{V}{4}.$$

We now split the proof by considering two cases.

In the general case, we define

$$\varphi_{\tau\underline{t}\mu} = \mathcal{D}_{\mathbf{w}}^{\underline{\sigma}_\mu}\varphi_{\tau\underline{t}},$$

so that we have

$$\widetilde{\gamma}_{\tau\underline{t}}^{(\underline{\sigma}_\mu s_\mu)} = \varphi_{\tau\underline{t}\mu}(\underline{\zeta}'_\mu) + \Lambda\delta_{\tau\underline{t}\mu}.$$

We need to check

$$\sup_{|z|=E}\max_{1\le\mu\le L'}|\varphi_{\tau\underline{t}\mu}(z\underline{\zeta}'_\mu)| \le e^{M_{\tau\underline{t}}} \quad\text{and}\quad \max_{1\le\mu\le L'}|\delta_{\tau\underline{t}\mu}| \le e^{M_{\tau\underline{t}}}.$$

Here

$$\varphi_{\tau\underline{t}\mu}(z\underline{\zeta}'_\mu) = \sum_{\kappa=0}^{\sigma_{\mu0}}\binom{\sigma_{\mu0}}{\kappa}\delta_{T_0^\sharp}(sz;\tau,\kappa)(t_m\beta_0)^{\sigma_{\mu0}-\kappa}\left(\prod_{i=1}^{m-1}(t_i+t_m\beta_i)^{\sigma_{\mu i}}\right)e^{(t_1\lambda_1+\cdots+t_m\lambda_m)sz}.$$

For $|z| = E$, $|t_i| \le T_i$ and $|s| \le (m+1)S_1$ we have

$$\log\left|e^{(t_1\lambda_1+\cdots+t_m\lambda_m)sz}\right| \le (m+1)S_1E\sum_{i=1}^m T_i|\lambda_i| \le m(m+1)U$$

and (recall Lemma 9.8)

$$\left|\delta_{T_0^\sharp}(sz;\tau,\kappa)\right| \le \kappa!e^{T_0+T_0^\sharp}\left(1+\frac{E(m+1)S_1}{T_0^\sharp}\right)^{T_0}.$$

Since $\kappa \le (m+1)S_0$ we have

$$\kappa! \le \big((m+1)S_0\big)^{(m+1)S_0} \le (B_2S_0)^{(m+1)S_0}.$$

Using the assumptions

$$E^{1/D} \le B_1, \quad \frac{S_1}{T_0^\sharp} \le B_1, \quad T_0^\sharp \le \log B_2 \le U \quad\text{and}\quad E \ge e,$$

we deduce, for $|z| = E$,

$$\log\left|\delta_{T_0^\sharp}(sz;\tau,\kappa)\right| \le T_0 + T_0^\sharp + 2(m+1)S_0\log B_2 + 3DT_0\log B_1 \le (2m+7)U.$$

Recall that $\mathrm{h}(\beta_j) \le \log B_2$, hence $|\beta_j| \le B_2^D$. Using only the assumption $T_i \le B_2^D$, we bound

$$\sum_{\kappa=0}^{\sigma_{\mu0}}\binom{\sigma_{\mu0}}{\kappa}|t_m\beta_0|^{\sigma_{\mu0}-\kappa}\prod_{i=1}^{m-1}|t_i+t_m\beta_i|^{\sigma_{\mu i}}$$

from above by

$$\Bigl(\max_{1\le i\le m} T_i\Bigr)^{(m+1)S_0}\Bigl(1+\max_{0\le j\le m}|\beta_j|\Bigr)^{(m+1)S_0} \le (2B_2)^{2(m+1)DS_0} \le e^{4(m+1)U}.$$

The estimate $|\delta_{\tau\underline{t}\mu}| \le e^{M_{\tau\underline{t}}}$ is proved in the same way: since

$$|\delta_{\tau\underline{t}\mu}| \le 2\bigl|s_\mu t_m \widetilde{\gamma}_{\tau\underline{t}}^{(\underline{\sigma}_\mu s_\mu)}\bigr|,$$

it suffices to use the inequalities

$$2|s_\mu t_m| \le 2(m+1)S_1T_m \le 2(m+1)U < e^U.$$

Consider now the homogeneous rational case. Define rational numbers $q_{\sigma\kappa}$ $(\sigma \ge \kappa \ge 0)$ by

$$\Delta(z;\sigma) = \sum_{\kappa=0}^{\sigma} q_{\sigma\kappa}z^\kappa.$$

Set

$$\varphi_{\tau\underline{t}\mu} = \frac{1}{\sigma_{\mu 0}!}\sum_{\kappa_1=0}^{\sigma_{\mu 1}}\cdots\sum_{\kappa_{m-1}=0}^{\sigma_{\mu,m-1}} q_{\sigma_{\mu 1}\kappa_1}\cdots q_{\sigma_{\mu,m-1}\kappa_{m-1}} b_m^{\|\underline{\kappa}\|-\sigma_{\mu 0}}\mathcal{D}_{\underline{w}}^{\underline{\kappa}}\varphi_{\tau\underline{t}},$$

where $\underline{\kappa}$ stands for $(\sigma_{\mu 0}, \kappa_1, \ldots, \kappa_{m-1}) \in \mathbb{N}^m$. We have

$$\varphi_{\tau\underline{t}\mu}(z\underline{\zeta}'_\mu) = \frac{1}{\sigma_{\mu 0}!}\delta_{T_0^\sharp}(sz;\tau,\sigma_{\mu 0})\left(\prod_{i=1}^{m-1}\Delta(t_ib_m - t_mb_i;\sigma_{\mu i})\right)e^{(t_1\lambda_1+\cdots+t_m\lambda_m)sz}.$$

Therefore

$$\widetilde{\gamma}_{\tau\underline{t}}^{(\underline{\sigma}_\mu s_\mu)} = \varphi_{\tau\underline{t}\mu}(\underline{\zeta}'_\mu) + \Lambda\delta_{\tau\underline{t}\mu}.$$

Let us check

$$\sup_{|z|=E}\max_{1\le\mu\le L'}|\varphi_{\tau\underline{t}\mu}(z\underline{\zeta}'_\mu)| \le e^{M_{\tau\underline{t}}}.$$

We have

$$\frac{(m+1)S_1}{T_0^\sharp} + 1 \le (m+1)B_1 + 1 \le B_1^{m+1},$$

$$1 + \frac{|t_ib_m - t_mb_i|}{(m+1)S_0} \le B_2 \qquad (1 \le i \le m-1),$$

and

$$e^{T_0+T_0^\sharp+(m+1)S_0}B_1^{(m+1)T_0}B_2^{(m+1)S_0} \le e^{(3m+5)U}.$$

From Lemma 9.11 we deduce, for $|z| = E$,

$$\frac{1}{\sigma_{\mu 0}!}\bigl|\delta_{T_0^\sharp}(sz;\tau,\sigma_{\mu 0})\bigr|\prod_{i=1}^{m-1}\bigl|\Delta(t_ib_m - t_mb_i;\sigma_{\mu i})\bigr| \le e^{(3m+5)U}.$$

The estimate $|\delta_{\tau\underline{t}\mu}| \le e^{M_{\tau\underline{t}}}$ also follows.

From Proposition 10.5 we deduce

$$\text{either } |\epsilon| \le e^{-V} \text{ or else } |\Delta| \le e^{-L'V/4}.$$

On the other hand, we claim that Liouville's inequality (Proposition 3.14) yields $|\Delta| > e^{-L'V/4}$. The arithmetic estimates are similar to the analytic estimates, apart from the fact that we need to take into account the denominator $\nu(T_0^\sharp)^{S_0}$ and the degree D. This is the place in the proof where we need the assumption $B_2 \geq e^{T_0^\sharp}$, and also $B_2 \geq T_i$ in the general case.

Hence $|\Lambda| > e^{-V}$. □

10.2.6 Second Proof of Theorem 9.1

Assume the hypotheses of Theorem 9.1 are satisfied. Without loss of generality we may assume also $\beta_m \neq 0$. Define $\beta_i' = -\beta_i/\beta_m$ $(0 \leq i \leq m)$ and $\Lambda' = -\Lambda/\beta_m$. Our goal is to deduce from Theorem 10.12 a lower bound for $|\Lambda'|$.

Let N be a sufficiently large integer which depends only on m. Define

$$B_1 = (E^*)^N, \quad B_2 = B^N,$$

$$T_0^\sharp = [\log B_2], \quad S_1 = \left[\frac{N^2 D \log B_2}{\log E}\right],$$

$$U = N^{2m+1} D^{m+2} (\log B_1)(\log B_2)(\log A_1) \cdots (\log A_m)(\log E)^{-m-1},$$

$$T_0 = \left[\frac{U}{D \log B_1}\right], \quad S_0 = \left[\frac{U}{D \log B_2}\right],$$

$$T_i = \left[\frac{U}{D S_1 \log A_i}\right] \quad (1 \leq i \leq m)$$

and

$$L = (T_0 + 1)(2T_1 + 1) \cdots (2T_m + 1).$$

Notice that each of the parameters $T_0^\sharp$, $T_0, \ldots, T_m$, S_0, S_1 and L is a (large) positive integer, because

$$D \log B \geq \log E, \quad D \log E^* \geq \log E \quad \text{and} \quad D \log A_i \geq \log E.$$

It is plain that the following hypotheses of Theorem 10.12 are satisfied:

$$D T_0 \log B_1 \leq U, \quad D S_0 \log B_2 \leq U, \quad D S_1 \max_{1 \leq i \leq m} T_i \log A_i \leq U.$$

Also, for sufficiently large N, we check

$$S_0 \geq 2(m+1) \max_{1 \leq i \leq m} T_i \quad \text{because} \quad \log A_i \geq \frac{1}{D} \log E,$$

and

$$(S_0 + 1)(2S_1 + 1) > (m+1)! 2^m T_0 \quad \text{because} \quad \log E^* \geq \frac{1}{D} \log E.$$

Define $V = c_1 U$ where c_1 is the constant occurring in Theorem 10.12. We plainly have (again for sufficiently large N)

$$V \le c_2 S_0 S_1 \log E.$$

From the estimates

$$L \ge \left(\frac{U}{D\log B_2}\right)^m \cdot \frac{U(\log E)^m}{N^{2m}D^{m+1}(\log B_1)(\log A_1)\cdots(\log A_m)}$$

and

$$2\binom{S_0+m-1}{m-1}\cdot\frac{V}{\log E} \le 2\left(\frac{U}{D\log B_2}\right)^m \cdot \frac{c_1 D\log B_2}{\log E},$$

using the definition of U, we deduce

$$L \ge 2\binom{S_0+m-1}{m-1}\cdot\frac{V}{\log E}.$$

We also deduce the estimates

$$B_1 \ge \frac{S_1}{T_0^\sharp} \quad \text{from the assumption} \quad E^* \ge \frac{D}{\log E},$$

$$B_1 \ge E^{1/D} \quad \text{from the assumption} \quad E^* \ge E^{1/D},$$

and

$$B_2 \ge e^{T_0^\sharp}.$$

In the general case, we have, for $1 \le i \le m$,

$$\begin{aligned} T_i &\le 4N^{2m}\cdot\frac{D\log E^*}{\log E}\prod_{j\ne i}\frac{D\log A_j}{\log E} \\ &\le 4N^{2m}B^m\log B \\ &\le B_2. \end{aligned}$$

In the homogeneous rational case, we have

$$\frac{T_i}{S_0} \le \frac{\log E}{D\log A_i} \qquad (1 \le i \le m),$$

hence the assumption

$$B \ge \frac{\log E}{D}\cdot\max_{1\le i<m}\left(\frac{|b_m|}{\log A_i}+\frac{|b_i|}{\log A_m}\right)$$

yields

$$B_2 \ge \frac{1}{S_0}\max_{1\le i<m}\left(|b_m|T_i+|b_i|T_m\right).$$

From Theorem 10.12 we deduce $|\Lambda'| > e^{-V}$, and the conclusion of Theorem 9.1 easily follows with $C(m) = N^{2m+4}$. □

10.3 Baker's Method with Auxiliary Function

Baker's method involving an auxiliary function has been developed in a collection of original papers (references are given in the comments on methods 1 and 2 in § 14.4.6). A few monographs include a description of this method: [W 1974], Chap. 8, [B 1975], Chap. 2 and 3, [L 1978], Chap. 8, 10, 11, [W 1979a], Lectures 4 and 5, [F 1982], Chap. 10, [Sp 1982], Chap. III and [FNe 1998], Chap. 4, § 1.

Here we outline the main points of this method.

One introduces an auxiliary function

$$F(\underline{z}) = P(z_0, e^{z_1}, \ldots, e^{z_m}),$$

where $P \in K[X, Y_1^{\pm 1}, \ldots, Y_m^{\pm 1}]$ is a nonzero polynomial with coefficients in the number field $K = \mathbb{Q}(\alpha_1, \ldots, \alpha_m, \beta_0, \ldots, \beta_m)$. This polynomial P is constructed by means of Dirichlet's box principle. To be more precise, P is not explicitly constructed, but the mere existence of P is sufficient for the proof. One requires

$$\mathcal{D}_{\underline{w}}^{\underline{\sigma}} P(s, s\lambda_1, \ldots, s\lambda_m) = 0$$

for $\underline{\sigma} \in \mathbb{N}^m$, $\|\underline{\sigma}\| \leq S_0$ and $s \in \mathbb{Z}$, $|s| \leq S_1$. Like in § 10.2.1,

$$\mathcal{D}_{\underline{w}}^{\underline{\sigma}} = \left(\frac{\partial}{\partial z_0} + \beta_0 \frac{\partial}{\partial z_m}\right)^{\sigma_{\mu 0}} \cdots \left(\frac{\partial}{\partial z_{m-1}} + \beta_{m-1} \frac{\partial}{\partial z_m}\right)^{\sigma_{\mu m-1}}.$$

To prove the existence of such a polynomial $P \neq 0$ of degree $\leq T_0$ in X and degree $\leq T_i$ in $Y_i^{\pm 1}$ amounts to show that a system with $\binom{S_0+m}{m}(2S_1+1)$ equations and $(T_0+1)(2T_1+1)\cdots(2T_m+1)$ unknowns has a nontrivial solution. The unknowns are nothing else than the coefficients of P. The equations are homogeneous and linear, and they have their coefficients in K (these coefficients are just entries of our interpolation matrix M in § 10.2.1). The existence of P is guaranteed as soon as $\binom{S_0+m}{m}(2S_1+1)$ is larger than the number $L = (T_0+1)(2T_1+1)\cdots(2T_m+1)$. Assuming the quotient of these two numbers not only is larger than 1, but is a little bit bigger, enables one to get a good control of the height of the coefficients of P in K. This control is necessary for checking several estimates occurring in the proof. We shall not tell more on this point, since we only give a sketch, but is suffices to say that Thue-Siegel's Lemma (see § 4.5) is the clue.

The goal of the extrapolation process is to prove that

$$\mathcal{D}_{\underline{w}}^{\underline{\sigma}'} P(s', s'\lambda_1, \ldots, s'\lambda_m) = 0$$

for further values of $\underline{\sigma}' \in \mathbb{N}^m$ and $s' \in \mathbb{Z}$, say $\|\underline{\sigma}'\| \leq S_0'$ and $|s'| \leq S_1'$. If we succeed to show that these equations hold with S_0' and S_1' sufficiently large, namely such that $\binom{S_0'+m}{m}(2S_1'+1)$ is somewhat bigger than L, then we shall be in a position to apply Proposition 10.9 with S_0, S_1 replaced by S_0', S_1'.

To begin with, assume for a while $\Lambda = 0$. Define, for $\underline{\sigma} \in \mathbb{N}^m$,

$$F_{\underline{\sigma}}(\underline{z}) = \mathcal{D}_{\underline{w}}^{\underline{\sigma}} P(z_0, e^{z_1}, \ldots, e^{z_m}),$$

where, as in § 10.2.1,

$$\mathcal{D}_{\underline{w}}^{\underline{\sigma}} = \prod_{i=0}^{m-1} \left(\frac{\partial}{\partial z_i} + \beta_i \frac{\partial}{\partial z_m} \right)^{\sigma_{\mu i}}.$$

A fundamental remark of Baker's is that, for $\|\underline{\sigma}'\| \leq S_0/2$, the function of a single variable

$$\varphi_{\underline{\sigma}'}(z) = F_{\underline{\sigma}'}(z, z\lambda_1, \ldots, z\lambda_m)$$

has a zero of multiplicity $\geq S_0/2$ at each point $s \in \mathbb{Z}$ with $|s| \leq S_1$.

We check this assertion as follows. Define

$$f_{\underline{\sigma}'}(z_0, \ldots, z_{m-1}) = F_{\underline{\sigma}'}(z_0, \ldots, z_{m-1}, \beta_0 z_0 + \cdots + \beta_{m-1} z_{m-1}).$$

Since $\Lambda = 0$, we have on one hand

$$\varphi_{\underline{\sigma}'}(z) = f_{\underline{\sigma}'}(z, z\lambda_1, \ldots, z\lambda_{m-1}),$$

and on the other hand, for $\underline{\kappa} \in \mathbb{N}^m$,

$$\mathcal{D}_{\underline{w}}^{\underline{\kappa}} F_{\underline{\sigma}'}(z_0, \ldots, z_{m-1}, \beta_0 z_0 + \cdots + \beta_{m-1} z_{m-1}) = \left(\frac{\partial}{\partial z_0} \right)^{\kappa_0} \cdots \left(\frac{\partial}{\partial z_{m-1}} \right)^{\kappa_{m-1}} f_{\underline{\sigma}'}(z_0, \ldots, z_{m-1}).$$

The chain rule for derivatives of the composite of functions gives

$$\left(\frac{d}{dz} \right)^{\sigma''} \varphi_{\underline{\sigma}'} = \sum_{\|\underline{\kappa}\| = \sigma''} \frac{\sigma''!}{\underline{\kappa}!} \lambda_1^{\kappa_1} \cdots \lambda_{m-1}^{\kappa_{m-1}} \varphi_{\underline{\sigma}' + \underline{\kappa}}.$$

By assumption

$$\varphi_{\underline{\sigma}' + \underline{\kappa}}(s) = 0 \quad \text{for} \quad \|\underline{\sigma}'\| \leq \frac{S_0}{2}, \quad \|\underline{\kappa}\| \leq \frac{S_0}{2} \quad \text{and} \quad |s| \leq S_1.$$

Hence for $\|\underline{\sigma}'\| \leq S_0/2$ we have

$$\left(\frac{d}{dz} \right)^{\sigma''} \varphi_{\underline{\sigma}'}(s) = 0 \quad \text{for} \quad 0 \leq \sigma'' \leq \frac{S_0}{2} \quad \text{and} \quad |s| \leq S_1,$$

which means that $\varphi_{\underline{\sigma}'}$ has a zero of multiplicity $\geq S_0/2$ at each point $s \in \mathbb{Z}$ with $|s| \leq S_1$.

From Schwarz' Lemma for a function of a single variable (for instance the case $n = 1$ of Proposition 4.7) we deduce an upper bound for $|\varphi_{\underline{\sigma}'}(s')|$ when $s' \in \mathbb{Z}$ satisfies $|s'| \leq S_1'$, and S_1' is larger than S_1.

We insist here that it is crucial to deal with a function $\varphi_{\underline{\sigma}'}(z)$ which depends only on a single variable. Proposition 4.7 would yield an upper bound for the maximum modulus of a function vanishing on a Cartesian product, but the points

$$(s, s\lambda_1, \ldots, s\lambda_m) \quad (s \in \mathbb{Z},\ |s| \leq S_1)$$

do not constitute a Cartesian product in $\mathbb{C}^{m+1}$. Due to the lack of a suitable Schwarz Lemma in several variables, one would not be able to perform a similar argument with the method of Chapters 6, 7 and 9 when $m \geq 3$.

The assumption $\Lambda = 0$ is good enough for the proof of Baker's transcendence results (Theorems 1.5 and 1.6), but not for the quantitative refinements. Assume now only that $|\Lambda|$ is small. One still deduces an upper bound for $|\varphi_{\underline{\sigma}'}(s')|$, but Schwarz' Lemma needs to be replaced by an interpolation formula. The point is that the function $\varphi_{\underline{\sigma}'}(z)$ (of a single variable) takes small values at the points $s \in \mathbb{Z},\ |s| \leq S_1$, and its derivatives of order $\leq S_0/2$ also. An interpolation formula provides an upper bound for the maximum modulus of such a function on a disc.

If the upper bound for $|\varphi_{\underline{\sigma}'}(s')|$ is sufficiently sharp, Liouville's estimate implies $\varphi_{\underline{\sigma}'}(s') = 0$, which was our goal.

It is not difficult to work out the details of this proof; one obtains in this way another proof of Theorem 9.1.

By the way, all the proofs given in Chap. 2 can also be worked out by means of an auxiliary function in place of an interpolation determinant, along the same lines. The method with an auxiliary function is in fact the older one, and there are many references for it, like [Si 1949], [G 1952], [Sch 1957], [L 1966], [W 1974], [B 1975], [W 1979a] and [W 1979b].

The proof of Schneider-Lang's criterion in Chap. 4 involved also an auxiliary function, but the construction was slightly different: we did not consider a system of equations, because we did not require the auxiliary function to have many zeroes. In place, we used the *universal* auxiliary function provided by Proposition 4.10. More precisely this function was constructed so that its first derivatives at the origin have small absolute values, and then from an interpolation formula (Lemma 4.13, involving several variables, but a single point) we deduced an upper bound for the maximum modulus of the function on a large disc.

The transcendence proofs in Chapters 6, 7, 9 and 10 could also be given by means of an auxiliary function like the one of Proposition 4.10 (see [W 1991a]), but we do not know how to work out the transcendence proofs of Chapters 6, 7 and 9 with an auxiliary function constructed to have many zeroes, as in Baker's extrapolation argument. There is no interpolation lemma, let alone Schwarz' Lemma, in several variables, which would be suitable for this purpose.

One of the main interests of Baker's extrapolation technique is the following: if one computes the value of $C(m)$ by means of the interpolation determinant method where there is no extrapolation, we find a much larger number than the one given by Proposition 9.18. Some explanation for this difference between the outputs of two methods will be given in § 14.4 (we shall see that it seems mainly due to a lack of symmetry in the multiplicity estimate). However, using an auxiliary function and an extrapolation, a much smaller value for $C(m)$ can be achieved (see Theorems 10.20, 10.21, 10.23 and 10.24).

The idea is the following. Starting from the system of relations $F_{\underline{\sigma}}(s) = 0$ ($\|\underline{\sigma}\| \leq S_0,\ |s| \leq S_1$), we have explained how to deduce $F_{\underline{\sigma}'}(s') = 0$ for $\|\underline{\sigma}'\| \leq S_0'$

and $|s'| \leq S_1'$, where $S_0' = S_0/2$, while S_1' is larger than S_1. It may be wise to repeat the argument and to get $F_{\underline{\sigma}'}(s') = 0$ for $\|\sigma'\| \leq S_0^{(j)}$ and $|s'| \leq S_1^{(j)}$, where $S_0^{(j)} = S_0/2^j$ and $S_1^{(j)}$ is larger than $S_1^{(j-1)}$. One cannot repeat this argument forever, since $S_0^{(j)}$ decreases [17]. But performing this extrapolation several times is better than just once if one is interested in getting a sharp estimate for $C(m)$.

The above mentioned smallest known numerical values for $C(m)$ are achieved by a slight modification of this argument. The extrapolation is performed not only on the integers $s \in \mathbb{Z}$, but also on some rational numbers, after H. M. Stark [St 1971]. Several attempts have been made, and the best results nowadays are obtained by using the numbers s/q with $s = 0, \pm 1, \pm 2, \ldots$, while q is a fixed prime, say $q = 2$. For this purpose a strong independence condition on $\alpha_1, \ldots, \alpha_m$ is introduced.

Definition. Dealing with $\beta_0+\beta_1\lambda_1+\cdots+\beta_m\lambda_m$, we shall say that *Kummer's condition is satisfied* if the field $K = \mathbb{Q}(\alpha_1, \ldots, \alpha_m, \beta_0, \ldots, \beta_m)$ satisfies

$$[K(\sqrt{\alpha_1}, \ldots, \sqrt{\alpha_m}) : K] = 2^m. \tag{10.13}$$

This condition implies that $\alpha_1, \ldots, \alpha_m$ generate a multiplicative subgroup of $K^\times$ of rank $\geq m-1$ (see Exercise 10.6).

Once a measure of linear independence is proved in the special case where (10.13) holds, a *final descent* enables one to remove it; see for instance [LoxV 1976], [B 1977], § 8, [L 1978], Chap. XI, § 5 and [W 1980].

Thanks to assumption (10.13), each relation $F_{\underline{\sigma}'}(s'/2) = 0$ can be decomposed into 2^m relations[18] (recall that the coefficients of P are in K). After performing a sufficiently large number of steps for this extrapolation, one finally applies the multiplicity estimate, which produces an algebraic subgroup G^* of $G = \mathbb{G}_a \times \mathbb{G}_m^m$. There are mainly three ways to conclude the proof.

(1) The simplest one, used in [B 1977], [W 1980] and [Mat 1998] for instance (see Theorems 10.19, 10.20 and 10.24), consists in pushing the extrapolation process sufficiently far so that G^* is trivial. The drawback of this approach is that one needs to assume $E^* \geq \log A_i$ for $1 \leq i \leq m-1$.

(2) In [PW 1988a] (see Theorem 10.21), one starts the proof by selecting, among the algebraic subgroups of G, an obstruction subgroup, say G^-, which is maximal among the G^* which could be produced by the zero estimate. The whole construction of the auxiliary function is performed on G/G^- in place of G. This is just what we did in § 10.2.1.

(3) The argument of [Wü 1988] and [BWü 1993] (see Theorem 10.23) is basically the same, but it is introduced in a slightly different way, involving an inductive

[17] Due to this fact, Baker's method always requires some kind of rather sharp zero estimate, even for the transcendence result; one cannot just use the fact that a nonzero function cannot have a zero of infinite order for instance. Compare with the zero estimate which is implicit in the proof given in § 4.6.

[18] In the extrapolation process, it is not the number of equations which increases, but the number of coefficients which decreases!

process starting with the algebraic subgroup G^* produced by the multiplicity estimate.

10.4 The State of the Art

It is not so easy to give a survey of known explicit measures of linear independence for logarithms of algebraic numbers, since different authors use different notation: normalization for the height, choice of parameters, assumptions differ from one text to another. We shall keep here the notation and assumptions of Theorem 9.1 and try to give an idea of the state of the art on this topic.

As before we refer to the *general case* for

$$\Lambda = \beta_0 + \beta_1\lambda_1 + \ldots + \beta_m\lambda_m$$

where β_i and $\alpha_j = e^{\lambda_j}$ are algebraic numbers in a number field of degree D, and to the *homogeneous rational case* when $\beta_0 = 0$ and $\beta_1, \ldots, \beta_m$ are rational integers; in such a situation we write b_i for β_i so that

$$\Lambda = b_1\lambda_1 + \ldots + b_m\lambda_m$$

Surveys dealing with measures of linear independence of logarithms are given in [B 1977], § 1, [FNe 1998], Chap. 4, § 1.1 and [Mat 1998].

We now discuss some of the recent refinements. We start with the dependence on the *main parameters*, namely B and A_j, next we look at D and E, and finally at m and the absolute constant. To finish with we quote a few recent estimates.

10.4.1 Dependence on B and A_j

The most natural parameters to measure the height of $\beta_0, \ldots, \beta_m$ are either

$$\mathrm{h}(1{:}\,\beta_0{:}\cdots{:}\,\beta_m) \quad \text{or} \quad \max\bigl\{\mathrm{h}(\beta_0), \ldots, \mathrm{h}(\beta_m)\bigr\}.$$

It does not make much difference to choose one or the other (cf. Exercise 3.3.a). In the homogeneous rational case, there is no harm to assume that the integers $b_1, \ldots, b_m$ are relatively prime, and then both quantities are nothing else than

$$\log \max\bigl\{|b_1|, \ldots, |b_m|\bigr\}.$$

Theorem 9.1 includes Fel'dman's estimate [F 1968], which, by Lemma 1.8, is *best possible* in terms of the maximal heights of the β_j:

$$|\Lambda| \geq \exp\Bigl\{-C \max\bigl\{1, \mathrm{h}(\beta_0), \ldots, \mathrm{h}(\beta_m)\bigr\}\Bigr\},$$

where C does not depend on B (it depends only on $m, \lambda_1, \ldots, \lambda_m$ and D).

Apart from the condition $B \geq \log \max_{1 \leq i \leq m} A_i$ which occurs in the general case, the final result is optimal also in terms of each A_i separately (see Exercise 10.5).

However one would expect that the product $(\log B)(\log A_1) \cdots (\log A_m)$ could be replaced by the sum $\log B + \log A_1 + \cdots + \log A_m$ — see Conjectures 1.11 and 14.25.

Let us now restrict the discussion to the homogeneous rational case until the end of § 10.4.1. For convenience define

$$B_0 = \max\{e, |b_1|, \ldots, |b_m|\}.$$

The choice of the constant e is not important, but one should not exclude the case where all b_i are ± 1. In fact one might assume that B_0 is *large*, otherwise Liouville's estimate is stronger than what we can prove via transcendence methods (see Remark 2 in § 7.1.1).

In spite of the fact that Fel'dman's result yields a best possible estimate in terms of B_0, namely B_0^{-C}, one can improve this lower bound by introducing a smaller parameter than B_0: this is what we did in Theorem 9.1 with the number

$$\max_{1 \leq j \leq m-1} \left\{ \frac{|b_m|}{\log A_j} + \frac{|b_j|}{\log A_m} \right\}. \tag{10.14}$$

Feld'man's polynomials are the main tool which enables one to replace B_0 by (10.14) The introduction of this quantity (10.14) has the following origin. As we already pointed out in § 1.2, the estimates we are studying have dramatic consequences in several diophantine problems, in particular to diophantine equations. Typically, one wishes to prove that some diophantine equation has no solution, apart from those which are already known. Assuming a contrario that there is a *nontrivial* solution, one produces a number

$$\Lambda = b_1\lambda_1 + \cdots + b_m\lambda_m$$

which satisfies $0 < |\Lambda| \leq e^{-\delta B_0}$ with some δ in the range $0 < \delta \leq 1$. Usually the number $\lambda_1, \ldots, \lambda_m$ (in $\mathcal{L}$) and δ (in $\mathbb{R}$) are explicitly known, while $b_1, \ldots, b_m$ (in $\mathbb{Z}$) depend on the exceptional solution we started from. From this information one wishes to deduce an upper bound for B_0, and usually this restricts the exceptional solution to belong to some finite set. Once B_0 will be bounded, the rest of the proof consists in checking that the given equation has no nontrivial solution in the finite set: it is a finite problem - which does not mean that it is always trivial! But we do not address this issue here: our concern is only to deduce from $0 < |\Lambda| \leq e^{-\delta B_0}$ an upper bound for B_0.

Any nontrivial lower bound for $|\Lambda|$ (like Theorem 1.9, which Gel'fond proved in fact for algebraic α's in place of rational a's) will do. For instance, from Fel'dman's estimate $|\Lambda| \geq B_0^{-C}$ with $C = C(m, \lambda_1, \ldots, \lambda_m)$ we deduce

$$\frac{B_0}{\log B_0} \leq \frac{C}{\delta}.$$

Before Fel'dman's result was available, in part IV of [B 1966], A. Baker proved, under the hypothesis $0 < |\Lambda| \leq e^{-\delta B_0}$, the upper bound

$$B_0 < \left(4^{m^2}\delta^{-1}D_+^{2m}\log A\right)^{(2m+1)^2},$$

where $A = \max\{4, A_1, \ldots, A_m\}$ and $D_+ = \max\{4, D\}$.

Such an estimate has been refined by N. I. Feld'man in [F 1968], and then by A. Baker in part II of [B 1972], who proved an estimate of the form

$$|\Lambda| > \left(\frac{\delta}{B_m}\right)^{C\log A_m} e^{-\delta B'} \quad \text{for any } \delta \text{ with } \quad 0 < \delta \leq \frac{1}{2}, \tag{10.15}$$

with

$$B' = \max\{2, |b_1|, \ldots, |b_{m-1}|\}, \quad B_m = \max\{2, |b_m|\}$$

and C depends only on $m, D, A_1, \ldots, A_{m-1}$. Our Corollary 9.24, is a refinement of (10.15).

Using (10.15) with $\delta = 1/B_0$ yields

$$|\Lambda| \geq e^{-C(\log B_0)(\log A_m)} \tag{10.16}$$

with another constant C which depends also on $m, D, A_1, \ldots, A_{m-1}$. We already remarked that in terms of either B_0 or A_m, the measure (10.16) is best possible.

Another consequence of (10.15) is the following: assume $b_m = -1$ and $0 < |\Lambda| \leq e^{-\epsilon B_0}$ with $0 < \epsilon \leq 1$. Taking $\delta = \epsilon/2$, one deduces from (10.15)

$$B \leq \kappa \log A_m$$

where κ depends only on $m, D, A_1, \ldots, A_{m-1}$ and ϵ. The estimate (10.16) would yield only $B \leq \kappa(\log A_m)(\log\log A_m)$; hence (10.15) produces a sharper conclusion than (10.16).

The next step goes back to [LoxVMW 1987]. One remarks that for any $\lambda > 0$, the minimum of the function $x \mapsto e^{-x}x^{\lambda}$ is obtained for $x = \lambda$, hence there is an optimal value for δ which minimizes the right hand side of (10.15). In fact it is slightly more efficient to look directly where this δ appears in Baker's proof and to optimize at this point. This is how we included this refinement into the final estimate for the homogeneous rational case by introducing the number (10.14) as in Theorem 9.1 (see also [W 1993], [Lau 1994], [LauMN 1995] and [Mat 1998] for instance).

Let us describe an example where the refinement from B_0 to (10.14) is relevant. The following result is Corollary 1 in [LauMN 1995]:

Theorem 10.17. *Let λ_1, λ_2 be two elements in $\mathcal{L}$ and b_1, b_2 two rational integers such that $\Lambda = b_1\lambda_1 + b_2\lambda_2$ is not zero. Define*

$$\alpha_1 = e^{\lambda_1}, \quad \alpha_2 = e^{\lambda_2} \quad \textit{and} \quad D = [\mathbb{Q}(\alpha_1, \alpha_2) : \mathbb{Q}].$$

Let A_1, A_2 and B be positive real numbers satisfying

$$\log A_i = \max\left\{h(\alpha_i), \frac{|\lambda_i|}{D}, \frac{1}{D}\right\} \qquad (i = 1, 2)$$

and

$$B \geq e, \quad B \geq \frac{|b_2|}{D\log A_1} + \frac{|b_1|}{D\log A_2}.$$

Then

$$|\Lambda| \geq \exp\left\{-CD^4(\log A_1)(\log A_2)(\log B)^2\right\}.$$

Theorem 10.17 follows from Theorem 9.1 with $m = 2$, $E = e$ and $E^* = B$ (the main result of [LauMN 1995] includes also a parameter E, but we do not need it here). The proof of [LauMN 1995] is very close to the proof in Chap. 7 of Theorem 7.1 (in the special case $m = 2$, so that only functions of a single variable are needed), but it uses Lemma 7.15 in the same way as we did in Chap. 9 (that means, more efficiently than the estimates in Chap. 7 where we were content with B_0).

Also [LauMN 1995] includes a very small numerical value for the absolute constant C: assuming the two algebraic numbers α_1, α_2 are multiplicatively independent and

$$B \geq e^{21/D},$$

then the conclusion of Theorem 10.17 holds with $C = 31$.

Using an idea of E. Bombieri in [Bo 1993] and [BoCoh 1997], worked out by Y. Bilu and Y. Bugeaud in [BilBu 2000], we deduce from Theorem 10.17 the following result of N. I. Fel'dman [F 1971] (see also [F 1982], Chap. 10, Th. 7.10, [Sp 1982], Chap. 3, Lemma 1.1 and [FNe 1998], Chap. 4, § 1.6, Theorem 4.18):

Corollary 10.18. *Given positive integers n and D, a real number δ in the range $0 < \delta \leq 1$ and elements $\lambda_1, \ldots, \lambda_n$ in $\mathcal{L}$, there exists a positive constant κ with the following property. Let λ be an element of $\mathcal{L}$ and let $b_1, \ldots, b_n$ be rational integers. Define $\alpha = e^\lambda$,*

$$\Lambda = b_1\lambda_1 + \cdots + b_n\lambda_n - \lambda,$$

$$\log A = \max\{e, h(\alpha), |\lambda|\} \quad \text{and} \quad B_0 = \max\{2, |b_1|, \ldots, |b_n|\}.$$

Assume $[\mathbb{Q}(\alpha) : \mathbb{Q}] \leq D$ *and*

$$0 < |\Lambda| \leq e^{-\delta B_0}.$$

Then

$$B_0 \leq \kappa \log A.$$

In case $n = 1$ one may use directly Theorem 10.17 with

$$b_2 = -1, \quad \lambda_2 = \lambda, \quad B = \max\left\{e, 1 + \frac{B_0}{\log A}\right\}$$

and deduce an upper bound for B.

Similarly, for $n \geq 2$ one might apply Corollary 9.24. But the point is that we want to deduce Corollary 10.18 from Theorem 10.17, (the latter involves only two logarithms).

Proof of Corollary 10.18. Given n, D, δ and $\lambda_1, \dots, \lambda_n$, select a sufficiently large integer M. The main condition which will be required for M is

$$M(\log M)^{-2} > 6C\delta^{-1}(D_0 D)^4(1 + n \log A_0)^2,$$

where C is the constant in Theorem 10.17,

$$D_0 = [\mathbb{Q}(\alpha_1, \dots, \alpha_n) : \mathbb{Q}], \quad \log A_0 = \max_{1 \leq i \leq n} \max\{e, h(\alpha_i), |\lambda_i|\}$$

and $\alpha_i = e^{\lambda_i}$ $(1 \leq i \leq n)$.

Assume now $b_1, \dots, b_m$ and λ are such that $\Lambda \neq 0$ and

$$B_0 > M^{2n+1}, \quad B_0 > M^{n+1} \log A.$$

Our goal is to prove $|\Lambda| > e^{-\delta B_0}$. This will plainly imply the conclusion of Corollary 10.18 with, say, $\kappa = M^{2n+1}$.

Denote by N the smallest integer satisfying

$$N^2 \geq B_0 M^{2n+1} \quad \text{and} \quad N \geq M^{n+1} \log A.$$

From our assumption on B_0 we deduce $B_0 \geq N$.

Using Dirichlet's box principle (see § 15.2), we deduce that there exist rational integers $p_0, \dots, p_n$ such that $1 \leq p_0 < M^n$ and

$$\max_{1 \leq i \leq n} \left| p_0 \frac{b_i}{N} - p_i \right| \leq \frac{1}{M}.$$

Let $r \in \mathbb{Z}$ be the integer in the range

$$\frac{N}{p_0} \leq r < \frac{N}{p_0} + 1.$$

We have for $1 \leq i \leq n$

$$|b_i - rp_i| \leq \frac{r}{M} + \frac{b_i}{r-1} \leq \frac{r}{M} + \frac{2B}{r}.$$

Since $p_0 \leq M^n$ and $N^2 \geq B_0 M^{2n+1}$ we have

$$\frac{B_0}{r} \leq \frac{B_0 p_0}{N} \leq \frac{B_0 M^n}{N} \leq \frac{N}{M^{n+1}} \leq \frac{N}{p_0 M} \leq \frac{r}{M},$$

hence

$$|b_i - rp_i| \leq \frac{3r}{M} \quad \text{and} \quad |p_i| \leq \frac{|b_i|}{r} + \frac{3}{M} \leq \frac{2B_0}{r}.$$

From $N > M^{n+1} \log A$ we deduce

$$\frac{r}{M} \geq \frac{N}{M^{n+1}} \geq \log A.$$

Define

$$\widetilde{\lambda}_1 = \sum_{i=1}^{n} p_i \lambda_i, \quad \widetilde{\lambda}_2 = \sum_{i=1}^{n} (b_i - r p_i)\lambda_i + \lambda,$$

so that $\Lambda = r\widetilde{\lambda}_1 + \widetilde{\lambda}_2$, and set

$$\log \widetilde{A}_1 = \frac{2B_0}{r} \cdot n \log A_0 \quad \text{and} \quad \log \widetilde{A}_2 = \frac{r}{M}\bigl(1 + 3n \log A_0\bigr).$$

Since $\log \widetilde{A}_1 \geq 1$ and $\log \widetilde{A}_2 \geq 2r/M$, we have

$$\frac{1}{\log \widetilde{A}_1} + \frac{r}{\log \widetilde{A}_2} \leq M.$$

Hence we may use Theorem 10.17 with

$$\lambda_1, \quad \lambda_2, \quad b_1, \quad b_2, \quad D, \quad A_1, \quad A_2, \quad B,$$

replaced respectively by

$$\widetilde{\lambda}_1, \quad \widetilde{\lambda}_2, \quad r, \quad 1, \quad D_0 D, \quad \widetilde{A}_1, \quad \widetilde{A}_2, \quad M.$$

We deduce

$$\begin{aligned} |\Lambda| &\geq \exp\bigl\{-C(D_0D)^4(\log \widetilde{A}_1)(\log \widetilde{A}_2)(\log M)^2\bigr\} \\ &\geq \exp\Bigl\{-C(D_0D)^4 \cdot \frac{2B_0}{r} \cdot (n \log A_0) \cdot \frac{r}{M} \cdot (1 + 3n \log A_0)(\log M)^2\Bigr\} \\ &> e^{-\delta B_0}. \end{aligned}$$

□

Among the applications of Corollary 10.18 is Fel'dman's improvement of Liouville's Theorem 1.1 (see for instance [Sp 1982], Chap. V, § 5 and [FNe 1998], Chap. 4, § 2.2):

- *For any algebraic number α of degree $d \geq 3$, there exists two positive numbers c and η, which are explicitly known, such that, for any $p/q \in \mathbb{Q}$,*

$$\left|\alpha - \frac{p}{q}\right| > \frac{c}{q^{d-\eta}}.$$

It is a remarkable fact that Gel'fond knew how to deduce such a result from Corollary 10.18, and he also knew how to derive effective measures of linear independence for two logarithms; but his measures involved B_0, and not (10.14) like in Theorem 10.17!

10.4.2 The Degree D

The degree D has not been considered as an important parameter at the early stage of the subject (apart for questions dealing with transcendence measures). The first papers who achieved a strong dependence in D involved Schneider's method for two logarithms [MiW 1978], thanks to the systematic use of Weil's absolute logarithmic height. However Baker's method also yields the best known estimate in this respect, namely D^{m+2} (see Theorem 10.20). One should insist that the dependence on D depends strongly on the choice of the height: we recall that here the parameters A_i and B are defined by means of Weil's absolute logarithmic height, as in Theorems 7.1 and 9.1.

10.4.3 The Parameter E

The idea of introducing the parameter E in [MiW 1978] arose from a work by T. N. Shorey [Sho 1974] who got very sharp estimates when the algebraic numbers α_i are close to 1. This happens in several applications, especially to some problems related to prime number theory. Two examples are Mignotte's paper *A note on the equation* $ax^n - by^n = c$, Acta Arith. **75** (1997), 287–295 and the paper by Bennett and de Weger *On the diophantine equation* $|ax^n - by^n| = 1$, Math. Comp. **67** (1998), 413–438.

This parameter has been incorporated in Baker's method in [W 1980], and used also in [PW 1988a] and [BlaGMMS 1990] (see Theorem 10.20 and 10.21). It is useful not only when the $|\log\alpha_i|$ are very small, but even when they are not too large. In particular it plays an important role for getting sharp transcendence measures [W 1978].

There is a limitation for E: for an algebraic number $\alpha \neq 1$ of degree $\leq D$, and for $A \geq e$ satisfying $\log A \geq \mathrm{h}(\alpha)$, Liouville's estimate (§ 3.5) yields

$$|\alpha - 1| \geq 2^{1-D}A^{-D},$$

hence any logarithm λ of α has $|\lambda| \geq (2A)^{-D}$. On the other hand an extreme case where E can be chosen quite large is given in Exercise 10.4.

This parameter turns out to be quite important in the work of E. Matveev [Mat 1998] for getting a sharp dependence on the number m of logarithms. Also we shall see examples in Chap. 14 where E plays a fundamental role for the dependence on the degree. For instance without the introduction of E the dependence on the degree D in the measure of simultaneous approximation for the two numbers $2^{\sqrt[3]{2}}$ and $2^{\sqrt[3]{4}}$ in § 14.1.3 would not be sharp enough to yield Gel'fond's result of algebraic independence of these two numbers in § 15.3.2. It is interesting to notice here that the p-adic analog of this algebraic independence result is not yet proved, and the main reason for that is the fact that E cannot be chosen large enough in the p-adic case. A p-adic analog of the parameter E has been obtained by Y. Bugeaud [*Linear forms in p-adic logarithms and the Diophantine equation* $\frac{x^n-1}{x-1} = y^q$, Math. Proc. Cambridge Phil. Soc., (1999) **127**, 373–381]; notice however that this improvement does not yield so spectacular results as in the Archimedean case and is useful only in some (very) particular cases.

10.4.4 The Number m of Logarithms

The number $C(m)$ occurring in the conclusion of Theorem 9.1 depends only on the number m of logarithms.

In the early 70's, the estimates obtained by Baker and Stark for instance (in connection with Gauss' class number problem, which they solved for imaginary quadratic fields with class number one and two) involved a function $C(m)$ which grows like the exponential of m^2 (one should say that the rest of the estimate was not as sharp as it is now in Theorem 9.1, but this is a different issue). An improvement of this value of $C(m)$ has been achieved by T. N. Shorey [Sho 1976], who got $C(m) = m^{cm}$ for some absolute constant c by introducing small steps in Baker's inductive argument. J. H. Loxton and A. J. van der Poorten [LoxV 1976] have shown that c can be replaced by $2 + \epsilon(m)$ with $\epsilon(m) \to 0$ as $m \to \infty$, and even by $1 + \epsilon(m)$ under Kummer's condition (10.13).

In [W 1980] (see Theorem 10.20), it was shown that in the final estimate one needs only $m^m c^m$ under Kummer's condition (10.13), $m^{2m} c^m$ without any Kummer's condition (here c is an absolute constant with $c > 1$). More precisely, the final descent costs $c^m m^m$ for the constant $C(m)$.

Using Matveev's arguments in [Mat 1998] (Theorem 10.24) one should be able to prove that the number $C(m)$ in Theorem 9.1 may be replaced by c^m under Kummer's condition. Hence in that case the main dependence on m will arise from the product $\log A_1 \cdots \log A_m$.

Also one expects Theorem 9.1 holds with $C(m) = c^m m^m$ without Kummer's condition. However one should be careful here: removing hypothesis (10.13) requires to measure the height of the coefficients with the parameter B_0 and not with (10.14). Furthermore, this final descent introduces $\max\{B_0, \log A_m\}$ in place of B_0 in the estimate.

One conjectures that Theorem 9.1 holds with $C(m) = c^m$ without any further assumption like (10.13), but this is still an open problem.

10.4.5 The Numerical Constant

Computation of the numerical value of the constants appearing in the final estimates are particularly significant for applications, and several works have been devoted to get sharp estimates. The first completely explicit measure of linear independence of logarithms is due to A. Schinzel [S 1967] for two logarithms and to A. Baker [B 1966] for m logarithms. Considerable improvements have been obtained later, motivated by the wide range of applications. Not all the arguments have been explained here. For instance Blaschke factors (see Exercise 4.3) have proved to be very efficient in this respect (see [MiW 1978] and [BWü 1993], § 18).

The smallest numerical constants so far occurring in the final estimate for two logarithms are given in [LauMN 1995] (an example is given in Theorem 10.17), for three logarithms in [BeBGMS 1997] (there is also some unpublished work by P. Voutier dealing with the case $m = 3$), and for m logarithms in [Mat 1998].

10.4.6 A Sample of Recent Estimates

We quote some linear independence measures from the literature. Each of the six theorems below refers to a text where a completely explicit estimate is provided. But the notation and assumptions (for $A_1, \dots, A_m, B$ for instance) differ from one text to another. Here we use the same notation for all of the six results, but we do not give a numerical value for the absolute constant C. We insist that the results we quote are only consequences of the original statements, and we recommend the interested reader to see the corresponding paper for a more precise result.

Let $m \geq 1$ be a positive integer, $\alpha_1, \dots, \alpha_m$ nonzero algebraic numbers, $\beta_0, \dots, \beta_m$ algebraic numbers, $\lambda_1, \dots, \lambda_m$ logarithms of $\alpha_1, \dots, \alpha_m$ respectively, D the degree over $\mathbb{Q}$ of the number field

$$\mathbb{Q}(\alpha_1, \dots, \alpha_m, \beta_0, \dots, \beta_m).$$

Assume that the number

$$\Lambda = \beta_0 + \beta_1\lambda_1 + \cdots + \beta_m\lambda_m$$

is nonzero. As before, in the homogeneous rational case where $\beta_0 = 0$ and $\beta_i \in \mathbb{Z}$, we write $\beta_i = b_i$ and

$$\Lambda = b_1\lambda_1 + \cdots + b_m\lambda_m.$$

Let $A_1, \dots, A_m, B, E$ and E^* be positive numbers with $E \geq e$ and

$$\log A_i = \max\left\{h(\alpha_i), \frac{E}{D}|\lambda_i|, \frac{\log E}{D}\right\} \qquad (1 \leq i \leq m).$$

Theorem 10.19 [B 1977]. *Assume $\lambda_i = \log \alpha_i$ $(1 \leq i \leq m)$, where the logarithms have their principal values. Assume also $A_i \geq e$. Define*

$$E = e \quad \textit{and} \quad E^* = \max\{e, \log A_1, \dots, \log A_{m-1}\}.$$

<u>In the general case</u>, assume B satisfies

$$\log B \geq \max_{0 \leq i \leq m} \mathrm{h}(\beta_i) \quad B \geq E^* \quad \textit{and} \quad B \geq \log A_m.$$

<u>In the homogeneous rational case</u>, define

$$B = \max\{2, |b_1|, \dots, |b_m|\}.$$

Then

$$|\Lambda| \geq \exp\left\{-C^m m^{200m} D^{200m} (\log A_1) \cdots (\log A_m)(\log B)(\log E^*)\right\}.$$

This result, which is a consequence of Theorem 9.1 (taking Proposition 9.18 into account), includes most of the known results in 1977, apart from some already quoted (in § 10.4.1) refinements in the homogeneous rational case, providing an upper bound for B under the assumption $|\Lambda| < e^{-\delta B}$ for some δ in the range $0 < \delta \le 1$.

Theorem 10.20* [W 1980]. *Define*

$$E^* = \max\{E, D, \log A_1, \dots, \log A_{m-1}\}$$

and assume

$$\log B \ge \max_{0 \le i \le m} \mathrm{h}(\beta_i), \quad B \ge E^* \quad \text{and} \quad B \ge \log A_m.$$

Then

$$|\Lambda| \ge \exp\left\{-C^m m^{2m} D^{m+2} (\log A_1) \cdots (\log A_m)(\log B)(\log E^*)(\log E)^{-m-1}\right\}.$$

This is almost a consequence of Theorem 9.1: only the constant $C(m)$ in Theorem 9.1 is replaced here by $C^m m^{2m}$.

Moreover, in the case where Kummer's condition (10.13) is satisfied, one may replace m^{2m} by m^m in the conclusion of Theorem 10.20.

Theorem 10.21* [PW 1988a]. *Define* $E^* = \max\{E, D\}$ *and assume*

$$\log A_j \ge \frac{m}{D} \log E, \quad (1 \le j \le m),$$

$$\log B \ge \max_{0 \le i \le m} \mathrm{h}(\beta_i) \quad \text{and} \quad B \ge \max\{E^*, \log A_1, \dots, \log A_m\}.$$

Then

$$|\Lambda| \ge \exp\left\{-C^m m^{2m} D^{m+2} (\log A_1) \cdots (\log A_m)(\log B)(\log E^*)(\log E)^{-m-1}\right\}.$$

This result is Theorem 2.1 of [PW 1988a] and refers to the general case (as does Theorem 10.20). Here also, when Kummer's condition (10.13) holds, one may replace m^{2m} by m^m.

The conclusions of Theorems 10.20 and 10.21 are the same, but the term $\log E^*$ is smaller in Theorem 10.21.

A similar statement has been proved by G. Wüstholz in [Wü 1988]: he removes the factor $\log E^*$ in the conclusion of Theorem 10.19 (assuming $\beta_0 = 0$).

A sharp estimate for the numerical value of the constant C in Theorem 10.21 is given in part I of [BlaGMMS 1990].

From now on we assume that we are in the homogeneous rational case.

A refinement of Theorem 10.21 is provided in Theorem 2.2 of [PW 1988a], but the dependence of the final constant in terms of m is not explicitly given. This refinement reads as follows:

- *Define* $E^* = \max\{E, D\}$ *and assume* $b_m \neq 0$,

$$B \geq e, \quad B \geq E^{1/D} \quad \textit{and} \quad B \geq \max_{1\leq j\leq m-1}\left\{\frac{|b_m|}{\log A_j} + \frac{|b_j|}{\log A_m}\right\}.$$

Then

$$|\Lambda| \geq \exp\bigl\{-C(m)D^{m+2}(\log A_1)\cdots(\log A_m)(\log B)(\log E^*)(\log E)^{-m-1}\bigr\},$$

where $C(m)$ *depends only on* m.

(This is a consequence of Theorem 9.1).

In part II of [BlaGMMS 1990], an explicit value for this constant $C(m)$ is computed, but only under additional restrictions: the authors assume

$$B \geq \max\{E^*, \log A_1, \ldots, \log A_m\},$$

and they also assume that Kummer's condition (10.13) is satisfied.

Theorem 10.22 [W 1993]. *Assume* $b_m \neq 0$,

$$E^* \geq E^{1/D}, \quad E^* \geq \frac{D}{\log E}$$

and

$$B \geq E^*, \quad B \geq \max_{1\leq j\leq m-1}\left\{\frac{|b_m|}{\log A_j} + \frac{|b_j|}{\log A_m}\right\}.$$

Then

$$|\Lambda| \geq \exp\bigl\{-C^m m^{3m} D^{m+2}(\log A_1)\cdots(\log A_m)(\log B)(\log E^*)(\log E)^{-m-1}\bigr\}.$$

A proof of Theorem 10.22 has been given in Chap. 9, using interpolation determinants. In [W 1993] the proof uses an auxiliary function, and for this reason the conclusion is (very marginally) weaker: either the stated lower bound for $|\Lambda|$ holds, or else

$$|\Lambda| \geq \exp\bigl\{-C^m m^{3m} D^2 \log A\bigr\},$$

where $A = \max\{A_1, \ldots, A_m\}$.

Theorem 10.23* [BWü 1993]. *Define*

$$E = e, \quad E^* = eD \quad \textit{and} \quad B = \max\{|b_1|, \ldots, |b_m|, e^{1/D}\}.$$

Then

$$|\Lambda| \geq \exp\bigl\{-C^m m^{2m} D^{m+2}(\log A_1)\cdots(\log A_m)(\log B)(\log E^*)\bigr\}.$$

Theorem 10.24* [Mat 1998]. *Assume* $b_m \neq 0$ *and* $E = e$. *Assume further that Kummer's condition (10.13) is satisfied. Define*

$$E^* = \max\{e, D, \log A_1, \ldots, \log A_{m-1}\}$$

and assume furthermore

$$B \geq e, \quad B \geq \max_{1 \leq j \leq m} \frac{|b_j| \log A_j}{\log A_m}.$$

Then

$$|\Lambda| \geq \exp\{-C^m D^{m+2} (\log A_1) \cdots (\log A_m)(\log B)(\log E^*)\}.$$

In [Y 1998], Yu Kunrui proves p-adic estimates which may be considered as ultrametric analogues to the results of [BWü 1993] and [Mat 1998].

Exercises

Exercise 10.1.

a) Let K be an algebraically closed field of zero characteristic and $\beta_1, \ldots, \beta_{m-1}$ elements of K such that $1, \beta_1, \ldots, \beta_{m-1}$ are linearly independent over $\mathbb{Q}$. On the ring $K[X_1^{\pm 1}, \ldots, X_m^{\pm 1}]$, introduce derivative operators by

$$\mathcal{D}_i = X_i \frac{\partial}{\partial X_i} + \beta_i X_m \frac{\partial}{\partial X_m} \quad (1 \leq i \leq m-1)$$

and, for $\underline{\sigma} \in \mathbb{N}^{m-1}$,

$$\mathcal{D}^{\underline{\sigma}} = \mathcal{D}_1^{\sigma_1} \cdots \mathcal{D}_{m-1}^{\sigma_{m-1}}.$$

Let $m \geq 2$ be an integer, $\alpha_1, \ldots, \alpha_m$ nonzero elements of K which generate a multiplicative subgroup of $K^\times$ of rank $\geq m-1$. Let T, S_0, S_1 be positive integers satisfying

$$S_0 \geq 2mT, \quad (S_0+1)(2S_1+1) > m!(m-1)!2T$$

and

$$S_0^{m-1}(2S_1+1) > m!(m-1)!(2T)^m.$$

For $\underline{t} \in \mathbb{Z}^m$, $\underline{\sigma} \in \mathbb{N}^{m-1}$ and $s \in \mathbb{Z}$, define $\gamma_{\underline{t}}^{(\underline{\sigma} s)} \in K$ as the value, at the point

$$(\alpha_1^s, \ldots, \alpha_m^s) \in (K^\times)^m,$$

of the polynomial

$$\mathcal{D}^{\underline{\sigma}}(X_1^{t_1} \cdots X_m^{t_m}) \in K[X_1^{\pm 1}, \ldots, X_m^{\pm 1}].$$

Consider the following matrix:

$$M = \left(\gamma_{\underline{t}}^{(\underline{\sigma} s)}\right)_{\substack{\underline{t} \\ (\underline{\sigma}, s)}}$$

where the index of rows $\underline{t}$ runs over the elements in $\mathbb{Z}^m$ with $|\underline{t}| \leq T$, while the index of columns $(\underline{\sigma}, s)$ runs over the elements of $\mathbb{N}^{m-1} \times \mathbb{Z}$ with $\|\underline{\sigma}\| \leq mS_0$ and $|s| \leq mS_1$.

Using Theorem 8.1, show that the matrix M has rank $(2T+1)^m$.

b) Let $\lambda_1, \ldots, \lambda_m, \beta_1, \ldots, \beta_{m-1}$ be complex numbers, with $m \geq 2$, $(\lambda_1, \ldots, \lambda_m) \neq (0, \ldots, 0)$ and $1, \beta_1, \ldots, \beta_{m-1}$ linearly independent over $\mathbb{Q}$. Assume

$$\lambda_m = \beta_1 \lambda_1 + \cdots + \beta_{m-1} \lambda_{m-1}.$$

Let $E \geq e$ be a real number and T, S_0, S_1, L be four integers, all greater than one, satisfying

$$L = (2T+1)^m, \quad S_0 \geq 2(m!)^2 T, \quad T \geq 8m^2$$

and

$$S_0^{m-1} S_1 > 2^m (m!)^2 T^m.$$

Using a) with $K = \mathbb{C}$ and $\alpha_i = e^{\lambda_i}$, show that there exists a polynomial f in $\mathbb{Z}[X_1^{\pm 1}, \dots, X_m^{\pm 1}, Y_1, \dots, Y_{m-1}]$ satisfying

$$\deg f \leq mL\big(m(T+1)S_1 + S_0\big), \quad \mathrm{L}(f) \leq L!(2T)^{mLS_0},$$

and

$$0 < |f(\alpha_1, \dots, \alpha_m, \beta_1, \dots, \beta_{m-1})| \leq \exp\left\{ -\frac{1}{3} L^{1+(1/(m-1))} \log E + L\big(mS_0 \log(TE) + c_0 S_0 + c_0 T S_1 E\big)\right\}$$

where

$$c_0 = \max\left\{m \log \max_{1 \leq i \leq m-1}(1+|\beta_i|)\,;\ 1 + m(|\lambda_1| + \cdots + |\lambda_m|)\right\}.$$

c) From b), using Lemmas 1.7 and 2.1, deduce Baker's Homogeneous Theorem 1.5.

Exercise 10.2.
a) Under the assumptions of Proposition 10.3, compute an explicit positive number ϵ such that, for each $\underline{\theta} \in (\mathbb{C}^\times)^m \times \mathbb{C}^m$ satisfying

$$\max_{1 \leq i \leq m} |\theta_i - \alpha_i| \leq \epsilon \quad \text{and} \quad \max_{0 \leq i \leq m-1} |\theta_{m+i} - \beta_i| \leq \epsilon,$$

there exists a polynomial f satisfying the conclusion of Proposition 10.3 with the number

$$f(\alpha_1, \dots, \alpha_m, \beta_0, \dots, \beta_{m-1})$$

replaced by $f(\underline{\theta})$.
b) Deduce an explicit measure of linear independence for logarithms of algebraic numbers.
c) Extend the result in a) including multiplicities and deduce an improved estimate for b).

Hint. *Compare with Exercise 15.4.*

Exercise 10.3. Let $\lambda_1, \dots, \lambda_m$ be complex numbers and $\beta_1, \dots, \beta_m$ algebraic numbers such that

$$\Lambda = \beta_1 \lambda_1 + \cdots + \beta_m \lambda_m \neq 0.$$

For $i = 1, \dots, n$ define $\alpha_i = e_i^{\lambda}$ and $A_i = \max |\lambda_i|$. Let D be the degree of a number field containing $\beta_1, \dots, \beta_m$. Define B by

$$\log B = \max_{1 \leq i \leq m} \mathrm{h}(\beta_i).$$

Assume that there exists an algebraic subgroup G^* of $\mathbb{G}_{\mathrm{m}}^m$, defined by equations of degree $\leq L$, and a positive integer $s \geq 1$, such that

$$(\alpha_1^s, \dots, \alpha_m^s) \in G^* \quad \text{and} \quad T_e(G^*) \subset \big\{\underline{z} \in \mathbb{C}^m\,;\ \beta_1 z_1 + \cdots + \beta_m z_m = 0\big\}.$$

Check

$$|\Lambda| \geq \exp\big\{ -mD\log B - mD\log(mL) - \log(m^2 As)\big\}.$$

Hint. *See Lemma 3.7 of* [PW 1988a].

Exercise 10.4. For any integer $m \geq 2$, show the existence of a number $c(m) > 0$ with the following property.

Let $\alpha_1, \ldots, \alpha_m$ be nonzero algebraic numbers in a number field of degree $\leq D$ and $b_1, \ldots, b_m$ rational integers such that $\alpha_1^{b_1} \cdots \alpha_m^{b_m} \neq 1$. Let A, B and κ be positive real numbers with $A \geq e$, $B \geq e$ and $0 < \kappa \leq 1$. Assume

$$|\alpha_i - 1| \leq A^{-D/\kappa} D \log A \quad \text{for} \quad 1 \leq i \leq m.$$

Assume further

$$B \geq \max\{2, D, \log A\} \quad \text{and} \quad B \geq \max_{1 \leq i \leq m} \frac{2|b_i|}{\log A}.$$

Furthermore, assume

$$A \geq \max_{1 \leq i \leq m} \mathrm{h}(\alpha_i).$$

Then

$$\left|\alpha_1^{b_1} \cdots \alpha_m^{b_m} - 1\right| \geq B^{-c(m)D\kappa^{-m}}.$$

Hint. *Using Exercise 1.1.b, check that the principal value λ_i of the logarithm of α_i satisfies, for $1 \leq i \leq m$,*

$$|\lambda_i| \leq \frac{e}{e-1}|\alpha_i - 1| \qquad (1 \leq i \leq m).$$

Use Theorem 9.1 with $E = A^{\kappa D}$ and $E^* = A^{\kappa}$.

Remark. See [Sho 1974] for arithmetical applications.

Exercise 10.5.
a) For any $\vartheta \in \mathbb{R}$, there are infinitely many $(p, q) \in \mathbb{Z}^2$ with $p > 0$, $q > 0$ such that

$$\left|\vartheta - \log \frac{p}{q}\right| \leq \frac{1}{pq}.$$

b) Fix $(b_1, b_2) \in \mathbb{Z}^2$ with $b_1 > 0$ and $b_2 < 0$. For $A > 2$, define

$$\Phi(A) = \min\left\{|a_1^{b_1} a_2^{b_2} - 1| \,;\, (a_1, a_2) \in \mathbb{Z}^2,\ 2 \leq a_i \leq A,\ a_1^{b_1} a_2^{b_2} \neq 1\right\}.$$

Check

$$\lim_{A \to \infty} \frac{\log \Phi(A)}{\log A} < 0.$$

Exercise 10.6. Let K be a number field and $\alpha_1, \ldots, \alpha_m$ elements in $K^\times$.
a) Check that the following two conditions are equivalent.

(i) $[K(\sqrt{\alpha_1}, \ldots, \sqrt{\alpha_m}) : K] = 2^m$.

(ii) For $(h_1, \ldots, h_m) \in \mathbb{Z}^m$ and $\gamma \in K^\times$, the relation $\alpha_1^{h_1} \cdots \alpha_m^{h_m} = \gamma^2$ implies that each of the integers $h_1, \ldots, h_m$ is even.

Hint. *The field $K(\sqrt{\alpha_1}, \ldots, \sqrt{\alpha_m})$ is an abelian extension of K whose Galois group is described by Kummer's theory; see for instance* [L 1993], *Chap. VI, § 8, Th. 8.1.*

b) Deduce that if Kummer's condition is satisfied for $\beta_0 + \beta_1\lambda_1 + \cdots + \beta_m\lambda_m$, then $\lambda_1, \ldots, \lambda_m$ are linearly independent over $\mathbb{Q}$.

11. Points Whose Coordinates are Logarithms of Algebraic Numbers

The main result of this chapter (Theorem 11.5) includes Baker's Theorems 1.5 and 1.6 (hence also Hermite-Lindemann's Theorem 1.2 as well as Gel'fond-Schneider's Theorem 1.4), the six exponentials Theorem 1.12, and much more (especially extensions of these results to several variables). It provides information on the distribution of elements of $\mathcal{L}^d$ into the vector space $\mathbb{C}^d$, and more generally on the distribution of elements of $\mathbb{C}^d$ whose coordinates are linear combinations of logarithms of algebraic numbers.

11.1 Introduction

11.1.1 The $\mathbb{Q}$-Vector Subspace $\mathcal{L}^d$ of $\mathbb{C}^d$

The set $\mathcal{L}^d$ is a $\mathbb{Q}$-vector subspace of $\mathbb{C}^d$:

$$\mathcal{L}^d = \left\{(\log\alpha_1, \dots, \log\alpha_d)\,;\, (\alpha_1, \dots, \alpha_d) \in (\overline{\mathbb{Q}}^\times)^d\right\}.$$

Our first goal is to study the intersection of $\mathcal{L}^d$ with a $\mathbb{C}$-vector subspace $\mathcal{V}$ of $\mathbb{C}^d$.

A preliminary remark is that if $\mathcal{V}$ is a vector subspace of $\mathbb{C}^d$ which contains a nonzero rational point, that means an element $\underline{b} = (b_1, \dots, b_d) \neq 0$ of $\mathbb{Q}^d$, then $\mathcal{V}$ contains $(b_1\lambda, \dots, b_d\lambda)$ for any $\lambda \in \mathcal{L}$; therefore in this case the $\mathbb{Q}$-vector space $\mathcal{V} \cap \mathcal{L}^d$ has infinite dimension.

Next, assume that $\mathcal{V}$ is a complex vector subspace of $\mathbb{C}^d$ which is *rational* over $\overline{\mathbb{Q}}$ and such that $\mathcal{V} \cap \mathbb{Q}^d = \{0\}$. One deduces from Baker's Theorem that $\mathcal{V} \cap \mathcal{L}^d = \{0\}$. In fact it is not difficult (see Exercise 1.5) to check that this statement is equivalent to Baker's homogeneous Theorem 1.5. This settles the problem in the case of a vector space which is rational over $\overline{\mathbb{Q}}$:

- *If $\mathcal{V}$ is is rational over $\overline{\mathbb{Q}}$ and $\mathcal{V} \cap \mathbb{Q}^d = \{0\}$, then $\mathcal{V} \cap \mathcal{L}^d = \{0\}$, and otherwise $\mathcal{V} \cap \mathcal{L}^d$ has infinite dimension over $\mathbb{Q}$.*

Consider now a vector subspace $\mathcal{V}$ of $\mathbb{C}^d$ which is not assumed to be rational over $\overline{\mathbb{Q}}$. Of course $\mathcal{V} \cap \mathcal{L}^d$ may contain nonzero elements even if $\mathcal{V} \cap \mathbb{Q}^d = \{0\}$: the most obvious example is a complex line $\mathbb{C}\underline{\lambda}$ spanned by an element $\underline{\lambda} = (\lambda_1, \dots, \lambda_d)$ in $\mathcal{L}^d$ with $\dim_{\mathbb{Q}}(\mathbb{Q}\lambda_1 + \cdots + \mathbb{Q}\lambda_d) \geq 2$.

In § 11.2 we deal with the special case $\dim_{\mathbb{C}}(\mathcal{V}) = 1$ and show that the six exponentials Theorem 1.12 is equivalent to the following result:

- *If $\mathcal{V}$ is a vector subspace of $\mathbb{C}^2$ such that* $\dim_{\mathbb{C}}(\mathcal{V}) = 1$ *and* $\mathcal{V} \cap \mathbb{Q}^2 = \{0\}$, *then*

$$\dim_{\mathbb{Q}}(\mathcal{V} \cap \mathcal{L}^2) \leq 2.$$

On the other hand, the four exponentials Conjecture 1.13 is equivalent to:

(?) *If $\mathcal{V}$ is a vector subspace of $\mathbb{C}^d$ such that* $\dim_{\mathbb{C}}(\mathcal{V}) = 1$ *and* $\mathcal{V} \cap \mathbb{Q}^d = \{0\}$, *then* $\dim_{\mathbb{Q}}(\mathcal{V} \cap \mathcal{L}^d) \leq 1$.

The hypotheses imply $d \geq 2$. Plainly, it would be sufficient to prove the conclusion in the case $d = 2$. One can establish this estimate for $d \geq 3$, but some extra hypothesis is needed (otherwise one would deduce the four exponentials Conjecture 1.13), namely that $\mathcal{V}$ is not contained in a subspace of $\mathbb{C}^d$ of dimension $< d$ rational over $\mathbb{Q}$:

- *Let $\mathcal{V}$ be a $\mathbb{C}$-vector subspace of $\mathbb{C}^d$ of dimension* 1. *Assume that $\mathbb{C}^d$ itself is the only subspace of $\mathbb{C}^d$ rational over $\mathbb{Q}$ which contains $\mathcal{V}$. If $d \geq 3$, then* $\dim_{\mathbb{Q}}(\mathcal{V} \cap \mathcal{L}^d) \leq 1$.

Again, this statement is equivalent to the six exponentials Theorem 1.12.

The situation concerning higher dimensional vector subspaces of $\mathbb{C}^d$ will be dealt with in § 11.5. We shall prove (Corollary 11.6) that *the condition $\mathcal{V} \cap \mathbb{Q}^d = \{0\}$ is necessary and sufficient for $\mathcal{V} \cap \mathcal{L}^d$ to be of finite dimension over $\mathbb{Q}$*. Moreover,

- *if $\mathcal{V} \cap \mathbb{Q}^d = \{0\}$, then*

$$\dim_{\mathbb{Q}}(\mathcal{V} \cap \mathcal{L}^d) \leq n(n+1)$$

  where $n = \dim_{\mathbb{C}}(\mathcal{V})$.

It is expected that this dimension is at most $n(n+1)/2$ and this would be best possible.

11.1.2 The $\mathbb{Q}$-Vector Subspace $\mathcal{L}_G$ of $\mathbb{C}^d$

A more general situation will be considered in § 11.3. Let d_0 and d_1 be two nonnegative rational integers with $d = d_0 + d_1 > 0$ (in fact the case $d_1 = 0$ will turn out not to be interesting). Consider the linear algebraic groups

$$G_0 = \mathbb{G}_{\mathrm{a}}^{d_0}, \quad G_1 = \mathbb{G}_{\mathrm{m}}^{d_1} \quad \text{and} \quad G = G_0 \times G_1.$$

The exponential map

$$\exp_G : \mathbb{C}^d \longrightarrow G(\mathbb{C}) = \mathbb{C}^{d_0} \times (\mathbb{C}^\times)^{d_1}$$

is a surjective morphism of groups, with kernel the group of *periods*:

$$\Omega_G = \{0\} \times (2i\pi\mathbb{Z})^{d_1} \subset \mathbb{C}^{d_0} \times \mathbb{C}^{d_1}.$$

The $\mathbb{Q}$-vector space

$$\mathcal{L}_G = \exp_G^{-1} G(\overline{\mathbb{Q}}) = \overline{\mathbb{Q}}^{d_0} \times \mathcal{L}^{d_1}$$

consists of all elements in $\mathbb{C}^d$ whose images under the exponential map of $G(\mathbb{C})$ are in the group of algebraic points of G over $\overline{\mathbb{Q}}$, namely in $\overline{\mathbb{Q}}^{d_0} \times (\overline{\mathbb{Q}}^{\times})^{d_1}$. An element in $\mathcal{L}_G$ can be written

$$(\beta_1, \ldots, \beta_{d_0}, \lambda_1, \ldots, \lambda_{d_1}),$$

where $\beta_1, \ldots, \beta_{d_0}$ are algebraic numbers, while $\lambda_1, \ldots, \lambda_{d_1}$ are logarithms of algebraic numbers. We now repeat the previous discussion in this more general framework.

The set $\mathcal{L}_G$ is a $\mathbb{Q}$-vector subspace of $\mathbb{C}^d$. Our second goal is to study the intersection of $\mathcal{L}_G$ with a $\mathbb{C}$-vector subspace $\mathcal{V}$ of $\mathbb{C}^d$.

A preliminary remark is that if $\mathcal{V}$ is a vector subspace of $\mathbb{C}^d$ which contains a point $\underline{b} = (0, \ldots, 0, b_1, \ldots, b_{d_1}) \neq 0$ in $\{0\} \times \mathbb{Q}^{d_1}$, then $\mathcal{V}$ contains $(0, \ldots, 0, b_1\lambda, \ldots, b_{d_1}\lambda)$ for any $\lambda \in \mathcal{L}$; therefore in this case the $\mathbb{Q}$-vector space $\mathcal{V} \cap \mathcal{L}_G$ has infinite dimension. Another circumstance where $\mathcal{V} \cap \mathcal{L}_G$ has infinite dimension over $\mathbb{Q}$ is when $\mathcal{V}$ contains a nonzero element $\underline{\beta} = (\beta_1, \ldots, \beta_{d_0}, 0, \ldots, 0)$ of $\overline{\mathbb{Q}}^{d_0} \times \{0\}$, since in this case $\mathcal{V}$ contains all elements of the form $(\beta_1\gamma, \ldots, \beta_{d_0}\gamma, 0, \ldots, 0)$ with $\gamma \in \overline{\mathbb{Q}}$.

Therefore it is natural to consider first the $\mathbb{C}$-subspaces $\mathcal{V}$ of $\mathbb{C}^d$ for which

$$\mathcal{V} \cap \left(\{0\} \times \mathbb{Q}^{d_1}\right) = \{0\} \quad \text{and} \quad \mathcal{V} \cap \left(\overline{\mathbb{Q}}^{d_0} \times \{0\}\right) = \{0\}. \tag{11.1}$$

Notice that this condition means exactly that no algebraic subgroup G^* of G of positive dimension has its tangent space contained in $\mathcal{V}$.

Any subspace $\mathcal{V}$ of $\mathbb{C}^d$ contains a unique maximal subspace of the form $E_0 \times E_1$, where E_0 is a subspace of $\mathbb{C}^{d_0}$ rational over $\overline{\mathbb{Q}}$ and E_1 is a subspace of $\mathbb{C}^{d_1}$ rational over $\mathbb{Q}$. *We denote this maximal subspace by* $\mathcal{V}_{\max}$. It is also defined as follows: $\mathcal{V}_{\max} = E_0 \times E_1$, where E_0 is the subspace of $\mathbb{C}^{d_0}$ which is spanned by $\mathcal{V} \cap \left(\overline{\mathbb{Q}}^{d_0} \times \{0\}\right)$, while E_1 is the subspace of $\mathbb{C}^{d_1}$ which is spanned by $\mathcal{V} \cap \left(\{0\} \times \mathbb{Q}^{d_1}\right)$. Hence

condition (11.1) holds if and only if $\mathcal{V}_{\max} = \{0\}$

Warning. *This subspace* $\mathcal{V}_{\max}$ *depends on* (d_0, d_1), *not only on* d.

For a complex vector subspace $\mathcal{V}$ of $\mathbb{C}^d$ which is *rational* over $\overline{\mathbb{Q}}$ and such that (11.1) holds, $\mathcal{V} \cap \mathcal{L}_G = \{0\}$, as shown by Baker's Theorem 1.6 (see Exercise 1.5).

Consider now a vector subspace $\mathcal{V}$ of $\mathbb{C}^d$ which is not assumed to be rational over $\overline{\mathbb{Q}}$. We shall prove (Corollary 11.6) that *(11.1) is a necessary and sufficient condition for* $\mathcal{V} \cap \mathcal{L}_G$ *to be of finite dimension over* $\mathbb{Q}$.

The next goal is to produce an upper bound for this dimension, when it is finite (that is under condition (11.1)). In order to include both the situation where the given space is rational over $\overline{\mathbb{Q}}$ and the general case, we introduce a complex vector subspace $\mathcal{W}$ of $\mathbb{C}^d$, rational over $\overline{\mathbb{Q}}$, of dimension say ℓ_0, which is contained into $\mathcal{V}$.

The case $\mathcal{W} = \{0\}$, $\ell_0 = 0$ will of course not be excluded, while the other extreme case $\mathcal{W} = \mathcal{V}$, $\ell_0 = \dim_{\mathbb{C}}(\mathcal{V})$ occurs when $\mathcal{V}$ itself is rational over $\overline{\mathbb{Q}}$ (this will be related with Baker's method). The upper bound for the dimension will be, roughly speaking,

$$\dim_{\mathbb{Q}}(\mathcal{V} \cap \mathcal{L}_G) \leq \frac{d_1(n-\ell_0)}{d-n}, \tag{11.2}$$

with $n = \dim_{\mathbb{C}}(\mathcal{V})$, but this will be shown to be true only under some extra conditions. One example where hypothesis (11.1) is sufficient for (11.2) to hold is when $\mathcal{V}$ is a hyperplane (that is $n = d - 1$ — see Corollary 11.6 in § 11.3).

In general, we shall require that d_0 and d_1 are *minimal* in the following sense:

(11.3) *If E_0 is a $\mathbb{C}$-vector subspace of $\mathbb{C}^{d_0}$ which is rational over $\overline{\mathbb{Q}}$ and E_1 is a $\mathbb{C}$-vector subspace of $\mathbb{C}^{d_1}$ which is rational over $\mathbb{Q}$, such that $\mathcal{V} \subset E_0 \times E_1$, then $E_0 = \mathbb{C}^{d_0}$ and $E_1 = \mathbb{C}^{d_1}$.*

This condition means that $\exp_G \mathcal{V}$ is Zariski dense in $G(\mathbb{C})$; this is certainly a natural condition, for otherwise one would replace G by the Zariski closure of $\exp_G \mathcal{V}$.

In § 11.2, we show that the estimate (11.2) is valid under condition (11.3) when $\mathcal{V}$ is a complex line (that is $n = 1$ — see Theorem 11.4).

Concerning condition (11.3), one may notice that there is a unique minimal subspace of $\mathbb{C}^d$ of the form $E_0 \times E_1$, with $E_0 \subset \mathbb{C}^{d_0}$ rational over $\overline{\mathbb{Q}}$ and $E_1 \subset \mathbb{C}^{d_1}$ rational over $\mathbb{Q}$, which contains $\mathcal{V}$. *We denote it by $\mathcal{V}_{\min}$*. Explicitly, $\mathcal{V}_{\min} = E_0 \times E_1$, where E_0 is the intersection of all hyperplanes of $\mathbb{C}^{d_0}$, rational over $\overline{\mathbb{Q}}$, which contain the projection of $\mathcal{V}$ onto $\mathbb{C}^{d_0}$, and similarly, E_1 is the intersection of the hyperplanes of $\mathbb{C}^{d_1}$ rational over $\mathbb{Q}$, which contain the projection of $\mathcal{V}$ onto $\mathbb{C}^{d_1}$. Since orthogonality $\mathcal{V} \mapsto \mathcal{V}^{\perp}$ for the scalar product $(\underline{z}, \underline{w}) \mapsto \underline{z}\underline{w}$, preserves rationality and reverses inclusions, we can write

$$(\mathcal{V}_{\min})^{\perp} = (\mathcal{V}^{\perp})_{\max}.$$

From the definition of $\mathcal{V}_{\min}$ we deduce:

condition (11.3) holds if and only if $\mathcal{V}_{\min} = \mathbb{C}^d$

On the other hand, if (11.3) is not satisfied, then one can apply the results conditional to (11.3) but with d replaced by $d_{\min} = \dim_{\mathbb{C}}(\mathcal{V}_{\min})$, with d_0 replaced by $\dim_{\mathbb{C}}(E_0)$ and d_1 by $\dim_{\mathbb{C}}(E_1)$, where $\mathcal{V}_{\min} = E_0 \times E_1$.

Remark. Assume $d \geq 2$. For a complex line $\mathcal{V}$, condition (11.1) means that $\mathcal{V}$ is not spanned by an element in $\overline{\mathbb{Q}}^{d_0} \times \{0\}$, nor by an element in $\{0\} \times \mathbb{Q}^{d_1}$. In dimension $n = 1$ condition (11.3) implies (11.1).

For a hyperplane $\mathcal{V}$, condition (11.3) means that $\mathcal{V}$ is not defined by an equation

$$\beta_1 z_1 + \cdots + \beta_{d_0} z_{d_0} = 0 \quad \text{for some} \quad (\beta_1, \ldots, \beta_{d_0}) \in \overline{\mathbb{Q}}^{d_0} \setminus \{0\},$$

nor by an equation

$$b_1 z_{d_0+1} + \cdots + b_{d_1} z_d = 0 \quad \text{for some} \quad (b_1, \ldots, b_{d_1}) \in \mathbb{Q}^{d_1} \setminus \{0\}.$$

In codimension $d - n = 1$ condition (11.1) implies (11.3).

The very general result we shall state in § 11.3 does not involve any condition like (11.1) nor (11.3). As we know, if we remove condition (11.1), we cannot expect $\mathcal{V} \cap \mathcal{L}_G$ to be of finite dimension; but we show that *most* elements in $\mathcal{V} \cap \mathcal{L}_G$ belong to a set $\mathcal{L}_{G^*}$ for some algebraic subgroup G^* of G, of positive dimension. This is the meaning of the *Linear Subgroup Theorem* 11.5.

11.1.3 The $\mathbb{Q}$-Vector Subspace $\widetilde{\mathcal{L}}^d$ of $\mathbb{C}^d$

Our last goal (§ 11.6) is to study the $\overline{\mathbb{Q}}$-vector space $\widetilde{\mathcal{L}}$ spanned by 1 and $\mathcal{L}$. As above, we start with a preliminary remark: *if $\mathcal{V}$ is a vector subspace of $\mathbb{C}^d$ such that $\mathcal{V} \cap \overline{\mathbb{Q}}^d \neq \{0\}$, say*

$$\underline{\beta} = (\beta_1, \ldots, \beta_d) \in \mathcal{V} \cap \overline{\mathbb{Q}}^d \setminus \{0\}$$

then $\mathcal{V}$ contains $(\beta_1\lambda, \ldots, \beta_d\lambda)$ for any $\lambda \in \widetilde{\mathcal{L}}$. Therefore in this case the $\overline{\mathbb{Q}}$-vector space $\mathcal{V} \cap \widetilde{\mathcal{L}}^d$ has infinite dimension. We shall prove conversely that *if $\mathcal{V} \cap \overline{\mathbb{Q}}^d = \{0\}$, then the dimension over $\overline{\mathbb{Q}}$ of $\mathcal{V} \cap \widetilde{\mathcal{L}}^d$ is finite.* Moreover, when this condition is fulfilled, we shall give an upper bound for this dimension in Corollary 11.15. In particular we have

$$\dim_{\overline{\mathbb{Q}}}\bigl(\mathcal{V} \cap \widetilde{\mathcal{L}}^d\bigr) \leq n(n+1)$$

where n is the dimension of $\mathcal{V}$. It is expected that this dimension is at most $n(n+1)/2$ and this would be best possible (Lemma 11.20).

11.2 One Parameter Subgroups

Here we restrict the discussion of the previous section § 11.1 to the special case where $\mathcal{V}$ has dimension 1.

Let, as before, d_0 and d_1 be two nonnegative integers with $d = d_0 + d_1 \geq 2$.

11.2.1 The Main Result in Dimension 1

We shall show that the following result is equivalent to the conjunction of the Theorem of Hermite-Lindemann 1.2, the Theorem of Gel'fond-Schneider 1.4 and the six exponentials Theorem 1.12.

Theorem 11.4. *Let $\mathcal{V}$ be a vector subspace of $\mathbb{C}^d$ of dimension* 1.
(1) The $\mathbb{Q}$-vector space $\mathcal{V} \cap \mathcal{L}_G$ has finite dimension if and only if (11.1) holds.
(2) Assume (11.3). Then

$$\dim_{\mathbb{Q}}\bigl(\mathcal{V} \cap \mathcal{L}_G\bigr) \leq \frac{d_1}{d-1}.$$

(3) Assume (11.1). Assume further that $\mathcal{V}$ is rational over $\overline{\mathbb{Q}}$. Then

$$\mathcal{V} \cap \mathcal{L}_G = \{0\}.$$

We now deduce corollaries from Theorem 11.4. As a matter of fact one can, conversely, deduce Theorem 11.4 from these special cases.

11.2.2 On Hermite-Lindemann's Theorem

Assume α is a nonzero algebraic number and β an algebraic number such that $e^\beta = \alpha$. Hence β is a logarithm of an algebraic number, i.e. β belongs to $\mathcal{L}$. We want to deduce $\beta = 0$.

Take $d_0 = d_1 = 1$ so that $d = 2$ and $\mathcal{L}_G = \overline{\mathbb{Q}} \times \mathcal{L}$. The complex line $\mathcal{V} = \mathbb{C}(1, 1)$ in $\mathbb{C}^2$ is rational over $\overline{\mathbb{Q}}$ and contains $(\beta, \beta) \in \mathcal{L}_G$. Notice that $\mathcal{V}$ is neither $\mathbb{C} \times \{0\}$ nor $\{0\} \times \mathbb{C}$; hence condition (11.1) holds. By part 3) of Theorem 11.4, we have $\mathcal{V} \cap \mathcal{L}_G = \{0\}$, which gives $\beta = 0$. □

11.2.3 On Schneider's Solution to Hilbert's Seventh Problem

Assume λ_1 and λ_2 are two elements of $\mathcal{L}$ and β is an algebraic number such that $\lambda_2 = \beta\lambda_1$. Assume $\lambda_1 \neq 0$. We want to prove $\beta \in \mathbb{Q}$.

We choose again $d_0 = d_1 = 1$ so that $d = 2$ and $\mathcal{L}_G = \overline{\mathbb{Q}} \times \mathcal{L}$. Now let $\mathcal{V}$ be the complex line in $\mathbb{C}^2$ of equation $z_2 = \lambda_1 z_1$. Notice that $\mathcal{V} \cap \mathcal{L}_G$ contains the points $(1, \lambda_1)$ and (β, λ_2). Clearly, $\mathcal{V} \neq \{0\} \times \mathbb{C}$. Since $\lambda_1 \neq 0$, we also have $\mathcal{V} \neq \mathbb{C} \times \{0\}$. Therefore (11.3) holds. From part 2) of Theorem 11.4 we deduce $\dim_{\mathbb{Q}}(\mathcal{V} \cap \mathcal{L}_G) \leq 1$. Hence $(1, \lambda_1)$ and (β, λ_2) are linearly dependent over $\mathbb{Q}$ in $\mathcal{L}_G$, which implies that β is rational. □

11.2.4 On Gel'fond's Solution to Hilbert's Seventh Problem

Again, λ_1 and λ_2 are two elements of $\mathcal{L}$ and β is an algebraic number such that $\lambda_2 = \beta\lambda_1$. Assume $\beta \notin \mathbb{Q}$. We want to prove $\lambda_1 = 0$.

Take $d_0 = 0$ and $d_1 = 2$, so that $d = 2$ and $\mathcal{L}_G = \mathcal{L}^2$. Let $\mathcal{V}$ be the complex line in $\mathbb{C}^2$ of equation $z_2 = \beta z_1$, that is $\mathcal{V} = \mathbb{C}(1, \beta)$. Condition (11.1) follows from the assumption $\beta \notin \mathbb{Q}$. From part 3) of Theorem 11.4 one deduces $\mathcal{V} \cap \mathcal{L}_G = \{0\}$. Now $(\lambda_1, \lambda_2) \in \mathcal{V} \cap \mathcal{L}_G$, hence $\lambda_1 = \lambda_2 = 0$. □

11.2.5 On the Six Exponentials Theorem 1.12

Let $x_1, \ldots, x_d$ be $\mathbb{Q}$-linearly independent complex numbers and $y_1, \ldots, y_\ell$ also $\mathbb{Q}$-linearly independent complex numbers. Assume that the $d\ell$ numbers $x_i y_j$ $(1 \leq i \leq d, 1 \leq j \leq \ell)$ all belong to $\mathcal{L}$. We want to deduce $d\ell \leq d + \ell$.

One may obviously assume $d \geq 2$ and $\ell \geq 2$, otherwise the conclusion is plain. Take $d_0 = 0$, so that $d_1 = d$ and $\mathcal{L}_G = \mathcal{L}^d$. Define $\mathcal{V}$ as the complex line in $\mathbb{C}^d$ spanned by the point $(x_1, \ldots, x_d)$. Condition (11.3) holds, since it is equivalent to the hypothesis of linear independence of $x_1, \ldots, x_d$ over $\mathbb{Q}$. Now $\mathcal{V} \cap \mathcal{L}_G$ contains the ℓ points $(x_1 y_j, \ldots, x_d y_j)$ $(1 \leq j \leq \ell)$, which are linearly independent over $\mathbb{Q}$. From part 2) of Theorem 11.4 one deduces

$$\ell \leq \frac{d}{d-1}.$$

□

Remark. Theorem 11.4 can be stated in an equivalent way as follows.

- *Let m and n be two positive integers, $\{x_1, \ldots, x_m\}$ and $\{y_1, \ldots, y_n\}$ two families of $\mathbb{Q}$-linearly independent complex numbers. Define*

$$\begin{aligned} E_1 &= \left\{e^{x_i y_j}\,,\ 1 \leq i \leq m,\ 1 \leq j \leq n\right\}, & \kappa_1 &= mn, \\ E_2 &= E_1 \cup \left\{x_1, \ldots, x_m\right\}, & \kappa_2 &= \kappa_1 + m, \\ E_3 &= E_2 \cup \left\{y_1, \ldots, y_n\right\}, & \kappa_3 &= \kappa_2 + n. \end{aligned}$$

Let $h \in \{1, 2, 3\}$. Assume $\kappa_h > m + n$. Then one at least of the κ_h elements of E_h is transcendental.

Extensions of this statement to algebraic independence are considered in § 15.3.3 and § 15.4.

11.3 Six Variants of the Main Result

Let d_0 and d_1 be two nonnegative integers with $d = d_0 + d_1 \geq 2$. Let Y be a subgroup of $\mathcal{L}_G$ of finite rank $\ell_1 > 0$ over $\mathbb{Z}$, let $\mathcal{W}$ be a vector subspace of $\mathbb{C}^d$, rational over $\overline{\mathbb{Q}}$, of dimension $\ell_0 \geq 0$ over $\mathbb{C}$ and let $\mathcal{V}$ be a vector subspace of $\mathbb{C}^d$ of dimension n over $\mathbb{C}$ which contains both Y and $\mathcal{W}$.

We want to show that ℓ_1 and ℓ_0 cannot be too large with respect to n and d, unless *most* elements in Y and $\mathcal{W}$ belong to a subspace $T_e(G^*)$ of $\mathbb{C}^d$, where G^* is a connected algebraic subgroup of G of positive dimension. The best estimate we can reach turns out to be

$$\ell_1(d-n) \leq d_1(n - \ell_0).$$

This is the expected conclusion in a *general* situation, and there are several equivalent formulations for this result.

11.3.1 The Main Result

In the next statement, G^* is a connected algebraic subgroup of G, defined over $\overline{\mathbb{Q}}$; for each such G^* we define

$$Y^* = Y \cap T_e(G^*), \quad \mathcal{V}^* = \mathcal{V} \cap T_e(G^*), \quad \mathcal{W}^* = \mathcal{W} \cap T_e(G^*)$$

and

$$d^* = \dim(G^*), \quad \ell_1^* = \mathrm{rank}_{\mathbb{Z}}(Y^*), \quad n^* = \dim_{\mathbb{C}}(\mathcal{V}^*), \quad \ell_0^* = \dim_{\mathbb{C}}(\mathcal{W}^*).$$

By Theorem 5.13, we may write $G^* = G_0^* \times G_1^*$ where G_0^* is an algebraic subgroup of G_0 and G_1^* is an algebraic subgroup of G_1. Define

$$d_0^* = \dim(G_0^*), \quad d_1^* = \dim(G_1^*),$$

so that $d^* = d_0^* + d_1^*$.

Further, we introduce

$$G_0' = \frac{G_0}{G_0^*}, \quad G_1' = \frac{G_1}{G_1^*}, \quad G' = \frac{G}{G^*} = G_0' \times G_1',$$

$$Y' = \frac{Y}{Y^*}, \quad \mathcal{V}' = \frac{\mathcal{V}}{\mathcal{V}^*}, \quad \mathcal{W}' = \frac{\mathcal{W}}{\mathcal{W}^*},$$

and

$$d_0' = \dim(G_0'), \quad d_1' = \dim(G_1'), \quad d' = \dim(G'),$$

$$\ell_1' = \mathrm{rank}_{\mathbb{Z}}(Y'), \quad n' = \dim_{\mathbb{C}}(\mathcal{V}'), \quad \ell_0' = \dim_{\mathbb{C}}(\mathcal{W}').$$

The relations

$$d_0 = d_0^* + d_0', \quad d_1 = d_1^* + d_1', \quad d = d^* + d',$$

$$\ell_1 = \ell_1^* + \ell_1', \quad n = n^* + n', \quad \ell_0 = \ell_0^* + \ell_0'$$

plainly hold.

Theorem 11.5 — The Linear Subgroup Theorem.
(1) Assume $d > n$. Then there exists a connected algebraic subgroup G^* of G such that

$$d' > \ell_0' \quad \text{and} \quad \frac{\ell_1' + d_1'}{d' - \ell_0'} \le \frac{d_1}{d-n}.$$

(1') Assume $\ell_1 > 0$. Then there is a G^* for which

$$(d_1^*, \ell_1^*) \neq (0,0) \quad \text{and} \quad \frac{d^* - \ell_0^*}{d_1^* + \ell_1^*} \le \frac{n - \ell_0}{\ell_1}.$$

(2) Assume $d > n$ and $\ell_1 > 0$. Assume further that for any G^* for which $Y^* \neq \{0\}$, we have

$$\frac{n^* - \ell_0^*}{\ell_1^*} \ge \frac{n - \ell_0}{\ell_1}.$$

Assume also that there is no G^* for which the three conditions $\ell_1' = 0$, $n' = \ell_0'$ and $d' > 0$ simultaneously hold. Then

$$d_1 > 0 \quad \text{and} \quad \ell_1(d-n) \le d_1(n - \ell_0).$$

(2') Assume $d > n$ and $\ell_1 > 0$. Assume further that for any G^* for which $d' > n'$, we have

$$\frac{d_1}{d-n} \leq \frac{d_1'}{d'-n'}.$$

Assume also that there is no G^* for which the three conditions $d_1^* = 0$, $d^* = n^*$ and $d^* > 0$ simultaneously hold. Then

$$n > \ell_0 \quad \text{and} \quad \ell_1(d-n) \leq d_1(n-\ell_0).$$

(3) Assume $\ell_1 > 0$. Then the family of G^* for which $\ell_1^* \neq 0$ and $(n^* - \ell_0^*)/\ell_1^*$ is minimal is not empty. Let G^* be such an element for which d^* is minimal. Then either $d^* = n^*$ or else

$$d_1^* > 0 \quad \text{and} \quad \frac{n-\ell_0}{\ell_1} \geq \frac{n^*-\ell_0^*}{\ell_1^*} \geq \frac{d^*-n^*}{d_1^*}.$$

(3') Assume $d > n$. Then the family of G^* for which $d' > n'$ and $d_1'/(d'-n')$ is minimal is not empty. Let G^* be such an element for which d' is minimal. Then either $\ell_1' = 0$ or else

$$n' > \ell_0' \quad \text{and} \quad \frac{d_1}{d-n} \geq \frac{d_1'}{d'-n'} \geq \frac{\ell_1'}{n'-\ell_0'}.$$

The proof of Theorem 11.5 will be given in § 11.7. Notice that there is a *duality* (explained in [Roy 1990]) which relates (1) and (1'), as well as (2) and (2') and also (3) and (3') (see § 11.7).

11.3.2 A Second Proof of Theorem 11.4

Here we deduce Theorem 11.4 from the case $n = 1$ of Theorem 11.5. We assume that $\mathcal{V}$ is a complex line in $\mathbb{C}^d$ for which (11.3) holds (this is no loss of generality). Under this assumption we prove that the number

$$\ell_1 = \dim_{\mathbb{Q}}(\mathcal{V} \cap \mathcal{L}_G)$$

is bounded by

$$\ell_1 \leq \frac{d_1(1-\ell_0)}{d-1} \quad \text{where} \quad \ell_0 = \begin{cases} 1 & \text{if } \mathcal{V} \text{ is rational over } \overline{\mathbb{Q}}, \\ 0 & \text{otherwise.} \end{cases}$$

Part (2) of Theorem 11.4 will follow from the case $\mathcal{W} = \{0\}$, $\ell_0 = 0$ and part (3) from the case $\mathcal{W} = \mathcal{V}$, $\ell_0 = 1$.

Since $n = 1$ and $d \geq 2$, part (1) of Theorem 11.5 shows the existence of $G^* \subset G$ such that $d' > \ell_0'$ and

$$(\ell_1' + d_1')(d-1) \leq d_1(d' - \ell_0').$$

Since $d' > 0$, we have $G^* \neq G$. Assumption (11.3) then gives $\mathcal{V} \not\subset T_e(G^*)$. Hence $n' = n = 1$, $\ell_0' = \ell_0$, $\ell_1' = \ell_1$ and we get

$$(\ell_1 + d_1')(d-1) \leq d_1(d' - \ell_0).$$

Since $d = d_1 + d_0$ and $d' = d_1' + d_0'$ with $d_0' \leq d_0 \leq 1$, we have

$$d_0'd_1 - d_0d_1' \leq d_0(d_1 - d_1') \leq d_1 - d_1'$$

and therefore

$$(d'-1)d_1 = (d_1' + d_0' - 1)d_1 \leq d_1'(d_1 + d_0 - 1) = d_1'(d-1).$$

Hence

$$\ell_1(d-1) \leq d_1d' - d_1'(d-1) - d_1\ell_0 \leq d_1(1 - \ell_0).$$

□

11.3.3 The $\mathbb{Q}$-Vector Space $\mathcal{V} \cap \mathcal{L}_G$

Corollary 11.6. *Let $\mathcal{V}$ be a vector subspace of $\mathbb{C}^d$. Then the $\mathbb{Q}$-vector space $\mathcal{V} \cap \mathcal{L}_G$ has finite dimension if and only if (11.1) holds. If this condition is satisfied, then*

$$\dim_{\mathbb{Q}}(\mathcal{V} \cap \mathcal{L}_G) \leq d_1(n - \ell_0)$$

where n denotes the dimension of $\mathcal{V}$ and ℓ_0 the dimension of the $\overline{\mathbb{Q}}$-vector space spanned by $\mathcal{V} \cap \overline{\mathbb{Q}}$.

This result includes not only the six exponentials Theorem 1.12, but also the five exponentials Theorem (Example 1 in § 11.3.3).

Proof. We first recall why (11.1) is a necessary condition for $\mathcal{V} \cap \mathcal{L}_G$ to have finite dimension: if G^* is an algebraic subgroup of G such that $T_e(G^*) \subset \mathcal{V}$, then $\mathcal{V} \cap \mathcal{L}_G \supset \mathcal{L}_{G^*}$. Further, if G^* has positive dimension over $\mathbb{C}$, then $\mathcal{L}_{G^*}$ has infinite dimension over $\mathbb{Q}$.

Conversely, if the dimension of the $\mathbb{Q}$-vector space $\mathcal{V} \cap \mathcal{L}_G$ is not finite, then part (3) of Theorem 11.5 shows that there exists a G^* such that $d^* = n^*$, hence such that $\mathcal{V} \supset T_e(G^*)$, and in this case (11.1) does not hold.

Assume now condition (11.1) holds (so that $d > n$) and denote by ℓ_1 the dimension of $\mathcal{V} \cap \mathcal{L}_G$ over $\mathbb{Q}$. Plainly we may assume $\ell_1 > 0$. We prove the estimate by induction on d. For $d = 2$ the result follows from Theorem 11.4.

We use conclusion (3') of Theorem 11.5: there exists an algebraic subgroup G^* of G such that $d' > n'$ and

$$\ell_1' \leq \frac{d_1}{d-n}(n' - \ell_0') \leq d_1(n' - \ell_0').$$

If $G^* = G$, then $d' = d$, $n' = n$, $\ell_0' = \ell_0$, $\ell_1' = \ell_1$ and the desired estimate $\ell_1 \leq d_1(n - \ell_0)$ holds.

Otherwise we have $d^* < d$, and $\mathcal{V}^* = \mathcal{V} \cap T_e(G^*)$ satisfies (11.1) as a subspace of $T_e(G^*)$. Therefore we may apply the inductive hypothesis:

$$\ell_1^* \leq d_1^*(n^* - \ell_0^*).$$

However we have $d_1^* \leq d_1$, $\ell_0 = \ell_0' + \ell_0^*$ and $n = n' + n^*$, so that

$$\ell_1 = \ell_1' + \ell_1^* \leq d_1(n' - \ell_0') + d_1^*(n^* - \ell_0^*) \leq d_1(n - \ell_0).$$

□

The following examples turn out to be special cases of Corollary 11.15 below.

Example 1. Here is the *Five Exponentials Theorem* of [W 1988], Corollary 2.2.

- *Let x_1, x_2 be two $\mathbb{Q}$-linearly independent complex numbers and y_1, y_2 be also two $\mathbb{Q}$-linearly independent complex numbers. Further let γ be a nonzero algebraic number. Then one at least of the five numbers*

$$e^{x_1y_1},\ e^{x_1y_2},\ e^{x_2y_1},\ e^{x_2y_2},\ e^{\gamma x_1/x_2}$$

is transcendental.

Proof. We deduce this result from Corollary 11.6 by taking $d_0 = 1$, $d_1 = 2$, while $\mathcal{V}$ is the hyperplane of $\mathbb{C}^3$ of equation

$$\gamma x_1 z_0 - x_2 z_1 + x_1 z_2 = 0$$

which contains the point $(1, 0, -\gamma)$ of $\overline{\mathbb{Q}}^3$. Since x_1, x_2 are $\mathbb{Q}$-linearly independent and $\gamma \neq 0$, $\mathcal{V}$ satisfies (11.1), hence $\dim_{\mathbb{Q}}(\mathcal{V} \cap \mathcal{L}_G) \leq 2$. The three points

$$(1, \gamma x_1/x_2, 0), \quad (0, x_1y_1, x_2y_1), \quad (0, x_1y_2, x_2y_2)$$

are $\mathbb{Q}$-linearly independent and in $\mathcal{V}$, therefore one at least of them does not belong to $\mathcal{L}_G = \overline{\mathbb{Q}} \times \mathcal{L}^2$. □

Without loss of generality we may set $x_2 = 1$, in which case the five exponentials become

$$e^{y_1},\ e^{y_2},\ e^{x_1y_1},\ e^{x_1y_2},\ e^{\gamma x_1}.$$

Therefore the five exponentials Theorem can be stated as follows:

- *Let λ_0 be a nonzero element of $\mathcal{L}$, λ_1, λ_2 two $\mathbb{Q}$-linearly independent elements of $\mathcal{L}$ and β a nonzero algebraic number. Then one at least of the two numbers*

$$e^{\beta\lambda_0\lambda_1},\ e^{\beta\lambda_0\lambda_2}$$

is transcendental.

Example 2. The following result was called the *strong six exponentials Theorem* in [W 1988], Corollary 2.1 and [W 1990], Corollary 2.3 – but it is now a special case of the more general Corollary 11.16 who deserves this name.

- *Let x_1, x_2 be two $\mathbb{Q}$-linearly independent complex numbers and y_1, y_2, y_3 be three $\mathbb{Q}$-linearly independent complex numbers. Further let β_{ij} $(i = 1, 2,\ j = 1, 2, 3)$ be six algebraic numbers. Assume that the six numbers*

$$e^{x_i y_j - \beta_{ij}}, \quad i = 1, 2,\ j = 1, 2, 3$$

are algebraic. Then

$$x_i y_j = \beta_{ij} \quad \textit{for} \quad i = 1, 2, \quad \textit{and} \quad j = 1, 2, 3.$$

This result clearly contains the five exponentials Theorem: take $\beta_{ij} = 0$ for $(i, j) \neq (1, 3)$, $\beta_{13} = \gamma$, $y_3 = \gamma/x_1$, and use Baker's Theorem to deduce the linear independence of x_1y_1, x_1y_2, γ.

Proof. By assumption the six numbers $\lambda_{ij} = x_i y_j - \beta_{ij}$ $(i = 1, 2, j = 1, 2, 3)$ are in $\mathcal{L}$. Put $d_0 = d_1 = 2$. The hyperplane $\mathcal{V}$ of $\mathbb{C}^4$ of equation

$$x_2(z_1 + z_3) = x_1(z_2 + z_4)$$

contains the points $(1, 0, -1, 0)$ and $(0, -1, 0, 1)$ of $\overline{\mathbb{Q}}^4$ as well as the three points

$$(\beta_{1j}, \beta_{2j}, \lambda_{1j}, \lambda_{2j}) \quad (j = 1, 2, 3)$$

of $\mathcal{L}_G = \overline{\mathbb{Q}}^2 \times \mathcal{L}^2$. These three points are $\mathbb{Q}$-linearly independent, because y_1, y_2, y_3 are $\mathbb{Q}$-linearly independent. From Corollary 11.6 we deduce that $\mathcal{V}$ does not satisfy (11.1). It follows that x_1, x_2 are $\overline{\mathbb{Q}}$-linearly dependent, so that $\gamma = x_2/x_1$ is an irrational algebraic number. The relations

$$\lambda_{2j} + \beta_{2j} = \gamma(\lambda_{1j} + \beta_{1j}) \quad (j = 1, 2, 3)$$

together with Baker's Theorem 1.6 imply $\lambda_{ij} = 0$ for $i = 1, 2$ and $j = 1, 2, 3$. □

11.3.4 Subspaces which are Rational over $\overline{\mathbb{Q}}$

A simple statement can be deduced from Theorem 11.5 in the case where $\mathcal{V}$ is rational over $\overline{\mathbb{Q}}$, that is $\ell_0 = n$ and $\mathcal{W} = \mathcal{V}$. We fix $d_0 \geq 0$ and $d_1 \geq 1$ and we set $d = d_0 + d_1$, as before.

Corollary 11.7. *Let $\mathcal{V}$ be a subspace of $\mathbb{C}^d$ which is rational over $\overline{\mathbb{Q}}$. Then*

$$\mathcal{V} \cap \mathcal{L}_G = \mathcal{V}_{\max} \cap \mathcal{L}_G.$$

Proof. For a subspace $\mathcal{V}$ of $\mathbb{C}^d$ which is rational over $\overline{\mathbb{Q}}$ and satisfies (11.1), we have $\mathcal{V} \cap \mathcal{L}_G = \{0\}$: this follows from Corollary 11.6 with $\ell_0 = n$.

For the general case, write

$$\mathcal{V}_{\max} = E_0 \times E_1, \quad d_0' = \dim_{\mathbb{C}}\left(\frac{\mathbb{C}^d}{E_0}\right), \quad d_1' = \dim_{\mathbb{C}}\left(\frac{\mathbb{C}^d}{E_1}\right).$$

Define $G' = \mathbb{G}_a^{d_0'} \times \mathbb{G}_m^{d_1'}$. Let $\mathbb{C}^{d_0} \to \mathbb{C}^{d_0'}$ be a surjective linear map, rational over $\overline{\mathbb{Q}}$, with kernel E_0 and let $\mathbb{C}^{d_1} \to \mathbb{C}^{d_1'}$ be a surjective linear map, rational over $\mathbb{Q}$, with kernel E_1. Denote by

$$\varphi\colon \mathbb{C}^{d_0} \times \mathbb{C}^{d_1} \to \mathbb{C}^{d_0'} \times \mathbb{C}^{d_1'}$$

their product, so that $\ker\varphi = \mathcal{V}_{\max}$. Notice that $\varphi(\mathcal{L}_G) = \mathcal{L}_{G'}$. From the definition of $\mathcal{V}_{\max}$ we deduce that condition (11.1) holds for the subspace $\mathcal{V}' = \varphi(\mathcal{V})$ of $\mathbb{C}^{d_0'} \times \mathbb{C}^{d_1'}$, hence $\mathcal{V}' \cap \mathcal{L}_{G'} = \{0\}$. Therefore

$$\mathcal{V} \cap \mathcal{L}_G \subset \varphi^{-1}\bigl(\mathcal{V}' \cap \mathcal{L}_{G'}\bigr) = \ker\varphi = \mathcal{V}_{\max}.$$

□

11.4 Linear Independence of Logarithms

There are different ways to recover Baker's Theorem 1.5 (homogeneous case) from Theorem 11.5. Two of them (§ 11.4.1) involve the special case $\ell_0 = n$, which means that $\mathcal{V}$ is rational over $\overline{\mathbb{Q}}$. They correspond to Gel'fond-Baker's method. Two others (§ 11.4.2) are *dual* (in the sense of § 13.7) of the previous ones and correspond to Schneider's method. We also deduce in § 11.4.3 Baker's nonhomogeneous Theorem 1.6 from Theorem 11.5.

We display values for the parameters d_0, d_1, ℓ_0, ℓ_1 and n for which

$$\ell_1(d-n) > d_1(n-\ell_0).$$

By Theorem 11.5, some degeneracy should take place, which will yield the desired result.

11.4.1 Gel'fond-Baker's Method

We give two proofs of Baker's Theorem 1.5.

Assume

$$\beta_1\lambda_1 + \cdots + \beta_m\lambda_m = 0 \tag{11.8}$$

where $\lambda_1, \ldots, \lambda_m$ are elements in $\mathcal{L}$, while $\beta_1, \ldots, \beta_m$ are algebraic numbers.

1 Set $\boxed{d_0 = 0, d_1 = m, \ell_0 = m-1, \ell_1 = 1, n = m-1}$

Assume $(\beta_1, \ldots, \beta_m) \neq (0, \ldots, 0)$ and consider the hyperplane $\mathcal{V}$ in $\mathbb{C}^m$ of equation

$$\beta_1 z_1 + \cdots + \beta_m z_m = 0$$

which is plainly defined over $\overline{\mathbb{Q}}$. Since $d_0 = 0$, $\mathcal{V}_{\max}$ is the maximal vector subspace of $\mathbb{C}^m$ which is rational over $\mathbb{Q}$. This is nothing else than the vector subspace of $\mathcal{V}$ spanned by

$$\mathcal{V} \cap \mathbb{Q}^m = \left\{\underline{b} \in \mathbb{Q}^m \, ; \, b_1\beta_1 + \cdots + b_m\beta_m = 0\right\}.$$

Therefore

$$\mathcal{V}_{\max} = \{0\} \iff \beta_1, \ldots, \beta_m \text{ are } \mathbb{Q}\text{-linearly independent.}$$

Assume now that in (11.8) the numbers $\beta_1, \ldots, \beta_m$ are $\mathbb{Q}$-linearly independent. Since $(\lambda_1, \ldots, \lambda_m) \in \mathcal{V} \cap \mathcal{L}^m$, we deduce from Corollary 11.7 $\lambda_1 = \cdots = \lambda_m = 0$ and Baker's Theorem 1.5 follows (see Lemma 1.7). □

2 Set $\boxed{d_0 = 1, d_1 = m, \ell_0 = m, \ell_1 = 1, n = m}$

Consider the hyperplane $\mathcal{V}$ in $\mathbb{C}^{m+1}$ of equation

$$z_0 + \beta_1 z_1 + \cdots + \beta_m z_m = 0.$$

Again $\mathcal{V}$ is defined over $\overline{\mathbb{Q}}$. Now $d_0 = 1$ and $d_1 = m$. Since $(1, 0, \ldots, 0) \notin \mathcal{V}$, $\mathcal{V}_{\max}$ is the vector subspace of $\mathcal{V}$ spanned by

$$\left\{(0, \underline{b}) \in \{0\} \times \mathbb{Q}^m \, ; \, b_1\beta_1 + \cdots + b_m\beta_m = 0\right\}.$$

Therefore we still have

$$\mathcal{V}_{\max} = \{0\} \iff \beta_1, \ldots, \beta_m \text{ are } \mathbb{Q}\text{-linearly independent.}$$

Since $(0, \lambda_1, \ldots, \lambda_m) \in \mathcal{V} \cap \mathcal{L}_G$, Corollary 11.7 yields $\lambda_1 = \cdots = \lambda_m = 0$, and Baker's Theorem 1.5 follows as before. □

11.4.2 Schneider's Method

Here are two other proofs of Baker's Theorem 1.5.

1' Set $\boxed{d_0 = m-1, d_1 = 1, \ell_0 = 0, \ell_1 = m, n = m-1}$

We shall prove, by induction on m, that if $\lambda_1, \ldots, \lambda_m$ are $\overline{\mathbb{Q}}$-linearly dependent elements of $\mathcal{L}$, then they are linearly dependent over $\mathbb{Q}$. Assume now $\lambda_1, \ldots, \lambda_m$ are $\overline{\mathbb{Q}}$-linearly dependent: they satisfy a relation (11.8) with $(\beta_1, \ldots, \beta_m) \neq 0$. Without loss of generality we may assume $\beta_m = -1$. Also, thanks to the induction hypothesis, we may assume that $\lambda_1, \ldots, \lambda_{m-1}$ are $\overline{\mathbb{Q}}$-linearly independent.

For $d_0 = m - 1$ and $d_1 = 1$, we have $d = m$ and $\mathcal{L}_G = \overline{\mathbb{Q}}^{m-1} \times \mathcal{L}$. Let $\mathcal{V}$ be the hyperplane in $\mathbb{C}^m$ of equation

$$z_1\lambda_1 + \cdots + z_{m-1}\lambda_{m-1} = z_m$$

and let $\mathcal{W} = \{0\}$. We have $\mathcal{V} \cap (\{0\} \times \mathbb{C}) = \{0\}$. Moreover, since $\lambda_1, \ldots, \lambda_{m-1}$ are $\overline{\mathbb{Q}}$-linearly independent, we also have $\mathcal{V} \cap (\overline{\mathbb{Q}}^{m-1} \times \{0\}) = \{0\}$. This shows that condition (11.1) is fulfilled.

Using Kronecker's symbol δ_{ij}, we check that the elements

$$\underline{\eta}_j = (\delta_{1j}, \ldots, \delta_{m-1,j}, \lambda_j) \quad (1 \le j \le m-1), \qquad \underline{\eta}_m = (\beta_1, \ldots, \beta_{m-1}, \lambda_m)$$

belong to $\mathcal{V} \cap \mathcal{L}_G$. From Corollary 11.6 one deduces that $\underline{\eta}_1, \ldots, \underline{\eta}_m$ are $\mathbb{Q}$-linearly dependent, hence $\lambda_1, \ldots, \lambda_m$ are $\mathbb{Q}$-linearly dependent. □

2' Set $d_0 = m,\ d_1 = 1,\ \ell_0 = 1,\ \ell_1 = m,\ n = m$

Let $\lambda_1, \ldots, \lambda_m$ be $\mathbb{Q}$-linearly independent elements of $\mathcal{L}$ which satisfy (11.8) for some $(\beta_1, \ldots, \beta_m) \neq 0$. We want to get a contradiction.

Since $d_0 = m$ and $d_1 = 1$, we have $d = m + 1$ and $\mathcal{L}_G = \overline{\mathbb{Q}}^m \times \mathcal{L}$. Consider the hyperplane $\mathcal{V}$ of equation $z_1\lambda_1 + \cdots + z_m\lambda_m = z_{m+1}$ in $\mathbb{C}^{m+1}$. Further, let $\mathcal{W}$ be the $\mathbb{C}$-vector space of dimension 1, rational over $\overline{\mathbb{Q}}$, spanned by $(\beta_1, \ldots, \beta_m, 0)$. The points

$$\underline{\eta}_j = (\delta_{1j}, \ldots, \delta_{mj}, \lambda_j) \quad (1 \le j \le m)$$

belong to $\mathcal{V} \cap \mathcal{L}_G$ and they are linearly independent over $\mathbb{C}$ (hence also over $\mathbb{Q}$). We use part (3') of Theorem 11.5 with $Y = \mathbb{Z}\underline{\eta}_1 + \cdots + \mathbb{Z}\underline{\eta}_m$. Consider the G^* for which $d' > n'$. They satisfy $d^* = n^*$, hence $T_e(G^*) \subset \mathcal{V}$. Since $\mathcal{V} \cap (\{0\} \times \mathbb{C}) = \{0\}$, we have $d_1^* = 0$ and $d_1' = 1$. Moreover $d' - n' = 1$, so that

$$\frac{d_1'}{d' - n'} = 1 = \frac{d_1}{d - n}.$$

An example of such a $T_e(G^*)$ is $\mathbb{C}(\beta_1, \ldots, \beta_m, 0)$, whose codimension d' is m. Consider such a G^* with d' minimal. Our example shows that $d' \le m$. The conclusion (3') of Theorem 11.5 yields $\ell_0' + \ell_1' \le n'$. Since $\lambda_1, \ldots, \lambda_m$ are $\mathbb{Q}$-linearly independent, $Y \cap (\mathbb{C}^m \times \{0\}) = \{0\}$, hence $\ell_1^* = 0$ and $\ell_1' = m$. We deduce $m \le m + \ell_0' \le n'$ and $d' = n' + 1 \ge m + 1$, which is the desired contradiction. □

11.4.3 Affine Linear Forms

Our goal is to show that any relation

$$\beta_0 + \beta_1\lambda_1 + \cdots + \beta_m\lambda_m = 0, \tag{11.9}$$

with $\lambda_1, \ldots, \lambda_m$ in $\mathcal{L}$ and $\beta_0, \ldots, \beta_m$ in $\overline{\mathbb{Q}}$ implies $\beta_0 = 0$. This will complete the proof of Baker's Theorem 1.6.

It will be convenient to argue by contradiction: assume $\beta_0 \neq 0$ and consider such a relation with m minimal. Therefore $\lambda_1, \dots, \lambda_m$ are $\overline{\mathbb{Q}}$-linearly independent and at the same time $\beta_1, \dots, \beta_m$ are $\mathbb{Q}$-linearly independent.

We give two proofs which are just slight modifications of proofs in §§ 11.4.1 and 11.4.2 respectively.

2 Set $\boxed{d_0 = 1, d_1 = m, \ell_0 = m, \ell_1 = 1, n = m}$

The hyperplane $\mathcal{V}$ in $\mathbb{C}^{m+1}$ of equation

$$\beta_0 z_0 + \beta_1 z_1 + \cdots + \beta_m z_m = 0$$

is rational over $\overline{\mathbb{Q}}$ and satisfies property (11.1), because $\beta_1, \dots, \beta_m$ are linearly independent over $\mathbb{Q}$. However it contains the nonzero element $(1, \lambda_1, \dots, \lambda_m)$ of $\mathcal{L}_G = \overline{\mathbb{Q}} \times \mathcal{L}^m$, which contradicts Corollary 11.7. □

Replacing z_0 by z_0/β_0 one might as well consider the hyperplane $z_0 + \beta_1 z_1 + \cdots + \beta_m z_m = 0$ and the point $(\beta_0, \lambda_1, \dots, \lambda_m)$.

2' Set $\boxed{d_0 = m, d_1 = 1, \ell_0 = 1, \ell_1 = m, n = m}$

Since $\lambda_1, \dots, \lambda_m$ are $\overline{\mathbb{Q}}$-linearly independent, the hyperplane $\mathcal{V}$ of $\mathbb{C}^{m+1}$ of equation $z_1\lambda_1 + \cdots + z_m\lambda_m = z_{m+1}$ satisfies condition (11.1). Let $\mathcal{W}$ be the $\mathbb{C}$-vector subspace of $\mathbb{C}^{m+1}$, rational over $\overline{\mathbb{Q}}$, of dimension 1, spanned by $(\beta_1, \dots, \beta_m, -\beta_0)$.

Since the points

$$\underline{\eta}_j = (\delta_{1j}, \dots, \delta_{mj}, \lambda_j) \in \overline{\mathbb{Q}}^m \times \mathcal{L} \quad (1 \le j \le m)$$

belong to $\mathcal{V} \cap \mathcal{L}_G$ and are linearly independent over $\mathbb{C}$, we deduce our contradiction from Corollary 11.6. □

Another description of the same method (involving the same values for d_0, d_1, ℓ_0, ℓ_1 and m) arises from the following change of variables

$$(z_1, \dots, z_{m+1}) \longmapsto (Z_0, Z_1, \dots, Z_m)$$

with

$$Z_0 = \beta_0 z_m, \quad Z_i = z_i + \beta_i z_m \quad (1 \le i \le m-1), \quad Z_m = z_{m+1}.$$

Assume $\beta_m = -1$, so that

$$\beta_0 + \beta_1\lambda_1 + \cdots + \beta_{m-1}\lambda_{m-1} = \lambda_m.$$

The hyperplane $\mathcal{V}$ in $\mathbb{C}^{m+1}$ of equation

$$Z_0 + \lambda_1 Z_1 + \cdots + \lambda_{m-1} Z_{m-1} = Z_m$$

satisfies (11.1) (recall that m is minimal for (11.9)) and contains the complex vector subspace $\mathcal{W}$ of $\mathbb{C}^{m+1}$ spanned by $(1, 0, \dots, 0, 1)$. Moreover $\mathcal{V} \cap \mathcal{L}_G$ contains the m points

$$\underline{\eta}_j = (0, \delta_{1j}, \dots, \delta_{m-1,j}, \lambda_j) \quad (1 \le j \le m-1)$$

and

$$\underline{\eta}_m = (\beta_0, \beta_1, \dots, \beta_{m-1}, \lambda_m).$$

11.4.4 Exponential Polynomials

One may analyze the underlying principle to each of the previous proofs by writing the functions and the points (as well as derivatives, if applicable):

[1] Assume $\beta_m = -1$; we work with m functions of $m-1$ complex variables:

$$e^{z_1}, \dots, e^{z_{m-1}}, e^{\beta_1 z_1 + \cdots + \beta_{m-1} z_{m-1}},$$

take the derivatives with respect to $m-1$ variables $z_1, \dots, z_{m-1}$ and consider the values at the points of the subgroup $\mathbb{Z}(\lambda_1, \dots, \lambda_{m-1})$ which has rank 1.

Due to the fact that these points span a $\mathbb{C}$-vector space of dimension 1, one may introduce Baker's method without mentioning several complex variables. Indeed, for $\underline{t} = (t_1, \dots, t_m) \in \mathbb{Z}^m$, consider the exponential monomial

$$\begin{aligned}\Phi_{\underline{t}}(z_1, \dots, z_{m-1}) &= e^{t_1 z_1 + \cdots + t_{m-1} z_{m-1} + t_m(\beta_1 z_1 + \cdots + \beta_{m-1} z_{m-1})} \\ &= e^{(t_1 + t_m \beta_1) z_1 + \cdots + (t_{m-1} + t_m \beta_{m-1}) z_{m-1}}.\end{aligned}$$

Let $\phi_{\underline{t}}$ be the restriction of $\Phi_{\underline{t}}$ to the complex line $\mathbb{C}(\lambda_1, \dots, \lambda_{m-1})$:

$$\phi_{\underline{t}}(z) = e^{z\left((t_1 + t_m\beta_1)\lambda_1 + \cdots + (t_{m-1} + t_m \beta_{m-1})\lambda_{m-1}\right)}.$$

For $\kappa \in \mathbb{N}$, we have the relations

$$\left(\frac{d}{dz}\right)^{\kappa} \phi_{\underline{t}}(z) = \sum_{\|\underline{\sigma}\| = \kappa} \frac{\kappa!}{\underline{\sigma}!} \lambda_1^{\sigma_1} \cdots \lambda_{m-1}^{\sigma_{m-1}} \mathcal{D}^{\underline{\sigma}} \Phi_{\underline{t}}(\lambda_1 z, \dots, \lambda_{m-1} z),$$

where $\underline{\sigma}$ runs over the elements $(\sigma_1, \dots, \sigma_{m-1})$ in $\mathbb{N}^{m-1}$ satisfying $\|\underline{\sigma}\| = \kappa$ and

$$\mathcal{D}^{\underline{\sigma}} = \left(\frac{\partial}{\partial z_1}\right)^{\sigma_1} \cdots \left(\frac{\partial}{\partial z_{m-1}}\right)^{\sigma_{m-1}}$$

Specializing $z = s \in \mathbb{Z}$, the number $(d/dz)^{\kappa} \phi_{\underline{t}}(s)$ is a polynomial in $\lambda_1, \dots, \lambda_{m-1}$, and this is not so nice since these numbers are transcendental. However the numbers $\mathcal{D}^{\underline{\sigma}} \Phi_{\underline{t}}(\lambda_1 s, \dots, \lambda_{m-1} s)$ are algebraic, and the main feature of Baker's method is to work with these algebraic numbers.

[2] Let us start with the homogeneous situation of § 11.4.1:

$$\beta_1 \lambda_1 + \cdots + \beta_m \lambda_m = 0.$$

We consider $m+1$ functions of m variables, namely

$$\beta_1 z_1 + \cdots + \beta_m z_m, \; e^{z_1}, \dots, e^{z_m}.$$

We differentiate them in all directions and evaluate their derivatives at the points of the set $\mathbb{Z}(\lambda_1, \dots, \lambda_m)$.

Let us assume $\beta_m = -1$ and perform the change of variables

$$-z_0 = \beta_1 z_1 + \cdots + \beta_{m-1} z_{m-1} - z_m.$$

Take the derivatives of the functions

$$z_0,\ e^{z_1}, \ldots, e^{z_{m-1}},\ e^{z_0+\beta_1 z_1+\cdots+\beta_{m-1}z_{m-1}},$$

with respect to the m variables $z_0, \ldots, z_{m-1}$ and consider their values at the points of the subgroup $\mathbb{Z}\big(\beta_0, \lambda_1, \ldots, \lambda_{m-1}\big)$ which has rank 1.

For $(\tau, \underline{t}) \in \mathbb{N} \times \mathbb{Z}^m$, define

$$\Phi_{\tau\underline{t}}(z_0, \ldots, z_{m-1}) = z_0^\tau e^{t_m z_0 + (t_1+t_m\beta_1)z_1+\cdots+(t_{m-1}+t_m\beta_{m-1})z_{m-1}}$$

and

$$\begin{aligned}\phi_{\tau\underline{t}}(z) &= \Phi_{\tau\underline{t}}(z, \lambda_1 z, \ldots, \lambda_{m-1}z)\\ &= z^\tau e^{z\left(t_m+(t_1+t_m\beta_1)\lambda_1+\cdots+(t_{m-1}+t_m\beta_{m-1})\lambda_{m-1}\right)}.\end{aligned}$$

The derivatives of $\phi_{\tau\underline{t}}$ can be written

$$\left(\frac{d}{dz}\right)^\kappa \phi_{\tau\underline{t}}(z) = \sum_{\|\underline{\sigma}\|=\kappa} \frac{\kappa!}{\underline{\sigma}!}\lambda_1^{\sigma_1}\cdots\lambda_{m-1}^{\sigma_{m-1}} \mathcal{D}^{\underline{\sigma}}\Phi_{\tau\underline{t}}(z, \lambda_1 z, \ldots, \lambda_{m-1}z),$$

for $\underline{\sigma} = (\sigma_0, \ldots, \sigma_{m-1}) \in \mathbb{N}^m$. For $s \in \mathbb{Z}$, the numbers

$$\mathcal{D}^{\underline{\sigma}}\Phi_{\tau\underline{t}}(s, \lambda_1 s, \ldots, \lambda_{m-1}s)$$

are algebraic, while $(d/dz)^\kappa \phi_{\tau\underline{t}}(s)$ is a polynomial in $\lambda_1, \ldots, \lambda_{m-1}$.

In the situation of § 11.4.3 with a linear form

$$\beta_0 z_0 + \beta_1 z_1 + \cdots + \beta_{m-1} z_{m-1} - z_m$$

which vanishes at $(1, \lambda_1, \ldots, \lambda_m)$, we consider the functions

$$z_0,\ e^{z_1}, \ldots, e^{z_{m-1}},\ e^{\beta_0 z_0+\beta_1 z_1+\cdots+\beta_{m-1}z_{m-1}},$$

so that we replace $\Phi_{\tau\underline{t}}(\underline{z})$ by

$$z_0^\tau e^{t_m\beta_0 z_0 + (t_1+t_m\beta_1)z_1+\cdots+(t_{m-1}+t_m\beta_{m-1})z_{m-1}}.$$

Notice that we do not recover the previous discussion of the homogeneous case if we just set $\beta_0 = 0$! See § 14.4.3 for a further discussion of this point.

[1'] Again assume $\beta_m = -1$. We deal with m functions of $m-1$ variables:

$$z_1, \ldots, z_{m-1},\ e^{\lambda_1 z_1+\cdots+\lambda_{m-1}z_{m-1}},$$

we take no derivative at all (this is why we speak of *Schneider's method*) and we consider the values at the points of the subgroup $\mathbb{Z}^{m-1} + \mathbb{Z}\big(\beta_1, \ldots, \beta_{m-1}\big)$ of $\mathbb{C}^{m-1}$.

$\boxed{2'}$ In the homogeneous situation of § 11.4.2 as well as in the affine case of § 11.4.1, we work with $m+1$ functions of m variables. We can

- either consider

$$z_1, \ldots, z_m, \; e^{\lambda_1 z_1 + \cdots + \lambda_m z_m};$$

  in this case we introduce a single derivative

$$\mathcal{D} = \frac{\beta_1 \partial}{\partial z_1} + \cdots + \beta_m \frac{\partial}{\partial z_m}$$

  and we take the values at the points in $\mathbb{Z}^m$.

- or else, assuming $\beta_m = -1$, deal with

$$z_0, z_1, \ldots, z_{m-1}, \; e^{z_0 + \lambda_1 z_1 + \cdots + \lambda_{m-1} z_{m-1}};$$

  then we consider one derivative $\partial/\partial z_0$, and we evaluate our functions at the points of the subgroup $\{0\} \times \mathbb{Z}^{m-1} + \mathbb{Z}(\beta_0, \beta_1, \ldots, \beta_{m-1})$ of $\mathbb{C}^{m-1}$.

As we saw in § 11.4.3, a simple change of variables passes from one option to the other.

11.4.5 Comparison

Methods $\boxed{1}$ and $\boxed{2}$ correspond to Baker's method. They involve several exponential functions (that is several factors $\mathbb{G}_{\mathrm{m}}$) and several derivatives, but one point $\underline{\eta}$ is sufficient (together with its multiples). In method $\boxed{1}$, there is no factor $\mathbb{G}_{\mathrm{a}}$ (since $d_0 = 0$), so that one works on a torus $\mathbb{G}_{\mathrm{m}}^m$.

In the case $m = 2$, method $\boxed{1}$ reduces to Gel'fond's solution of Hilbert's seventh problem.

Methods $\boxed{1'}$ and $\boxed{2'}$ are *dual* (see § 13.7) of $\boxed{1}$ and $\boxed{2}$ respectively. They involve only one multiplicative factor, but several additive factors $\mathbb{G}_{\mathrm{a}}$. Also several independent points of $G(\overline{\mathbb{Q}})$ are needed (not just a rank one subgroup). In $\boxed{1'}$ there is no derivative (and in $\boxed{2'}$ there is just a single direction for multiplicities).

When one restricts to only two logarithms, $\boxed{1'}$ is Schneider's solution to Hilbert's seventh problem.

In the nonhomogeneous case (11.9) with $m = 1$, methods $\boxed{2}$ and $\boxed{2'}$ coincide: they just reduce to the proof of Lindemann-Weierstraß' Theorem by means of Gel'fond's method, involving one additive factor, one multiplicative factor, one derivative and one point:

$$d_0 = d_1 = \ell_0 = \ell_1 = 1.$$

It may seem pointless to produce several proofs for the same result. However it will turn out that our variants lead to a few differences in the diophantine estimates which are obtained as quantitative refinements of the transcendence result. We refer to § 14.4 for a detailed discussion of this matter.

11.5 Complex Toruses

By the case $d_0 = 0$, $d_1 = d$ of Corollary 11.6, for a $\mathbb{C}$-vector subspace $\mathcal{V}$ of $\mathbb{C}^d$, the $\mathbb{Q}$-vector space $\mathcal{V} \cap \mathcal{L}^d$ has finite dimension if and only if $\mathcal{V} \cap \mathbb{Q}^d = \{0\}$. This point of view was initiated by M. Emsalem [E 1987] and his estimates were refined by D. Roy.

For d and n positive integers satisfying $0 \le n < d$, define

$$\Psi(n,d) = \begin{cases} 1 & \text{for } n = 1,\ d \ge 3, \\ n(n-1)+1 & \text{for } 2 \le n \le d-3, \\ n(n-1)+2 & \text{for } n = d-2 \ge 2, \\ n(n+1) & \text{for } n = d-1 \end{cases}$$

and $\Psi(0,d) = 0$ for $d > 0$.

Here is Th. 5 of [Roy 1992b], which we shall deduce from Theorem 11.5 applied to the torus $G = \mathbb{G}_{\mathrm{m}}^d$.

For $\mathcal{V} \subset \mathbb{C}^d$, recall that $\mathcal{V}_{\min}$ is the least vector subspace of $\mathbb{C}^d$ which is rational over $\mathbb{Q}$ and contains $\mathcal{V}$, while $d_{\min}$ is the dimension of $\mathcal{V}_{\min}$.

Corollary 11.10. *Let $\mathcal{V}$ be a vector subspace of $\mathbb{C}^d$ of dimension n such that $\mathcal{V} \cap \mathbb{Q}^d = \{0\}$. Then*

$$\dim_{\mathbb{Q}}\big(\mathcal{V} \cap \mathcal{L}^d\big) \le \Psi(n, d_{\min}).$$

Since $\Psi(1,2) = 2$ and $\Psi(1,d) = 1$ for $d \ge 3$, taking $n = 1$ and either $d = 2$ or $d = 3$, we deduce from Corollary 11.10 the six exponentials Theorem 1.12.

Before going into the proof of Corollary 11.10, we state a property of the function Ψ, whose proof is left as an exercise (Exercise 11.4).

Lemma 11.11. *For $0 < n < d$ we have*

$$\Psi(n,d) \ge \left[\frac{nd}{d-n}\right]. \tag{11.12}$$

Moreover, if $n = n' + n^*$ and $d = d' + d^*$ with $0 < n' < d'$ and $0 < n^* < d^*$, then

$$\Psi(n',d') + \Psi(n^*,d^*) \le \Psi(n,d). \tag{11.13}$$

For the proof of Corollary 11.10 we shall apply only the following consequence of Theorem 11.5.

Corollary 11.14. *Let $\mathcal{V}$ be a vector subspace of $\mathbb{C}^d$ of dimension $n < d$. Among the subspaces S of $\mathbb{C}^d$ which are rational over $\mathbb{Q}$ and distinct from $\mathbb{C}^d$, we select one for which the quantity n'/d' is minimal, with*

$$n' = \dim_{\mathbb{C}}\left(\frac{\mathcal{V}}{\mathcal{V} \cap S}\right), \qquad d' = \dim_{\mathbb{C}}\left(\frac{\mathbb{C}^d}{S}\right).$$

Then the dimension ℓ' of the $\mathbb{Q}$-vector space $\mathcal{V}\cap\mathcal{L}^d/\mathcal{V}\cap S\cap\mathcal{L}^d$ is finite and satisfies

$$\frac{\ell'}{d'+\ell'}\le\frac{n'}{d'}\le\frac{n}{d}.$$

Proof of Corollary 11.14. Corollary 11.14 would immediately follow from part (3') of Theorem 11.5 if we were asking S to have minimal d'. Since we did not include this condition, we shall proceed by induction on d and we split the proof in three cases.

The first case is when the only subspace S which satisfies the assumptions is $S=\{0\}$. We use part (3') of Theorem 11.5 with $d_0=0$, $d_1=d$, $\ell_0=0$: the only G^* for which $n'/d'\le n/d$ is $G^*=\{e\}$. Hence $d'=d$, $n'=n$, and the dimension ℓ of the $\mathbb{Q}$-vector space $\mathcal{V}\cap\mathcal{L}^d$ is finite and bounded above by $dn/(d-n)$.

The second case is when the selected subspace S has positive dimension. We use the induction hypothesis with $\mathbb{C}^d$, $\mathcal{V}$, S replaced respectively by $\mathbb{C}^d/S$, $\mathcal{V}/(\mathcal{V}\cap S)$ and $\{0\}$. We need to check that the subspace $\{0\}$ of $\mathbb{C}^d/S$ satisfies the assumption with respect to $\mathcal{V}/(\mathcal{V}\cap S)$. A subspace of $\mathbb{C}^d/S$, rational over $\mathbb{Q}$, can be written S'/S where S' is a subspace of $\mathbb{C}^d$, rational over $\mathbb{Q}$, containing S. The required inequality

$$\frac{\dim_{\mathbb{C}}\big((\mathcal{V}/\mathcal{V}\cap S)/(\mathcal{V}\cap S'/\mathcal{V}\cap S)\big)}{\dim_{\mathbb{C}}\big((\mathbb{C}^d/S)/(S'/S)\big)}\ge\frac{\dim_{\mathbb{C}}(\mathcal{V}/\mathcal{V}\cap S)}{\dim_{\mathbb{C}}(\mathbb{C}^d/S)}$$

follows from our assumption that n'/d' is minimal, namely

$$\frac{\dim_{\mathbb{C}}(\mathcal{V}/\mathcal{V}\cap S')}{\dim_{\mathbb{C}}(\mathbb{C}^d/S')}\ge\frac{\dim_{\mathbb{C}}(\mathcal{V}/\mathcal{V}\cap S)}{\dim_{\mathbb{C}}(\mathbb{C}^d/S)}.$$

Hence in this second case we may use the induction hypothesis and the desired result follows at once.

Finally we need to consider the case where $S=\{0\}$, but there exists a nonzero subspace, say S^*, for which $n'/d'=n/d$, where $d'=d-d^*$, $n'=n-n^*$, d^* is the dimension of S^* and n^* the dimension of $\mathcal{V}\cap S^*$.

From the induction hypothesis with $\mathbb{C}^d$, $\mathcal{V}$, S replaced respectively by $\mathbb{C}^d/S^*$, $\mathcal{V}/(\mathcal{V}\cap S^*)$ and $\{0\}$, we deduce (like in the second case) that the dimension ℓ' of $(\mathcal{V}\cap\mathcal{L}^d)/\big(\mathcal{V}\cap S^*\cap\mathcal{L}^d\big)$ is bounded by

$$\ell'\le\frac{d'n'}{d'-n'}=\frac{d'n}{d-n}.$$

Next we use the induction hypothesis with $\mathbb{C}^d$, $\mathcal{V}$, S replaced respectively by S^*, $\mathcal{V}\cap S^*$ and $\{0\}$. We need to check the hypothesis: if S' is a rational subspace of $\mathbb{C}^d$ contained in S^* and distinct from S^*, then we have by assumption

$$\frac{n-\dim_{\mathbb{C}}(\mathcal{V}\cap S')}{d-\dim_{\mathbb{C}}(S')}\ge\frac{n}{d}$$

and this implies

$$\frac{n^* - \dim_{\mathbb{C}}(\mathcal{V} \cap S')}{d^* - \dim_{\mathbb{C}}(S')} \geq \frac{n^*}{d^*}.$$

Hence the dimension ℓ^* of $\mathcal{V} \cap S^* \cap \mathcal{L}^d$ is bounded by

$$\ell^* \leq \frac{d^* n^*}{d^* - n^*} = \frac{(d - d')n}{d - n}.$$

Therefore the number $\ell = \dim_{\mathbb{Q}}(\mathcal{V} \cap \mathcal{L}^d) = \ell^* + \ell'$ satisfies the desired upper bound:

$$\ell \leq \frac{dn}{d - n}.$$

□

Proof of Corollary 11.10.

We first notice that the assumption $\mathcal{V} \cap \mathbb{Q}^d = \{0\}$ is nothing else that (11.1).

Without loss of generality we may assume $d_{\min} = d$: otherwise we just replace $\mathbb{C}^d$ by $\mathcal{V}_{\min}$, which is a subspace of $\mathbb{C}^d$ rational over $\mathbb{Q}$ of dimension $d_{\min}$ containing $\mathcal{V}$. When (11.3) holds, the result we want to prove can be stated as follows:

Let $\mathcal{V}$ be a vector subspace of $\mathbb{C}^d$ of dimension n such that $\mathcal{V} \cap \mathbb{Q}^d = \{0\}$. Assume that the only subspace of $\mathbb{C}^d$ which is rational over $\mathbb{Q}$ and contains $\mathcal{V}$ is $\mathbb{C}^d$ itself. Then $\dim_{\mathbb{Q}}(\mathcal{V} \cap \mathcal{L}^d) \leq \Psi(n, d)$.

The proof is by induction on d. The case $d = 1$ is trivial. Assume $d \geq 2$ and define

$$Y = \mathcal{V} \cap \mathcal{L}^d, \quad \ell = \dim_{\mathbb{Q}}(Y).$$

We use Corollary 11.14: there exists a subspace S of $\mathbb{C}^d$, rational over $\mathbb{Q}$ of codimension $d' > 0$ in $\mathbb{C}^d$ such that, if we set

$$n' = \dim_{\mathbb{C}}\left(\frac{\mathcal{V}}{\mathcal{V} \cap S}\right) \quad \text{and} \quad \ell' = \dim_{\mathbb{Q}}\left(\frac{Y \cap \mathcal{L}^d}{Y \cap S \cap \mathcal{L}^d}\right),$$

then

$$\ell' \leq \frac{n'd'}{d' - n'}.$$

If $S = \{0\}$, then $\ell' = \ell$, $n' = n$, $d' = d$, we obtain $\ell \leq \psi(n, d)$, and in this case the result is proved.

If $\dim(S) > 0$, we may apply the induction hypothesis to $Y \cap S$ and deduce that the numbers

$$d^* = \dim_{\mathbb{C}}(S), \quad n^* = \dim_{\mathbb{C}}(\mathcal{V} \cap S), \quad \ell^* = \ell - \ell' = \dim_{\mathbb{Q}}(Y \cap S \cap \mathcal{L}^d),$$

satisfy

$$\ell^* \leq \frac{n^* d^*}{d^* - n^*},$$

hence $\ell^* \leq \psi(n^*, d^*)$. We conclude

$$\ell = \ell' + \ell^* \leq \psi(n', d') + \psi(n^*, d^*) \leq \psi(n, d).$$

This completes the proof of Corollary 11.10. □

Consider now the special case of vector subspaces of $\mathbb{C}^d$ which are rational over $\overline{\mathbb{Q}}$. Baker's homogeneous Theorem 1.5 says that for such a space $\mathcal{W}$,

$$\mathcal{W} \cap \mathcal{L}^d = \bigcup_V V \cap \mathcal{L}^d,$$

where V ranges over the vector subspaces of $\mathbb{C}^d$ which are rational over $\mathbb{Q}$ and contained in $\mathcal{W}$ (see Exercise 1.5). Baker's nonhomogeneous Theorem 1.6 can be stated in the same way, replacing $\mathcal{W}$ by a linear affine subvariety of $\mathbb{C}^d$ defined over $\overline{\mathbb{Q}}$.

Conjecture 1.15 on algebraic independence of logarithms can be stated as follows (see [Roy 1995]):

Any affine algebraic subvariety $\mathfrak{V}$ of $\mathbb{C}^d$ defined over $\overline{\mathbb{Q}}$ satisfies the following property:

$$\mathfrak{V} \cap \mathcal{L}^d = \bigcup_V V \cap \mathcal{L}^d, \tag{a.i.}$$

where V ranges over the vector subspaces of $\mathbb{C}^d$ which are rational over $\mathbb{Q}$ and contained in $\mathfrak{V}$.

The first examples of nonlinear algebraic varieties for which property (a.i.) holds are given by D. Roy in [Roy 1995]. Let $\mathcal{V}$ be a $\mathbb{C}$-vector space of finite dimension equipped with a $\mathbb{Q}$-structure (see Exercise 1.4.c). For each integer $k \geq 1$, the external product $\bigwedge^k \mathcal{V}$ has a natural $\mathbb{Q}$-structure ([Roy 1995], § 1). Denote by $G(k, \mathcal{V})$ the image of the map

$$\begin{array}{ccc} \mathcal{V}^k & \longrightarrow & \bigwedge^k \mathcal{V} \\ (v_1, \ldots, v_k) & \longmapsto & v_1 \wedge \cdots \wedge v_k \end{array}.$$

Hence $G(k, \mathcal{V})$ is an algebraic subvariety of $\bigwedge^k \mathcal{V}$ defined over $\mathbb{Q}$: it is the affine cone over the Grassmannian of subspaces of $\mathcal{V}$ of dimension k. By [Roy 1995],

- *If property (a.i.) holds for $G(2, \mathbb{C}^4)$, then it also holds for $G(k, \mathcal{V})$ for any $\mathcal{V}$ and any $k \geq 1$.*

It remains to deal with $G(2, \mathbb{C}^4)$, which amounts to prove that property (a.i.) holds for the hypersurface

$$x_1y_1 + x_2y_2 + x_3y_3 = 0$$

in $\mathbb{C}^6$ (see Exercise 12.12). Compare with the four exponentials conjecture which raises the question of proving property (a.i.) for the hypersurface

$$x_1y_1 + x_2y_2 = 0$$

in $\mathbb{C}^4$ (see Exercise 1.8).

Further examples involving tensor products or symmetric products have been worked out by S. Fischler. The general context is that of an affine algebraic group, defined over $\mathbb{Q}$, acting on a vector space W over $\mathbb{C}$ endowed with a $\mathbb{Q}$-structure. The action is given by a representation ϱ, which is a morphism, defined over $\mathbb{Q}$, between the algebraic groups G and $\mathrm{GL}(W)$. Let X be an orbit for the action of G over W. From Conjecture 1.15 one deduces that, if X is not of maximal dimension among the orbits of ϱ, then any element of $X(L)$ belongs to a hyperplane of W defined over $\mathbb{Q}$.

A simple example of one of Fischler's results is the following.

- *Let k and m be two positive integers with $km > k+m$. Let $L = a_1X_1 + \ldots + a_mX_m$ be a linear form in m variables with complex coefficients. Assume that the coefficients $\lambda_{\underline{i}}$ ($\|\underline{i}\| = k$) of the polynomial*

$$L^k = \sum_{\|\underline{i}\|=k} \lambda_{\underline{i}} \underline{X}^{\underline{i}}$$

are in $\mathcal{L}$. Then these coefficients are $\mathbb{Q}$-linearly dependent.

11.6 Linear Combinations of Logarithms with Algebraic Coefficients

We extend the results of § 11.5 in two directions. On one hand one replaces $\mathcal{L}$ by $\mathbb{Q}+\mathcal{L}$: we consider points in $\mathbb{C}^d$ with coordinates $(b_i+\lambda_i)_{1\le i\le d}$, where b_i are rational numbers and λ_i are logarithms of algebraic numbers. On the other hand we replace the $\mathbb{Q}$-vector space $\mathcal{L}$ by the $\overline{\mathbb{Q}}$-vector space $\widetilde{\mathcal{L}}$: we consider points in $\mathbb{C}^d$ whose coordinates are linear combinations of 1 and logarithms of algebraic numbers.

We denote by $(\mathbb{K}, \boldsymbol{L})$ either $(\mathbb{Q}, \mathbb{Q}+\mathcal{L})$ or else $(\overline{\mathbb{Q}}, \widetilde{\mathcal{L}})$. One could also take $(\mathbb{Q}, \mathcal{L})$: one would just recover previous results. In any case $\boldsymbol{L}$ is a $\mathbb{K}$-vector subspace of $\mathbb{C}$.

11.6.1 Results and Conjectures

Following [Roy 1992b], we deduce from Theorem 11.5 the next result (compare with Corollary 11.10 above).

Corollary 11.15. *Let $\mathcal{V}$ be a vector subspace of $\mathbb{C}^d$ of dimension n such that $\mathcal{V} \cap \mathbb{K}^d = \{0\}$; denote by $d_{\min}$ the dimension of the least vector subspace of $\mathbb{C}^d$ which is rational over $\mathbb{K}$ and contains $\mathcal{V}$. Then*

$$\dim_{\mathbb{K}}\left(\mathcal{V} \cap \boldsymbol{L}^d\right) \le \Psi(n, d_{\min}).$$

One deduces from Corollary 11.15 the strong six exponentials Theorem of D. Roy [Roy 1992b], Corollary 2 § 4 p. 38:

Corollary 11.16 — Strong Six Exponentials Theorem. *Let $x_1, \ldots, x_d$ be $\overline{\mathbb{Q}}$-linearly independent complex numbers and $y_1, \ldots, y_\ell$ be also $\overline{\mathbb{Q}}$-linearly independent complex numbers. Assume $d\ell > d + \ell$. Then one at least of the $d\ell$ numbers $x_i y_j$ $(1 \leq i \leq d,\ 1 \leq j \leq \ell)$ does not belong to $\widetilde{\mathcal{L}}$.*

In other terms, for $d = 2$ and $\ell = 3$ the six numbers $e^{x_i y_j}$ cannot all be of the form

$$e^{\beta_0}\alpha_1^{\beta_1} \cdots \alpha_s^{\beta_s}.$$

Remark. One deduces the six exponentials Theorem 1.12 as well as the five exponentials Theorem (Example 1 of § 11.3.3) from Corollary 11.16 using the following fact: if the d numbers $x_i y_1$ $(1 \leq i \leq d)$ are logarithms of algebraic numbers with $x_1, \ldots, x_d$ linearly independent over $\mathbb{Q}$ and $y_1 \neq 0$, then Baker's Theorem shows that $x_1, \ldots, x_d$ are also linearly independent over $\overline{\mathbb{Q}}$.

One expects that Corollary 11.16 also holds in the limit case $d\ell = d + \ell$, i.e. $d = \ell = 2$:

Conjecture 11.17 — Strong Four Exponentials Conjecture. *Let x_1, x_2 be two $\overline{\mathbb{Q}}$-linearly independent complex numbers and y_1, y_2 be also two $\overline{\mathbb{Q}}$-linearly independent complex numbers. Then at least one of the four numbers x_1y_1, x_1y_2, x_2y_1, x_2y_2 does not belong to $\widetilde{\mathcal{L}}$.*

Again, using Gel'fond-Schneider's Theorem 1.4, it is easy to check that Conjecture 11.17 includes the four exponentials Conjecture 1.13.

As pointed out by G. Diaz (see [Di 1997a]), the strong four exponentials Conjecture implies the following refinement of the Hermite-Lindemann's Theorem, which is an open problem:

(?) *For any $\lambda \in \mathcal{L}$ with $\lambda \neq 0$, the number $|\lambda|$ is transcendental.*

In other terms, for any nonzero complex number $u \in \mathbb{C}$, if $|u|$ is algebraic, then e^u should be transcendental. This follows from Conjecture 11.17 by taking

$$\lambda = e^u, \quad x_1 = 1, \quad x_2 = \lambda, \quad y_1 = 1, \quad y_2 = \overline{\lambda}.$$

In the same way, the following conjectural refinement of the Gel'fond-Schneider's Theorem is also a consequence of the strong four exponentials Conjecture:

(?) *Let $\lambda \in \mathcal{L}$, $\lambda \neq 0$ and let $u \in \mathbb{C}$ with $|u| \in \overline{\mathbb{Q}}$. If the number $e^{u\lambda}$ is algebraic, then either $u \in \mathbb{Q}$, or else $u\lambda/\overline{\lambda} \in \mathbb{Q}$.*

(Compare with Exercise 1.8)

11.6.2 Proof

The proof of Corollary 11.15 (following [Roy 1992b]) will rest on two preliminary results.

Let k be a number field of degree m. We choose a basis $(\omega_1, \ldots, \omega_m)$ of k as a $\mathbb{Q}$-vector space. The mapping

$$\begin{array}{ccc} \mathbb{Q}^m & \longrightarrow & k = \mathbb{Q}\omega_1 + \cdots + \mathbb{Q}\omega_m \\ (a_1, \ldots, a_m) & \longmapsto & a_1\omega_1 + \cdots + a_m\omega_m \end{array}$$

is an isomorphism of $\mathbb{Q}$-vector spaces from $\mathbb{Q}^m$ to k.

Denote by $\widetilde{\mathcal{L}}_k$ the k-vector subspace of $\mathbb{C}$ spanned by 1 and $\mathcal{L}$: this is the set of numbers of the form

$$\beta + \omega_1\lambda_1 + \cdots + \omega_m\lambda_m$$

where β runs over the elements of k and $\lambda_1, \ldots, \lambda_m$ over the elements of $\mathcal{L}$.

The linear map

$$\begin{array}{cccc} \Upsilon: & \mathbb{C}^d \times (\mathbb{C}^d)^m & \longrightarrow & \mathbb{C}^d \\ & (\underline{x}, \underline{y}_1, \ldots, \underline{y}_m) & \longmapsto & \underline{x} + \omega_1\underline{y}_1 + \cdots + \omega_m\underline{y}_m \end{array}$$

will play a fundamental role. We define

$$\underline{d}_0 = d, \quad \underline{d}_1 = dm, \quad \underline{d} = \underline{d}_0 + \underline{d}_1 = (m+1)d$$

and we introduce the algebraic group $G = G_0 \times G_1$ with $G_0 = \mathbb{G}_{\mathrm{a}}^{\underline{d}_0}$, $G_1 = \mathbb{G}_{\mathrm{m}}^{\underline{d}_1}$.

Lemma 11.18.
1.– The kernel of Υ is the $\mathbb{C}$-vector subspace of $\mathbb{C}^{\underline{d}}$, rational over k, of dimension md, defined by the equation

$$\underline{x} + \omega_1\underline{y}_1 + \cdots + \omega_m\underline{y}_m = 0.$$

2.– Let $\mathcal{U}$ be a $\mathbb{C}$-vector subspace of $\mathbb{C}^d$ and let $\mathcal{V} = \Upsilon^{-1}(\mathcal{U})$. Then

$$\dim_{\mathbb{C}}\left(\frac{\mathbb{C}^{\underline{d}}}{\mathcal{V}}\right) = \dim_{\mathbb{C}}\left(\frac{\mathbb{C}^d}{\mathcal{U}}\right).$$

Hence $\dim_{\mathbb{C}}(\mathcal{V}) = \dim_{\mathbb{C}}(\mathcal{U}) + md$. *Moreover, if $\mathcal{U} \cap k^d = \{0\}$, then*

$$\mathcal{V} \cap \left(\{0\} \times \mathbb{Q}^{\underline{d}_1}\right) = \{0\} \quad \textit{and} \quad \mathcal{V} \cap \left(\overline{\mathbb{Q}}^{\underline{d}_0} \times \{0\}\right) = \{0\}.$$

3.– Let $G^* = G_0^* \times G_1^*$ be a connected algebraic subgroup G^* of G, defined over $\overline{\mathbb{Q}}$. The tangent space $S = T_e(G^*)$ is a subspace of $\mathbb{C}^{\underline{d}}$ and $T = \Upsilon(S)$ is a subspace of $\mathbb{C}^d$ which is rational over $\overline{\mathbb{Q}}$. Define

$$\underline{d}_1' = \dim\left(\frac{G_1}{G_1^*}\right) \quad \textit{and} \quad d' = \dim_{\mathbb{C}}\left(\frac{\mathbb{C}^d}{T}\right).$$

Then $\underline{d}_1' \geq md'$. Moreover if $\mathcal{U}$ is a $\mathbb{C}$-vector subspace of $\mathbb{C}^d$ and if we set

$$\underline{d}' = \dim\left(\frac{G}{G^*}\right), \quad n' = \dim_{\mathbb{C}}\left(\frac{\mathcal{U}}{\mathcal{U}\cap T}\right),$$

$$\mathcal{V} = \Upsilon^{-1}(\mathcal{U}), \quad \underline{n}' = \dim_{\mathbb{C}}\left(\frac{\mathcal{V}}{\mathcal{V}\cap S}\right),$$

we have $\underline{d}' - \underline{n}' = d' - n'$.

4.– The restriction of Υ *to* $k^{\underline{d}_0} \times \mathcal{L}^{\underline{d}_1}$ *defines an isomorphism of* $\mathbb{Q}$*-vector spaces:*

$$\upsilon: k^{\underline{d}_0} \times \mathcal{L}^{\underline{d}_1} \longrightarrow \widetilde{\mathcal{L}}_k^d.$$

Therefore, if Z *is a* k*-vector subspace of* $\widetilde{\mathcal{L}}_k^d$ *and if we set* $Y = \upsilon^{-1}(Z)$*, we have* $\dim_{\mathbb{Q}}(Y) = m\dim_k(Z)$.

Proof of Lemma 11.18. The first property is trivial. The second one follows from the fact that the spaces $\mathbb{C}^{\underline{d}}/\mathcal{V}$ and $\mathbb{C}^d/\mathcal{U}$ are isomorphic.

Before going further, we make two remarks.

a) If k is a field, K an extension of k and if $\underline{x}_1, \dots, \underline{x}_n$ are k-linearly independent elements in k^n, then in K^n the elements $\underline{x}_1, \dots, \underline{x}_n$ are K-linearly independent (the rank of a matrix does not change under scalar extension).

b) If Z is a k-vector space of finite dimension, then Z is a $\mathbb{Q}$-vector space of finite dimension $m\dim_k(Z)$.

We now prove the estimate $\underline{d}_1' \geq md'$ in part 3 of Lemma 11.18. Let us define

$$S = T_e(G^*) \subset \mathbb{C}^{\underline{d}} \quad \text{and} \quad S_1 = T_e(G_1^*) \subset \mathbb{C}^{\underline{d}_1}.$$

The subspace S_1 of $\mathbb{C}^{\underline{d}_1}$ is rational over $\mathbb{Q}$; hence its codimension in $\mathbb{C}^{\underline{d}_1}$, namely $\underline{d}_1'$, is the same as the codimension in $\{0\}\times\mathbb{Q}^{\underline{d}_1}$ of the $\mathbb{Q}$-vector space $E = S\cap\left(\{0\}\times\mathbb{Q}^{\underline{d}_1}\right)$. Since Υ induces an $\mathbb{Q}$-isomorphism between $\{0\}\times\mathbb{Q}^{\underline{d}_1}$ and k^d, the $\mathbb{Q}$-vector subspace $\Upsilon(E)$ of k^d is isomorphic to E. Moreover $\Upsilon(E)$ is contained in the $\mathbb{Q}$-vector space $T\cap k^d$. Hence

$$\dim_{\mathbb{Q}}(E) \leq \dim_{\mathbb{Q}}(T\cap k^d).$$

Since $T\cap k^d$ is a k-vector space, we have $\dim_{\mathbb{Q}}(T\cap k^d) = m\dim_k(T\cap k^d)$. From remark a) above we deduce $\dim_k(T\cap k^d) \leq \dim_{\mathbb{C}}(T)$. Finally we conclude $\dim_{\mathbb{Q}}(E) \leq m\dim_{\mathbb{C}}(T)$, which means $dm - \dim_{\mathbb{Q}}(E) \geq m\dim_{\mathbb{C}}(\mathbb{C}^d/T)$, hence $\underline{d}_1' \geq md'$.

We deduce $\underline{d}' - \underline{n}' = d' - n'$ as follows: since $\mathcal{V} \supset \ker\Upsilon$, using property 2 for $\mathcal{V}+S$ yields $\Upsilon^{-1}(\mathcal{U}+T) = \mathcal{V}+S$. Hence

$$\dim_{\mathbb{C}}\left(\frac{\mathbb{C}^{\underline{d}}}{\mathcal{V}+S}\right) = \dim_{\mathbb{C}}\left(\frac{\mathbb{C}^d}{\mathcal{U}+T}\right)$$

with $(\mathcal{V}+S)/S \simeq \mathcal{V}/\mathcal{V}\cap S$ and $(\mathcal{U}+T)/T \simeq \mathcal{U}/\mathcal{U}\cap T$.

Finally the fact that υ is surjective is a consequence of the very definition of $\widetilde{\mathcal{L}}_k$, while the injectivity uses Baker's Theorem: if $\lambda_1, \dots, \lambda_m$ belong to $\mathcal{L}$ and if β belongs to k, then the relation $\beta + \omega_1\lambda_1 + \cdots + \omega_m\lambda_m = 0$ implies $\beta = 0$ and $\lambda_1 = \cdots = \lambda_m = 0$. □

The main tool for the proof of Corollary 11.15 is the following (see [Roy 1992b], Th. 4).

Proposition 11.19. *Let $\mathcal{U}$ be a vector subspace of $\mathbb{C}^d$ of dimension $n < d$. Among the subspaces of T which are rational over $\mathbb{K}$ and distinct from $\mathbb{C}^d$, we select one for which the quantity n'/d' is minimal, with*

$$n' = \dim_{\mathbb{C}}\left(\frac{\mathcal{U}}{\mathcal{U}\cap T}\right), \qquad d' = \dim_{\mathbb{C}}\left(\frac{\mathbb{C}^d}{T}\right).$$

Then the dimension ℓ' of the $\mathbb{K}$-vector space $\mathcal{U}\cap \boldsymbol{L}^d/\mathcal{U}\cap T\cap \boldsymbol{L}^d$ is finite and satisfies

$$\frac{\ell'}{d'+\ell'} \le \frac{n'}{d'} \le \frac{n}{d}.$$

In particular if the inequality $n'/d' \ge n/d$ holds for any $\mathbb{K}$-rational subspace T, then the hypothesis holds with $T = \{0\}$. In this case we deduce that the dimension ℓ of the $\mathbb{K}$-vector space $\mathcal{U}\cap \boldsymbol{L}^d$ is finite and satisfies $\ell d \le n(\ell + d)$.

Proof of Proposition 11.19. The statement of Proposition 11.19 is obtained from Corollary 11.14 by replacing $\mathbb{Q}$, $\mathcal{L}$, $\mathcal{V}$ and S by $\mathbb{K}$, $\boldsymbol{L}$, $\mathcal{U}$ and T respectively. The proof of Corollary 11.14 was by induction on d, and the general case was a formal consequence of the so-called first case, where $S = \{0\}$ is the only subspace of $\mathbb{C}^d$, rational over $\mathbb{Q}$ and distinct from $\mathbb{C}^d$, for which $n'/d' \le n/d$. In exactly the same way, for the proof of Proposition 11.19, we may restrict to the case where the space $T = \{0\}$ is the only one which satisfies the assumptions. Therefore we assume that for any $\mathbb{K}$-rational subspace T of dimension d^* in the range $0 < d^* < d$,

$$\frac{\dim_{\mathbb{C}}\big(\mathcal{U}/\mathcal{U}\cap T\big)}{\dim_{\mathbb{C}}(\mathbb{C}^d/T)} > \frac{d^*}{d}.$$

This implies $\mathcal{U}\cap \mathbb{K}^d = \{0\}$.

Let Z_1 be a $\mathbb{K}$-vector subspace of $\mathcal{U}\cap \boldsymbol{L}^d$ of finite dimension ℓ. We want to prove the upper bound $\ell(d-n) \le nd$. Choose a basis of Z_1 over $\mathbb{K}$. The elements of this basis belong to $\boldsymbol{L}$: they are linear combinations $\beta_0 + \beta_1\lambda_1 + \cdots + \beta_s\lambda_s$ with algebraic coefficients β_i of elements in $\mathcal{L}$; these coefficients β_i generate a number field k (in the case $(\mathbb{K}, \boldsymbol{L}) = (\mathbb{Q}, \mathbb{Q}+\mathcal{L})$ we have $k = \mathbb{Q}$). Let Z be the k-vector subspace of $\widetilde{\mathcal{L}}_k$ which is spanned by our selected basis of Z_1. Then $\dim_k(Z) = \dim_{\mathbb{K}}(Z_1) = \ell$.

Thanks to Lemma 11.18, we may apply part (3') of Theorem 11.5 with d_0, d_1, d, n and ℓ_0 replaced respectively by

$$\underline{d}_0 = d, \quad \underline{d}_1 = dm, \quad \underline{d} = \underline{d}_0 + \underline{d}_1 = (m+1)d, \quad \underline{n} = n + md, \quad \underline{\ell}_0 = md,$$

and with

$$Y = \upsilon^{-1}(Z), \qquad \mathcal{W} = \ker\Upsilon \subset \mathcal{V} = \Upsilon^{-1}(\mathcal{U}) \subset \mathbb{C}^{\underline{d}}.$$

The rank of Y is at least $\underline{\ell}_1 = m\ell$. Since $\underline{d} > \underline{n}$, we deduce $\underline{\ell}_1(\underline{d} - \underline{n}) \le \underline{d}_1(\underline{n} - \underline{\ell}_0)$, which yields the desired inequality $\ell(d - n) \le dn$. □

Proof of Corollary 11.15. The proof is now exactly the same as the proof of Corollary 11.10, using Proposition 11.19 in place of Corollary 11.14 and $\mathbb{K}$, L, $\mathcal{U}$, T in place of $\mathbb{Q}$, $\mathcal{L}$, $\mathcal{V}$, S respectively. □

11.6.3 Are we Far from the Truth?

As shown by D. Roy [Roy 1992b], § 5 Th. 6, the estimates in Corollaries 11.10 and 11.15 are optimal up to a coefficient 1/2. More precisely, set

$$\phi(n,d) = \begin{cases} \dfrac{n(n-1)}{2} + 1 & \text{for } 1 \le n \le d-2, \\ \dfrac{n(n+1)}{2} & \text{for } n = d-1. \end{cases}$$

Lemma 11.20. *Let n and d be two positive integers with $1 \le n < d$. There exists a vector subspace $\mathcal{V}$ of $\mathbb{C}^d$, of dimension n, which is not contained in any subspace of $\mathbb{C}^d$ rational over $\mathbb{Q}$ distinct from $\mathbb{C}^d$, and such that*
$\mathcal{V} \cap \overline{\mathbb{Q}}^d = \{0\}$, $\dim_{\mathbb{Q}}(\mathcal{V} \cap \mathcal{L}^d) \ge \phi(n,d)$ *and* $\dim_{\overline{\mathbb{Q}}}(\mathcal{V} \cap \widetilde{\mathcal{L}}^d) \ge \phi(n,d)$.

Proof. The proof of Lemma 11.20 uses the following consequence of Baker's Theorem (see Exercise 1.5):

- *Let $y_1, \dots, y_\ell$ be $\mathbb{Q}$-linearly independent elements in $\mathcal{L}^d$. Then $y_1, \dots, y_\ell$ are $\overline{\mathbb{Q}}$-linearly independent.*

Hence for any subspace $\mathcal{V}$ of $\mathbb{C}^d$ we have

$$\dim_{\overline{\mathbb{Q}}}(\mathcal{V} \cap \widetilde{\mathcal{L}}^d) \ge \dim_{\mathbb{Q}}(\mathcal{V} \cap \mathcal{L}^d).$$

The problem now is to construct $\mathcal{V}$ with $\dim_{\mathbb{Q}}(\mathcal{V} \cap \mathcal{L}^d) \ge \phi(n,d)$. Let $\lambda_1, \dots, \lambda_d$ be $\mathbb{Q}$-linearly independent elements of $\mathcal{L}$.
a) If $n = 1$ we take for $\mathcal{V}$ the complex line $\mathbb{C}(\lambda_1, \dots, \lambda_d)$ in $\mathbb{C}^d$, so that $\dim_{\mathbb{Q}}(\mathcal{V} \cap \mathcal{L}^d) \ge 1$.
b) For $n = d-1$ we take for $\mathcal{V}$ the hyperplane of equation $\lambda_1 z_1 + \dots + \lambda_d z_d = 0$ in $\mathbb{C}^d$, which contains the $d(d-1)/2$ points $\lambda_i \underline{e}_j - \lambda_j \underline{e}_i$ of $\mathcal{L}^d$ $(1 \le i < j \le d)$ (here $(\underline{e}_1, \dots, \underline{e}_d)$ denotes the canonical basis in $\mathbb{C}^d$).
c) Finally, for $2 \le n \le d-2$ we take $\mathcal{V} = \mathcal{V}_1 \times \mathcal{V}_2$ where $\mathcal{V}_1$ is the line $\mathbb{C}(\lambda_1, \dots, \lambda_{d-n})$ in $\mathbb{C}^{d-n}$ while $\mathcal{V}_2$ is the hyperplane $\lambda_{d-n+1} z_1 + \dots + \lambda_d z_n = 0$ in $\mathbb{C}^n$. Then $\mathcal{V}_1$ contains a nonzero element in $\mathcal{L}^d$; moreover $\mathcal{V}_2$ contains $n(n-1)/2$ elements of $\mathcal{L}^d$ which are $\mathbb{Q}$-linearly independent. Therefore $\mathcal{V}$ contains $1 + n(n-1)/2$ elements of $\mathcal{L}$ which are $\mathbb{Q}$-linearly independent. □

Remark. Assuming Conjecture 1.15 on algebraic independence of logarithms of algebraic numbers, Lemma 11.20 is optimal; see Exercise 12.8.

11.7 Proof of the Linear Subgroup Theorem

We split the proof of Theorem 11.5 in three parts. In the first one we apply the transcendence method and produce a variant of part (1) (it seems weaker than (1), but it will turn out to be equivalent). Next we establish the equivalence between the six statements. Finally we complete the proof by taking possible periods into account. It may be worthwhile to point out that the same method works more generally for commutative algebraic groups (and not only for the linear ones).

11.7.1 The Transcendence Argument

We first prove the following statement (compare with Theorem 4.1 of [W 1988]):

Theorem 11.21. *Let d_0 and d_1 be two nonnegative integers with $d = d_0 + d_1 \geq 2$. Let $\mathcal{V}$ be a complex subspace of $\mathbb{C}^d$ of dimension n with $1 \leq n < d$, let $\mathcal{W}$ be a $\mathbb{C}$-vector subspace of $\mathbb{C}^d$, rational over $\overline{\mathbb{Q}}$, of dimension $\ell_0 \geq 0$, contained in $\mathcal{V}$ and let Y be a subgroup of $\mathcal{L}_G$, of finite rank $\ell_1 > 0$ over $\mathbb{Z}$, also contained in $\mathcal{V}$. Consider the subgroup $\Gamma = \exp_G(Y)$ of $G(\overline{\mathbb{Q}})$. Then there exists a connected algebraic subgroup $G^* = G_0^* \times G_1^*$ of G, defined over $\overline{\mathbb{Q}}$, such that*

$$d' > \ell_0' \quad \textit{and} \quad \frac{\lambda' + d_1'}{d' - \ell_0'} \leq \frac{d_1}{d - n}$$

where

$$d' = \dim\left(\frac{G}{G^*}\right), \quad \lambda' = \operatorname{rank}_{\mathbb{Z}}\left(\frac{\Gamma}{\Gamma \cap G^*(\mathbb{C})}\right),$$

$$d_1' = \dim\left(\frac{G_1}{G_1^*}\right), \quad \ell_0' = \dim_{\mathbb{C}}\left(\frac{\mathcal{W}}{\mathcal{W} \cap T_e(G^*)}\right).$$

Proof.

Step 1. Preliminary Reduction

It will be convenient (at the end of the proof) to assume $d_0 \leq n$. So let us start by proving the conclusion in the alternative case.

Assume $d_0 > n$. Denote by π_0 the projection of $\mathbb{C}^d$ onto $\mathbb{C}^{d_0}$ with kernel $\{0\} \times \mathbb{C}^{d_1}$. The complex vector subspace of $\mathbb{C}^{d_0}$ spanned by $\pi_0(Y \cup \mathcal{W})$ is contained in $\pi_0(\mathcal{V})$, hence has dimension $\leq n < d_0$. Therefore $Y \cup \mathcal{W} \subset T_e(G^*)$ where $G^* = G_0^* \times G_1$, and G_0^* is an algebraic subgroup of G_0 of codimension $d_0' > 0$. In this case we trivially get the conclusion with $d_1' = \ell_0' = \lambda' = 0$.

Step 2. Introducing the Parameters

We introduce positive integers T_0, T_1, S_0, S_1 and we define

$$L = \binom{T_0 + d_0}{d_0}(2T_1 + 1)^{d_1}.$$

We denote by $c_1, \ldots, c_{11}$ positive integers which depend only on the data of Theorem 11.21 but do not depend on T_0, T_1, S_0, S_1.

Step 3. Construction of the Matrix M

In the ring $\mathbb{C}[G] = \mathbb{C}[\underline{X}, \underline{Y}^{\pm 1}]$ (see § 5.1), consider the monomials

$$\underline{X}^{\underline{\tau}}\underline{Y}^{\underline{t}} = X_1^{\tau_1} \cdots X_{d_0}^{\tau_{d_0}} Y_1^{t_1} \cdots Y_{d_1}^{t_{d_1}},$$

where $(\underline{\tau}, \underline{t})$ runs over the set of all L elements in $\mathbb{N}^{d_0} \times \mathbb{Z}^{d_1}$ which satisfy

$$\|\underline{\tau}\| \le T_0, \quad |\underline{t}| \le T_1.$$

We fix an ordering of these L elements and we define entire functions of d variables by

$$\underline{z}^{\underline{\tau}} e^{\underline{t}\underline{z}} = z_1^{\tau_1} \cdots z_{d_0}^{\tau_{d_0}} e^{t_1 z_{d_0+1} + \cdots + t_{d_1} z_d}.$$

Let $\boldsymbol{w} = (\underline{w}_1, \ldots, \underline{w}_{\ell_0})$ be a basis of $\mathcal{W}$ over $\mathbb{C}$ contained in $\overline{\mathbb{Q}}^{d\ell_0}$. The coordinates of $\underline{w}_k$ $(1 \le k \le \ell_0)$ will be written

$$\underline{w}_k = (\beta_{1k}, \ldots, \beta_{dk}).$$

Recall the notation, for $\underline{\sigma} \in \mathbb{N}^{\ell_0}$,

$$\mathcal{D}_{\boldsymbol{w}}^{\underline{\sigma}} = \mathcal{D}_{\underline{w}_1}^{\sigma_1} \cdots \mathcal{D}_{\underline{w}_{\ell_0}}^{\sigma_{\ell_0}},$$

with

$$\mathcal{D}_{\underline{w}_k} = \beta_{1k} \frac{\partial}{\partial z_1} + \cdots + \beta_{dk} \frac{\partial}{\partial z_d}.$$

Further, let $\underline{\eta}_1, \ldots, \underline{\eta}_{\ell_1}$ be $\mathbb{Z}$-linearly independent elements of Y. We shall denote the coordinates of $\underline{\eta}_j$ $(1 \le j \le \ell_1)$ by

$$\underline{\eta}_j = (\beta_{1,\ell_0+j}, \ldots, \beta_{d_0,\ell_0+j}, \lambda_{1j}, \ldots, \lambda_{d_1,j}).$$

We set $\alpha_{ij} = e^{\lambda_{ij}}$ $(1 \le i \le d_1,\ 1 \le j \le \ell_1)$ and we denote by K a number field which contains all the $d_1\ell_1 + d_1\ell_0 + d_0\ell_1 + d_0\ell_0 = (d_0 + d_1)(\ell_0 + \ell_1)$ algebraic numbers

$$\alpha_{ij}, \qquad \beta_{hk}, \qquad \beta_{d_0+i,k}, \qquad \beta_{h,\ell_0+j}$$

for

$$1 \le i \le d_1, \quad 1 \le j \le \ell_1, \quad 1 \le k \le \ell_0, \quad 1 \le h \le d_0.$$

Notice that the points

$$\underline{\gamma}_j = \exp_G(\underline{\eta}_j) = (\beta_{1,\ell_0+j}, \ldots, \beta_{d_0,\ell_0+j}, \alpha_{1j}, \ldots, \alpha_{d_1,j}) \quad (1 \le j \le \ell_1)$$

are in $G(K) = K^{d_0} \times (K^\times)^{d_1}$.

For each $(\underline{\sigma}, \underline{s}) \in \mathbb{N}^{\ell_0} \times \mathbb{Z}^{\ell_1}$ satisfying

$$\|\underline{\sigma}\| \le S_0, \quad |s_j| \le S_1 \qquad (1 \le j \le \ell_1),$$

we introduce the number

$$\gamma_{\underline{\tau t}}^{\underline{\sigma s}} = \mathcal{D}_{\underline{w}}^{\underline{\sigma}}\big(\underline{z}^{\underline{\tau}} e^{\underline{tz}}\big)(s_1 \underline{\eta}_1 + \cdots + s_{\ell_1} \underline{\eta}_{\ell_1})$$

which lies in the number field K. We also choose an ordering for these tuples $(\underline{\sigma}, \underline{s})$ and we build the matrix

$$M = \Big(\gamma_{\underline{\tau t}}^{\underline{\sigma s}}\Big)_{\substack{(\underline{\tau}, \underline{t}) \\ (\underline{\sigma}, \underline{s})}}.$$

Our goal is to show that M has rank $< L$. This will allow us to apply the multiplicity estimate (Theorem 8.1) which will in turn produce the subgroup G^* of G.

Our goal is trivially achieved if the number of columns of M is less than L. Otherwise, let Δ be the determinant of a square $L \times L$ matrix extracted from M.

Step 4. Upper Bound for $|\Delta|$

The determinant $F(\zeta)$ of the matrix

$$\Big(\mathcal{D}_{\underline{w}}^{\underline{\sigma}}\big(\underline{z}^{\underline{\tau}} e^{\underline{tz}}\big)\big(\zeta(s_1 \underline{\eta}_1 + \cdots + s_{\ell_1} \underline{\eta}_{\ell_1})\big)\Big)_{\substack{(\underline{\tau}, \underline{t}) \\ (\underline{\sigma}, \underline{s})}}$$

is a function of a complex variable ζ which has a zero of multiplicity

$$\geq c_1 L^{1+(1/n)} - c_2 L S_0$$

at the origin (the correcting term $c_2 L S_0$ arises from $\mathcal{D}_{\underline{w}}^{\underline{\sigma}}$; compare with Lemma 9.2 which settles the case $\ell_0 = 1$ and with Lemma 10.6; further explicit estimates will be provided in Lemma 13.4). Moreover on the disc $|\zeta| \leq e$ we have

$$\log |F(\zeta)| \leq c_3 L\big(S_0 \log(T_0 T_1) + T_0 \log(S_0 S_1) + T_1 S_1\big).$$

Again, explicit estimates for a more general situation will be provided in Lemma 13.9.

Since $\Delta = F(1)$, from Schwarz' Lemma 6.1 we conclude

$$\log |\Delta| \leq -c_4 L^{1+(1/n)} + c_5 L\big(S_0 \log(T_0 T_1) + T_0 \log(S_0 S_1) + T_1 S_1\big).$$

Step 5. Lower Bound for $|\Delta|$

From Liouville's estimate (see Exercise 3.8 for the case $\ell_0 = 0$ and Proposition 13.8 for the general case) we deduce that either $\Delta = 0$ or else

$$\log |\Delta| \geq -c_6 L\Big(S_0 \log(T_0 T_1) + T_0 \log(S_0 S_1) + T_1 S_1\Big).$$

Therefore if our parameters satisfy

$$S_0 \log(T_0 T_1) + T_0 \log(S_0 S_1) + T_1 S_1 \leq c_7 L^{1/n} \tag{11.22}$$

for a suitable (sufficiently small) $c_7 > 0$, we shall deduce from steps 4 and 5 that $\Delta = 0$. We assume that this condition (11.22) is fulfilled and we shall confirm it in step 7 thanks to a suitable choice of the parameters.

Step 6. Using the Multiplicity Estimate

Since $\Delta = 0$ it follows that M has rank $< L$. Therefore there exist complex numbers $c_{\tau t}$, not all of which are zero, such that

$$\sum_{\underline{\tau}} \sum_{\underline{t}} c_{\underline{\tau}\underline{t}} \gamma_{\underline{\tau}\underline{t}}^{\underline{\sigma}\underline{s}} = 0$$

for all $(\underline{\sigma}, \underline{s})$ as above. This means that the polynomial

$$P = \sum_{\underline{\tau}} \sum_{\underline{t}} c_{\underline{\tau}\underline{t}} \underline{X}^{\underline{\tau}} \underline{Y}^{\underline{t}} \in \mathbb{C}[\underline{X}, \underline{Y}^{\pm 1}]$$

yields a function

$$F(\underline{z}) = P(z_1, \ldots, z_{d_0}, e^{z_{d_0+1}}, \ldots, e^{z_d})$$

which satisfies

$$\mathcal{D}_{\underline{w}}^{\underline{\sigma}} F(s_1 \underline{\eta}_1 + \cdots + s_{\ell_1} \underline{\eta}_{\ell_1}) = 0$$

for all $(\underline{\sigma}, \underline{s}) \in \mathbb{N}^{\ell_0} \times \mathbb{Z}^{\ell_1}$ with $\|\underline{\sigma}\| \leq S_0$ and $|s_j| \leq S_1 \quad (1 \leq j \leq \ell_1)$.

Using Theorem 8.1 with $G^+ = G$, $G^- = 0$, $D_0 = T_0$, $D_1 = \cdots = D_{d_1} = T_1$ and

$$\Sigma = \left\{ s_1 \underline{\gamma}_1 + \cdots + s_{\ell_1} \underline{\gamma}_{\ell_1} \, ; \, s \in \mathbb{Z}^{\ell_1}, \, |s_j| \leq S_1 \, (1 \leq j \leq \ell_1) \right\},$$

we obtain the existence of a subgroup G^* of G of dimension $< d$ such that, if we set

$$\ell_0' = \dim_{\mathbb{C}} \left(\frac{\mathcal{W}}{\mathcal{W} \cap T_e(G^*)} \right),$$

then

$$\binom{S_0 + \ell_0'}{\ell_0'} \operatorname{Card} \left(\frac{\Sigma + G^*}{G^*} \right) \mathcal{H}(G^*; T_0; T_1) \leq \mathcal{H}(G; T_0; T_1).$$

Therefore we deduce

$$S_0^{\ell_0'} S_1^{\lambda'} \leq c_8 T_0^{d_0'} T_1^{d_1'}. \tag{11.23}$$

Step 7. Choice of Parameters

Define $\mu = d_1/(d-n)$ and $\nu = d+n$. We choose for S_1 a sufficiently large integer and we define

$$S_0 = \left[S_1^{\mu} (\log S_1)^{\nu} \right], \qquad T_0 = \left[S_0 / \log S_1 \right], \qquad T_1 = \left[S_0 / S_1 \right].$$

From step 1 we deduce $\mu \geq 1$. We check (11.22): indeed we have

$$S_0 \log(T_0 T_1) + T_0 \log(S_0 S_1) + T_1 S_1 \leq c_9 S_1^{\mu} (\log S_1)^{\nu+1}$$

and

$$L \geq \frac{1}{d_0!} S_1^{d\mu - d_1} (\log S_1)^{d\nu - d_0}.$$

This explains the choice of μ. As far as ν is concerned, any $\nu > (d_0 + n)/(d - n)$ is admissible.

Therefore (11.23) holds. We first replace T_0 and T_1 by their values in terms of S_0 and S_1:

$$S_1^{\lambda'+d_1'} \le c_{10} S_0^{d'-\ell_0'} (\log S_1)^{-d_0'}$$

and then in terms of S_1 only:

$$S_1^{\lambda'+d_1'} (\log S_1)^{d_0'} \le c_{11} S_1^{(d'-\ell_0')\mu} (\log S_1)^{(d'-\ell_0')\nu}.$$

The inequality $\lambda' + d_1' \le (d' - \ell_0')\mu$ plainly follows. Further, if $\lambda' + d_1' = (d' - \ell_0')\mu$, then $d_0' \le (d' - \ell_0')\nu$.

Hence $d' \ge \ell_0'$, and equality $d' = \ell_0'$ would imply $\lambda' = d_1' = 0$ and $d_0' = 0$. But this is not compatible with the condition $d' > 0$. □

11.7.2 Equivalence between Six Statements

Following [Roy 1992a], we show that if any of the six statements (1) to (3') from Theorem 11.5 holds for any $(d_0, d_1, Y, \mathcal{V}, \mathcal{W})$, then so do the five others.

The following simple fact will be useful: for positive real numbers a, b, c, d, we have

$$\min\left\{\frac{a}{b}, \frac{c}{d}\right\} \le \frac{a+c}{b+d} \le \max\left\{\frac{a}{b}, \frac{c}{d}\right\}.$$

(1) ⇒ (2)

Assume (1). Assume also that the hypotheses of (2) are satisfied. Among the G^* for which $d' > n'$, select one (and call it G^*) for which $d_1'/(d' - n')$ is minimal. From (1) with G replaced by $G' = G/G^*$ we deduce that there exists a quotient $G'' = G/G^{**}$ of G', where G^{**} is an algebraic subgroup of G containing G^*, such that, if we set

$$G'' = G_0'' \times G_1'', \quad d'' = \dim(G''), \quad d_1'' = \dim(G_1''),$$

$$Y'' = \frac{Y}{Y \cap T_e(G^{**})}, \quad \mathcal{V}'' = \frac{\mathcal{V}}{\mathcal{V} \cap T_e(G^{**})}, \quad \mathcal{W}'' = \frac{\mathcal{W}}{\mathcal{W} \cap T_e(G^{**})}$$

and

$$\ell_1'' = \mathrm{rank}_{\mathbb{Z}}(Y''), \quad n'' = \dim_{\mathbb{C}}(\mathcal{V}''), \quad \ell_0'' = \dim_{\mathbb{C}}(\mathcal{W}''),$$

then

$$d'' > \ell_0'' \quad \text{and} \quad \frac{\ell_1'' + d_1''}{d'' - \ell_0''} \le \frac{d_1'}{d' - n'}.$$

Since G'' is a quotient of G, from the choice of G^* we deduce

$$\frac{d_1'}{d' - n'} \le \frac{d_1''}{d'' - n''} \quad \text{if} \quad d'' > n''.$$

This inequality together with

$$\frac{\ell_1'' + d_1''}{(d'' - n'') - (\ell_0'' - n'')} \le \frac{d_1'}{d' - n'}$$

imply

$$\frac{\ell_1''}{n'' - \ell_0''} \le \frac{d_1'}{d' - n'} \quad \text{if} \quad d'' > n'' \quad \text{and} \quad n'' > \ell_0''.$$

The last inequality plainly holds also if $d'' = n''$. Further, we have $n'' > \ell_0''$. Indeed otherwise we would get $d'' > \ell_0''$, hence $\ell_1'' = 0$, which has been excluded in the assumptions of (2).

Again from the choice of G^* we deduce

$$\frac{d_1'}{d' - n'} \le \frac{d_1}{d - n}.$$

Hence

$$\frac{\ell_1''}{n'' - \ell_0''} \le \frac{d_1}{d - n}.$$

Using the relations

$$\ell_1'' = \ell_1 - \ell_1^{**}, \quad n'' = n - n^{**} \quad \text{and} \quad \ell_0'' = \ell_0 - \ell_0^{**}$$

we can write the last inequality:

$$\frac{\ell_1 - \ell_1^{**}}{(n - n^{**}) - (\ell_0 - \ell_0^{**})} \le \frac{d_1}{d - n}.$$

However from the assumptions in (2) we deduce

$$\ell_1(n^{**} - \ell_0^{**}) \ge \ell_1^{**}(n - \ell_0).$$

Since $n'' > \ell_0''$ we have $\mathcal{W}'' \neq \mathcal{V}''$, hence $\mathcal{W} \neq \mathcal{V}$ and $n > \ell_0$. Therefore the last inequality yields

$$\frac{\ell_1}{n - \ell_0} \le \frac{\ell_1 - \ell_1^{**}}{(n - \ell_0) - (n^{**} - \ell_0^{**})}.$$

The desired inequality

$$\frac{\ell_1}{n - \ell_0} \le \frac{d_1}{d - n}$$

follows.

(2) $\Rightarrow$ (3)

The first inequality

$$\frac{n - \ell_0}{\ell_1} \ge \frac{n^* - \ell_0^*}{\ell_1^*}$$

is plain. Replacing G by G^*, we may assume that for any $G^* \neq G$ for which $\ell_1^* \neq 0$, we have

$$\frac{n^* - \ell_0^*}{\ell_1^*} > \frac{n - \ell_0}{\ell_1}.$$

If $\ell_1' = 0$, then $\ell_1 = \ell_1^*$ and $n' < \ell_0'$. From (2) we deduce

$$\ell_1(d-n) \le d_1(n-\ell_0).$$

(3) ⇒ (1)'

This is a consequence of the following remark: if $\ell_1^* \neq 0$ and $d_1^* \neq 0$, then

$$\frac{d^*-\ell_0^*}{d_1^*+\ell_1^*} \le \max\left\{\frac{d^*-n^*}{d_1^*}, \frac{n^*-\ell_0^*}{\ell_1^*}\right\}.$$

(1)' ⇒ (2)' ⇒ (3)' ⇒ (1)

One can just repeat the same arguments as before, permuting

$$\begin{array}{cccc} d_1' & n'-\ell_0' & \ell_1' & d'-n' \\ \updownarrow & \updownarrow & \updownarrow & \updownarrow \\ n^*-\ell_0^* & d_1^* & d^*-n^* & \ell_1^* \end{array}$$

respectively. However there is a much more elegant solution in [Roy 1992a] (see also [Roy 1992b]) involving a category and its opposite.

11.7.3 Taking Periods into Account

Theorem 11.21 does not look as sharp as statement (1) in Theorem 11.5, because $\lambda' \le \ell_1'$: the difference is the rank of $Y' \cap \ker \exp_{G'}$. In order to keep track of the periods and to get rid of the discrepancy between λ' and ℓ_1', we shall use the following Lemma:

Lemma 11.24. *Under the assumptions of Theorem 11.5, denote by Ω_G the kernel of* $\exp_G$ *in* $\mathbb{C}^{d_0} \times \mathbb{C}^{d_1}$. *Define also*

$$\Gamma = \exp_G Y, \quad \lambda = \mathrm{rank}_{\mathbb{Z}}(\Gamma), \quad \kappa = \mathrm{rank}_{\mathbb{Z}}(Y \cap \Omega_G),$$

so that $\lambda = \ell_1 - \kappa$. Then there exists a connected algebraic subgroup G^* of G such that, if we define G', Y', d_0', d_1', d', n', ℓ_1', ℓ_0' as in § 11.3.1, and also

$$\Omega_{G'} = \ker \exp_{G'}, \quad \kappa' = \mathrm{rank}_{\mathbb{Z}}(Y' \cap \Omega_{G'}),$$

then we have

$$d_0' = d_0, \quad d_1' = d_1 - \kappa, \quad d' = d - \kappa, \quad n' = n - \kappa,$$

$$\ell_1' \le \ell_1 - \kappa, \quad \ell_0 - \kappa \le \ell_0' \le \ell_0, \quad \ell_0' + \kappa \le n, \quad \kappa' = 0.$$

Proof. From $\Omega_G = \{0\} \times (2i\pi\mathbb{Z})^{d_1}$ we deduce that elements in Ω_G which are linearly independent over $\mathbb{Z}$ are also linearly independent over $\mathbb{C}$. The $\mathbb{C}$-vector subspace of $\mathbb{C}^d$ spanned by Ω_G can be written $T_e(G^*)$ where $G^* = \{0\} \times G_1^*$ satisfies the required properties. □

The conclusion of Theorem 11.21 involves an upper bound where the left hand side depends on a quotient of G and the right hand side is $d_1/(d-n)$. Now if there is a quotient of G which has a corresponding value of $d_1/(d-n)$ smaller than the initial value associated to G, then one gets a sharper estimate by applying Theorem 11.21 to this quotient rather than to G. An example is given by using Lemma 11.24: it produces a quotient where $d_1/(d-n)$ is replaced by $\tilde{d}_1/(d-n)$, where $\tilde{d}_1 = d_1 - \kappa$.

We now observe that in the left hand side of the conclusion of Theorem 11.21, the quantity $\lambda' + d_1'$ can be written $\ell_1' + \tilde{d}_1'$, where $\tilde{d}_1' = d_1' - \kappa'$. Therefore, with this notation, we deduce from Theorem 11.21 together with Lemma 11.24 the following statement (which is Theorem 4.1 of [W 1988]), where the assumptions are those of Theorem 11.5:

($\tilde{1}$) *Assume $d > n$. Then there exists a connected algebraic subgroup G^* of G such that*

$$d' > \ell_0' \quad \textit{and} \quad \frac{\ell_1' + \tilde{d}_1'}{d' - \ell_0'} \le \frac{\tilde{d}_1}{d-n}.$$

Repeating the proof of (1) $\Rightarrow$ (2) with d_1 replaced by $\tilde{d}_1$, and we deduce

($\tilde{2}$) *Assume $d > n$ and $\ell_1 > 0$. Assume further that for any G^* for which $Y^* \neq \{0\}$, we have*

$$\frac{n^* - \ell_0^*}{\ell_1^*} \ge \frac{n - \ell_0}{\ell_1}.$$

Assume also that there is no G^* for which the three conditions $\ell_1' = 0$, $n' = \ell_0'$ and $d' > 0$ simultaneously hold. Then

$$d_1 > 0 \quad \textit{and} \quad \ell_1(d-n) \le \tilde{d}_1(n - \ell_0).$$

Now $\tilde{d}_1 \le d_1$, therefore ($\tilde{2}$) is stronger than (2). Since ($\tilde{1}$) holds, it follows that ($\tilde{2}$) and (2) also hold, and from § 11.7.2 we deduce finally that all properties (3), (1'), (2'), (3'), (1) also hold. □

Exercises

Exercise 11.1. Assume, in Theorem 11.5, that K is a subfield of $\overline{\mathbb{Q}}$ such that $\mathcal{W}$ is rational over K and $\exp_G(Y) \subset G(K)$. Show that in the conclusion one can restrict to algebraic subgroups G^* of G which are defined over K.

Exercise 11.2. Let $\mathcal{V}$ be a vector subspace of $\mathbb{C}^d = \mathbb{C}^{d_0} \times \mathbb{C}^{d_1}$ satisfying (11.1). Show that there exists a hyperplane H of $\mathbb{C}^d$ containing $\mathcal{V}$ and satisfying (11.1).

Hint. *Write $\mathcal{V}$ as intersection of hyperplanes $H_1, \dots, H_m$, where m is the codimension of $\mathcal{V}$. For $1 \le i \le m$ let L_i be a nonzero linear form whose kernel is H_i. Consider the set of complex tuples $(t_1, \dots, t_m)$ for which the kernel H of $t_1L_1 + \cdots + t_mL_m$ does not satisfy (11.1).*

Exercise 11.3. *(Six exponentials Theorem in several variables)* For X and Y subsets of $\mathbb{C}^n$, denote by XY the set of scalar products xy $(x \in X, y \in Y)$.

Let $\mathcal{U}$ and $\mathcal{V}$ be two $\overline{\mathbb{Q}}$-vector spaces of $\mathbb{C}^n$ and X, Y two $\mathbb{Q}$-vector subspaces of $\mathbb{C}^n$ of dimensions

$$\dim_{\overline{\mathbb{Q}}}(\mathcal{U}) = d_0 \geq 0, \quad \dim_{\overline{\mathbb{Q}}}(\mathcal{V}) = \ell_0 \geq 0, \quad \dim_{\mathbb{Q}}(X) = d_1 \geq 1, \quad \dim_{\mathbb{Q}}(Y) = \ell_1 \geq 1.$$

Assume

$$\mathcal{U}\mathcal{V} \subset \overline{\mathbb{Q}}, \quad \mathcal{U}Y \subset \overline{\mathbb{Q}}, \quad X\mathcal{V} \subset \overline{\mathbb{Q}}, \quad XY \subset \mathcal{L}.$$

Deduce from part (3') of Theorem 11.5 the existence of decompositions as direct sums of vector subspaces

$$\mathcal{U} = \mathcal{U}^* \oplus \mathcal{U}', \quad \mathcal{V} = \mathcal{V}^* \oplus \mathcal{V}', \quad X = X^* \oplus X', \quad Y = Y^* \oplus Y',$$

where

$$\mathcal{U}'\mathcal{V}^* = \{0\}, \quad \mathcal{U}'Y^* = \{0\}, \quad X'\mathcal{V}^* = \{0\}, \quad X'Y^* = \{0\},$$

and such that

$$n' < d_0' + d_1' \quad \text{and} \quad (n' - \ell_0')d_1' \geq \ell_1'(d_0' + d_1' - n'),$$

where n' is the dimension of the $\mathbb{C}$-vector subspace of $\mathbb{C}^n$ spanned by $\mathcal{U}' \cap X'$, and where

$$d_0' = \dim_{\overline{\mathbb{Q}}}(\mathcal{U}'), \quad \ell_0' = \dim_{\overline{\mathbb{Q}}}(\mathcal{V}'), \quad d_1' = \dim_{\mathbb{Q}}(X'), \quad \ell_1' = \dim_{\mathbb{Q}}(Y').$$

Hint. *For the case $d_0 = \ell_0 = 0$, see Theorem 1 of* [W 1981].

Exercise 11.4. Prove Lemma 11.11.

Hint. *The proof of (11.12) is easy. For (11.13), first check*

$$n(n-1) + 1 \leq \Psi(n, d) \leq n(n+1)$$

and deduce the desired estimate when $n'n^* \geq n' + n^*$. By symmetry, assume now $n^* = 1$ and $n = n' + 1$. In the case $d^* \geq 3$, show $\Psi(1, d^*) = 1$ and

$$\Psi(n', d') + \Psi(1, d^*) \leq n'(n'+1) + 1 \leq \Psi(n, d).$$

Next assume $d^* = 2$. In order to prove

$$\Psi(n', d') + 2 \leq \Psi(n'+1, d'+2),$$

consider several cases:
a) If $n' = 1$ check

$$\Psi(n', d') = \begin{cases} 1 & \text{for } d' \geq 3, \\ 2 & \text{for } d' = 2, \end{cases} \qquad \text{and} \qquad \Psi(2, d'+2) = \begin{cases} 3 & \text{for } d' \geq 3, \\ 4 & \text{for } d' = 2. \end{cases}$$

Assume now $n' \geq 2$.
b) Assume $d' \geq n' + 2$. From $\Psi(n', d') \leq n'(n'-1) + 2$ deduce

$$\Psi(n', d') + 2 \leq n'(n'-1) + 4 = (n-1)(n-2) + 4 \leq n(n-1) + 1 = \Psi(n, d'+2).$$

c) The last case is $n' = d' - 1 \geq 2$. In this case show $\Psi(n', d') = n'(n'+1)$ and

$$\Psi(n', d') + 2 = n'(n'+1) + 2 = n(n-1) + 2 = \Psi(n, d).$$

Exercise 11.5.
a) Show that the function Ψ of § 11.5 is the smallest arithmetic function satisfying both properties (11.12) and (11.13) of Lemma 11.11.
b) Show also that another equivalent definition of $\Psi(n, d)$ is

$$\Psi(n,d) = \begin{cases} \Psi(1, d-n) + \Psi(n-1, n) & \text{if } n \geq 2 \text{ and } d - n \geq 2, \\ [nd/(d-n)] & \text{if } n = 1 \text{ or } d - n = 1. \end{cases}$$

Hint. *Check first*

$$\Psi(n,d) \geq \left[\frac{nd}{d-n}\right] \qquad \text{and} \qquad \Psi(n,d) \geq \Psi(n', d') + \Psi(n^*, d^*)$$

for $n = n' + n^*$ *and* $d = d' + d^*$. *Deduce*

$$\Psi(n,d) \geq \max\left\{\left[\frac{n_1 d_1}{d_1 - n_1}\right] + \cdots + \left[\frac{n_k d_k}{d_k - n_k}\right]\right\}.$$

c) Show that this function Ψ is also defined by

$$\Psi(n,d) = \max\left\{\left[\frac{n_1 d_1}{d_1 - n_1}\right] + \cdots + \left[\frac{n_k d_k}{d_k - n_k}\right]\right\}$$

when $(n_1, \dots, n_k, d_1, \dots, d_k)$ runs over the finite set of tuples with $0 < n_i < d_i$ $(1 \leq i \leq k)$ and $n_1 + \cdots + n_k = n$, $d_1 + \cdots + d_k = d$.

Hint. *Check, for* $d = 1$ *and for* $d - n = 1$, $\Psi(n, d) = [nd/(d-n)]$. *Assume now* $d \geq 2$ *and* $d - n \geq 2$. *Check*

$$\Psi(n,d) = \left[\frac{n'd'}{d'-n'}\right] + \left[\frac{n^*d^*}{d^*-n^*}\right]$$

with $n' = 1$, $n^* = n - 1$, $d' = d - n$, $d^* = n$, *because*

$$\left[\frac{n^*d^*}{d^*-n^*}\right] = n(n-1) = \Psi(n-1, n).$$

Deduce the conclusion (if one wishes one may restrict to $k \in \{1, 2\}$*).*

d) Compare with [Roy 1992b].

Exercise 11.6. Let $\mathcal{V}$ be a vector subspace of $\mathbb{C}^d$ and Y a subgroup of $\mathcal{V} \cap \mathcal{L}^d$ of rank ℓ. For a subspace S of $\mathbb{C}^d$, rational over $\mathbb{Q}$, define

$$\begin{array}{lll} & \mathcal{V}^* = \mathcal{V} \cap S, & Y^* = Y \cap S, \\ d^* = \dim_{\mathbb{C}}(S), & n^* = \dim_{\mathbb{C}}(\mathcal{V}^*), & \ell^* = \dim_{\mathbb{Q}}(Y^*), \\ S' = \dfrac{\mathbb{C}^d}{S}, & \mathcal{V}' = \dfrac{\mathcal{V}}{\mathcal{V}^*}, & Y' = \dfrac{Y}{Y^*}, \\ d' = \dim_{\mathbb{C}}(S'), & n' = \dim_{\mathbb{C}}(\mathcal{V}'), & \ell' = \dim_{\mathbb{Q}}(Y'), \end{array}$$

so that $d = d^* + d'$, $n = n^* + n'$ and $\ell = \ell^* + \ell'$.

Check that the following statements are equivalent to Corollary 11.14.
(1) Assume $d > n$. Show the existence of a subspace S of $\mathbb{C}^d$, rational over $\mathbb{Q}$, of codimension $d' \geq 1$ such that

$$\frac{\ell' + d'}{d'} \leq \frac{d}{d-n}.$$

(1') Assume $Y \neq \{0\}$. Show the existence of a subspace S of $\mathbb{C}^d$, rational over $\mathbb{Q}$, for which

$$(d^*, \ell^*) \neq (0, 0) \quad \text{and} \quad \frac{d^*}{d^* + \ell^*} \leq \frac{n}{\ell}.$$

(2) Assume $d > n$ and $Y \neq \{0\}$. Assume further that for any S for which $Y \cap S \neq \{0\}$, we have

$$\frac{n^*}{\ell^*} \geq \frac{n}{\ell}.$$

Assume furthermore that there is no $S \neq \mathbb{C}^d$ which contains Y. Check

$$\ell(d-n) \leq dn.$$

(2') Assume $d > n$ and $Y \neq \{0\}$. Assume further that for any $S \neq \mathbb{C}^d$, we have

$$\frac{n'}{d'} \leq \frac{n}{d}.$$

Assume furthermore that there is no $S \neq \{0\}$ contained in $\mathcal{V}$. Check

$$\ell(d-n) \leq dn.$$

(3) Assume $Y \neq \{0\}$. Then the family of S for which $\ell^* \neq 0$ and n^*/ℓ^* is minimal is not empty. Let S be such an element for which d^* is minimal. Assume $d^* > n^*$. Check

$$\frac{n}{\ell} \geq \frac{n^*}{\ell^*} \geq \frac{d^* - n^*}{d^*}.$$

Exercise 11.7. Let F be any subfield of $\overline{\mathbb{Q}}$. Define $\mathcal{L}_F$ as the F-vector space spanned by 1 and $\mathcal{L}$. Extend the results of § 11.6, as well as Exercise 11.6, replacing $\overline{\mathbb{Q}}$ by F. For instance prove:

- *Let $\mathcal{U}$ be a vector subspace of $\mathbb{C}^d$ of dimension $n < d$. Among the subspaces of T which are rational over $\overline{\mathbb{Q}}$ and distinct from $\mathbb{C}^d$, we select one for which the quantity n'/d' is minimal, with*

$$n' = \dim_{\mathbb{C}} \left(\frac{\mathcal{U}}{\mathcal{U} \cap T} \right), \qquad d' = \dim_{\mathbb{C}} \left(\frac{\mathbb{C}^d}{T} \right).$$

Then the dimension ℓ' of the F-vector space $\mathcal{U} \cap \mathcal{L}_F^d / \mathcal{U} \cap T \cap \mathcal{L}_F^d$ is finite and satisfies

$$\frac{\ell'}{d' + \ell'} \leq \frac{n'}{d'} \leq \frac{n}{d}.$$

Hint. *See* [Roy 1992b], *Remark (i) p.37.*

Exercise 11.8.
a) Deduce from Corollary 11.16 the following result. Let $x_1, \ldots, x_d$ be $\mathbb{Q}$-linearly independent complex numbers and $y_1, \ldots, y_\ell$ be also $\mathbb{Q}$-linearly independent complex numbers. Assume $d\ell > d+\ell$. Then one at least of the $d\ell$ numbers $x_i y_j$ $(1 \le i \le d, 1 \le j \le \ell)$ does not belong to $\mathbb{Q} + \mathcal{L}$.

b) Deduce from the five exponentials Theorem of § 11.3.3:

- *Let λ_0 and λ_1 be nonzero elements of $\mathcal{L}$ and β a nonzero algebraic number. Then one at least of the two numbers*

$$e^{\beta\lambda_0\lambda_1}, \quad e^{(\beta\lambda_0)^2\lambda_1}$$

is transcendental.

c) Deduce that one at least of the two numbers

$$2^{\log 2}, \quad 2^{(\log 2)^2}$$

is transcendental.

Exercise 11.9. Let λ_{ij} $(0 \le i \le n, 1 \le j \le m)$ be elements in $\mathcal{L}$ and $t_1, \ldots, t_n$ complex numbers. Assume $m > n(n+1)$. Assume also

$$\sum_{i=1}^{n} t_i \lambda_{ij} = \lambda_{0j} \quad (1 \le j \le m).$$

a) If the m elements

$$(\lambda_{1j}, \ldots, \lambda_{nj}) \quad (1 \le j \le m)$$

in $\mathcal{L}^n$ are $\mathbb{Q}$-linearly independent, then the numbers $1, t_1, \ldots, t_n$ are $\mathbb{Q}$-linearly dependent.

b) If the mn numbers

$$\lambda_{ij} \quad (1 \le i \le n, \ 1 \le j \le m)$$

are $\mathbb{Q}$-linearly independent, then the numbers $t_1, \ldots, t_n$ are are all rational.

Hint. *See* [W 1981], *Corollary 1.2.*

Exercise 11.10. Let λ_{ij} $(1 \le i \le n, 0 \le j \le m)$ be elements in $\mathcal{L}$ with $m > n^2$. Assume $\lambda_{10}, \ldots, \lambda_{n0}$ are $\mathbb{Q}$-linearly independent. Assume also the m elements

$$(\lambda_{1j}, \ldots, \lambda_{nj}) \quad (1 \le j \le m)$$

in $\mathcal{L}^n$ are $\mathbb{Q}$-linearly independent. Show that one at least of the numbers

$$\exp\left(\sum_{i=1}^{n} \lambda_{i0}\lambda_{ij}\right) \quad (1 \le j \le m)$$

is transcendental

Hint. *(See* [W 1990]*). Consider the hyperplane $\mathcal{V}$ of equation*

$$\lambda_{10}(z_{n+1} - z_1) + \cdots + \lambda_{n0}(z_{2n} - z_n) = z_{2n+1}$$

in $\mathbb{C}^{2n+1}$. Using Baker's Theorem, check that $\mathcal{V}$ satisfies (11.1) for $d_0 = n$ and $d_1 = n+1$. Check also $\dim_{\overline{\mathbb{Q}}}\big(\mathcal{V} \cap \overline{\mathbb{Q}}^{2n+1}\big) \ge n$. *Define*

$$\lambda_{0j} = \sum_{i=1}^{n} \lambda_{i0}\lambda_{ij} \quad (1 \le j \le m).$$

Deduce that one at least of the $m + n$ points

$$\big(1, 0, \ldots, 0, -\lambda_{i0}\big) \quad (1 \le i \le n)$$

$$\big(0, \ldots, 0, \lambda_{1j}, \ldots, \lambda_{nj}, \lambda_{0j}\big) \quad (1 \le j \le m)$$

does not belong to $\mathcal{L}_G = \overline{\mathbb{Q}}^n \times \mathcal{L}^{n+1}$.

Exercise 11.11.
a) Let G be a subgroup of $\mathbb{R}^d$. Show that the following properties are equivalent.

(i) There exists a finitely generated subgroup of G which is dense in $\mathbb{R}^d$.
(ii) For each hyperplane $\mathcal{V}$ of $\mathbb{R}^d$, the lower bound

$$\mathrm{rk}_{\mathbb{Z}} \left(\frac{G + \mathcal{V}}{\mathcal{V}} \right) \ge 2$$

holds.
(iii) For each vector subspace $\mathcal{V}$ of $\mathbb{R}^d$ with $\mathcal{V} \ne \mathbb{R}^d$, the lower bound

$$\mathrm{rk}_{\mathbb{Z}} \left(\frac{G + \mathcal{V}}{\mathcal{V}} \right) > \dim_{\mathbb{R}} \left(\frac{\mathbb{R}^d}{\mathcal{V}} \right)$$

holds.

b) Let ℓ and d be positive integers with $\ell > d^2 - d + 1$. Let α_{ij}, $(1 \le i \le d, 1 \le j \le \ell)$ be multiplicatively independent positive real algebraic numbers. Denote by $\mathbb{R}_+^*$ the multiplicative group of positive real numbers, and by Γ the multiplicative subgroup of $(\mathbb{R}_+^*)^d$ which is spanned by $\underline{\alpha}_1, \ldots, \underline{\alpha}_\ell$, with $\underline{\alpha}_j = (\alpha_{1j}, \ldots, \alpha_{dj})$:

$$\Gamma = \left\{ \left(\prod_{j=1}^{\ell} \alpha_{1j}^{s_j}, \ldots, \prod_{j=1}^{\ell} \alpha_{dj}^{s_j} \right) ; \underline{s} = (s_1, \ldots, s_\ell) \in \mathbb{Z}^\ell \right\}.$$

Prove that Γ is dense in $(\mathbb{R}_+^*)^d$.

Hint. *Let Y be the subgroup of $\mathbb{R}^d$ which is spanned by $\underline{\lambda}_1, \ldots, \underline{\lambda}_\ell$, with*

$$\underline{\lambda}_j = (\log \alpha_{1j}, \ldots, \log \alpha_{dj}), \quad (1 \le j \le \ell).$$

Show that for each hyperplane $\mathcal{V}$ of $\mathbb{R}^d$,

$$\mathrm{rk}_{\mathbb{Z}} \left(\frac{Y + \mathcal{V}}{\mathcal{V}} \right) \ge \ell - d(d-1).$$

Deduce from a) that Y is dense in $\mathbb{R}^d$, and conclude.

12. Lower Bounds for the Rank of Matrices

Hermite-Lindemann's Theorem, Gel'fond-Schneider's Theorem and the four exponentials Conjecture which have been stated in Chap. 1 can be phrased in terms of rank of 2×2 matrices, respectively

$$\begin{pmatrix} \beta_0 & \beta_1 \\ \beta_2 & \lambda \end{pmatrix}, \quad \begin{pmatrix} \beta_0 & \beta_1 \\ \lambda_0 & \lambda_1 \end{pmatrix}, \quad \begin{pmatrix} \lambda_1 & \lambda_2 \\ \lambda_3 & \lambda_4 \end{pmatrix},$$

where $\beta_0, \beta_1, \beta_2$ are algebraic numbers and $\lambda, \lambda_0, \ldots, \lambda_4$ are elements of $\mathcal{L} = \exp^{-1}(\overline{\mathbb{Q}}^\times)$. In this chapter we study the rank of a matrix whose entries are either algebraic numbers, or else logarithms of algebraic numbers, and more generally whose entries are linear combinations with algebraic coefficients of logarithms of algebraic numbers.

We first study matrices whose entries are linear polynomials (§ 12.1). This will enable us to introduce the *structural rank* (see Chap. 1 § 1.4) which will be our main tool for studying (in § 12.4) the situation from a conjectural point of view. By Conjecture 1.15 on algebraic independence of logarithms of algebraic numbers:

(?) *$\mathbb{Q}$-linearly independent elements of $\mathcal{L}$ are algebraically independent.*

As explained in Chap. 1, it is not yet known that the transcendence degree over $\mathbb{Q}$ of the field $\mathbb{Q}(\mathcal{L})$ is at least 2. However a few partial results are known, and they deal with the rank of matrices whose entries are either in $\mathcal{L}$, or else in the $\mathbb{Q}$-vector space

$$\mathbb{Q} + \mathcal{L} = \{b + \lambda \; ; \; b \in \mathbb{Q}, \; \lambda \in \mathcal{L}\}$$

spanned by 1 and $\mathcal{L}$, or, more generally, in the $\overline{\mathbb{Q}}$-vector space $\widetilde{\mathcal{L}}$ (already considered in Chap. 11) spanned by 1 and $\mathcal{L}$.

Conjecture 1.15 provides a simple (conjectural) description for the rank of a matrix whose entries are in $\mathcal{L}$: it should be equal to the structural rank. It turns out that the Linear Subgroup Theorem (Theorem 11.5) enables one to show: *the rank of a matrix in* $\mathrm{Mat}_{d\times\ell}(\mathcal{L})$ *is at least half its structural rank.* Moreover, as noticed by D. Roy, in order to solve the problem of algebraic independence of logarithms, it would be sufficient to show that the rank of a matrix in $\mathrm{Mat}_{d\times\ell}(\mathbb{Q} + \mathcal{L})$ is equal to the structural rank.

Theorem 1.16 provides a lower bound for the rank of a $d \times \ell$ matrix $\mathsf{M} = (\lambda_{ij})$ with entries in $\mathcal{L}$, namely $\mathrm{rank}(\mathsf{M}) \geq d\ell/(d+\ell)$. This estimate is valid under a rather strong hypothesis: one assumes that there is no nontrivial relation

$$\sum_{i=1}^{d}\sum_{j=1}^{\ell} t_i s_j \lambda_{ij} = 0$$

with rational integer coefficients $t_1, \ldots, t_d$ and $s_1, \ldots, s_\ell$. This is a simple but rather strong assumption: for instance no entry λ_{ij} of M is allowed to be zero!

In § 12.2 we prove similar lower bounds for the rank under much weaker assumptions. We also consider matrices (like in § 1.5)

$$\mathsf{M} = \underbrace{\begin{pmatrix} \mathsf{B}_0 & \mathsf{B}_1 \\ \mathsf{B}_2 & \mathsf{L} \end{pmatrix}}_{\ell_0 \quad \ell_1} \begin{matrix} \}d_0 \\ \}d_1 \end{matrix}$$

where $\mathsf{B}_0, \mathsf{B}_1, \mathsf{B}_2$ have algebraic entries, while the entries of L are in $\mathcal{L}$. From Theorem 11.5 we deduce (see Theorem 12.19) that if M has rank

$$\operatorname{rank}(\mathsf{M}) < \frac{d_1\ell_1 + d_1\ell_0 + d_0\ell_1}{d_1 + \ell_1},$$

then, after linear combinations of rows and columns, one gets a matrix with many zeroes.

In § 12.3 we deal with matrices whose entries are linear combinations, with algebraic coefficients, of logarithms of algebraic numbers.

The main references for all this chapter are [Roy 1990] and [Roy 1989].

In the last section (§ 12.5) we consider quadratic relations between logarithms of algebraic numbers, following [RoyW 1997a] and [RoyW 1997b].

12.1 Entries are Linear Polynomials

In all this section, K is a field and k a subfield of K.

12.1.1 k-Equivalent Matrices

Two $d \times \ell$ matrices M and N with entries in K are *k-equivalent* if there exist two regular matrices $\mathsf{P} \in \mathrm{GL}_d(k)$ and $\mathsf{Q} \in \mathrm{GL}_\ell(k)$ such that $\mathsf{N} = \mathsf{PMQ}$. In this case the rank of M is the same as the rank of N. We shall use $\mathbb{Q}$-equivalence when dealing with matrices with entries either in $\mathcal{L}$ (in § 12.2) or in $\mathbb{Q} + \mathcal{L}$ (in § 12.3), because $\mathcal{L}$ is a $\mathbb{Q}$-vector space, and for a similar reason we shall use $\overline{\mathbb{Q}}$-equivalence when dealing with matrices with entries in $\widetilde{\mathcal{L}}$ (in § 12.3).

If a matrix $\mathsf{M} \in \mathrm{Mat}_{d\times\ell}(K)$ is k-equivalent to a matrix which can be written by block as

$$\underbrace{\begin{pmatrix} \mathsf{A} & \mathsf{B} \\ \mathsf{C} & 0 \end{pmatrix}}_{\ell' \quad \ell^*} \begin{matrix} \}d^* \\ \}d' \end{matrix} \tag{12.1}$$

then the rank of M is bounded above by $d^* + \ell'$.

The connection with the previous chapter is given by elementary considerations of linear algebra. Given $\mathsf{M} \in \mathrm{Mat}_{d\times\ell}(K)$, denote by Y the k-vector subspace of K^d spanned by the ℓ column vectors of M. Further, let S be a vector subspace of K^d, rational over k, of dimension d^* and codimension d'. Select a basis $(\underline{e}_1, \dots, \underline{e}_d)$ of k^d with $\underline{e}_i \in S$ for $1 \le i \le d^*$, and denote by $\mathsf{P} \in \mathrm{GL}_d(k)$ the transition matrix from the standard basis of k^d to $(\underline{e}_1, \dots, \underline{e}_d)$. In the same way, define $\ell^* = \dim_k(Y \cap S)$ and let $\mathsf{Q} \in \mathrm{GL}_\ell(k)$ be the matrix associated with a linear automorphism of k^ℓ which maps the last ℓ^* elements of the standard basis of k^ℓ onto elements of $Y \cap S$. Then PMQ has the shape (12.1) with $\ell' = \ell - \ell^*$.

12.1.2 Vector Spaces Spanned by Algebraically Independent Elements

Here is the key result (Proposition 1 of [Roy 1990]).

Proposition 12.2. *Let $\mathcal{E}$ be a k-vector subspace of K which is spanned over k by a family (finite or not) of elements of K which are algebraically independent over k. Then any matrix* M *with entries in $\mathcal{E}$ is k-equivalent to a matrix of the form* $\begin{pmatrix} \mathsf{A} & \mathsf{B} \\ \mathsf{C} & 0 \end{pmatrix}$ *where* A *is either the zero-size matrix or else a regular square matrix.*

Remark. It is convenient not to exclude the following three trivial examples:
– If the matrix M is null, we take $d' = d$ and $\ell' = 0$, so that A has size 0×0;
– If the rows of M are linearly independent over K, we take $d' = \ell' = 0$, hence the size of C is $0 \times \ell$, which means that M is k-equivalent to a matrix (A, B) where A is invertible;
– If the columns of M are K-linearly independent, we take $d' = d$ and $\ell' = \ell$, which means that B has size $d \times 0$: M is k-equivalent to a matrix $\begin{pmatrix} \mathsf{A} \\ \mathsf{C} \end{pmatrix}$ where A is a square matrix with maximal rank.

We shall give two proofs of Proposition 12.2 (for the second one only we shall assume that the field k has infinitely many elements).

For the first proof, we need the following elementary result.

Lemma 12.3. *Let $\mathcal{E}$ be a k-vector subspace of K. The following properties are equivalent.*

(i) The vector space $\mathcal{E}$ is spanned over k by a family of elements in K which are algebraically independent over k.

(ii) Elements in $\mathcal{E}$ which are k-linearly independent are also algebraically independent over k.

(iii) If $\mathcal{E}'$ is a k-vector subspace of $\mathcal{E}$ and x an element of $\mathcal{E}$ which does not belong to $\mathcal{E}'$, then x is transcendental over $k(\mathcal{E}')$.

Proof of Lemma 12.3.
(i) $\Rightarrow$ (ii)

Let $x_1, \dots, x_m$ be k-linearly independent elements of $\mathcal{E}$. Let B be a basis of $\mathcal{E}$ over k consisting of elements in K which are algebraically independent over k. We write each x_i $(1 \le i \le m)$ as a linear combination, with coefficients in k, of elements of B. This involves only finitely many elements of B, say $y_1, \dots, y_n$, which are algebraically independent over k:

$$x_i = \sum_{j=1}^{n} a_{ij} y_j \quad (1 \le i \le m).$$

The matrix $\left(a_{ij}\right)_{\substack{1\le i\le m\\ 1\le j\le n}}$ has rank m. From linear algebra it follows that there is a subset $\{z_1, \dots, z_{n-m}\}$ of $\{y_1, \dots, y_n\}$ such that

$$k(y_1, \dots, y_n) = k(x_1, \dots, x_m, z_1, \dots, z_{n-m}).$$

Therefore $x_1, \dots, x_m$ are algebraically independent over k.

(ii) $\Rightarrow$ (iii)

Let $x \in \mathcal{E}$ be algebraic over $k(\mathcal{E}')$. There exist $y_1, \dots, y_n$ in $\mathcal{E}'$, linearly independent over k, such that x is algebraic over $k(y_1, \dots, y_n)$. Then $y_1, \dots, y_n, x$ are algebraically dependent over k, and since they belong to $\mathcal{E}$ it follows from (ii) that they are linearly dependent over k. Since $y_1, \dots, y_n$ are linearly independent over k, we conclude $x \in ky_1 + \dots + ky_n \subset \mathcal{E}'$.

(iii) $\Rightarrow$ (i)

Let B be a basis of $\mathcal{E}$ over k. We shall prove that any set $\{y_1, \dots, y_n\}$ of k-linearly independent elements in B consists of k-algebraically independent elements. We prove this result by induction on n. For $n = 1$ we use assumption (iii) with $\mathcal{E}' = \{0\}$: since $y_1 \ne 0$, we have $y_1 \notin \mathcal{E}'$, hence y_1 is transcendental over k.

Assume the result holds for $n-1$ with $n \ge 2$. Let $y_1, \dots, y_n$ be k-linearly independent elements of B. Consider the vector subspace $\mathcal{E}'$ of $\mathcal{E}$ over k spanned by $y_1, \dots, y_{n-1}$. From the induction hypothesis we deduce that $y_1, \dots, y_{n-1}$ are algebraically independent over k. Since $y_n \notin \mathcal{E}'$ we deduce from (iii) that y_n is transcendental over the field $k(y_1, \dots, y_{n-1})$. Hence $y_1, \dots, y_n$ are algebraically independent over k. □

First proof of Proposition 12.2. Let $\mathcal{E}_0$ be the k-vector subspace of $\mathcal{E}$ spanned by the entries of M. Let us warm up by looking at the situation where the dimension n of $\mathcal{E}_0$ is 1 or 2.

For $n = 1$ we can write $\mathsf{M} = \mathsf{N}x$ where N has entries in k and $x \in \mathcal{E}$, $x \ne 0$. Let r be the rank of N. Let P and Q be regular square matrices with entries in k such that $\mathsf{PNQ} = \begin{pmatrix} \mathsf{I}_r & 0 \\ 0 & 0 \end{pmatrix}$. Then $\mathsf{PMQ} = \begin{pmatrix} \mathsf{I}_r x & 0 \\ 0 & 0 \end{pmatrix}$. We deduce the conclusion of Proposition 12.2 with $\mathsf{A} = \mathsf{I}_r x$, $\mathsf{B} = 0$, $\mathsf{C} = 0$.

Next consider the case $n = 2$. Write $\mathsf{M} = \mathsf{M}_1x_1 + \mathsf{M}_2x_2$ with x_1 and x_2 in $\mathcal{E}$ linearly (hence, by Lemma 12.3, also algebraically) independent over k, while M_1 and M_2 are matrices with entries in k. Denote by r_1 the rank of M_1. There exist regular matrices P_1 and Q_1 with entries in k such that $\mathsf{P}_1\mathsf{M}_1\mathsf{Q}_1 = \begin{pmatrix} \mathsf{I}_{r_1} & 0 \\ 0 & 0 \end{pmatrix}$. Denote by $\mathsf{A}_2, \mathsf{B}_2, \mathsf{C}_2, \mathsf{D}_2$ the matrices with entries in k (where A_2 is a square $r_1 \times r_1$ matrix) such that $\mathsf{P}_1\mathsf{M}_2\mathsf{Q}_1 = \begin{pmatrix} \mathsf{A}_2 & \mathsf{B}_2 \\ \mathsf{C}_2 & \mathsf{D}_2 \end{pmatrix}$. Hence

$$\mathsf{P}_1\mathsf{M}\mathsf{Q}_1 = \begin{pmatrix} \mathsf{I}_{r_1}x_1 + \mathsf{A}_2x_2 & \mathsf{B}_2x_2 \\ \mathsf{C}_2x_2 & \mathsf{D}_2x_2 \end{pmatrix}.$$

Let now P_2 and Q_2 be regular matrices with entries in k such that $\mathsf{P}_2\mathsf{D}_2\mathsf{Q}_2 = \begin{pmatrix} \mathsf{I}_{r_2} & 0 \\ 0 & 0 \end{pmatrix}$, where r_2 is the rank of D_2. Then

$$\begin{pmatrix} \mathsf{I}_{r_1} & 0 \\ 0 & \mathsf{P}_2 \end{pmatrix} \mathsf{P}_1\mathsf{M}\mathsf{Q}_1 \begin{pmatrix} \mathsf{I}_{r_1} & 0 \\ 0 & \mathsf{Q}_2 \end{pmatrix} = \begin{pmatrix} \mathsf{I}_{r_1}x_1 + \mathsf{A}_2x_2 & \mathsf{B}'_2x_2 & \mathsf{B}''_2x_2 \\ \mathsf{C}'_2x_2 & \mathsf{I}_{r_2}x_2 & 0 \\ \mathsf{C}''_2x_2 & 0 & 0 \end{pmatrix}$$

where $\mathsf{B}'_2, \mathsf{B}''_2, \mathsf{C}'_2, \mathsf{C}''_2$ have entries in k. We now set

$$\mathsf{A} = \begin{pmatrix} \mathsf{I}_{r_1}x_1 + \mathsf{A}_2x_2 & \mathsf{B}'_2x_2 \\ \mathsf{C}'_2x_2 & \mathsf{I}_{r_2}x_2 \end{pmatrix}.$$

The determinant of A is a polynomial in x_1 and x_2 and the coefficient of $x_1^{r_1}x_2^{r_2}$ is 1. Hence this determinant is not zero.

After having considered the cases $n = 1$ and $n = 2$, here is the general case. We proceed by induction on n. Assume $n \geq 1$. Then $\mathcal{E}_0$ contains a nonzero element x. Let $\mathcal{E}_1$ be a subspace of $\mathcal{E}_0$ such that $\mathcal{E}_0 = \mathcal{E}_1 \oplus kx$. Write $\mathsf{M} = x\mathsf{N} + \mathsf{M}_1$ with N having entries in k, while M_1 has entries in $\mathcal{E}_1$. If r denotes the rank of N, there exist matrices $\mathsf{P} \in \mathrm{GL}_d(k)$ and $\mathsf{Q} \in \mathrm{GL}_\ell(k)$ such that

$$\mathsf{P}\mathsf{N}\mathsf{Q} = \begin{pmatrix} \mathsf{I}_r & 0 \\ 0 & 0 \end{pmatrix}.$$

Then

$$\mathsf{P}\mathsf{M}\mathsf{Q} = \begin{pmatrix} x\mathsf{I}_r + \mathsf{A}_1 & \mathsf{B}_1 \\ \mathsf{C}_1 & \mathsf{D}_1 \end{pmatrix},$$

with A_1, B_1, C_1 and D_1 having their entries in $\mathcal{E}_1$. Use the induction hypothesis for D_1: there exist $\mathsf{P}' \in \mathrm{GL}_{d-r}(k)$ and $\mathsf{Q}' \in \mathrm{GL}_{\ell-r}(k)$ such that $\mathsf{P}'\mathsf{D}_1\mathsf{Q}' = \begin{pmatrix} \mathsf{A}' & \mathsf{B}' \\ \mathsf{C}' & 0 \end{pmatrix}$, where A' is a regular matrix with entries in $\mathcal{E}_1$. Then

$$\begin{pmatrix} \mathsf{I}_r & 0 \\ 0 & \mathsf{P}' \end{pmatrix} \begin{pmatrix} x\mathsf{I}_r + \mathsf{A}_1 & \mathsf{B}_1 \\ \mathsf{C}_1 & \mathsf{D}_1 \end{pmatrix} \begin{pmatrix} \mathsf{I}_r & 0 \\ 0 & \mathsf{Q}' \end{pmatrix} = \begin{pmatrix} \mathsf{A} & \mathsf{B} \\ \mathsf{C} & 0 \end{pmatrix}$$

with

$$\mathsf{A} = \begin{pmatrix} x\mathsf{I}_r + \mathsf{A}_1 & \mathsf{B}'' \\ \mathsf{C}'' & \mathsf{A}' \end{pmatrix},$$

the entries of the matrices B'' and C'' being in $\mathcal{E}_1$. The determinant of A is a polynomial in x with coefficients in $k(\mathcal{E}_1)$, whose term of highest degree is $x^r \det \mathsf{A}'$. Since x does not belong to $\mathcal{E}_1$, by Lemma 12.3 x is transcendental over $k(\mathcal{E}_1)$, and since A' is regular, we conclude $\det \mathsf{A} \neq 0$. □

This proof of Proposition 12.2 yields a more general result where the x_i are not assumed to be algebraically independent ([Roy 1989], Lemme).

Proposition 12.4. *Let $x_1, \ldots, x_n$ be k-linearly independent elements of K. For $0 \leq m \leq n$ denote by $\mathcal{E}_m = kx_1 + \cdots + kx_m$ the k-vector subspace of K spanned by $x_1, \ldots, x_m$. Then any $d \times \ell$ matrix with entries in $\mathcal{E}_n$ is k-equivalent to a block matrix*

$$\begin{pmatrix} \mathsf{M}_{11} & \cdots & \mathsf{M}_{1n} & \mathsf{M}_{1,n+1} \\ \vdots & \ddots & \vdots & \vdots \\ \mathsf{M}_{n1} & \cdots & \mathsf{M}_{nn} & \mathsf{M}_{n,n+1} \\ \mathsf{M}_{n+1,1} & \cdots & \mathsf{M}_{n+1,n} & 0 \end{pmatrix}$$

with the following properties:

- *For $1 \leq i \leq n$, M_{ii} is a square $r_i \times r_i$ matrix with $r_i \geq 0$, the diagonal elements belong to $\mathcal{E}_{n-i+1}$ and not to $\mathcal{E}_{n-i}$, the other elements belong to $\mathcal{E}_{n-i}$.*
- *For $1 \leq i < j \leq n+1$, the entries of the matrices M_{ij} and M_{ji} are in $\mathcal{E}_{n-i}$.*

Proposition 12.2 again follows from the fact that the determinant of the matrix

$$\begin{pmatrix} \mathsf{M}_{11} & \cdots & \mathsf{M}_{1n} \\ \vdots & \ddots & \vdots \\ \mathsf{M}_{n1} & \cdots & \mathsf{M}_{nn} \end{pmatrix}$$

is a nonzero polynomial in $x_1, \ldots, x_n$, since the coefficient of the monomial $x_n^{r_1} \cdots x_1^{r_n}$ is not zero.

Proof of Proposition 12.4. We repeat the proof of Proposition 12.2: write $\mathsf{M} = x_n\mathsf{A} + \mathsf{N}$ where A is a matrix of rank $r_1 - 1$ with entries in k while N has entries in $\mathcal{E}_{n-1}$. There exist matrices $\mathsf{P} \in \mathrm{GL}_d(k)$ and $\mathsf{Q} \in \mathrm{GL}_\ell(k)$ with $\mathsf{PAQ} = \begin{pmatrix} \mathsf{I}_{r_1} & 0 \\ 0 & 0 \end{pmatrix}$. Then

$$\mathsf{PMQ} = \begin{pmatrix} \mathsf{M}_{11} & \mathsf{N}_1 \\ \mathsf{N}_2 & \mathsf{M}' \end{pmatrix},$$

where M_{11} is a $r_1 \times r_1$ matrix whose diagonal has entries in $\mathcal{E}_n \setminus \mathcal{E}_{n-1}$, while the other entries are in $\mathcal{E}_{n-1}$. The entries of matrices N_1, N_2 and M' are in $\mathcal{E}_{n-1}$. If $n = 1$ Proposition 12.4 follows. If $n \geq 2$ we use the induction hypothesis for M'. □

Before starting the second proof, let us come back to the initial problem. Denote by $k[X_1, \ldots, X_n]$ the ring of polynomials in n variables over k and consider a $d \times \ell$ matrix $\mathsf{M} = \mathsf{M}_1 X_1 + \cdots + \mathsf{M}_n X_n$, where each M_i has entries in k. We keep the same symbol M_i to denote the associated linear map from k^ℓ to k^d in the canonical bases. We shall assume k is infinite, and we shall specialize X_i in k. It is therefore natural to introduce the k-vector subspace T of $\mathrm{Hom}_k(k^\ell, k^d)$ which is spanned by $\mathsf{M}_1, \ldots, \mathsf{M}_n$. Consider in T a matrix N of maximal rank. After a change of bases, we may assume $\mathsf{N} = \begin{pmatrix} \mathsf{I}_r & 0 \\ 0 & 0 \end{pmatrix}$, where r is the rank of N. We shall prove (this is the main point) that in these new bases, we have $\mathsf{M}_i = \begin{pmatrix} \mathsf{A}_i & \mathsf{B}_i \\ \mathsf{C}_i & 0 \end{pmatrix}$, where A_i is a square $r \times r$ matrix. Then

$$\mathsf{M} = \begin{pmatrix} \mathsf{A} & \mathsf{B} \\ \mathsf{C} & 0 \end{pmatrix},$$

with

$$\mathsf{A} = \sum_{i=1}^{n} \mathsf{A}_i X_i, \quad \mathsf{B} = \sum_{i=1}^{n} \mathsf{B}_i X_i, \quad \mathsf{C} = \sum_{i=1}^{n} \mathsf{C}_i X_i.$$

Then A is a regular matrix (one of its specializations is I_r).

It remains to check that each matrix M_i can be written as we have claimed. Hence we need to prove $\mathsf{M}_i \cdot x \in k^r \times \{0\}^{n-r}$ for $x \in \{0\}^r \times k^{n-r}$, which means $\mathsf{M}_i(\ker \mathsf{N}) \subset \mathrm{Im}\mathsf{N}$.

The following result is Proposition 3 of [Roy 1990] (see also Lemma 3.2 of [Roy 1992c]).

Proposition 12.5. *Let k be a field with infinitely many elements, $\mathcal{U}$ and $\mathcal{V}$ vector spaces of finite dimension over k and T a vector subspace of $\mathrm{Hom}_k(\mathcal{U}, \mathcal{V})$. Let θ be an element in T of maximal rank. Then for any $\xi \in T$ we have $\xi(\ker\theta) \subset \mathrm{Im}\theta$.*

Proof of Proposition 12.5. Fix $\theta \in T$ of maximal rank, $\xi \in T$ and $u \in \ker\theta$. Let $\mathcal{W}$ be a vector subspace of $\mathcal{U}$ such that $\mathcal{U} = \mathcal{W} \oplus \ker\theta$ and let $(u_1, \ldots, u_r)$ be a basis of $\mathcal{W}$, so that $\theta(u_1), \ldots, \theta(u_r)$ is a basis of $\mathrm{Im}\theta$. For any $a \in k$, the linear map $\theta + a\xi$ has rank $\leq r$, hence

$$(\theta + a\xi)(u_1), \ldots, (\theta + a\xi)(u_r), (\theta + a\xi)(u)$$

are k-linearly dependent. Since $u \in \ker\theta$ we have $(\theta + a\xi)(u) = a\xi(u)$. It follows that for any $a \in k$, $a \neq 0$, the following $r + 1$ elements

$$(\theta + a\xi)(u_1), \ldots, (\theta + a\xi)(u_r), \xi(u),$$

which we denote by $\mu_1^{(a)}, \ldots, \mu_{r+1}^{(a)}$, are also k-linearly dependent. Let us check that this holds true also for $a = 0$. Indeed, in a basis $(e_1, \ldots, e_d)$ of $\mathcal{V}$, the $(r + 1) \times d$ matrix associated with $\mu_1^{(a)}, \ldots, \mu_{r+1}^{(a)}$ has rank $\leq r$. Each of the $\binom{d}{r+1}$ determinants of $(r + 1) \times (r + 1)$ submatrices is zero. Consider any one of them: it is a polynomial in a, which vanishes for any $a \in k^\times$. Since k is infinite, it also vanishes at 0.

It follows that the vectors $\theta(u_1), \ldots, \theta(u_r), \xi(u)$ are k-linearly dependent. Therefore $\xi(u)$ is a linear combination of $\theta(u_1), \ldots, \theta(u_r)$, hence belongs to $\mathrm{Im}\theta$. □

Proof of Proposition 12.2 when the field k is infinite.

Write the matrix M as $\mathsf{M}_1 x_1 + \cdots + \mathsf{M}_n x_n$ where $x_1, \ldots, x_n$ are algebraically independent over k. Denote by T the subspace of $\mathrm{Mat}_{d\times\ell}(k)$ which is spanned by $\mathsf{M}_1, \ldots, \mathsf{M}_n$, and by N an element in T of maximal rank, say r. Let $\mathsf{P} \in \mathrm{GL}_d(k)$ and $\mathsf{Q} \in \mathrm{GL}_\ell(k)$ be regular matrices such that $\mathsf{PNQ} = \begin{pmatrix} \mathsf{I}_r & 0 \\ 0 & 0 \end{pmatrix}$. From Proposition 12.5 we deduce that each of the matrices $\mathsf{PM}_i\mathsf{Q}$ is of the form $\begin{pmatrix} \mathsf{A}_i & \mathsf{B}_i \\ \mathsf{C}_i & 0 \end{pmatrix}$, where A_i is a square $r \times r$ matrix. Then

$$\mathsf{PMQ} = \begin{pmatrix} \mathsf{A} & \mathsf{B} \\ \mathsf{C} & 0 \end{pmatrix},$$

with

$$\mathsf{A} = \sum_{i=1}^n \mathsf{A}_i x_i, \quad \mathsf{B} = \sum_{i=1}^n \mathsf{B}_i x_i, \quad \mathsf{C} = \sum_{i=1}^n \mathsf{C}_i x_i.$$

The determinant of A is a polynomial in $x_1, \ldots, x_n$. Since there exists $u \in k^n$ such that $\mathsf{N} = \mathsf{M}_1 u_1 + \cdots + \mathsf{M}_n u_n$, this polynomial does not vanish at the point u. Since $x_1, \ldots, x_n$ are algebraically independent over k, we deduce $\det \mathsf{A} \neq 0$, which shows that A is a regular matrix. □

12.1.3 Property $\left(\begin{smallmatrix} A & B \\ C & 0 \end{smallmatrix}\right)$

Let $\mathcal{E}$ be a k-vector subspace of K. We denote by $\left(\begin{smallmatrix} A & B \\ C & 0 \end{smallmatrix}\right)$ the following property:

> *Any nonzero matrix* M *with entries in* $\mathcal{E}$ *is k-equivalent to a matrix* $\begin{pmatrix} \mathsf{A} & \mathsf{B} \\ \mathsf{C} & 0 \end{pmatrix}$ *where* A *is a regular square matrix.*

This is a property for the triple $(k, K, \mathcal{E})$, but we shall often simply say only that $\mathcal{E}$ satisfies $\left(\begin{smallmatrix} A & B \\ C & 0 \end{smallmatrix}\right)$.

Remark. By Proposition 12.2, any k-vector space spanned by algebraically independent elements over k satisfies Property $\left(\begin{smallmatrix} A & B \\ C & 0 \end{smallmatrix}\right)$. Moreover

> *If* $\mathcal{E}_0$ *is a k-vector subspace of K which is spanned by k-algebraically independent elements and if* $\mathcal{E}_0 \cap k = \{0\}$, *then* $\mathcal{E} = k + \mathcal{E}_0$ *satisfies Property* $\left(\begin{smallmatrix} A & B \\ C & 0 \end{smallmatrix}\right)$.

Indeed, as a k-vector space, $\mathcal{E}$ is isomorphic to the subspace $\mathcal{E}' = kX + \mathcal{E}_0$ of $K(X)$, and Property $\left(\begin{smallmatrix} A & B \\ C & 0 \end{smallmatrix}\right)$ holds for the triple $(k, K(X), \mathcal{E}')$ by Proposition 12.2.

The next result uses the definition of K-vector subspace of K^d, rational over k, when K is a field and k a subfield (see Exercise 1.4).

Proposition 12.6. *Let $\mathcal{E}$ be a k-vector subspace of K satisfying Property* $\left(\begin{smallmatrix} A & B \\ C & 0 \end{smallmatrix}\right)$. *Let $\mathcal{V}$ be a K-vector subspace of K^d of dimension $n < d$ and let Y be a k-vector space of finite dimension contained in $\mathcal{V} \cap \mathcal{E}^d$. Then there exists a K-vector subspace S of K^d, rational over k, such that, if we write*

$$d' = \dim_K \left(\frac{K^d}{S} \right) \quad \text{and} \quad \ell' = \dim_k \left(\frac{Y}{Y \cap S} \right),$$

then

$$\ell' \leq d - d' \leq n.$$

In order to compare with Corollary 11.14, one can write the conclusion as

$$\frac{\ell'}{d' + \ell'} \leq \frac{d - d'}{d} \leq \frac{n}{d}.$$

Proof. If $Y = \{0\}$, we take $S = \{0\}$. Assume $Y \neq \{0\}$. Let $\underline{\eta}_1, \ldots, \underline{\eta}_\ell$ be a basis of Y over k. Consider the matrix M whose j-th column consists of the components y_{ij} of $\underline{\eta}_j$ in the canonical basis of K^d. From condition $\left(\begin{smallmatrix} A & B \\ C & 0 \end{smallmatrix}\right)$ we obtain two matrices

$$\mathsf{P} = \left(p_{si}\right)_{1 \leq s, i \leq d} \in \mathrm{GL}_d(k) \quad \text{and} \quad \mathsf{Q} = \left(q_{jt}\right)_{1 \leq j, t \leq \ell} \in \mathrm{GL}_\ell(k)$$

such that $\mathsf{PMQ} = \begin{pmatrix} \mathsf{A} & \mathsf{B} \\ \mathsf{C} & 0 \end{pmatrix}$ where A is a regular $r \times r$ matrix. Since the columns of M span a K-vector space of dimension $\leq n$, we have $1 \leq r \leq n < d$. The coefficients of PMQ are

$$m_{st} = \sum_{i=1}^{d} \sum_{j=1}^{\ell} p_{si} y_{ij} q_{jt}, \qquad 1 \leq s \leq d, \; 1 \leq t \leq \ell$$

and we have

$$m_{st} = 0 \qquad \text{for} \quad r < s \leq d \quad \text{and} \quad r < t \leq \ell.$$

Let S denote the subspace of K^d intersection of hyperplanes

$$\sum_{i=1}^{d} p_{si} z_i = 0, \qquad (r < s \leq d).$$

Its codimension d' in K^d is $d - r$. Define $\underline{\theta}_1, \ldots, \underline{\theta}_\ell$ in Y by

$$\underline{\theta}_t = \sum_{j=1}^{\ell} q_{jt} \underline{\eta}_j, \qquad (1 \leq t \leq \ell).$$

Since Q is regular, these elements provide another basis of Y over k and the $\ell - r$ elements $\underline{\theta}_{r+1}, \ldots, \underline{\theta}_\ell$ belong to S. Hence $\dim_k(Y \cap S) \geq \ell - r$ and $\ell' = \dim_k(Y/Y \cap S) \leq r$. □

Proposition 12.7. *Let $\mathcal{E}$ be a k-vector subspace of K satisfying Property $\left(\begin{smallmatrix} A & B \\ C & 0 \end{smallmatrix}\right)$. Let $\mathcal{V}$ be a K-vector subspace of K^d of dimension n such that $\mathcal{V} \cap k^d = \{0\}$. Then the k-vector subspace $\mathcal{V} \cap \mathcal{E}^d$ has dimension $\leq n(n+1)/2$.*

Proof. For $d = 1$ we have $\mathcal{V} = \{0\}$ and $\mathcal{V} \cap \mathcal{E} = \{0\}$. Assume now $d \geq 2$. By induction on d we shall prove the following assertion: *for any $r < d$, if $\mathcal{V}_1$ is a k-vector subspace of K^r such that $\mathcal{V}_1 \cap k^r = \{0\}$, then the k-vector space $\mathcal{V}_1 \cap \mathcal{E}^r$ has finite dimension $\leq r(r-1)/2$.*

Take ℓ elements in $\mathcal{V} \cap \mathcal{E}^d$ which are linearly independent over k, and consider the $d \times \ell$ matrix M whose columns are given by the coordinates of these elements. The rank of M is $\leq n < d$. From Property $\left(\begin{smallmatrix} A & B \\ C & 0 \end{smallmatrix}\right)$ it follows that M is k-equivalent to a matrix $\begin{pmatrix} \mathsf{A} & \mathsf{B} \\ \mathsf{C} & 0 \end{pmatrix}$, where A is a regular $r \times r$ matrix. The rank of M is $\geq r$, hence $r \leq n < d$. Put $\ell_1 = \ell - r$, so that B is a $r \times \ell_1$ matrix. Let $\mathcal{V}_1$ be the k-vector space spanned by the columns of B in K^r. Since $\mathcal{V}$ contains $\mathcal{V}_1 \times \{0\}^{d-r}$, we have $\mathcal{V}_1 \cap k^r = \{0\}$. The columns of M are k-linearly independent, hence the same holds for $\begin{pmatrix} \mathsf{A} & \mathsf{B} \\ \mathsf{C} & 0 \end{pmatrix}$, and also for B. Therefore we can use the induction hypothesis: $\ell_1 \leq r(r-1)/2$. We deduce

$$\ell \leq r + \frac{1}{2}r(r-1) \leq n + \frac{1}{2}n(n-1) = \frac{1}{2}n(n+1)\cdot$$

Therefore the conclusion of Proposition 12.7 follows from the induction hypothesis. Moreover, since $n \leq d-1$, we find $\ell \leq d(d-1)/2$, which completes the inductive argument. □

12.1.4 Structural Rank

Following [Roy 1989] and [Roy 1995], we now define the *structural rank with respect to k* of a matrix M whose entries are in K. Consider the k-vector subspace $\mathcal{E}$ of K spanned by the entries of M. Choose an injective morphism φ of $\mathcal{E}$ into a k-vector space $kX_1 + \cdots + kX_n$. The image $\varphi(\mathsf{M})$ of M is a matrix whose entries are in the field $k(X_1, \ldots, X_n)$ of rational fractions. We shall check that its rank does not depend on the choice of φ. This will be the structural rank of M, which will be denoted by $r_{\mathrm{str}}(\mathsf{M})$.

Lemma 12.8. *Let M be a matrix with entries in K. Choose a basis $(e_1, \ldots, e_n)$ of the k-vector subspace of K spanned by the entries of M. Write*

$$\mathsf{M} = \mathsf{M}_1 e_1 + \cdots + \mathsf{M}_n e_n,$$

where $\mathsf{M}_1, \ldots, \mathsf{M}_n$ are matrices of the same size as M with entries in k. Then the rank of the matrix $\mathsf{M}_1 X_1 + \cdots + \mathsf{M}_n X_n$ does not depend on the choice of the basis $(e_1, \ldots, e_n)$.

Proof. Assume $(x_1, \dots, x_n)$ and $(y_1, \dots, y_n)$ are two families of k-algebraically independent elements of K and $\mathsf{M}_1, \dots, \mathsf{M}_n$ be matrices with entries in k. Then the rank of the matrix $\mathsf{M}_1 x_1 + \cdots + \mathsf{M}_n x_n$ is the same as the rank of $\mathsf{M}_1 y_1 + \cdots + \mathsf{M}_n y_n$. In particular for $K = k(X_1, \dots, X_n)$, if $(Y_1, \dots, Y_n)$ is a basis of the k-vector space $kX_1 + \cdots + kX_n$, then the two matrices $\mathsf{M}_1 X_1 + \cdots + \mathsf{M}_n X_n$ and $\mathsf{M}_1 Y_1 + \cdots + \mathsf{M}_n Y_n$ have the same rank. Lemma 12.8 easily follows. □

Examples.
1. If the k-vector space $\mathcal{E}$ is spanned by elements of K which are algebraically independent over k, then the rank of M is the same as its structural rank with respect to k.
2. If $\mathcal{E}$ has dimension 1, then again the rank of any matrix is the same as the structural rank with respect to k. For instance the structural rank with respect to K is nothing else than the rank of M.
3. The rank of M is always bounded above by its structural rank. Here is an example with a strict inequality. Let $P \in K[X]$ be the determinant of a square $d \times d$ matrix $\mathsf{M}_0 + \mathsf{M}_1 X$, where M_0 and M_1 have entries in k, and let $\theta \in K$ be a root of P with $\theta \notin k$. The matrix $\mathsf{M}_0 + \mathsf{M}_1\theta$ has rank $< d$ (its determinant is $P(\theta) = 0$) but its structural rank with respect to k is d (because the determinant of $\mathsf{M}_0 + \mathsf{M}_1 X$ is $P(X) \neq 0$). For instance[19] the matrix $\begin{pmatrix} \sqrt{2} & 2 \\ 1 & \sqrt{2} \end{pmatrix}$ has rank 1 but structural rank 2 with respect to $\mathbb{Q}$. This shows that the structural rank depends on the field k.

Remark. If M *is a* $r \times r$ *matrix with entries in a k-vector subspace of K which satisfies Property* $(\begin{smallmatrix} A & B \\ C & 0 \end{smallmatrix})$, *then the rank of* M *is at least r, while its structural rank with respect to k is at most* $2r$. Indeed, if $(x_1, \dots, x_n)$ is a basis of the k-vector space spanned by the entries of M, so that $M = M_1 x_1 + \cdots + M_n x_n$ where each M_i has entries in k, and if $M_1 X_1 + \cdots + M_n X_n$ is equivalent over k to a matrix (12.1), then the matrix $\begin{pmatrix} \mathsf{A} \\ \mathsf{C} \end{pmatrix}$ has rank r, while $\begin{pmatrix} \mathsf{B} \\ 0 \end{pmatrix}$ has rank $\leq r$. Therefore

> *If* $\mathcal{E}$ *is a k-vector subspace of K which satisfies Property* $(\begin{smallmatrix} A & B \\ C & 0 \end{smallmatrix})$, *the rank of any matrix with coefficients in* $\mathcal{E}$ *is at least half its structural rank with respect to k.*

The determinant of a matrix whose entries are linear forms is a homogeneous polynomial. In order to deal with nonhomogeneous polynomials, we consider matrices whose entries are linear polynomials (i.e. polynomials of total degree ≤ 1).

Lemma 12.9. *Let* M *be a matrix with entries in* K, $\mathcal{E}$ *the k-vector subspace of K spanned by the entries of* M, *and* φ *an injective linear map from* $\mathcal{E}$ *into a k-vector space* $k + kX_1 + \cdots + kX_n$. *Then the rank of the matrix* $\varphi(\mathsf{M})$ *is the same as the structural rank of* M *with respect to k.*

---

[19] On the other hand, by Lemma 12.16, the rank of any matrix M with entries in the field $\mathbb{Q}(\sqrt{2})$ is at least half the structural rank of M with respect to $\mathbb{Q}$.

Proof of Lemma 12.9. Compose φ with the injective linear mapping

$$\begin{array}{ccc} k + kX_1 + \cdots + kX_n & \longrightarrow & kY_0 + kY_1 + \cdots + kY_n \\ a_0 + a_1X_1 + \cdots + a_nX_n & \longmapsto & a_0Y_0 + a_1Y_1 + \cdots + a_nY_n \end{array}$$

and use Lemma 12.8. □

12.1.5 Any Polynomial is the Determinant of a Matrix

Again, the results of this section are due to D. Roy [Roy 1990](see also [Roy 1995]§ 3.1).

We first show that any polynomial in variables $X_1, \ldots, X_n$ is the determinant of a matrix whose entries are linear polynomials in $1, X_1, \ldots, X_n$.

Proposition 12.10. *For any $P \in k[X_1, \ldots, X_n]$ there exists a square matrix with entries in the k-vector space $k + kX_1 + \cdots + kX_n$ whose determinant is P.*

The proof involves two lemmas.

Lemma 12.11. *Let* M *be a matrix whose entries are bilinear forms*

$$\mathsf{M} = \left(\sum_{s=0}^{S} \sum_{t=0}^{T} m_{ijst} X_s Y_t \right)_{\substack{1 \le i \le d \\ 1 \le j \le \ell}}.$$

There exist a matrix A *whose entries are linear forms in $X_0, \ldots, X_S$ and a matrix* B *whose entries are linear forms in $Y_0, \ldots, Y_T$ such that* $\mathsf{M} = \mathsf{AB}$.

Proof of Lemma 12.11. Here is one solution (plainly, there is no unicity). Write

$$\mathsf{M} = \mathsf{M}_0 X_0 + \cdots + \mathsf{M}_S X_S$$

with

$$\mathsf{M}_s = \left(\sum_{t=0}^{T} m_{ijst} Y_t \right)_{\substack{1 \le i \le d \\ 1 \le j \le \ell}}, \qquad (0 \le s \le S).$$

Take for A the $d \times dS$ matrix

$$\mathsf{A} = \left(X_0 \mathsf{I}_d, \ldots, X_S \mathsf{I}_d \right)$$

and for B the $dS \times \ell$ matrix

$$\mathsf{B} = \begin{pmatrix} \mathsf{M}_0 \\ \vdots \\ \mathsf{M}_S \end{pmatrix}.$$

□

Lemma 12.12. *The determinant of a product* AB *of a $d \times \ell$ matrix* A *by a $\ell \times d$ matrix* B *is the determinant of the $(d + \ell) \times (d + \ell)$ matrix written as blocks*

$$\begin{pmatrix} \mathsf{I}_\ell & \mathsf{B} \\ -\mathsf{A} & 0 \end{pmatrix}.$$

Proof of Lemma 12.12. Multiply on the left the matrix $\begin{pmatrix} \mathsf{I}_\ell & \mathsf{B} \\ -\mathsf{A} & 0 \end{pmatrix}$ by the determinant 1 matrix $\begin{pmatrix} \mathsf{I}_\ell & 0 \\ \mathsf{A} & \mathsf{I}_d \end{pmatrix}$. This will not change the determinant, and the product is $\begin{pmatrix} \mathsf{I}_\ell & \mathsf{B} \\ 0 & \mathsf{AB} \end{pmatrix}$, whose determinant is det(AB). □

Proof of Proposition 12.10. If P has degree 1 the result is trivial. We first explain the argument by considering a polynomial P of total degree 2. Write P as $L_0 + L_1X_1 + \cdots + L_nX_n$ where each L_i is a polynomial of degree ≤ 1, which means that each L_i lies in $k + kX_1 + \cdots + kX_n$. Then P is the determinant of the $(n+2) \times (n+2)$ matrix

$$\begin{pmatrix} & & & 1 \\ & & & X_1 \\ & \mathsf{I}_{n+1} & & \vdots \\ & & & X_n \\ -L_0 & \cdots & -L_n & 0 \end{pmatrix}$$

Consider now the general case. For any $m \geq 1$, denote by $\mathcal{E}_m$ the k-vector subspace of $k[X_1, \ldots, X_n]$ consisting of all polynomials of degree $\leq m$. For instance $\mathcal{E}_1 = k + kX_1 + \cdots + kX_n$. Let M be a matrix with entries in $\mathcal{E}_m$ with $m \geq 2$. Write $\mathsf{M}_0 + \mathsf{M}_1X_1 + \cdots + \mathsf{M}_nX_n$, where each M_i has entries in $\mathcal{E}_{m-1}$. Thanks to Lemma 12.11 (with $S = T = n$, $Y_s = X_s$ for $0 \leq s \leq n$), we may write M = AB where the entries of A are in $\mathcal{E}_{m-1}$ while the entries of B lie in $\mathcal{E}_1$. By Lemma 12.12, M has the same determinant as the square matrix $\begin{pmatrix} \mathsf{I}_\ell & \mathsf{B} \\ -\mathsf{A} & 0 \end{pmatrix}$ whose entries are in $\mathcal{E}_{m-1}$. By induction on m this implies that any square matrix M with entries in $\mathcal{E}_m$ has the same determinant as a square matrix whose entries are in $\mathcal{E}_1$. In particular any polynomial (which is the determinant of a 1×1 matrix with entries in $\mathcal{E}_m$, when m is any upper bound for the total degree) is the determinant of a square matrix with entries in $\mathcal{E}_1$. □

Remark. The determinant of a square $d \times d$ matrix whose entries are homogeneous linear forms in $kX_0 + \cdots + kX_n$ is a homogeneous polynomial of degree d. But not all homogeneous polynomials occur as such determinants (see Exercise 12.4).

Proposition 12.13. *For a subspace $\mathcal{E}$ of K over k containing k the four following properties are equivalent:*

(*i*) *There exists a basis $(x_i)_{i\in I}$ of $\mathcal{E}$ over k with $0 \in I$, $x_0 = 1$ and $\{x_i\,;\, i \in I,\ i \neq 0\}$ algebraically independent over k.*

(*ii*) *If $x_1, \ldots, x_n$ are elements in $\mathcal{E}$ such that $1, x_1, \ldots, x_n$ are linearly independent over k, then $x_1, \ldots, x_n$ are algebraically independent over k.*

(*iii*) *For any tuple $(x_0, \ldots, x_n)$ which consists of k-linearly independent elements of $\mathcal{E}$ and for any nonzero homogeneous polynomial $P \in k[X_0, \ldots, X_n]$, the number $P(x_0, \ldots, x_n)$ is not zero.*

(*iv*) *Any matrix* M *with entries in* $\mathcal{E}$ *has a rank equal to its structural rank with respect to* k.

Proof of Proposition 12.13. Implication $(iii) \Rightarrow (ii)$ is easy (take $x_0 = 1$), $(ii) \Rightarrow (i)$ is plain, while $(i) \Rightarrow (iv)$ follows from Lemma 12.9.

It remains to check $(iv) \Rightarrow (iii)$. Assume (iii) does not hold: there exist $x_0, \dots, x_n$ in $\mathcal{E}$ which are k-linearly independent such that, if we set $y_i = x_i/x_0$, then $y_1, \dots, y_n$ are algebraically dependent over k. Let $P \in k[X_1, \dots, X_n]$ be a nonzero polynomial such that $P(y_1, \dots, y_n) = 0$. Proposition 12.10 shows that there exists a square matrix $\mathsf{M}_0 + \mathsf{M}_1 X_1 + \cdots + \mathsf{M}_n X_n$ with entries in $k + kX_1 + \cdots + kX_n$ whose determinant is P. Hence the determinant of the matrix $\mathsf{M} = \mathsf{M}_0 + \mathsf{M}_1 y_1 + \cdots + \mathsf{M}_n y_n$ is zero. Since $1, y_1, \dots, y_n$ are k-linearly independent, there exists an injective k-linear map φ from $k + ky_1 + \cdots + ky_n$ into $k + kX_1 + \cdots + kX_n$ which maps y_i onto X_i and is the identity on k. The determinant of $\varphi(\mathsf{M})$ is not zero, hence the structural rank of M with respect to k is strictly larger than the rank of M and condition (iv) does not hold. □

We apply Proposition 12.13 to a k-vector subspace of K of the form $k + \mathcal{E}_0$ where $k \cap \mathcal{E}_0 = \{0\}$ so that the sum $k + \mathcal{E}_0$ is direct.

Corollary 12.14. *Let* $\mathcal{E}_0$ *be a* k*-vector subspace of* K *such that* $k \cap \mathcal{E}_0 = \{0\}$. *Then the two following properties are equivalent.*

(*i*) *Elements of* $\mathcal{E}_0$ *which are* k*-linearly independent are algebraically independent over* k.
(*ii*) *The rank of any matrix* M *with coefficients in* $k + \mathcal{E}_0$ *is equal to the structural rank of* M *with respect to* k.

12.1.6 Conclusion

One can summarize the preceding properties as follows. Consider a triple $(k, K, \mathcal{E}_0)$, where k is a field, K an extension of k and $\mathcal{E}_0$ a k-vector subspace of K satisfying $\mathcal{E}_0 \cap k = \{0\}$. Define $\mathcal{E} = k + \mathcal{E}_0$.

$$\begin{array}{ccc}
\boxed{x_i \text{ a.i.}} & \Longleftrightarrow & \boxed{r = r_{\mathrm{str}}} \\
 & & \Downarrow \\
\boxed{\ell \le n(n+1)/2} & \Leftarrow & \boxed{\begin{pmatrix} A & B \\ C & 0 \end{pmatrix}} \\
\Downarrow & & \Downarrow \\
\boxed{\ell \le n(n+1)} & & \boxed{r \ge r_{\mathrm{str}}/2}
\end{array}$$

In the first row, the left hand side means that $\mathcal{E}_0$ is spanned as a k-vector space by elements of K which are algebraically independent over k. The right hand side means that the rank of any matrix with entries in $\mathcal{E}$ is equal to its structural rank with respect to k. The fact that these two properties are equivalent is Corollary 12.14.

In the intermediate row, the right hand side is Property $\binom{A\ B}{C\ 0}$ for the k-vector space $\mathcal{E}$, the left hand side means that for any K-vector subspace $\mathcal{V}$ of K^d satisfying $\mathcal{V} \cap k^d = \{0\}$, the dimension ℓ of $\mathcal{V} \cap \mathcal{E}^d$ over k is finite and $\leq n(n+1)/2$. The implication from right to left is Proposition 12.7.

Property $\binom{A\ B}{C\ 0}$ is a consequence of $r = r_{\mathrm{str}}$ and it implies $r \geq r_{\mathrm{str}}/2$ as pointed out in § 12.1.5. Lemmas 12.15 and 12.16 below show that the converse implications do not hold: it is not possible to go from one row to the row above it.

For arithmetic applications (see § 12.2 and 12.3), two cases are most important:
(1) Take $k = \mathbb{Q}$, $\mathcal{E}_0$ is the $\mathbb{Q}$-vector space $\mathcal{L}$ of logarithms of algebraic numbers, $\mathcal{E} = \mathbb{Q}+\mathcal{L}$ (see §§ 12.2 and 12.3). The condition $\mathcal{E}_0 \cap \overline{\mathbb{Q}} = \{0\}$ is Hermite-Lindemann's Theorem.
(2) Choose $k = \overline{\mathbb{Q}}$, $\mathcal{E}_0$ is the $\overline{\mathbb{Q}}$-vector space of homogeneous linear combinations of elements of $\mathcal{L}$ with coefficients in $\overline{\mathbb{Q}}$, while $\mathcal{E} = \widetilde{\mathcal{L}}$ (see § 12.3). The condition $\mathcal{E}_0 \cap \overline{\mathbb{Q}} = \{0\}$ follows from Baker's Theorem.

By Conjecture 1.15, one *conjectures* that the properties on the top line are satisfied in both cases. As we shall see below, the bottom line corresponds to the known facts so far. It would already be interesting to prove the results corresponding to the intermediate row: this would include for instance a solution of the four exponentials Conjecture, but this would not enable us to infer that the field $\mathbb{Q}(\mathcal{L})$ has transcendence degree ≥ 2 over $\mathbb{Q}$.

We complete this section with two examples (Proposition 1 of [Roy 1989] and Theorem 3.4 of [Roy 1995]) whose proofs are left as exercises (12.9 and 12.10).

Lemma 12.15*. *Let $x \in K$ be transcendental over k. There exists a k-vector subspace $\mathcal{E}$ of $k[x]$ of infinite dimension which satisfies Property $\binom{A\ B}{C\ 0}$.*

Lemma 12.16*. *Let $\mathcal{E}$ be a k-vector subspace of K of dimension ≤ 3. Then the rank of any matrix M with entries in $\mathcal{E}$ is at least half its structural rank with respect to k.*

Here is a motivation for Lemma 12.16 (cf. Proposition 2 of [Roy 1989]). Let x and u be two elements in K such that u, ux, ux^2 are k-linearly independent. Denote by $\mathcal{E}$ the k-vector space $ku + kux + kux^2$. Then $\dim_k(\mathcal{E}) = 3$, and therefore Lemma 12.16 shows that the rank of any matrix M with entries in $\mathcal{E}$ is at least half its structural rank with respect to k. However $\mathcal{E}$ does not satisfy a property like the four exponentials Conjecture: the line $\mathcal{V} = K(1, x)$ in K^2 (which is also the hyperplane of equation $z_2 = xz_1$) satisfies $\mathcal{V} \cap k^2 = \{0\}$ (because $x \notin k$), while $\mathcal{V} \cap \mathcal{E}^2$ contains the two points (u, ux) and (ux, ux^2) which are k-linearly independent (because $x \notin k$). In particular this triple $(k, K, \mathcal{E})$ does not satisfy Property $\binom{A\ B}{C\ 0}$.

12.2 Entries are Logarithms of Algebraic Numbers

In Chap. 11 we investigated the relationships between the dimension ℓ of $\mathbb{Q}$-vector space $\mathcal{V} \cap \mathcal{L}^d$ and the dimension n of $\mathcal{V}$, when $\mathcal{V}$ is a complex vector subspace of $\mathbb{C}^d$. Taking a basis over $\mathbb{Q}$ of $\mathcal{V} \cap \mathcal{L}^d$ gives rise to a $d \times \ell$ matrix M whose entries are logarithms of algebraic numbers. The question studied in Chap. 11 amounts to investigate the rank n of M in terms of d and ℓ.

12.2.1 A Consequence of the Linear Subgroup Theorem

From Corollary 11.14 we deduce a lower bound for the rank of matrices with entries in $\mathcal{L}$ as follows (see [W 1981], Th. 2.1).

Theorem 12.17. *Any $d \times \ell$ matrix M of rank n with entries in $\mathcal{L}$ is $\mathbb{Q}$-equivalent to a block matrix (12.1) where the matrix C has size $d' \times \ell'$ with $d' > 0$ and*

$$n \geq \frac{d\ell'}{d' + \ell'}.$$

We insist that $d' > 0$, but (with the notation of (12.1)) we allow $\ell^* = 0$. In particular if the conclusion holds with $\ell' = \ell$ then $n \geq \ell d/(\ell + d)$. For instance if a matrix with entries in $\mathcal{L}$ has its $d\ell$ entries linearly independent over $\mathbb{Q}$, then its rank is $\geq \ell d/(\ell + d)$.

Proof of Theorem 12.17. In (12.1) we allow $\ell' = \ell$. Hence we may, without loss of generality, assume $n < \ell d/(\ell + d)$. Therefore the assumption $n < d$ of Corollary 11.14 is satisfied.

Denote by $\underline{\eta}_1, \dots, \underline{\eta}_\ell$ the columns of M in $\mathbb{C}^d$, by $\mathcal{V}$ the complex vector subspace of $\mathbb{C}^d$ which they span and by Y the $\mathbb{Q}$-vector space which they span in $\mathcal{L}^d$. The dimension n of $\mathcal{V}$ is nothing else than the rank of M. From Corollary 11.14 we deduce that there exists a complex vector subspace S of $\mathbb{C}^d$, rational over $\mathbb{Q}$, of dimension d^* and codimension $d' > 0$, such that the dimension ℓ' of the $\mathbb{Q}$-vector space $Y/Y \cap S$ satisfies

$$\frac{\ell'}{d' + \ell'} \leq \frac{n'}{d'} \leq \frac{n}{d}.$$

Let $(\underline{e}_1, \dots, \underline{e}_{d^*})$ be a basis of S which we complete into a basis $(\underline{e}_1, \dots, \underline{e}_d)$ of $\mathbb{C}^d$. We denote by $\mathsf{P} = \bigl(p_{hi}\bigr)_{1 \leq h,i \leq d}$ the transition matrix, so that, for $\underline{z} = (z_1, \dots, z_d) \in \mathbb{C}^d$,

$$\underline{z} = \sum_{h=1}^{d} \sum_{i=1}^{d} p_{hi} z_i \underline{e}_h.$$

Notice that $\underline{z} \in S$ if and only if

$$\sum_{i=1}^{d} p_{hi} z_i = 0 \quad \text{for} \quad d^* < h \leq d.$$

Next we take a basis $(\underline{\theta}_1, \ldots, \underline{\theta}_\ell)$ of Y over $\mathbb{Q}$ such that $(\underline{\theta}_{\ell'+1}, \ldots, \underline{\theta}_\ell)$ is a basis of $Y \cap S$. Let $\mathsf{Q} = (q_{jt})_{1 \leq j,t \leq \ell}$ the transition matrix between the two bases of Y:

$$\underline{\theta}_t = \sum_{j=1}^{\ell} q_{jt} \underline{\eta}_j \qquad (1 \leq t \leq \ell).$$

For $\ell' < t \leq \ell$,

$$\underline{\theta}_t = \left(\sum_{j=1}^{\ell} q_{jt} \lambda_{ij} \right)_{1 \leq i \leq d}$$

is in S, hence for $d^* < h \leq d$ we have,

$$\sum_{i=1}^{d} p_{hi} \sum_{j=1}^{\ell} q_{jt} \lambda_{ij} = 0,$$

which means that PMQ has the form (12.1). □

Proof of Theorem 1.16. Let $\mathsf{M} = (\lambda_{ij})$ be a $d \times \ell$ matrix with entries in $\mathcal{L}$. Assume, for any $\underline{t} = (t_1, \ldots, t_d) \in \mathbb{Z}^d \setminus \{0\}$ and any $\underline{s} = (s_1, \ldots, s_\ell) \in \mathbb{Z}^\ell \setminus \{0\}$,

$$\sum_{i=1}^{d} \sum_{j=1}^{\ell} t_i s_j \log \alpha_{ij} \neq 0.$$

We use Theorem 12.17: since no matrix of the form PMQ can have a vanishing entry, and since $d' > 0$, in (12.1) we have $\ell^* = 0$, which means $\ell' = \ell$. Hence the rank of M is at least $d\ell/(d + \ell)$. □

12.2.2 Rank and Structural Rank

From Theorem 12.17 we deduce:

Corollary 12.18. *Any matrix with entries in $\mathcal{L}$ has rank at least half its structural rank with respect to $\mathbb{Q}$.*

Proof. Without loss of generality we may assume $r_{\text{str}} = \ell = d$ (just take a square submatrix of maximal structural rank). From Theorem 12.17 we deduce that the rank of M is bounded from below by

$$\operatorname{rank}(\mathsf{M}) \geq \frac{d\ell'}{d' + \ell'}$$

while the structural rank is bounded from above by

$$r_{\text{str}}(\mathsf{M}) \leq \ell' + d - d'.$$

Since $r_{\text{str}}(\mathsf{M}) = d$, we deduce $d' \leq \ell'$ and $\ell'/(d' + \ell') \geq 1/2$. □

12.2.3 A Further Consequence of the Linear Subgroup Theorem

Here is a reformulation of Theorem 11.5 part (1) in terms of matrices.

Theorem 12.19. *Let*

$$\mathsf{M} = \begin{pmatrix} \mathsf{B}_0 & \mathsf{B}_1 \\ \mathsf{B}_2 & \mathsf{L} \end{pmatrix} \begin{matrix} \}d_0 \\ \}d_1 \end{matrix}, \qquad \text{column blocks } \underbrace{\quad}_{\ell_0}\ \underbrace{\quad}_{\ell_1}$$

be a $d \times \ell$ matrix of rank n with $1 \leq n < d$, where B_0, B_1, B_2 are matrices with entries in $\overline{\mathbb{Q}}$, while the entries of L are in $\mathcal{L}$. Then there exist regular matrices

$$\mathsf{P} = \begin{pmatrix} \mathsf{P}_0 & 0 \\ 0 & \mathsf{P}_1 \end{pmatrix} \quad \textit{and} \quad \mathsf{Q} = \begin{pmatrix} \mathsf{Q}_0 & 0 \\ 0 & \mathsf{Q}_1 \end{pmatrix}$$

where

$$\mathsf{P}_0 \in \mathrm{GL}_{d_0}(\overline{\mathbb{Q}}), \quad \mathsf{P}_1 \in \mathrm{GL}_{d_1}(\mathbb{Q}),$$
$$\mathsf{Q}_0 \in \mathrm{GL}_{\ell_0}(\overline{\mathbb{Q}}), \quad \mathsf{Q}_1 \in \mathrm{GL}_{\ell_1}(\mathbb{Q}),$$

such that

$$\mathsf{PMQ} = \begin{pmatrix} \mathsf{P}_0\mathsf{B}_0\mathsf{Q}_0 & \mathsf{P}_0\mathsf{B}_1\mathsf{Q}_1 \\ \mathsf{P}_1\mathsf{B}_2\mathsf{Q}_0 & \mathsf{P}_1\mathsf{L}\mathsf{Q}_1 \end{pmatrix}$$

with

$$\mathsf{P}_0\mathsf{B}_0\mathsf{Q}_0 = \begin{pmatrix} \mathsf{B}_{00} & \mathsf{B}_{01} \\ \mathsf{B}_{02} & 0 \end{pmatrix} \qquad \mathsf{P}_0\mathsf{B}_1\mathsf{Q}_1 = \begin{pmatrix} \mathsf{B}_{10} & \mathsf{B}_{11} \\ \mathsf{B}_{12} & 0 \end{pmatrix} \begin{matrix} \}d_0^* \\ \}d_0' \end{matrix}$$

$$\mathsf{P}_1\mathsf{B}_2\mathsf{Q}_0 = \begin{pmatrix} \mathsf{B}_{20} & \mathsf{B}_{21} \\ \mathsf{B}_{22} & 0 \end{pmatrix} \qquad \mathsf{P}_1\mathsf{L}\mathsf{Q}_1 = \begin{pmatrix} \mathsf{L}_0 & \mathsf{L}_1 \\ \mathsf{L}_2 & 0 \end{pmatrix} \begin{matrix} \}d_1^* \\ \}d_1' \end{matrix}$$

$$\text{column blocks: } \underbrace{\quad}_{\ell_0'}\ \underbrace{\quad}_{\ell_0^*} \qquad \underbrace{\quad}_{\ell_1'}\ \underbrace{\quad}_{\ell_1^*}$$

and $d' = d_0' + d_1' > \ell_0'$,

$$(\ell_1' + d_1')(d - n) \leq d_1(d' - \ell_0').$$

As for Theorem 12.17, we insist that $d' > 0$, but we allow $\ell_0^* = 0$ or/and $\ell_1^* = 0$. Examples involving matrices like

$$\begin{pmatrix} 1 & 1 & 0 & 0 \\ 0 & \gamma x_1/x_2 & x_1y_1 & x_1y_2 \\ -\gamma & 0 & x_2y_1 & x_2y_2 \end{pmatrix} \quad \text{and} \quad \begin{pmatrix} 1 & 0 & \beta_{11} & \beta_{12} & \beta_{13} \\ 0 & -1 & \beta_{21} & \beta_{22} & \beta_{23} \\ -1 & 0 & \lambda_{11} & \lambda_{12} & \lambda_{13} \\ 0 & 1 & \lambda_{21} & \lambda_{22} & \lambda_{23} \end{pmatrix}$$

occur in § 11.3.3.

12.3 Entries are Linear Combinations of Logarithms

We now translate some of the results of § 11.6 in terms of matrices whose entries are in $\mathbb{Q}+\mathcal{L}$ or more generally in $\widetilde{\mathcal{L}}$.

12.3.1 A Consequence of the Linear Subgroup Theorem

The following result is nothing else than Proposition 11.19.

Theorem 12.20. *Denote by* $(\mathbb{K}, \boldsymbol{L})$ *either* $(\mathbb{Q}, \mathbb{Q}+\mathcal{L})$ *or else* $(\overline{\mathbb{Q}}, \widetilde{\mathcal{L}})$. *Let* M *be a* $d \times \ell$ *matrix (with* $d > 0$*) of rank* n *with entries in* $\boldsymbol{L}$. *Then* M *is* $\mathbb{K}$*-equivalent to a block matrix (12.1) where the matrix* C *has size* $d' \times \ell'$ *with* $d' > 0$ *and*

$$n \geq \frac{d\ell'}{d'+\ell'}.$$

One deduces the analog of Theorem 1.16 for $\widetilde{\mathcal{L}}$.

Corollary 12.21. *Let* $\mathsf{M} = \left(\Lambda_{ij}\right)_{\substack{1\le i\le d\\ 1\le j\le \ell}}$ *be a* $d \times \ell$ *matrix with entries in* $\widetilde{\mathcal{L}}$. *Assume, for any* $\underline{\tau} = (\tau_1, \ldots, \tau_d) \in \overline{\mathbb{Q}}^d \setminus \{0\}$ *and any* $\underline{\sigma} = (\sigma_1, \ldots, \sigma_\ell) \in \overline{\mathbb{Q}}^\ell \setminus \{0\}$,

$$\sum_{i=1}^{d}\sum_{j=1}^{\ell} \tau_i\sigma_j\Lambda_{ij} \neq 0.$$

Then the rank of M *is at least* $d\ell/(d+\ell)$.

12.3.2 Rank and Structural Rank

For a matrix with entries in $\mathcal{L}$ or in $\mathbb{Q}+\mathcal{L}$, it is natural to consider its structural rank with respect to $\mathbb{Q}$, since $\mathcal{L}$ is a $\mathbb{Q}$-vector space. Now $\widetilde{\mathcal{L}}$ is a $\overline{\mathbb{Q}}$-vector space; hence for a matrix with coefficients in $\widetilde{\mathcal{L}}$, one should rather consider its structural rank with respect to $\overline{\mathbb{Q}}$. But $\mathcal{L} \subset \mathbb{Q}+\mathcal{L} \subset \widetilde{\mathcal{L}}$. Fortunately, *for a matrix* M *with entries in* $\mathbb{Q}+\mathcal{L}$, *its structural rank with respect to* $\mathbb{Q}$ *is the same as its structural rank with respect to* $\overline{\mathbb{Q}}$. Indeed, let $1, \lambda_1, \ldots, \lambda_m$ is a basis over $\mathbb{Q}$ of the subspace of $\mathcal{L}$ spanned by 1 and the entries of M. Write

$$\mathsf{M} = \mathsf{M}_0 + \mathsf{M}_1\lambda_1 + \cdots + \mathsf{M}_m\lambda_m$$

where each M_i is in $\mathrm{Mat}_{d\times\ell}(\mathbb{Q})$. From Baker's Theorem, it follows that $1, \lambda_1, \ldots, \lambda_m$ is also a basis over $\overline{\mathbb{Q}}$ of the subspace of $\widetilde{\mathcal{L}}$ spanned by 1 and the entries of M. Hence both structural ranks are just the rank of the $d \times \ell$ matrix

$$\mathsf{M}_0 + \mathsf{M}_1X_1 + \cdots + \mathsf{M}_mX_m$$

with entries in the field $\mathbb{Q}(X_1, \ldots, X_m)$.

Therefore, dealing with matrices with entries in $\mathbb{Q} + \mathcal{L}$, we shall not specify the field with respect to which we consider the structural rank[20].

The next statement follows from Theorem 12.20 exactly as Corollary 12.18 from Theorem 12.17.

Corollary 12.22. *Any matrix with entries in $\widetilde{\mathcal{L}}$ has rank at least half its structural rank.*

Also:

Corollary 12.23. *The rank of any matrix with entries in $\mathbb{Q} + \mathcal{L}$ is at least half its structural rank.*

Therefore the properties stated in the bottom line in the diagram of § 12.1.7 are satisfied for $(k, \mathcal{E})$ either $(\mathbb{Q}, \mathbb{Q} + \mathcal{L})$ or $(\overline{\mathbb{Q}}, \widetilde{\mathcal{L}})$: the left hand side follows from Corollary 11.15 and the right hand side from Corollaries 12.22 and 12.23.

Lemma 12.15 shows that one cannot deduce from Corollary 12.23 that the transcendence degree over $\mathbb{Q}$ of the field $\mathbb{Q}(\mathcal{L})$ is at least 2. Moreover, Lemma 12.16 shows that one cannot deduce either the four exponentials Conjecture.

The strong six exponentials Theorem (Corollary 11.16) follows from Corollary 12.22:

Corollary 12.24. *Let M be a $d \times \ell$ matrix with entries in $\widetilde{\mathcal{L}}$ whose rows are $\overline{\mathbb{Q}}$-linearly independent in $\widetilde{\mathcal{L}}^{\ell}$ and whose columns are $\overline{\mathbb{Q}}$-linearly independent in $\widetilde{\mathcal{L}}^{d}$. If $\ell d > \ell + d$, then the rank of M is at least 2.*

One can deduce the following result either from Corollary 12.23 or from Corollary 12.24 :

- *If M is a $d \times \ell$ matrix with entries in $\mathbb{Q} + \mathcal{L}$ whose rows are $\mathbb{Q}$-linearly independent and whose columns are also $\mathbb{Q}$-linearly independent, if $\ell d > \ell + d$, then the rank of M is at least 2.*

[20] If $\mathsf{M} \in \mathrm{Mat}_{d\times\ell}(\widetilde{\mathcal{L}})$ does not belong to $\mathrm{Mat}_{d\times\ell}(\mathbb{Q} + \mathcal{L})$, it is implicit that we consider its structural rank with respect to $\overline{\mathbb{Q}}$, not to $\mathbb{Q}$.

12.4 Assuming the Conjecture of Algebraic Independence of Logarithms

In this last section we discuss consequences of Conjecture 1.15 on algebraic independence of logarithms of algebraic numbers.

12.4.1 The $\mathbb{Q}$-Vector Space $\mathcal{L}$

One conjectures that Property $\binom{A\ B}{C\ 0}$ holds for the triple $(\mathbb{Q}, \mathbb{C}, \mathcal{L})$. Using Proposition 12.14, one would deduce:

(?) *For any complex vector subspace $\mathcal{V}$ of $\mathbb{C}^d$ of dimension n such that $\mathcal{V} \cap \mathbb{Q}^d = \{0\}$, the $\mathbb{Q}$-vector subspace $\mathcal{V} \cap \mathcal{L}^d$ has dimension $\le n(n+1)/2$.*

Clearly this would solve the four exponentials Conjecture. On the other hand Lemma 12.15 shows that Property $\binom{A\ B}{C\ 0}$ would not be sufficient to deduce that there exist two algebraically independent logarithms of algebraic numbers.

Corollary 12.14 shows that the two following statements are equivalent:

(?) *The rank of any matrix in $\mathrm{Mat}_{d\times\ell}(\mathcal{L})$ is equal to its structural rank.*
(?) *If $\lambda_0, \ldots, \lambda_n$ are $\mathbb{Q}$-linearly independent elements in $\mathcal{L}$, then the numbers $\lambda_1/\lambda_0, \ldots, \lambda_n/\lambda_0$ are algebraically independent over $\mathbb{Q}$.*

12.4.2 The $\mathbb{Q}$-Vector Space $\mathbb{Q} + \mathcal{L}$

From Corollary 12.14 one deduces that Conjecture 1.15 is equivalent to the following statement

(?) *Any matrix*

$$\left(b_{ij} + \lambda_{ij}\right)_{\substack{1\le i\le d\\ 1\le j\le \ell}}$$

with $b_{ij} \in \mathbb{Q}$ and $\lambda_{ij} \in \mathcal{L}$ has a rank equal to its structural rank.

12.4.3 The $\overline{\mathbb{Q}}$-Vector Space $\widetilde{\mathcal{L}}$

By Conjecture 1.15 the rank of any matrix in $\mathrm{Mat}_{d\times\ell}(\widetilde{\mathcal{L}})$ should be equal to its structural rank. This would imply that Property $\binom{A\ B}{C\ 0}$ also holds for the triple $(\overline{\mathbb{Q}}, \mathbb{C}, \widetilde{\mathcal{L}})$. Hence from Proposition 12.7 one would deduce:

(?) *For any complex vector subspace $\mathcal{V}$ of $\mathbb{C}^d$ of dimension n such that $\mathcal{V} \cap \overline{\mathbb{Q}}^d = \{0\}$, the $\overline{\mathbb{Q}}$-vector subspace $\mathcal{V} \cap \widetilde{\mathcal{L}}^d$ has dimension $\le n(n+1)/2$.*

This result includes the strong four exponentials Conjecture 11.17. A weaker statement is the so-called *Strong Five Exponentials Conjecture* of [W 1988]:

(?) *Let x_1, x_2 be two $\mathbb{Q}$-linearly independent complex numbers and y_1, y_2 be also two $\mathbb{Q}$-linearly independent complex numbers. Further let β_{ij} $(i = 1, 2,\ j = 1, 2)$, γ_1 and γ_2 be six algebraic numbers with $\gamma_1 \neq 0$. Assume that the five numbers*

$$e^{x_1y_1-\beta_{11}},\ e^{x_1y_2-\beta_{12}},\ e^{x_2y_1-\beta_{21}},\ e^{x_2y_2-\beta_{22}},\ e^{(\gamma_1x_1/x_2)-\gamma_2}$$

are algebraic. Then

$$x_iy_j = \beta_{ij} \quad \textit{for} \quad i = 1, 2, \quad \textit{and} \quad j = 1, 2 \quad \textit{and} \quad \gamma_1x_1 = \gamma_2x_2.$$

12.5 Quadratic Relations

In this section, $k \subset K$ are two fields and $\mathcal{E}$ is a k-subspace of K.

In the next proposition, we assume that any nonzero matrix with coefficients in $\mathcal{E}$ has rank larger than half the structural rank, with strict inequality. For $k = \mathbb{Q}$, $K = \mathbb{C}$ and $\mathcal{E} = \mathcal{L}$ this condition is not yet known: we have no strict inequality so far. However we shall see (Theorem 15.30) that this property is satisfied for any $\mathbb{Q}$-vector subspace of $\mathcal{L}$ spanned by elements in a field of transcendence degree 1. It is unlikely that any such space of dimension ≥ 2 exists, but it is a challenge to prove that there is none.

For such a vector space $\mathcal{E}$, we prove a property akin to the assertion (a.i) of § 11.5 for all quadratic hypersurfaces of K^n (i.e. affine hypersurfaces which are defined by a homogeneous polynomial of degree 2).

The proof rests on an explicit representation of a Clifford algebra, following [RoyW 1997a] and [RoyW 1997b].

Proposition 12.25. *Assume the rank of any nonzero matrix M with entries in $\mathcal{E}$ satisfies* $\operatorname{rank}(\mathsf{M}) > (1/2)r_{\mathrm{str}}(\mathsf{M})$, *where $r_{\mathrm{str}}(\mathsf{M})$ is the structural rank of M with respect to k. Let $Q \in k[X_1, \dots, X_n]$ be a nonzero homogeneous polynomial of degree 2. Denote by $Z(Q)$ the hypersurface $Q(\underline{x}) = 0$ in K^n. Then $Z(Q) \cap \mathcal{E}^n$ is the union of $E \cap \mathcal{E}^n$, where E ranges over the vector subspaces of K^n, rational over k, contained in $Z(Q)$.*

Roughly speaking, the conclusion means that the only $\underline{x} \in \mathcal{E}^n$ which satisfy $Q(\underline{x}) = 0$ are the trivial ones. For instance any $\underline{x} = (x_1, \dots, x_n) \in \mathcal{E}^n$ with $x_1, \dots, x_n$ linearly independent over k has $Q(\underline{x}) \neq 0$.

Let us start with an easy case: take $n = 4$ and

$$Q(\underline{X}) = X_1X_4 - X_2X_3.$$

We use the hypothesis of Proposition 12.25 for the matrix

$$\mathsf{M} = \begin{pmatrix} x_1 & x_2 \\ x_3 & x_4 \end{pmatrix}.$$

For $\underline{x} \in \mathcal{E}^4 \cap Z(Q) \setminus \{0\}$, this matrix M has rank 1, hence structural rank 1 also. The conclusion easily follows (see Exercise 1.8).

The next lemma reduces the proof of Proposition 12.25 to the special case $n = 2m$ and

$$Q = X_1Y_1 + \cdots + X_mY_m.$$

Lemma 12.26. *Let $\mathcal{E}$ be a k-vector subspace of K. The two following assertions are equivalent.*

(i) For any $m \geq 1$, the hypersurface $\mathcal{Z}$ of K^{2m} of equation

$$x_1y_1 + \cdots + x_my_m = 0$$

satisfies:

$$\mathcal{Z} \cap \mathcal{E}^m = \bigcup_{E \subset \mathcal{Z}} E \cap \mathcal{E}^m,$$

where E ranges over the vector subspaces of K^m, rational over k, contained in $\mathcal{Z}$.

(ii) For any $n \geq 1$ and any nonzero homogeneous quadratic polynomial $Q \in k[X_1, \ldots, X_n]$,

$$Z(Q) \cap \mathcal{E}^n = \bigcup_{E \subset Z(Q)} E \cap \mathcal{E}^n,$$

where E ranges over the vector subspaces of K^n, rational over k, contained in $Z(Q)$.

Proof of Lemma 12.26. Obviously (i) is a consequence of (ii) with $n = 2m$ and $Q = X_1Y_1 + \cdots + X_mY_m$.

Conversely, assume (i). Let $Q \in k[X_1, \ldots, X_n]$ be a nonzero homogeneous polynomial of degree 2, and let $\underline{u} \in Z(Q) \cap \mathcal{E}^n$.

We choose one way of writing

$$Q(\underline{X}) = \sum_{i=1}^{n}\sum_{j=1}^{n} a_{ij}X_iX_j$$

with $a_{ij} \in k$. Define a k-linear map

$$\begin{array}{cccc} \varphi: & K^n & \longrightarrow & K^{2n} \\ & (z_1, \ldots, z_n) & \longmapsto & \left(z_1, \ldots, z_n, \sum_{j=1}^n a_{1j}z_j, \ldots, \sum_{j=1}^n a_{nj}z_j\right) \end{array}$$

and set $\underline{v} = \varphi(\underline{u})$. From $\underline{u} \in Z(Q) \cap \mathcal{E}^n$ we deduce $\underline{v} \in \mathcal{Z} \cap \mathcal{E}^{2n}$, where $\mathcal{Z}$ is the hypersurface of K^{2n} of equation $x_1y_1 + \cdots + x_ny_n = 0$. Now let F be the minimal vector subspace of K^{2n}, rational over k, containing $\underline{v}$. Thanks to (i) with $m = n$ we know that F is contained in $\mathcal{Z}$. Define $E = \varphi^{-1}(F)$. Then E is a vector subspace of K^n, rational over k, containing $\underline{u}$, and

$$E = \varphi^{-1}(F) \subset \varphi^{-1}(\mathcal{Z}) = Z(Q).$$

□

The next result occurs in [RoyW 1997a]. Define inductively, for each $m \geq 1$, two K-linear mappings φ_m and ψ_m of K^{2m} into $\mathrm{Mat}_{2^{m-1}\times 2^{m-1}}(K)$ as follows. For $m = 1$, define

$$\varphi_1(x_1, y_1) = (x_1), \quad \psi_1(x_1, y_1) = (y_1),$$

and, for $m \geq 1$, setting

$$\underline{v} = (x_1, \ldots, x_m, y_1, \ldots, y_m), \quad \underline{v}' = (x_1, \ldots, x_{m+1}, y_1, \ldots, y_{m+1}),$$

define

$$\varphi_{m+1}(\underline{v}') = \begin{pmatrix} x_{m+1}\mathrm{I}_{2^{m-1}} & \psi_m(\underline{v}) \\ -\varphi_m(\underline{v}) & y_{m+1}\mathrm{I}_{2^{m-1}} \end{pmatrix}$$

and

$$\psi_{m+1}(\underline{v}') = \begin{pmatrix} y_{m+1}\mathrm{I}_{2^{m-1}} & -\psi_m(\underline{v}) \\ \varphi_m(\underline{v}) & x_{m+1}\mathrm{I}_{2^{m-1}} \end{pmatrix}.$$

Lemma 12.27. *The following properties hold.*

(1) For $m \geq 1$,

$$\varphi_m(\underline{v})\psi_m(\underline{v}) = \psi_m(\underline{v})\varphi_m(\underline{v}) = (x_1y_1 + \cdots + x_my_m)\mathrm{I}_{2^{m-1}}.$$

(2) For $m \geq 2$,

$$\det \varphi_m(\underline{v}) = \det \psi_m(\underline{v}) = (x_1y_1 + \cdots + x_my_m)^{2^{m-2}}.$$

(3) Both mappings $\varphi_m(\underline{v})$ and $\psi_m(\underline{v})$ are injective. For $m \geq 2$ and $\underline{v} \neq 0$, the rank of each of the two matrices $\varphi_m(\underline{v})$ and $\psi_m(\underline{v})$ is either 2^{m-1} or 2^{m-2}.

Proof of Lemma 12.27. Statement (1) is clear by induction (and products of bloc matrices). Since $\det \varphi_m(\underline{v})$ and $\det \psi_m(\underline{v})$ are homogeneous polynomials of degree 2^{m-1}, and since the coefficient of $(x_my_m)^{2^{m-2}}$ is 1, property (2) follows. Injectivity as well as the lower bound for the ranks of the matrices in (3) are plain by induction on m. The upper bound for the ranks then follows from $\varphi_m(\underline{v})\psi_m(\underline{v}) = 0$ when $x_1y_1 + \cdots + x_my_m = 0$. $\square$

Proof of Proposition 12.25. Using Lemma 12.26, we may assume $n = 2m$ and $Q = X_1Y_1 + \cdots + X_mY_m$. Let $\mathcal{Z} = \mathcal{Z}(Q)$, let $\underline{v} = (\underline{x}, \underline{y}) \in \mathcal{E}^{2m} \cap \mathcal{Z}$ with $\underline{v} \neq 0$ and let $(e_1, \ldots, e_s) \in \mathcal{E}^s$ be a basis of the k-vector space spanned by $x_1, \ldots, x_m, y_1, \ldots, y_m$. Write

$$x_i = \sum_{\sigma=1}^{s} a_{i\sigma}e_\sigma, \quad y_i = \sum_{\sigma=1}^{s} b_{i\sigma}e_\sigma \qquad (1 \leq i \leq m)$$

with $a_{i\sigma}$ and $b_{i\sigma}$ in k. For the proof, we can use either the map φ_m, or ψ_m, or else

$$\theta_m = \begin{pmatrix} 0 & \psi_m \\ \varphi_m & 0 \end{pmatrix},$$

as we wish. Let us use φ_m. Since $\underline{v} \in Z(Q)$ and $\underline{v} \neq 0$, the rank of the matrix $\varphi_m(\underline{v})$ is 2^{m-2}. Using the hypothesis on $\mathcal{E}$ with Lemma 12.27, we deduce that the structural rank of this matrix is $< 2^{m-1}$. This structural rank is nothing else than the rank of the matrix $\varphi_m\big(\underline{w}(\underline{T})\big)$, where

$$\underline{w}(\underline{T}) = \big(\xi_1(\underline{T}), \dots, \xi_m(\underline{T}), \eta_1(\underline{T}), \dots, \eta_m(\underline{T})\big)$$

with

$$\xi_i(\underline{T}) = \sum_{\sigma=1}^{s} a_{i\sigma} T_\sigma, \quad \eta_i(\underline{T}) = \sum_{\sigma=1}^{s} b_{i\sigma} T_\sigma \qquad (1 \le i \le m).$$

Therefore

$$\sum_{i=1}^{m} \xi_i(\underline{T})\eta_i(\underline{T}) = 0$$

in $k[\underline{T}] = k[T_1, \dots, T_s]$. The image of K^s under the linear map $\underline{t} \mapsto \underline{w}(\underline{t})$ is a vector subspace of K^{2m}, rational over k, which is contained in the hypersurface $\mathcal{Z}$ and contains $\underline{v}$. This completes the proof of Proposition 12.25. □

Remark. The linear map

$$\begin{array}{rccc} \theta_m: & K^{2m} & \longrightarrow & \mathrm{Mat}_{2^m \times 2^m}(K) \\ & \underline{v} & \longmapsto & \begin{pmatrix} 0 & \psi_m(\underline{v}) \\ \varphi_m(\underline{v}) & 0 \end{pmatrix} \end{array}$$

is injective and satisfies

$$\theta_m(\underline{v})^2 = Q(\underline{v}) \mathrm{I}_{2^m}$$

for any $\underline{v} \in K^{2m}$, where Q is the quadratic form $X_1Y_1 + \cdots + X_mY_m$. This shows that $\mathrm{Mat}_{2^m \times 2^m}(K)$ is the Clifford algebra attached to the quadratic form $X_1Y_1 + \cdots + X_mY_m$ (see [L 1993], Chap. 19, S 4, [RoyW 1997a], § 7 and [RoyW 1997b], § 10).

Exercises

Exercise 12.1. Let $\mathrm{M} \in \mathrm{Mat}_{d\times\ell}(\mathcal{L})$ be a $d \times \ell$ matrix with entries in $\mathcal{L}$. Denote by $\varphi: \mathbb{C}^\ell \to \mathbb{C}^d$ the associated linear map (in the canonical bases).

a) Check that the smallest vector subspace of $\mathbb{C}^d$ which is rational over $\overline{\mathbb{Q}}$ and contains the image of φ is in fact rational over $\mathbb{Q}$.

b) Check that the smallest vector subspace of $\mathbb{C}^\ell \times \mathbb{C}^d$ which is rational over $\overline{\mathbb{Q}}$ and contains the graph $\mathcal{G}(\varphi)$ of φ, namely

$$\mathcal{G}(\varphi) = \big\{\big(\underline{z}, \varphi(\underline{z})\big) \; ; \; \underline{z} \in \mathbb{C}^\ell\big\} \subset \mathbb{C}^\ell \times \mathbb{C}^d$$

is of the form $\mathbb{C}^\ell \times V$, where V is a subspace of $\mathbb{C}^d$ which is rational over $\mathbb{Q}$.

c) Check that the largest vector subspace of $\mathbb{C}^\ell$ which is rational over $\overline{\mathbb{Q}}$ and is contained in the kernel of φ is rational over $\mathbb{Q}$.

Exercise 12.2. In the field $k(X_1, \dots, X_n)$ of rational fractions in n unknowns over k, denote by $\mathcal{E}$ the k-vector subspace spanned by $1, X_1, \dots, X_n$. Let $Y_1, \dots, Y_m$ be elements in $\mathcal{E}$ which are algebraically dependent over k. Deduce from Lemma 12.3 that $1, Y_1, \dots, Y_m$ are k-linearly dependent.

Exercise 12.3. Any polynomial in $\mathbb{Z}[X_1, \dots, X_n]$ is the determinant of a matrix whose entries are in $\mathbb{Z} + \mathbb{Z}X_1 + \cdots + \mathbb{Z}X_n$.

Exercise 12.4. Check that the polynomial $X_0X_1+X_2X_3+X_4X_5$ cannot be written as $AD-BC$ with A, B, C, D linear forms in $X_0, \dots, X_5$.

Exercise 12.5. Let $k \subset K$ be two fields. Assume k is infinite.
a) Let $x_1, \dots, x_n$ be elements in K. Show that the following properties are equivalent.

(i) *The elements $x_1, \dots, x_n$ are algebraically independent over k.*
(ii) *The rank of any matrix* M *with entries in the k-vector space $k + kX_1 + \cdots + kX_n$ equals the rank of the matrix* M′ *with coefficients in the k-vector space $k + kx_1 + \cdots + kx_n$ which is derived from* M *by specializing X_i in x_i for each $i = 1, \dots, n$.*

b) Let $x_0, \dots, x_n$ be elements in K. Show that the following properties are equivalent.

(i) *For any nonzero homogeneous polynomial $P \in k[X_0, \dots, X_n]$,*

$$P(x_0, \dots, x_n) \neq 0.$$

(ii) *The rank of any matrix* M *with entries in the k-vector space $kX_0 + \cdots + kX_n$ is equal to the rank of the matrix* M′ *with coefficients in the k-vector space $kx_0 + kx_1 + \cdots + kx_n$ which is obtained by specializing X_i in x_i for each $i = 0, \dots, n$.*

Exercise 12.6. Let K be a field and k a subfield.
a) Let $\mathcal{E}_0$ be a k-subspace of K containing k. Check that the following conditions are equivalent.

(i) *There exists a basis B of $\mathcal{E}_0$ over k such that, for any distinct elements $x_0, \dots, x_m$ in B and for any nonzero homogeneous polynomial P in the ring $k[X_0, \dots, X_m]$, we have $P(x_0, \dots, x_m) \neq 0$.*
(ii) *For any tuple $(x_0, \dots, x_m)$ of k-linearly independent elements in $\mathcal{E}_0$ and any nonzero homogeneous polynomial $P \in k[X_0, \dots, X_m]$, $P(x_0, \dots, x_m) \neq 0$.*
(iii) *For any nonzero homogeneous polynomial $P \in k[X_0, \dots, X_n]$, we have*

$$\mathcal{E}_0^{n+1} \cap Z(P) = \bigcup_{\mathcal{V} \subset Z(P)} \mathcal{E}_0^{n+1} \cap \mathcal{V},$$

where $\mathcal{V}$ runs over the set of K-vector subspaces of K^{n+1}, rational over k, and contained in

$$Z(P) = \{\underline{x} \in K^{n+1};\ P(\underline{x}) = 0\}.$$

b) Let $\mathcal{E}$ be a k-subspace of K. Check that the conditions (i), (ii) and (iii) of Lemma 12.3 are equivalent to the following one

(iv) *For any nonzero polynomial $P \in k[X_1, \dots, X_n]$, we have*

$$\mathcal{E}^n \cap Z(P) = \bigcup_{\mathcal{V} \subset Z(P)} \mathcal{E}^n \cap \mathcal{V},$$

where $\mathcal{V}$ runs over the set of K-vector subspaces of K^n, rational over k, and contained in

$$Z(P) = \{\underline{x} \in K^n \,;\, P(\underline{x}) = 0\}.$$

Hint. *Compare with Exercise 1.8.a. See also* [Roy 1995].

Exercise 12.7. Let K be a field, k a subfield of K and $\mathcal{E}$ a k-vector subspace of K spanned by elements of K which are algebraically independent over k. Let M be a $d \times \ell$ matrix with coefficients in $\mathcal{E}$. Denote by $\mathcal{V}$ the K-vector subspace of K^d spanned by the columns of M. Assume $\mathcal{V} \cap k^d = \{0\}$. Show that the columns of M span a k-vector space of dimension at most $d(d-1)/2$.

Exercise 12.8. Under the assumptions from Proposition 12.7, check the estimate

$$\dim_k \mathcal{V} \cap \mathcal{E}^d \le \phi(n, d)$$

involving the function $\phi(n, d)$ from § 11.6.3.

Exercise 12.9. Prove Lemma 12.15.

Hint. *(Following* [Roy 1989], *Proposition 1). Let $(p_n)_{n\ge1}$ be a sequence of elements in $\mathbb{Q}[x]$, where p_n has degree $t_n > 0$, such that $t_{n+1} \ge 2t_n$ for any $n \ge 1$. Set $\mathcal{E}_n = \mathbb{Q}p_1 + \cdots + \mathbb{Q}p_n$ for $n \ge 1$ and*

$$\mathcal{E} = \bigcup_{n\ge1} \mathcal{E}_n.$$

Exercise 12.10. Prove Lemma 12.16. More precisely, under the assumptions of Lemma 12.16, if the ℓ columns of M are k-linearly independent, and if the same holds for the d rows, then the rank of M is $\ge (d+\ell)/4$.

Hint. *Apply Proposition 12.4 – see* [Roy 1995], *Theorem 3.4.*

Exercise 12.11 (D. Roy). Let k be a field of zero characteristic, $\overline{k}$ an algebraic closure of k, K an extension of $\overline{k}$ and $\mathcal{E}$ a subspace of K spanned by elements in K which are algebraically independent over k. Further, let $\mathcal{V}$ be a K-vector subspace of K^n spanned by elements of $\mathcal{E}^n$. Show that $\mathcal{V} \cap \overline{k}^n$ is contained in the maximal subspace of $\mathcal{V}$ rational over k.

Hint. *Check $\overline{k} \cap k(\mathcal{E}) = k$. Deduce that any automorphism of $\overline{k}$ can be extended to an automorphism of $\overline{k}(\mathcal{E})$ which fixes $k(\mathcal{E})$.*

Exercise 12.12. Here is Conjecture 2.6 of [Roy 1995]:

(?) *For any 4×4 skew-symmetric matrix* M *with entries in $\mathcal{L}$ and rank ≤ 2, either the rows of* M *are linearly dependent over $\mathbb{Q}$, or the columns space of* M *contains a nonzero element of $\mathbb{Q}^4$.*

Check that this statement is a consequence of Conjecture 1.15 on algebraic independence of logarithms of algebraic numbers, and that it contains the four exponentials Conjecture 1.13.

Hint. *A* 4×4 *skew-symmetric matrix*

$$\begin{pmatrix} 0 & x_{12} & x_{13} & x_{14} \\ -x_{12} & 0 & x_{23} & x_{24} \\ -x_{13} & -x_{23} & 0 & x_{34} \\ -x_{14} & -x_{24} & -x_{34} & 0 \end{pmatrix}$$

has rank ≤ 2 *if and only if*

$$x_{12}x_{34} - x_{13}x_{24} + x_{14}x_{23} = 0.$$

Exercise 12.13. Let $\mathcal{V}$ be a real vector space of positive dimension n and Y a finitely generated subgroup of $\mathcal{V}$ of rank ℓ. Assume no subgroup of Y of rank $n+1$ is dense in $\mathcal{V}$. Show that there exists a subgroup of Y of rank $\geq \ell - n + 1$ which is not dense in $\mathcal{V}$.

Hint. *Use Proposition 12.5 and see* [Roy 1992c], *Lemma 3.3.*

Exercise 12.14 (É. Reyssat). Show that the assumption that k has infinitely many elements cannot be removed from Proposition 12.5.

Exercise 12.15. Let K be a field of zero characteristic and $m \geq 1$ a positive integer. Denote by $\mathcal{Z}$ the set of zeros in K^{2m} of the quadratic polynomial $X_1Y_1 + \cdots + X_mY_m$. Recall the map $\theta_m \colon K^{2m} \to \mathrm{Mat}_{N\times N}(K)$ of § 12.5, with $N = 2^m$, which satisfies

$$\theta_m(\underline{x}, \underline{y})^2 = (x_1y_1 + \cdots + x_my_m)\mathrm{I}_N.$$

Let $\underline{w} = (\underline{x}, \underline{y}) \in \mathcal{Z}$. Denote by X the $\mathbb{Q}$-vector subspace of K^N spanned by the column vectors of $\theta_m(\underline{w})$. Assume that there exists a vector subspace U of K^N, defined over $\mathbb{Q}$, such that

$$\dim_{\mathbb{Q}}\left(\frac{X}{X\cap U}\right) < \dim_K\left(\frac{K^N}{U}\right).$$

Show that there exists a vector subspace of K^{2m}, defined over $\mathbb{Q}$, which contains $\underline{w}$ and is contained in $\mathcal{Z}$.

Hint. *Define*

$$d' = \dim_K\left(\frac{K^N}{U}\right) \quad \textit{and } \ell' = \dim_{\mathbb{Q}}\left(\frac{X}{X\cap U}\right).$$

Show that there exist two matrices P *and* Q *in* $\mathrm{GL}_N(\mathbb{Q})$ *such that*

$$\mathsf{P}\theta_m(\underline{w})\mathsf{Q} = \begin{pmatrix} \mathsf{A} & \mathsf{B} \\ \mathsf{C} & 0 \end{pmatrix} \begin{matrix} \}d^* \\ \}d' \end{matrix}$$
$$\qquad \underbrace{\quad}_{\ell'} \ \underbrace{\quad}_{\ell^*}$$

Define $\mathcal{E}$ *as the set of* $\underline{v} \in K^{2m}$ *such that there exist matrices* $\mathsf{A}(\underline{v})$, $\mathsf{B}(\underline{v})$ *and* $\mathsf{C}(\underline{v})$ *for which*

$$\mathsf{P}\theta_m(\underline{v})\mathsf{Q} = \begin{pmatrix} \mathsf{A}(\underline{v}) & \mathsf{B}(\underline{v}) \\ \mathsf{C}(\underline{v}) & 0 \end{pmatrix} \begin{matrix} \}d^* \\ \}d' \end{matrix}$$
$$\qquad \underbrace{\quad}_{\ell'} \ \underbrace{\quad}_{\ell^*}$$

Check that $\mathcal{E}$ *is a vector subspace of* K^{2m} *which satisfies the desired properties. (See also* [RoyW 1997b], *Lemma 10.3).*

13. A Quantitative Version of the Linear Subgroup Theorem

The main result of this chapter (Theorem 13.1) is a quantitative version of Theorem 11.5.

Let $G = \mathbb{G}_{\mathrm{a}}^{d_0} \times \mathbb{G}_{\mathrm{m}}^{d_1}$ be a commutative linear algebraic group over $\overline{\mathbb{Q}}$ of dimension $d = d_0 + d_1 > 0$. Denote by

$$\begin{array}{rccc} \exp_G\colon & \mathbb{C}^d & \longrightarrow & G(\mathbb{C}) = \mathbb{C}^{d_0} \times (\mathbb{C}^\times)^{d_1} \\ & (z_1, \ldots, z_d) & \longmapsto & \left(z_1, \ldots, z_{d_0}, e^{z_{d_0+1}}, \ldots, e^{z_d}\right) \end{array}$$

its exponential map. There are two kinds of interesting points in $\mathbb{C}^d$ related to the field $\overline{\mathbb{Q}}$ of algebraic numbers: the points $\underline{w} = (\beta_1, \ldots, \beta_d)$ in $\overline{\mathbb{Q}}^d$ whose coordinates are algebraic numbers, and the points

$$\underline{\eta} = (\beta_1, \ldots, \beta_{d_0}, \lambda_1, \ldots, \lambda_{d_1}) \in \overline{\mathbb{Q}}^{d_0} \times \mathcal{L}^{d_1}$$

whose images under $\exp_G$ lie in $G(\overline{\mathbb{Q}}) = \overline{\mathbb{Q}}^{d_0} \times (\overline{\mathbb{Q}}^\times)^{d_1}$. According to Hermite-Lindemann's Theorem 1.2, the intersection of the two sets is $\overline{\mathbb{Q}}^{d_0} \times \{0\}$.

The Linear Subgroup Theorem 11.5 provides information on the dimension of the vector subspace of $\mathbb{C}^d$ spanned by points

$$\underline{w}_1, \ldots, \underline{w}_{\ell_0}, \underline{\eta}_1, \ldots, \underline{\eta}_{\ell_1},$$

when $\underline{w}_1, \ldots, \underline{w}_{\ell_0}$ belong to $\overline{\mathbb{Q}}^d$ and $\underline{\eta}_1, \ldots, \underline{\eta}_{\ell_1}$ to $\overline{\mathbb{Q}}^{d_0} \times \mathcal{L}^{d_1}$.

Our quantitative version is as follows: let r be the lower bound for the dimension which is provided by the Linear Subgroup Theorem and let

$$\underline{w}'_1, \ldots, \underline{w}'_{\ell_0}, \underline{\eta}'_1, \ldots, \underline{\eta}'_{\ell_1}$$

be points in a subspace of $\mathbb{C}^d$ of dimension $< r$ over $\mathbb{C}$. Then we refine the conclusion

$$(\underline{w}_1, \ldots, \underline{w}_{\ell_0}, \underline{\eta}_1, \ldots, \underline{\eta}_{\ell_1}) \neq (\underline{w}'_1, \ldots, \underline{w}'_{\ell_0}, \underline{\eta}'_1, \ldots, \underline{\eta}'_{\ell_1})$$

of the Linear Subgroup Theorem by giving a lower bound for

$$\max\left\{ \max_{1 \le k \le \ell_0} |\underline{w}_k - \underline{w}'_k|, \max_{1 \le j \le \ell_1} |\underline{\eta}_j - \underline{\eta}'_j| \right\}.$$

This lower bound will be completely explicit. We describe briefly this result here – the exact statement (Theorem 13.1) is given in § 13.1.

The data involve algebraic numbers, namely the coordinates of $\underline{w}_k$ $(1 \leq k \leq \ell_0)$ in $\overline{\mathbb{Q}}^d$ and the coordinates of $\exp_G(\underline{\eta}_j)$ $(1 \leq j \leq \ell_1)$ in $\overline{\mathbb{Q}}^{d_0} \times (\overline{\mathbb{Q}}^\times)^{d_1}$. If we write

$$\underline{w}_k = (\beta_{1k}, \ldots, \beta_{dk}) \quad (1 \leq k \leq \ell_0)$$

and

$$\underline{\eta}_j = (\beta_{1,\ell_0+j}, \ldots, \beta_{d_0,\ell_0+j}, \lambda_{1j}, \ldots, \lambda_{d_1 j}) \quad (1 \leq j \leq \ell_1),$$

then all β_{ij} are algebraic as well as all $\alpha_{ij} = e^{\lambda_{ij}}$. We denote by D an upper bound for the degree of a number field generated by these $d\ell$ algebraic numbers, where $\ell = \ell_0 + \ell_1$.

The height of these algebraic numbers will be measured by parameters B_1, B_2 and A_{ij} $(1 \leq i \leq d_1,\ 1 \leq j \leq \ell_1)$. The parameter A_{ij} takes care of the algebraic number α_{ij}, the parameter B_1 is related to the heights of the projections of $\exp_G(\underline{\eta}_j)$ on $\overline{\mathbb{Q}}^{d_0}$, while B_2 is related to the heights of the coordinates of $\underline{w}_1, \ldots, \underline{w}_{\ell_0}$ in $\overline{\mathbb{Q}}^d$.

There is also a parameter E, which often in applications will be chosen as e, but which will enable us to reach sharper estimates when it is chosen larger; this will be possible when the numbers $|\lambda_{ij}|$ are comparatively not too large with respect to $\log A_{ij}$.

There are further parameters r_1, r_2 and r_3 which can be described as ranks of matrices and satisfy $r = r_1 + r_2 + r_3$: while r is the rank of the full $d \times \ell$ matrix M' whose columns are the components of

$$\underline{w}'_1, \ldots, \underline{w}'_{\ell_0}, \underline{\eta}'_1, \ldots, \underline{\eta}'_{\ell_1},$$

the number r_3 is the rank of the $d_1 \times \ell_1$ matrix $\left(\eta'_{d_0+i,j}\right)_{\substack{1\leq i\leq d_1\\ 1\leq j\leq \ell_1}}$ which approximates $\left(\lambda_{ij}\right)_{\substack{1\leq i\leq d_1\\ 1\leq j\leq \ell_1}}$, while $r_1 + r_3$ and $r_2 + r_3$ are related to the ranks of the matrices composed of the last ℓ_1 columns (resp. the last d_1 rows) of M'.

The main tools for the proof (§ 13.4) are: Philippon's multiplicity estimate (Theorem 8.1) on one hand, and analytic estimates (§ 13.2) on the other hand. Of course Liouville's Lemma will also be needed; we apply it to produce a lower bound for the absolute value of a determinant involving exponential polynomials (§ 13.3).

In § 13.5 we investigate which optimal result could be deduced from Th. 13.1. At the same time we show how to choose the parameters $T_0, T_1, \ldots, T_{d_1}$ and $S_0, S_1, \ldots, S_{\ell_1}$ in most cases. We shall see in the next chapter that this optimal value is reached (up to constants) in a number of cases.

Theorem 13.1 (which is a variant of the main result in [W 1997a]) includes a lot of diophantine estimates; some examples will be given in the next chapter. It is not the final word on this topic, even within the present limitations of the theory: it is possible to refine it by introducing Fel'dman's polynomials. In § 13.6 we suggest how such a refinement could be performed.

13.1 The Main Result

Let $d_0 \geq 0$, $d_1 \geq 1$, $\ell_0 \geq 0$, $\ell_1 \geq 1$, $r_1 \geq 0$, $r_2 \geq 0$, $r_3 \geq 1$ be rational integers. Define $d = d_0 + d_1$, $\ell = \ell_0 + \ell_1$ and $r = r_1 + r_2 + r_3$.

Let K be a number field of degree $\leq D$ over $\mathbb{Q}$. We consider the linear algebraic groups

$$G_0 = \mathbb{G}_{\mathrm{a}}^{d_0}, \qquad G_1 = \mathbb{G}_{\mathrm{m}}^{d_1}, \qquad G = G_0 \times G_1.$$

Let G_1^- and G_1^+ be connected algebraic subgroups of G_1, defined over K, with $G_1^- \subset G_1^+$. Define

$$G^- = \{0\} \times G_1^-, \qquad G^+ = G_0 \times G_1^+.$$

Assume that the dimension d^+ of G^+ is positive.

Let $\underline{w}_1, \ldots, \underline{w}_{\ell_0}$ be elements of K^d, with

$$\underline{w}_k = (\beta_{1k}, \ldots, \beta_{dk}) \quad (1 \leq k \leq \ell_0).$$

The complex vector subspace $\mathbb{C}\underline{w}_1 + \cdots + \mathbb{C}\underline{w}_{\ell_0}$ of $\mathbb{C}^d = T_e(G)$ they span will be denoted by $\mathcal{W}$. We assume $\mathcal{W} \subset T_e(G^+)$.

Let $\underline{\eta}_1, \ldots, \underline{\eta}_{\ell_1}$ be elements of $K^{d_0} \times \mathcal{L}^{d_1}$, with

$$\underline{\eta}_j = (\beta_{1,\ell_0+j}, \ldots, \beta_{d_0,\ell_0+j}, \lambda_{1j}, \ldots, \lambda_{d_1 j}) \quad (1 \leq j \leq \ell_1).$$

For $1 \leq i \leq d_1$ and $1 \leq j \leq \ell_1$ define

$$\alpha_{ij} = e^{\lambda_{ij}}$$

and assume $\alpha_{ij} \in K^\times$. Hence for $1 \leq j \leq \ell_1$ the point

$$\underline{\gamma}_j = \exp_G \underline{\eta}_j = (\beta_{1,\ell_0+j}, \ldots, \beta_{d_0,\ell_0+j}, \alpha_{1j}, \ldots, \alpha_{d_1 j})$$

belongs to $G(K) = K^{d_0} \times (K^\times)^{d_1}$; we assume that in fact $\underline{\eta}_j$ belongs to $T_e(G^+)$ so that $\underline{\gamma}_j \in G^+(K)$.

Let $\underline{w}'_1, \ldots, \underline{w}'_{\ell_0}, \underline{\eta}'_1, \ldots, \underline{\eta}'_{\ell_1}$ be elements of $\mathbb{C}^d$. Define

$$\mathcal{W}' = \mathbb{C}\underline{w}'_1 + \cdots + \mathbb{C}\underline{w}'_{\ell_0}, \qquad \mathcal{X}' = \mathbb{C}\underline{\eta}'_1 + \cdots + \mathbb{C}\underline{\eta}'_{\ell_1}.$$

Denote by π the projection $\mathbb{C}^d \longrightarrow \mathbb{C}^d / T_e(G^-)$ and by π_1 the projection $\mathbb{C}^d \longrightarrow \mathbb{C}^{d_1} / T_e(G_1^-)$ with kernel $\mathbb{C}^{d_0} \times T_e(G_1^-)$. Assume

$$r = \dim_{\mathbb{C}}\bigl(\pi(\mathcal{W}' + \mathcal{X}')\bigr), \qquad r_3 = \dim_{\mathbb{C}}\bigl(\pi_1(\mathcal{X}')\bigr)$$

and

$$r_1 + r_3 \geq \dim_{\mathbb{C}}\bigl(\pi(\mathcal{X}')\bigr), \qquad r_2 + r_3 \geq \dim_{\mathbb{C}}\bigl(\pi_1(\mathcal{W}' + \mathcal{X}')\bigr).$$

Let B_1 and B_2 be two real numbers which are > 1 and satisfy

$$\mathrm{h}(1 : \beta_{h,\ell_0+1} : \cdots : \beta_{h\ell}) \leq \log B_1 \quad (1 \leq h \leq d_0)$$

and

$$\mathrm{h}(1{:}\,\beta_{d_0+1,k}{:}\cdots{:}\,\beta_{dk}) \le \log B_2 \quad (1 \le k \le \ell_0).$$

Moreover, assume

either

$$\mathrm{h}(1{:}\,\beta_{h1}{:}\cdots{:}\,\beta_{h\ell}) \le \log B_1 \quad (1 \le h \le d_0)$$

or

$$\mathrm{h}(1{:}\,\beta_{1k}{:}\cdots{:}\,\beta_{dk}) \le \log B_2 \quad (1 \le k \le \ell_0).$$

Let $A_{ij} > 1$ be real numbers satisfying

$$\log A_{ij} \ge \max\left\{\mathrm{h}(\alpha_{ij}),\ \frac{1}{D}\right\} \qquad (1 \le i \le d_1,\ 1 \le j \le \ell_1).$$

Let U, V, E be positive real numbers with $E \ge e$ and $T_0, T_1, \ldots, T_{d_1}, S_0, S_1, \ldots, S_{\ell_1}$ nonnegative integers. Assume

$$U \ge \max\Bigl\{20rD\log(2dD),\ r_3\log E,\ DT_0\log B_1,\ DS_0\log B_2,\ D\sum_{i=1}^{d_1}\sum_{j=1}^{\ell_1} T_iS_j\log A_{ij}\Bigr\}$$

and

$$V = (12d^+ + 9)U.$$

Define $T^* = T_1 + \cdots + T_{d_1}$, $S^* = S_1 + \cdots + S_{\ell_1}$ and assume

$$B_1^D \ge E, \qquad B_1 \ge d^+S^*, \qquad B_2^D \ge E, \qquad B_2 \ge T^*;$$

assume moreover $T^* \ge 1$, $S^* \ge 1$ and

$$\text{either} \quad B_1 \ge d^+S^* + \frac{d^+S_0}{T^*} \quad \text{or} \quad B_2 \ge T^* + \frac{T_0}{d^+S^*}.$$

Let $\mathcal{T}_1$ be a subset of $\mathbb{Z}^{d_1}$ consisting of tuples $\underline{t}$ for which $|t_i| \le T_i$ for $1 \le i \le d_1$ and

$$\underline{y}^{\underline{t}} = 1$$

for any $\underline{y} \in G_1^-$.

Define $\mathcal{T} = \mathcal{T}_0 \times \mathcal{T}_1 \subset \mathbb{N}^{d_0} \times \mathbb{Z}^{d_1}$, where $\mathcal{T}_0$ is the subset of $\mathbb{N}^{d_0}$ consisting of all tuples $\underline{\tau}$ for which $\|\underline{\tau}\| \le T_0$. Denote by $H(G^+;\mathcal{T})$ the dimension of the $\mathbb{C}$-vector space of polynomial maps $G^+(\mathbb{C}) \to \mathbb{C}$ which are given by polynomials in $\mathbb{C}[G^+]$ of the form

$$\sum_{\|\underline{\tau}\|\le T_0}\sum_{\underline{t}\in\mathcal{T}_1} c_{\underline{\tau}\underline{t}}\underline{X}^{\underline{\tau}}\underline{Y}^{\underline{t}}.$$

For instance if $G^+ = G$ then

$$H(G;\mathcal{T}) = \binom{T_0+d_0}{d_0}\mathrm{Card}\mathcal{T}_1,$$

while for $G_1^- = \{1\}$ and

$$\mathcal{T}_1 = \left\{\underline{t} \in \mathbb{Z}^{d_1}\,;\ |t_i| \le T_i\ (1 \le i \le d_1)\right\},$$

we have

$$H(G^+;\mathcal{T}) = H(G^+;T_0,\underline{T}) = \binom{T_0+d_0}{d_0} H(G_1^+;\underline{T})$$

with the notation of Chap. 8 and $\underline{T} = (T_1,\ldots,T_{d_1})$.

Further, let $\mathcal{S}_1$ be a subset of $\mathbb{Z}^{\ell_1}$ consisting of tuples $\underline{s}$ for which $|s_j| \le S_j$ for $1 \le j \le \ell_1$. Assume

$$\left|\sum_{i=1}^{d_1}\sum_{j=1}^{\ell_1} t_i s_j \lambda_{ij}\right| \le \frac{U}{E}$$

for any $\underline{t} \in \mathcal{T}_1$ and any $\underline{s} \in \mathcal{S}_1$.

We assume

$$H(G^+;\mathcal{T}) \ge 2\binom{T_0+r_1}{r_1}\binom{d^+S_0+r_2}{r_2}\left(\frac{V}{\log E}\right)^{r_3}.$$

Finally we denote by Σ the following subset of $G^+(K)$:

$$\Sigma = \left\{s_1\underline{\gamma}_1 + \cdots + s_{\ell_1}\underline{\gamma}_{\ell_1}\,;\ \underline{s} \in \mathcal{S}_1\right\}.$$

Theorem 13.1. *Assume*

$$\max_{1\le k\le \ell_0} |\underline{w}_k - \underline{w}_k'| \le e^{-V} \quad \textit{and} \quad \max_{1\le j\le \ell_1} |\underline{\eta}_j - \underline{\eta}_j'| \le e^{-V}.$$

Then there exists a connected algebraic subgroup G^* of G^+ of dimension $< d^+$, which contains G^-, which is incompletely defined in G^+ by polynomials of multidegree $\le (T_0,\underline{T})$, such that, if we set

$$\ell_0^b = \dim_{\mathbb{C}}\left(\frac{W + T_e(G^*)}{T_e(G^*)}\right),$$

then

$$\binom{S_0+\ell_0^b}{\ell_0^b}\mathrm{Card}\left(\frac{\Sigma + G^*}{G^*}\right)\mathcal{H}(G^*;T_0,\underline{T}) \le \mathcal{H}(G^+;T_0,\underline{T}).$$

Remark. The vector columns in $\mathbb{C}^d$ of the matrix

$$\mathsf{M} = \begin{pmatrix} \mathsf{B}_0 & \mathsf{B}_1 \\ \mathsf{B}_2 & \mathsf{L} \end{pmatrix} = \begin{pmatrix} \beta_{11} & \cdots & \beta_{1\ell_0} & \beta_{1,\ell_0+1} & \cdots & \beta_{1\ell} \\ \vdots & \ddots & \vdots & \vdots & \ddots & \vdots \\ \beta_{d_01} & \cdots & \beta_{d_0\ell_0} & \beta_{d_0,\ell_0+1} & \cdots & \beta_{d_0\ell} \\ \beta_{d_0+1,1} & \cdots & \beta_{d_0+1,\ell_0} & \lambda_{11} & \cdots & \lambda_{1\ell_1} \\ \vdots & \ddots & \vdots & \vdots & \ddots & \vdots \\ \beta_{d1} & \cdots & \beta_{d\ell_0} & \lambda_{d_11} & \cdots & \lambda_{d_1\ell_1} \end{pmatrix}$$

are $\underline{w}_1, \ldots, \underline{w}_{\ell_0}, \underline{\eta}_1, \ldots, \underline{\eta}_{\ell_1}$. Notice that the conditions on the parameters B_1 and B_2 are the following:

- either B_1 is an upper bound for the height of projective points corresponding to the rows of the matrix $(\mathsf{B}_0\ \mathsf{B}_1)$ and B_2 is an upper bound corresponding to the columns of the matrix (B_2)
- or B_1 is an upper bound for the height of projective points corresponding to the rows of the matrix (B_1) and B_2 is an upper bound corresponding to the columns of the matrix $\begin{pmatrix}\mathsf{B}_0\\ \mathsf{B}_2\end{pmatrix}$.

13.2 Analytic Estimates

In this section we prove some analytic estimates which hold for analytic functions; since not all data from § 13.1 are involved, we shall repeat the piece of notation we need.

13.2.1 Analytic Upper Bound for a Determinant

We give here an extension of Propositions 9.13 and 10.5 (compare with Proposition 5.1 of [W 1997a]).

Let $d_0 \geq 0$, $d_1 > 0$, $\ell_0 \geq 0$ be rational integers. Put $d = d_0 + d_1$. Let $\mathcal{W}$ and $\mathcal{X}$ be two subspaces of $\mathbb{C}^d$, $\mathcal{U}_0$ a subspace of $\mathbb{C}^{d_0}$ and $\mathcal{U}_1$ a subspace of $\mathbb{C}^{d_1}$. Define $\mathcal{U} = \mathcal{U}_0 \times \mathcal{U}_1$. Denote by π the linear projection $\mathbb{C}^d \longrightarrow \mathbb{C}^d/\mathcal{U}$ and by π_1 the projection $\mathbb{C}^d \longrightarrow \mathbb{C}^{d_1}/\mathcal{U}_1$ with kernel $\mathbb{C}^{d_0} \times \mathcal{U}_1$. Define

$$r = \dim_{\mathbb{C}}\bigl(\pi(\mathcal{W} + \mathcal{X})\bigr), \quad r_3 = \dim_{\mathbb{C}}\bigl(\pi_1(\mathcal{X})\bigr).$$

We assume $r_3 \geq 1$. Let $r_1 \geq 0$ and $r_2 \geq 0$ be rational integers satisfying $r = r_1 + r_2 + r_3$ and

$$\dim_{\mathbb{C}}\bigl(\pi(\mathcal{X})\bigr) - \dim_{\mathbb{C}}\bigl(\pi_1(\mathcal{X})\bigr) \leq r_1 \leq \dim_{\mathbb{C}}\bigl(\pi(\mathcal{W} + \mathcal{X})\bigr) - \dim_{\mathbb{C}}\bigl(\pi_1(\mathcal{W} + \mathcal{X})\bigr).$$

Notice that $\dim_{\mathbb{C}}\bigl(\pi(\mathcal{X})\bigr) - \dim_{\mathbb{C}}\bigl(\pi_1(\mathcal{X})\bigr)$ is the dimension of

$$\pi(\mathcal{X}) \cap \left(\frac{\mathbb{C}^{d_0}}{\mathcal{U}_0} \times \{0\}\right).$$

Let $L > 0$, $T_0 \geq 0$ and $S_0 \geq 0$ be rational integers, $\varphi_1, \ldots, \varphi_L$ entire functions in $\mathbb{C}^d$, $\underline{\zeta}_1, \ldots, \underline{\zeta}_L$ elements of $\mathcal{X}$, $\underline{w}_1, \ldots, \underline{w}_{\ell_0}$ be elements of $\mathcal{W}$ and $\underline{\sigma}_1, \ldots, \underline{\sigma}_L$ elements of $\mathbb{N}^{\ell_0}$ satisfying $\|\underline{\sigma}_\mu\| \leq S_0$ $(1 \leq \mu \leq L)$. Assume

$$\varphi_\lambda(\underline{z} + \underline{u}) = \varphi_\lambda(\underline{z})$$

for $1 \le \lambda \le L$, $\underline{u} \in \mathcal{U}$ and $\underline{z} \in \mathbb{C}^d$. We define $\boldsymbol{w} = (\underline{w}_1, \dots, \underline{w}_{\ell_0}) \in \mathcal{W}^{\ell_0}$ and we take derivatives:

$$\mathcal{D}_{\boldsymbol{w}}^{\underline{\sigma}} = \mathcal{D}_{\underline{w}_1}^{\sigma_1} \cdots \mathcal{D}_{\underline{w}_{\ell_0}}^{\sigma_{\ell_0}}.$$

For $1 \le \lambda \le L$ and $1 \le \mu \le L$, let $\delta_{\lambda\mu}$ be a complex number. For $1 \le \lambda \le L$ and $\underline{\tau} \in \mathbb{N}^{d_0}$ with $\|\underline{\tau}\| \le T_0$, let $\psi_{\lambda\underline{\tau}}$ be an entire function in $\mathbb{C}^{d_1}$. We assume, for $1 \le \lambda \le L$,

$$\varphi_\lambda(\underline{z}) = \sum_{\|\underline{\tau}\| \le T_0} z_1^{\tau_1} \cdots z_{d_0}^{\tau_{d_0}} \psi_{\lambda\underline{\tau}}(z_{d_0+1}, \dots, z_d).$$

Let $V, E, M_1, \dots, M_L$ be positive numbers and ϵ a complex number satisfying

$$E \ge 1, \quad |\epsilon| \le e^{-V},$$

$$|\delta_{\lambda\mu}| \le e^{M_\lambda}, \quad \sup_{\substack{z \in \mathbb{C} \\ |z| = E}} \left| \left(\mathcal{D}_{\boldsymbol{w}}^{\underline{\sigma}_\mu} \varphi_\lambda\right)(z \underline{\zeta}_\mu) \right| \le e^{M_\lambda} \qquad (1 \le \lambda, \mu \le L).$$

Define

$$\widetilde{V} = V + r_3 T_0 \log E + (r_3+1) S_0 \log E + r_3(r_3+1) \log E$$

and assume

$$L \ge \frac{2}{(r_3+1)!} \binom{T_0 + r_1}{r_1} \binom{S_0 + r_2}{r_2} \left(\frac{\widetilde{V}}{\log E}\right)^{r_3} \cdot \frac{\widetilde{V}}{V}.$$

Proposition 13.2. *The determinant Δ of the $L \times L$ matrix*

$$\left(\mathcal{D}_{\boldsymbol{w}}^{\underline{\sigma}_\mu} \varphi_\lambda(\underline{\zeta}_\mu) + \epsilon \delta_{\lambda\mu}\right)_{1 \le \lambda, \mu \le L}$$

has absolute value bounded by

$$\log|\Delta| \le -\frac{1}{2} L V + L \log(2L) + M_1 + \cdots + M_L.$$

This Proposition 13.2 will be the main (analytic) tool providing an upper bound for the absolute value of an arithmetic determinant Δ_{ar} occurring in § 13.3.2. A suitable value for M_λ will be computed in § 13.4.2.

13.2.2 A Combinatorial Lemma

The lower bound we shall produce, for the order of vanishing at the origin of the interpolation determinant, will depend on the following combinatorial lemma (compare with Lemmas 6.5 and 7.3):

Lemma 13.3. *Let s be a nonnegative integer, $L, K_1, \dots, K_s$ be positive integers and $I_0, I_1, \dots, I_s$ be a partition of the set $\{1, \dots, d\}$, with* $\mathrm{Card} I_\sigma = i_\sigma$ $(0 \le \sigma \le s)$, *where $i_0 > 0$. Define Θ_L as the minimum of $\|\underline{\kappa}_1\| + \cdots + \|\underline{\kappa}_L\|$ for $\underline{\kappa}_1, \dots, \underline{\kappa}_L$ pairwise distinct elements of $\mathbb{N}^d$ satisfying*

$$\max_{1\le\lambda\le L}\sum_{i\in I_\sigma}\kappa_{\lambda i}\le K_\sigma \qquad (1\le\sigma\le s).$$

Put $K = K_1 + \cdots + K_s$ and define M by

$$M^{i_0} = \left(\frac{i_0}{i_0+1}\right)^{i_0}\cdot\frac{i_0!L}{\prod_{\sigma=1}^{s}\binom{K_\sigma+i_\sigma}{i_\sigma}}.$$

Then

$$\Theta_L \ge LM - i_0L(K+i_0+1).$$

Proof. For any nonnegative integer a, define

$$N_a = \mathrm{Card}\left\{\underline{\kappa}\in\mathbb{N}^d;\ \|\underline{\kappa}\| = a \ \text{ and } \sum_{i\in I_\sigma}\kappa_i\le K_\sigma\ (1\le\sigma\le s)\right\}.$$

From this definition and the definition of Θ_L it follows that if A is a positive integer such that

$$\sum_{a=0}^{A}N_a\le L,\qquad\text{then}\qquad \Theta_L\ge\sum_{a=0}^{A}aN_a.$$

We claim that for any $a\ge 0$,

$$N_a\le\binom{a+i_0-1}{i_0-1}\cdot\prod_{\sigma=1}^{s}\binom{K_\sigma+i_\sigma}{i_\sigma}.$$

Indeed, once the coordinates κ_i of $\underline{\kappa}$ for $i\in I_\sigma$ $(1\le\sigma\le s)$ are chosen, the number of $(\kappa_i)_{i\in I_0}$ which will give a $\underline{\kappa}\in\mathbb{N}^d$ with $\|\underline{\kappa}\| = a$ is

$$\binom{a-\sum_{i\notin I_0}\kappa_i+i_0-1}{i_0-1}.$$

Therefore

$$N_a = \sum_{c_0=0}^{a}\sum_{c_1=0}^{K_1}\cdots\sum_{c_s=0}^{K_s}\prod_{\sigma=0}^{s}\binom{c_\sigma+i_\sigma-1}{i_\sigma-1},$$

where $(c_0, c_1, \dots, c_s)$ is restricted to the condition $c_0+\cdots+c_s = a$. For $c_0\le a$ we have

$$\binom{c_0+i_0-1}{i_0-1}\le\binom{a+i_0-1}{i_0-1},$$

while, for $1\le\sigma\le s$, we have

$$\sum_{c_\sigma=0}^{K_\sigma}\binom{c_\sigma+i_\sigma-1}{i_\sigma-1} = \binom{K_\sigma+i_\sigma}{i_\sigma}.$$

Our claim readily follows.

We use the same argument to show that for $a\ge K$, we have

$$N_a \geq \binom{a-K+i_0-1}{i_0-1} \cdot \prod_{\sigma=1}^{s} \binom{K_\sigma+i_\sigma}{i_\sigma}.$$

Indeed, for $a \geq K$, we have

$$\binom{c_0+i_0-1}{i_0-1} \geq \binom{a-K+i_0-1}{i_0-1},$$

because $c_0+\cdots+c_s = a$ and $c_1+\cdots+c_s \leq K$. The claimed lower bound for N_a plainly follows.

We define A by the condition

$$\sum_{a=0}^{A} N_a \leq L < \sum_{a=0}^{A+1} N_a.$$

Since

$$\sum_{a=0}^{A+1} \binom{a+i_0-1}{i_0-1} = \binom{A+i_0+1}{i_0},$$

we deduce from the upper bound for N_a

$$L < \binom{A+i_0+1}{i_0} \cdot \prod_{\sigma=1}^{s} \binom{K_\sigma+i_\sigma}{i_\sigma} \leq \frac{(A+i_0+1)^{i_0}}{i_0!} \cdot \prod_{\sigma=1}^{s} \binom{K_\sigma+i_\sigma}{i_\sigma};$$

hence

$$(A+i_0+1)^{i_0} \prod_{\sigma=1}^{s} \binom{K_\sigma+i_\sigma}{i_\sigma} > i_0! L.$$

Using the definition of M, we can write

$$A+i_0+1 > \left(1+\frac{1}{i_0}\right) M.$$

On the other hand, from the lower bound for N_a, we deduce immediately

$$\Theta_L \geq \sum_{a=K}^{A} a \binom{a-K+i_0-1}{i_0-1} \cdot \prod_{\sigma=1}^{s} \binom{K_\sigma+i_\sigma}{i_\sigma}.$$

We now check the formula

$$\sum_{a=K}^{A} a \binom{a-K+i_0-1}{i_0-1} = \frac{i_0 A + K}{i_0+1} \binom{A-K+i_0}{i_0}.$$

Indeed, we have

$$\sum_{a=K}^{A} a \binom{a-K+i_0-1}{i_0-1} = \sum_{\alpha=0}^{A-K} (\alpha+K) \binom{\alpha+i_0-1}{i_0-1}.$$

Since

$$\alpha\binom{\alpha+i_0-1}{i_0-1}=i_0\binom{\alpha+i_0-1}{i_0},$$

we have

$$\sum_{\alpha=0}^{A-K}\alpha\binom{\alpha+i_0-1}{i_0-1}=i_0\binom{A-K+i_0}{i_0+1}=i_0\cdot\frac{A-K}{i_0+1}\binom{A-K+i_0}{i_0};$$

on the other hand

$$\sum_{\alpha=0}^{A-K}K\binom{\alpha+i_0-1}{i_0-1}=K\binom{A-K+i_0}{i_0}.$$

The desired formula follows from

$$\frac{i_0}{i_0+1}(A-K)+K=\frac{i_0A+K}{i_0+1}.$$

We use the lower bound

$$\binom{A-K+i_0}{i_0}\geq\frac{(A-K)^{i_0}}{i_0!}$$

and deduce

$$\begin{aligned}\Theta_L&\geq\frac{1}{(i_0+1)!}(i_0A+K)(A-K)^{i_0}\cdot\prod_{\sigma=1}^{s}\binom{K_\sigma+i_\sigma}{i_\sigma}\\&\geq\frac{i_0A+K}{i_0+1}\cdot\left(\frac{i_0(A-K)}{(i_0+1)M}\right)^{i_0}\cdot L;\end{aligned}$$

we now use the weak estimate $i_0A+K\geq i_0(A-K)$ and obtain, thanks to the lower bound $i_0(A+i_0+1)>(i_0+1)M$,

$$\Theta_L\geq LM\left(1-\frac{i_0(K+i_0+1)}{(i_0+1)M}\right)^{i_0+1}.$$

Finally we notice that Lemma 13.3 holds trivially if $M\leq i_0(K+i_0+1)$ and we use, with $x=i_0(K+i_0+1)/(i_0+1)M$ and $t=i_0+1$, the estimate

$$(1-x)^t\geq 1-tx$$

which holds for $t\geq 1$ and $0\leq x\leq 1$. □

13.2.3 Lower Bound for the Order of Vanishing of the Interpolation Determinant

Let L' be an integer in the range $0 \leq L' \leq L$ and let I be a subset of $\{1, \dots, L\}$ with L' elements. Define an entire function Δ_I on $\mathbb{C}$ by

$$\Delta_I(z) = \det\Big(d_{\lambda\mu}(z)\Big)_{1\leq\lambda,\mu\leq L}$$

with

$$d_{\lambda\mu}(z) = \begin{cases} (\mathcal{D}_{\underline{w}}^{\underline{\sigma}_\mu}\varphi_\lambda)(z\underline{\zeta}_\mu) & \text{for } \lambda \in I, \\ \delta_{\lambda\mu} & \text{for } \lambda \notin I. \end{cases}$$

We now give a lower bound for the order of vanishing of $\Delta_I(z)$ at the origin which generalizes Lemmas 6.4, 9.2 and 10.6.

Lemma 13.4. *The function Δ_I has a zero at the origin of multiplicity at least*

$$\Theta_{L'} - L'S_0,$$

where $\Theta_{L'}$ satisfies

$$\frac{1}{L'}\Theta_{L'} \geq \frac{r_3}{r_3+1}\left(\frac{r_3!L'}{\binom{T_0+r_1}{r_1}\binom{S_0+r_2}{r_2}}\right)^{1/r_3} - r_3(T_0+S_0+r_3+1).$$

Proof. Assume first $\mathcal{U}_0 = \{0\}$ and $\mathcal{U}_1 = \{0\}$. In this case π is the identity and π_1 is the canonical linear projection $\mathbb{C}^d \longrightarrow \mathbb{C}^{d_1}$ with kernel $\mathbb{C}^{d_0} \times \{0\}$.

Let $\underline{e}_1, \dots, \underline{e}_{r_3}$ be r_3 elements in $\mathcal{X}$ such that $\big(\pi_1(\underline{e}_1), \dots, \pi_1(\underline{e}_{r_3})\big)$ is a basis of $\pi_1(\mathcal{X})$. From the assumption

$$\dim_{\mathbb{C}}\big(\mathcal{X} \cap \ker \pi_1\big) \leq r_1 \leq \dim_{\mathbb{C}}\big((\mathcal{W}+\mathcal{X}) \cap \ker \pi_1\big),$$

we deduce that there exists elements $\underline{e}_{r_3+1}, \dots, \underline{e}_{r_1+r_3}$ in $(\mathcal{W}+\mathcal{X}) \cap \ker \pi_1$ such that the vector space $\mathcal{V}$ spanned by $\underline{e}_1, \dots, \underline{e}_{r_1+r_3}$ contains $\mathcal{X}$. We complete first into a basis $(\underline{e}_1, \dots, \underline{e}_r)$ of $\mathcal{W}+\mathcal{X}$, and then into a basis $(\underline{e}_1, \dots, \underline{e}_d)$ of $\mathbb{C}^d$. We denote by $Z_1, \dots, Z_d$ the homogeneous linear forms in $(z_1, \dots, z_d)$ satisfying

$$\underline{z} = Z_1\underline{e}_1 + \cdots + Z_d\underline{e}_d.$$

For $\underline{\kappa} = (\kappa_1, \dots, \kappa_d) \in \mathbb{N}^d$, $\underline{Z}^{\underline{\kappa}}$ denotes the polynomial $Z_1^{\kappa_1}\cdots Z_d^{\kappa_d}$ which is homogeneous of degree $\|\underline{\kappa}\|$.

For $1 \leq \lambda \leq L$ consider the Taylor expansion at the origin of the function $\underline{Z} \mapsto \varphi_\lambda(Z_1\underline{e}_1 + \cdots + Z_d\underline{e}_d)$:

$$\varphi_\lambda(Z_1\underline{e}_1 + \cdots + Z_d\underline{e}_d) = \sum_{\underline{\kappa}_\lambda \in \mathbb{N}^d} c_{\lambda\underline{\kappa}_\lambda} \underline{Z}^{\underline{\kappa}_\lambda}.$$

The determinant $\Delta_I = \Delta_{I,\varphi_1,\ldots,\varphi_L}$ associated with the functions $\varphi_1, \ldots, \varphi_L$ can be written

$$\Delta_{I,\varphi_1,\ldots,\varphi_L} = \sum_{\underline{\kappa}_1 \in \mathbb{N}^d} \cdots \sum_{\underline{\kappa}_L \in \mathbb{N}^d} c_{1\underline{\kappa}_1} \cdots c_{L\underline{\kappa}_L} \Delta_{I,\underline{Z}^{\underline{\kappa}_1},\ldots,\underline{Z}^{\underline{\kappa}_L}}$$

where $\Delta_{I,\underline{Z}^{\underline{\kappa}_1},\ldots,\underline{Z}^{\underline{\kappa}_L}}$ denotes the corresponding determinant associated with the functions $\varphi_\lambda(\underline{z}) = \underline{Z}^{\underline{\kappa}_\lambda}$ $(1 \le \lambda \le L)$. Therefore we may assume without loss of generality $\varphi_\lambda(\underline{z}) = \underline{Z}^{\underline{\kappa}_\lambda}$, for some $\underline{\kappa}_\lambda \in \mathbb{N}^d$.

Since, for $\underline{\kappa} \in \mathbb{N}^d$ and $\underline{\sigma} \in \mathbb{N}^{\ell_0}$, $\mathcal{D}_{\underline{w}}^{\underline{\sigma}}(\underline{Z}^{\underline{\kappa}})$ is either 0 or a homogeneous polynomial in $z_1, \ldots, z_d$ of degree $\|\underline{\kappa}\| - \|\underline{\sigma}\|$, if we multiply each entry in a row of index $\lambda \in I$ by z^{S_0}, we get a common factor $z^{\|\underline{\kappa}_\lambda\|}$ in all elements of this row. Hence, if Δ_I does not vanish identically, then it has a zero of multiplicity $\ge \max\{0, \|\underline{\kappa}_1\| + \cdots + \|\underline{\kappa}_{L'}\| - L'S_0\}$. It remains to produce a lower bound for $\|\underline{\kappa}_1\| + \cdots + \|\underline{\kappa}_{L'}\|$ when Δ_I does not vanish identically.

Consider one row with index $\lambda \in I$ of the matrix whose determinant is $\Delta_I(z)$:

$$(d_{\lambda 1} \ \cdots \ d_{\lambda L})$$

with

$$d_{\lambda\mu} = d_{\lambda\mu}(z) = \mathcal{D}_{\underline{w}}^{\underline{\sigma}_\mu}\big(\underline{Z}^{\underline{\kappa}_\lambda}\big)(z\underline{\zeta}_\mu) \qquad (1 \le \mu \le L).$$

By construction $\underline{e}_{r_3+1}, \ldots, \underline{e}_{r_1+r_3}$ belong to $\ker \pi_1$; it follows that the linear forms $Z_{r_3+1}, \ldots, Z_{r_1+r_3}$ do not depend on the d_1 variables $z_{d_0+1}, \ldots, z_d$. Hence, as a function of $z_1, \ldots, z_{d_0}$, $\underline{Z}^{\underline{\kappa}_\lambda}$ is a homogeneous polynomial of degree

$$\ge \sum_{i=r_3+1}^{r_1+r_3} \kappa_{\lambda i}.$$

Since $\|\underline{\tau}_\mu\| \le T_0$, if our row does not consist only of zeroes, then we have

$$\sum_{i=r_3+1}^{r_1+r_3} \kappa_{\lambda i} \le T_0.$$

Since $\underline{\zeta}_\mu \in \mathcal{X} \subset \mathcal{V}$, the linear forms $Z_{r_1+r_3+1}, \ldots, Z_r$ vanish at $\underline{\zeta}_\mu$. Hence for

$$\sum_{i=r_1+r_3+1}^{r} \kappa_{\lambda i} > \|\underline{\sigma}\|,$$

we have

$$\mathcal{D}_{\underline{w}}^{\underline{\sigma}}\big(\underline{Z}^{\underline{\kappa}_\lambda}\big)(z\underline{\zeta}_\mu) = 0.$$

Therefore, if

$$\sum_{i=r_1+r_3+1}^{r} \kappa_{\lambda i} > S_0,$$

then $d_{\lambda 1} = \cdots = d_{\lambda L} = 0$.

Since $\underline{w}_k \in \mathcal{W} \subset \mathcal{W} + \mathcal{X}$ and $\underline{\zeta}_\mu \in \mathcal{X} \subset \mathcal{W} + \mathcal{X}$, if one component $\kappa_{\lambda i}$ of $\underline{\kappa}_\lambda$ with $r < i \leq d$ is not zero, then again $d_{\lambda\mu} = 0$ for $1 \leq \mu \leq L$.

Define $I_0 = \{1, \ldots, r_3\}$, $I_1 = \{r_3 + 1, \ldots, r_3 + r_1\}$, $I_2 = \{r_3 + r_1 + 1, \ldots, r\}$, $I_3 = \{r + 1, \ldots, d\}$. Let $\mathcal{K}$ denote the set of $\underline{\kappa} \in \mathbb{N}^d$ for which

$$\sum_{i \in I_1} \kappa_i \leq T_0, \quad \sum_{i \in I_2} \kappa_i \leq S_0 \quad \text{and} \quad \kappa_i = 0 \quad \text{for } i \in I_3.$$

From the above remarks, it follows that Δ_I vanishes, unless $\underline{\kappa}_1, \ldots, \underline{\kappa}_{L'}$ are distinct elements in $\mathcal{K}$. We use Lemma 13.3 with $s = 3$, $K_1 = T_0$, $K_2 = S_0$, $K_3 = 0$, $i_0 = r_3$, $i_1 = r_1$, $i_2 = r_2$, $i_3 = d - r$ and L replaced by L': for $\underline{\kappa}_1, \ldots, \underline{\kappa}_L$ distinct elements in $\mathcal{K}$, we have

$$\|\underline{\kappa}_1\| + \cdots + \|\underline{\kappa}_{L'}\| \geq \Theta_{L'}.$$

This completes the proof of Lemma 13.4 in case $\mathcal{U} = \{0\}$.

For the general case, we repeat the argument in the proof of Lemma 10.6. Define

$$\widetilde{d}_0 = \dim_{\mathbb{C}}\left(\frac{\mathbb{C}^{d_0}}{\mathcal{U}_0}\right), \quad \widetilde{d}_1 = \dim_{\mathbb{C}}\left(\frac{\mathbb{C}^{d_1}}{\mathcal{U}_1}\right), \quad \widetilde{d} = \widetilde{d}_0 + \widetilde{d}_1 = \dim_{\mathbb{C}}\left(\frac{\mathbb{C}^{d}}{\mathcal{U}}\right),$$

choose bases of $\mathbb{C}^{d_0}/\mathcal{U}_0$ and $\mathbb{C}^{d_1}/\mathcal{U}_1$ giving isomorphisms $\iota_0 : \mathbb{C}^{d_0}/\mathcal{U}_0 \to \mathbb{C}^{\widetilde{d}_0}$ and $\iota_1 : \mathbb{C}^{d_1}/\mathcal{U}_0 \to \mathbb{C}^{\widetilde{d}_1}$ and denote by $\widetilde{\pi} : \mathbb{C}^d \to \mathbb{C}^{\widetilde{d}}$ the composition of $\iota = \iota_0 \times \iota_1$ with π. The relations $\varphi_\lambda(\underline{z} + \underline{u}) = \varphi_\lambda(\underline{z})$ mean that for $1 \leq \lambda \leq L$ there exists a unique entire function $\widetilde{\varphi}_\lambda : \mathbb{C}^{\widetilde{d}} \to \mathbb{C}$ such that $\widetilde{\varphi}_\lambda \circ \widetilde{\pi} = \varphi_\lambda$. Define

$$\widetilde{\underline{w}}_k = \widetilde{\pi}(\underline{w}_k) \quad (1 \leq k \leq \ell_0).$$

Clearly we have for $\underline{\sigma} \in \mathbb{N}^{\ell_0}$

$$\mathcal{D}_{\underline{w}}^{\underline{\sigma}} \varphi_\lambda = \left(\mathcal{D}_{\widetilde{\underline{w}}}^{\underline{\sigma}} \widetilde{\varphi}_\lambda\right) \circ \widetilde{\pi}.$$

Hence for $\lambda \in I$ and $1 \leq \mu \leq L$,

$$d_{\lambda\mu}(z) = \mathcal{D}_{\widetilde{\underline{w}}}^{\underline{\sigma}_\mu}(z\widetilde{\underline{\xi}}_\mu)$$

where

$$\widetilde{\underline{\xi}}_\mu = \widetilde{\pi}(\underline{\zeta}_\mu) \quad (1 \leq \mu \leq L).$$

Applying the special case of Lemma 13.4 (where $\widetilde{\mathcal{U}} = \{0\}$) to $\mathbb{C}^{\widetilde{d}}$ with

$$\widetilde{\mathcal{W}} = \widetilde{\pi}(\mathcal{W}) \quad \text{and} \quad \widetilde{\mathcal{X}} = \widetilde{\pi}(\mathcal{X})$$

completes the proof of Lemma 13.4 in the general case. □

13.2.4 Upper Bound for the Interpolation Determinant

We now extend Lemma 7.7 in order to include derivatives:

Corollary 13.5. *Under the assumptions of Lemma 13.4, for* $1 \le \lambda \le L$, *let* M_λ *be a positive real number for which*

$$\log \sup_{|z|=E} \max_{1\le\mu\le L} |d_{\lambda\mu}(z)| \le M_\lambda;$$

then

$$\log|\Delta_I(1)| \le -\Theta_{L'}\log E + L'S_0\log E + \log(L!) + M_1 + \cdots + M_L.$$

Proof. We use Schwarz' Lemma (Lemma 6.3) together with Lemma 13.4:

$$\log|\Delta_I(1)| \le -(\Theta_{L'} - L'S_0)\log E + \log\sup_{|z|=E}|\Delta_I(z)|.$$

For $|z| = E$ we plainly have

$$\log|\Delta_I(z)| \le \log(L!) + M_1 + \cdots + M_L.$$

□

13.2.5 Proof of Proposition 13.2

Proof. From Corollary 13.5 and the lower bound for $\Theta_{L'}$ in Lemma 13.4, we deduce that the hypotheses of Lemma 7.6 are satisfied with $r = r_3$ and with χ_0, χ_1 and χ_3 defined by

$$\left(\frac{r_3+1}{r_3}\cdot\frac{\chi_0}{\log E}\right)^{r_3}\binom{T_0+r_1}{r_1}\binom{S_0+r_2}{r_2} = r_3!,$$

$$\chi_1 = \widetilde{V}, \qquad \chi_2 = \log(L!) + M_1 + \cdots + M_L.$$

The assumption on L implies

$$\frac{r_3^{r_3}}{(r_3+1)^{r_3+1}}\cdot\frac{\widetilde{V}^{r_3+1}}{\chi_0^{r_3}} \le \frac{1}{2}LV.$$

□

13.3 Exponential Polynomials

For $\underline{\tau} = (\tau_1, \ldots, \tau_{d_0}) \in \mathbb{N}^{d_0}$ and $\underline{t} = (t_1, \ldots, t_{d_1}) \in \mathbb{Z}^{d_1}$, consider the exponential monomial in d variables

$$\underline{z}^{\underline{\tau}} e^{\underline{tz}} = z_1^{\tau_1} \cdots z_{d_0}^{\tau_{d_0}} e^{t_1 z_{d_0+1} + \cdots + t_{d_1} z_d}.$$

In order to take derivatives, we introduce $\boldsymbol{w} = (\underline{w}_1, \ldots, \underline{w}_{\ell_0}) \in (\mathbb{C}^d)^{\ell_0}$ and $\underline{\sigma} = (\sigma_1, \ldots, \sigma_{\ell_0}) \in \mathbb{N}^{\ell_0}$. Let $\underline{\eta}_1, \ldots, \underline{\eta}_{\ell_1}$ be elements of $\mathbb{C}^d$. For $\underline{s} = (s_1, \ldots, s_{\ell_1}) \in \mathbb{Z}^{\ell_1}$, the number

$$\mathcal{D}_{\boldsymbol{w}}^{\underline{\sigma}}\big(\underline{z}^{\underline{\tau}} e^{\underline{tz}}\big)(s_1 \underline{\eta}_1 + \cdots + s_{\ell_1} \underline{\eta}_{\ell_1})$$

is the value of a polynomial at a point given by the coordinates of

$$\exp_G \underline{\eta}_j \in G(\mathbb{C}) = \mathbb{C}^{d_0} \times (\mathbb{C}^\times)^{d_1} \quad (1 \le j \le \ell_1).$$

Our first goal (§ 13.3.1) is to introduce these polynomials.

Write

$$\underline{w}_k = (\beta_{1k}, \ldots, \beta_{dk}) \quad (1 \le k \le \ell_0),$$

$$\underline{\eta}_j = (\beta_{1,\ell_0+j}, \ldots, \beta_{d_0,\ell_0+j}, \lambda_{1j}, \ldots, \lambda_{d_1 j}) \quad (1 \le j \le \ell_1)$$

and for $1 \le i \le d_1,\ 1 \le j \le \ell_1$, define

$$\alpha_{ij} = e^{\lambda_{ij}}$$

so that

$$\exp_G \underline{\eta}_j = (\beta_{1,\ell_0+j}, \ldots, \beta_{d_0,\ell_0+j}, \alpha_{1j}, \ldots, \alpha_{d_1 j}).$$

Further, set

$$\underline{\eta}_{\underline{s}} = s_1 \underline{\eta}_1 + \cdots + s_{\ell_1} \underline{\eta}_{\ell_1}.$$

Then we will get

$$\mathcal{D}_{\boldsymbol{w}}^{\underline{\sigma}}\big(\underline{z}^{\underline{\tau}} e^{\underline{tz}}\big)(\underline{\eta}_{\underline{s}}) = P_{\underline{\tau t}}^{(\underline{\sigma s})}(\underline{\beta}) \prod_{i=1}^{d_1} \prod_{j=1}^{\ell_1} \alpha_{ij}^{t_i s_j}$$

where $P_{\underline{\tau t}}^{(\underline{\sigma s})}$ is a polynomial in $d_0\ell_0 + d_1\ell_0 + d_0\ell_1$ variables.

Next we put these numbers into a square matrix and we express again its determinant Δ_{ar} as the value of a polynomial Δ (§ 13.3.2). In the special case where all the numbers α and β are algebraic, Liouville's estimate will enable us to produce a lower bound for the absolute value of this determinant, assuming it is not zero.

13.3.1 Algebraic Expression for $\mathcal{D}_{\boldsymbol{w}}^{\underline{\sigma}}\big(\underline{z}^{\underline{\tau}} e^{\underline{tz}}\big)(\underline{\eta}_{\underline{s}})$

We introduce $d(\ell + 1)$ variables, named as follows:

$$\begin{pmatrix} \mathbb{W} & \mathbb{X} & \underline{\mathrm{X}} \\ & \mathbb{Y} & \underline{\mathrm{Y}} \end{pmatrix} = \begin{pmatrix} \mathrm{W}_{hk} & \mathrm{X}_{hj} & \mathrm{X}_h \\ \underbrace{\mathrm{W}_{d_0+i,k}}_{1\le k\le \ell_0} & \underbrace{\mathrm{Y}_{ij}}_{1\le j\le \ell_1} & \mathrm{Y}_i \end{pmatrix} \quad \begin{matrix} \} & 1 \le h \le d_0 \\ \} & 1 \le i \le d_1. \end{matrix}$$

Hence $\mathbb{Z}[\mathbb{W}, \mathbb{X}, \mathbb{Y}]$ is a ring of polynomials in $d\ell$ variables, while $\mathbb{Z}[\mathbb{W}, \underline{\mathrm{X}}, \underline{\mathrm{Y}}]$ is a ring of polynomials in $d(\ell_0 + 1)$ variables.

For $\underline{\tau} = (\tau_1, \dots, \tau_{d_0}) \in \mathbb{N}^{d_0}$ and $\underline{t} = (t_1, \dots, t_{d_1}) \in \mathbb{Z}^{d_1}$, we write

$$\underline{\mathrm{X}}^{\underline{\tau}} = \mathrm{X}_1^{\tau_1} \cdots \mathrm{X}_{d_0}^{\tau_{d_0}}, \qquad \underline{\mathrm{Y}}^{\underline{t}} = \mathrm{Y}_1^{t_1} \cdots \mathrm{Y}_{d_1}^{t_{d_1}}.$$

For $1 \le k \le \ell_0$, write $\underline{\mathrm{W}}_k = (\mathrm{W}_{1k}, \dots, \mathrm{W}_{dk})$ and define a derivative operator $\mathcal{D}_{\underline{\mathrm{W}}_k}$ on the field of rational functions in $d(\ell_0 + 1)$ variables $\mathbb{Q}(\mathbb{W}, \underline{\mathrm{X}}, \underline{\mathrm{Y}})$ as follows:

$$\mathcal{D}_{\underline{\mathrm{W}}_k} = \sum_{h=1}^{d_0} \mathrm{W}_{hk} \frac{\partial}{\partial \mathrm{X}_h} + \sum_{i=1}^{d_1} \mathrm{W}_{d_0+i,k} \mathrm{Y}_i \frac{\partial}{\partial \mathrm{Y}_i}.$$

For $\underline{\sigma} = (\sigma_1, \dots, \sigma_{\ell_0}) \in \mathbb{N}^{\ell_0}$, define

$$\mathcal{D}_{\mathbb{W}}^{\underline{\sigma}} = \mathcal{D}_{\underline{\mathrm{W}}_1}^{\sigma_1} \cdots \mathcal{D}_{\underline{\mathrm{W}}_{\ell_0}}^{\sigma_{\ell_0}}.$$

Further, for $\underline{s} = (s_1, \dots, s_{\ell_1}) \in \mathbb{Z}^{\ell_1}$, we put

$$\underline{Z}_{\underline{s}} = \left(\sum_{j=1}^{\ell_1} s_j \mathrm{X}_{1j}, \dots, \sum_{j=1}^{\ell_1} s_j \mathrm{X}_{d_0 j}, \prod_{j=1}^{\ell_1} \mathrm{Y}_{1j}^{s_j}, \dots, \prod_{j=1}^{\ell_1} \mathrm{Y}_{d_1 j}^{s_j} \right).$$

We assume that $\underline{\tau}$, $\underline{t}$, $\underline{\sigma}$ and $\underline{s}$ satisfy

$$\|\underline{\tau}\| \le T_0, \qquad \|\underline{\sigma}\| \le S_0,$$

$$\|\underline{t}\| \le T^* \quad \text{and} \quad \|\underline{s}\| \le S^*.$$

Lemma 13.6. *For each $\underline{\tau} \in \mathbb{N}^{d_0}$, $\underline{t} \in \mathbb{Z}^{d_1}$, $\underline{\sigma} \in \mathbb{N}^{\ell_0}$ and $\underline{s} \in \mathbb{Z}^{\ell_1}$, there exists a polynomial $P_{\underline{\tau t}}^{(\underline{\sigma s})}$ in the ring $\mathbb{Z}[\mathbb{W}, \mathbb{X}]$, such that*

$$\mathcal{D}_{\mathbb{W}}^{\underline{\sigma}} (\underline{\mathrm{X}}^{\underline{\tau}} \underline{\mathrm{Y}}^{\underline{t}})_{(\underline{\mathrm{X}}, \underline{\mathrm{Y}}) = \underline{Z}_{\underline{s}}} = P_{\underline{\tau t}}^{(\underline{\sigma s})} (\mathbb{W}, \mathbb{X}) \prod_{i=1}^{d_1} \prod_{j=1}^{\ell_1} \mathrm{Y}_{ij}^{t_i s_j}.$$

This polynomial $P_{\underline{\tau t}}^{(\underline{\sigma s})}$ is
– for $1 \le k \le \ell_0$, homogeneous of degree σ_k in the d variables $\mathrm{W}_{1k}, \dots, \mathrm{W}_{dk}$;
– for $1 \le h \le d_0$, homogeneous of degree τ_h in the ℓ variables $\mathrm{W}_{h1}, \dots, \mathrm{W}_{h\ell_0}$, $\mathrm{X}_{h1}, \dots, \mathrm{X}_{h\ell_1}$;
moreover, the length of $P_{\underline{\tau t}}^{(\underline{\sigma s})}$ is bounded from above by

$$L\big(P_{\underline{\tau t}}^{(\underline{\sigma s})}\big) \le \sum_{m=0}^{\min\{T_0, S_0\}} m! \binom{T_0}{m} \binom{S_0}{m} (S^*)^{T_0 - m} (T^*)^{S_0 - m}.$$

Proof. Define the polynomial $Q_{\underline{\tau}\underline{t}}^{(\underline{\sigma})}(\mathbb{W},\underline{\mathrm{X}})$ in $\mathbb{Z}[\mathbb{W},\underline{\mathrm{X}}]$ by

$$Q_{\underline{\tau}\underline{t}}^{(\underline{\sigma})}(\mathbb{W},\underline{\mathrm{X}})\underline{\mathrm{Y}}^{\underline{t}} = \mathcal{D}_{\mathbb{W}}^{\underline{\sigma}}(\underline{\mathrm{X}}^{\underline{\tau}}\underline{\mathrm{Y}}^{\underline{t}}).$$

By induction on $\|\underline{\sigma}\|$, it is easily checked that this polynomial $Q_{\underline{\tau}\underline{t}}^{(\underline{\sigma})}$ is homogeneous of degree σ_k in the d variables $\mathrm{W}_{1k},\dots,\mathrm{W}_{dk}$ $(1\le k\le \ell_0)$, and homogeneous of degree τ_h in the ℓ_0+1 variables $\mathrm{W}_{h1},\dots,\mathrm{W}_{h\ell_0}$, X_h $(1\le h\le d_0)$. One obtains the polynomial $P_{\underline{\tau}\underline{t}}^{(\underline{\sigma}\underline{s})}$ by substituting, in $Q_{\underline{\tau}\underline{t}}^{(\underline{\sigma})}$,

$$\sum_{j=1}^{\ell_1} s_j\mathrm{X}_{hj} \quad\text{to the variable}\quad \mathrm{X}_h \qquad (1\le h\le d_0).$$

We now estimate the length. Since the upper bound we announced is a nondecreasing function of each of the parameters S_0, T_0, S^* and T^*, there is no loss of generality to assume $S_0=\|\underline{\sigma}\|$, $T_0=\|\underline{\tau}\|$, $S^*=\|\underline{s}\|$ and $T^*=\|\underline{t}\|$. For each $\underline{\sigma}$, $\underline{s}$, $\underline{\tau}$ and $\underline{t}$, we can write

$$P_{\underline{\tau}\underline{t}}^{(\underline{\sigma}\underline{s})} = \sum_{\underline{\alpha}}\sum_{\underline{\beta}} \underline{t}^{\underline{\alpha}}\underline{s}^{\underline{\beta}} p_{\underline{\alpha}\underline{\beta}}, \qquad (\underline{\alpha}\in\mathbb{N}^{d_1},\quad \underline{\beta}\in\mathbb{N}^{\ell_1}),$$

where $p_{\underline{\alpha}\underline{\beta}}\in\mathbb{Z}[\mathbb{W},\mathbb{X}]$ are polynomials with nonnegative coefficients. It easily follows that if we set

$$\boldsymbol{t}=(|t_1|,\dots,|t_{d_1}|)\in\mathbb{N}^{d_1} \qquad\text{and}\qquad \boldsymbol{s}=(|s_1|,\dots,|s_{\ell_1}|)\in\mathbb{N}^{\ell_1},$$

then we have

$$L\big(P_{\underline{\tau}\underline{t}}^{(\underline{\sigma}\underline{s})}\big) \le \sum_{\underline{\alpha}}\sum_{\underline{\beta}} \boldsymbol{t}^{\underline{\alpha}}\boldsymbol{s}^{\underline{\beta}} L(p_{\underline{\alpha}\underline{\beta}}).$$

For a polynomial with nonnegative coefficients, the length is the value of the polynomial where one substitutes 1 to each variable. Denote by $(\mathbf{1},\mathbf{1},\mathbf{1})$ the corresponding point in $\mathbb{C}^{d_0\ell_0}\times\mathbb{C}^{d_1\ell_0}\times\mathbb{C}^{d_0\ell_1}$. We get

$$\begin{aligned} L\big(P_{\underline{\tau}\underline{t}}^{(\underline{\sigma}\underline{s})}\big) &\le P_{\underline{\tau}\underline{t}}^{(\underline{\sigma}\underline{s})}(\mathbf{1},\mathbf{1},\mathbf{1}) \\ &\le \left(\sum_{h=1}^{d_0}\frac{\partial}{\partial \mathrm{X}_h} + \sum_{i=1}^{d_1}\mathrm{Y}_i\frac{\partial}{\partial \mathrm{Y}_i}\right)^{S_0} \big(\underline{\mathrm{X}}^{\underline{\tau}}\underline{\mathrm{Y}}^{\boldsymbol{t}}\big)_{\substack{\mathrm{X}_1=\cdots=\mathrm{X}_{d_0}=S^* \\ \mathrm{Y}_1=\cdots=\mathrm{Y}_{d_1}=1}}. \end{aligned}$$

For $f\in\mathbb{C}[\underline{\mathrm{X}},\underline{\mathrm{Y}}]$, we have

$$\frac{d}{dz}f(z,\dots,z,e^z,\dots,e^z)\big|_{z=S^*} = \left(\sum_{h=1}^{d_0}\frac{\partial}{\partial \mathrm{X}_h} + \sum_{i=1}^{d_1}\mathrm{Y}_i\frac{\partial}{\partial \mathrm{Y}_i}\right) f\,\Big|_{\substack{\mathrm{X}_1=\cdots=\mathrm{X}_{d_0}=S^* \\ \mathrm{Y}_1=\cdots=\mathrm{Y}_{d_1}=e^{S^*}}}.$$

Therefore

$$\begin{aligned} L\big(P_{\underline{\tau}\underline{t}}^{(\underline{\sigma}\underline{s})}\big) &\le e^{-T^*S^*}\left(\frac{d}{dz}\right)^{S_0}\big(z^{T_0}e^{T^*z}\big)\big|_{z=S^*} \\ &\le \sum_{m=0}^{\min\{T_0,S_0\}} m!\binom{T_0}{m}\binom{S_0}{m}(S^*)^{T_0-m}(T^*)^{S_0-m}. \end{aligned}$$

□

Remark. Using the bound $m!\binom{N}{m} \le N^m$ for $N \ge 0$ – with the usual conventions, namely

$$N^m = 1 \quad \text{for} \quad N = m = 0,$$
$$\binom{N}{m} = 0 \quad \text{unless} \quad 0 \le m \le N,$$
$$\binom{N}{0} = 1 \quad \text{for} \quad N \ge 1,$$

we deduce

$$L\big(P_{\underline{\tau}\underline{t}}^{(\underline{\sigma}\underline{s})}\big) \le (T^*)^{S_0}(S^*)^{T_0} \min\left\{\left(1+\frac{S_0}{T^*S^*}\right)^{T_0}, \left(1+\frac{T_0}{T^*S^*}\right)^{S_0}\right\}.$$

A special case of Lemma 13.6 is Lemma 4.9.

13.3.2 The Arithmetic Determinant

In this section we use the notation of § 13.1 and we assume that the conditions stated there are fulfilled.

a) The Polynomial Δ

Let L be nonnegative integers. For $1 \le \lambda, \mu \le L$, let

$$\underline{\tau}_\lambda = (\tau_{1\lambda}, \ldots, \tau_{d_0\lambda}) \in \mathbb{N}^{d_0}, \qquad \underline{t}_\lambda = (t_{1\lambda}, \ldots, t_{d_1\lambda}) \in \mathbb{Z}^{d_1},$$

$$\underline{\sigma}_\mu = (\sigma_{1\mu}, \ldots, \sigma_{\ell_0\mu}) \in \mathbb{N}^{\ell_0}, \quad \underline{s}_\mu = (s_{1\mu}, \ldots, s_{\ell_1\mu}) \in \mathbb{Z}^{\ell_1}$$

satisfy

$$\|\underline{\tau}_\lambda\| \le T_0, \qquad |t_{i\lambda}| \le T_i, \qquad (1 \le i \le d_1),$$

$$\|\underline{\sigma}_\mu\| \le d^+S_0, \qquad |s_{j\mu}| \le d^+S_j, \quad (1 \le j \le \ell_1).$$

Define

$$\vartheta_{\lambda\mu} = \mathcal{D}_{\underline{\mathbb{W}}}^{\underline{\sigma}_\mu} \underline{\mathbb{X}}^{\underline{\tau}_\lambda} \underline{\mathbb{Y}}^{\underline{t}_\lambda}\Big|_{(\underline{\mathbb{X}},\underline{\mathbb{Y}})=Z_{\underline{s}_\mu}}$$
$$= P_{\underline{\tau}_\lambda\underline{t}_\lambda}^{(\underline{\sigma}_\mu\underline{s}_\mu)}(\mathbb{W}, \mathbb{X}) \prod_{i=1}^{d_1}\prod_{j=1}^{\ell_1} \mathbb{Y}_{ij}^{t_{i\lambda}s_{j\mu}}.$$

Recall the notation $T^* = T_1 + \cdots + T_{d_1}$ and $S^* = S_1 + \cdots + S_{\ell_1}$.

Lemma 13.7. *The determinant* Δ *of the* $L \times L$ *matrix*

$$\big(\vartheta_{\lambda\mu}\big)_{1\le\lambda,\mu\le L}$$

is a polynomial with integer coefficients in $\mathbb{W}$, $\mathbb{X}$, $\mathbb{Y}^{\pm 1}$, *of total degree* $\le d^+LS_0$ *in the* $d\ell_0$ *variables* $\mathbb{W}_{ik}$ $(1 \le i \le d,\ 1 \le k \le \ell_0)$,

$\le LT_0$ in the $d_0\ell$ variables W_{hk}, X_{hj} $(1 \le h \le d_0,\ 1 \le k \le \ell_0,\ 1 \le j \le \ell_1)$, $\le d^+LT_iS_j$ in each of the variables Y_{ij} and $1/\mathrm{Y}_{ij}$ $(1 \le i \le d_1,\ 1 \le j \le \ell_1)$. The length of this polynomial is bounded from above by

$$L!e^{LU_0},$$

where

$$U_0 = T_0\log(d^+S^*) + d^+S_0\log T^* + \min\left\{T_0\log\left(1+\frac{S_0}{T^*S^*}\right), d^+S_0\log\left(1+\frac{T_0}{d^+T^*S^*}\right)\right\}.$$

Proof. We develop the determinant

$$\Delta = \sum_{\{\varphi\}} \epsilon(\varphi)\prod_{\mu=1}^{L}\left(P^{(\underline{\sigma}_\mu\underline{s}_\mu)}_{\underline{\tau}_{\varphi(\mu)}\underline{t}_{\varphi(\mu)}}(\mathbb{W},\mathbb{X})\prod_{i=1}^{d_1}\prod_{j=1}^{\ell_1}\mathrm{Y}_{ij}^{t_{i,\varphi(\mu)}s_{j\mu}}\right),$$

where φ runs over the set of bijective maps from the set $\{1,\dots,L\}$ onto itself and $\epsilon(\varphi) = \pm 1$. The length which we want to estimate is bounded by

$$\sum_{\{\varphi\}}\prod_{\mu=1}^{L} L\left(P^{\underline{\sigma}_\mu\underline{s}_\mu}_{\underline{\tau}_{\varphi(\mu)}\underline{t}_{\varphi(\mu)}}\right).$$

The number of terms in the sum is $L!$ and therefore the desired bound follows from Lemma 13.6 and the remark at the end of § 13.3.1 (with $S_0, S_1, \dots, S_{\ell_1}$ replaced by $d^+S_0, d^+S_1, \dots, d^+S_{\ell_1}$ respectively). □

b) Liouville's Estimate for the Arithmetic Determinant

We substitute to $\mathbb{W}$, $\mathbb{X}$ and $\mathbb{Y}$ algebraic numbers.

Define

$$\theta^{\underline{\sigma}\underline{s}}_{\underline{\tau}\underline{t}} = \mathcal{D}^{\underline{\sigma}}_{\underline{w}}\left(\underline{z}^{\underline{\tau}}e^{\underline{t}\underline{z}}\right)(\underline{\eta}_{\underline{s}})$$

and

$$\theta_{\lambda\mu} = \theta^{\underline{\sigma}_\mu\underline{s}_\mu}_{\underline{\tau}_\lambda\underline{t}_\lambda}.$$

Proposition 13.8. *Assume that the determinant Δ_{ar} of the $L\times L$ matrix*

$$\left(\theta_{\lambda\mu}\right)_{1\le\lambda,\mu\le L}$$

is not zero. Then

$$-\frac{1}{L}\log|\Delta_{\mathrm{ar}}| \le (D-1)(U_0+\log L) + U_1 + 2U_2$$

with

$$U_1 = DT_0 \log B_1 + d^+ DS_0 \log B_2,$$

$$U_2 = d^+ D \sum_{i=1}^{d_1} \sum_{j=1}^{\ell_1} T_i S_j \log A_{ij}.$$

Proof. The number $\theta_{\lambda\mu}$ is the value of the polynomial $\vartheta_{\lambda\mu}$ at the point

$$\mathrm{W}_{hk} = \beta_{hk}\ (1 \le h \le d_0,\ 1 \le k \le \ell_0),\quad \mathrm{X}_{hj} = \beta_{h,\ell_0+j}\ (1 \le h \le d_0,\ 1 \le j \le \ell_1),$$
$$\mathrm{W}_{d_0+i,k} = \beta_{d_0+i,k}\ (1 \le i \le d_1,\ 1 \le k \le \ell_0),\quad \mathrm{Y}_{ij} = \alpha_{ij}\ (1 \le i \le d_1,\ 1 \le j \le \ell_1),$$

hence Δ_{ar} is the value at the same point of the determinant Δ from Lemma 13.7.

We use Proposition 3.14 (Liouville's inequality); in the case where

$$\mathrm{h}(1{:}\, \beta_{h1}{:} \cdots {:}\, \beta_{h\ell}) \le \log B_1 \quad (1 \le h \le d_0)$$

one uses the fact that for $1 \le h \le d_0$ the degree of the polynomial Δ in the variables $\mathrm{W}_{h1}, \ldots, \mathrm{W}_{h\ell_0}, \mathrm{X}_{h1}, \ldots, \mathrm{X}_{h\ell_1}$ is at most LT_0, while for $1 \le k \le \ell_0$ the degree in $\mathrm{W}_{d_0+1,k}, \ldots, \mathrm{W}_{dk}$ is at most d^+LS_0. In the other case where

$$\mathrm{h}(1{:}\, \beta_{1k}{:} \cdots {:}\, \beta_{dk}) \le \log B_2 \quad (1 \le k \le \ell_0)$$

we use the fact that for $1 \le k \le \ell_0$, Δ has degree $\le d^+LS_0$ in $\mathrm{W}_{1k}, \ldots, \mathrm{W}_{dk}$ and for $1 \le h \le d_0$, it has degree $\le LT_0$ in $\mathrm{X}_{h1}, \ldots, \mathrm{X}_{h\ell_1}$. □

13.4 Proof of Theorem 13.1

We assume that the hypotheses of Theorem 13.1 are satisfied. We first give the main arguments, taking for granted some technical estimates; next we check these estimates.

13.4.1 Sketch of the Proof

We choose for L the least integer for which

$$L \ge 2\binom{T_0 + r_1}{r_1}\binom{d^+S_0 + r_2}{r_2}\left(\frac{V}{\log E}\right)^{r_3}.$$

From the hypotheses in § 13.1 we deduce

$$L \le H(G^+; \mathcal{T}).$$

By the definition of $H(G^+; \mathcal{T})$, there exist monomials in $\mathbb{C}[G] = \mathbb{C}[\underline{X}, \underline{Y}^{\pm 1}]$, say

$$\underline{X}^{\underline{\tau}_\lambda} \underline{Y}^{\underline{t}_\lambda} = X_1^{\tau_{1\lambda}} \cdots X_{d_0}^{\tau_{d_0\lambda}} Y_1^{t_{1\lambda}} \cdots Y_{d_1}^{t_{d_1\lambda}} \qquad (1 \le \lambda \le L),$$

with $\underline{\tau}_\lambda \in \mathbb{N}^{d_0}$, $\|\underline{\tau}_\lambda\| \le T_0$, $\underline{t}_\lambda \in \mathcal{T}_1$, such that the restrictions to $G^+(\mathbb{C})$ of the associated mappings

$$\begin{array}{rcl} G(\mathbb{C}) = \mathbb{C}^{d_0} \times (\mathbb{C}^\times)^{d_1} & \longrightarrow & \mathbb{C} \\ (\underline{u}, \underline{v}) & \longmapsto & \underline{u}^{\underline{\tau}_\lambda} \underline{v}^{\underline{t}_\lambda} \end{array}$$

are linearly independent.

Our goal is to show that the matrix

$$M = \left(\mathcal{D}_{\boldsymbol{w}}^{\underline{\sigma}} \big(\underline{z}^{\underline{\tau}_\lambda} e^{\underline{t}_\lambda \underline{z}} \big) (\underline{\eta}_{\underline{s}}) \right)_{\substack{1 \le \lambda \le L \\ (\underline{\sigma}, \underline{s})}}$$

has rank $< L$; the index of rows is λ with $1 \le \lambda \le L$, while the index of columns is $(\underline{\sigma}, \underline{s})$ with $\underline{\sigma} \in \mathbb{N}^{\ell_0}$, $\|\underline{\sigma}\| \le d^+ S_0$ and $\underline{s} \in \mathcal{S}_1[d^+]$. Recall the notation

$$\mathcal{S}_1[d^+] = \left\{ \underline{s}^{(1)} + \cdots + \underline{s}^{(d^+)} \, ; \, \underline{s}^{(i)} \in \mathcal{S}_1 \ (1 \le i \le d^+) \right\}.$$

This will allow us to check the hypotheses of Philippon's multiplicity estimate (Chap. 10): indeed a vanishing nontrivial linear combination between the rows gives a nonzero element of $\mathbb{C}[G]$ of multidegree $\le (T_0, \ldots, T_{d_1})$ which does not vanish identically on G^+ but vanishes to order $> d^+ S_0$ with respect to $\mathcal{W}$ at each point of $\Sigma[d^+] + G^-$ (recall the definition of $\mathcal{T}_1$) with

$$\begin{aligned} \Sigma[d^+] &= \left\{ \underline{\gamma}^{(1)} + \cdots + \underline{\gamma}^{(d^+)} \, ; \, \underline{\gamma}^{(i)} \in \Sigma \ (1 \le i \le d^+) \right\} \\ &= \left\{ s_1 \underline{\gamma}_1 + \cdots + s_{\ell_1} \underline{\gamma}_{\ell_1} \, ; \, (s_1, \ldots, s_{\ell_1}) \in \mathcal{S}_1[d^+] \right\}. \end{aligned}$$

Theorem 8.1 then gives the conclusion of Theorem 13.1.

We select L elements $(\underline{\sigma}_\mu, \underline{s}_\mu) \in \mathbb{N}^{\ell_0} \times \mathcal{S}_1$ with $\|\underline{\sigma}_\mu\| \le S_0$ and we consider the corresponding determinant (see Proposition 13.8)

$$\Delta_{\mathrm{ar}} = \det\Big(\theta_{\lambda\mu} \Big)_{1 \le \lambda, \mu \le L}.$$

We want to prove $\Delta_{\mathrm{ar}} = 0$.

We shall check below that the assumptions of Proposition 13.2 are satisfied with $\boldsymbol{w}$, $\mathcal{W}$ and $\mathcal{X}$ replaced by $\boldsymbol{w}'$, $\mathcal{W}'$ and $\mathcal{X}'$ respectively, and with $\mathcal{U}_0 = \{0\}$, $\mathcal{U}_1 = T_e(G_1^-)$,

$$\psi_{\lambda\underline{\tau}}(z_{d_0+1}, \ldots, z_d) = \begin{cases} e^{\underline{t}_\lambda \underline{z}} = e^{t_{1\lambda} z_{d_0+1} + \cdots + t_{d_1\lambda} z_d} & \text{for } \underline{\tau} = \underline{\tau}_\lambda, \\ 0 & \text{otherwise} \end{cases},$$

$$\varphi_\lambda(\underline{z}) = \underline{z}^{\underline{\tau}_\lambda} e^{\underline{t}_\lambda \underline{z}} = z_1^{\tau_{1\lambda}} \cdots z_{d_0}^{\tau_{d_0\lambda}} \psi_{\lambda\underline{\tau}_\lambda}(z_{d_0+1}, \ldots, z_d),$$

$$\underline{\zeta}_\mu = \underline{\eta}'_{\underline{s}_\mu}, \quad \epsilon = e^{-V},$$

$$M_\lambda = 2(d^+ + 1)U + (d^+ + 1)\frac{U}{D} + 1,$$

while $\delta_{\lambda\mu}$ is defined by

$$\theta_{\lambda\mu} = \mathcal{D}_{\boldsymbol{w}'}^{\underline{\sigma}_\mu} \big(\underline{z}^{\underline{\tau}_\lambda} e^{\underline{t}_\lambda \underline{z}} \big) (\underline{\eta}'_{\underline{s}_\mu}) + \epsilon \delta_{\lambda\mu}.$$

Hence

$$\frac{1}{L}\log|\Delta_{\rm ar}| \le -\frac{1}{2}V + 2(d^+ + 1)U + (d^+ + 1)\frac{U}{D} + 2 + \log L.$$

From the estimates

$$U_0 \le (d^+ + 1)\frac{U}{D}, \qquad U_1 \le (d^+ + 1)U, \qquad U_2 \le d^+U,$$

it follows that the conclusion of Proposition 13.8 gives the lower bound

$$\frac{1}{L}\log|\Delta_{\rm ar}| \ge -2(2d^+ + 1)U - (D-1)\log L + (d^+ + 1)\frac{U}{D}\cdot$$

The first technical estimate below implies $e^2L^D < e^{U/2}$; the condition $V \ge cU$ with $c = 12d^+ + 9$ shows that this lower bound for $|\Delta_{\rm ar}|$ is not compatible with the previous upper bound ; hence the assumption of Proposition 13.8 is not satisfied, which means $\Delta_{\rm ar} = 0$. This is what we wanted to prove.

13.4.2 First Technical Estimate

We prove now:

$$e^2L^D < e^{U/2}.$$

Using the lower bound

$$2\binom{T_0 + r_1}{r_1}\binom{d^+S_0 + r_2}{r_2}\left(\frac{V}{\log E}\right)^{r_3} \ge 2(12d^+ + 9) > 1$$

we derive

$$L < 3\binom{T_0 + r_1}{r_1}\binom{d^+S_0 + r_2}{r_2}\left(\frac{V}{\log E}\right)^{r_3}.$$

From the estimates $\log E \ge 1$ and $\max\{T_0, S_0, r_1, r_2\} \le U$, we deduce

$$\begin{aligned} L &< 3(T_0 + r_1)^{r_1}(d^+S_0 + r_2)^{r_2}\left(\frac{V}{\log E}\right)^{r_3} \\ &< 3(12d^+ + 9)^r U^r. \end{aligned}$$

Using the assumption $U \ge 20rD\log(2dD)$ we obtain $U^{rD} < e^{U/4}$ and $9e^2(4d+3) < e^{U/4rD}$ (for this last upper bound one may notice that the assumptions of § 13.1 cannot be satisfied with $d = 1$). The first technical estimate follows.

13.4.3 Second Technical Estimate

We check the upper bounds

$$e^2d^+T^*S^*(T_0 + d^+S_0 + d^+T^*S^*) < e^U \quad \text{and} \quad T_0 + d^+S_0 + d^+T^*S^*E < e^V$$

which will be needed later. We bound firstly $T_0 + d^+S_0 + d^+T^*S^*$ by $(2d + 1)U$, secondly T^*S^* by U, and next $e^2d^+(2d + 1)U^2$ by e^U, using again the lower bound $U \ge 20\log(2d)$. This proves the first estimate, while the second is proved in the same way using the inequality $U > \log E$.

13.4.4 Upper Bound for $\widetilde{V}$

We check

$$\widetilde{V} \leq \left(1 + \frac{r_3}{4}\right) V.$$

Indeed, from $V = cU$ with $c = 12d^+ + 9$, using the assumption

$$\max\{T_0 \log E,\ S_0 \log E,\ r_3 \log E\} \leq U,$$

we deduce

$$\begin{aligned}\widetilde{V} &= V + r_3 T_0 \log E + (r_3 + 1)d^+ S_0 \log E + r_3(r_3 + 1) \log E \\ &\leq V + (d^+ r_3 + d^+ + 2r_3 + 1)U \leq \left(1 + \frac{d^+ r_3 + d^+ + 2r_3 + 1}{c}\right) V.\end{aligned}$$

Since

$$\frac{d^+ r_3 + d^+ + 2r_3 + 1}{c} \leq \frac{r_3}{4} \quad \text{and} \quad \left(1 + \frac{r_3}{4}\right)^{r_3+1} < (r_3 + 1)!,$$

we deduce

$$\frac{2}{(r_3 + 1)!} \left(\frac{\widetilde{V}}{V}\right)^{r_3+1} < 2.$$

13.4.5 Estimate for M_λ

We check now that the assumptions of Proposition 13.2 are satisfied with

$$M_\lambda = 2(d^+ + 1)U + (d^+ + 1)\frac{U}{D} + 1.$$

Recall that a complex algebraic number β of degree $\leq D$ and absolute logarithmic height $\leq \log B$ satisfies $|\beta| \leq B^D$.

For $1 \leq k \leq \ell_0$ write $\underline{w}'_k = (w'_{1k}, \ldots, w'_{dk})$. From the assumptions of Theorem 13.1 we derive, for $1 \leq h \leq d_0$ and $1 \leq k \leq \ell_0$,

$$|w'_{hk} - \beta_{hk}| \leq e^{-V} \qquad \text{and} \qquad |\beta_{hk}| \leq \max\{B_1^D, B_2^D\},$$

hence

$$|w'_{hk}| \leq \max\{B_1^D, B_2^D\}(1 + e^{-V}),$$

because $B_i \geq 1$; from the estimate $\log(1 + t) \leq t$ (for $t \geq 0$), we obtain

$$\log |w'_{hk}| \leq D \log \max\{B_1, B_2\} + e^{-V}.$$

We also write $\underline{\eta}'_j = (\eta'_{1j}, \ldots, \eta'_{dj})$ for $1 \leq j \leq \ell_1$. The same argument gives, for $1 \leq h \leq d_0$, $1 \leq i \leq d_1$ and $1 \leq j \leq \ell_1$:

$$\log |\eta'_{hj}| \leq D \log B_1 + e^{-V}, \qquad \log |w'_{d_0+i,k}| \leq D \log B_2 + e^{-V}$$

and

$$|\eta'_{d_0+i,j}| \le |\lambda_{ij}| + e^{-V}.$$

Fix λ and μ in the range $1 \le \lambda \le L$, $1 \le \mu \le L$, and define $F_{\lambda\mu}\colon \mathbb{C} \to \mathbb{C}$ by

$$F_{\lambda\mu}(z) = \mathcal{D}_{\boldsymbol{w}'}^{\underline{\sigma}_\mu}\left(\underline{z}^{\underline{\tau}_\lambda} e^{\underline{t}_\lambda \underline{z}}\right)(z\underline{\eta}'_{\underline{s}_\mu}).$$

Lemma 13.9. *We have*

$$\log \sup_{|z|=E} |F_{\lambda\mu}(z)| \le U_0 + U_1 + U_3 + T_0 \log E + 1,$$

where

$$U_3 = E\left|\sum_{i=1}^{d_1}\sum_{j=1}^{\ell_1} t_i s_j \lambda_{ij}\right|.$$

Proof. We can write

$$F_{\lambda\mu}(z) = P_{\underline{\tau t}}^{(\sigma s)}(\boldsymbol{w}', z\boldsymbol{x}') \prod_{i=1}^{d_1}\prod_{j=1}^{\ell_1} e^{t_i s_j \eta'_{d_0+i,j} z}$$

where $\boldsymbol{w}'$ denotes the point in $\mathbb{C}^{d\ell_0}$ with coordinates (w'_{ik}), $\boldsymbol{x}'$ denotes the point in $\mathbb{C}^{d_0\ell_1}$ with coordinates (η'_{hj}) and $P_{\underline{\tau t}}^{(\sigma s)}$ is the polynomial of Proposition 13.6, in the variables W_{ik}, X_{hj}, with length bounded by e^{U_0} and total degree in $X_{h1}, \ldots, X_{h\ell_1}$ at most T_0. Hence

$$\log\left|P_{\underline{\tau t}}^{(\sigma s)}(\boldsymbol{w}', z\boldsymbol{x}')\right| \le U_0 + U_1 + T_0 \log E + (T_0 + d^+ S_0)e^{-V}.$$

Finally for $|z| \le E$ we have

$$\begin{aligned}\log\left|\prod_{i=1}^{d_1}\prod_{j=1}^{\ell_1} e^{t_i s_j \eta'_{d_0+i,j} z}\right| &\le \left|\sum_{i=1}^{d_1}\sum_{j=1}^{\ell_1} t_i s_j \eta'_{d_0+i,j} z\right| \\ &\le E\left|\sum_{i=1}^{d_1}\sum_{j=1}^{\ell_1} t_i s_j \lambda_{ij}\right| + d^+T^*S^*Ee^{-V} \\ &\le U_3 + d^+T^*S^*Ee^{-V}\end{aligned}$$

By the second technical estimate, we may bound

$$(T_0 + d^+S_0 + d^+T^*S^*E)e^{-V} \quad \text{by} \quad 1.$$

□

For the proof of the next analytic lemma, we need a simple auxiliary result which is useful to estimate differences.

Lemma 13.10. *Let $\nu_1, \ldots, \nu_\ell$ be positive integers and P a polynomial with complex coefficients in $\nu_1 + \cdots + \nu_\ell$ variables X_{ij} $(1 \le j \le \nu_i,\ 1 \le i \le \ell)$. Assume P has total degree N. Assume also that for any $i = 1, \ldots, \ell$, P has total degree $\le L_i$ with respect to the ν_i variables $X_{i1}, \ldots, X_{i\nu_i}$. Let a_{ij}, b_{ij} be complex numbers and $A_1, \ldots, A_\ell$, ϵ positive real numbers such that*

$$\max\{1, |a_{ij}|, |b_{ij}|\} \le A_i \quad \text{and} \quad |a_{ij} - b_{ij}| \le \epsilon A_i$$

for $1 \le j \le \nu_i$ and $1 \le i \le \ell$. Then

$$|P(\underline{a}) - P(\underline{b})| \le \epsilon N L(P) A_1^{L_1} \cdots A_\ell^{L_\ell}.$$

Proof. For any integer $k \ge 0$, the identity

$$a^k - b^k = (a - b)(a^{k-1} + \cdots + b^{k-1})$$

gives the estimate

$$|a^k - b^k| \le k|a - b| \max\{|a|, |b|\}^{k-1}.$$

By induction on m, one deduces, for $k_1, \ldots, k_m$ nonnegative integers,

$$|a_1^{k_1} \cdots a_m^{k_m} - b_1^{k_1} \cdots b_m^{k_m}| \le \sum_{\nu=1}^{m} k_\nu |a_\nu - b_\nu| \max\{|a_\nu|, |b_\nu|\}^{k_\nu - 1} \prod_{\substack{\mu \ne \nu \\ 1 \le \mu \le m}} \max\{|a_\mu|, |b_\mu|\}^{k_\mu}.$$

This proves Lemma 13.10 when P is a monomial; the general case easily follows. □

Remark. The same proof yields a homogeneous version of Lemma 13.10: for each $i \in \{1, \ldots, \ell\}$ for which P is homogeneous of degree L_i with respect to the ν_i variables $X_{i1}, \ldots, X_{i\nu_i}$, one may replace

$$\max\{1, |a_{ij}|, |b_{ij}|\} \le A_i \quad \text{by} \quad \max\{|a_{ij}|, |b_{ij}|\} \le A_i.$$

For instance if $P \in \mathbb{C}[X_0, \ldots, X_m]$ is homogeneous of degree N, then

$$|P(a_0, \ldots, a_m) - P(b_0, \ldots, b_m)| \le N L(P) \Big(\max_{0 \le i \le m} |a_i - b_i|\Big) \Big(\max_{0 \le i \le m} \max\{|a_i|, |b_i|\}\Big)^{N-1}.$$

Lemma 13.11. *The number*

$$\delta_{\lambda\mu} = e^V \left(\mathcal{D}_{\underline{w}}^{\underline{\sigma}_\mu} \left(\underline{z}^{\underline{\tau}_\lambda} e^{\underline{t}_\lambda \underline{z}} \right)(\underline{\eta}_{\underline{s}_\mu}) - \mathcal{D}_{\underline{w}'}^{\underline{\sigma}_\mu} \left(\underline{z}^{\underline{\tau}_\lambda} e^{\underline{t}_\lambda \underline{z}} \right)(\underline{\eta}'_{\underline{s}_\mu}) \right)$$

has absolute value bounded by

$$\log\left|\delta_{\lambda\mu}\right| \le U_0 + U_1 + \frac{U_3}{E} + U.$$

Proof. For simplicity we write $\underline{\tau}$, $\underline{t}$, $\underline{\sigma}$ and $\underline{s}$ in place of $\underline{\tau}_\lambda$, $\underline{t}_\lambda$, $\underline{\sigma}_\mu$ and $\underline{s}_\mu$. We use Exercise 1.1.a:

$$|e^z - e^{z'}| \le |z - z'| \max\{|e^z|, |e^{z'}|\}$$

with

$$z = \sum_{i=1}^{d_1}\sum_{j=1}^{\ell_1} t_i s_j \lambda_{ij}, \quad z' = \sum_{i=1}^{d_1}\sum_{j=1}^{\ell_1} t_i s_j \eta'_{d_0+i,j}.$$

For these values of z and z' we have

$$e^z = \prod_{i=1}^{d_1}\prod_{j=1}^{\ell_1} \alpha_{ij}^{t_i s_j}, \quad e^{z'} = \prod_{i=1}^{d_1}\prod_{j=1}^{\ell_1} e^{t_i s_j \eta'_{d_0+i,j}},$$

and

$$|z - z'| \le d^+ T^* S^* e^{-V}, \quad |e^z| \le \exp\left(\frac{U_3}{E}\right),$$

$$|e^{z'}| \le \exp\left(\frac{U_3}{E} + d^+ T^* S^* e^{-V}\right).$$

Hence

$$|e^z - e^{z'}| \le d^+ T^* S^* e^{-V} \max\{1, |e^z|, |e^{z'}|\}$$

with

$$\log\max\{1, |e^z|, |e^{z'}|\} \le \frac{U_3}{E} + d^+ T^* S^* e^{-V}.$$

We use Lemma 13.10 for the polynomial

$$P_{\underline{\tau}\underline{t}}^{(\underline{\sigma}\underline{s})}(\mathbb{W}, \mathbb{X})Z.$$

The number $e^{-V}\delta_{\lambda\mu}$ which we want to estimate is the absolute value of the difference of the values of this polynomial at two different points, namely the points with coordinates

$$W_{hk} = \beta_{hk}, \quad X_{hj} = \beta_{h,\ell_0+j}, \quad Z = \prod_{i=1}^{d_1}\prod_{j=1}^{\ell_1} \alpha_{ij}^{t_i s_j}$$

and

$$W_{hk} = w'_{hk}, \quad X_{hj} = \eta'_{hj}, \quad Z = \prod_{i=1}^{d_1}\prod_{j=1}^{\ell_1} e^{t_i s_j \eta'_{d_0+i,j}}$$

respectively. We use Lemma 13.10 with $\epsilon = d^+ T^* S^* e^{-V}$. The total degree N is bounded by $T_0 + d^+ S_0 + 1$ and the length by e^{U_0}. We deduce

$$\log\left|\delta_{\lambda\mu}\right| \le \log(d^+ T^* S^*) + \log(T_0 + d^+ S_0 + 1) + U_0 + U_1 + \frac{U_3}{E} + (T_0 + d^+ S_0 + d^+ T^* S^*)e^{-V}.$$

From the technical estimates we deduce

$$ed^+T^*S^*(T_0 + d^+S_0 + 1) \leq e^U, \quad (T_0 + d^+S_0 + d^+T^*S^*)e^{-V} \leq 1$$

and the conclusion of Lemma 13.11 readily follows. □

From the estimates

$$U_0 \leq (d^+ + 1)\frac{U}{D}, \quad U_1 \leq (d^+ + 1)U,$$

and

$$U_3 \leq d^+U, \quad T_0 \log E \leq DT_0 \log B_1 \leq U$$

we conclude that

$$M_\lambda = (d^+ + 1)\left(2 + \frac{1}{D}\right)U + 1$$

is admissible, which is what we wanted to check. □

13.5 Directions for Use

In this section we explain how to use Theorem 13.1. For simplicity we shall consider only the simpler situation where $G^- = \{e\}$, $G^+ = G$ (see § 14.4.5.for comments on the relevance of G^- and G^+).

We also assume that $\mathcal{T}_1$ is the full set of tuples $(t_1, \ldots, t_{d_1})$ for which $|t_i| \leq T_i$ $(1 \leq i \leq d_1)$, and similarly for $\mathcal{S}_1$. Even when $G^- = \{e\}$, it may be useful to take for $\mathcal{T}_1$ and $\mathcal{S}_1$ smaller subsets (for instance in connection with Matveev's trick - see § 9.3), but we shall no insist further on this point.

The statement of Theorem 13.1 involves a number of parameters. In most applications (examples already occurred earlier in this book, and further applications are to come), the parameters A_{ij}, B_1, B_2 and E are imposed by the data. For instance, when proving a measure of linear independence for logarithms of algebraic numbers, that means a lower bound for the absolute value of nonzero numbers of the form

$$\beta_0 + \beta_1\lambda_1 + \cdots + \beta_m\lambda_m,$$

the parameters A_{ij} measure the height of the algebraic numbers e^{λ_i}, B_1 and B_2 are related to the heights of the coefficients $\beta_0, \ldots, \beta_m$, and E is connected with $|\lambda_j|$.

On the other hand we have to choose the parameters $T_0, T_1, \ldots, T_{d_1}$ and $S_0, S_1, \ldots, S_{\ell_1}$, and the main goal of this section is to provide a few tips for a good choice; here, *good* means that we are looking for a sharp conclusion, which means that U and V should be as small as possible.

In many applications, the most restrictive conditions on the parameters are imposed by the fact that we wish to avoid the trivial case where the conclusion of Theorem 13.1 holds with $G^* = \{e\}$. We deal with this issue here.

13.5.1 Optimal Value for U

We prove that, under the assumptions of Theorem 13.1, if the parameter U is sufficiently small, then the conclusion is satisfied with $G^* = \{e\}$. This indicates, to a certain extent, a limit of application of Theorem 13.1: one can deduce a nontrivial conclusion only by taking U large enough.

Proposition 13.12. *Under the assumptions of Theorem 13.1, suppose*

$$\mathcal{T}_1 = \{\underline{t} \in \mathbb{Z}^{d_1}\,;\ |t_i| \le T_i\ (1 \le i \le d_1)\},$$
$$\mathcal{S}_1 = \{\underline{s} \in \mathbb{Z}^{\ell_1}\,;\ |s_j| \le S_j\ (1 \le j \le \ell_1)\}$$

and $G_1^- = \{1\}$, $G_1^+ = G_1$. *Suppose also* $T_i \ge 1$ *for* $1 \le i \le d_1$, $S_j \ge 1$ *for* $1 \le j \le \ell_1$, *and moreover* $T_0 \ge 1$ *if* $d_0 > 0$, $S_0 \ge 1$ *if* $\ell_0 > 0$. *Assume further that the numbers*

$$u = d_1\ell_1 + d_0\ell_1 + d_1\ell_0 - r(d_1 + \ell_1),$$

$$\delta = d_1\ell_1 + d_0\ell_1 + d_1\ell_0 - (r - r_3)(d_1 + \ell_1),$$

$$b_1 = d_0\ell_1 - r_1(d_1 + \ell_1),$$

$$b_2 = d_1\ell_0 - r_2(d_1 + \ell_1)$$

are ≥ 0. *If the conclusion of Theorem 13.1 is not satisfied for the trivial subgroup* $G^* = \{e\}$, *then the parameter* U *satisfies*

$$U^u \ge c_0 D^\delta (\log B_1)^{b_1} (\log B_2)^{b_2} \left(\prod_{i=1}^{d_1} \prod_{j=1}^{\ell_1} \log A_{ij} \right) (\log E)^{-r_3(d_1+\ell_1)}$$

where c_0 *depends only on* d_0, d_1, ℓ_0, ℓ_1, r_1, r_2, r_3.

Proof.
Define

$$L = \binom{T_0 + d_0}{d_0}(2T_1 + 1)\cdots(2T_{d_1} + 1) = \binom{T_0 + d_0}{d_0}\mathrm{Card}\mathcal{T}_1$$

and

$$N = \binom{S_0 + \ell_0}{\ell_0}(2S_1 + 1)\cdots(2S_{\ell_1} + 1) = \binom{S_0 + \ell_0}{\ell_0}\mathrm{Card}\mathcal{S}_1$$

which represent (to a certain extent) the *number of unknowns* and the *number of equations* in the multiplicity estimate. We have

$$L \le (1 + \eta_1)\frac{2^{d_1}}{d_0!} T_0^{d_0} T_1 \cdots T_{d_1} \quad \text{and} \quad N \le (1 + \eta_2)\frac{2^{\ell_1}}{\ell_0!} S_0^{\ell_0} S_1 \cdots S_{\ell_1},$$

with

$$1+\eta_1 = \left(1+\frac{d_0}{T_0}\right)^{d_0}\left(1+\frac{1}{2T_1}\right)\cdots\left(1+\frac{1}{2T_{d_1}}\right)$$

and

$$1+\eta_2 = \left(1+\frac{\ell_0}{S_0}\right)^{\ell_0}\left(1+\frac{1}{2S_1}\right)\cdots\left(1+\frac{1}{2S_{\ell_1}}\right).$$

Since the conclusion of Theorem 13.1 is not satisfied for the trivial subgroup G^* of dimension 0, we have

$$N \ge c_{ZE}L \quad \text{with} \quad c_{ZE} = d!.$$

The *analytic condition* implies

$$L \ge c_{AN}T_0^{r_1}S_0^{r_2}\left(\frac{V}{\log E}\right)^{r_3} \quad \text{with} \quad c_{AN} = \frac{2d^{r_2}}{r_1!r_2!},$$

while the *arithmetic condition* requires

$$V \ge c_{AR}U, \qquad \text{with} \qquad c_{AR} = 12d+9.$$

Hence the arithmetic plus analytic constraints are

$$L \ge c_{AN}(c_{AR})^{r_3}T_0^{r_1}S_0^{r_2}\left(\frac{U}{\log E}\right)^{r_3}.$$

From the upper bounds for L and N one deduces

$$L^{\ell_1}N^{d_1} \le (1+\eta_1)^{\ell_1}(1+\eta_2)^{d_1}\left(\frac{T_0^{d_0}}{d_0!}\right)^{\ell_1}\left(\frac{S_0^{\ell_0}}{\ell_0!}\right)^{d_1}4^{d_1\ell_1}\prod_{i=1}^{d_1}\prod_{j=1}^{\ell_1}T_iS_j.$$

We estimate the left hand side:

$$L^{\ell_1}N^{d_1} \ge (c_{ZE})^{d_1}L^{d_1+\ell_1} \ge (c_{ZE})^{d_1}\left(c_{AN}(c_{AR})^{r_3}\right)^{d_1+\ell_1}\left(T_0^{r_1}S_0^{r_2}\left(\frac{U}{\log E}\right)^{r_3}\right)^{d_1+\ell_1}.$$

Therefore we have

$$(1+\eta_1)^{\ell_1}(1+\eta_2)^{d_1}T_0^{d_0\ell_1}S_0^{d_1\ell_0}\prod_{i=1}^{d_1}\prod_{j=1}^{\ell_1}T_iS_j \ge$$

$$\frac{d_0!^{\ell_1}\ell_0!^{d_1}}{4^{d_1\ell_1}}(c_{ZE})^{d_1}\left(c_{AN}(c_{AR})^{r_3}\right)^{d_1+\ell_1}\left(T_0^{r_1}S_0^{r_2}\left(\frac{U}{\log E}\right)^{r_3}\right)^{d_1+\ell_1}.$$

We use the bounds

$$T_0 \le \frac{U}{D\log B_1}, \quad S_0 \le \frac{U}{D\log B_2}, \quad T_iS_j \le \frac{U}{D\log A_{ij}},$$

together with $u \geq 0$, $\delta \geq 0$, $b_1 \geq 0$, $b_2 \geq 0$ and we find the given condition on U. □

Remark. The proof yields also an estimate for c_0. For simplicity we replace η_1 and η_2 by 0 (in all applications these numbers are pretty small, in any case < 1). We also assume

$$T_i S_j \leq \frac{U}{d_1 \ell_1 D \log A_{ij}}.$$

Then

$$c_0 \geq (d_0!)^{\ell_1}(\ell_0!)^{d_1} \cdot \left(\frac{d_1\ell_1}{4}\right)^{d_1\ell_1} (c_{ZE})^{d_1}\left(c_{AN}(c_{AR})^{r_3}\right)^{d_1+\ell_1} \tag{13.13}$$

$$\geq \frac{(d_0!)^{\ell_1}(\ell_0!)^{d_1}(d_1\ell_1)^{d_1\ell_1} d!^{d_1}}{4^{d_1\ell_1}} \left(\frac{2d^{r_2}(12d+9)^{r_3}}{r_1!r_2!}\right)^{d_1+\ell_1}.$$

This is only a lower bound for the value of the constant which can be expected using Theorem 13.1. The actual value involves extra terms, coming for instance from the need to choose the parameters $T_0, \dots, T_{d_1}, S_0, \dots, S_{\ell_1}$ as integers, and taking into account the difference between a real number and its integral part. On the other hand if one wishes to get a sharp numerical constant, it is usually advisable to repeat the proof introducing a few minor refinements at some places. Such tricks can be found in the literature on this subject. The paper [LauMN 1995] (which deals only with homogeneous linear combinations of two logarithms only) contains several such refinements. Just to name one of them, the use of Blaschke factors (see Exercise 4.3), which was introduced in [MiW 1978], easily yields sharper numerical values for the final constants.

13.5.2 Choice of the Parameters

Proposition 13.12 gives a limit of application to Theorem 13.1. When applying Theorem 13.1, one wishes to come as close as possible to this limit. We explain how to select the parameters for this purpose.

We shall denote by c_1 a *constant* which depends only on $d_0, d_1, \ell_0, \ell_1, r_1, r_2, r_3$.

We assume, as we may without loss of generality,

$$\log A_{ij} \leq \frac{\log A_{i1} \log A_{1j}}{\log A_{11}}$$

for $1 \leq i \leq d_1$, $1 \leq j \leq \ell_1$ (a permutation reduces to this case).

Remark. From the proof of Proposition 13.12 it is easy to see that the optimal value of U cannot be reached unless the rank of the matrix $\left(\log A_{ij}\right)$ is 1.

We first define U by the following condition involving u, δ, b_1, b_2:

$$U^u = c_0 D^\delta (\log B_1)^{b_1} (\log B_2)^{b_2} \cdot \left(\prod_{i=1}^{d_1} \log A_{i1}\right)^{\ell_1} \left(\prod_{j=1}^{\ell_1} \log A_{1j}\right)^{d_1} (\log A_{11})^{-d_1\ell_1} (\log E)^{-r_3(d_1+\ell_1)}.$$

The definition of S_0 and T_0 is easy:

$$T_0 = \left[\frac{U}{D \log B_1}\right], \qquad S_0 = \left[\frac{U}{D \log B_2}\right].$$

We shall select a positive integer S_1 below and then define

$$S_j = \left[S_1 \frac{\log A_{11}}{\log A_{1j}}\right] \quad (1 \le j \le \ell_1),$$

$$T_i = \left[\frac{U}{d_1 \ell_1 D S_1 \log A_{i1}}\right] \quad (1 \le i \le d_1).$$

This enables us to check

$$\sum_{i=1}^{d_1} \sum_{j=1}^{\ell_1} T_i S_j \log A_{ij} \le \frac{U}{D}.$$

Now we define S_1 in terms of U in such a way that the condition $N \ge c_{ZE} L$ of § 13.5.1 is satisfied. For simplicity we replace

$$N \quad \text{by} \quad \frac{2^{\ell_1}}{\ell_0!} \left(\frac{U}{D \log B_2}\right)^{\ell_0} \frac{(S_1 \log A_{11})^{\ell_1}}{\log A_{11} \cdots \log A_{1\ell_1}}$$

and

$$L \quad \text{by} \quad \frac{2^{d_1}}{d_0!} \left(\frac{U}{D \log B_1}\right)^{d_0} \frac{U^{d_1}}{(d_1 \ell_1 D S_1)^{d_1} \log A_{11} \cdots \log A_{d_1 1}}.$$

Accordingly we choose the least positive integer for which

$$S_1^{d_1+\ell_1} \ge c_1 \frac{U^{d-\ell_0} (\log B_2)^{\ell_0} \log A_{11} \cdots \log A_{1\ell_1}}{D^{d-\ell_0} (\log B_1)^{d_0} (\log A_{11})^{\ell_1} \log A_{11} \cdots \log A_{d_1 1}}$$

with some c_1 satisfying

$$c_1 \ge \frac{\ell_0! 2^{d_1-\ell_1}}{d_0! (d_1 \ell_1)^{d_1}} \cdot c_{ZE}$$

(see Exercise 13.7). This is essentially how our parameters will be chosen in the next chapter. Notice that this choice leads a value for $T_1^{d_1+\ell_1}$ which is close to

$$c_1' \frac{U^{\ell-d_0} (\log B_1)^{d_0} \log A_{11} \cdots \log A_{d_1 1}}{D^{\ell-d_0} (\log B_2)^{\ell_0} (\log A_{11})^{d_1} \log A_{11} \cdots \log A_{1\ell_1}}$$

where

$$c_1' = \frac{d_0! 2^{\ell_1-d_1}}{\ell_0! (d_1 \ell_1)^{\ell_1}} \cdot \frac{1}{c_{ZE}}.$$

A good approximation for L is then

$$\frac{2^{d_1}}{d_0!(d_1\ell_1)^{d_1}}\cdot\frac{U^d}{D^d(\log B_1)^{d_0}S_1^{d_1}\log A_{11}\cdots\log A_{d_11}},$$

which is close to

$$c_{AN}(c_{AR})^{r_3}\frac{U^r}{(D\log B_1)^{r_1}(D\log B_2)^{r_2}}.$$

13.5.3 On the Conditions $T_i \geq 1$ and $S_j \geq 1$

Recall that the letters $T_0, T_1, \ldots, T_{d_1}$ and $S_0, S_1, \ldots, S_{\ell_1}$ in Theorem 13.1 denote positive integers. A general principle is that if the choice suggested in § 13.5.1 yields a value for some of the parameters T_i or S_j less than one, then one should omit the corresponding factor. The condition $S_j \geq 1$ amounts to

$$c_1U^{d-\ell_0}(\log B_2)^{\ell_0}(\log A_{11})^{d_1}\log A_{11}\cdots\log A_{1\ell_1}\geq \\ D^{d-\ell_0}(\log B_1)^{d_0}(\log A_{1j})^{d_1+\ell_1}\log A_{11}\cdots\log A_{d_11}$$

while the condition $T_i \geq 1$ is essentially

$$c_1'U^{\ell-d_0}(\log B_1)^{d_0}(\log A_{11})^{\ell_1}\log A_{11}\cdots\log A_{d_11}\geq \\ D^{\ell-d_0}(\log B_2)^{\ell_0}(\log A_{i1})^{d_1+\ell_1}\log A_{11}\cdots\log A_{1\ell_1}.$$

We shall use this remark in §§ 14.1.2, 14.2.2, 14.3.1 and 14.3.2.

13.6 Introducing Fel'dman's Polynomials

As mentioned in the introduction of the present chapter, we did not include Fel'dman's polynomials in our proof. Therefore Theorem 13.1 does not contain the refined measures of linear independence of logarithms given in Theorem 9.1 for instance. We explain here one way of including such a refinement in Theorem 13.1.

In this section we assume that the $d_0\times\ell_0$ matrix B_0 has a special shape:

$$\beta_{hk}=\delta_{hk}\qquad(1\leq h\leq d_0,\quad 1\leq k\leq \ell_0).$$

(This assumption is satisfied, by agreement, when either $d_0=0$ or $\ell_0=0$). The specialization $\widetilde{P}_{\underline{\tau t}}^{(\sigma s)}$ at

$$\mathrm{W}_{hk}=\delta_{hk}\qquad(1\leq h\leq d_0,\quad 1\leq k\leq \ell_0)$$

of the polynomial $P_{\underline{\tau t}}^{(\sigma s)}$ of Lemma 13.6 has a simple closed form[21], namely

[21] Compare with Lemma 4.9.

$$\widetilde{P}^{(\sigma s)}_{\underline{\tau} t} = \sum_{\underline{\kappa}} \frac{\underline{\sigma}!\underline{\tau}!}{\underline{\kappa}!(\underline{\sigma}-\underline{\kappa})!(\underline{\tau}-\underline{\kappa})!} (\underline{s\mathrm{X}})^{\underline{\tau}-\underline{\kappa}} (\underline{t\mathrm{W}})^{\underline{\sigma}-\underline{\kappa}},$$

where $\underline{\kappa}$ runs over the tuples $(\kappa_1, \kappa_2, \dots)$ of rational integers with

$$0 \le \kappa_h \le \min\{\sigma_h, \tau_h\}, \qquad \kappa_h = 0 \quad \text{for} \quad h \ge \min\{d_0, \ell_0\}.$$

For simplicity we have written

$$(\underline{s\mathrm{X}})^{\underline{\tau}-\underline{\kappa}} = \left(\sum_{j=1}^{\ell_1} s_j \underline{\mathrm{X}}_j\right)^{\underline{\tau}-\underline{\kappa}} = \prod_{h=1}^{d_0} \left(\sum_{j=1}^{\ell_1} s_j \mathrm{X}_{hj}\right)^{\tau_h-\kappa_h}$$

and similarly

$$(\underline{t\mathrm{W}})^{\underline{\sigma}-\underline{\kappa}} = \left(\sum_{i=1}^{d_1} t_i \underline{\mathrm{W}}_{d_0+i}\right)^{\underline{\sigma}-\underline{\kappa}} = \prod_{k=1}^{\ell_0} \left(\sum_{i=1}^{d_1} t_i \mathrm{W}_{d_0+i,k}\right)^{\sigma_k-\kappa_k}$$

It is sometimes useful to perform a change of basis, as we already saw in Chapters 9 and 10, where we used Fel'dman's Δ polynomials. We are going to do the same in the general context.

To start with, consider

$$\frac{\underline{\tau}!}{(\underline{\tau}-\underline{\kappa})!} (\underline{s\mathrm{X}})^{\underline{\tau}-\underline{\kappa}}.$$

This term arises from the specialization of

$$\mathcal{D}^{\underline{\kappa}}_{\underline{w}}(\underline{z}^{\underline{\tau}})$$

at

$$\underline{s\mathrm{X}} = \left(\sum_{j=1}^{\ell_1} s_j \mathrm{X}_{1j}, \dots, \sum_{j=1}^{\ell_1} s_j \mathrm{X}_{d_0 j}\right).$$

One may replace $\underline{z}^{\underline{\tau}} = z_1^{\tau_1} \cdots z_{d_0}^{\tau_{d_0}}$ by other polynomials, which may have better arithmetic properties. So let us choose any basis

$$\delta^{(1)}(\underline{z};\tau) \qquad (1 \le \tau \le M) \quad \text{with} \quad M = \binom{T_0 + d_0}{d_0}$$

of the space of polynomials in d_0 variables with coefficients in the number field K and total degree $\le T_0$. For $\underline{\kappa} \in \mathbb{N}^{d_0}$, define

$$\delta^{(1)}(\underline{z};\tau,\underline{\kappa}) = \left(\frac{\partial}{\partial z_1}\right)^{\kappa_1} \cdots \left(\frac{\partial}{\partial z_{d_0}}\right)^{\kappa_{d_0}} \delta^{(1)}(\underline{z};\tau).$$

Then in the expression for $\widetilde{P}^{(\sigma s)}_{\underline{\tau} t}$ one may replace

$$\frac{\underline{\tau}!}{(\underline{\tau}-\underline{\kappa})!} (\underline{s\mathrm{X}})^{\underline{\tau}-\underline{\kappa}}$$

by

$$\delta^{(1)}(\underline{s}\mathrm{X}; \tau, \underline{\kappa}).$$

The consequence on the matrix

$$M = \left(\mathcal{D}_{\underline{w}}^{\underline{\sigma}}\left(\underline{z}^{\underline{\tau}} e^{\underline{t}\underline{z}}\right)(\underline{s}\underline{\eta})\right)_{\substack{(\underline{\tau}\underline{t})\\(\underline{\sigma}\underline{s})}}$$

is described as follows. Define $\widetilde{M}$ to be the matrix of the same size as M whose entries are

$$\sum_{\underline{\kappa}} \frac{\underline{\sigma}!}{\underline{\kappa}!(\underline{\sigma}-\underline{\kappa})!} \delta^{(1)}(\underline{s}\underline{\beta}; \tau, \underline{\kappa})\left(\underline{t}\underline{\beta}\right)^{\underline{\sigma}-\underline{\kappa}} \prod_{i=1}^{d_1}\prod_{j=1}^{\ell_1} \alpha_{ij}^{t_i s_j}.$$

Here, $\underline{s}\underline{\beta}$ is nothing but

$$\left(\sum_{j=1}^{\ell_1} s_j \beta_{1,\ell_0+j}, \ldots, \sum_{j=1}^{\ell_1} s_j \beta_{d_0,\ell_0+j}\right) \in K^{d_0}$$

while

$$\underline{t}\underline{\beta} = \left(\sum_{i=1}^{d_1} t_i \underline{\beta}_{d_0+i,1}, \ldots \sum_{i=1}^{d_1} t_i \underline{\beta}_{d_0+i,\ell_0}\right) \in K^{\ell_0},$$

so that

$$\left(\underline{t}\underline{\beta}\right)^{\underline{\sigma}-\underline{\kappa}} = \left(\sum_{i=1}^{d_1} t_i \underline{\beta}_{d_0+i}\right)^{\underline{\sigma}-\underline{\kappa}} = \prod_{k=1}^{\ell_0}\left(\sum_{i=1}^{d_1} t_i \beta_{d_0+i,k}\right)^{\sigma_k-\kappa_k}.$$

We choose any ordering for the tuples $(\underline{\tau}, \underline{t})$ so that the tuples with the same $\underline{t}$ are consecutive. Denote by

$$Q^{(1)} = \left(q_{\tau\underline{\tau}}^{(1)}\right) \in \mathrm{GL}_M(\mathbb{C})$$

the transition matrix such that

$$\delta^{(1)}(\underline{z}; \tau) = \sum_{\|\underline{\tau}\| \le T_0} q_{\tau\underline{\tau}}^{(1)} \underline{z}^{\underline{\tau}} \qquad (1 \le \tau \le M).$$

Denote by $\widetilde{Q}^{(1)}$ the matrix in $\mathrm{GL}_L(\mathbb{C})$ (where $L = M(2T_1+1)\cdots(2T_{d_1}+1)$) which is diagonal by blocs with

$$\widetilde{Q}^{(1)} = \mathrm{diag}(Q^{(1)}, \ldots, Q^{(1)}).$$

Then

$$\widetilde{Q}^{(1)} M = \widetilde{M}.$$

The change of basis we just performed turns out to be specially useful when the entries of the matrix B_1 are rational integers. In the same way, when B_2 has rational integers entries, it may be useful to perform a similar change of basis in order to avoid a too strong condition on the parameter B_2 of Theorem 13.1. For this, consider any basis

$$\delta^{(2)}(\underline{z};\sigma) \qquad 1 \le \sigma \le \binom{S_0+\ell_0}{\ell_0}$$

of the space of polynomials in ℓ_0 variables with coefficients in K and total degree $\le S_0$. For $\underline{\kappa} \in \mathbb{N}^{\ell_0}$, define

$$\delta^{(2)}(\underline{z};\sigma,\underline{\kappa}) = \left(\frac{\partial}{\partial z_1}\right)^{\kappa_1}\cdots\left(\frac{\partial}{\partial z_{\ell_0}}\right)^{\kappa_{\ell_0}}\delta^{(1)}(\underline{z};\sigma).$$

Denote by

$$Q^{(2)} = \left(q^{(2)}_{\underline{\sigma}\sigma}\right)$$

the (transposed of the) transition matrix such that

$$\delta^{(2)}(\underline{z};\sigma) = \sum_{\|\underline{\sigma}\|\le S_0} q^{(2)}_{\underline{\sigma}\sigma}\underline{z}^{\underline{\sigma}} \qquad 1 \le \sigma \le \binom{S_0+\ell_0}{\ell_0}$$

and by $\widetilde{Q}^{(2)}$ the matrix $\mathrm{diag}(Q^{(2)},\dots,Q^{(2)})$. Then the entries of

$$\widetilde{Q}^{(1)}M\widetilde{Q}^{(2)}$$

are

$$\sum_{\underline{\kappa}}\frac{1}{\underline{\kappa}!}\delta^{(1)}(\underline{s}\underline{\beta};\tau,\underline{\kappa})\delta^{(2)}(\underline{t}\underline{\beta};\sigma,\underline{\kappa})\prod_{i=1}^{d_1}\prod_{j=1}^{\ell_1}\alpha_{ij}^{t_is_j}.$$

This means that we have replaced the polynomials $\widetilde{P}^{(\underline{\sigma}\underline{s})}_{\underline{\tau}\underline{t}}$ by

$$\sum_{\underline{\kappa}}\frac{1}{\underline{\kappa}!}\delta^{(1)}(\underline{s}\underline{X};\tau,\underline{\kappa})\delta^{(2)}(\underline{t}\underline{W};\sigma,\underline{\kappa}).$$

This polynomial depends on the choices of bases $\delta^{(1)}$ and $\delta^{(2)}$. Of course if we take the canonical bases then we find $\widetilde{P}^{(\underline{\sigma}\underline{s})}_{\underline{\tau}\underline{t}}$ again (the transition matrices $Q^{(1)}$, $Q^{(2)}$, $\widetilde{Q}^{(1)}$, $\widetilde{Q}^{(2)}$ are then identity matrices), so we do not lose any generality with this modification.

Remark. When the entries of the matrix $(\mathsf{B}_0,\mathsf{B}_1)$ are rational integers, one may take for instance for $\delta^{(1)}(\underline{z};\tau)$ the basis

$$\prod_{h=1}^{d_0}\Delta(z_h;\tau_h) \qquad (\|\underline{\tau}\| \le T_0),$$

or some variants of these like

$$\prod_{h=1}^{d_0}\Delta(z_h+\tau'_h;T'_0)^{\tau''_h} \quad\text{or}\quad \prod_{h=1}^{d_0}\delta_{T'_0}(z_h;\tau_h)$$

(see Lemma 9.8).

In a symmetric way, if the entries of $\binom{\mathsf{B}_0}{\mathsf{B}_2}$ are in $\mathbb{Z}$, one may also take for $\delta^{(2)}(\underline{z};\tau)$ products of Delta polynomials.

The choice of such polynomials should be done in order to sharpen the arithmetic estimates and to get rid (at least in some cases) of the conditions $DT_0 \log S_1 \leq U$ (using $\delta^{(1)}$) and $DS_0 \log T_1 \leq U$ (using $\delta^{(2)}$) of Theorem 13.1. In the context of measures of linear independence of logarithms of algebraic numbers, this enables one to weaken the assumptions on E^*, as we have seen in Chap. 9 and Chap. 10. However this process has its own limitation: for instance in the nonhomogeneous rational case ($\beta_i \in \mathbb{Z}$ for $0 \leq i \leq m$), the available estimates are not yet strong enough to enable us to omit the assumption $B \geq \max \log A_j$ in Theorem 9.1 (see Exercise 9.4).

13.7 Duality: the Fourier-Borel Transform

We already mentioned several times the existence of a *duality*[22] between the solution by Gel'fond on one hand, Schneider on the other hand, of Hilbert's seventh problem. Let us consider this matter more thoroughly.

13.7.1 Duality between Gel'fond-Baker's Method and Schneider's Method

Start for simplicity with a *homogeneous* linear combination of logarithms of algebraic numbers with algebraic coefficients:

$$\Lambda = \beta_1\lambda_1 + \cdots + \beta_{m-1}\lambda_{m-1} - \lambda_m,$$

where β_i and $\alpha_i = e^{\lambda_i}$ $(1 \leq i \leq m)$ are algebraic numbers.

The method we described in Chap. 10 (Gel'fond-Baker) involves the functions

$$e^{z_1}, \ldots, e^{z_{m-1}}, e^{\beta_1 z_1 + \cdots + \beta_{m-1} z_{m-1}}$$

which satisfy differential equations with algebraic coefficients The proof required a multiplicity estimate on $\mathbb{G}_{\mathrm{m}}^m$ related to an exponential sum

$$\Phi_G^{(1)}(\underline{z}) = \sum_{\underline{t}} c_{\underline{t}} \prod_{i=1}^{m-1} e^{(t_i + t_m\beta_i)z_i} = \sum_{\underline{t}} c_{\underline{t}} e^{\underline{x}_{\underline{t}}\underline{z}},$$

with

$$\underline{x}_{\underline{t}} = (t_1 + t_m\beta_1, \ldots, t_{m-1} + t_m\beta_{m-1}) \in \mathbb{C}^{m-1}.$$

[22] This duality is certainly different from the one introduced by D. Roy in [Roy 1992a] and [Roy 1992b] related to a category and its opposite — see § 11.7.2.

Here, $\underline{t}$ runs over a finite subset of $\mathbb{Z}^m$ and $c_{\underline{t}}$ are algebraic numbers. We take derivatives $(\partial/\partial z_1)^{\sigma_1} \cdots (\partial/\partial z_{m-1})^{\sigma_{m-1}}$ for $\underline{\sigma} \in \mathbb{N}^{m-1}$. If $\Lambda = 0$, the value of this derivative of $\Phi_G^{(1)}$ at a point $s\underline{\lambda} = (s\lambda_1, \ldots, s\lambda_{m-1})$ $(s \in \mathbb{Z})$ is an algebraic number:

$$\mathcal{D}^{\underline{\sigma}}\Phi_G^{(1)}(s\underline{\lambda}) = \sum_{\underline{t}} c_{\underline{t}} \prod_{i=1}^{m-1} (t_i + t_m\beta_i)^{\sigma_i} \cdot \prod_{i=1}^{m} \alpha_i^{t_i s}. \tag{13.14}$$

If Λ is not zero, then equality (13.14) does not hold anymore, but the difference between both sides of (13.14) has a small absolute value if $|\Lambda|$ is small.

In case $m = 2$, this is akin to Gel'fond's solution of Hilbert's seventh problem [G 1934]. He used a similar approach later (see references for instance in [G 1952]) when he established the first effective estimates. Baker's work [B 1966] extends this method to the general case $m \geq 2$. The proof, without any refinement nor extrapolation, yields, under the notation of Theorem 9.1,

$$|\Lambda| \geq \exp\left\{-C(m)D^{m^2}(\log B)^{m^2-m} \log A_1 \cdots \log A_m (\log E)^{-m^2+1}\right\}$$

(see § 14.4.1).

The main new feature of Baker's method, as opposed to Gel'fond's method, is to exploit the fact that the points lie on a complex line. This enabled Baker to use an extrapolation formula which holds only for functions of a single complex variable. By means of an induction he increases progressively the number of zeroes of his auxiliary function, and accordingly he increases the number of equations which are satisfied by the numbers $c_{\underline{t}}$. Finally this yields an estimate

$$|\Lambda| \geq \exp\left\{-C(m)D^{m+2}(\log B)^2 \log A_1 \cdots \log A_m\right\}.$$

The best known results so far by this method have been described in § 10.4: substituting $(\log B)(\log E^*)$ to $(\log B)^2$ is achieved by replacing $\Phi_G^{(1)}$ by a function of m variables

$$\Phi_G^{(2)}(\underline{z}) = \sum_{\tau} \sum_{\underline{t}} c_{\tau\underline{t}} z_0^{\tau} \prod_{i=1}^{m-1} e^{(t_i + t_m\beta_i)z_i},$$

where $(\tau, \underline{t})$ runs over a finite subset of $\mathbb{N} \times \mathbb{Z}^m$. If $\Lambda = 0$, the derivative of order $\underline{\sigma} = (\sigma_0, \ldots, \sigma_{m-1}) \in \mathbb{N}^n$ at the point $s\underline{\lambda} = (s, s\lambda_1, \ldots, s\lambda_{m-1})$ $(s \in \mathbb{Z})$ is

$$\mathcal{D}^{\underline{\sigma}}\Phi_G^{(2)}(s\underline{\lambda}) = \sum_{\tau,\underline{t}} c_{\tau\underline{t}} \frac{\tau!}{(\tau - \sigma_0)!} s^{\tau-\sigma_0} \prod_{i=1}^{m-1} (t_i + t_m\beta_i)^{\sigma_i} \cdot \prod_{i=1}^{m} \alpha_i^{\lambda_i s}. \tag{13.15}$$

This is what Baker, Fel'dman, Stark and others did (with an auxiliary function), and also what we did (with an interpolation determinant) in Chap. 10.

In Chapters 7 and 9 we used a quite different approach. Transpose (13.14) and consider numbers

$$\sum_{\underline{\sigma}} \sum_{s} c_{\underline{\sigma}s} \prod_{i=1}^{m-1} (t_i + t_m\beta_i)^{\sigma_i} \cdot \prod_{i=1}^{m} \alpha_i^{t_i s} \tag{13.16}$$

with some algebraic coefficients $c_{\underline{\sigma}s}$, where $(\underline{\sigma}, s)$ runs over a finite subset of $\mathbb{N}^{m-1} \times \mathbb{Z}$. For $\underline{t} \in \mathbb{Z}^m$ the value $\Phi_S^{(1)}(\underline{x}_{\underline{t}})$ of the exponential polynomial

$$\Phi_S^{(1)}(\underline{z}) = \sum_{\underline{\sigma}} \sum_{s} c_{\underline{\sigma}s} \prod_{i=1}^{m-1} \left(z_i^{\sigma_i} \alpha_i^{sz_i}\right)$$

at the point $\underline{x}_{\underline{t}} \in \mathbb{C}^{m-1}$, is close to the algebraic number (13.16) if $|\Lambda|$ is small (and is equal if Λ is zero).

For $m = 2$ this is just the starting point of Schneider's solution to Hilbert's seventh problem [Sch 1934], which has been developed later for getting sharp measures of linear independence for two logarithms of algebraic numbers in [MiW 1978], [Lau 1994], [LauMN 1995]. The dependence in B is $(\log B)^2$, and this is quite comparable to what can be achieved by Gel'fond's method alone. The zero estimate is simpler for Schneider's method because no derivative is involved. The zero estimate used in [MiW 1978] is due to D. W. Masser (see also [Ma 1981b]). The improvement in [Lau 1994] arises mainly from using an interpolation determinant in place of an auxiliary function, while [LauMN 1995] rests on an improved zero estimate of Nesterenko's (Proposition 2.12).

This method has been extended to cover the case $m \geq 2$ in [W 1991b], [W 1992] and [W 1993].

In Chap. 7 we used this approach (for $m \geq 2$), which is the exact dual to the Gel'fond-Baker's method. If you compare the notation of Chap. 7 with (13.16), you will notice a permutation between $(\tau, \underline{t})$ and $(\underline{\sigma}, s)$, which is of course due to this duality.

In this duality, the analog of Baker's refinement (related to the fact that his points were on a complex line) is that, apart from a single exponential factor

$$\left(\alpha_1^{z_1} \cdots \alpha_{m-1}^{z_{m-1}}\right)^s = \exp(s\underline{\lambda z}),$$

we have only polynomial functions (see Chap. 9). This yields an estimate involving $(\log B)^2$, like in [MiW 1978], [Lau 1994], [LauMN 1995] for the case $m = 2$. In Chap. 9, in order to replace $(\log B)^2$ by $(\log B)(\log E^*)$, we transposed (13.15). The identity

$$\left(\frac{d}{dz}\right)^{\sigma} \left(z^{\tau}\right|_{z=s} = \left(\frac{d}{dz}\right)^{\tau} \left(z^{\sigma} e^{sz}\right|_{z=0}$$

(for σ, τ in $\mathbb{N}$ and s in $\mathbb{Z}$) suggests to consider a complex function of m variables

$$\Phi_S^{(2)}(\underline{z}) = \sum_{\underline{\sigma}} \sum_{s} c_{\underline{\sigma}s} z_0^{\sigma_0} e^{sz_0} \prod_{i=1}^{m-1} \left(z_i^{\sigma_i} \alpha_i^{z_i s}\right),$$

(where $(\underline{\sigma}, s)$ runs over a finite subset of $\mathbb{N}^m \times \mathbb{Z}$) and to introduce

$$\left(\frac{\partial}{\partial z_0}\right)^{\tau} \Phi_S^{(2)}(0, t_1 + t_m\beta_1, \ldots, t_{m-1} + t_m\beta_{m-1})$$

for $\tau \in \mathbb{N}$ and $\underline{t} \in \mathbb{Z}^m$.

This duality between the two main classical methods extends to the very general setting of Theorem 13.1. Consider the matrix M of § 13.1 together with its transpose:

$$\mathsf{M} = \begin{pmatrix} \mathsf{B}_0 & \mathsf{B}_1 \\ \mathsf{B}_2 & \mathsf{L} \end{pmatrix} \qquad {}^t\mathsf{M} = \begin{pmatrix} {}^t\mathsf{B}_0 & {}^t\mathsf{B}_2 \\ {}^t\mathsf{B}_1 & {}^t\mathsf{L} \end{pmatrix}$$

This transposition exchanges (d_0, d_1) and (ℓ_0, ℓ_1) on one hand, permutes r_1 and r_2 on the other hand, while n, r and r_3 are invariant. The quantities

$$\ell_0 d_1 + \ell_1 d_0, \quad d_1 \ell_1, \quad d_1 + \ell_1$$

are invariant; so the inequality

$$\ell_0 d_1 + \ell_1 d_0 + \ell_1 d_1 \leq n(\ell_1 + d_1)$$

(which occurred several times in many disguises in Chap. 11) is invariant by transposition. This duality establishes a correspondence between methods [1] and [1'] on one hand, [2] and [2'] on the other, for proving Baker's Theorem (see § 11.4.5 and § 14.4).

The most remarkable fact is the effect of such a transposition on the matrix $\boldsymbol{M}$ (§ 13.4.1) which gave rise to the interpolation determinant Δ_{ar}: the parameters

$$U, \quad D, \quad E, \quad n, \quad r, \quad r_3, \quad u, \quad \delta$$

are left invariant, while the following parameters are exchanged:

$$\begin{array}{ccccccc} B_1 & A_{ij} & d_0 & d_1 & d & r_1 & b_1 \\ \updownarrow & \updownarrow & \updownarrow & \updownarrow & \updownarrow & \updownarrow & \updownarrow \\ B_2 & A_{ji} & \ell_0 & \ell_1 & \ell & r_2 & b_2 \end{array}$$

Some proofs are *auto-dual* (to a certain extent): this is the case for the Theorem of Hermite-Lindemann and for the six exponentials Theorem. One of the main point in considering this duality is that it enables us to mix the methods: the arithmetic estimate is invariant by transposition, but this is not exactly true for the analytic estimate, and this is not true at all for the multiplicity estimate. So there is no obstruction to use in the same proof the zero estimate related to Schneider's method, say, and the analytic upper bound arising from Gel'fond-Baker's method. This shows that in spite of the rigidity of the method (alluded to in Lang's book [L 1966][23]), there are many possible variants. Unfortunately, so far, they do not seem to lead to substantially different conclusions!

By means of the *Fourier-Borel transform* (see [LelGru 1986], Chap. 8 and [W 1991a]) we now explain why this correspondence holds for the analytic estimates of § 13.2 as well as for the arithmetic estimates of § 13.3 (e.g. Lemma 13.6: see Exercise 13.3). This will explain why the conclusion of Proposition 13.12 is essentially invariant under transposition: indeed, it is invariant as far as the dependence in D,

[23] "The few examples which one has now do suggest an absolute fantastic rigidity in the entire theory." [L 1966], Chap. 6, Historical Note.

B_1, B_2, A_{ij} and E is concerned, but it is not invariant if we consider the value of c_0 in terms of d, ℓ, d_1 and ℓ_1 given by (13.13). Clearly, the lack of symmetry occurs in the term $(c_{ZE})^{d_1}$ only. It would be nice to have a symmetric multiplicity estimate.

13.7.2 Fourier-Borel Transform

Recall that $\mathcal{A}_n$ denotes the $\mathbb{C}$-vector space of entire functions in $\mathbb{C}^n$. Further, let $\mathcal{A}'_n$ denote the space of $\mathbb{C}$-linear maps $\eta\colon \mathcal{A}_n \to \mathbb{C}$ which are *bounded* as follows: there exist two positive constants C and R (depending on η) such that, for any $F \in \mathcal{A}_n$,

$$|\eta(F)| \le C|F|_R \tag{13.17}$$

The elements of $\mathcal{A}'_n$ are called *analytic functionals* . Each $\eta \in \mathcal{A}'_n$ is determined by its *moments* $\eta(\underline{z}^{\underline{\kappa}})$, $\underline{\kappa} \in \mathbb{N}^n$, as shown by the following lemma:

Lemma 13.18. *Let $\eta \in \mathcal{A}'_n$ and $F \in \mathcal{A}_n$ be given. Consider the Taylor expansion at the origin of F:*

$$F(\underline{z}) = \sum_{\underline{\kappa}\in\mathbb{N}^n} a_{\underline{\kappa}} \underline{z}^{\underline{\kappa}}.$$

Then the series

$$\sum_{\underline{\kappa}\in\mathbb{N}^n} a_{\underline{\kappa}} \eta(\underline{z}^{\underline{\kappa}})$$

converges absolutely and the sum is $\eta(F)$.

Proof. From (13.17) follows

$$|\eta(\underline{z}^{\underline{\kappa}})| \le C R^{\|\underline{\kappa}\|},$$

with some positive constants C and R independent of $\underline{\kappa} \in \mathbb{N}^n$. Fix $R_1 > R$. Using Cauchy's inequalities we deduce

$$|a_{\underline{\kappa}}| \le |F|_{R_1} R_1^{-\|\underline{\kappa}\|}.$$

Hence

$$\left|a_{\underline{\kappa}}\eta(\underline{z}^{\underline{\kappa}})\right| \le C|F|_{R_1}\left(\frac{R}{R_1}\right)^{\|\underline{\kappa}\|}.$$

Fix $\epsilon > 0$. Let N be a sufficiently large integer, so that

$$C(2+N)^n|F|_{R_1}\left(\frac{R}{R_1}\right)^N < \epsilon.$$

Define

$$G(\underline{z}) = F(\underline{z}) - \sum_{\|\underline{\kappa}\|\le N} a_{\underline{\kappa}} \underline{z}^{\underline{\kappa}}.$$

On one hand we have

$$|G|_{R_1} \leq |F|_{R_1} + \sum_{\|\underline{\kappa}\| \leq N} |a_{\underline{\kappa}}| R_1^{\|\underline{\kappa}\|} \leq (2+N)^n |F|_{R_1}.$$

On the other hand G has a zero at $\underline{z} = 0$ of multiplicity $\geq N$, and Schwarz' Lemma (Lemma 2.4 in one variable suffices) gives

$$|G|_R \leq \left(\frac{R}{R_1}\right)^N |G|_{R_1}.$$

Therefore we deduce from (13.17)

$$\left| \eta(F) - \sum_{\|\underline{\kappa}\| \leq N} a_{\underline{\kappa}} \eta(\underline{z}^{\underline{\kappa}}) \right| = \left|\eta(G)\right| \leq C|G|_R < \epsilon.$$

□

It follows from Lemma 13.18 that η is also determined by its values

$$\mathcal{F}_\eta(\underline{\zeta}) = \eta(e^{\underline{\zeta}\underline{z}})$$

on the functions $\underline{z} \mapsto e^{\underline{\zeta}\underline{z}}$, for $\underline{\zeta}$ running over $\mathbb{C}^n$:

$$\mathcal{F}_\eta(\underline{\zeta}) = \sum_{\underline{\kappa} \in \mathbb{N}^n} \frac{1}{\underline{\kappa}!} \eta(\underline{z}^{\underline{\kappa}}) \underline{\zeta}^{\underline{\kappa}}.$$

By Lemma 13.18, this series $\sum_{\underline{\kappa} \in \mathbb{N}^n} \eta(\underline{z}^{\underline{\kappa}}) \underline{\zeta}^{\underline{\kappa}} / \underline{\kappa}!$ converges absolutely, and condition (13.17) shows that $\mathcal{F}_\eta$ is an entire function *of finite exponential type*:

Definition. An entire function $\Phi \in \mathcal{A}_n$ is *of finite exponential type* if there exist two positive constants C_1 and C_2 such that

$$|\Phi(\underline{\zeta})| \leq C_1 e^{C_2 |\underline{\zeta}|}$$

for any $\underline{\zeta} \in \mathbb{C}^n$. We denote by $\mathcal{A}_n^0$ the vector space of entire functions of finite exponential type.

Notice that any $f \in \mathcal{A}_n^0$ has order of growth ≤ 1.

Definition. The *Fourier-Borel Transform* of $\eta \in \mathcal{A}'_n$ is $\mathcal{F}_\eta \in \mathcal{A}_n^0$.

Conversely, to an entire function Φ in $\mathcal{A}_n^0$ we associate an analytic functional $\eta \in \mathcal{A}'_n$ such that $\Phi = \mathcal{F}_\eta$ as follows:

Lemma 13.19. *Let Φ be an entire function of finite exponential type. Let*

$$\Phi(\underline{\zeta}) = \sum_{\underline{\kappa} \in \mathbb{N}^n} a_{\underline{\kappa}} \frac{\underline{z}^{\underline{\kappa}}}{\underline{\kappa}!}$$

be its Taylor expansion at the origin. Then for any entire function

$$F(\underline{z}) = \sum_{\kappa \in \mathbb{N}^n} b_{\underline{\kappa}} \underline{z}^{\underline{\kappa}} \in \mathcal{A}_n$$

the series

$$\mathcal{H}_\Phi(F) = \sum_{\kappa \in \mathbb{N}^n} a_{\underline{\kappa}} b_{\underline{\kappa}}$$

is absolutely convergent and the mapping

$$F \longmapsto \mathcal{H}_\Phi(F)$$

belongs to $\mathcal{A}_n'$.

Moreover

$$\eta \longmapsto \mathcal{F}_\eta \quad \text{and} \quad \Phi \longmapsto \mathcal{H}_\Phi$$

define inverse bijections between $\mathcal{A}_n'$ *and* $\mathcal{A}_n^0$:

$$\Phi = \mathcal{F}_\eta \iff \eta = \mathcal{H}_\Phi.$$

Proof. We use Cauchy's inequality twice: first for F, with $R_1 > 0$:

$$|b_{\underline{\kappa}}| \leq |F|_{R_1} R_1^{-\|\underline{\kappa}\|},$$

and then for Φ, with $\varrho > 0$, using the assumption $\Phi \in \mathcal{A}_n^0$:

$$|a_{\underline{\kappa}}| \leq \underline{\kappa}! |\Phi|_\varrho \varrho^{-\|\underline{\kappa}\|} \leq \underline{\kappa}! C_1 e^{C_2 \varrho} \varrho^{-\|\underline{\kappa}\|}.$$

Therefore

$$\left|a_{\underline{\kappa}} b_{\underline{\kappa}}\right| \leq \underline{\kappa}! C_1 e^{C_2 \varrho} |F|_{R_1} (R_1 \varrho)^{-\|\underline{\kappa}\|}.$$

This estimate holds for any $R_1 > 0$, $\varrho > 0$ and $\underline{\kappa} \in \mathbb{N}^n$. Fix $R_1 > C_2$ and choose $\varrho = \|\underline{\kappa}\|/C_2$. For any $\underline{\kappa} \in \mathbb{N}^n$, we get

$$\left|a_{\underline{\kappa}} b_{\underline{\kappa}}\right| \leq u_{\underline{\kappa}} |F|_{R_1}$$

with

$$u_{\underline{\kappa}} = C_1 \left(\frac{C_2 e}{R_1}\right)^{\|\underline{\kappa}\|} \frac{\underline{\kappa}!}{\|\underline{\kappa}\|^{\|\underline{\kappa}\|}}.$$

The series $\sum_{\underline{\kappa}} u_{\underline{\kappa}}$ is convergent. Let c be its sum. We obtain

$$\left|\mathcal{H}_\Phi(F)\right| \leq \sum_{\kappa \in \mathbb{N}^n} \left|a_{\underline{\kappa}} b_{\underline{\kappa}}\right| \leq c |F|_{R_1},$$

which proves $\mathcal{H}_\Phi \in \mathcal{A}_n'$.

From the definition of $\mathcal{H}_\Phi$ we deduce

$$\mathcal{H}_\Phi(\underline{z}^{\underline{\kappa}}) = \mathcal{D}^{\underline{\kappa}} \Phi(0) \quad \text{for any } \underline{\kappa} \in \mathbb{N}^n$$

and therefore

$$\Phi(\underline{\zeta}) = \sum_{\underline{\kappa} \in \mathbb{N}^n} \frac{1}{\underline{\kappa}!} \mathcal{H}_\Phi(\underline{z}^{\underline{\kappa}}) \underline{\zeta}^{\underline{\kappa}}.$$

Let $\eta \in \mathcal{A}'_n$; put $\Phi = \mathcal{F}_\eta$. For any $\underline{\kappa} \in \mathbb{N}^n$ we have

$$\mathcal{H}_\Phi(\underline{z}^{\underline{\kappa}}) = \mathcal{D}^{\underline{\kappa}} \Phi(0) = \mathcal{D}^{\underline{\kappa}} \mathcal{F}_\eta(0) = \eta(\underline{z}^{\underline{\kappa}}),$$

hence $\mathcal{H}_\Phi = \eta$.

Conversely, let $\Phi \in \mathcal{A}_n^0$; put $\eta = \mathcal{H}_\Phi$. Then

$$\mathcal{F}_\eta(\underline{\zeta}) = \sum_{\underline{\kappa} \in \mathbb{N}^n} \frac{1}{\underline{\kappa}!} \eta(\underline{z}^{\underline{\kappa}}) \underline{\zeta}^{\underline{\kappa}} = \sum_{\underline{\kappa} \in \mathbb{N}^n} \frac{1}{\underline{\kappa}!} \mathcal{H}_\Phi(\underline{z}^{\underline{\kappa}}) \underline{\zeta}^{\underline{\kappa}} = \Phi(\underline{\zeta}).$$

□

Example 1.
1. For $\underline{z}_0 \in \mathbb{C}^n$, denote by $\delta_{\underline{z}_0} \in \mathcal{A}'_n$ the analytic functional $F \mapsto F(\underline{z}_0)$. Then

$$\mathcal{F}_{\delta_{\underline{z}_0}}(\underline{\zeta}) = e^{\underline{z}_0 \underline{\zeta}}.$$

Example 2. For $\eta \in \mathcal{A}'_n$ and $1 \le i \le n$, denote by $\eta \circ \partial/\partial z_i$ the functional

$$F \mapsto \eta\left(\frac{\partial}{\partial z_i} F\right).$$

Then

$$\mathcal{F}_{\eta \circ \partial/\partial z_i}(\underline{\zeta}) = \eta\left(\frac{\partial}{\partial z_i} e^{\underline{\zeta}\underline{z}}\right) = \eta\left(\zeta_i e^{\underline{\zeta}\underline{z}}\right) = \zeta_i \eta\left(e^{\underline{\zeta}\underline{z}}\right) = \zeta_i \mathcal{F}_\eta(\underline{\zeta}).$$

Of course an equivalent statement is

$$\mathcal{H}_{\zeta_i \Phi} = \mathcal{H}_\Phi \circ \frac{\partial}{\partial z_i}$$

for any $\Phi \in \mathcal{A}_n^0$.

Example 3. For any $\eta \in \mathcal{A}'_n$ and for $1 \le i \le n$, denote by $\eta \circ z_i$ the functional

$$F \mapsto \eta(z_i F).$$

Then

$$\mathcal{F}_{\eta \circ z_i} = \frac{\partial}{\partial \zeta_i} \mathcal{F}_\eta.$$

This relation can be written

$$\mathcal{H}_\Phi \circ z_i = \mathcal{H}_{\frac{\partial}{\partial \zeta_i} \Phi}$$

for any $\Phi \in \mathcal{A}_n^0$, which is proved as follows: for $\Phi = \mathcal{F}_\eta$ (so that $\eta = \mathcal{H}_\Phi$) and $\underline{\kappa} \in \mathbb{N}^n$,

$$\eta(\underline{z}^{\underline{\kappa}}) = \mathcal{D}^{\underline{\kappa}}\Phi(0) \quad \text{and} \quad \eta(z_i\underline{z}^{\underline{\kappa}}) = \mathcal{D}^{\underline{\kappa}}\left(\frac{\partial}{\partial \zeta_i}\Phi\right)(0),$$

hence

$$\eta \circ z_i = \mathcal{H}_{\frac{\partial}{\partial \zeta_i}\Phi}.$$

4. Let $\pi: \mathbb{C}^d \to \mathbb{C}^n$ be a linear map and ${}^{\mathrm{t}}\pi: \mathbb{C}^n \to \mathbb{C}^d$ its transpose. Define

$$\begin{array}{ccc} \mathcal{A}'_n & \longrightarrow & \mathcal{A}'_d \\ \eta & \longmapsto & \pi^*\eta \end{array}$$

by

$$\pi^*\eta(F) = \eta(F \circ \pi) \quad \text{for} \quad F \in \mathcal{A}_n.$$

Then

$$\mathcal{F}_{\pi^*\eta} = \mathcal{F}_\eta \circ {}^{\mathrm{t}}\pi.$$

In other terms, for $\Phi \in \mathcal{A}_d^0$

$$\mathcal{H}_{\Phi \circ {}^{\mathrm{t}}\pi} = \mathcal{H}_\Phi(F \circ \pi) \quad \text{for} \quad F \in \mathcal{A}_n.$$

It suffices to check this equality when $F(\underline{z}) = e^{\underline{\zeta}\underline{z}}$ for $\underline{\zeta} \in \mathbb{C}^n$:

$$\mathcal{H}_{\Phi \circ {}^{\mathrm{t}}\pi}(e^{\underline{\zeta}\underline{z}}) = \Phi \circ {}^{\mathrm{t}}\pi(\underline{\zeta}) = \mathcal{H}_\Phi\big(\underline{\xi} \mapsto e^{{}^{\mathrm{t}}\pi(\underline{\zeta})\underline{\xi}}\big) = \mathcal{H}_\Phi\big(\underline{\xi} \mapsto e^{\underline{\zeta}\pi(\underline{\xi})}\big).$$

Putting these examples together one deduces:

Lemma 13.20.
a) Let $\underline{u}_1, \dots, \underline{u}_{d_0}$ be elements of $\mathbb{C}^n$, $\underline{\tau} \in \mathbb{N}^{d_0}$ and $\mu \in \mathcal{A}'_n$. Define $\eta \in \mathcal{A}'_n$ by

$$\eta(F) = \mu\big((\boldsymbol{u}\underline{z})^{\underline{\tau}}F\big) \quad \textit{for} \quad F \in \mathcal{A}_n,$$

where

$$(\boldsymbol{u}\underline{z})^{\underline{\tau}} = (\underline{u}_1\underline{z})^{\tau_1} \cdots (\underline{u}_{d_0}\underline{z})^{\tau_{d_0}}.$$

Then

$$\mathcal{F}_\eta = \mathcal{D}_{\boldsymbol{u}}^{\underline{\tau}}\mathcal{F}_\mu.$$

b) Let $\underline{w}_1, \dots \underline{w}_{\ell_0}$ be elements of $\mathbb{C}^n$, $\underline{\sigma} \in \mathbb{N}^{\ell_0}$ and $\eta \in \mathcal{A}'_n$. Define

$$\mu = \eta \circ \mathcal{D}_{\boldsymbol{w}}^{\underline{\sigma}} \in \mathcal{A}'_n.$$

Then

$$\mathcal{F}_\mu(\underline{\zeta}) = (\boldsymbol{w}\underline{\zeta})^{\underline{\sigma}}\mathcal{F}_\eta(\underline{\zeta}).$$

Corollary 13.21. *Let $\underline{w}_1, \dots, \underline{w}_{\ell_0}, \underline{u}_1, \dots, \underline{u}_{d_0}, \underline{x}$ and $\underline{y}$ in $\mathbb{C}^n$, $\underline{\tau} \in \mathbb{N}^{d_0}$ and $\underline{\sigma} \in \mathbb{N}^{\ell_0}$. For $\underline{z} \in \mathbb{C}^d$, write $(\boldsymbol{u}\underline{z})^{\underline{\tau}}$ for $(\underline{u}_1\underline{z})^{\tau_1} \cdots (\underline{u}_{d_0}\underline{z})^{\tau_{d_0}}$.*
a) The Fourier-Borel transform of the analytic functional $\eta \in \mathcal{A}'_n$ defined by

$$\eta(F) = \mathcal{D}_{\boldsymbol{w}}^{\underline{\sigma}}\big((\boldsymbol{u}\underline{z})^{\underline{\tau}}F(\underline{z})\big|_{\underline{z}=\underline{y}}$$

is

$$\mathcal{F}_\eta(\underline{\zeta}) = \mathcal{D}_{\underline{u}}^{\underline{\tau}}\big((\underline{w}\underline{z})^{\underline{\sigma}} e^{\underline{y}\underline{z}}\big|_{\underline{z}=\underline{\zeta}}.$$

b) We have

$$\mathcal{D}_{\underline{w}}^{\underline{\sigma}}\big((\underline{u}\underline{z})^{\underline{\tau}} e^{\underline{x}\underline{z}}\big)\big|_{\underline{z}=\underline{y}} = \mathcal{D}_{\underline{u}}^{\underline{\tau}}\big((\underline{w}\underline{z})^{\underline{\sigma}} e^{\underline{y}\underline{z}}\big)\big|_{\underline{z}=\underline{x}}.$$

Proof. (See [W 1991a], lemme 7.6.)

Since $\mathcal{F}_\eta(\underline{x}) = \eta(e^{\underline{x}\underline{z}})$, part b) follows from part a).

Recall the notation $\delta_{\underline{y}}$ for the functional $F \mapsto F(\underline{y})$. Define $\mu \in \mathcal{A}'_n$ by $\mu = \delta_{\underline{y}} \circ \mathcal{D}_{\underline{w}}^{\underline{\sigma}}$. By part b) of Lemma 13.20, we have

$$\mathcal{F}_\mu(\underline{\zeta}) = \mathcal{F}_{\delta_{\underline{y}}\circ\mathcal{D}_{\underline{w}}^{\underline{\sigma}}}(\underline{\zeta}) = (\underline{w}\underline{\zeta})^{\underline{\sigma}}\mathcal{F}_{\delta_{\underline{y}}}(\underline{\zeta}) = (\underline{w}\underline{\zeta})^{\underline{\sigma}} e^{\underline{\zeta}\underline{y}}.$$

On the other hand, since

$$\eta(F) = \mu\big((\underline{u}\underline{z})^{\underline{\tau}} F\big)$$

we deduce from part a) of Lemma 13.20

$$\mathcal{F}_\eta(\underline{\zeta}) = \mathcal{D}_{\underline{u}}^{\underline{\tau}}\big((\underline{w}\underline{z})^{\underline{\sigma}} e^{\underline{y}\underline{z}}\big|_{\underline{z}=\underline{\zeta}}.$$

□

Remark. It follows from Corollary 13.21 that the functions $\Phi_G^{(1)}$ and $\Phi_G^{(2)}$ in Gel'fond's method (as sketched in § 13.7.1) are the Fourier-Borel transforms of the analytic functionals

$$\eta_G^{(1)} : F \longmapsto \sum_{\underline{t}} c_{\underline{t}} F(\underline{x}_{\underline{t}})$$

and

$$\eta_G^{(2)} : F \longmapsto \sum_{\tau}\sum_{\underline{t}} c_{\tau\underline{t}} \left(\frac{\partial}{\partial z_0}\right)^{\tau} F(\underline{x}_{\underline{t}})$$

respectively, while the function $\Phi_S^{(1)}$ and $\Phi_S^{(2)}$ in Schneider's method are the Fourier-Borel transforms of the analytic functionals

$$F \longmapsto \sum_{\underline{\sigma}}\sum_{s} c_{\underline{\sigma}s}\mathcal{D}^{\underline{\sigma}} F(s\underline{\lambda})$$

with

$$\underline{\sigma} = (\sigma_1, \dots, \sigma_{m-1}) \in \mathbb{N}^{m-1}, \quad \underline{\lambda} = (\lambda_1, \dots, \lambda_{m-1}) \in \mathbb{C}^{m-1} \quad \text{for} \quad \Phi_S^{(1)},$$

$$\underline{\sigma} = (\sigma_0, \dots, \sigma_{m-1}) \in \mathbb{N}^{m}, \quad \underline{\lambda} = (1, \lambda_1, \dots, \lambda_{m-1}) \in \mathbb{C}^{m} \quad \text{for} \quad \Phi_S^{(2)}.$$

Part b) of Corollary 13.21 explains why the matrices M which occurred in the transcendence proofs of Chapters 9 and 10 are transposed one of the other. This is why the results reached by the two methods are so closed. In § 14.4 we shall go further and compare the outputs of both methods in the light of Proposition 13.12.

Exercises

Exercise 13.1. Under the assumptions of § 13.1, denote by

$$\mathsf{M}' = \begin{pmatrix} \mathsf{B}'_0 & \mathsf{B}'_1 \\ \mathsf{B}'_2 & \mathsf{L}' \end{pmatrix}$$

the $d \times \ell$ matrix whose column vectors are respectively

$$\underline{w}'_1, \dots, \underline{w}'_{\ell_0}, \underline{\eta}'_1, \dots, \underline{\eta}'_{\ell_1}.$$

a) Check

$$r = \operatorname{rank}(\mathsf{M}') \quad \text{and} \quad r_3 = \operatorname{rank}(\mathsf{L}').$$

b) Define ϱ_1 and ϱ_2 by

$$\varrho_1 + r_3 = \operatorname{rk}\begin{pmatrix} \mathsf{B}'_1 \\ \mathsf{L}' \end{pmatrix} \qquad \text{and} \qquad \varrho_2 + r_3 = \operatorname{rk}\bigl(\mathsf{B}'_2,\ \mathsf{L}'\bigr).$$

Check

$$\varrho_1 + r_3 = \dim_{\mathbb{C}}(\mathcal{X}') \quad \text{and} \quad \varrho_2 + r_3 = \dim_{\mathbb{C}}\bigl(\pi_1(\mathcal{W}' + \mathcal{X}')\bigr).$$

Deduce

$$\varrho_1 \le r_1 \le r - r_3 - \varrho_2.$$

c) Check

$$\dim_{\mathbb{C}}\bigl(\mathcal{X}' \cap \ker \pi_1\bigr) = \varrho_1 \quad \text{and} \quad \dim_{\mathbb{C}}\bigl((\mathcal{W}' + \mathcal{X}') \cap \ker \pi_1\bigr) = r - r_3 - \varrho_2.$$

Deduce that $r = \varrho_1 + \varrho_2 + r_3$ if and only if $(\mathcal{W}' + \mathcal{X}') \cap \ker \pi_1 \subset \mathcal{X}'$.

Exercise 13.2. Check the following explicit formula for the polynomial $P_{\underline{\tau t}}^{(\underline{\sigma s})}$ of Lemma 13.6:

$$P_{\underline{\tau t}}^{(\underline{\sigma s})} = \underline{\tau}!\underline{\sigma}! \sum_{\underline{\kappa}} \left(\prod_{h=1}^{d_0} \prod_{k=1}^{\ell_0} \frac{W_{hk}^{\kappa_{hk}}}{\kappa_{hk}!} \right) \left(\prod_{i=1}^{d_1} \prod_{k=1}^{\ell_0} \frac{(t_i W_{d_0+i,k})^{\kappa_{d_0+i,k}}}{\kappa_{d_0+i,k}!} \right) \left(\prod_{h=1}^{d_0} \prod_{j=1}^{\ell_1} \frac{(s_j X_{hj})^{\kappa_{h,\ell_0+j}}}{\kappa_{h,\ell_0+j}!} \right),$$

where $\underline{\kappa}$ runs over the set of elements in $\mathbb{N}^{d\ell - d_1\ell_1}$ which satisfy

$$\sum_{k=1}^{\ell_0} \kappa_{hk} + \sum_{j=1}^{\ell_1} \kappa_{h,\ell_0+j} = \tau_h, \qquad (1 \le h \le d_0),$$

and

$$\sum_{h=1}^{d_0} \kappa_{hk} + \sum_{i=1}^{d_1} \kappa_{d_0+i,k} = \sigma_k, \qquad (1 \le k \le \ell_0).$$

Exercise 13.3.
a) The polynomial $P_{\underline{\tau t}}^{(\underline{\sigma s})}$ of Lemma 13.6 lies in the ring $\mathbb{Z}[\mathbb{W}, \mathbb{X}]$ involving $d_1\ell_1 + d_0\ell_1 + d_1\ell_0$ variables. Write these variables in matrices

$$\mathbb{W} = \begin{pmatrix} W_{11} & \cdots & W_{1\ell_0} \\ \vdots & \ddots & \vdots \\ W_{d1} & \cdots & W_{d\ell_0} \end{pmatrix} \quad \text{and} \quad \mathbb{X} = \begin{pmatrix} X_{11} & \cdots & X_{1\ell_1} \\ \vdots & \ddots & \vdots \\ X_{d_01} & \cdots & X_{d_0\ell_1} \end{pmatrix}$$

and denote by $\mathbb{W}'$, $\mathbb{X}'$ the transposed matrices:

$$\mathbb{W}' = \begin{pmatrix} W_{11} & \cdots & W_{d1} \\ \vdots & \ddots & \vdots \\ W_{1\ell_0} & \cdots & W_{d\ell_0} \end{pmatrix} \quad \text{and} \quad \mathbb{X}' = \begin{pmatrix} X_{11} & \cdots & X_{d_0 1} \\ \vdots & \ddots & \vdots \\ X_{1\ell_1} & \cdots & X_{d_0\ell_1} \end{pmatrix}.$$

Check

$$P_{\underline{\tau t}}^{(\underline{\sigma s})}(\mathbb{W}, \mathbb{X}) = P_{\underline{\sigma s}}^{(\underline{\tau t})}(\mathbb{W}', \mathbb{X}').$$

Hint. *Use Exercise 13.2.*

b) Deduce from Corollary 13.21 an analytic proof of a).

Hint. *Define*

$$\underline{\eta} = \sum_{j=1}^{\ell_1} s_j(x_{1j}, \ldots, x_{d_0 j}, y_{1j}, \ldots, y_{d_1 j}) \in \mathbb{C}^d.$$

The value of $P_{\underline{\tau t}}^{(\underline{\sigma s})}$ at a point $\boldsymbol{w} = (w_{ik})$, $\boldsymbol{x} = (x_{hj})$ is

$$\mathcal{D}_{\boldsymbol{w}}^{\underline{\sigma}}\left(\underline{z}^{\underline{\tau}} e^{\underline{tz}}\right)(\underline{\eta}) \prod_{i=1}^{d_1} \prod_{j=1}^{\ell_1} e^{-t_i s_j y_{ij}}.$$

Introduce variables $\zeta_1, \ldots, \zeta_\ell$. Define $\boldsymbol{v} = (\underline{v}_1, \ldots, \underline{v}_{d_0})$ where

$$\underline{v}_h = (w_{h1}, \ldots, w_{h\ell_0}, x_{h1}, \ldots, x_{h\ell_1}) \in \mathbb{C}^\ell.$$

Put also

$$\underline{\xi} = \sum_{i=1}^{d_1} t_i(w_{d_0+i,1}, \ldots, w_{d_0+i,\ell_0}, y_{i1}, \ldots, y_{i\ell_1}) \in \mathbb{C}^\ell.$$

The formula we want to check reads

$$\mathcal{D}_{\boldsymbol{w}}^{\underline{\sigma}}\left(\underline{z}^{\underline{\tau}} e^{\underline{tz}}\right)(\underline{\eta}) = \mathcal{D}_{\boldsymbol{v}}^{\underline{\tau}}\left(\underline{\zeta}^{\underline{\sigma}} e^{\underline{s\zeta}}\right)(\underline{\xi}).$$

Using Corollary 13.21 with $n = d$,

$$u_{hk} = \delta_{hk} \quad (1 \le h \le d_0,\ 1 \le k \le d), \qquad \underline{x} = (0, \ldots, 0, t_1, \ldots, t_{d_1}),$$

check

$$\mathcal{D}_{\boldsymbol{w}}^{\underline{\sigma}}\left(\underline{z}^{\underline{\tau}} e^{\underline{tz}}\right)(\underline{\eta}) = \left(\frac{\partial}{\partial z_1}\right)^{\tau_1} \cdots \left(\frac{\partial}{\partial z_{d_0}}\right)^{\tau_{d_0}}$$

$$\prod_{k=1}^{\ell_0}\left(\sum_{h=1}^{d_0} w_{hk} z_h + \sum_{i=1}^{d_1} w_{d_0+i,k} t_i\right)^{\sigma_k} \prod_{h=1}^{d_0}\prod_{j=1}^{\ell_1} e^{s_j x_{hj} z_h} \cdot \prod_{i=1}^{d_1}\prod_{j=1}^{\ell_1} e^{t_i s_j y_{ij}} \Bigg|_{z_1 = \cdots = z_{d_0} = 0}.$$

Conclude the proof by means of the formula

$$\left(\frac{\partial}{\partial z_1}\right)^{\tau_1} \cdots \left(\frac{\partial}{\partial z_{d_0}}\right)^{\tau_{d_0}} F(\boldsymbol{v}\underline{z} + \underline{\zeta}) \Bigg|_{z_1 = \cdots = z_{d_0} = 0} = \mathcal{D}_{\boldsymbol{v}}^{\underline{\sigma}} F(\underline{\zeta}).$$

Exercise 13.4. In Proposition 13.8, assume

$$L = \binom{T_0 + d_0}{d_0} \prod_{i=1}^{d_1} (2T_i + 1),$$

which means that the set $\{(\underline{\tau}_\lambda, \underline{t}_\lambda),\ 1 \le \lambda \le L\}$ is

$$\big\{(\underline{\tau}, \underline{t})\,;\ \|\underline{\tau}\| \le T_0,\ |t_i| \le T_i,\ (1 \le i \le d_1)\big\}.$$

Show that, in the conclusion of Proposition 13.8, the quantity

$$2U_2 = 2d^+ D \sum_{i=1}^{d_1} \sum_{j=1}^{\ell_1} T_i S_j \log A_{ij}$$

can be replaced by

$$d^+ D \sum_{i=1}^{d_1} \sum_{j=1}^{\ell_1} \frac{2T_i(T_i + 1)}{2T_i + 1} \cdot S_j \log A_{ij}.$$

Hint. *See the hint of Exercise 3.8.*

Exercise 13.5. Show that D can be replaced by $D/2$ in all estimates when the number field K is not contained into $\mathbb{R}$.

Hint. *Use Exercise 3.5.*

Exercise 13.6. With the notation of Lemma 13.3, let $a > 0$ be such that

$$(i_0!)^{1/i_0} \left(1 - \frac{i_0 + 1}{a}\right) \ge \frac{i_0 + 1}{e}.$$

(For instance $a = 8i_0^2$ will do.) Assume

$$L \ge \frac{a^{i_0}}{i_0!} (K + i_0 + 1)^{i_0} \cdot \prod_{\sigma=1}^{s} \binom{K_\sigma + i_\sigma}{i_\sigma}.$$

Check

$$\Theta_L^{i_0} \ge \left(\frac{i_0}{e}\right)^{i_0} \cdot \left(\frac{L^{i_0+1}}{\prod_{\sigma=1}^{s} \binom{K_\sigma + i_\sigma}{i_\sigma}}\right).$$

Hint. *Check*

$$M - \frac{i_0}{e} \cdot \left(\frac{L}{\prod_{\sigma=1}^{s} \binom{K_\sigma + i_\sigma}{i_\sigma}}\right)^{1/i_0} \ge \left(\frac{L}{\prod_{\sigma=1}^{s} \binom{K_\sigma + i_\sigma}{i_\sigma}}\right)^{1/i_0} \cdot \left(\frac{i_0}{i_0 + 1} \cdot (i_0!)^{1/i_0} - \frac{i_0}{e}\right)$$
$$\ge a i_0 (K + i_0 + 1) \left(\frac{1}{i_0 + 1} - \frac{1}{e (i_0!)^{1/i_0}}\right),$$

with

$$\frac{1}{i_0+1}-\frac{1}{e(i_0!)^{1/i_0}}\geq\frac{1}{a}.$$

Exercise 13.7. Check that an admissible value for the constant c_1 in the definition of S_1 from § 13.5.2 is

$$c_1=(1+\eta_1)(1+\eta_3)\frac{2^{d_1}}{d_0!}\cdot\frac{\ell_0!}{2^{\ell_1}}\cdot\frac{1}{(d_1\ell_1)^{d_1}}\cdot c_{ZE}$$

provided that $\eta_3>0$ satisfies

$$(1+\eta_3)^{-1}\leq\left(1-\frac{D\log B_1}{2U}\right)^{\ell_0}\prod_{j=1}^{\ell_1}\left(1-\frac{\log A_{1j}}{2S_1\log A_{11}}\right).$$

Assume further $S_j\geq 1$ and $T_0\geq d_0$. Check the inequality $\eta_1\leq 2^{d_0}(3/2)^{d_1}$ and show that $\eta_3=2^{\ell}-1$ and

$$c_1=\frac{2^{d_0}3^{d_1}\ell_0!c_{ZE}}{d_0!(d_1\ell_1)^{d_1}}$$

are admissible values.

Exercise 13.8.
a) From Theorem 13.1 deduce Theorem 7.10.
b) State and prove a generalization of Theorem 13.1 which includes Fel'dman's polynomials (see § 13.6) and implies Theorem 9.1.

Exercise 13.9.
a) The Fourier-Borel transform (in a single variable) of

$$F\mapsto\int_0^1 F(z)dz\quad\text{is}\quad\zeta\mapsto\frac{e^{\zeta}-1}{\zeta}.$$

b) Given a real valued function φ on $\mathbb{R}$ (continuous with compact support, for simplicity), the Fourier-Borel transform of the functional

$$F\mapsto\int_{-\infty}^{+\infty}\varphi(z)F(iz)dz$$

is nothing else than the Fourier transform of φ, namely

$$\zeta\mapsto\int_{-\infty}^{+\infty}\varphi(z)e^{iz\zeta}dz.$$

Exercise 13.10. For $\Phi\in\mathcal{A}_n^0$ with Taylor expansion

$$\Phi(\underline{\zeta})=\sum_{\underline{\kappa}\in\mathbb{N}^n}a_{\underline{\kappa}}\underline{\zeta}^{\underline{\kappa}},$$

define the Laplace transform $\widetilde{\Phi}$ of Φ by

$$\widetilde{\Phi}(\underline{z})=\sum_{\underline{\kappa}\in\mathbb{N}^n}a_{\underline{\kappa}}\underline{z}^{-\underline{\kappa}-\underline{1}},$$

where

$$\underline{z}^{-\underline{\kappa}-\underline{1}} = z_1^{-\kappa_1-1} \cdots z_n^{-\kappa_n-1}.$$

a) Check that $\widetilde{\Phi}$ is analytic in a domain of $\mathbb{C}^n$ which contains a closed set of the form

$$\big\{\underline{z} \in \mathbb{C}^n\ ;\ |z_j| \geq R \ \text{ for } 1 \leq j \leq n\big\}.$$

b) Using Cauchy's formula, derive, for $F \in \mathcal{A}_n$,

$$\mathcal{H}_\Phi(F) = \frac{1}{(2i\pi)^n} \int_{|z_1|=R} \cdots \int_{|z_n|=R} F(\underline{z})\widetilde{\Phi}(\underline{z})d\underline{z}.$$

c) Using the special case $F(\underline{z}) = e^{\underline{\zeta}\underline{z}}$ with $\underline{\zeta} \in \mathbb{C}^n$, deduce the inversion formula for Laplace's transform

$$\Phi(\underline{\zeta}) = \frac{1}{(2i\pi)^n} \int_{|z_1|=R} \cdots \int_{|z_n|=R} e^{\underline{\zeta}\underline{z}}\widetilde{\Phi}(\underline{z})d\underline{z}.$$

Hint. *See* [LelGru 1986], *Chap. 8.*

14. Applications to Diophantine Approximation

The purpose of the present chapter is to apply the main result of diophantine approximation of the previous chapter (Th. 13.1).

The first two sections are devoted to diophantine approximation in dimension 1 (we apply Theorem 13.1 with $r = 1$). We derive two lower bounds for numbers of the form

$$\sum_{i=1}^{m}\sum_{j=1}^{n}|\lambda_{ij} - \beta_j\lambda_i| \quad \text{and} \quad \sum_{i=1}^{m}\sum_{j=1}^{n}|\lambda_{ij} - \beta_j\beta_i'|$$

respectively, where β_j and β_i' are algebraic numbers, while λ_i and λ_{ij} are logarithms of algebraic numbers.

In § 14.3 we give examples involving several variables. In the next section (§ 14.4) we come back to the question of measures of linear independence of logarithms of algebraic numbers: the effective version of the Theorem of the linear subgroup includes such estimates in a number of ways, and we compare the results and the methods.

In the present chapter we shall use the following definition:

Definition. Let $\underline{\theta} = (\theta_1, \ldots, \theta_m)$ be a m-tuple of complex numbers. Following [RoyW 1997b], we say that a function $\varphi\colon \mathbb{N} \times \mathbb{R}_{>0} \to \mathbb{R}_{>0} \cup \{\infty\}$ is a *simultaneous*
[24] *approximation measure for* $\underline{\theta}$ if there exist a positive integer D_0 together with a real number $h_0 \geq 1$ such that, for any integer $D \geq D_0$, any real number $h \geq h_0$ and any m-tuple $\underline{\gamma} = (\gamma_1, \ldots, \gamma_m)$ of algebraic numbers satisfying

$$[\mathbb{Q}(\underline{\gamma})\colon \mathbb{Q}] \leq D \quad \text{and} \quad \max_{1\leq i\leq m} \mathrm{h}(\gamma_i) \leq h,$$

we have

$$\max_{1\leq i\leq m} |\theta_i - \gamma_i| \geq \exp\bigl\{-\varphi(D, h)\bigr\}.$$

[24] Notice that the meaning of "simultaneous approximation" here is not the usual one related with *rational* approximation measure to a tuple of real numbers: in the latter case the control is on a common denominator of the approximations. For algebraic approximations in higher degree, one could also introduce the projective height of the point $(1\colon \gamma_1\colon \cdots\colon \gamma_m)$, but for our purpose this is not relevant. It would only change some constants (as shown by Exercise 3.3.a), which is not our main concern here.

The main results of this chapter (namely Theorems 14.1, 14.6, 14.17 and 14.20) deal with simultaneous approximation of logarithms of algebraic numbers. Let us fix some notation.

Let m and n be two positive integers. We denote by

$$\mathsf{L}_{m,n} = \left(\lambda_{ij}\right)_{\substack{1\le i\le m\\ 1\le j\le n}}$$

a $m \times n$ matrix with entries in $\mathcal{L}$. We consider a number field K of degree $D = [K:\mathbb{Q}]$ such that the algebraic numbers $\alpha_{ij} = e^{\lambda_{ij}}$ belong to $K^\times$. Let A_{ij} $(1 \le i \le m,$ $1 \le j \le n)$, E be positive real numbers satisfying, for $1 \le i \le m$ and $1 \le j \le n$, the following conditions:

$$\mathrm{h}(\alpha_{ij}) \le \log A_{ij}, \quad |\lambda_{ij}| \le \frac{D}{E}\log A_{ij} \quad \text{and} \quad 1 \le \log E \le D\log A_{ij}.$$

We assume further that the $m \times n$ matrix $\left(\log A_{ij}\right)_{\substack{1\le i\le m\\ 1\le j\le n}}$ has rank 1, which can be written

$$(\log A_{ij})(\log A_{11}) = (\log A_{i1})(\log A_{1j})$$

for $1 \le i \le m$ and $1 \le j \le n$.

We need some independence condition on these λ_{ij}. For simplicity we introduce the following definition: given a positive real number U, we shall say that *the matrix* L_{mn} *satisfies the linear independence condition for* U if, for any $\underline{t} \in \mathbb{Z}^m \setminus \{0\}$ and any $\underline{s} \in \mathbb{Z}^n \setminus \{0\}$ with $|\underline{t}| \le U$ and $|\underline{s}| \le U$,

$$\sum_{i=1}^{m}\sum_{j=1}^{n} t_i s_j \lambda_{ij} \ne 0.$$

We shall say also that L_{mn} satisfies the *linear independence condition* if, for any $U > 0$, it satisfies the linear independence condition for U. This means that for any nonzero tuple $\underline{t} = (t_1, \dots, t_m)$ in $\mathbb{Z}^m$ and any nonzero tuple $\underline{s} = (s_1, \dots, s_n)$ in $\mathbb{Z}^n$, we have

$$\sum_{i=1}^{m}\sum_{j=1}^{n} t_i s_j \lambda_{ij} \ne 0.$$

14.1 A Quantitative Refinement to Gel'fond-Schneider's Theorem

By the Theorem of Gel'fond-Schneider, if λ_{ij} are elements of $\mathcal{L}$, not all zero, $\lambda_1, \dots, \lambda_m$ elements of $\mathcal{L}$ and $\beta_1, \dots, \beta_n$ algebraic numbers, not all rational, then the $(m+1) \times (n+1)$ matrix

$$\begin{pmatrix} 1 & \beta_1 & \cdots & \beta_n \\ \lambda_1 & & & \\ \vdots & & \lambda_{ij} & \\ \lambda_m & & & \end{pmatrix}$$

has rank ≥ 2. We show that such a matrix cannot be too close to a rank 1 matrix. In order to get a sharper result when m and n are large, some independence condition is clearly necessary.

14.1.1 A Lower Bound for $\sum_{i=1}^{m}\sum_{j=1}^{n}\left|\log\alpha_{ij} - \beta_j \log\alpha_i\right|$

Our first result deals with the simultaneous approximation of numbers $\alpha_i^{\beta_j}$ by algebraic numbers. Since we work with a $m \times (n+1)$ matrix

$$\mathsf{L}_{m,n+1} = \left(\lambda_{ij}\right)_{\substack{1\leq i\leq m\\ 0\leq j\leq n}}$$

with entries in $\mathcal{L}$, we complete the notation of the introduction as follows: we write λ_i, α_i, A_i for λ_{i0}, α_{i0} and A_{i0} respectively, so that

$$\mathrm{h}(\alpha_i) \leq \log A_i, \quad |\lambda_i| \leq \frac{D}{E}\log A_i \quad \text{and} \quad \log E \leq D\log A_i.$$

The number field K contains the algebraic numbers $\alpha_i = \alpha_{i0} = e^{\lambda_i}$ and we have

$$(\log A_{ij})(\log A_1) = (\log A_i)(\log A_{1j})$$

for $1 \leq i \leq m$ and $0 \leq j \leq n$.

Theorem 14.1. *There exists a positive constant c, which depends only on m and n, with the following property. Let $\beta_1, \ldots, \beta_n$ be algebraic numbers in K and B a positive real number satisfying*

$$B \geq e \quad \textit{and} \quad \mathrm{h}(1:\beta_1:\cdots:\beta_n) \leq \log B.$$

Define

$$U_1^{mn} = D^{(m+1)(n+1)}(\log B)^{n+1}\left(\prod_{i=1}^{m}\prod_{j=0}^{n}\log A_{ij}\right)(\log E)^{-m-n-1}$$

and assume that the matrix $\mathsf{L}_{m,n+1}$ satisfies the linear independence condition for $(cU_1)^2$. Assume further

$$\log E \leq D\log B \leq U_1, \quad B \geq D$$

and

$$D\log A_{ij} \leq B$$

for $1 \leq i \leq m$ and $0 \leq j \leq n$. Then

$$\sum_{i=1}^{m}\sum_{j=1}^{n}\left|\lambda_{ij}-\beta_j\lambda_i\right| \geq e^{-cU_1}.$$

Remark 1. To give an explicit numerical value for the constant c is not so important, but an efficient way of stating that it is effectively computable and that it depends only on m and n is to produce an explicit admissible value. We shall check the result with

$$c = 2^{32}m^4n^2(2m)^m.$$

Remark 2. One may replace the linear independence condition on the matrix $\mathsf{L}_{m,n+1}$ by the following hypotheses:

for any m-tuple $\underline{t} = (t_1, \ldots, t_m) \in \mathbb{Z}^m \setminus \{0\}$ satisfying $|t_i| \leq (cU_1)^2$ for $1 \leq i \leq m$, we have

$$t_1\lambda_1 + \cdots + t_m\lambda_m \neq 0$$

and

for any $n+1$-tuple $\underline{s} = (s_1, \ldots, s_{n+1}) \in \mathbb{Z}^{n+1} \setminus \{0\}$ satisfying $|s_j| \leq (cU_1)^2$ for $1 \leq j \leq n+1$, we have

$$s_1\beta_1 + \cdots + s_n\beta_n \neq s_{n+1}.$$

Plainly, the first condition is satisfied if the numbers $\lambda_1, \ldots, \lambda_m$ are linearly independent over $\mathbb{Q}$, and the second if the numbers $1, \beta_1, \ldots, \beta_n$ are linearly independent over $\mathbb{Q}$.

Remark 3. The conclusion of Theorem 14.1 corresponds to the optimal value for U given by Proposition 13.12, with

$$u = mn, \quad \delta = mn + m + n + 1, \quad b_1 = n + 1 \quad b_2 = 0.$$

14.1.2 Proof of Theorem 14.1

Let us start by checking that there is no loss of generality to assume

$$D(\log B)(\log A_1)\cdots(\log A_m) \geq (\log A_i)^m \log E \tag{14.2}$$

for $1 \leq i \leq m$. The proof is based on the remark in § 13.5.3: if, for some i, say $i = m$, this condition is not satisfied, then one should just omit the value $i = m$ in the statement and prove a lower bound for

$$\sum_{i=1}^{m-1}\sum_{j=1}^{n}\left|\lambda_{ij}-\beta_j\lambda_i\right|.$$

More precisely, assume (without loss of generality) $A_1 \leq \cdots \leq A_m$. Since $D \log B \geq \log E$, there is at least one integer m' in the range $1 \leq m' \leq m$ for which

$$D(\log B)\prod_{\iota=1}^{m'} \log A_\iota \geq (\log A_{m'})^{m'} \log E.$$

Let m' be the maximal element with this property. In the case $m' = m$ there is nothing to prove. Otherwise from the definition of m' and the assumption $A_1 \leq \cdots \leq A_m$ we deduce

$$D(\log B)\prod_{\iota=1}^{m'} \log A_\iota < (\log A_i)^{m'} \log E$$

for $i = m'+1$, hence also for any $i = m'+1, \ldots, m$. Assuming the result holds for m replaced by m', we have

$$\sum_{\iota=1}^{m'}\sum_{j=1}^{n}\left|\lambda_{\iota j} - \beta_j \lambda_\iota\right| \geq e^{-c'V_1},$$

where c' depends only on m' and n, and where V_1 is defined by

$$V_1^{m'n} = D^{(m'+1)(n+1)}(\log B)^{n+1}\left(\prod_{\iota=1}^{m'}\prod_{j=0}^{n} \log A_{\iota j}\right)(\log E)^{-m'-n-1}.$$

We have

$$\left(\frac{D\log B}{\log E}\right)^{m-m'}\prod_{\iota=1}^{m'}(\log A_\iota)^{m-m'} < \prod_{i=m'+1}^{m}(\log A_i)^{m'},$$

hence

$$\left(\frac{D\log B}{\log E}\right)^{(m-m')(n+1)}\prod_{\iota=1}^{m'}\prod_{j=0}^{n}(\log A_{\iota j})^{m-m'} < \prod_{i=m'+1}^{m}\prod_{j=0}^{n}(\log A_{ij})^{m'},$$

which yields $V_1 < U_1$.

We use the same argument and check that we may assume without loss of generality

$$D^{m+1}(\log B)\prod_{i=1}^{m}\big((\log A_{i0})\cdots(\log A_{in})\big) \geq \left(\prod_{i=1}^{m} \log A_{ij}\right)^{n}(\log E)^{m+1} \qquad (14.3)$$

for $0 \leq j \leq n$. For $n = 1$, (14.3) holds because $D\log B \geq \log E$ and $D\log A_{ij} \geq \log E$. Next, if condition (14.3) is not satisfied for, say, $j = n$, then one should work with n replaced by $n-1$ and consider rather

$$\sum_{i=1}^{m}\sum_{j=1}^{n-1}\left|\lambda_{ij} - \beta_j\lambda_i\right|.$$

Under condition (14.3) we shall obtain a sharper numerical value for c, namely

$$c = 2^{32}m^4n^2(2m)^{m/n}.$$

There is a symmetry in the statement of Theorem 14.1 which we shall break in the proof. We may apply Theorem 13.1 either with

$$d_0 = 1, \quad d_1 = m, \quad \ell_0 = 0, \quad \ell_1 = n+1,$$

or else with

$$d_0 = 0, \quad d_1 = n+1, \quad \ell_0 = 1, \quad \ell_1 = m.$$

Here we choose the former solution: we set $G = \mathbb{G}_{\mathrm{a}} \times \mathbb{G}_{\mathrm{m}}^m$ and $d = m+1$. We also take $G^+ = G$, $G^- = \{e\}$, $r = r_3 = 1$, $r_1 = r_2 = 0$. It will be convenient to define $\beta_{n+1} = 1$, $A_{1,n+1} = A_{10}$ and to use also the notation, for $1 \le i \le m$,

$$\lambda_{i,n+1} = \lambda_i, \quad \alpha_{i,n+1} = \alpha_i.$$

For $1 \le j \le n+1$, define $\underline{\eta}_j \in K \times \mathcal{L}^m$ and $\underline{\gamma}_j \in G(K) = K \times (K^\times)^m$ as follows:

$$\underline{\eta}_j = (\beta_j, \lambda_{1j}, \ldots, \lambda_{mj}) \quad \text{and} \quad \underline{\gamma}_j = \exp_G \underline{\eta}_j = (\beta_j, \alpha_{1j}, \ldots, \alpha_{mj}).$$

Hence

$$\underline{\eta}_{n+1} = (1, \lambda_1, \ldots, \lambda_m) \quad \text{and} \quad \underline{\gamma}_{n+1} = (1, \alpha_1, \ldots, \alpha_m).$$

Next put $\underline{\eta}'_j = \beta_j \underline{\eta}_{n+1}$ $(1 \le j \le n+1)$. The vector space

$$\mathcal{X}' = \mathbb{C}\underline{\eta}'_1 + \cdots + \mathbb{C}\underline{\eta}'_{n+1} = \mathbb{C}\underline{\eta}_{n+1}$$

has dimension $r = 1$. Since $\ell_0 = 0$, we have $\mathcal{W} = 0$, $\mathcal{W}' = 0$, we take $S_0 = 0$, and the parameter B_2 (which will play no role) can be selected as

$$B_2 = E^{1/D} + T_1 + \cdots + T_m.$$

We are going now to introduce parameters $T_0, T_1, \ldots, T_m, S_1, \ldots, S_{n+1}$. Because of the assumption $(m+1)(S_1 + \cdots + S_{n+1}) \le B_1$ (which we shall check later) of Theorem 13.1, we define $B_1 = B^{c_1}$ with some constant $c_1 \ge 1$ which will be explicitly given.

We shall define below two positive real numbers S and U. Instead of giving the values now, we explain where they come from. We first define T_0 as follows:

$$T_0 = \left[\frac{U}{D \log B_1}\right],$$

so that the estimate

$$DT_0 \log B_1 \le U$$

is satisfied.

Next we set

$$S_j = \left[S \frac{\log A_{10}}{\log A_{1j}}\right] \quad (1 \le j \le n+1).$$

Notice that $S_{n+1} = [S]$. For $1 \le i \le m$ and $1 \le j \le n+1$ we have

$$S_j \log A_{ij} \le S \log A_{i0}.$$

Therefore if we define $T_1, \ldots, T_m$ by

$$T_i = \left[\frac{U}{2mnDS\log A_{i0}}\right] \quad (1 \le i \le m),$$

then the condition

$$D\sum_{i=1}^{m}\sum_{j=1}^{n+1} T_i S_j \log A_{ij} \le U$$

is fulfilled.

This shows that our parameters $T_0, T_1, \ldots, T_m, S_1, \ldots, S_{n+1}$ will be known as soon as S and U are chosen. We take

$$\mathcal{S} = \left\{\underline{s} \in \mathbb{Z}^{n+1}\,;\ |s_j| \le S_j\ (1 \le j \le n+1)\right\}$$

and

$$\mathcal{T} = \left\{\underline{t} \in \mathbb{Z}^{m}\,;\ |t_i| \le T_i\ (1 \le i \le m)\right\}.$$

There are two main conditions which will help us to fix S and U. The first one arises from the hypothesis

$$(T_0+1)(2T_1+1)\cdots(2T_m+1) \ge \frac{2V}{\log E}$$

of Theorem 13.1, where $V \le 33mU$. Using the lower bounds $2[x]+1 \ge [x]+1 \ge x$ which hold for any $x \ge 0$ (such lousy estimates will occur repeatedly during the proof), one checks

$$(T_0+1)(2T_1+1)\cdots(2T_m+1) \ge \frac{U^{m+1}}{(2mn)^m D^{m+1} S^m (\log B_1)(\log A_{10})\cdots(\log A_{m0})}.$$

This explains the first main condition relating U and S that we require, namely:

$$U^m \ge 66m(2mn)^m D^{m+1} S^m (\log B_1)(\log A_{10})\cdots(\log A_{m0})(\log E)^{-1}.$$

Our second main condition will enable us to check

$$(2S_1+1)\cdots(2S_{n+1}+1) > (m+1)!2^m T_0 T_1 \cdots T_m.$$

Otherwise the conclusion of Theorem 13.1 would be trivial by taking for G^* the trivial subgroup $\{e\}$.

Since $(m+1)! \le 2m^m$,

$$T_0 T_1 \cdots T_m \le \frac{U^{m+1}}{(2mn)^m D^{m+1} S^m (\log B_1)(\log A_{10})\cdots(\log A_{m0})}$$

and

$$(2S_1+1)\cdots(2S_{n+1}+1) > S^{n+1}\frac{(\log A_{10})^n}{(\log A_{11})\cdots(\log A_{1n})},$$

our second main condition relating U and S will be:

$$n^m D^{m+1} S^{m+n+1} (\log B_1)(\log A_{10})^n \prod_{i=1}^m \log A_{i0} \geq 2U^{m+1} \prod_{j=1}^n \log A_{1j}.$$

This is why we define S and U as follows:

$$S^{mn} = c_2 D^{m+1} (\log B_1) \left(\prod_{i=1}^m \prod_{j=0}^n \log A_{ij} \right) \prod_{i=1}^m (\log A_{i0})^{-n} \cdot (\log E)^{-m-1}$$

and

$$U^{mn} = c_3 D^{(m+1)(n+1)} (\log B_1)^{n+1} \left(\prod_{i=1}^m \prod_{j=0}^n \log A_{ij} \right) (\log E)^{-m-n-1}$$

with positive constants c_2 and c_3 which should satisfy

$$c_3 \geq c_2 (66m)^n (2mn)^{mn}$$

and

$$c_2^{m+n+1} n^{m^2 n} \geq c_3^{m+1} 2^{mn}.$$

We define

$$c_4 = 33(2m)^{m+1} n^m, \quad c_3 = c_4^n c_2, \quad c_2 = 33^{mn+1} 2^{m^2+3mn} m^{m^2+2mn+1} n^m$$

so that

$$c_2 \geq c_4^{m+1} 2^m n^{-m^2},$$

and

$$c_3 = 33^{mn+n+1} 2^{m^2+4mn+n+1} m^{m^2+3mn+n+1} n^{m(n+1)} \leq \left(66m^3 n\right)^{mn} (2m)^{m^2}.$$

We now check the condition $(m+1)(S_1 + \cdots + S_{n+1}) \leq B_1$. We need to bound S_j from above. From the definitions of S and the inequality $S_j \log A_{ij} \leq S \log A_{i0}$ we deduce

$$S_j^{mn} \leq c_2 D^{m+1} (\log B_1) \left(\prod_{i=1}^m \frac{(\log A_{i0}) \cdots (\log A_{in})}{(\log A_{ij})^n} \right) (\log E)^{-m-1}.$$

Using the estimates $\log E \leq D \log A_{ij} \leq B$ we deduce

$$\frac{(\log A_{i1}) \cdots (\log A_{in})}{(\log A_{ij})^n} \leq \left(\frac{B}{\log E} \right)^{n-1} \quad \text{and} \quad \log A_{i0} \leq \frac{B}{D},$$

hence

$$S_j^{mn} \leq c_2 D (\log B_1) B^{mn} (\log E)^{-mn-1}.$$

We now use the assumptions $D \leq B$ and $E \geq e$ with $\log B_1 = c_1 \log B$ and we find

$$S_j^{mn} \leq c_1 c_2 B^{mn+1} \log B.$$

Therefore we only need to check

$$\big((m+1)(n+1)\big)^{mn} c_1 c_2 B^{mn+1} \log B \le B^{c_1 mn},$$

and since we have assumed $B \ge e$ we deduce that this inequality holds with $c_1 = 20m^2 n$. The inequality $U \ge 20D\log(4mD)$ also plainly follows from $B \ge D$ and $U_1 \ge D \log B$ since $c_3 c_1^{n+1} > (2^{10} m^3 n)^{mn}$.

By the preceding choices, we have $U = (c_3 c_1^{n+1})^{1/mn} U_1$,

$$T_i = \left[\left(\frac{c_4 D (\log B_1)(\log A_{10}) \cdots (\log A_{m0})}{\log E} \right)^{1/m} \frac{1}{2mn \log A_{i0}} \right]$$

and $S_j = [\tilde{S}_j]$ where

$$\tilde{S}_j^{mn} = c_2 D^{m+1} (\log B_1) \left(\prod_{i=1}^{m} (\log A_{i0}) \cdots (\log A_{in})(\log A_{ij})^{-n} \right) (\log E)^{-m-1}.$$

Assume now that the conclusion of Theorem 14.1 does not hold. Then

$$\max_{1 \le j \le n+1} |\underline{\eta}_j - \underline{\eta}'_j| < e^{-cU_1}.$$

We wish to check the hypotheses of Theorem 13.1. Hence we want to deduce

$$\max_{1 \le j \le n+1} |\underline{\eta}_j - \underline{\eta}'_j| < e^{-V}.$$

If we get a contradiction with some value of the parameter U, then we shall deduce the desired result for any c satisfying

$$c \ge 33m \big(c_3 c_1^{n+1}\big)^{1/mn}.$$

The estimate

$$c_1^{n+1} \le (400n)^{mn}$$

shows that the value

$$c \le 2^{32} m^4 n^2 (2m)^{m/n}$$

is admissible.

Hence we have checked that if the conclusion of Theorem 14.1 does not hold, then all the assumptions of Theorem 13.1 are satisfied for

$$\Sigma = \Big\{ \big(s_1\beta_1 + \cdots + s_{n+1}\beta_{n+1}, \alpha_{11}^{s_1} \cdots \alpha_{1,n+1}^{s_{n+1}}, \ldots, \alpha_{m1}^{s_1} \cdots \alpha_{m,n+1}^{s_{n+1}} \big) ; \\ (s_1, \ldots, s_{n+1}) \in \mathcal{S} \Big\} \subset G(K).$$

Therefore there exists a connected algebraic subgroup G^* of G, which is incompletely defined by polynomials of degrees $\le (T_0, T_1, \ldots, T_m) = (T_0, \underline{T})$, such that

$$\operatorname{Card}\left(\frac{\Sigma + G^*}{G^*} \right) \mathcal{H}(G^*; T_0, \underline{T}) \le (m+1)! 2^m T_0 T_1 \cdots T_m.$$

Notice that this inequality would be trivial if $\mathcal{H}(G^*; T_0, \underline{T}) = 0$, that is if one T_i would vanish. However the inequality $T_0 \geq 1$ follows from the lower bounds

$$U_1 \geq D \log B, \quad c_3 > c_1^{mn} \quad \text{and} \quad U \geq c_1 U_1 \geq D \log B_1,$$

while condition (14.2) implies $T_i \geq 1$ for $1 \leq i \leq m$, since $c_1 c_4 > (2mn)^m$.

From the inequality

$$(m+1)! 2^m T_0 T_1 \cdots T_m < (2S_1 + 1) \cdots (2S_{n+1} + 1)$$

which we have checked earlier, and since $\mathcal{H}(G^*; T_0, \underline{T}) \geq 1$, we deduce

$$\operatorname{Card}\left(\frac{\Sigma + G^*}{G^*}\right) < (2S_1 + 1) \cdots (2S_{n+1} + 1),$$

hence there exist $\underline{\gamma}' \neq \underline{\gamma}''$ in Σ such that $\underline{\gamma} = \underline{\gamma}' - \underline{\gamma}'' \in G^*$.

Here comes the final descent. We write $G^* = G_0^* \times G_1^*$, where G_0^* is an algebraic subgroup of $G_0 = \mathbb{G}_a$, hence G_0^* is either 0 or $\mathbb{G}_a$, while G_1^* is a connected algebraic subgroup of $G_1 = \mathbb{G}_m^m$ which is incompletely defined by polynomials of degrees $\leq (T_1, \ldots, T_m)$.

We first check $G_0^* = \mathbb{G}_a$, which means $G_0^* \neq 0$. For this it is sufficient to show that any relation

$$s_1 \beta_1 + \cdots + s_{n+1} \beta_{n+1} = 0$$

with $\underline{s} \in \mathbb{Z}^{n+1}$ satisfying $|s_j| \leq 2S_j \quad (1 \leq j \leq n+1)$ implies $\underline{s} = 0$. Using our assumption

$$\sum_{i=1}^{m} \sum_{j=1}^{n+1} |\lambda_{ij} - \beta_j \lambda_i| \leq e^{-33mU},$$

we deduce, for $1 \leq i \leq m$,

$$\left| \sum_{j=1}^{n+1} s_j \lambda_{ij} \right| \leq B_1 e^{-33mU}.$$

In order to deduce from Liouville's inequality (Exercise 3.7.b) that the left hand side vanishes for at least one index i in the range $1 \leq i \leq m$, it suffices to check

$$B_1 e^{-33mU} < 2^{-D} \prod_{j=1}^{n+1} e^{-D|s_j| \mathrm{h}(\alpha_{ij})}.$$

From (14.2) we deduce $\log E \leq D \log B$. Since

$$\left(\frac{U}{DS}\right)^m = \frac{c_4 D \log B_1}{\log E} \prod_{i=1}^{m} \log A_{i0} \geq c_1 c_4 \prod_{i=1}^{m} \log A_{i0}$$

and $c_1 c_4 (8m)^m > n^m$, there is at least one i with $1 \leq i \leq m$ for which

$$\frac{U}{DS} \geq (c_1 c_4)^{1/m} \log A_{i0} \geq \frac{n}{8m} \log A_{i0}.$$

We bound first $\log B_1 + D\log 2$ by mU (recall $c_3 > c_1^{mn}$), next $S_j \log A_{ij}$ by $S\log A_{i0}$, and $n+1$ by $2n$, and we deduce

$$33mU > \log B_1 + D\log 2 + 2D\sum_{j=1}^{n+1} S_j \log A_{ij}.$$

Therefore we may conclude

$$\sum_{j=1}^{n+1} s_j \lambda_{ij} = 0.$$

We want to use the condition of linear independence of λ_{ij}; we need to check $2S_j \leq (cU_1)^2$. In fact we shall need later the estimate $2S_j \leq cU$ which is much stronger since $U < cU_1$. Indeed, since $S_j \log A_{ij} \leq S\log A_{i0}$ for $1 \leq j \leq n+1$, we have

$$\left(\frac{U}{DS_j}\right)^m \geq \frac{c_4 D\log B_1}{\log E}\prod_{i=1}^{m}\log A_{ij}.$$

We deduce from (14.2) the lower bound $D\log B \geq \log E$; we also use the assumptions $D\log A_{ij} \geq \log E \geq 1$ and $\log B_1 = c_1 \log B$. We obtain

$$S_j \leq (c_1 c_4)^{1/m} U.$$

This completes the proof of our claim $G_0^* = \mathbb{G}_{\mathrm{a}}$.

Since $G^* \neq G$, it follows that $G_1^* \neq G_1$. Let Σ_1 be the projection of Σ on $\mathbb{G}_{\mathrm{m}}^m$. For $\underline{s} \in \mathbb{Z}^{n+1}$, define

$$\underline{\gamma}(\underline{s}) = \left(\alpha_{11}^{s_1}\cdots\alpha_{1,n+1}^{s_{n+1}}, \ldots, \alpha_{m1}^{s_1}\cdots\alpha_{m,n+1}^{s_{n+1}}\right) \in (K^\times)^m,$$

so that

$$\Sigma_1 = \left\{\underline{\gamma}(\underline{s})\,;\ \underline{s} \in \mathcal{S}\right\}.$$

Define $\mathcal{E}$ as the set of $\underline{s} \in \mathbb{Z}^{n+1}$ such that $|s_j| \leq 2S_j$ $(1 \leq j \leq n+1)$ and $\underline{\gamma}(\underline{s}) \in G_1^*$. Then, by Lemma 7.8, we have

$$\operatorname{Card}\left(\frac{\Sigma + G^*}{G^*}\right) = \operatorname{Card}\left(\frac{\Sigma_1 + G_1^*}{G_1^*}\right) \geq (2S_1+1)\cdots(2S_{n+1}+1)\big(\operatorname{Card}\mathcal{E}\big)^{-1}.$$

We deduce

$$(2S_1+1)\cdots(2S_{n+1}+1)\big(\operatorname{Card}\mathcal{E}\big)^{-1}\mathcal{H}(G^*; T_0, \underline{T}) \leq (m+1)!2^m T_0 T_1 \cdots T_m.$$

Since G_1^* is incompletely defined by polynomials of degrees $\leq (T_1, \ldots, T_m)$ and $G_1^* \neq G_1$, it follows that $T_e(G_1^*)$ is contained in some hyperplane $t_1 z_1 + \cdots + t_m z_m = 0$ of $\mathbb{C}^m = T_e(G_1)$ where $\underline{t} \in \mathbb{Z}^m \setminus \{0\}$ satisfies $|t_i| \leq T_i$ $(1 \leq i \leq m)$. For each such hyperplane and each $\underline{s} \in \mathcal{E}$, we have

$$\prod_{i=1}^{m}\prod_{j=1}^{n+1}\alpha_{ij}^{t_i s_j} = 1.$$

Hence the number

$$k(\underline{t}, \underline{s}) := \frac{1}{2i\pi} \sum_{i=1}^{m} \sum_{j=1}^{n+1} t_i s_j \lambda_{ij}$$

is a rational integer. For $\underline{s} \in \mathcal{E}$, we have $|s_j| \le 2S_j \quad (1 \le j \le n+1)$. From the assumption $|\lambda_j|E \le D \log A_{ij}$ we deduce the upper bound

$$|k(\underline{t}, \underline{s})| \le \frac{D}{\pi E} \sum_{i=1}^{m} \sum_{j=1}^{n+1} T_i S_j \log A_{ij} \le \frac{U}{\pi e} < U.$$

We split the proof in three cases:

(i) Assume $\mathcal{E}$ contains two $\mathbb{Q}$-linearly independent elements, say $\underline{s}'$ and $\underline{s}''$. Let $\underline{t} \in \mathbb{Z}^m \setminus \{0\}$ with $|t_i| \le T_i$ be such that the hyperplane $t_1 z_1 + \cdots + t_m z_m = 0$ contains $T_e(G_1^*)$. Define

$$\underline{s} = k(\underline{t}, \underline{s}')\underline{s}'' - k(\underline{t}, \underline{s}'')\underline{s}' \in \mathbb{Z}^{n+1} \setminus \{0\}.$$

Then we have

$$\sum_{i=1}^{m} \sum_{j=1}^{n+1} t_i s_j \lambda_{ij} = 0.$$

Since $|t_i| \le T_i \le U < cU_1$ and $|s_j| \le 2US_j \le (cU_1)^2$, we get a contradiction with the assumption that the matrix $\mathsf{L}_{m,n+1}$ satisfies the condition of linear independence for $(cU_1)^2$.

(ii) Assume G_1^* has codimension ≥ 2 in G_1. In this case there exist two linearly independent elements $\underline{t}'$ and $\underline{t}''$ in $\mathbb{Z}^m \setminus \{0\}$ with $|t_i'| \le T_i$ and $|t_i''| \le T_i$ such that $T_e(G_1^*)$ is contained in the intersection of the two corresponding hyperplanes. Let $\underline{s} \in \mathcal{E}$. Define

$$\underline{t} = k(\underline{t}', \underline{s})\underline{t}'' - k(\underline{t}'', \underline{s})\underline{t}'.$$

Then we have

$$\sum_{i=1}^{m} \sum_{j=1}^{n+1} t_i s_j \lambda_{ij} = 0.$$

Since $|t_i| \le 2UT_i \le (cU_1)^2$ and $|s_j| \le 2S_j \le cU_1$, we get again a contradiction with the assumption of linear independence of the matrix $\mathsf{L}_{m,n+1}$.

(iii) In the remaining case, we have $\mathrm{Card}\mathcal{E} \le \max_{1\le j\le n+1}(2S_j + 1)$ and G_1^* has codimension 1. By Proposition 5.14 we have

$$\mathcal{H}(G^*; T_0, \underline{T}) \ge m!2^{m-1} T_0 T_1 \cdots T_m \left(\max_{1\le i\le m} T_i \right)^{-1}.$$

Therefore we get

$$(2S_1 + 1) \cdots (2S_{n+1} + 1) \le 2(m+1) \left(\max_{1\le i\le m} T_i \right) \left(\max_{1\le j\le n+1} (2S_j + 1) \right).$$

We bound the left hand side from below:

$$
\begin{aligned}
(2S_1+1)\cdots(2S_{n+1}+1) &\geq S^{n+1}\frac{(\log A_{10})^n}{(\log A_{11})\cdots(\log A_{1n})} \\
&\geq \frac{(m+1)!2^m}{(2mn)^m}\left(\frac{U}{DS}\right)^m \frac{U}{D\log B_1}\prod_{i=1}^m (\log A_{i0})^{-1} \\
&\geq \frac{2c_4U}{n^m \log E}.
\end{aligned}
$$

Now only we use (14.3) and the inequality $c_1c_2 > 1$ to deduce $S_j \geq 1$. Hence $2S_j + 1 \leq 3S_j$. Since

$$
T_iS_j \leq \frac{U}{2mnD\log A_{ij}} \leq \frac{U}{2mn\log E}
$$

we again derive a contradiction from the estimate $2mnc_4 > 3(m+1)n^m$. □

14.1.3 Simultaneous Approximation for $\alpha_s^{\beta_j\beta'_r}$

We derive from Theorem 14.1 a simultaneous approximation measure for numbers of the form $\alpha_s^{\beta_j\beta'_r}$ when α_s, β_j and β'_r are algebraic numbers ($0 \leq j \leq n$, $1 \leq r \leq p$, $1 \leq s \leq q$).

Corollary 14.4. *Let $\beta_0, \ldots, \beta_n$ be $\mathbb{Q}$-linearly independent algebraic numbers, $\beta'_1, \ldots, \beta'_p$ also $\mathbb{Q}$-linearly independent algebraic numbers and $\lambda_1, \ldots, \lambda_q$ be $\mathbb{Q}$-linearly independent elements of $\mathcal{L}$. There exists a constant $c > 0$ such that the function*

$$
\varphi(D,h) = cD^{(n+1)(pq+1)/npq}h^{1+(1/n)}(\log h + \log D)^{-1/n}
$$

is a simultaneous approximation measure for the $(n+1)pq$ numbers

$$
e^{\beta_j\beta'_r\lambda_s} \quad (0 \leq j \leq n,\ 1 \leq r \leq p,\ 1 \leq s \leq q).
$$

Example. (Compare with [RoyW 1997b], Th. 2.1). Let β be an algebraic number of degree d and λ a nonzero element of $\mathcal{L}$. Then there exists a positive number $c = c(\beta, \lambda)$ such that

$$
cD^{(d+1)/(d-1)}h^{d/(d-1)}(\log h + \log D)^{-1/(d-1)}
$$

is a simultaneous approximation measure for the $d-1$ numbers

$$
e^{\beta\lambda}, \ldots, e^{\beta^{d-1}\lambda}.
$$

This follows from Corollary 14.4 by taking $n = d-1$, $p = d$, $q = 1$,

$$
\beta_j = \beta^j \quad (0 \leq j \leq n), \qquad \beta'_r = \beta^{r-1} \quad (1 \leq r \leq p) \quad \text{and} \quad \lambda_1 = \lambda.
$$

Proof. We deduce Corollary 14.4 from Theorem 14.1 with $m = pq$ (and the index i, $1 \le i \le m$ is replaced here by (r, s) with $1 \le r \le p$, $1 \le s \le q$).

Replacing if necessary β_j by β_j/β_0 and β'_r by $\beta_0\beta'_r$, there is no loss of generality to assume $\beta_0 = 1$.

Assume γ_{rsj} are algebraic numbers of height $\mathrm{h}(\gamma_{rsj}) \le \log A$ $(0 \le j \le n$, $1 \le r \le p$, $1 \le s \le q)$ in a field of degree $\le D$ with

$$\max_{\substack{0\le j\le n\\ 1\le r\le p\\ 1\le s\le q}} |e^{\beta_j\beta'_r\lambda_s} - \gamma_{rsj}| \le e^{-cU}$$

where

$$U^{npq} = D^{(n+1)(pq+1)}(\log A)^{(n+1)pq}\left(\log\log A + \log D\right)^{-pq}.$$

Define λ_{rsj} as the logarithm of γ_{rsj} which is close to $\beta_j\beta'_r\lambda_s$ and take

$$\lambda_{rs} = \lambda_{rs0}, \quad \log A_{rsj} = \log A \ge c_0, \quad B = (D\log A)^{c_0}, \quad E = (D\log A)^{1/c_0}$$

where c_0 is a sufficiently large constant (independent of the γ_{rsj}).

For $\underline{t} = (t_{rs}) \in \mathbb{Z}^{pq}$ and $T \ge 2$ with $0 < |\underline{t}| \le T$, the lower bound

$$\left|\sum_{r=1}^{p}\sum_{s=1}^{q} t_{rs}\beta'_r\lambda_s\right| \ge T^{-c_0}$$

holds (see for instance Theorem 9.1). Hence, if there is such a $\underline{t} \neq 0$ with $T = (cU)^2$ for which

$$\sum_{r=1}^{p}\sum_{s=1}^{q} t_{rs}\lambda_{rs} = 0,$$

then the conclusion of Corollary 14.4 plainly follows. Otherwise, the required linear independence condition on the numbers λ_{rs} is satisfied, and one may apply Theorem 14.1. This completes the proof of Corollary 14.4. □

Remark. This proof is dual of the proof given in [RoyW 1997b] (see the remark at the end of § 6 p. 407 of [RoyW 1997b]).

14.1.4 Simultaneous Approximation for y_j and $e^{x_i y_j}$

It will be useful to introduce the following definition.

Definition. A n-tuple $(\theta_1, \dots, \theta_n)$ of complex numbers satisfies a *linear independence measure condition* if for any $\epsilon > 0$ there exists $S_0 > 0$ such that, for any $S \ge S_0$ and any $\underline{s} \in \mathbb{Z}^n$ satisfying $0 < |\underline{s}| \le S$, we have

$$|s_1\theta_1 + \cdots + s_n\theta_n| \ge e^{-S^\epsilon}.$$

In particular such a tuple consists of $\mathbb{Q}$-linearly independent numbers. This linear independence measure condition will occur as an hypothesis in several results below.

Often, it would be possible to weaken this condition (for instance asking it holds only for a restricted range of ϵ, for instance $\epsilon > \kappa$ for some explicit $\kappa > 0$). In the results of diophantine approximation, such an hypothesis cannot be completely avoided (see Exercise 15.12). On the other hand, when it occurs in statements of algebraic independence (see Chap. 15), this assumption is known as "a technical hypothesis" and one expects that it is superfluous.

Example. Any tuple $(\Lambda_1, \ldots, \Lambda_n)$ of $\mathbb{Q}$-linearly independent elements in $\widetilde{\mathcal{L}}$ satisfies a linear independence measure condition. Indeed, if

$$\Lambda_i = \gamma_{i0} + \sum_{j=1}^{m} \gamma_{ij}\lambda_j$$

with $\gamma_{ij} \in \overline{\mathbb{Q}}$ $(1 \le i \le n, 0 \le j \le m)$ and with $\lambda_j \in \mathcal{L}$ $(1 \le j \le m)$, then

$$\sum_{i=1}^{n} s_i\Lambda_i = \beta_0 + \sum_{j=1}^{m} \beta_j\lambda_j,$$

where

$$\beta_j = \sum_{i=1}^{n} s_i\gamma_{ij} \qquad (0 \le j \le m).$$

From Lemma 3.7 we deduce $\mathrm{h}(\beta_j) \le c_1 \log S$ where c_1 does not depend on $\underline{s}$, hence Theorem 9.1 gives

$$|s_1\Lambda_1 + \cdots + s_n\Lambda_n| \ge S^{-c_2},$$

a much stronger estimate than what is actually needed.

Here is another example: if $\theta_1, \ldots, \theta_n$ are $\mathbb{Q}$-linearly independent elements in the $\overline{\mathbb{Q}}$-vector space spanned by $\exp(\overline{\mathbb{Q}})$, which means

$$\theta_i = \sum_{j=1}^{m} \beta_{ij}e^{\gamma_j} \quad (1 \le i \le n),$$

with algebraic γ's and β's, then the n-tuple $(\theta_1, \ldots, \theta_n)$ satisfies a linear independence measure condition. This result follows from a quantitative refinement to Lindemann-Weierstraß' Theorem (works of D. Morduhai-Boltovskoi, C. L. Siegel, K. Mahler,...); see for instance [Sh 1989], Chap. XIII § 3 Th. I. Further examples of tuples satisfying a linear independence measure condition can be deduced from the results of [Sh 1989], Chap. XIII § 3 and § 5 (see also [FNe 1998], Chap. V § 5).

Corollary 14.5. *Let $m \ge 1$ and $k \ge 2$ be positive integers, $(x_1, \ldots, x_m)$ be a m-tuple of complex numbers satisfying a linear independence measure condition, and $(y_1, \ldots, y_k)$ be a k-tuple of complex numbers satisfying a linear independence measure condition. There exists a constant $c > 0$ such that a simultaneous approximation measure for the $k + km$ numbers*

$$y_j, \quad e^{x_iy_j} \qquad (1 \le i \le m,\ 1 \le j \le k)$$

is

$$\varphi(D,h) = cD^{1+\frac{m+k}{m(k-1)}} h^{1+\frac{1}{k-1}} (h + \log D)^{\frac{1}{m}+\frac{1}{m(k-1)}} (\log h + \log D)^{-\frac{1}{m}-\frac{1}{k-1}-\frac{1}{m(k-1)}}.$$

Proof. Let $\gamma_1, \ldots, \gamma_k$ and γ_{ij} $(1 \leq i \leq m,\ 1 \leq j \leq k)$ be algebraic numbers in a field of degree $\leq D$ and heights $\leq h$. Our goal is to produce a lower bound for

$$\sum_{j=1}^{k} \left(|y_j - \gamma_j| + \sum_{i=1}^{m} |x_i y_j - \gamma_{ij}| \right).$$

Since $y_1 \neq 0$ and $x_i y_j \neq 0$, there is no loss of generality to assume $\gamma_1 \neq 0$ and $\gamma_{ij} \neq 0$. For $1 \leq i \leq m$ and $1 \leq j \leq k$, let $\log \gamma_{ij}$ be the value of the logarithm of γ_{ij} which is closer to $x_i y_j$.

In Theorem 14.1, set

$$n = k-1, \quad \beta_j = \frac{\gamma_{j+1}}{\gamma_1} \quad (1 \leq j \leq n), \qquad \lambda_i = \log \gamma_{i1} \quad (1 \leq i \leq m)$$

and

$$\lambda_{ij} = \log \gamma_{i,j+1} \quad (1 \leq j \leq n,\ 1 \leq i \leq m).$$

Next we choose

$$A_i = A_{ij} = e^h, \quad B = c_0 D e^h, \quad E = \frac{1}{c_0} Dh,$$

where c_0 is a sufficiently large positive number which depends only on $x_1, \ldots, x_m$ and $y_1, \ldots, y_k$. □

14.2 A Quantitative Refinement to Hermite-Lindemann's Theorem

By Hermite-Lindemann's Theorem, for $m \geq 1$ and $n \geq 1$, when $\beta_1, \ldots, \beta_n$, $\beta'_1, \ldots, \beta'_m$ are algebraic numbers and λ_{ij} are in $\mathcal{L}$ and not all zero, the matrix

$$\begin{pmatrix} 1 & \beta_1 & \cdots & \beta_n \\ \beta'_1 & & & \\ \vdots & & \lambda_{ij} & \\ \beta'_m & & & \end{pmatrix}$$

has rank ≥ 2. Theorem 14.6 below provides a lower bound for one at least of the 2×2 minors.

14.2.1 A Lower Bound for $\sum_{i=1}^{m}\sum_{j=1}^{n}|\log\alpha_{ij}-\beta_j\beta'_i|$

Our second main result deals with the simultaneous approximation of numbers $e^{\beta_j\beta'_i}$ by algebraic numbers. Recall the notation for $m, n, \mathsf{L}_{mn}, K, D, \lambda_{ij}, A_{ij}$ and E already given at the end of the introduction.

Theorem 14.6. *There exists a positive constant c, which depends only on m and n, with the following property. Let $\beta_1,\dots,\beta_n,\beta'_1,\dots,\beta'_m$ be algebraic numbers in K. Let B and B' be positive real numbers satisfying the following conditions:*

$$B\ge e,\quad B'\ge e,\quad B\ge D\log B',\quad B'\ge D\log B,$$

$$\mathrm{h}(1{:}\,\beta_1{:}\cdots{:}\,\beta_n)\le\log B\quad\textit{and}\quad \mathrm{h}(1{:}\,\beta'_1{:}\cdots{:}\,\beta'_m)\le\log B'.$$

Define

$$U_2^{mn}=D^{mn+m+n}(\log B)^n(\log B')^m\left(\prod_{i=1}^{m}\prod_{j=1}^{n}\log A_{ij}\right)(\log E)^{-m-n}$$

and assume that the matrix L_{mn} satisfies the linear independence condition for $(cU_2)^2$. Assume further

$$\log E\le D\log B\le U_2,\quad D\log B'\le U_2,\quad D\log A_{ij}\le B\quad\textit{and}\quad D\log A_{ij}\le B'$$

for $1\le i\le m$ and $1\le j\le n$. Then

$$\sum_{i=1}^{m}\sum_{j=1}^{n}|\lambda_{ij}-\beta_j\beta'_i|\ge e^{-cU_2}.$$

Remark 1. We shall check the conclusion with

$$c=2^{23}m^3n^2(2m)^{m/n}.$$

This is the only part of the result which is not symmetric by replacing

$$m,\ n,\ \beta_1,\dots,\beta_n,\ \beta'_1,\dots,\beta'_m,\ B,\ B'$$

and the matrix L_{mn} respectively by

$$n,\ m,\ \beta'_1,\dots,\beta'_m,\ \beta_1,\dots,\beta_n,\ B',\ B$$

and the transposed matrix ${}^t\mathsf{L}_{mn}$.

Remark 2. One may replace the linear independence condition on the matrix L_{mn} by the following hypotheses:

for any $\underline{t}\in\mathbb{Z}^m\setminus\{0\}$ satisfying $|t_i|\le(cU_2)^2$ for $1\le i\le m$, we have

$$t_1\beta'_1+\cdots+t_m\beta'_m\ne0$$

and

for any $\underline{s} \in \mathbb{Z}^n \setminus \{0\}$ *satisfying* $|s_j| \leq (cU_2)^2$ *for* $1 \leq j \leq n$, *we have*

$$s_1\beta_1 + \cdots + s_n\beta_n \neq 0.$$

Remark 3. Up to the numerical values of the constants, the estimate is the best one can expect from Theorem 13.1 since, with the notation of Proposition 13.12, we have

$$u = mn, \quad \delta = mn + m + n, \quad b_1 = n \quad b_2 = m.$$

14.2.2 Proof of Theorem 14.6

Thanks to the remark in § 13.5.3, we may assume without loss of generality

$$D(\log B)(\log A_{11})\cdots(\log A_{m1}) \geq (\log A_{i1})^m \log E \tag{14.7}$$

for $1 \leq i \leq m$ and

$$D(\log B')(\log A_{11})\cdots(\log A_{1n}) \geq (\log A_{1j})^n \log E \tag{14.8}$$

for $1 \leq j \leq n$.

We apply Theorem 13.1 with $d_0 = 1$, $d_1 = m$, $d = m+1$, $G = G^+ = G_0 \times G_1$, $G_0 = \mathbb{G}_{\mathrm{a}}$, $G_1 = \mathbb{G}_{\mathrm{m}}^m$, $G^- = \{e\}$, $\ell_0 = 1$, $\ell_1 = n$, $r = r_3 = 1$, $r_1 = r_2 = 0$. Define

$$\underline{w}_1 = (1, \beta_1', \ldots, \beta_m') \in K^{m+1}, \quad \mathcal{W} = K\underline{w}_1$$

and, for $1 \leq j \leq n$,

$$\underline{\eta}_j = (\beta_j, \lambda_{1j}, \ldots, \lambda_{mj}) \in K \times \mathcal{L}^m,$$

$$\underline{\gamma}_j = \exp_G \underline{\eta}_j = (\beta_j, \alpha_{1j}, \ldots, \alpha_{mj}) \in G(K) = K \times (K^\times)^m.$$

Next put $\underline{w}_1' = \beta_1\underline{w}_1$ and, for $1 \leq j \leq n$,

$$\underline{\eta}_j' = \beta_j\underline{w}_1 = (\beta_j, \beta_1'\beta_j, \ldots, \beta_m'\beta_j) \in K^{m+1}.$$

The vector spaces

$$\mathcal{W}' = \mathbb{C}\underline{w}_1' \quad \text{and} \quad \mathcal{X}' = \mathbb{C}\underline{\eta}_1' + \cdots + \mathbb{C}\underline{\eta}_n'$$

are both equal to $\mathbb{C}\underline{w}_1$.

We define $B_1 = B^{c_1}$ and $B_2 = B^{c_1'}$ where $c_1 > 1$ and $c_1' > 1$ are two real numbers which we are going to fix later. We introduce three positive numbers c_2, c_3, c_4 related by $c_3 = c_4^n c_2^m$ and we define S and U by

$$S^n = \frac{c_2 D \log B_2}{\log E}\left(\prod_{j=1}^n \frac{\log A_{1j}}{\log A_{11}}\right)$$

and

$$U^{mn} = c_3 D^{mn+m+n} (\log B_1)^n (\log B_2)^m \left(\prod_{i=1}^{m} \prod_{j=1}^{n} \log A_{ij} \right) (\log E)^{-m-n}$$

so that

$$\left(\frac{U}{DS}\right)^m = \frac{c_4 D \log B_1}{\log E} \cdot (\log A_{11}) \cdots (\log A_{m1}).$$

Then define

$$T_0 = \left[\frac{U}{D\log B_1}\right], \quad S_0 = \left[\frac{U}{D\log B_2}\right],$$

$$T_i = \left[\frac{U}{mnDS\log A_{i1}}\right] \quad (1 \le i \le m),$$

$$S_j = \left[S\frac{\log A_{11}}{\log A_{1j}}\right] \quad (1 \le j \le n),$$

so that the condition

$$D\sum_{i=1}^{m}\sum_{j=1}^{n} T_i S_j \log A_{ij} \le U$$

is satisfied.

We now want to check

$$(T_0+1)(2T_1+1)\cdots(2T_m+1) \ge \frac{2V}{\log E}$$

where $V \le 33mU$. The left hand side is bounded from below by

$$(T_0+1)(2T_1+1)\cdots(2T_m+1) \ge \frac{U^{m+1}}{(mn)^m D^{m+1} S^m (\log B_1)(\log A_{11})\cdots(\log A_{m1})}.$$

Therefore the condition we need reads

$$U^m \ge 33m(mn)^m D^{m+1} S^m (\log B_1)(\log A_{11})\cdots(\log A_{m1})(\log E)^{-1}$$

and it will hold as soon as

$$c_3 \ge (33m^{m+1}n^m)^n c_2^m.$$

Next we show

$$(S_0+1)(2S_1+1)\cdots(2S_n+1) > (m+1)!2^m T_0 T_1 \cdots T_m$$

On one hand we have

$$(S_0+1)(2S_1+1)\cdots(2S_n+1) > \frac{U}{D\log B_2} \cdot S^n (\log A_{11})^n \prod_{j=1}^{n} (\log A_{1j})^{-1}.$$

On the other hand $(m+1)!2^m \le m^m 2^{m+1}$ and

$$T_0T_1\cdots T_m \le \frac{U^{m+1}}{(mn)^m D^{m+1}S^m(\log B_1)(\log A_{11})\cdots(\log A_{m1})}.$$

Hence the required condition

$$n^m S^{m+n} D^m (\log A_{11})^n (\log B_1)\prod_{i=1}^{m}\log A_{i1} \ge 2^{m+1}U^m(\log B_2)\prod_{j=1}^{n}\log A_{1j}$$

will follow if

$$n^{mn}c_2^{m+n} \ge 2^{(m+1)n}c_3.$$

This explains the following choice:

$$c_4 = 33m^{m+1}n^m, \quad c_2 = 33(2m)^{m+1}, \quad c_3 = c_4^n c_2^m.$$

We now check that the inequalities

$$\log E \le D\log B_1, \quad \log E \le D\log B_2$$

and

$$B_1 \ge (m+1)S^* + (m+1)\frac{S_0}{T^*}, \quad B_2 \ge T^* + \frac{T_0}{dS^*}$$

are satisfied if one chooses $c_1 = 12mn$ and $c_1' = 8mn$. Recall that $T^* = T_1 + \cdots + T_m$ and $S^* = S_1 + \cdots + S_n$. Since

$$S_j^n \le c_2\frac{D\log B_2}{\log E}\cdot\frac{(\log A_{11})\cdots(\log A_{1n})}{(\log A_{1j})^n}$$

we have

$$S_j \le (c_2 D(\log B_2)(B')^{n-1})^{1/n} \le (c_2c_1')^{1/n}B'(\log B')^{1/n}$$

and

$$(m+1)S^* + (m+1)\cdot\frac{S_0}{T^*} \le \big((m+1)(n+1)c_1' + 2m(m+1)(c_1')^{-1}\big)c_2^{1/n}B^2$$

because $\log B' \le B$. Further we have

$$5mn(c_1'c_2)^{1/n}B^2 \le B^{c_1}$$

because $B \ge e$. Similarly, from the upper bound

$$T_i^m \le c_4\frac{D\log B_1}{\log E}\cdot\frac{(\log A_{11})\cdots(\log A_{m1})}{(mn\log A_{i1})^m},$$

we deduce

$$T_i \le \frac{c_4^{1/m}}{mn}B'(\log B_1)^{1/m}$$

and therefore

$$T^* \le \frac{1}{n}c_4^{1/m}(B')^{c_1'}(\log B_1)^{1/m} \le B_2$$

because $\log B \leq B'$ and

$$\frac{1}{n}(c_4c_1)^{1/m} \leq (B')^{c_1'-1-1/m}.$$

The connection between U and U_2 is

$$U^{mn} = c_3 c_1^n (c_1')^m U_2^{mn}.$$

If the conclusion of Theorem 14.6 does not hold, then we deduce

$$|\underline{w}_1 - \underline{w}_1'| \leq e^{-cU_1} \quad \text{and} \quad \max_{1\leq j\leq n} |\underline{\eta}_j - \underline{\eta}_j'| < e^{-cU_1}.$$

The hypotheses of Theorem 13.1 will follow if we take

$$c \geq 33mc_3^{1/mn}c_1^{1/m}(c_1')^{1/n}.$$

Since $c_1^{1/m} \leq 12n$, $(c_1')^{1/n} \leq 8m$ and

$$c_3 \leq (2m)^{m^2}(2\cdot 33^2mn)^{mn},$$

we shall obtain the desired result with

$$c = 2^{23}m^3n^2(2m)^{m/n}.$$

We have now checked all hypotheses of Theorem 13.1. Hence we obtain an algebraic subgroup $G^* = G_0^* \times G_1^*$ of G such that

$$\binom{S_0 + \ell_0^b}{\ell_0^b}\operatorname{Card}\left(\frac{\Sigma + G^*}{G^*}\right)\mathcal{H}(G^*;\, T_0, \underline{T}) \leq (m+1)!2^m T_0T_1\cdots T_m$$

where

$$\ell_0^b = \dim_{\mathbb{C}}\left(\frac{\mathcal{W} + T_e(G^*)}{T_e(G^*)}\right) \in \{0, 1\}.$$

Inequalities (14.7) and (14.8) imply $T_i \geq 1$ for $1 \leq i \leq m$ and $S_j \geq 1$ for $1 \leq j \leq n$ respectively. The condition $T_0 > 0$ follows from $U > D\log B_1$ and inequality $S_0 > 0$ follows from $U > D\log B_2$.

In particular we have $\mathcal{H}(G^*;\, T_0, \underline{T}) \geq 1$. Therefore

$$\binom{S_0 + \ell_0^b}{\ell_0^b}\operatorname{Card}\left(\frac{\Sigma + G^*}{G^*}\right) < (S_0+1)(2S_1+1)\cdots(2S_n+1).$$

We check $\ell_0^b = 1$. Otherwise one would have $\mathcal{W} \subset T_e(G^*)$, which means $(1, \beta_1', \ldots, \beta_m') \in T_e(G^*)$. In this case $G_0^* = \mathbb{G}_a$, hence $G_1^* \neq G_1$. Let $t_1z_1 + \cdots + t_mz_m = 0$ be an equation of an hyperplane in $T_e(G_1) = \mathbb{C}^m$ containing $T_e(G_1^*)$, where $\underline{t} \in \mathbb{Z}^m \setminus \{0\}$ satisfies $|t_i| \leq T_i$ $(1 \leq i \leq m)$. We get a relation

$$t_1\beta_1' + \cdots + t_m\beta_m' = 0,$$

and Liouville's inequality easily yields a contradiction with the hypothesis of linear independence on the matrix L_{mn}.

Therefore $\ell_0^b = 1$, and we obtain

$$\operatorname{Card}\left(\frac{\Sigma + G^*}{G^*}\right) < (2S_1 + 1) \cdots (2S_n + 1).$$

Next we check $G_0^* = G_0$. Otherwise we would have $G_0^* = \{0\}$, and if Σ_0 denotes the projection of Σ on G_0, then

$$\operatorname{Card}\left(\frac{\Sigma + G^*}{G^*}\right) = \operatorname{Card}\left(\frac{\Sigma_0 + G_0^*}{G_0^*}\right) = \operatorname{Card}\Sigma_0 = (2S_1 + 1) \cdots (2S_n + 1)$$

because the elements $s_1\beta_1 + \cdots + s_n\beta_n$ are pairwise distinct (again this follows from Liouville's inequality together with the hypothesis of linear independence on the matrix L_{mn}). Hence $G_0^* = \mathbb{G}_a$.

Let Σ_1 denotes the projection of Σ on G_1. For each $\underline{\gamma}' \neq \underline{\gamma}''$ in Σ_1 for which $\underline{\gamma}' - \underline{\gamma}'' \in G_1^*$, and for each hyperplane of $T_e(G^*)$ containing $T_e(G_1^*)$ of equation $t_1 z_1 + \cdots + t_m z_m = 0$, we get a relation

$$\prod_{i=1}^{m} \prod_{j=1}^{n} \alpha_{ij}^{t_i s_j} = 1.$$

Using the hypothesis of linear independence of λ_{ij} together with Liouville's inequality and the estimates $mnT_iS_j \leq U$, $2U^2 \leq c^2 U_1^2$, we deduce

$$\operatorname{Card}\left(\frac{\Sigma + G^*}{G^*}\right) = \operatorname{Card}\left(\frac{\Sigma_1 + G_1^*}{G_1^*}\right) \geq \frac{(2S_1 + 1) \cdots (2S_n + 1)}{\max\limits_{1 \leq j \leq n} (2S_j + 1)}$$

and

$$\mathcal{H}(G^*; T_0, \underline{T}) \geq m! 2^{m-1} \frac{T_0 T_1 \cdots T_m}{\max\limits_{1 \leq i \leq m} T_i}.$$

Therefore

$$(S_0 + 1)(2S_1 + 1) \cdots (2S_n + 1) \leq 2(m + 1) \left(\max_{1 \leq i \leq m} T_i\right) \left(\max_{1 \leq j \leq n} (2S_j + 1)\right).$$

On the other hand combining the inequalities

$$(S_0 + 1)(2S_1 + 1) \cdots (2S_n + 1) \geq \frac{c_2 U}{\log E}$$

and

$$2(m + 1)(2S_j + 1)T_i \leq 6(m + 1)T_i S_j \leq 6(m + 1)\frac{U}{mnD \log A_{ij}} \leq \frac{12U}{n \log E}$$

with $c_2 > 12/n$, we deduce

$$2(m + 1)(2S_j + 1)T_i < (S_0 + 1)(2S_1 + 1) \cdots (2S_n + 1)$$

for $1 \leq i \leq m$ and $1 \leq j \leq n$, which is not compatible with the previous estimate. This contradiction concludes the proof of Theorem 14.6. □

14.2.3 Lower Bound for $\sum_{j=1}^{n} |\lambda_j - \beta_j|$

The next statement is the special case $m = 1$ of Theorem 14.6. It provides a refinement to Theorem 8.1 of [RoyW 1997b] (here we introduce several parameters $A_1, \ldots, A_n$ instead of the single $A = \max_{1\le j\le n} A_j$).

Corollary 14.9. *Let n be a positive integer. There exists a positive constant c with the following property. Let $\alpha_1, \ldots, \alpha_n$ and $\beta_1, \ldots, \beta_n$ be algebraic numbers, let D be the degree of the number field they generate, and let $A_1, \ldots, A_n$, A, B, B', E be real numbers which satisfy*

$$B \ge e, \quad B' \ge e, \quad A = \max_{1\le j\le n} A_j,$$

$$\mathrm{h}(\alpha_j) \le \log A_j \quad (1 \le j \le n) \qquad \text{and} \quad \mathrm{h}(1\colon \beta_1 \colon \cdots \colon \beta_n) \le \log B.$$

For $1 \le j \le n$, assume that the number α_j is nonzero, choose $\lambda_j \in \mathcal{L}$ such that $e^{\lambda_j} = \alpha_j$ and assume

$$|\lambda_j| \le \frac{D}{E} \log A_j.$$

Let U be a positive real number satisfying

$$U \ge D^{2+(1/n)}(\log B)\big((\log B')(\log A_1)\cdots(\log A_n)\big)^{1/n}(\log E)^{-1-(1/n)};$$

$$U \ge D^2(\log B)(\log A)(\log E)^{-1-(1/n)}.$$

Further, assume

$$1 \le \log E \le D\log A_j \le B, \quad \log B' \le D\log A,$$

$$B' \ge D\log A, \quad U \ge D\log B,$$

$$\log E \le D\log B \le B', \quad \text{and} \quad \log E \le D\log B' \le B.$$

Furthermore, assume

$$s_1\beta_1 + \cdots + s_n\beta_n \ne 0$$

for any $\underline{s} \in \mathbb{Z}^n \setminus \{0\}$ with

$$0 < |\underline{s}| \le (cU)^2.$$

Then, we have

$$\sum_{j=1}^{n} |\lambda_j - \beta_j| \ge e^{-cU}$$

where

$$c = 2^{24}n^2.$$

Proof. We deduce Corollary 14.9 from Theorem 14.1 by taking $m = 1$, $\beta_1' = 1$. □

We deduce a diophantine approximation estimate related with Schanuel's Conjecture (Theorem 2.5 of [RoyW 1997b]).

Corollary 14.10. *Let $x_1, \dots, x_n$ be complex numbers which satisfy a linear independence measure condition. There exists a positive constant $c = c(n, x_1, \dots, x_n)$ such that the function*

$$\varphi(D, h) = cD^{2+(1/n)} h(h + \log D)(\log h + \log D)^{-1}$$

is a simultaneous approximation measure for the $2n$ numbers

$$x_1, \dots, x_n, e^{x_1}, \dots, e^{x_n}.$$

Here is an estimate of simultaneous diophantine approximation related to the Lindemann-Weierstraß' Theorem:

Corollary 14.11. *Let $\beta_1, \dots, \beta_m$ be $\mathbb{Q}$-linearly independent algebraic numbers. There exists a positive constant $c = c(\beta_1, \dots, \beta_m)$ such that the function*

$$\varphi(D, h) = cD^{1+(1/m)} h(\log h + D \log D)(\log h + \log D)^{-1}$$

is a simultaneous approximation measure for the numbers $e^{\beta_1}, \dots, e^{\beta_m}$.

Finally we deduce from Corollary 14.9 a statement on simultaneous approximation of logarithms of algebraic numbers:

Corollary 14.12. *Let $\alpha_1, \dots, \alpha_m$ be nonzero algebraic numbers. For $1 \le i \le m$, let λ_i be a determination of the logarithm of α_i. Assume the numbers $\lambda_1, \dots, \lambda_m$ are $\mathbb{Q}$-linearly independent. Then there exists a positive constant $c = c(\lambda_1, \dots, \lambda_m)$ such that*

$$\varphi(D, h) = cD^{2+(1/m)}(h + \log D)(\log h + \log D)^{1/m}(\log D)^{-1-(1/m)}$$

is a simultaneous approximation measure for the numbers $\lambda_1, \dots, \lambda_m$.

Proof. We permute m and n and take

$$E = D, \qquad B' = Dh, \qquad B = De^h.$$

Recall that without loss of generality we may assume $D \ge 2$. □

Remark. The following estimate, due to N.I. Feld'man ([F 1982], Th. 7.7 Chap. 7 § 5) is stronger when $h > D$ (the point is that our proof of Theorem 14.6, hence of Corollary 14.9 and then of Corollary 14.12, does not involve Feld'man's polynomials; see Exercise 14.5):

(14.13*) *Let $\lambda_1, \dots, \lambda_m$ be $\mathbb{Q}$-linearly independent logarithm of algebraic numbers. There exists a positive constant $c = c(\lambda_1, \dots, \lambda_m)$ such that*

$$cD^{2+(1/m)}(h + \log D)(\log D)^{-1}$$

is a simultaneous approximation measure for the numbers $\lambda_1, \ldots, \lambda_m$.

All constants are easy to compute explicitly. Here is an example (see [NeW 1996] - see also Exercise 14.3):

- *Let* β *be an algebraic number and* λ *a logarithm of an algebraic number. Define* $\alpha = e^{\lambda}$, $K = \mathbb{Q}(\alpha, \beta)$ *and* $D = [K : \mathbb{Q}]$. *Let* A, B *and* E *be positive real numbers satisfying* $E \geq e$,

$$\log A \geq \max\bigl(h(\alpha),\ D^{-1}\log E,\ D^{-1}|\beta|E\bigr)$$

and

$$\log B \geq h(\beta) + \log_+ \log A + \log D + \log E,$$

where $\log_+ x = \log\max(1, x)$. *Then*

$$|\beta - \lambda| \geq \exp\Bigl(-105500 \cdot D^2(\log A)(\log B)\bigl(D\log D + \log E\bigr)(\log E)^{-2}\Bigr).$$

14.2.4 Simultaneous Approximation for x_i, y_j and $e^{x_i y_j}$

Corollary 14.14. *Let* $m \geq 1$ *and* $n \geq 1$ *be positive integers,* $(x_1, \ldots, x_m)$ *a* m-*tuple of complex numbers satisfying a linear independence measure condition and* $(y_1, \ldots, y_n)$ *a* n-*tuple of complex numbers satisfying also a linear independence measure condition. There exists a constant* $c > 0$ *such that a simultaneous approximation measure for the* $m + n + mn$ *numbers*

$$x_i,\quad y_j,\quad e^{x_i y_j} \qquad (1 \leq i \leq m,\ 1 \leq j \leq n)$$

is

$$\varphi(D, h) = cD^{1+\frac{m+n}{mn}} h(h + \log D)^{\frac{m+n}{mn}} (\log h + \log D)^{-\frac{m+n}{mn}}.$$

For $D > e^h$, the measure is simply

$$c' D^{1+\frac{m+n}{mn}} h$$

with another constant c'.

Proof. Let $\beta_1, \ldots, \beta_n, \beta'_1, \ldots, \beta'_m$ and γ_{ij} $(1 \leq i \leq m,\ 1 \leq j \leq n)$ be algebraic numbers in a field of degree $\leq D$ and heights $\leq h$. Our goal is to produce a lower bound for

$$\sum_{i=1}^{m} |x_i - \beta'_i| + \sum_{j=1}^{n} |y_j - \beta_j| + \sum_{i=1}^{m}\sum_{j=1}^{n} |x_i y_j - \gamma_{ij}|.$$

There is no loss of generality to assume $\gamma_{ij} \neq 0$. For $1 \leq i \leq m$ and $1 \leq j \leq n$, let λ_{ij} be the value of the logarithm of γ_{ij} which is closer to $x_i y_j$.

In Theorem 14.6, set

$$A_{ij} = e^h, \quad B = (De^h)^{c_0}, \quad E = \frac{1}{c_0} Dh,$$

where c_0 is a sufficiently large positive number which depends only on $x_1, \ldots, x_m$ and $y_1, \ldots, y_n$. □

14.2.5 Simultaneous Approximation for $\log \alpha_1, \log \alpha_2, \alpha_1^\beta$ and α_2^β

In this section we consider the special case $m = n = 2$ of Theorem 14.6.

Corollary 14.15. *Let K be a number field of degree D, β, β_1', β_2' be elements of K, $\lambda_1, \lambda_2\ \lambda_1', \lambda_2'$ elements in $\mathcal{L}$ such that the algebraic numbers*

$$\alpha_1 = e^{\lambda_1}, \ \alpha_2 = e^{\lambda_2}, \ \alpha_1' = e^{\lambda_1'}, \ \alpha_2' = e^{\lambda_2'}$$

are in K. Assume λ_1, λ_2 are linearly independent over $\mathbb{Q}$ and β is irrational. Let $B \geq e$ and $B' \geq e$ be real numbers with

$$\mathrm{h}(\beta) \leq \log B, \quad \mathrm{h}(1 : \beta_1' : \beta_2') \leq \log B'.$$

Let A_1, A_2, A_1', A_2' be positive numbers, all $\geq e^2$, and E a real number $\geq e$, which satisfy

$$(\log A_1)(\log A_2') = (\log A_2)(\log A_1')$$

and, for $i = 1, 2$,

$$\mathrm{h}(\alpha_i) \leq \log A_i, \quad \mathrm{h}(\alpha_i') \leq \log A_i',$$

and

$$|\lambda_i| \leq \frac{D}{E} \log A_i, \quad |\lambda_i'| \leq \frac{D}{E} \log A_i'.$$

Assume

$$\log E \leq D \log A_i \leq \min\{B, B'\}, \quad \log E \leq D \log A_i' \leq \min\{B, B'\},$$

$$\log E \leq D \log B', \quad \log B' \leq B, \quad \log B \leq B'$$

and

$$\log E \leq D(\log B) \cdot \frac{\log A_1}{\log A_2}, \quad \log E \leq D(\log B) \cdot \frac{\log A_2}{\log A_1}.$$

Define

$$U = D^2 (\log B)^{1/2} (\log B')^{1/2} \big((\log A_1)(\log A_2)(\log A_1')(\log A_2') \big)^{1/4} (\log E)^{-1}.$$

Then

$$|\lambda_1 - \beta_1'| + |\lambda_2 - \beta_2'| + |\beta\lambda_1 - \lambda_1'| + |\beta\lambda_2 - \lambda_2'| > \exp\{-2^{30} U\}.$$

Remark. By Corollary 14.15, a matrix of the form

$$\begin{pmatrix} 1 & \beta'_1 & \beta'_2 \\ 1 & \lambda_1 & \lambda_2 \\ \beta & \lambda'_1 & \lambda'_2 \end{pmatrix}$$

cannot be too close to a rank 1 matrix.

Example 14.16. Let λ_1, λ_2 be two elements of $\mathcal{L}$ which are linearly independent over $\mathbb{Q}$ and let θ be a complex irrational number which satisfies a linear independence measure condition. Then there exists a constant $c > 0$ such that the function

$$\varphi(D, h) = cD^2(h + \log D)h^{1/2}(\log D)^{-1}$$

is a simultaneous approximation measure for the five numbers $\lambda_1, \lambda_2, \theta, e^{\theta\lambda_1}, e^{\theta\lambda_2}$.

This follows from Corollary 14.15 by considering the rank 1 matrix

$$\begin{pmatrix} 1 & 1 & \theta \\ \lambda_1 & \lambda_1 & \theta\lambda_1 \\ \lambda_2 & \lambda_2 & \theta\lambda_2 \end{pmatrix}.$$

Several applications of Theorem 14.6 with $m = n = 2$ are given in Exercise 14.4.

14.3 Simultaneous Approximation in Higher Dimension

In the two previous sections, we were dealing with rank one matrices. We consider now higher values for r. In § 14.3.1 we take $d_0 = \ell_0 = 0$, while in § 14.3.2 we study the opposite extreme case where $d_0 = \ell_0 = r$.

14.3.1 Simultaneous Approximation of Logarithms by Complex Numbers

By Theorem 12.17, a $m \times n$ matrix L_{mn} with entries in $\mathcal{L}$ which satisfies the linear independence condition has rank $\geq mn/(m+n)$. We produce a quantitative refinement to this statement, namely a lower bound for the distance between a matrix with entries in $\mathcal{L}$ and a matrix (with complex entries) of given rank $r < mn/(m+n)$.

Theorem 14.17. *For any pair (m, n) of positive integers, there exists a positive constant c with the following property. Let L_{mn} be a $m \times n$ matrix with entries in $\mathcal{L}$. Recall the notation for K, D, A_{ij} and E in the introduction of this chapter. Further, let r be a real number in the range*

$$0 \leq r < \frac{mn}{m+n}$$

and let U_3 be a real number satisfying

$$U_3 \geq D \log D$$

and

$$U_3^{mn-r(m+n)} \geq D^{mn}\left(\prod_{i=1}^{m}\prod_{j=1}^{n}\log A_{ij}\right)(\log E)^{-r(m+n)}.$$

Assume further that L_{mn} *satisfies the linear independence condition for* $(cU_3)^2$.

Then for any $m \times n$ *complex matrix*

$$\mathsf{M} = \left(x_{ij}\right)_{\substack{1\leq i\leq m\\ 1\leq j\leq n}}$$

of rank r *we have*

$$\sum_{i=1}^{m}\sum_{j=1}^{n}\left|\lambda_{ij} - x_{ij}\right| \geq e^{-cU_3}.$$

Proof. From the assumptions $D\log A_{ij} \geq \log E$ we deduce

$$U_3^{mn-r(m+n)} \geq (\log E)^{mn}(\log E)^{-r(m+n)},$$

hence $U_3 \geq \log E$.

We first show that we may assume, without loss of generality,

$$U_3^r(\log A_{11})\cdots(\log A_{m1}) \geq (\log A_{i1})^m(\log E)^r \quad (1 \leq i \leq m).$$

We proceed by induction on m. Assume

$$U_3^r(\log A_{11})\cdots(\log A_{m1}) < (\log A_{m1})^m(\log E)^r.$$

Taking into account the definition of U_3, we deduce

$$D^{rn}\prod_{i=1}^{m}(\log A_{i1})^{n-r}\cdot\prod_{j=1}^{n}(\log A_{1j})^r < (\log A_{m1})^{mn-r(m+n)}(\log E)^{rn}(\log A_{11})^{rn}.$$

One deduces that the number $V > 0$ defined by

$$V^{mn-r(m+n)-n+r} = D^{(m-1)n}\prod_{i=1}^{m}\prod_{j=1}^{n}(\log A_{ij})(\log E)^{-r(m+n)+r}$$

satisfies $V < U$, and this allows us to apply the remark in § 13.5.3.

For the same reason we may assume

$$U_3^r(\log A_{11})\cdots(\log A_{1n}) \geq (\log A_{1j})^n(\log E)^r \quad (1 \leq j \leq n).$$

From the condition on U_3 one deduces

$$U_3^{m-r}(\log E)^r(\log A_{11})^m \geq D^m(\log A_{11})\cdots(\log A_{m1})(\log A_{1j})^m$$

for $1 \leq j \leq n$.

We start with the easy case where all entries x_{ij} of M are zero: in this special case Liouville's inequality (Exercise 3.7.a) gives

$$\sum_{i=1}^{m}\sum_{j=1}^{n}|\lambda_{ij}| \geq 2^{-D} \max_{\substack{1\le i\le m\\ 1\le j\le n}} A_{ij}^{-D}.$$

Next we remark that we may, without loss of generality, replace the number r by the rank of the matrix M.

Let c_0 be a sufficiently large integer. How large it should be can be explicitly written in terms of m and n only.

We shall apply Theorem 13.1 with $d_0 = \ell_0 = 0$, $d = d_1 = m$, $G = G^+ = \mathbb{G}_{\mathrm{m}}^m$, $G^- = \{e\}$, $\ell = \ell_1 = n$, $r_3 = r$, $r_1 = r_2 = 0$,

$$\underline{\eta}_j = (\lambda_{ij})_{1\le i\le m}, \qquad \underline{\eta}'_j = (x_{ij})_{1\le i\le m} \qquad (1 \le j \le n).$$

Since $d_0 = \ell_0 = 0$ we set $T_0 = S_0 = 0$. Therefore the parameters B_1 and B_2 will play no role, but for completeness we set

$$B_1 = B_2 = mnE\Big(D \max_{\substack{1\le i\le m\\ 1\le j\le n}} A_{ij}\Big)^{mn}.$$

Let U be a real number satisfying

$$U \geq c_0^{mn(r+1)} U_3.$$

Define a real number S by

$$S^m = \frac{U^{m-r}(\log E)^r}{c_0^{r+1} D^m (\log A_{11}) \cdots (\log A_{m1})}.$$

Next, let $V = c_0 U$,

$$T_i = \left[\frac{U}{mnDS\log A_{i1}}\right] \qquad (1 \le i \le m)$$

and

$$S_j = \left[S\cdot\frac{\log A_{11}}{\log A_{1j}}\right] \qquad (1 \le j \le n).$$

From the preliminary reduction we infer $S_j \geq 1$ for $1 \le j \le n$ and $T_i \geq 1$ for $1 \le i \le m$. From the definitions of U and S one deduces

$$D\sum_{i=1}^{m}\sum_{j=1}^{n} T_i S_j \log A_{ij} < U,$$

$$(2T_1+1)\cdots(2T_m+1) > 2\left(\frac{V}{\log E}\right)^r$$

and

$$m!2^m T_1\cdots T_m < (2S_1+1)\cdots(2S_n+1).$$

We take

$$\mathcal{T}_1 = \mathbb{Z}^m[\underline{T}], \quad \mathcal{S}_1 = \mathbb{Z}^n[\underline{S}],$$

so that

$$H(G;\mathcal{T}_1)=\mathrm{Card}(\mathcal{T}_1)=(2T_1+1)\cdots(2T_m+1)$$

and

$$\mathrm{Card}(\mathcal{S}_1)=(2S_1+1)\cdots(2S_n+1).$$

Define

$$\Sigma=\left\{\left(\alpha_{11}^{s_1}\cdots\alpha_{1n}^{s_n},\ldots,\alpha_{m1}^{s_1}\cdots\alpha_{mn}^{s_n}\right)\in(K^\times)^m\ ;\ \underline{s}\in\mathbb{Z}^n[\underline{S}]\right\}.$$

Assume that the conclusion of Theorem 14.17 does not hold for $c=c_0^{mn(r+1)+1}$. Then the hypotheses of Theorem 13.1 are satisfied, and we deduce that there exists a connected algebraic subgroup G^* of G, which is incompletely defined by polynomials of multidegree $\le \underline{T}$, such that

$$\mathrm{Card}\left(\frac{\Sigma+G^*}{G^*}\right)\mathcal{H}(G^*;\,\underline{T})\le m!2^mT_1\cdots T_m.$$

Here $\underline{T}$ stands for the m-tuple $(T_1,\ldots,T_m)$.

Since $\mathcal{H}(G^*;\,\underline{T})\ge 1$ we deduce

$$\mathrm{Card}\left(\frac{\Sigma+G^*}{G^*}\right)<(2S_1+1)\cdots(2S_n+1).$$

Hence $\Sigma[2]\cap G^*(K)\ne\{e\}$. We deduce that there exist $\underline{s}\in\mathbb{Z}^n[2\underline{S}]\setminus\{0\}$ and $\underline{t}\in\mathbb{Z}^m[\underline{T}]\setminus\{0\}$ with

$$\sum_{i=1}^m\sum_{j=1}^n t_is_j\lambda_{ij}\in 2\pi\sqrt{-1}\mathbb{Z}.$$

Let us check, by contradiction, that G^* has codimension 1. We already know $G^*\ne G$. If the codimension of G^* were ≥ 2, we would have two linearly independent elements $\underline{t}'$ and $\underline{t}''$ in $\mathbb{Z}^m[\underline{T}]$ such that the two numbers

$$a'=\frac{1}{2\pi\sqrt{-1}}\sum_{i=1}^m\sum_{j=1}^n t_i's_j\lambda_{ij}\quad\text{and}\quad a''=\frac{1}{2\pi\sqrt{-1}}\sum_{i=1}^m\sum_{j=1}^n t_i''s_j\lambda_{ij}$$

are in $\mathbb{Z}$. Notice that

$$\max\{|a'|,\,|a''|\}\le\frac{1}{\pi E}D\sum_{i=1}^m\sum_{j=1}^n T_iS_j\log A_{ij}<\frac{1}{2}U\cdot$$

We eliminate $2\pi\sqrt{-1}$: set $\underline{t}=a''\underline{t}'-a'\underline{t}''$, so that

$$\sum_{i=1}^m\sum_{j=1}^n t_is_j\lambda_{ij}=0$$

and

$$0<|\underline{t}|\le U\max_{1\le i\le m}T_i<U^2.$$

This is not compatible with our hypothesis that the matrix L_{mn} satisfies the linear independence condition for $(cU_3)^2$.

Hence G^* has codimension 1 in G. Therefore

$$\mathcal{H}(G^*;\underline{T}) \geq \frac{2^{m-1}T_1\cdots T_m}{\max_{1\leq i\leq m} T_i}.$$

A similar argument shows that any $\underline{s}'$, $\underline{s}''$ in $\mathbb{Z}^n[2\underline{S}]$ for which

$$\sum_{i=1}^m\sum_{j=1}^n t_i s'_j\lambda_{ij} \in 2\pi\sqrt{-1}\mathbb{Z} \quad \text{and} \quad \sum_{i=1}^m\sum_{j=1}^n t_i s''_j\lambda_{ij} \in 2\pi\sqrt{-1}\mathbb{Z}$$

are linearly dependent over $\mathbb{Z}$. From Lemma 7.8 we deduce

$$\operatorname{Card}\left(\frac{\Sigma+G^*}{G^*}\right) \geq \frac{(2S_1+1)\cdots(2S_n+1)}{\max_{1\leq j\leq n}(2S_j+1)}.$$

Therefore

$$(2S_1+1)\cdots(2S_n+1) \leq 2\left(\max_{1\leq i\leq m} T_i\right)\left(\max_{1\leq j\leq n}(2S_j+1)\right).$$

Since

$$(2S_1+1)\cdots(2S_n+1) > \left(\frac{V}{\log E}\right)^r$$

and

$$T_iS_j \leq \frac{U}{D\log A_{ij}} \leq \frac{U}{\log E},$$

we get a contradiction. This completes the proof of Theorem 14.17. □

Remark. The linear independence condition on L_{mn} can be much weakened. For instance when $A_{ij} = A$ is independent on i, j (hence $T_1 = \cdots = T_m$ and $S_1 = \cdots = S_n$) the assumption which is needed in the proof is the following:

For any algebraic subgroup G^* of $G = \mathbb{G}_{\mathrm{m}}^m$, we have

$$\operatorname{Card}\left(\frac{\Sigma+G^*}{G^*}\right) \geq (2S+1)^{nm^\flat/m}$$

where $m^\flat = \dim(G/G^*)$.

For instance this condition is satisfied if we assume that the subgroup Γ of $G(K)$ generated by

$$(\alpha_{1j},\ldots,\alpha_{mj}) \in (K^\times)^m \quad (1\leq j\leq n)$$

satisfies

$$\operatorname{rank}_{\mathbb{Z}}\left(\frac{\Gamma}{\Gamma\cap G^*(K)}\right) \geq \frac{n}{m}\dim(G/G^*)$$

for any algebraic subgroup G^* of G.

Theorem 14.17 yields the following statement, which extends Theorem 2.3 of [RoyW 1997b] to higher dimension (replacing rank one matrices by matrices of arbitrary ranks, which amounts to deal with several variables instead of just one):

Corollary 14.18. *Let m, n and r be positive rational integers with $mn > r(m+n)$. Define*

$$\kappa = \frac{mn}{mn - r(m+n)}.$$

Let $\mathsf{M} = (x_{ij})_{\substack{1\le i\le m\\ 1\le j\le n}} \in \mathrm{Mat}_{m\times n}(\mathbb{C})$ *satisfying the following technical condition: for any sufficiently large integers T and S, and any $\underline{t} \in \mathbb{Z}^d \setminus \{0\}$, $\underline{s} \in \mathbb{Z}^\ell \setminus \{0\}$ satisfying $|\underline{t}| \le T$ and $|\underline{s}| \le S$, we have*

$$\left|\sum_{i=1}^{m}\sum_{j=1}^{n} t_i s_j x_{ij}\right| \ge \exp\{-c(TS)^{1/5}\}. \tag{14.19}$$

Then there exists a positive constant c such that

$$c(Dh)^\kappa \bigl(\log h + \log D\bigr)^{1-\kappa}$$

is a simultaneous approximation measure for the mn numbers $e^{x_{ij}}$, $(1 \le i \le m$, $1 \le j \le n)$.

One should not pay too much attention to the exponent $1/5$ in the technical hypothesis (14.19): one could weaken this assumption by replacing $1/5$ by a slightly larger constant; but one cannot completely omit such a condition (see Exercise 14.7).

Proof. We apply Theorem 14.17. Since x_{ij} are fixed, the condition

$$h \ge \max_{\substack{1\le i\le m\\ 1\le j\le n}} \frac{E}{D}|\lambda_{ij}|$$

is satisfied with $E = (Dh)^{1/2}$. Also we may assume that the linear independence condition on the matrix L_{mn} is satisfied, because for $|\underline{t}| \le (cU_3)^2$ and $|\underline{s}| \le (cU_3)^2$, applying (14.19) with $T = S = (cU_3)^2$ gives (if the conclusion of Corollary 14.18 does not hold)

$$\left|\sum_{i=1}^{m}\sum_{j=1}^{n} t_i s_j \lambda_{ij}\right| \ge \exp\{-2c(TS)^{1/5}\} > 0.$$

□

Remark. Let us consider the case $r = 1$. Corollary 14.18 is a quantitative refinement to the six exponentials Theorem (see [MiW 1977]). In this case the assumption (14.19) can be written more simply: writing $x_{ij} = u_i v_j$ with u_i and v_j in $\mathbb{C}$ (see Exercice 1.9), we need only to assume that each of the tuples $(u_1, \ldots, u_m)$ and $(v_1, \ldots, v_n)$ satisfies a linear independence measure condition. See [RoyW 1997b], Theorem 2.3. See also [RoyW 1997b], Theorem 2.4 for another equivalent formulation in terms of lower bounds for 2×2 minors in a matrix whose entries are logarithms of algebraic numbers.

14.3.2 Simultaneous Approximation of Logarithms by Algebraic Numbers

Let L_{mn} be a matrix with entries λ_{ij} in $\mathcal{L}$. We approximate simultaneously the numbers λ_{ij} by algebraic numbers β_{ij}. If r denotes the rank of the matrix

$$\mathsf{B} = \left(\beta_{ij}\right)_{\substack{1\le i\le m\\ 1\le j\le n}},$$

then the matrix

$$\begin{pmatrix} \mathsf{B} & \mathsf{B} \\ \mathsf{B} & \mathsf{L}_{mn} \end{pmatrix}$$

is close to the rank r matrix

$$\begin{pmatrix} \mathsf{B} & \mathsf{B} \\ \mathsf{B} & \mathsf{B} \end{pmatrix}.$$

Theorem 14.20. *There exists a constant $c > 0$ which depends only on n and m with the following property. Let L_{mn} be a matrix with entries in $\mathcal{L}$ and let K, D, A_{ij} $(1 \le i \le m,\ 1 \le j \le n)$ and E satisfy the conditions stated in the introduction. Further, let $\mathsf{B} = \left(\beta_{ij}\right)_{\substack{1\le i\le m\\ 1\le j\le n}}$ be a rank r matrix with entries in the number field K. Let B be a positive real numbers satisfying, for $1 \le i \le m$ and $1 \le j \le n$, the following conditions:*

$$\mathrm{h}(\beta_{ij}) \le \log B, \qquad B \ge \log A_{ij}, \quad B \ge D \quad \text{and} \quad B \ge e.$$

Let U_4 be a positive real number satisfying

$$U_4^{mn} \ge D^{mn+r(m+n)}(\log B)^{r(m+n)}\left(\prod_{i=1}^{m}\prod_{j=1}^{n}\log A_{ij}\right)(\log E)^{-r(m+n)}.$$

Assume L_{mn} satisfies the linear independence condition for $(cU_4)^2$. Assume furthermore

$$\log E \le D\log B \le U_4.$$

Then

$$\sum_{i=1}^{m}\sum_{j=1}^{n}\left|\lambda_{ij} - \beta_{ij}\right| \ge e^{-cU_4}.$$

Remark. In case $r = 1$, Theorem 14.20 is nothing else than the special case of Theorem 14.6 where $B = B'$ (cf. Exercise 1.9).

Proof of Theorem 14.20. Since $D\log B \ge \log E$, there is no loss of generality to assume

$$(D\log B)^r(\log A_{11})\cdots(\log A_{m1}) \ge (\log A_{i1})^m(\log E)^r \quad (1 \le i \le m)$$

and

$$(D\log B)^r(\log A_{11})\cdots(\log A_{1n}) \geq (\log A_{1j})^n(\log E)^r \quad (1\leq j\leq n).$$

We are going to apply Theorem 13.1 with

$$d_0=\ell_0=r,\quad d_1=m,\quad \ell_1=n,$$

$$G=G^+=\mathbb{G}_{\mathrm{a}}^r\times\mathbb{G}_{\mathrm{m}}^m,\quad G^-=\{e\},\quad r_3=r,\quad r_1=r_2=0.$$

Since B has rank r we may assume that the matrix $\left(\beta_{ij}\right)_{1\leq i,j\leq r}$ is regular. Define

$$\underline{w}_k=\underline{w}'_k=\left(\beta_{1k},\ldots,\beta_{rk},\beta_{1k},\ldots,\beta_{mk}\right)\in K^{r+m}\quad (1\leq k\leq r),$$

$$\underline{\eta}_j=\left(\beta_{1j},\ldots,\beta_{rj},\lambda_{1j},\ldots,\lambda_{mj}\right)\in K^r\times\mathcal{L}^m,\quad (1\leq j\leq n)$$

and

$$\underline{\eta}'_j=\left(\beta_{1j},\ldots,\beta_{rj},\beta_{1j},\ldots,\beta_{mj}\right)\quad (1\leq j\leq n).$$

Further, let

$$\gamma_j=\exp_G(\underline{\eta}_j)=\left(\beta_{1j},\ldots,\beta_{rj},\alpha_{1j},\ldots,\alpha_{mj}\right)\in K^r\times(K^\times)^m\quad (1\leq j\leq n).$$

Next define

$$B_1=B_2=B^{c_0},\quad U=c_0^{6m}U_4\quad\text{and}\quad V=c_0U.$$

Set

$$T_0=S_0=\left[\frac{U}{c_0D\log B}\right].$$

Notice that the hypotheses of Theorem 14.20 imply $T_0\geq c_0$ and $S_0\geq c_0$. Define $T_1,\ldots,T_m,S_1,\ldots,S_n$ by

$$T_i=\left[c_0^2\left(\frac{D\log B}{\log E}\right)^{r/m}\cdot\frac{\left((\log A_{11})\cdots(\log A_{m1})\right)^{1/m}}{\log A_{i1}}\right]\quad (1\leq i\leq m)$$

and

$$S_j=\left[c_0^{4m}\left(\frac{D\log B}{\log E}\right)^{r/n}\cdot\frac{\left((\log A_{11})\cdots(\log A_{1n})\right)^{1/n}}{\log A_{1j}}\right]\quad (1\leq j\leq n).$$

Notice that we have $T_i\geq c_0$ and $S_j\geq c_0$. We take

$$\mathcal{T}_1=\mathbb{Z}^m[\underline{T}],\quad \mathcal{S}_1=\mathbb{Z}^n[\underline{S}]$$

and

$$\Sigma=\left\{\gamma_1^{s_1}\cdots\gamma_m^{s_m}\ ;\ \underline{s}\in\mathbb{Z}^n[\underline{S}]\right\}\subset G(K)=K^r\times(K^\times)^m.$$

One easily checks

$$D\sum_{i=1}^m\sum_{j=1}^n T_iS_j\log A_{ij}<U,$$

$$\binom{T_0+r}{r}(2T_1+1)\cdots(2T_m+1) \geq 2\left(\frac{V}{\log E}\right)^r$$

and

$$\binom{S_0+r}{r}(2S_1+1)\cdots(2S_n+1) \geq c_0 T_0^r T_1 \cdots T_m.$$

The hypotheses

$$U \geq D\log D, \quad U \geq \log E, \quad B \geq S_j + \frac{S_0}{T^*}$$

of Theorem 13.1 are satisfied because

$$U \geq D\log B, \quad U \geq D\log A_{ij} \geq \log E, \quad B \geq D \quad \text{and} \quad B \geq \log A_{ij}.$$

Hence we may apply Theorem 13.1 and deduce the existence of an algebraic subgroup G^* of G satisfying

$$\binom{S_0+\ell_0^\flat}{\ell_0^\flat} M^\flat \mathcal{H}(G^*;\, T_0, \underline{T}) \leq \frac{(m+r)!}{r!} 2^m T_0^r T_1 \cdots T_m,$$

where

$$M^\flat = \operatorname{Card}\left(\frac{\Sigma + G^*}{G^*}\right)$$

and

$$\ell_0^\flat = \dim_{\mathbb{C}}\left(\frac{\mathcal{W} + T_e(G^*)}{T_e(G^*)}\right).$$

Define

$$d_0^* = \dim(G_0^*), \quad d_1^* = \dim(G_1^*), \quad d_0^\flat = \dim(G_0/G_0^*), \quad d_1^\flat = \dim(G_1/G_1^*)$$

and $d^* = d_0^* + d_1^* = \dim(G^*)$, $d^\flat = d_0^\flat + d_1^\flat = \dim(G/G^*)$.

We first notice that $\ell_0^\flat \geq d_0^\flat$. Indeed the restriction to $\mathcal{W}$ of the projection $\mathbb{C}^{r+m} \to \mathbb{C}^r$ on the first r components is surjective. Hence the restriction to $\mathcal{W}^\flat$ of the projection $\mathbb{C}^{d^\flat} \to \mathbb{C}^{d_0^\flat}$ is also surjective, hence the dimension $\ell_0^\flat$ of $\mathcal{W}^\flat$ is at least $d_0^\flat$.

Therefore we obtain $M^\flat < (2S_1+1)\cdots(2S_n+1)$. From the linear independence condition on L_{mn} one deduces, like in the proof of Theorem 14.17, that G_1^* has codimension 1 and that

$$M^\flat \geq \frac{(2S_1+1)\cdots(2S_n+1)}{\max_{1\leq j\leq n}(2S_j+1)}.$$

The final contradiction is reached, again, as in the proof of Theorem 14.17. □

The next statement combines the special cases of Theorems 14.17 and 14.20 where A_{ij} is independent of i and j.

Corollary 14.21. *Let m, n and r be positive rational integers. Define*

$$\theta = \frac{r(m+n)}{mn}.$$

There exists a positive constant c with the following property. Let B be a $m \times n$ matrix of rank $\leq r$ with coefficients β_{ij} in a number field K. For $1 \leq i \leq m$ and $1 \leq j \leq n$, let λ_{ij} be a complex number such that the number $\alpha_{ij} = e^{\lambda_{ij}}$ belongs to $K^\times$ and such that the matrix $\mathsf{L} = (\lambda_{ij})_{\substack{1\leq i\leq m\\ 1\leq j\leq n}}$ satisfies the linear independence condition. Define $D = [K : \mathbb{Q}]$. Let h_1, h_2 and E be positive real numbers satisfying the following conditions:

$$h_1 \geq \mathrm{h}(\alpha_{ij}), \quad h_1 \geq \frac{E}{D}|\lambda_{ij}|, \quad h_1 \geq \frac{\log E}{D}$$

and

$$h_2 \geq \mathrm{h}(\beta_{ij}), \quad h_2 \geq \log h_1 \quad h_2 \geq \log D, \quad h_2 \geq \frac{1}{D}\log E, \quad E \geq e$$

for $1 \leq i \leq m$ and $1 \leq j \leq n$. Then

$$\sum_{i=1}^{m}\sum_{j=1}^{n}\left|\lambda_{ij} - \beta_{ij}\right| \geq e^{-c\Phi}$$

where

$$\Phi = \begin{cases} Dh_1(Dh_2)^\theta(\log E)^{-\theta} & \text{if } \dfrac{Dh_1}{\log E} \geq \left(\dfrac{Dh_2}{\log E}\right)^{1-\theta}, \\[2ex] \max\left\{D\log D\,;\; (Dh_1)^{1/(1-\theta)}(\log E)^{-\theta/(1-\theta)}\right\} & \text{if } \dfrac{Dh_1}{\log E} < \left(\dfrac{Dh_2}{\log E}\right)^{1-\theta}. \end{cases}$$

Proof. From Theorem 14.20 with $A_{ij} = e^{h_1}$ $(1 \leq i \leq m, 1 \leq j \leq n)$ and $B = e^{2h_2}$ we deduce that the conclusion of Corollary 14.21 holds with Φ replaced by

$$\max\left\{D^{1+\theta}h_1h_2^\theta(\log E)^{-\theta}\;,\; Dh_2\right\}.$$

This proves the desired result in the case

$$\frac{Dh_1}{\log E} \geq \left(\frac{Dh_2}{\log E}\right)^{1-\theta}.$$

Assume now

$$\frac{Dh_1}{\log E} < \left(\frac{Dh_2}{\log E}\right)^{1-\theta}.$$

In this case we have $\Phi < Dh_2$, hence a further argument is necessary.

Since $Dh_1 \geq \log E$ and $Dh_2 \geq \log E$ we deduce $\theta < 1$. We apply Theorem 14.17 with $A_{ij} = e^{h_1}$ $(1 \leq i \leq m, 1 \leq j \leq n)$ and $U_3 = \Phi$. Thanks to the definition of Φ we have

$$\Phi \geq D \log D, \quad \Phi \geq \log E \quad \text{and} \quad \Phi^{m-r}(\log E)^r \geq (Dh_1)^m.$$

□

From Corollary 14.21 one deduces the following variant of Theorem 10.1 in [RoyW 1997b]:

Corollary 14.22. *Let m and n be positive integers and let $\mathsf{L} = (\lambda_{ij})_{\substack{1\le i\le m\\ 1\le j\le n}}$ be a $m \times n$ matrix of rank r with coefficients in $\mathcal{L}$ which satisfies the linear independence condition. Set $\kappa = (1/m) + (1/n)$. Then, there exists a positive constant c such that the function*

$$\varphi(D, h) = cD^{r\kappa+1}(h + \log D)^{r\kappa}(\log D)^{-r\kappa}$$

is a simultaneous approximation measure for the mn numbers λ_{ij} $(1 \leq i \leq m,$ $1 \leq j \leq n)$.

Remark. Using Dirichlet's box principle (see § 15.2.1), it is easy to check that any simultaneous approximation measure $\varphi(D, h)$ is bounded from below by $c(D)h$, where $c(D)$ depends only on D and on the given tuple. Therefore, under the assumptions of Corollary 14.22, one deduces $r\kappa \geq 1$. This is nothing else than Theorem 1.16.

Proof of Corollary 14.22. We apply Theorem 14.20 with h_1 a sufficiently large constant, $E = D$, $h_2 = h + \log(Dh_1)$ and $\theta = r\kappa$. □

The assumption in Corollaries 14.21 and 14.22 that the matrix L satisfies the linear independence condition is clearly too strong. On one hand, according to Theorem 14.20, it suffices to assume the linear independence condition for $(c\phi)^2$ in Corollary 14.21 and for $\big(c\varphi(D, h)\big)^2$ in Corollary 14.22. On the other hand, on the qualitative side, Corollary 12.18 involving the structural rank is stronger than Theorem 1.16 which assumes a linear independence condition (see also the remark after the proof of Theorem 14.17). While Corollary 14.22 is a quantitative sharpening to Theorem 1.16, the next result is a quantitative sharpening to Corollary 12.18.

Theorem 14.23. *Let $\lambda_1, \ldots, \lambda_n$ be elements of $\mathcal{L}$ and let M be a matrix with coefficients in the $\mathbb{Q}$-vector space spanned by $\lambda_1, \ldots, \lambda_n$. Assume*

$$\operatorname{rank}(\mathsf{M}) \leq \frac{1}{2} r_{\mathrm{str}}(\mathsf{M}).$$

Then there exists a positive constant c which depends only on $\lambda_1, \ldots, \lambda_n$ and M such that

$$cD^{3/2}(h + \log D)\big(h + D(\log D)^{-1}\big)^{1/2}(\log D)^{-1/2}$$

is a simultaneous approximation measure for $(\lambda_1, \ldots, \lambda_n)$.

In other terms, under the assumptions of Theorem 14.23, a simultaneous approximation measure for $(\lambda_1, \ldots, \lambda_n)$ is

$$\varphi(D,h) = \begin{cases} cD^2(h+\log D)(\log D)^{-1} & \text{if } h \le D(\log D)^{-1}, \\ cD^{3/2}h^{3/2}(\log D)^{-1/2} & \text{if } h \ge D(\log D)^{-1}. \end{cases}$$

Proof. Without loss of generality we may assume that M is a square $m \times m$ matrix of structural rank m. By Corollary 12.18 the assumption $\operatorname{rank}(\mathsf{M}) \le (1/2)r_{\rm str}(\mathsf{M})$ means that M has rank $m/2$.

Since M has coefficients in the $\mathbb{Q}$-vector space spanned by $\lambda_1, \ldots, \lambda_n$, starting from algebraic approximations to $\lambda_1, \ldots, \lambda_n$, one deduces algebraic approximations β_{ij} to the coefficients λ_{ij} of M. We repeat the proof of Theorem 14.20 with $n = m = 2r$ (hence $G_0 = \mathbb{G}_{\rm a}^{m/2}$ and $G_1 = \mathbb{G}_{\rm m}^m$), taking $E = D$, $B = De^h$ while A_{ij} are constants, but now U is defined by

$$U = c_0^{6m} D^{3/2}(h+\log D)\big(h + D(\log D)^{-1}\big)^{1/2}(\log D)^{-1/2}.$$

The proof of Theorem 14.20 would require only

$$U \ge c_0^{6m} D^2(h+\log D)(\log D)^{-1},$$

and indeed this condition will be also sufficient here for almost all the proof, apart from the very end of it.

Define

$$T_0 = S_0 = \left[\frac{U}{c_0 D(h+\log D)}\right],$$

$$T_i = \left[c_0^2\left(\frac{D(h+\log D)}{\log D}\right)^{1/2}\right] \qquad (1 \le i \le m)$$

and

$$S_j = \left[c_0^{4m}\left(\frac{D(h+\log D)}{\log D}\right)^{1/2}\right] \qquad (1 \le j \le n).$$

Define also $\underline{w}_1, \ldots, \underline{w}_r$ in K^{r+m}, $\underline{\eta}_1, \ldots, \underline{\eta}_m$ in $K^r \times \mathcal{L}^m$, $\underline{\eta}'_1, \ldots, \underline{\eta}'_m$ in $\mathbb{C}^{r+m}$, $\underline{\gamma}_1, \ldots, \underline{\gamma}_m$ in $G(K) = K^r \times (K^\times)^m$ and $\Sigma \subset G(K)$ as in the proof of Theorem 14.20. Again by Theorem 13.1 we deduce the existence of an algebraic subgroup G^* of G which satisfies

$$\binom{S_0 + \ell_0^\flat}{\ell_0^\flat} M^\flat \le \frac{(m+r)!}{r!} 2^{d_1^\flat} T_0^{d_0^\flat} T_1^{d_1^\flat},$$

where

$$\ell_0^\flat = \dim_{\mathbb{C}}(\mathcal{W}^\flat), \quad M^\flat = \operatorname{Card}(\Sigma^\flat),$$

$$\mathcal{W}^\flat = \frac{\mathcal{W} + T_e(G^*)}{T_e(G^*)}, \quad \Sigma^\flat = \frac{\Sigma + G^*}{G^*}$$

and

$$d_0^\flat = \dim(G_0^\flat), \quad d_1^\flat = \dim(G_1^\flat), \quad G_0^\flat = \frac{G_0}{G_0^*}, \quad G_1^\flat = \frac{G_1}{G_1^*}.$$

We are looking firstly for a lower bound for $\ell_0^\flat$, and secondly for a lower bound for $M^\flat$.

We already know from the proof of Theorem 14.20 that $\ell_0^\flat \geq d_0^\flat$. However here we need more.

Set

$$G^\flat = \frac{G}{G^*} = G_0^\flat \times G_1^\flat$$

and $d^\flat = d_0^\flat + d_1^\flat = \dim(G^\flat)$. Denote by

$$\pi_0 : \mathbb{C}^{r+m} \longrightarrow \mathbb{C}^r \quad \text{and} \quad \pi_0^\flat : T_e(G^\flat) \longrightarrow T_e(G_0^\flat)$$

the projections with kernels $\{0\} \times \mathbb{C}^m$ and $T_e(G_1^\flat)$ respectively, and by

$$g_0 : \mathbb{C}^r \longrightarrow T_e(G_0^\flat) \quad \text{and} \quad g : \mathbb{C}^{r+m} \longrightarrow T_e(G^\flat)$$

the projections whose kernels are $T_e(G_0^*)$ and $T_e(G^*)$ respectively. Since the diagram

$$\begin{array}{cccc}
\mathcal{W} \subset & \mathbb{C}^{r+m} & \xrightarrow{\ \pi_0\ } & \mathbb{C}^r \\
& \Big\downarrow g & & \Big\downarrow g_0 \\
\mathcal{W}^\flat \subset & T_e(G^\flat) & \xrightarrow[\ \pi_0^\flat\ ]{} & T_e(G_0^\flat)
\end{array}$$

commutes, since $g(\mathcal{W}) = \mathcal{W}^\flat$ and since $g_0 \circ \pi_0(\mathcal{W}) = T_e(G_0^\flat)$, we have

$$\ell_0^\flat = d_0^\flat + \dim_{\mathbb{C}}\big(\mathcal{W}^\flat \cap \ker(\pi_0^\flat)\big).$$

We shall derive below a lower bound for $\mathcal{W}^\flat \cap \ker(\pi_0^\flat)$, but we first estimate $M^\flat$ from below.

Denote by Y the $\mathbb{Q}$-vector space spanned by the vector columns $\underline{y}_1, \dots, \underline{y}_m$ of M in $\mathbb{C}^m$, set

$$Y^\flat = \frac{Y + T_e(G_1^*)}{T_e(G_1^*)} \subset T_e(G_1^\flat)$$

and denote by $\ell_1^\flat$ the dimension of $Y^\flat$ over $\mathbb{Q}$. Since M is $\mathbb{Q}$-equivalent to a matrix

$$\begin{pmatrix} \underset{}{\mathrm{A}} & \mathrm{B} \\ \underbrace{\mathrm{C}}_{\ell_1^\flat} & \underbrace{\mathrm{0}}_{m-\ell_1^\flat} \end{pmatrix} \begin{matrix} \}m - d_1^\flat \\ \}d_1^\flat \end{matrix}$$

we have

$$m = r_{\mathrm{str}}(\mathsf{M}) \le m - d_1^\flat + \ell_1^\flat,$$

which implies $\ell_1^\flat \ge d_1^\flat$.

Denote by Φ the subgroup of $\underline{\varphi} \in \mathbb{Z}^m$ for which $\underline{y}^{\underline{\varphi}} = 1$ for any $\underline{y} \in G_1^*$. Since G_1^* has codimension $d_1^\flat$ in G_1 and is incompletely defined in G_1 by polynomials of degrees $\le T_1$, there is a basis $\underline{\varphi}_1, \ldots, \underline{\varphi}_{d_1^\flat}$ of Φ with $\underline{\varphi}_j \in \mathbb{Z}^m[T_1]$ $(1 \le j \le d_1^\flat)$. The linear map

$$\begin{array}{rccc} g_1: & \mathbb{C}^m & \longrightarrow & \mathbb{C}^{d_1^\flat} \\ & \underline{z} & \longmapsto & \underline{\varphi}_j \underline{z} \end{array}$$

is surjective with kernel $T_e(G_1^*)$. We identify $T_e(G_1^\flat)$ with $\mathbb{C}^{d_1^\flat}$, so that $g = g_0 \times g_1$. The kernel $\Omega^\flat \subset T_e(G^\flat)$ of the exponential map of $G^\flat$ is $\Omega^\flat = \{0\} \times \Omega_1^\flat$, where $\Omega_1^\flat = (2i\pi\mathbb{Z})^{d_1^\flat}$. For $\underline{y} \in \mathbb{C}^m$ we have

$$\exp_{G_1}(\underline{y}) \in G_1^*(\mathbb{C}) \iff g_1(\underline{y}) \in \Omega_1^\flat.$$

For simplicity of notation, we permute (if necessary) the column vectors of M so that $g_1(\underline{y}_1), \ldots, g_1(\underline{y}_{\ell_1^\flat})$ are $\mathbb{Q}$-linearly independent in $Y^\flat$. Moreover we may assume that for an index κ in the range $0 \le \kappa \le \ell_1^\flat$, we have

$$\sum_{j=1}^{\ell_1^\flat - \kappa} s_j \underline{\gamma}_j \notin G^*(K)$$

for any $\underline{s} \in \mathbb{Z}^{\ell_1^\flat - \kappa}[S_1] \setminus \{0\}$, while for $1 \le k \le \kappa$ there exists $\underline{s}_k \in \mathbb{Z}^{\ell_1^\flat - \kappa + 1}[S_1] \setminus \{0\}$ such that

$$\sum_{j=1}^{\ell_1^\flat - \kappa} s_{jk} \underline{\gamma}_j + s_{\ell_1^\flat - \kappa + k, k} \underline{\gamma}_{\ell_1^\flat - \kappa + k} \in G^*(K).$$

From Lemma 7.8 we deduce

$$M^\flat \ge S_1^{\ell_1^\flat - \kappa}.$$

Therefore we have

$$S_0^{\ell_0^\flat} S_1^{\ell_1^\flat - \kappa} \le c' T_0^{d_0^\flat} T_1^{d_1^\flat}$$

with some constant c' depending only on m.

Let us check

$$\dim_{\mathbb{C}}\left(\mathcal{W}^\flat \cap \ker(\pi_0^\flat)\right) \ge \kappa.$$

For $1 \le k \le \kappa$ define

$$\underline{u}_k = \sum_{j=1}^{\ell_1^\flat - \kappa} s_{jk} g(\underline{\eta}_j) + s_{\ell_1^\flat - \kappa + k, k} g(\underline{\eta}_{\ell_1^\flat - \kappa + k})$$

and

$$\underline{v}_k = \sum_{j=1}^{\ell_1^\flat - \kappa} s_{jk} g(\underline{w}_j) + s_{\ell_1^\flat - \kappa + k, k} g(\underline{w}_{\ell_1^\flat - \kappa + k})$$

in $T_e(G^\flat)$. Since $\exp_{G^\flat}(\underline{u}_k) = e$ we have $\underline{u}_k \in \Omega^\flat$. Notice that $\Omega^\flat \subset \ker(\pi_0^\flat)$. Since $\pi_0(\underline{\eta}_j) = \pi_0(\underline{w}_j)$ we also have $\underline{v}_k \in \ker(\pi_0^\flat)$, hence

$$\underline{v}_k \in \mathcal{W}^\flat \cap \ker(\pi_0^\flat) \quad \text{for } 1 \le k \le \kappa.$$

We want to check that $\underline{v}_1, \dots, \underline{v}_\kappa$ are $\mathbb{C}$-linearly independent. We consider the $\kappa \times d_1^\flat$ matrix whose column vectors are $g_1(\underline{u}_1), \dots, g_1(\underline{u}_\kappa)$. The entries of this matrix are in $2i\pi\mathbb{Z}$. Since $\underline{y}_1, \dots, \underline{y}_{\ell_1^\flat}$ are $\mathbb{Q}$-linearly independent, it follows that the rank of this matrix is κ. Let A be a square regular $\kappa \times \kappa$ submatrix; denote by B the corresponding matrix obtained by replacing $\underline{u}_j$ by $\underline{v}_j$. Then

$$\left|\det(\mathsf{A}) - \det(\mathsf{B})\right| \le c''(TS)^m e^{-U} < 2\pi.$$

Since $\det(\mathsf{A})$ is a nonzero integral multiple of 2π, we deduce $\det(\mathsf{B}) \neq 0$, hence $\underline{v}_1, \dots, \underline{v}_\kappa$ are $\mathbb{C}$-linearly independent in $\mathbb{C}^{d^\flat}$.

This completes the proof of the claim $\dim_{\mathbb{C}}\left(\mathcal{W}^\flat \cap \ker(\pi_0^\flat)\right) \ge \kappa$, and therefore $\ell_0^\flat \ge d_0^\flat + \kappa$.

Now comes the extra condition for which we needed to take U so large: since

$$U \ge c_0^{6m} D^{3/2}(h + \log D)^{3/2}(\log D)^{-1},$$

we have

$$S_0 \ge S_1,$$

and we deduce that the inequalities $\ell_1^\flat \ge d_1^\flat$, $\ell_0^\flat \ge d_0^\flat + \kappa$ and

$$S_0^{\ell_0^\flat} S_1^{\ell_1^\flat - \kappa} \le c' T_0^{d_0^\flat} T_1^{d_1^\flat}$$

are not compatible. This completes the proof of Theorem 14.23. □

Remark. In the special case where $\lambda_1, \dots, \lambda_n$ are all real numbers, the proof simplifies notably and produces the simultaneous approximation measure

$$cD^2(h + \log D)(\log D)^{-1}$$

(cf. [RoyW 1997b], remark pp. 423–424). Indeed in this special case we have $\kappa = 0$ and therefore the condition $S_0 \ge S_1$ is not needed.

Using Proposition 12.25, one deduces from Theorem 14.23 a refinement of Theorem 2.8 in [RoyW 1997b]:

Corollary 14.24. *Let $n \ge 2$ be an integer and $\lambda_1, \dots, \lambda_n$ be $\mathbb{Q}$-linearly independent elements of $\mathcal{L}$. Assume that there exists a nonzero homogeneous polynomial Q in*

$\mathbb{Q}[X_1, \ldots, X_n]$ of degree 2, such that $Q(\lambda_1, \ldots, \lambda_n) = 0$. Then there exists a positive constant c such that the function

$$cD^2(h + \log D)^{3/2}(\log D)^{-3/2}$$

is a simultaneous approximation measure for the n-tuple $(\lambda_1, \ldots, \lambda_n)$.

14.4 Measures of Linear Independence of Logarithms (Again)

We investigate, by means of Proposition 13.12, the best possible results one may expect from Theorem 13.1 for the problem of measures of linear independence of logarithms of algebraic numbers, using anyone of the methods described in § 11.4. We take $G^- = \{0\}$ and $G^+ = G$ in Theorem 13.1, in order to apply Proposition 13.12. Next (§ 14.4.5) we explain why other choices for G^- and G^+ may be needed. Finally (§ 14.4.6) we provide further historical comments on the subject.

Proposition 13.12 indicates some limit for the range of application of Theorem 13.1. The estimates which follow are, to a certain extent, the best possible ones which can be reached by applying Theorem 13.1 - it does not mean that they follow actually from Theorem 13.1! To prove the corresponding estimates require some more work, which was done in Chapters 7, 9 and 10 for some cases. As we shall see, it turns out that the results which have been achieved in these chapters are very close to the limit indicated by Proposition 13.12. Even the numbers c_0 below (which depend only on the number m of logarithms) are not very far from (i.e. not much smaller than) the values which actually can be achieved.

14.4.1 Homogeneous Linear Forms: Gel'fond-Baker's Method

Start with a linear form $\beta_1 X_1 + \cdots + \beta_m X_m$ with algebraic coefficients and consider a point $(\lambda_1, \ldots, \lambda_m)$ in $\mathcal{L}^m$ where it does not vanish. Set

$$\Lambda = \beta_1\lambda_1 + \cdots + \beta_m\lambda_m,$$

as in Theorems 7.1 and 9.1 for instance. Assume $\beta_m = -1$.

$\boxed{1}$ The hyperplane $\mathcal{W}$ in $\mathbb{C}^m$ of equation

$$\beta_1 z_1 + \cdots + \beta_{m-1} z_{m-1} = z_m$$

is rational over $\overline{\mathbb{Q}}$ and contains the point

$$\underline{\eta}'_1 = (\lambda_1, \ldots, \lambda_{m-1}, \lambda_m + \Lambda)$$

The number $|\Lambda|$ estimates the distance between $\underline{\eta}'_1$ and the point

$$\underline{\eta}_1 = (\lambda_1, \ldots, \lambda_{m-1}, \lambda_m) \in \mathcal{L}^m.$$

Notice that the functions

$$e^{z_1}, \ldots, e^{z_m}$$

take algebraic values at the point $\underline{\eta}_1$ (and therefore also at the points $s\underline{\eta}_1$ for $s \in \mathbb{Z}$).

The restriction to $\mathcal{W}$ of these functions give rise to the m functions of $m-1$ variables

$$e^{z_1}, \ldots, e^{z_{m-1}}, e^{\beta_1 z_1 + \cdots + \beta_{m-1} z_{m-1}},$$

which satisfy differential equations with algebraic coefficients. This amounts to take a basis of $\mathcal{W}$, namely

$$\underline{w}_k = (\underline{e}_k, \beta_k) \quad (1 \le k \le m-1),$$

where $\underline{e}_1, \ldots, \underline{e}_{m-1}$ is the canonical basis of $\mathbb{C}^{m-1}$. Let us build the $m \times m$ matrix whose column vectors are the coordinates of $\underline{w}_1, \ldots, \underline{w}_{m-1}, \underline{\eta}_1$ in $\mathbb{C}^m$:

$$\mathsf{B}_2 = \begin{pmatrix} & \mathsf{I}_{m-1} & \\ \beta_1 & \cdots & \beta_{m-1} \end{pmatrix}, \qquad \mathsf{L} = \begin{pmatrix} \lambda_1 \\ \vdots \\ \lambda_m \end{pmatrix}$$

and

$$\mathsf{M} = \begin{pmatrix} \mathsf{B}_2 & \mathsf{L} \end{pmatrix} = \begin{pmatrix} & & & \lambda_1 \\ & \mathsf{I}_{m-1} & & \vdots \\ & & & \lambda_{m-1} \\ \beta_1 & \cdots & \beta_{m-1} & \lambda_m \end{pmatrix}.$$

The main point in introducing M is that its determinant is just $-\Lambda$.

In connection with Chap. 13, define $d_0 = 0$, $d = d_1 = m$, $G = \mathbb{G}_{\mathrm{m}}^m$, $\ell_0 = m-1$, $\ell_1 = 1$, $\ell = m$, $r_1 = 0$, $r_2 = m-2$, $r_3 = 1$, $r = m-1$. Hence $\underline{w}_1, \ldots, \underline{w}_{m-1}$ are in K^m, $\underline{\eta}_1$ in $\mathcal{L}^m$ and

$$\underline{\gamma}_1 = \exp_G \underline{\eta}_1 = (\alpha_1, \ldots, \alpha_m) \in (K^\times)^m.$$

The matrix M' which approximates M (see Exercise 13.1) is

$$\mathsf{M}' = \begin{pmatrix} \mathsf{B}_2 & \mathsf{L}' \end{pmatrix}$$

where L' is the same column matrix as L apart from the entry λ_m which is replaced by $\lambda_m + \Lambda$. Notice that the determinant of M' is zero.

These facts have been the basis of the proof of homogeneous Baker's Theorem in § 10.1.1.

With the notation of Proposition 13.12, we have

$$u = 1, \quad \delta = m+2, \quad b_1 = 0, \quad b_2 = 2.$$

This explains how to reach the estimate

$$|\Lambda| \ge \exp\{-c_0 D^{m+2} (\log B)^2 (\log A_1) \cdots (\log A_m) (\log E)^{-m-1}\}$$

with

$$c_0 = \frac{m!^{2m}}{4^m}\left(\frac{2m^{m-2}(12m+9)}{(m-2)!}\right)^{m+1}.$$

Notice that the right hand side is larger than $(m!)^m m^{m^2}$.

Remark. The main point in Baker's method, compared with Gel'fond's one, is the fact that $r_3 = 1$. In particular this enables Baker to use Schwarz' lemma for functions of a single variable, and this tool enables him to extrapolate (see § 10.3). It is also possible to extrapolate with interpolation determinants when $r_3 = 1$, but this is not so easy as with an auxiliary function.

However one could work with what may be called "Gel'fond's method in several variables", that means using only $r = r_3 = m$ and $r_1 = r_2 = 0$. In this case one gets $u = 1$, $\delta = m^2$, $b_2 = m(m-1)$, and the estimate reads

$$|\Lambda| \geq \exp\{-cD^{m^2}(\log B)^{m^2-m}(\log A_1)\cdots(\log A_m)(\log E)^{-m^2+1}\}.$$

2 Like in §§ 10.1.2 and 10.2.1, we start with the functions

$$z_0, e^{z_1}, \ldots, e^{z_m}$$

and the hyperplane $\mathcal{W}$ in $\mathbb{C}^{m+1}$ of equation

$$\beta_1 z_1 + \cdots + \beta_{m-1} z_{m-1} = z_m.$$

The number $|\Lambda|$ estimates the distance of the point

$$\underline{\eta}_1 = (1, \lambda_1, \ldots, \lambda_{m-1}, \lambda_m)$$

from $\mathcal{W}$, because

$$\underline{\eta}_1' = (1, \lambda_1, \ldots, \lambda_{m-1}, \lambda_m + \Lambda)$$

belongs to $\mathcal{W}$.

Take $d_0=1$, $d_1 = m$, hence $G = \mathbb{G}_a \times \mathbb{G}_m^m$, $d = m+1$; then put $\ell_1 = 1$, $\ell_0 = m$, $\ell = m+1$, $r_1 = 0$, $r_2 = m-1$, $r_3 = 1$, $r = m$. Let $\underline{w}_1, \ldots, \underline{w}_m$ in $\overline{\mathbb{Q}}^{m+1}$ and $\underline{\eta}_1$ in $\overline{\mathbb{Q}} \times \mathcal{L}^m$ be the column vectors of the $(m+1)\times(m+1)$ matrix

$$\mathsf{M} = \begin{pmatrix} \mathsf{B}_0 & \mathsf{B}_1 \\ \mathsf{B}_2 & \mathsf{L} \end{pmatrix}$$

with

$$\mathsf{B}_0 = \begin{pmatrix} 1 & 0 & \cdots & 0 \end{pmatrix} \quad \mathsf{B}_1 = \begin{pmatrix} 1 \end{pmatrix}$$

$$\mathsf{B}_2 = \begin{pmatrix} 0 & & & \\ \vdots & & \mathsf{I}_{m-1} & \\ 0 & \beta_1 & \cdots & \beta_{m-1} \end{pmatrix} \quad \mathsf{L} = \begin{pmatrix} \lambda_1 \\ \vdots \\ \lambda_m \end{pmatrix}.$$

Introduce also

$$\mathsf{M}' = \begin{pmatrix} \mathsf{B}_0 & \mathsf{B}_1 \\ \mathsf{B}_2 & \mathsf{L}' \end{pmatrix} = \begin{pmatrix} & & & & 1 \\ & & & & \lambda_1 \\ & & \mathsf{I}_m & & \vdots \\ & & & & \lambda_{m-1} \\ 0 & \beta_1 & \cdots & \beta_{m-1} & \lambda_m + \Lambda \end{pmatrix}.$$

Then

$$u = 1, \quad \delta = m + 2, \quad b_1 = b_2 = 1,$$

which yields

$$|\Lambda| \geq \exp\{-c_0 D^{m+2} (\log B)(\log A_1) \cdots (\log A_m)(\log E^*)(\log E)^{-m-1}\}$$

with

$$c_0 = \frac{m!^m m^m (m+1)!^m}{4^m} \left(\frac{2(m+1)^{m-1}(12m+21)}{(m-1)!} \right)^{m+1}.$$

Once more, the right hand side is $> m!^m m^{m^2}$. If we compute the value of $C(m)$ in Theorem 9.1 by means of the arguments in § 10.2, one should not expect to reach the sharp estimate of Proposition 9.18.

The parameters A_i, B, E^* and E are the same as in Theorem 9.1. However, if one does not uses Fel'dman polynomials, one needs to assume $E^* \geq \log B$ (cf. Exercise 14.5).

Remark. A variant of this method (see § 11.4.1) can be worked out starting with the matrix

$$\begin{pmatrix} \beta_1 & \cdots & \beta_m & 0 \\ & & & \lambda_1 \\ & \mathsf{I}_m & & \vdots \\ & & & \lambda_m \end{pmatrix}$$

in place of M.

14.4.2 Homogeneous Linear Forms: Schneider's Method

$\boxed{1'}$ The dual of method $\boxed{1}$ involves the matrix

$$\mathsf{M} = \begin{pmatrix} & & & \beta_1 \\ & \mathsf{I}_{m-1} & & \vdots \\ & & & \beta_{m-1} \\ \lambda_1 & \cdots & \lambda_{m-1} & \lambda_m \end{pmatrix}$$

which is just the transposed of the matrix from method $\boxed{1}$. The parameters are now $d_0 = m - 1$, $d_1 = 1$, hence $d = m$, and $\ell_0 = 0$, $\ell_1 = \ell = m$. This means that we consider the functions

$$z_1, \ldots, z_{m-1}, e^{z_m}$$

and the hyperplane $\mathcal{V}$ in $\mathbb{C}^m$ of equation

$$\lambda_1 z_1 + \cdots + \lambda_{m-1} z_{m-1} = z_m.$$

The m column vectors of M are the coordinates in $\mathbb{C}^m$ of the points

$$\underline{\eta}_j = (\underline{e}_j, \lambda_j) \quad (1 \le j \le m-1) \quad \text{and} \quad \underline{\eta}_m = (\beta_1, \ldots, \beta_{m-1}, \lambda_m)$$

(where $\underline{e}_1, \ldots, \underline{e}_{m-1}$ is, as usual, the canonical basis of $\mathbb{C}^{m-1}$). For $1 \le j \le m$ we have $\underline{\eta}_j \in \overline{\mathbb{Q}}^{m-1} \times \mathcal{L} = \mathcal{L}_G$. Moreover $\underline{\eta}_1, \ldots, \underline{\eta}_{m-1}$ belong to $\mathcal{V}$, as does

$$\underline{\eta}'_m = (\beta_1, \ldots, \beta_{m-1}, \lambda_m + \Lambda).$$

Here again we introduce a matrix M′, with zero determinant, which differs from M only because λ_m in M is replaced by $\lambda_m + \Lambda$. Hence the distance between the two matrices M and M′ is just $|\Lambda|$.

This yields $r_3 = 1$, $r = m-1$, $r_1 = m-2$, $r_2 = 0$, hence $u = 1$, $\delta = m+2$, $b_1 = 2$, $b_2 = 0$, and the value of U is the same as in method [1], apart from the constant c_0 which is now:

$$c_0 = \frac{(m-1)!^m m! m^m}{4^m} \left(\frac{2(12m+9)}{(m-2)!} \right)^{m+1}.$$

Now the right hand side is $> 4(6m^3)^{m+1}$. This corresponds to the method which is described in Chap. 7 of [W 1992].

[2’] Finally the proof given in § 9.2 is the dual of [2] and requires $d_0 = m$, $d_1 = 1$, $d = m+1$, $\ell_0 = 1$, $\ell_1 = m$. We consider the functions

$$z_0, z_1, \ldots, z_{m-1}, e^{z_m}$$

and the hyperplane $\mathcal{V}$ in $\mathbb{C}^{m+1}$ of equation

$$\lambda_1 z_1 + \cdots + \lambda_{m-1} z_{m-1} = z_m.$$

The first column vector of the matrix

$$\mathsf{M} = \begin{pmatrix} \mathsf{B}_0 & \mathsf{B}_1 \\ \mathsf{B}_2 & \mathsf{L} \end{pmatrix} = \begin{pmatrix} & & & & 0 \\ & & & & \beta_1 \\ & & \mathsf{I}_m & & \vdots \\ & & & & \beta_{m-1} \\ 1 & \lambda_1 & \cdots & \lambda_{m-1} & \lambda_m \end{pmatrix}$$

is given by the coordinates in $\mathbb{C}^{m+1}$ of the point

$$\underline{w}_1 = (1, 0, \ldots, 0, 1)$$

which lies in $\overline{\mathbb{Q}}^{m+1}$, while the m last column vectors are the coordinates in $\mathbb{C}^{m+1}$ of the points

$$\underline{\eta}_j = (0, \underline{e}_j, \lambda_j) \quad (1 \le j \le m-1), \qquad \underline{\eta}_m = (0, \beta_1, \dots, \beta_{m-1}, \lambda_m)$$

where $\underline{e}_1, \dots, \underline{e}_{m-1}$ is again the canonical basis of $\mathbb{C}^{m-1}$. Since $\mathcal{L}_G = \overline{\mathbb{Q}}^m \times \mathcal{L}$, for $1 \le j \le m-1$ we have $\underline{\eta}_j \in \mathcal{V} \cap \mathcal{L}_G$. Moreover

$$\underline{\eta}'_m = (0, \beta_1, \dots, \beta_{m-1}, \lambda_m + \Lambda) \in \mathcal{V}.$$

We can apply Proposition 13.12 with

$$\mathsf{M}' = \begin{pmatrix} \mathsf{B}_0 & \mathsf{B}_1 \\ & \\ \mathsf{B}_2 & \mathsf{L}' \end{pmatrix}, \qquad \mathsf{L}' = (\,\lambda_1 \quad \cdots \quad \lambda_{m-1} \quad \lambda_m + \Lambda\,),$$

$r_3 = 1$, $r = m$, $r_1 = m-1$ and $r_2 = 0$. Again the value of U is the same as in method $\boxed{2}$, apart from the value of c_0 which is now:

$$c_0 = \frac{m!^m m^m (m+1)!}{4^m} \left(\frac{2(12m+21)}{(m-1)!} \right)^{m+1}.$$

When m is large the right hand side is not less than $43(6m^3)^{m+1}$.

Remark. Once more, a variant of this method (compare with § 11.4.2) can be deduced from the following remark: the determinant of the matrix

$$\begin{pmatrix} \mathsf{B}_0 & \mathsf{B}_1 \\ & \\ \mathsf{B}_2 & \mathsf{L} \end{pmatrix} = \begin{pmatrix} \beta_1 & & & \\ \vdots & & \mathsf{I}_m & \\ \beta_m & & & \\ 0 & \lambda_1 & \cdots & \lambda_m \end{pmatrix}$$

is $(-1)^{m+1}(\beta_1\lambda_1 + \cdots + \beta_m\lambda_m)$.

14.4.3 Affine Linear Forms: Gel'fond-Baker's Method

Consider now a linear combination of 1 and logarithms of algebraic numbers with algebraic coefficients

$$\Lambda = \beta_0 + \beta_1\lambda_1 + \cdots + \beta_{m-1}\lambda_{m-1} + \beta_m\lambda_m$$

with, say, $\beta_m = -1$.

$\boxed{2}$ We modify the homogeneous method by considering the hyperplane $\mathcal{W}$ of equation

$$\beta_0 z_0 + \beta_1 z_1 + \cdots + \beta_{m-1} z_{m-1} = z_m$$

in $\mathbb{C}^{m+1}$. Take $d_0 = 1$, $d_1 = m$, $d = m+1$, $G = \mathbb{G}_a \times \mathbb{G}_m^m$, $\ell_0 = m$, $\ell_1 = 1$, $\ell = m+1$, $r_1 = 0$, $r_2 = m-1$, $r_3 = 1$, $r = m$. Let $\underline{w}_1, \dots, \underline{w}_m, \underline{\eta}_1$ be the elements of $\mathbb{C}^{m+1}$ whose coordinates are the column vectors of the $(m+1) \times (m+1)$ matrix

$$\mathsf{M} = \begin{pmatrix} & & & 1 \\ & & & \lambda_1 \\ & \mathsf{I}_m & & \vdots \\ & & & \lambda_{m-1} \\ \beta_0 & \cdots & \beta_{m-1} & \lambda_m \end{pmatrix}.$$

Hence

$$\underline{\gamma}_1 = \exp_G \underline{\eta}_1 = (\beta_0, \alpha_1, \ldots, \alpha_m) \in \overline{\mathbb{Q}} \times (\overline{\mathbb{Q}}^\times)^m.$$

With the notation of Proposition 13.12, we have

$$u = 1, \quad \delta = m + 2, \quad b_1 = b_2 = 1,$$

which gives

$$|\Lambda| \geq \exp\{-c_0 D^{m+2} (\log B)(\log A_1) \cdots (\log A_m)(\log E^*)(\log E)^{-m-1}\}$$

where

$$c_0 = \frac{(m!m(m+1)!)^m}{4^m} \left(\frac{2(m+1)^{m-1}(12m+21)}{(m-1)!} \right)^{m+1}.$$

The right hand side is $> m!^m m^{m^2}$.

Other square matrices than M have determinant $\pm\Lambda$. An example is

$$\widetilde{\mathsf{M}} = \begin{pmatrix} \beta_1 & \cdots & \beta_m & -\beta_0 \\ & & & \lambda_1 \\ & \mathsf{I}_m & & \vdots \\ & & & \lambda_m \end{pmatrix}.$$

Here there is no need to assume $\beta_m = -1$ in the definition of Λ. The choice of the starting matrix may influence the final estimate: in $\widetilde{\mathsf{M}}$, the algebraic numbers $\beta_1, \ldots, \beta_m$ arise in the upper left corner, which is the matrix B_0 of § 13.1. Therefore their height will be taken care of either by the parameter B_1 or B_2, as we wish. While in M the same algebraic numbers arise in B_2, so we have no choice: the height is controlled by B_2.

A slight modification enables us to include also the coefficient β_0 into B_0: consider the $(m+2) \times (m+2)$ matrix

$$\begin{pmatrix} 1 & 0 & \cdots & 0 & 1 \\ \beta_0 & \beta_1 & \cdots & \beta_m & 0 \\ 0 & 1 & \cdots & 0 & \lambda_1 \\ \vdots & \vdots & \ddots & \vdots & \vdots \\ 0 & 0 & \cdots & 1 & \lambda_m \end{pmatrix}$$

associated to the parameters $d_0 = 2$, $d_1 = m$, $\ell_0 = m+1$, $\ell_1 = m$. The number Λ is considered as the value, at the point $(1, 0, \lambda_1, \ldots, \lambda_m)$, of the linear form

$$\beta_0 z_{-1} - z_0 + \beta_1 z_1 + \cdots + \beta_m z_m.$$

This is related with the improvement, by N. Hirata-Kohno [Hir 1991], of the estimate in [PW 1988c] giving measures of linear independence of logarithms on commutative algebraic groups.

14.4.4 Affine Linear Forms: Schneider's Method

[2'] Consider the hyperplane of equation

$$z_0 + \lambda_1 z_1 + \cdots + \lambda_{m-1} z_{m-1} = z_m$$

in $\mathbb{C}^{m+1}$ and transpose the matrix M occurring in [2] :

$$ {}^{t}\mathsf{M} = \begin{pmatrix} & & & \beta_0 \\ & & & \beta_1 \\ & \mathsf{I}_m & & \vdots \\ & & & \beta_{m-1} \\ 1 & \lambda_1 \ \cdots & \lambda_{m-1} & \lambda_m \end{pmatrix} $$

with $d_0 = m, d_1 = 1, d = m+1, G = \mathbb{G}_{\mathrm{a}}^m \times \mathbb{G}_{\mathrm{m}}, \ell_0 = 1, \ell_1 = m, \ell = m+1, r_1 = m-1,$ $r_2 = 0, r_3 = 1, r = m.$

With the notation of Proposition 13.12, we still have

$$u = 1, \quad \delta = m+2, \quad b_1 = b_2 = 1,$$

which again leads to the estimate (under the condition $E^* \geq \log B$) from Chap. 9:

$$|\Lambda| \geq \exp\{-c_0 D^{m+2} (\log B)(\log A_1) \cdots (\log A_m)(\log E^*)(\log E)^{-m-1}\}$$

where

$$c_0 = \frac{(m!)^m (m+1)! m^{m+1}}{4^m} \left(\frac{2(12m+21)}{(m-1)!} \right)^{m+1}.$$

For large values of m the right hand side is $> 7(6m^3)^{m+1}$.

A variant of this method involves the transposed of the matrix $\widetilde{\mathsf{M}}$ in [2] :

$$ {}^{t}\widetilde{\mathsf{M}} = \begin{pmatrix} \beta_1 & & & \\ \vdots & & \mathsf{I}_m & \\ \beta_m & & & \\ -\beta_0 & \lambda_1 & \cdots & \lambda_m \end{pmatrix}. $$

It is interesting to compare methods [2] and [2'] when $m = 1$, namely for $\lambda - \beta_0$ (related to the Hermite-Lindemann's Theorem). The lack of symmetry suggests to replace β_0 by $\beta_0'\beta_0''$ and to consider for instance either the matrix

$$ \begin{pmatrix} & & & \beta_0' \\ & & & \beta_1 \\ & \mathsf{I}_m & & \vdots \\ & & & \beta_{m-1} \\ \beta_0'' & \lambda_1 \ \cdots & \lambda_{m-1} & \lambda_m \end{pmatrix} $$

or its transpose. Unfortunately it seems that one does not reach anything more than with the two trivial decompositions $(\beta_0', \beta_0'') = (\beta_0, 1)$ and $(\beta_0', \beta_0'') = (1, \beta_0)$.

14.4.5 The Subgroups G^+ and G^-

Proposition 13.12 involves only the algebraic subgroup $G^* = \{0\}$ of G. However the conclusion of Theorem 13.1 introduces an algebraic subgroup G^* of G, $G^* \neq G$, which may have a positive dimension, and we need to take it into account. The idea for applying Theorem 13.1 is to use it first with $G^+ = G$ and $G^- = \{0\}$, and to consider the possible G^* which may appear in the conclusion. Such a G^* is an *obstruction subgroup* for the given situation. We consider the "worst" one. Depending on the case, it will be such an obstruction subgroup of minimal or maximal dimension. Next we repeat the proof starting with either $G^- = \{0\}$, $G^+ = G^*$ or else $G^- = G^*$, $G^+ = G$. The former situation occurred in Chap. 9, the latter in Chap. 10. We explain here what happened.

Looking for a lower bound for the modulus of

$$\Lambda = \beta_0 + \beta_1\lambda_1 + \cdots + \beta_{m-1}\lambda_{m-1} - \lambda_m,$$

method $\boxed{2'}$ of § 14.4.4 involves the algebraic group $G = \mathbb{G}_{\mathrm{a}}^m \times \mathbb{G}_{\mathrm{m}}$, the hyperplane $\mathcal{V}$ of equation

$$z_0 + \lambda_1 z_1 + \cdots + \lambda_{m-1} z_{m-1} = z_m$$

in $\mathbb{C}^{m+1}$ and the subgroup $\mathbb{Z}\underline{\eta}_1 + \cdots + \mathbb{Z}\underline{\eta}_m$ of $\overline{\mathbb{Q}}^m \times \mathcal{L}$, where

$$\underline{\eta}_j = (0, \underline{e}_j, \lambda_j) \quad (1 \le j \le m), \qquad \underline{\eta}_m = (\beta_0, \beta_1, \ldots, \beta_{m-1}, \lambda_m).$$

Notice that $\mathcal{V}$ contains $\underline{\eta}_1, \ldots, \underline{\eta}_{m-1}$ as well as

$$\underline{\eta}'_m = (\beta_0, \beta_1, \ldots, \beta_{m-1}, \lambda_m + \Lambda).$$

Denote by Y the subgroup $\mathbb{Z}^{m-1} + \mathbb{Z}(\beta_1, \ldots, \beta_{m-1})$ of $\mathbb{C}^{m-1}$. An obstruction subgroup in this case is an algebraic subgroup $G^* = G_0^* \times G_1^*$ of G, of dimension $\le m$, where the algebraic subgroup G_0^* of $G_0 = \mathbb{G}_{\mathrm{a}}^m$ is nothing else than a vector subspace of $\mathbb{C}^m$. The projection onto $\{0\} \times \mathbb{C}^{m-1}$ associates to G_0^* a vector subspace of $\mathbb{C}^{m-1}$ which contains "many" points

$$(s_1 + s_m\beta_1, \ldots, s_{m-1} + s_m\beta_{m-1}) \in Y \quad (|s_j| \le S_j,\ 1 \le j \le m).$$

The existence of such a vector subspace of $\mathbb{C}^{m-1}$ is bad for the multiplicity estimate, but it is good for the transcendence proof because we wish to use the information provided by as many such points as possible. Hence we repeat the transcendence argument (i.e. we apply Theorem 13.1), taking for G^+ such an obstruction subgroup G^* of G of minimal dimension. This is what we did implicitly in Chap. 9. In fact, we had $G = \mathbb{G}_{\mathrm{a}}^m \times \mathbb{G}_{\mathrm{m}}$ and $G^+ = G_0^+ \times \mathbb{G}_{\mathrm{m}}$; because of this special and simple situation, a change of basis enabled us to work directly with $\mathbb{G}_{\mathrm{a}}^n \times \mathbb{G}_{\mathrm{m}}$, where $n = \dim(G_0^+)$. This is why G^+ did not appear explicitly in Chap. 9.

In Chap. 10 the situation is different. Starting with the same Λ, method $\boxed{2}$ of § 14.4.3 involves the linear algebraic group $G = \mathbb{G}_{\mathrm{a}} \times \mathbb{G}_{\mathrm{m}}^m$ and the hyperplane $\mathcal{W}$ of $\mathbb{C}^{m+1}$ of equation

$$z_m = \beta_0 z_0 + \beta_1 z_1 + \cdots + \beta_{m-1} z_{m-1},$$

which contains the point

$$(1, \lambda_1, \ldots, \lambda_{m-1}, \lambda_m + \Lambda)$$

close to

$$(1, \lambda_1, \ldots, \lambda_{m-1}, \lambda_m) \in \overline{\mathbb{Q}} \times \mathcal{L}^m$$

if $|\Lambda|$ is small. In this situation, Theorem 13.1 yields an obstruction subgroup G^* of G such that $T_e(G^*)$ is contained in $\mathcal{W}$ and $\mathcal{H}(G^*; \underline{T})$ is "small". This last bit of information means that we don't have many independent monomials at our disposal with G^*. On may expect that G/G^* will provide more monomials. Indeed, we take for G^- such an obstruction subgroup G^* of maximal dimension and we repeat the argument, using Theorem 13.1 with $G^+ = G$. The details have been given in Chap. 10.

Of course in the real life we do not repeat the construction: we immediately start with the right choice of G^+ and G^-, and the above initial construction just corresponds to the easiest situation where $G^- = \{0\}$ and $G^+ = G$.

14.4.6 Further Historical Comments

We described in § 11.4 several methods for proving Baker's Theorems 1.5 and 1.6, and we have just seen that they yield effective measures of linear independence. We now describe their developments and compare their respective merit. A general comment before we consider each method separately: because of applications (especially to solving explicitly diophantine equations), a special attention has been paid in published papers to the quality of the numerical estimates. The number of variables of the analytic functions occurring in the proof is one of the main limitation for getting small absolute constants. There is a discrepancy between proofs involving a single variable and proofs which require more than one variable. Part of the explanation is that complex analysis in one variable is better understood than in higher dimension. Therefore the estimates for $|\beta_1\lambda_1 + \beta_2\lambda_2|$ involving a dependence $(\log B)^2$ on the height B of β_1 and β_2 involve quite small numerical absolute constants, because the proof requires only exponential functions in a single variable [LauMN 1995]. Using two variables, one can either get estimate for the same linear combination in two logarithms with only $\log B$, or else get lower bounds for linear combination of three logarithms with $(\log B)^2$. Since both proofs involve the same number of variables, one should not be surprised that the numerical estimates one gets involve constants of comparable size.

Method 1

This method was initiated by A. O. Gel'fond for his proof of the transcendence of α^β in 1934, and soon after for proving effective measures of linear independence for two logarithms. In [S 1967], A. Schinzel produced the first explicit estimates and gave several arithmetic applications.

For several logarithms, this is also the method which enabled A. Baker in his first paper on this topic ([B 1966], I) to prove his homogeneous Theorem 1.5.

Using [1], one gets a lower bound for a measure of homogeneous linear independence of two logarithms with a dependence on the height B of the coefficients which is $\exp\{-c(\log B)^2\}$. This is the best estimate which can be achieved so far with this method. This explains why Gel'fond could not do better.

On the other hand, as pointed out in (10.14), one may use Fel'dman's polynomial and replace B by

$$\max_{1\leq j\leq m-1}\left\{\frac{|b_m|}{\log A_j}+\frac{|b_j|}{\log A_m}\right\}.$$

This is done in [Sp 1982], Chap. III.

This method has not been widely used for proving measures of homogeneous independence for several logarithms. It should be expected that essentially the same estimates can be achieved as with method [1'], apart from the numerical constant (which is the strong point of [1']).

Method [2]

Method [2] (and variants of the same) have been described in Chap. 10. This is certainly the method which has been the more widely used in papers dealing with "lower bounds for linear forms in logarithms" or with "logarithmic forms", [B 1966], [F 1968], [B 1972], [Sho 1974], [B 1975], Chap. 2, [LoxV 1976], [Sho 1976], [T 1976], [B 1977], [V 1977], [L 1978] (Chap. VIII, X and XI), [W 1980], [Lox 1986], [Wü 1988], [PW 1988a], [PW 1988b], [Y 1989], [BlaGMMS 1990], [BWü 1993], [BeBGMS 1997], [Mat 1998], [Y 1998] and [FNe 1998], Chap. 4, § 2.

This method has been extended to commutative algebraic groups in [Wü 1988], [PW 1988c], and [Hir 1991] (see also [D 1995] for explicit estimates in the elliptic case).

The surveys by A. Baker in [B 1977] and Fel'dman and Nesterenko in Chap. 4 § 1 of [FNe 1998] consider almost exclusively method [2]. Also the methods described in [L 1978] are only variants of [1] and [2].

Method [1']

Chapters 6, 7 and 9 described method [1'] (see also Chap. 7 and 9 of [W 1992]). This method was initiated in [MiW 1978] for studying linear combinations of two logarithms. The numerical estimates of [MiW 1978] have been improved in [Lau 1994] and [LauMN 1995]. The sharpest known numerical explicit measures of linear independence for two or three logarithms (which occur in many applications) all involve method [1'] : for three logarithms, see [BeBGMS 1997] (there are also unpublished manuscript by P. Voutier). The paper [BeBGMS 1997] deals with $b_1\log\alpha_1+b_2\log\alpha_2+b_3\log\alpha_3$ where $b_1,b_2,b_3,\ \alpha_1,\alpha_2,\alpha_3$ are positive rational numbers (in connection with Catalan's Conjecture), while Voutier's papers consider the more general situation where $\alpha_1,\alpha_2,\alpha_3$ are algebraic numbers. The numerical estimate in [LauMN 1995] for two logarithms is so sharp that one may sometimes use it for three or more logarithms, by grouping terms; this corresponds to a degenerate linear combination, and P. Voutier provides a systematic treatment of this degenerate case for any number of logarithms.

It should be pointed out that this method does not extend (so far) to commutative algebraic groups, apart from the elliptic case with complex multiplication (see [Y 1985]).

Method [2']

Method [2'], which was used in Chap. 9, is dual of [2]. It was introduced in [W 1979a], Chap. 6, for giving a new proof of Baker's qualitative result. It has been worked out in a quantitative form when $\beta_0 = 0$ in [W 1991b] and [W 1993], and for the general case in [W 1992], Chap. 11.

***p*-Adic Estimates**

Several authors considered p-adic measures of linear independence of logarithms of algebraic numbers, including

- K. Mahler, A. O. Gel'fond, and A. Schinzel and A. Brumer using method [1].
- J. H. Coates, V. G. Sprindžuk, R. M. Kaufman, A. J. van der Poorten, J. H. Loxton, and later Yu Kunrui, by means of method [2]. Surveys with references on this topic are included in [V 1977] and [Y 1989]. Yu Kunrui found an efficient way of avoiding the assumption that the α_i are close to 1 modulo p. In [Y 1998]-I he extended the work of Baker and Wüstholz [BWü 1993] to the p-adic case, and in [Y 1998]-II he did the same for the paper of Matveev [Mat 1998].
- Dong Pingping [Dpp 1995], with method [1'] for an arbitrary number of logarithms, and Y. Bugeaud and M. Laurent [BuLau 1996] for linear combinations of two logarithms only (p-adic analog of the main result of [LauMN 1995]).

Measures of Simultaneous Approximation

At an early stage of the theory, K. Ramachandra [R 1969b] obtained a comparatively sharp lower bound by considering several linear forms. He was using method [2]. His result was improved later in [Lox 1986] and [PW 1988b], again with method [2]. Further, J. H. Loxton gave arithmetic applications. Dong Pingping's p-adic result in [Dpp 1995] includes lower bound for simultaneous linear forms in logarithms.

Open Problems

We propose a simple but far reaching conjectural measure of linear independence for logarithms of algebraic numbers. There is no need to distinguish between general case, homogeneous case, rational case or whatever.

Conjecture 14.25. *There exist two positive absolute constants c_1 and c_2 with the following property. Let $\lambda_1, \dots, \lambda_m$ be logarithms of algebraic numbers with $\alpha_i = e^{\lambda_i}$ $(1 \le i \le m)$, let $\beta_0, \dots, \beta_m$ be algebraic numbers, D the degree of the number field $\mathbb{Q}(\alpha_1, \dots, \alpha_m, \beta_0, \dots, \beta_m)$ and finally let $h \ge 1/D$ satisfy*

$$h \geq \max_{1\leq i\leq m} h(\alpha_i), \quad h \geq \frac{1}{D}\max_{1\leq i\leq m} |\lambda_i| \quad \textit{and} \quad h \geq \max_{0\leq j\leq m} h(\beta_j).$$

1) Assume that the number

$$\Lambda = \beta_0 + \beta_1\lambda_1 + \cdots + \beta_m\lambda_m$$

is nonzero. Then

$$|\Lambda| \geq \exp\bigl\{-c_1 m D^2 h\bigr\}.$$

2) Assume $\lambda_1, \ldots, \lambda_m$ are linearly independent over $\mathbb{Q}$. Then

$$\sum_{i=1}^{m} |\lambda_i - \beta_i| \geq \exp\bigl\{-c_2 m D^{1+(1/m)} h\bigr\}.$$

In the special case $D = 1$ and $\beta_0 = 0$, Conjecture 1.11 is more precise than part 1 of Conjecture 14.25. On the other hand for $D = 1$, $m = 1$, $\beta_0 \neq 0$, both parts 1 and 2 of Conjecture 14.25 reduce to an open problem of Mahler [M 1967]:

(?) *Does there exist an absolute constant $c > 0$ such that, for any positive rational integers a and b,*

$$|e^b - a| \geq a^{-c}?$$

If $|e^b - a|$ is small, then b and $\log a$ are of the same order of magnitude, hence one can replace $a^{-c} = e^{-c\log a}$ in the right hand side by e^{-cb}. For the same reason, since $|e^b - a|/a = |e^{b-\log a} - 1|$ is close to $|b - \log a|$, one can replace $|e^b - a|$ in the left hand side by $|b - \log a|$ (replacing at the same time c by $c + 1$ in the right hand side).

The best known estimate in this direction is due to Mahler [M 1967]:

$$|e^b - a| \geq a^{-c\log\log a}$$

and

$$|b - \log a| \geq b^{-cb}$$

for $a \geq 3$. K. Mahler found a sharp explicit numerical value for c, namely $c = 33$ (for both estimates), provided that a is sufficiently large. A refinement is due to F. Wielonsky [Wi 1999]: for sufficiently large a, these estimates hold with $c = 20$.

In Chap. 15 we shall see that part 2 of Conjecture 14.25, dealing with a simultaneous approximation measure for logarithms of algebraic numbers, would imply results of *algebraic independence* for logarithms of algebraic numbers.

Exercises

Exercise 14.1.
a) Deduce from Theorem 14.1 the following measure of linear independence of two logarithms:

Let β be an algebraic number and λ_1, λ_2 elements of $\mathcal{L}$. Define $\alpha_i = e^{\lambda_i}$ and $D = [\mathbb{Q}(\alpha_1, \alpha_2, \beta) : \mathbb{Q}]$. Let A_1, A_2, B and E be positive real numbers satisfying

$$B \geq e, \quad B \geq D, \quad E \geq e, \quad B \geq E^{1/D}, \quad B \geq D \log A_i \geq \log E,$$

$$\mathrm{h}(\alpha_i) \leq \log A_i, \quad |\lambda_i| \leq \frac{D}{E} \log A_i \quad \text{and} \quad \mathrm{h}(\beta) \leq \log B$$

for $i = 1$ and $i = 2$. If $\beta\lambda_1 \neq \lambda_2$, then

$$|\lambda_2 - \beta\lambda_1| \geq \exp\left\{ - 2^{30} D^4 (\log B)^2 (\log A_1)(\log A_2)(\log E)^{-3} \right\}.$$

Hint. *Let*

$$U = 2^{30} D^4 (\log B)^2 (\log A_1)(\log A_2)(\log E)^{-3}.$$

If there exists $(s_1, s_2) \in \mathbb{Z}^2 \setminus \{0\}$ such that $s_1\lambda_1 = s_2\lambda_2$ and $|s_j| \leq U^2$, then apply Liouville's estimate (Exercise 3.7.a and Proposition 3.14) to deduce

$$|\lambda_2| \geq 2^{-D} A_2^{-D} \quad \text{and} \quad |s_2 - \beta s_1| \geq (2U^2)^{-D} B^{-D}.$$

b) We have shown in Chap. 9 how to improve $(\log B)^2$ to $(\log B) \log\log B$, and even to $\log B$. Use the same method and improve Theorem 14.1.
c) Produce a dual (in the sense of § 13.7) proof of Theorem 14.1 and compare the results.

Exercise 14.2. Deduce from Theorem 14.1 the following result:

Let $(\theta_1, \ldots, \theta_m)$ be a m-tuple of $\mathbb{Q}$-linearly independent complex numbers satisfying a linear independence measure condition. Let $\beta_0, \ldots, \beta_n$ be $\mathbb{Q}$-linearly independent algebraic numbers. There exists a constant $c > 0$ such that

$$\varphi(D, h) = c D^{(m+1)(n+1)/mn} h^{1+(1/n)} (\log h + \log D)^{-1/n}$$

is a simultaneous approximation measure for the $m(n+1)$ numbers

$$e^{\beta_j \theta_i} \quad (0 \leq j \leq n, \ 1 \leq i \leq m).$$

Hint. *Replacing if necessary β_j by β_j/β_0 and θ_i by $\theta_i\beta_0$, one may assume $\beta_0 = 1$. Define $U = \varphi(D, h)$. Assume $|e^{\beta_j\theta_i} - \alpha_{ij}| \leq e^{-U}$ for some nonzero algebraic numbers α_{ij}. Define $\lambda_{ij} \in \mathcal{L}$ by the conditions $e^{\lambda_{ij}} = \alpha_{ij}$ and $|\lambda_{ij} - \beta_j\theta_i| \leq e^{-2U/3}$. Further, let $\lambda_i = \lambda_{i0}$,*

$$A_{ij} = e^h, \quad E = (Dh)^{1/c_0}, \quad B = (Dh)^{c_0},$$

for some suitable constant $c_0 > 1$.

Let β be an algebraic number of degree d, a a nonzero complex number and $\log a$ a nonzero logarithm of a such that $(1, \log a)$ satisfies a linear independence measure condition. Write a^z for $e^{z \log a}$. Deduce that there exists $c > 0$ such that

$$cD^{(d+1)/(d-1)}h^{d/(d-1)}(\log h + \log D)^{-1/(d-1)}$$

is a simultaneous approximation measure for the d numbers

$$a,\ a^{\beta}, \dots, a^{\beta^{d-1}}.$$

Hint. *Take* $n = d - 1$, $m = d$,

$$\beta_j = \beta^j \quad (0 \le j \le n), \qquad \theta_i = \beta^{i-1} \log a \quad (1 \le i \le m).$$

(Compare with [RoyW 1997b], *Th. 2.1).*

Exercise 14.3.
a) Deduce from Theorem 14.6 a lower bound for $|\lambda - \beta|$ as follows:

Let β be an algebraic number and λ a nonzero element of $\mathcal{L}$. Define $\alpha = e^{\lambda}$ and $D = [\mathbb{Q}(\alpha, \beta) : \mathbb{Q}]$. Let A, B and E be positive real numbers satisfying

$$B \ge e, \quad B \ge D, \quad E \ge e, \quad B \ge E^{1/D}, \quad B \ge D \log A \ge \log E,$$

$$\mathrm{h}(\alpha) \le \log A, \quad |\lambda| \le \frac{D}{E} \log A \quad \textit{and} \quad \mathrm{h}(\beta) \le \log B$$

Then

$$|\lambda - \beta| \ge \exp\big\{ - 2^{30} D^3 (\log A)(\log B)(\log\log A + \log D)(\log E)^{-3} \big\}.$$

b) Compare this result with [NeW 1996] and with the special case $m = 1$ of Theorem 9.1 (see Remark 2 in § 9.4.1); deduce an improvement of Theorem 14.1.

Exercise 14.4. In this exercise, we say that a function $\phi \colon \mathbb{N} \times \mathbb{R}^m_{>0} \to \mathbb{R}_{>0} \cup \{\infty\}$ is a *simultaneous approximation measure for* $\underline{\theta}$ if there exist a positive integer D_0 together with a real number $h_0 \ge 1$ such that, for any integer $D \ge D_0$, any real numbers $h_i \ge h_0$ $(1 \le i \le m)$ and any m-tuple $\underline{\gamma} = (\gamma_1, \dots, \gamma_m)$ of algebraic numbers satisfying

$$[\mathbb{Q}(\underline{\gamma}) : \mathbb{Q}] \le D \quad \text{and} \quad \mathrm{h}(\gamma_i) \le h_i \ (1 \le i \le m),$$

we have

$$\max_{1 \le i \le m} |\theta_i - \gamma_i| \ge \exp\big\{ - \phi(D; h_1, \dots, h_m) \big\}.$$

When $\phi(D; h_1, \dots, h_m)$ depends only on D and $h = \max\{h_1, \dots, h_m)$, the function $\varphi \colon \mathbb{N} \times \mathbb{R}_{>0} \to \mathbb{R}_{>0} \cup \{\infty\}$ defined by $\varphi(D, h) = \phi(D; h, \dots, h)$ is a simultaneous approximation measure for $\underline{\theta}$, by the earlier definition in the introduction of this chapter.
Deduce from Theorem 14.6 with $m = n = 2$ the following results.
a) Let r be a nonzero rational number. There exists a constant $c = c(r)$ such that the function

$$\phi(D; h_1, h_2, h_3) = cD^2\big(h_1 + \log(Dh_2h_3)\big)(h_1 + h_2)^{1/2}(h_2 + h_3)^{1/2}\big(\log(Dh_2)\big)^{-1}$$

is a simultaneous approximation measure for the three numbers $e, e^{e^r}, e^{e^{2r}}$.

Hint. *The rank one matrix here is*

$$\begin{pmatrix} 1 & 1 & e^r \\ 1 & 1 & e^r \\ e^r & e^r & e^{2r} \end{pmatrix}.$$

b) Let λ be a nonzero element of $\mathcal{L}$ and let β be a nonzero algebraic number. There exists a constant $c = c(\lambda, \beta)$ such that the function

$$cD^2\big(h_1 + \log(Dh_2h_3)\big)h_2^{1/2}h_3^{1/2}(\log D)^{-1}$$

is a simultaneous approximation measure for the three numbers λ, e^β and $e^{\lambda^2/\beta}$.

Remark. An example is π, e and e^{π^2}.

Hint. *Here the rank one matrix is*

$$\begin{pmatrix} 1 & \beta & \lambda \\ 1 & \beta & \lambda \\ \dfrac{\lambda}{\beta} & \lambda & \dfrac{\lambda^2}{\beta} \end{pmatrix}.$$

c) Assuming one could avoid the assumption that the matrix $(\log A_{ij})$ has rank 1 in Theorem 14.6, show that in place of

$$cD^2h(h + \log D)(\log D)^{-1},$$

one would obtain the following simultaneous approximation measure for the numbers λ, e^β and $e^{\lambda^2/\beta}$:

$$cD^2h^{1/2}(h + \log D)(\log D)^{-1}.$$

d) Let λ be a nonzero element of $\mathcal{L}$. There exists a constant $c = c(\lambda) > 0$ such that

$$\phi(D; h_1, h_2, h_3) = cD^2\big(h_1 + \log(Dh_2h_3)\big)h_2^{1/2}(h_2 + h_3)^{1/2}(\log D)^{-1}$$

is a simultaneous approximation measure for the numbers λ, e^{λ^2} and e^{λ^3}.

Hint. *Consider the matrix*

$$\begin{pmatrix} 1 & \lambda & \lambda^2 \\ 1 & \lambda & \lambda^2 \\ \lambda & \lambda^2 & \lambda^3 \end{pmatrix}$$

which has rank 1.

e) Let β be an irrational quadratic number and let λ be a nonzero logarithm of an algebraic number. There exists a positive constant $c = c(\beta, \lambda)$ such that

$$\phi(D; h_1, h_2) = c\max\left\{Dh_1,\ D^2h_2\big(h_1 + \log(Dh_2)\big)^{1/2}\big(\log(Dh_1h_2)\big)^{1/2}\big(\log(Dh_2)\big)^{-1}\right\}$$

is a simultaneous approximation measure for the two numbers λ and $e^{\beta\lambda}$. In particular for $h_1 = h_2$ the measure is

$$cD^2h(h + \log D)^{1/2}(\log h + \log D)^{-1/2}.$$

Hint. *Observe that the matrix*

$$\begin{pmatrix} 1 & \lambda & \beta\lambda \\ 1 & \lambda & \beta\lambda \\ \beta & \beta\lambda & \beta^2\lambda \end{pmatrix}$$

has rank 1. *Compare with* [RoyW 1997b], *Th. 2.7.*

f) Let $\lambda_1, \lambda_2, \lambda_3, \lambda_4$ be nonzero elements of $\mathcal{L}$ such that both numbers λ_1/λ_2 and λ_1/λ_3 are irrational. Assume $\lambda_1\lambda_4 = \lambda_2\lambda_3$. Then a simultaneous approximation measure for the four numbers $\lambda_1, \lambda_2, \lambda_3, \lambda_4$ is

$$cD^2(h + \log D)(\log D)^{-1}.$$

Exercise 14.5. Using Feld'man's Delta polynomials (see § 13.6), improve:

a) Theorem 14.1, in the special case where $\beta_1, \dots, \beta_n$ are rational integers.
b) Theorem 14.6, when either one or both of the tuples of algebraic numbers $(\beta_1, \dots, \beta_n)$ and $(\beta'_1, \dots, \beta'_m)$ consist of rational integers.

Deduce a refinement of Corollary 14.12 which contains the result (14.13) of Fel'dman concerning the simultaneous approximation measure for logarithms of algebraic numbers.

Exercise 14.6. a) Using Theorem 14.6, show that a simultaneous approximation measure for the $p + q$ numbers

$$\lambda_1, \dots, \lambda_p, \quad e^{\beta_1}, \dots, e^{\beta_q}$$

when $\lambda_1, \dots, \lambda_p$ are $\mathbb{Q}$-linearly independent in $\mathcal{L}$ and $\beta_1, \dots, \beta_q$ are $\mathbb{Q}$-linearly independent in $\overline{\mathbb{Q}}$ is

$$cD^{2+\kappa}h^{q\kappa}(h + \log D)(\log h + \log D)^{\kappa}(\log D)^{-1-\kappa},$$

where $\kappa = 1/(p+q)$ and where $c > 0$ depends only on $\lambda_1, \dots, \lambda_p, \beta_1, \dots, \beta_q$.

Hint. *Choose for instance $n = 1$ and $m = p + q$.*

b) Deduce that, for $\lambda \in \mathcal{L} \setminus \{0\}$ and $\beta \in \overline{\mathbb{Q}} \setminus \{0\}$, a simultaneous approximation measure for λ and e^β is

$$c(\lambda, \beta)D^{5/2}h^{1/2}(h + \log D)(\log h + \log D)^{1/2}(\log D)^{-3/2}.$$

c) Show that for any $\beta \in \overline{\mathbb{Q}} \setminus \{0\}$, a simultaneous approximation measure for π and e^β is

$$c(\beta)D^2h^{1/2}(h + \log D)(\log h + \log D)^{1/2}.$$

Exercise 14.7. Check that Corollary 14.18 does not hold without condition (14.19).

Hint. *If $(p_0, \dots, p_m) \in \mathbb{Z}^m$ with $p_0 \neq 0$ and $(q_0, \dots, q_n) \in \mathbb{Z}^n$ with $q_0 \neq 0$ satisfy*

$$\max_{1 \le i \le m} \left| x_i - \frac{p_i}{p_0} \right| \le \epsilon \quad \textit{and} \quad \max_{1 \le j \le n} \left| y_j - \frac{q_j}{q_0} \log 2 \right| \le \epsilon$$

with $0 < \epsilon \le 1$, then

$$\max_{\substack{1 \le i \le m \\ 1 \le j \le n}} \left| e^{x_i y_j} - 2^{p_i q_j / p_0 q_0} \right| \le c\epsilon$$

where

$$c = \max_{\substack{1 \le i \le m \\ 1 \le j \le n}} (|x_i| + |y_j| + 1)e^{(|x_i|+1)(|y_j|+1)}.$$

Exercise 14.8. Deduce from Theorem 9.1 the approximation measures displayed in table 14.26 (with the definition of simultaneous approximation measure given in the introduction, but here $m = 1$ and $\theta \in \mathbb{C}$). The numbers

c_1 and c_2 are absolute constants,
$c_3(\lambda)$ depends on $\lambda \in \mathcal{L} \setminus \{0\}$,
$c_4(\beta)$ depends on $\beta \in \overline{\mathbb{Q}}^{\times}$,
$c_5(\lambda)$ depends on $\lambda \in \mathcal{L} \setminus 2i\pi\mathbb{Q}$,
$c_6(\lambda_1, \lambda_2)$ depends on λ_1, λ_2 which are $\mathbb{Q}$-linearly independent in $\mathcal{L}$,
$c_7(\beta, \lambda)$ depends on $\beta \in \overline{\mathbb{Q}} \setminus \mathbb{Q}$ and $\lambda \in \mathcal{L} \setminus \{0\}$.

Table 14.26. Approximation measures (Exercise 14.8).

θ	$\varphi(D, h)$
π	$c_1 D^2(h + \log D) \log D$
e^{π}	$c_2 D^3 h(\log h + \log D) \log D$
λ	$c_3(\lambda) D^3(h + \log D)(\log D)^{-1}$
e^{β}	$c_4(\beta) D^3 h$
λ/π	$c_5(\lambda) D^3(h + \log D)(\log D)$
λ_1/λ_2	$c_6(\lambda_1, \lambda_2) D^4(h + \log D)(\log D)^{-2}$
$e^{\beta\lambda}$	$c_7(\beta, \lambda) D^4 h(\log h + \log D)(\log D)^{-2}$

15. Algebraic Independence

Liouville's inequality which has been used many times so far (namely Lemma 2.1) involves *polynomial approximations* : a tuple $\underline{\theta} = (\theta_1, \ldots, \theta_m)$ of complex numbers for which there are polynomials $f \in \mathbb{Z}[X_1, \ldots, X_m]$ such that $|f(\underline{\theta})|$ is small but not zero contains at least one transcendental element. The converse is also true, and this yields a transcendence criterion. A similar statement holds for *algebraic approximations* to a complex number: a number $\theta \in \mathbb{C}$ is transcendental if and only if there are algebraic numbers γ such that $|\theta - \gamma|$ is small but not zero. One deduces that numbers $\theta_1, \ldots, \theta_m$ belonging to a field of transcendence degree 1 admit good simultaneous approximations by algebraic numbers $\gamma_1, \ldots, \gamma_m$, where the quality of the approximation, namely the number $\max_{1 \le i \le m} |\theta_i - \gamma_i|$, is controlled in terms of the degree $[\mathbb{Q}(\gamma_1, \ldots, \gamma_m) : \mathbb{Q}]$.

We explain these results in § 15.1, and we prove them in § 15.2. Several applications are given to algebraic independence results in § 15.3: Lindemann-Weierstraß' Theorem, values of the exponential function in a single variable or in several variables. We complete this chapter with further conjectures, especially concerning large transcendence degree (§ 15.4) and further results (§ 15.5).

15.1 Criteria: Irrationality, Transcendence, Algebraic Independence

Given a certain tuple $\underline{\theta} = (\theta_1, \ldots, \theta_m)$ of complex numbers, one wishes to produce a lower bound for the transcendence degree of the field $K = \mathbb{Q}(\underline{\theta})$ over $\mathbb{Q}$. We denote this transcendence degree by $\mathrm{trdeg}_{\mathbb{Q}} K$.

Proofs are not included in this section: they are postponed to § 15.2.

15.1.1 Irrationality Criterion, Linear Independence and Simultaneous Rational Approximation

As a warm up, we start by proving a criterion which ensures that one element at least in some set of real numbers is irrational.

Lemma 15.1. *Let $\vartheta_1, \ldots, \vartheta_m$ be real numbers. The following assertions are equivalent.*

(i) One at least of the numbers $\vartheta_1, \ldots, \vartheta_m$ is irrational, that is

$$\mathbb{Q}(\vartheta_1, \ldots, \vartheta_m) \neq \mathbb{Q}.$$

(ii) For any $\epsilon > 0$ there exist rational integers $p_1, \ldots, p_m, q$ in $\mathbb{Z}$ with $q > 0$ such that

$$0 < \max_{1 \le i \le m} \left| \vartheta_i - \frac{p_i}{q} \right| \le \frac{\epsilon}{q}.$$

(iii) There exist infinitely many tuples $(p_1, \ldots, p_m, q)$ in $\mathbb{Z}^{m+1}$ with $q > 0$ such that

$$0 < \max_{1 \le i \le m} \left| \vartheta_i - \frac{p_i}{q} \right| \le q^{-1-(1/m)}.$$

(iv) For any integer $Q > 1$ there exists a tuple $(p_1, \ldots, p_m, q)$ in $\mathbb{Z}^{m+1}$ with $1 \le q < Q^m$ such that

$$0 < \max_{1 \le i \le m} |q\vartheta_i - p_i| \le \frac{1}{Q}.$$

Remark 1. In the case $m = 1$ these conditions are also equivalent to a refined estimate in (iii), viz. $0 < |\vartheta - p/q| < 1/\sqrt{5}q^2$. This is a well known result from diophantine approximation due to A. Hurwitz (see for instance [Sc 1980], Chap. I § 2 Th. 2.F).

Remark 2. There is a gap between the exponent of q in (ii) which is -1 and in (iii) which is $-1 - (1/m)$. We shall discuss this matter later (§ 15.1.2).

Definition. A *measure of irrationality* of an irrational real number ϑ is a mapping $\psi\colon \mathbb{N} \to \mathbb{R}_{>0}$ such that, for any $(p, q) \in \mathbb{Z}^2$ with sufficiently large $q > 0$, say $q \ge q_0(\vartheta)$,

$$\left| \vartheta - \frac{p}{q} \right| \ge \frac{1}{\psi(q)}.$$

By Hurwitz' result quoted in Remark 1, any measure of irrationality ψ of an irrational real number satisfies

$$\psi(q) \ge \sqrt{5}q^2 \quad \text{for any } q \ge q_0(\vartheta).$$

Of course one could define *the* measure of irrationality ψ of ϑ to be

$$\psi(q) = \frac{q}{\|q\vartheta\|},$$

where $\|\cdot\|$ denotes the distance to the nearest integer. However one often prefers to work with increasing functions, so that one may restrict the condition in the definition of ψ to relatively prime integers (p, q). For any $q_0 > 0$, an increasing irrationality measure for ϑ is

$$\psi(q) = \max_{1 \le q_1 \le q} \frac{q_1}{\|q_1 \vartheta\|}.$$

A real number is a Liouville number (see (§ 3.5.3) if and only if, for any $\kappa > 0$, the function defined for $q \ge 2$ by $q \mapsto q^\kappa$ is not a measure of irrationality of ϑ.

On the opposite, for any $\epsilon > 0$ there exists a set of real numbers of Lebesgue's measure 0 such that, for any $\vartheta \in \mathbb{R}$ outside this set, there exists $c(\vartheta) > 0$ such that $c(\vartheta)q^{-2-\epsilon}$ is a measure of irrationality of ϑ (see for instance [Sc 1980], Chap. III, § 3).

In the definition of measure of irrationality we assumed that q is sufficiently large; this condition may be omitted, but it is convenient, for instance when the measure involves quantities like $\log q$ or $\log\log q$, to know that q is at least e or e^e, say.

Remark. From Lemma 15.1 one easily deduces the following statement:

- Let $\underline{\vartheta} = (\vartheta_1, \dots, \vartheta_m) \in \mathbb{R}^m$ be a m-tuple of real numbers. Denote by $r + 1$ the dimension of the $\mathbb{Q}$-vector space spanned by $1, \vartheta_1, \dots, \vartheta_m$ and assume $r \ge 1$ (which means that one at least of $\vartheta_1, \dots, \vartheta_m$ is irrational). Then there exist two constants $c > 0$ et $Q_0 > 0$ such that, for any integer $Q \ge Q_0$, there exists a tuple $(p_1, \dots, p_m, q)$ in $\mathbb{Z}^{m+1}$ with $1 \le q < Q^r$ such that

$$0 < \max_{1 \le i \le m} \left| q\vartheta_i - p_i \right| \le \frac{c}{Q}.$$

Therefore, given a tuple $\underline{\vartheta} \in \mathbb{R}^m$ and a positive real number k, if one can prove that there exists $c > 0$ such that, for any $(p_1, \dots, p_m, q) \in \mathbb{Z}^{m+1}$ with sufficiently large q,

$$\max_{1 \le i \le m} \left| \vartheta_i - \frac{p_i}{q} \right| > cq^{-1-(1/k)},$$

then

$$\dim_{\mathbb{Q}}\left(\mathbb{Q} + \mathbb{Q}\vartheta_1 + \cdots + \mathbb{Q}\vartheta_m\right) \ge 1 + k.$$

This sufficient condition for linear independence is not necessary: the set Θ of numbers

$$\sum_{n=0}^{\infty} \epsilon_n 2^{-n!} \quad \text{with} \quad \epsilon_n \in \{-1, +1\} \quad \text{for any} \quad n \ge 1$$

contains uncountably many numbers, hence spans a $\mathbb{Q}$-vector space of infinite dimension; moreover for any m-tuple $\underline{\vartheta} \in \Theta^m$, truncating the series with $0 \le n \le N$ produces good simultaneous rational approximations with $q = 2^{N!}$.

On the other hand (see [Sc 1980], Chap. III, § 3), for almost all tuples $\underline{\vartheta} \in \mathbb{R}^m$, for any ϵ there exists $c = c(\underline{\vartheta}, \epsilon) > 0$ such that

$$\max_{1 \le i \le m} \left| \vartheta_i - \frac{p_i}{q} \right| > cq^{-1-(1/m)-\epsilon}$$

for any $(p_1, \dots, p_m, q) \in \mathbb{Z}^{m+1}$ with $q > 1$.

Our main concern in this chapter is to study a similar situation where *rational approximations* to a tuple $(\vartheta_1, \ldots, \vartheta_m)$ of real numbers is replaced by *algebraic approximations* to a tuple $(\theta_1, \ldots, \theta_m)$ of complex numbers. At the same time, *linear independence* is replaced by *algebraic independence*. In place of rational numbers p_i/q we shall consider algebraic numbers γ_i, and the role of the dimension of the $\mathbb{Q}$-vector space spanned by $1, \vartheta_1, \ldots, \vartheta_m$ will be played (at least conjecturally) by the transcendence degree of the field $\mathbb{Q}(\theta_1, \ldots, \theta_m)$ over $\mathbb{Q}$.

15.1.2 Transcendence Criterion: Polynomial Approximation

Our next goal is to extend Lemma 15.1 and get a criterion for transcendence. When dealing with a single complex number θ, we may replace the condition of rational approximation $\theta - p/q$ either by a condition of algebraic approximation, considering $|\theta - \gamma|$ with algebraic γ's, or else replace the degree 1 polynomial $qX - p$ by a polynomial of arbitrary degree. In the first case, replacing p/q by an algebraic number, one deals with algebraic points, that is in dimension 0, while in the second case we deal with hypersurfaces, that is in codimension 1. Of course in a space of dimension 1 (which was the case for rational approximation) there is no difference, but in higher dimensional space there is a big difference. One may expect that intermediate situations are also relevant, and indeed this is the case. But we postpone this discussion to § 15.5 and for the time being we consider only the two extreme cases.

It turns out that the codimension one case is much easier. So we start with polynomial approximation . The next result is a transcendence criterion . We already stated part of it (namely $(ii) \Rightarrow (i)$) as Lemma 2.1, which was proved in § 3.5.

Proposition 15.2. *Let* $\underline{\theta} = (\theta_1, \ldots, \theta_m)$ *be a m-tuple of complex numbers. The following assertions are equivalent.*
(i) *One at least of the numbers* $\theta_1, \ldots, \theta_m$ *is transcendental, that is*

$$\operatorname{trdeg}_{\mathbb{Q}}\mathbb{Q}(\underline{\theta}) \geq 1.$$

(ii) *For any* $\kappa > 0$ *there exist a positive integer* T *and a polynomial* $f \in \mathbb{Z}[X_1, \ldots, X_m]$ *such that* $\deg f \leq T$, $\mathrm{H}(f) \leq e^T$ *and*

$$0 < |f(\underline{\theta})| \leq e^{-\kappa T}.$$

(iii) *For any* $\kappa < 1/2$ *there exists a positive integer* T_0 *such that, for any* $T \geq T_0$ *there is a polynomial* $f \in \mathbb{Z}[X_1, \ldots, X_m]$ *satisfying* $\deg f \leq T$, $\mathrm{H}(f) \leq e^T$ *and*

$$0 < |f(\underline{\theta})| \leq e^{-\kappa T^2}.$$

(iv) *For any* $H \geq 1$ *and* $D \geq 1$ *there exists a polynomial* $f \in \mathbb{Z}[X_1, \ldots, X_m]$ *of total degree* $\leq D$ *and usual height* $\mathrm{H}(f) \leq H$ *such that*

$$0 < |f(\underline{\theta})| \leq \sqrt{2}(1 + |\underline{\theta}|)^D H^{-(D-1)/2}.$$

Remark. There is a big gap between (ii) and (iii). First of all, (iii) claims the existence of a *dense* sequence of polynomials (for each T there is a polynomial f), while in (ii) the sequence of polynomials may be lacunary (there exist T and f). Moreover the quality of approximation given by (iii) is much better than in (ii). In fact (iv) is more precise than (iii), and there are many intermediate statements between (iv) and (ii) which are of course also equivalent. The fact that there are many such variants is due to the occurrence of two parameters, the degree and the height. We chose (iii) for simplicity of comparison with (ii).

This gap between (iii) and (ii) is a lucky event: in order to prove the transcendence of one at least among the numbers $\theta_1, \ldots, \theta_m$, it is sufficient to produce a sequence of polynomials which enables one to check (ii). By (iii), not only such a sequence does exist, but indeed there are sequences of polynomials satisfying much stronger requirements.

Finally we notice that the gap between (iii) and (ii) in Proposition 15.2 occurs at the second level of the exponential, while in Lemma 15.1 it occurred only at the first level.

15.1.3 Transcendence Measures and Measures of Algebraic Approximation

Let us make a small digression. Assume that we have proved that property (ii) in Proposition 15.2 holds for a certain tuple $(\theta_1, \ldots, \theta_m)$. Then one knows that (iii) also holds, and one may be tempted to feel that our proof of the weaker assertion (ii) has given all its juice. Often, this is not the case. Indeed (iii) asserts the existence of a polynomial f with the given property, but the proof of $(i) \Rightarrow (iii)$ rests on Dirichlet's box principle, and essentially nothing more is known about f.

On the opposite, it is often possible to construct an explicit sequence of polynomials which enables one to check (ii). In the previous chapters we gave many such examples by means of either interpolation determinants or auxiliary functions[25].

Such an explicit sequence of polynomial approximations may turn out to be useful to produce a quantitative refinement to assertion (i) of Proposition 15.2, namely a measure of simultaneous approximation for $\underline{\theta} = (\theta_1, \ldots, \theta_m)$ (a similar phenomenon related to Lemma 15.1 is described in Exercise 15.2).

Proposition 15.3. *Let $\underline{\theta} = (\theta_1, \ldots, \theta_m) \in \mathbb{C}^m$ be a m-tuple of complex numbers, $\underline{\gamma} = (\gamma_1, \ldots, \gamma_m) \in \overline{\mathbb{Q}}^m$ a m-tuple of algebraic numbers and $f \in \mathbb{Z}[X_1, \ldots, X_m]$ a polynomial such that $f(\underline{\gamma}) \neq 0$. Define D, L, d, μ and ϵ by*

$$D = [\mathbb{Q}(\underline{\gamma}) : \mathbb{Q}], \quad \mu = D\mathrm{h}(1 : \gamma_1 : \cdots : \gamma_m),$$

$$d = \deg f, \quad L = \mathrm{L}(f) \quad \textit{and} \quad \epsilon = \frac{1}{2} L^{-D} e^{-d\mu}.$$

[25] In spite of the fact that the construction of auxiliary functions also rests on Dirichlet's box principle, the resulting polynomials carry essentially the same amount of information as one gets from alternants or interpolation determinants.

Assume $|f(\underline{\theta})| \leq L\epsilon$. *Then*

$$|\underline{\theta} - \underline{\gamma}| \geq \frac{\epsilon}{d(1+|\underline{\theta}|)^{d-1}}.$$

Hence, if we know explicitly polynomials $f \in \mathbb{Z}[X]$ for which $|f(\theta)|$ is small, then in order to obtain a measure of approximation for θ it is sufficient to check it for the roots of these approximating polynomials f. Explicit polynomials f are very useful if we can also get further information on their zeroes. A simple case (Exercise 15.5) occurs when we can produce a lower bound for $|f(\theta)|$: this ensures that f does not vanish in a small neighborhood of θ.

An example of application of Proposition 15.3 to a measure of linear independence of logarithms is given in Exercise 15.4. We come back to this question in § 15.5.2.

Definition. Given a transcendental complex number θ, a *transcendence measure* for θ is a mapping $\Phi\colon \mathbb{N} \times \mathbb{R}_{>0} \to \mathbb{R}_{>0}$ such that, for any sufficiently large positive integer D, any sufficiently positive real number H and any nonzero polynomial $f \in \mathbb{Z}[X]$ of degree $\leq D$ and usual height $\mathrm{H}(f) \leq H$, we have

$$|f(\theta)| \geq \exp\{-\Phi(D, H)\}.$$

For any fixed $D \geq 1$ and $H \geq 1$, the set of algebraic numbers of degree $\leq D$ and usual height $\leq H$ is finite. Hence the number

$$\Phi_\theta(D, H) := \max\bigl\{-\log|f(\theta)|\,;\ f \in \mathbb{Z}[X],\ \deg f \leq D,\ \mathrm{H}(f) \leq H\bigr\}$$

is well defined, and Φ is a transcendence measure for θ if and only if there exist D_0 and H_0 such that $\Phi_\theta(D, H) \leq \Phi(D, H)$ for all $D \geq D_0$ and $H \geq H_0$.

Remark 1. If Φ is a transcendence measure for θ, then for any sufficiently large D, the mapping $q \mapsto \psi(q) = q\exp\{\Phi(D, q)\}$ is an irrationality measure for θ.

Remark 2. From Proposition 15.2 one deduces the lower bound

$$\Phi(D, H) \geq \frac{D-1}{2}\log H - D\log\bigl(1+|\theta|\bigr) - \frac{1}{2}\log 2.$$

Remark 3. We insist that D is an upper bound for the degree of f, and is not assumed to be the exact degree (similarly for H). This remark shows that the condition *D and H are sufficiently large* is not restrictive.

It is customary to define transcendence measure, as we did, with the usual height. It will be more convenient for our purpose to use Mahler's measure for the next definition.

Definition. Given a transcendental complex number θ, a *measure of algebraic approximation* for θ is a mapping $\psi\colon \mathbb{N}\times\mathbb{R}_{>0}\to\mathbb{R}_{>0}$ such that there exists $c>0$ with the following property: for any $D\in\mathbb{N}$ and $\mu\in\mathbb{R}$ with $D\geq c$ and $\mu\geq cD$, and for any algebraic number γ of degree $[\mathbb{Q}(\gamma)\colon\mathbb{Q}]\leq D$ and Mahler's measure $\mathrm{M}(\gamma)\leq e^{\mu}$, we have

$$|\theta-\gamma|\geq\exp\{-\psi(D,\mu)\}.$$

Remark. The estimates are sensitive to the choice of height, because the inequalities (3.12) relating h and $\log H$ for instance involve the exact degree d of α, and not an upper bound for d. By (3.12), if an algebraic number γ has degree $d\leq D$ and absolute logarithmic height $\leq h$, then its usual height is bounded by $\mathrm{H}(\gamma)\leq 2^d e^{dh}\leq 2^D e^{Dh}$. But on the other side if we know an upper bound $d\leq D$ for the degree and $\mathrm{H}(\gamma)\leq H$ for the usual height, one gets only the estimate

$$\mathrm{h}(\gamma)\leq\frac{1}{d}\log H+\frac{1}{2d}\log(d+1)\leq\log H+\frac{1}{2d}\log(d+1)$$

for the absolute logarithmic height (here the lower bound $d\geq 1$ is used, not the upper bound $d\leq D$). This is why it is of the utmost importance to produce (when possible) not only upper bounds, but also lower bounds for the degrees of algebraic approximations. When no lower bound is available for the degree, it makes a difference to phrase the results in terms of the absolute logarithmic height or else in terms of the logarithm of Mahler's measure. We choose the latter for a reason which will appear in § 15.4 (in connection with large transcendence degree).

We have required the conditions *D and μ/D are sufficiently large* by analogy with the conditions *D and h are sufficiently large* which appeared in Chap. 14. One might relax the condition on μ and require only that μ is sufficiently large (recall Lehmer's Problem in § 3.6.2: Mahler's measure should not be too small for nonzero algebraic numbers which are not roots of unity; diophantine approximation by roots of unity are not excluded here!)

Notice that if $\psi(D,\mu)$ is a measure of algebraic approximation for θ, then for any fixed sufficiently large $D>0$ the mapping $q\mapsto\exp\bigl\{\psi(D,\log q)\bigr\}$ is a measure of irrationality for θ .

A classical problem is, given a transcendental number θ, to produce a transcendence measure (that is an admissible function Φ) as well as a measure of algebraic approximation (that is an an admissible function ψ). See for instance [FNe 1998], Chap. 2. Close connections between transcendence measures and measures of algebraic approximation have been established by N. I. Fel'dman as soon as 1951 (see Chap. 7 § 1 Lemma 1.7 in [F 1982]; see also Exercise 15.14).

To begin with, the easy part: starting from an algebraic approximation γ to θ, the minimal polynomial $f\in\mathbb{Z}[X]$ produces a polynomial approximation to θ (see (15.12) below). Therefore, using only the information provided by irreducible polynomials in the definition of Φ, one deduces:

Lemma 15.4. *Let $\theta \in \mathbb{C}$ be a transcendental number. There exists a constant $c > 0$ such that, if $\Phi(D, H)$ is a transcendence measure for θ, then the function*

$$\psi(D, \mu) = \Phi\big(D, 2^D e^\mu\big) + \mu + cD$$

is a measure of algebraic approximation for θ.

The other direction is not so easy: if f is a polynomial approximation to θ, then a root γ of f at minimal distance of θ is a good candidate for an algebraic approximation to θ. However two difficulties occur: the first one is that several roots of f may contribute to the smallness of $|f(\theta)|$, the second one is that f may have a high multiplicity of zero at γ. With respect to the first one, it is useful to use Liouville's inequality and to produce a lower bound for the distance between two roots of f; but such an estimate is not always sharp enough and one cannot always avoid considering several roots of f. For the second one, one remarks that if f vanishes at γ with multiplicity say k, then $|f(\theta)|$ may be compared with $|\theta - \gamma|^k$ rather than with $|\theta - \gamma|$. However in this case the degree of γ is at most $(1/k) \deg f$, and moreover the first k derivatives of f at θ also are small; so some extra information is available.

Lemma 15.5. *For any transcendental number $\theta \in \mathbb{C}$ there exists a constant $c > 0$ with the following property. Let $\psi(D, \mu)$ be a measure of algebraic approximation for θ, which satisfies the two following conditions:*
(i) for any sufficiently large μ the mapping $D \mapsto \psi(D, \mu)$ is non-decreasing,
(ii) For any $k \geq 1$ and any sufficiently large d and μ, $k\psi(d, \mu) \leq \psi(kd, k\mu)$.
Then the function

$$\Phi(D, H) = \psi\big(D, \log(DH)\big) + 2D \log H + 3D \log D$$

is a transcendence measure for θ.

Remark 1. An error term $D \log H$ in the conclusion of Lemma 15.5 cannot be omitted (see Exercise 15.6). On the other hand it is not known whether the error term $3D \log D$ can be avoided.

Remark 2. Further properties (invariance under finite extension) of transcendence measures and measures of algebraic approximation are given in Proposition 15.19 and Exercise 15.7.

15.1.4 Transcendence Criterion: Algebraic Approximation to a Single Number

We start with the case $m = 1$ of Proposition 15.2. Conditions (ii), (iii) and (iv) involve a polynomial f such that $|f(\theta)|$ is small. Considering a root γ of f which is at minimal distance of θ, one may expect to be able to replace the corresponding assertion by the requirement that there exists an algebraic number γ which is close to θ.

Indeed, Lemmas 15.4 and 15.5 relate the problem of finding an *algebraic approximation* to θ (i.e. $\gamma \in \overline{\mathbb{Q}}$ such that $|\theta - \gamma|$ is small) and the problem of finding *a polynomial approximation* (that is $f \in \mathbb{Z}[X]$ such that $|f(\theta)|$ is small).

It turns out that the transcendence criterion involving algebraic approximations is not exactly the analog one might expect at first glance to Proposition 15.2: there exists a complex number θ such that, for any $\epsilon > 0$, for infinitely many integers $T > 0$, and for any algebraic number γ satisfying $[\mathbb{Q}(\gamma):\mathbb{Q}] \le T$ and $\mathrm{H}(\gamma) \le e^T$, we have

$$|\theta - \gamma| \ge e^{-T^{1+\epsilon}}$$

(see Exercise 15.8).

In view of such examples, when considering algebraic approximations, one cannot ask a condition as strong as the analog of property (iii) in Proposition 15.2: the sequence of algebraic approximations may be lacunary.

Theorem 15.6. *Let θ be a complex number. The following assertions are equivalent.*
(i) θ is transcendental.
(ii) For any $c > 0$ there exist a positive integer T and an algebraic number γ such that $[\mathbb{Q}(\gamma):\mathbb{Q}] \le T$, $\mathrm{H}(\gamma) \le e^T$ and

$$0 < |\theta - \gamma| \le e^{-cT}.$$

(iii) For any sufficiently large positive number c_0 and any sequences $(D_\nu)_{\nu\ge1}$ of positive integers and $(\mu_\nu)_{\nu\ge1}$ of real numbers satisfying

$$c_0 \le D_\nu \le D_{\nu+1} \le 2D_\nu, \quad c_0 D_\nu \le \mu_\nu \le \mu_{\nu+1} \le 2\mu_\nu \qquad (\nu \ge 1)$$

and

$$\lim_{\nu\to\infty} \mu_\nu = \infty,$$

for infinitely many ν there exists an algebraic number γ such that

$$\frac{1}{c_0} D_\nu \le [\mathbb{Q}(\gamma):\mathbb{Q}] \le D_\nu, \quad \mathrm{M}(\gamma) \le e^{\mu_\nu}$$

and

$$0 < |\theta - \gamma| \le e^{-D_\nu \mu_\nu / c_0}.$$

Remark 1. In this book, for proving that some complex numbers are transcendental, we used mainly the transcendence Criterion 15.2 involving polynomial approximation (for instance in Chapters 2 and 6). In fact the implication $(ii) \Rightarrow (i)$ of Theorem 15.6 is also a very useful tool to prove transcendence results. See for instance Chap. 2 of [FNe 1998].

Remark 2. As pointed out earlier, one strong difference between Proposition 15.2 and Theorem 15.6 is that, in the latter one, condition (iii) involves only infinitely many ν's, and not all sufficiently large ν's. Another important difference is that no statement like condition (iv) of Proposition 15.2 is known for algebraic approximation. There are variants of the equivalent conditions in Theorem 15.6 which are not immediate consequences of (iii) (see Exercise 15.10).

Remark 3. From Theorem 15.6 we deduce a lower bound for any measure of algebraic approximation ψ of a complex transcendental number θ: if $(D_\nu)_{\nu\geq 1}$ and $(\mu_\nu)_{\nu\geq 1}$ are two sequences satisfying the conditions in (iii), then

$$\limsup_{\nu\to\infty} \frac{1}{D_\nu \mu_\nu} \psi(D_\nu, \mu_\nu) > 0.$$

For instance any measure of algebraic approximation for a complex transcendental number of the form

$$\psi(D, \mu) = \kappa D^a (D^b + \mu^c)$$

with constants κ, a, b and c has $c \geq 1$ and $a + b \geq 1 + (b/c)$.

The dependence on the degree plays a fundamental role in this chapter. This is somehow a recent feature: earlier authors did not pay so much attention on the degree in their estimates as they did for the height. Their first goal was to get sharp irrationality measures, and then a natural extension is to consider approximation by algebraic numbers of bounded degree. It turns out that it is also useful to consider, for instance, approximation by algebraic numbers of bounded (absolute logarithmic) height. However one cannot expect too good algebraic approximations by numbers of bounded degree in general, unless one deals with Liouville-like numbers.

Remark 4. The set of Liouville numbers is uncountable, but nevertheless it is a rather small subset of $\mathbb{R}$: for instance it has Lebesgue's measure 0 ([Sc 1980], Chap. III, § 3). It is not so surprising that the study of rational approximation is in general not sufficient to decide whether a number is transcendental or not. For instance, if one wishes to prove that a complex number is not quadratic, one might expect that approximation by quadratic numbers should come into the picture.

One could be tempted to dub *Generalized Liouville Number* a complex number θ which admits algebraic approximations which are good enough so that Liouville's estimate

$$|\alpha - \beta| \geq 2^{-D+1}\mathrm{M}(\alpha)^{-D}\mathrm{M}(\beta)^{-D} \quad \text{for} \quad \alpha \neq \beta \quad \text{and} \quad D = [\mathbb{Q}(\alpha, \beta) : \mathbb{Q}]$$

is sufficient to prove that θ is transcendental. However Theorem 15.6 tells us that any transcendental complex number satisfies this condition!

Here is a better definition for *Liouville number in an extended sense*, as suggested by M. Laurent in [Lau 1999]: it is a complex number θ for which there exists a sequence $(\gamma_\nu)_{\nu\geq 1}$ of algebraic numbers such that $\gamma_\nu \neq \theta$ for $\nu \geq 1$ and

$$\frac{1}{[\mathbb{Q}(\gamma_\nu) : \mathbb{Q}] \cdot \log \mathrm{M}(\gamma_\nu)} \log |\theta - \gamma_\nu| \to -\infty \quad \text{as} \quad \nu \to \infty.$$

In other words such a number θ has much better algebraic approximations than those furnished by condition (iii) of Theorem 15.6.

An early result on this topic is due to E. Wirsing [Wir 1961] (see also [Sc 1980], Chap. VIII, Th. 3B).

Let ϑ be a real transcendental number. There exists a positive constant $c = c(\vartheta)$ such that, for any positive integer D, there exist infinitely many $\gamma \in \overline{\mathbb{Q}}$ satisfying $[\mathbb{Q}(\gamma)\colon \mathbb{Q}] \leq D$ and

$$|\vartheta - \gamma| \leq c\mathrm{H}(\gamma)^{-(D+3)/2}.$$

A similar statement holds for a complex number θ with $(D+3)/2$ replaced by $(D/4)+1$.

A refinement has been achieved by H. Davenport and W. M. Schmidt ([Sc 1980], Chap. VIII, Th. 3A) for *quadratic* approximations: for $D = 2$ they get the conclusion with the exponent -3 (and this is best possible: see [Sc 1980], Chap. VIII, Th. 2A). More precisely:

- *Let ϑ be a real number which is neither rational nor quadratic. There exist infinitely many rational or real quadratic γ such that*

$$|\vartheta - \gamma| \leq 18 \max\{1\,,\ |\vartheta|^2\}\mathrm{H}(\gamma)^{-3}.$$

In [Wir 1961], E. Wirsing conjectured that his result should hold with the exponent $(D+3)/2$ replaced by $D+1$ (or at least for $D+1-\epsilon$). This is known only for $D = 1$ (by Dirichlet's Theorem) and for $D = 2$ (by Davenport-Schmidt). In spite of some improvements of this exponent by V. Bernik and K. Tishchenko for small values of D, this problem is still open for $D \geq 3$.

Here we are not concerned with the best possible value for the exponent. In [LauRoy 1999b], M. Laurent and D. Roy provided an extra information, namely the lower bound for the degree of the approximant γ which occurred in condition (iii) of Theorem 15.6 (further related recent results are due to Y. Bugeaud and O. Theulié).

Corollary 15.7. *There exist absolute constants c_1, c_2 and c_3 with the following property. For any complex transcendental number θ and for any integer $D \geq c_1$, there exist infinitely many $\gamma \in \overline{\mathbb{Q}}$ satisfying*

$$c_2 D \leq [\mathbb{Q}(\gamma)\colon \mathbb{Q}] \leq D \quad \textit{and} \quad |\theta - \gamma| \leq \mathrm{M}(\gamma)^{-c_3 D}.$$

From Corollary 1, § 2 of [LauRoy 1999b], one deduces that

$$c_1 = 26, \quad c_2 = 10^{-3}, \quad c_3 = 2 \cdot 10^{-3}$$

are admissible values.

Thanks to the lower bound for the degree, up to numerical constants the same result holds with $\mathrm{M}(\gamma)$ replaced by $\mathrm{H}(\gamma)$. An instructive example is proposed by M. Laurent in Exercise 15.9.

One deduces Corollary 15.7 from the implication $(iii) \to (i)$ in Theorem 15.6 by choosing for $(D_\nu)_{\nu \ge 1}$ a constant sequence $D_\nu = D$ and, say, $\mu_\nu = \nu$. If one chooses instead $D_\nu = \nu$ and μ_ν / D_ν constant, one deduces Th. 3.2 in [RoyW 1997a]:

Corollary 15.8. *There exist absolute constants c_1 and c_2 with the following property: for any complex transcendental number θ, for any real number $h \ge c_1$ and for infinitely many positive integers d, there exists an algebraic number $\gamma \in \overline{\mathbb{Q}}$ of degree d satisfying*

$$\mathrm{h}(\gamma) \le h \quad \textit{and} \quad |\theta - \gamma| \le e^{-c_2 d^2 h}.$$

By Corollary 2 in [LauRoy 1999b], one can take $c_1 = 2000$, $c_2 = 2 \cdot 10^{-6}$.

15.1.5 Algebraic Independence: Simultaneous Diophantine Approximation

In § 15.1.4, we considered a single number θ, while Proposition 15.2 involved a tuple $\underline{\theta} = (\theta_1, \dots, \theta_m)$. Assume now that $\underline{\theta}$ is a m-tuple of complex numbers such that the field $\mathbb{Q}(\underline{\theta})$ has transcendence degree 1 over $\mathbb{Q}$. Let $\{\theta_0\}$ be a transcendence basis. Then each θ_i $(1 \le i \le m)$ is algebraic over $\mathbb{Q}(\theta_0)$. For $1 \le i \le m$, let $g_i \in \mathbb{Z}[X, Y]$ be a nonzero polynomial such that $g_i(\theta_0, \theta_i) = 0$. By Theorem 15.6, $(i) \Rightarrow (iii)$, θ_0 admits good algebraic approximations γ_0. If $\gamma_0 \in \overline{\mathbb{Q}}$ is sufficiently close to θ_0, then the polynomial $g_i(\gamma_0, Y) \in \overline{\mathbb{Q}}[Y]$ does not vanish, it has a root γ_i which is close to θ_i, and this produces an algebraic approximation to θ_i (if some θ_i is algebraic, then $\gamma_i = \theta_i$ is a perfect approximation!). Hence the m-tuple $\underline{\gamma} = (\gamma_1, \dots, \gamma_m)$ is a simultaneous algebraic approximation to $\underline{\theta}$ in the sense that one has a good control of the degree $[\mathbb{Q}(\underline{\gamma}) : \mathbb{Q}]$.

Proposition 15.9. *Let $\underline{\theta} = (\theta_1, \dots, \theta_m)$ be a m-tuple of complex numbers such that*

$$\mathrm{trdeg}_{\mathbb{Q}} \mathbb{Q}(\underline{\theta}) = 1.$$

There exists a constant $c > 0$ with the following property. Let $(D_\nu)_{\nu \ge 1}$ and $(\mu_\nu)_{\nu \ge 1}$ be two sequences of real numbers satisfying

$$c \le D_\nu \le D_{\nu+1} \le 2D_\nu \quad \textit{and} \quad cD_\nu \le \mu_\nu \le \mu_{\nu+1} \le 2\mu_\nu \qquad (\nu \ge 1).$$

Assume also that the sequence $(\mu_\nu)_{\nu \ge 1}$ is unbounded. Then for infinitely many ν there exists a m-tuple $\underline{\gamma} = (\gamma_1, \dots, \gamma_m)$ of algebraic numbers satisfying

$$\frac{1}{c} D_\nu \le [\mathbb{Q}(\underline{\gamma}) : \mathbb{Q}] \le D_\nu, \quad [\mathbb{Q}(\underline{\gamma}) : \mathbb{Q}] \max_{1 \le i \le m} \mathrm{h}(\gamma_i) \le \mu_\nu$$

and

$$\max_{1 \le i \le m} |\theta_i - \gamma_i| \le e^{-D_\nu \mu_\nu / c}.$$

An explicit value for c follows from Theorem 1 in [LauRoy 1999b].

Notice that we do not need to impose the condition $\max_{1\le i\le m} |\theta_i - \gamma_i| > 0$ since we assumed not all of the θ_i to be algebraic. One of the main difficulties in applying the transcendence criteria 15.2 or 15.6 is to check the nonvanishing condition for either $|f(\theta_1, \ldots, \theta_m)|$ or $|\theta - \gamma|$. This is the place where the zero estimate is required in transcendence proofs. But Proposition 15.9 does not involve such a condition.

Proposition 15.9 is a convenient tool for proving that the transcendence degree of some field $\mathbb{Q}(\theta_1, \ldots, \theta_m)$ is at least 2.

Definition. In this chapter we shall say that a function $\psi: \mathbb{N} \times \mathbb{R}_{>0} \to \mathbb{R}_{>0}$ is a *measure of simultaneous approximation* for a tuple $(\theta_1, \ldots, \theta_m) \in \mathbb{C}^m$ if there exists $c > 0$ such that, for any positive integer D, any positive real number μ, and any tuple $(\gamma_1, \ldots, \gamma_m)$ of algebraic numbers satisfying $D \ge c$, $\mu \ge cD$,

$$[\mathbb{Q}(\gamma_1, \ldots, \gamma_m) : \mathbb{Q}] \le D \quad \text{and} \quad [\mathbb{Q}(\gamma_1, \ldots, \gamma_m) : \mathbb{Q}] \max_{1\le i\le m} \mathrm{h}(\gamma_i) \le \mu$$

the inequality

$$\max_{1\le i\le m} |\theta_i - \gamma_i| \ge \exp\{-\psi(D, \mu)\}$$

holds.

Let us compare with the definition introduced in Chap. 14 involving a function $\varphi(D, h)$, where h is an upper bound for $\max_{1\le i\le m} \mathrm{h}(\gamma_i)$.

- *Let $\psi(D, \mu)$ be a measure of simultaneous approximation for the tuple $(\theta_1, \ldots, \theta_m)$. Then the function $\varphi(D, h) = \psi(D, Dh)$ satisfies the condition of the introduction of Chap. 14, namely:* for $D \ge D_0$, $h \ge h_0$ and $\underline{\gamma} \in \overline{\mathbb{Q}}^m$ with

  $$[\mathbb{Q}(\underline{\gamma}) : \mathbb{Q}] \le D \quad \text{and} \quad \max_{1\le i\le m} \mathrm{h}(\gamma_i) \le h,$$

  we have

  $$\max_{1\le i\le m} |\theta_i - \gamma_i| \ge \exp\bigl\{-\varphi(D, h)\bigr\}.$$

- Conversely, *let $\varphi(D, h)$ be a function which satisfies the condition in the introduction of Chap. 14. Assume that the inequality*

  $$\varphi(D_1, h_1) \le \varphi(D_2, h_2)$$

  holds for any D_1, D_2, h_1 and h_2 satisfying

  $$D_i \ge D_0, \quad h_i \ge h_0 \ (i = 1, 2), \qquad D_2 \ge D_1 \quad \text{and} \quad D_2 h_2 \ge D_1 h_1.$$

  Then the function $\psi(D, \mu) = \varphi(D, D_0\mu/D)$ is a measure of simultaneous approximation for the tuple $(\theta_1, \ldots, \theta_m)$.

Indeed, let $D \ge D_0$, $\mu \ge h_0 D$ and $\underline{\gamma} \in \overline{\mathbb{Q}}^m$ satisfy

$$[\mathbb{Q}(\underline{\gamma}) : \mathbb{Q}] \le D \quad \text{and} \quad [\mathbb{Q}(\underline{\gamma}) : \mathbb{Q}] \max_{1\le i\le m} \mathrm{h}(\gamma_i) \le \mu.$$

Define

$$D' = [\mathbb{Q}(\underline{\gamma})\colon \mathbb{Q}], \quad D'' = \max\{D', D_0\} \quad \text{and} \quad h = \frac{\mu}{D'}\cdot$$

We have

$$D'' \geq D_0, \quad h \geq \frac{h_0 D}{D'} \geq h_0,$$

$$[\mathbb{Q}(\underline{\gamma})\colon \mathbb{Q}] \leq D'' \quad \text{and} \quad \max_{1\leq i\leq m} \mathrm{h}(\gamma_i) \leq h.$$

Therefore

$$\max_{1\leq i\leq m} |\theta_i - \gamma_i| \geq \exp\bigl\{-\varphi(D'', h)\bigr\}.$$

Since $h = \mu/D'$ and $D'' \leq D_0 D'$, the assumption on φ with $D_1 = D''$, $D_2 = D$, $h_1 = h$ and $h_2 = D_0\mu/D$ yields

$$\varphi(D'', h) \leq \varphi\left(D, \frac{D_0\mu}{D}\right)\cdot$$

□

It is easy to check that any of the functions $\varphi(D, h)$ occurring in Chap. 14 satisfies the condition $\varphi(D_1, h_1) \leq \varphi(D_2, h_2)$ for $D_2 \geq D_1$ and $D_2 h_2 \geq D_1 h_1$, hence the function $\psi(D, \mu) = \varphi(D, D_0\mu/D)$ is a measure of simultaneous approximation for the corresponding tuple. Moreover in each single example one can take for D_0 either 1 or 2. So there is no harm to work with $\psi(D, \mu)$ in place of $\varphi(D, h)$.

Therefore one deduces from Proposition 15.9:

Corollary 15.10. *Let $\underline{\theta}$ be a m-tuple of complex numbers and ψ a measure of simultaneous approximation. Let $(D_\nu)_{\nu\geq 1}$ and $(\mu_\nu)_{\nu\geq 1}$ be sequences satisfying the conditions of Proposition 15.9 above, namely*

$$c \leq D_\nu \leq D_{\nu+1} \leq 2D_\nu \quad \text{and} \quad cD_\nu \leq \mu_\nu \leq \mu_{\nu+1} \leq 2\mu_\nu \qquad (\nu \geq 1)$$

and the sequence $(\mu_\nu)_{\nu\geq 1}$ is unbounded. Assume

$$\lim_{\nu\to\infty} \frac{1}{D_\nu \mu_\nu} \psi(D_\nu, \mu_\nu) = 0.$$

Then

$$\mathrm{trdeg}_{\mathbb{Q}} \mathbb{Q}(\underline{\theta}) \geq 2.$$

Taking either for $(D_\nu)_{\nu\geq 1}$ or else for $(\mu_\nu/D_\nu)_{\nu\geq 1}$ a (sufficiently large) constant sequence, we deduce from Corollary 15.10 that any measure of simultaneous approximation ψ of a m-tuple of complex numbers $\underline{\theta}$ for which

$$\mathrm{trdeg}_{\mathbb{Q}} \mathbb{Q}(\underline{\theta}) = 1$$

satisfies

$$\liminf_{D\to\infty} \frac{1}{D} \limsup_{\mu\to\infty} \frac{1}{\mu} \psi(D, \mu) > 0$$

and

$$\liminf_{h\to\infty} \frac{1}{h} \limsup_{D\to\infty} \frac{1}{D^2}\psi(D, Dh) > 0.$$

For any $m \geq 2$ one can show (see Exercise 15.12) the existence of m-tuples $\underline{\theta}$ of algebraically independent numbers which satisfy the conclusion of Proposition 15.9. Hence Corollary 15.10 is not a *criterion* for transcendence degree ≤ 1: it is a sufficient condition, but not a necessary one.

Proposition 15.9 has many interesting applications. For instance, assuming part 2 of Conjecture 14.25, Corollary 15.10 shows that two linearly independent logarithms of algebraic numbers are algebraically independent.

We propose a few such examples in § 15.3. Further applications of the present method are given in [RoyW 1997a] and [RoyW 1997b].

15.2 From Simultaneous Approximation to Algebraic Independence

The main purpose of this section is to provide complete proofs of the results stated in § 15.1.

15.2.1 Proof of the Irrationality Criterion

Proof of Lemma 15.1. The easiest implication is $(iii) \Rightarrow (ii)$: given $\epsilon > 0$, we have $q^{-1-(1/m)} < \epsilon/q$ as soon as $q > \epsilon^{-m}$.

For the proof of $(ii) \Rightarrow (i)$, we assume that (i) does not hold and we write $\vartheta_i = a_i/b$ with $a_1, \dots, a_m, b$ in $\mathbb{Z}$ and $b > 0$. Then for any $\epsilon < 1/b$, the condition

$$\max_{1\leq i\leq m} \left|\vartheta_i - \frac{p_i}{q}\right| \leq \frac{\epsilon}{q}.$$

implies $|a_i q - bp_i| < 1$ for $1 \leq i \leq m$, hence $a_i q - bp_i = 0$ and $p_i/q = a_i/b = \vartheta_i$ $(1 \leq i \leq m)$. This shows that (ii) does not hold either.

We prove now the implication $(i) \Rightarrow (iv)$.

Consider the mapping $q \mapsto \xi_q$ from the finite set $\{0, 1, \dots, Q^m\}$ to the cube $\mathcal{C} = [0, 1]^m$ in $\mathbb{R}^m$ which sends q to $\xi_q = \big(\{q\vartheta_1\}, \dots, \{q\vartheta_m\}\big)$, where $\{x\}$ denotes the fractional part of $x \in \mathbb{R}$:

$$x = [x] + \{x\}, \qquad [x] \in \mathbb{Z}, \quad 0 \leq \{x\} < 1.$$

We decompose the cube $\mathcal{C}$ into Q^m cubes

$$\mathcal{C}_{i_1\dots i_m} = \prod_{\nu=1}^{m} \left[\frac{i_\nu}{Q}, \frac{i_\nu+1}{Q}\right] \subset \mathcal{C} \qquad (0 \leq i_\nu \leq Q-1,\ 1 \leq \nu \leq m).$$

By Dirichlet's box principle, there exist two integers $q_1 \neq q_2$ in the interval $[0, Q^m]$ such that ξ_{q_1} and ξ_{q_2} belong to the same cube $\mathcal{C}_{i_1\dots i_m}$. Then, taking $q = |q_1 - q_2|$, we have $1 \le q < Q^m$, and if p_i is an integer which is at minimal distance of $q\vartheta_i$ $(1 \le i \le m)$, then $(p_1, \ldots, p_m, q)$ is a solution to (iv). Assumption (i) is used only to check that not all $q\vartheta_i - p_i$ are zero.

Finally we prove the implication $(iv) \Rightarrow (iii)$. Let $\big(p_1^{(\nu)}, \ldots, p_m^{(\nu)}, q_\nu\big)$ $(1 \le \nu \le N)$, be a given finite set of tuples in $\mathbb{Z}^{m+1}$ with $q_\nu > 0$ such that, for $1 \le \nu \le N$, the number

$$\eta_\nu = \max_{1 \le i \le m} \big|q_\nu \vartheta_i - p_i^{(\nu)}\big|$$

satisfies $0 < \eta_\nu < 1/2$. Notice that in this case, for $1 \le i \le m$ and $1 \le \nu \le N$, $p_i^{(\nu)}$ is the only integer in the interval $\big(q_\nu\vartheta_i - (1/2), q_\nu\vartheta_i + (1/2)\big)$; hence the tuple $\big(p_1^{(\nu)}, \ldots, p_m^{(\nu)}, q_\nu\big)$ is completely determined by q_ν.

Let Q be an integer satisfying $Q > 1/\eta_\nu$ for every $\nu = 1, \ldots, N$, so that $Q > 2$. From (iv) we deduce that there exists a tuple $(p_1, \ldots, p_m, q)$ satisfying $1 \le q < Q^m$ and

$$0 < \max_{1 \le i \le m} |q\vartheta_i - p_i| \le Q^{-1}.$$

Since $Q^{-1} < \min_{1 \le \nu \le N} \eta_\nu$, we deduce $(p_1, \ldots, p_m, q) \ne \big(p_1^{(\nu)}, \ldots, p_m^{(\nu)}, q_\nu\big)$ for $1 \le \nu \le N$. Hence $q \ne q_\nu$. Finally we have also $Q^{-1} \le q^{-1/m}$. □

Remark. The proof of $(i) \Rightarrow (iv)$ shows that for <u>any</u> tuple $(\vartheta_1, \ldots, \vartheta_m)$ of real numbers and any integer $Q > 1$, there exist rational integers $q, p_1, \ldots, p_m$ with $1 \le q < Q^m$ such that

$$\max_{1 \le i \le m} |q\vartheta_i - p_i| \le \frac{1}{Q}\cdot$$

Assumption (i) was used only to check that the left hand side is not zero. See [Sc 1980], Chap. II § 1 Th. 1.A

Notice also that by using Lemma 4.11 with

$$\mu = m, \quad \nu = m+1, \quad U = 2, \quad \ell = Q^{m+1}, \quad X = Q^m,$$

$$v_{ij} = \delta_{ij} \quad (1 \le i, j \le m)$$

(Kronecker's diagonal symbol) and

$$v_{m+1,j} = \{\vartheta_j\} \quad (1 \le j \le m),$$

one obtains a slightly weaker result, namely with

$$1 \le q \le Q^m \quad \text{and} \quad \max_{1 \le i \le m} |q\vartheta_i - p_i| \le \frac{2}{Q}\cdot$$

15.2.2 Dirichlet's Box Principle

Lemma 15.11. *Let $\underline{\theta} = (\theta_1, \ldots, \theta_m)$ be a m-tuple of complex numbers, H and D positive integers. There exists a nonzero polynomial $f \in \mathbb{Z}[X_1, \ldots, X_m]$, of total degree $\le D$ and usual height $\mathrm{H}(f) \le H$, such that*

$$|f(\underline{\theta})| \le cH^{-\frac{1}{2}\binom{D+m}{m}+1},$$

where

$$c = \sqrt{2} \sum_{i_1+\cdots+i_m \le D} |\theta_1^{i_1} \cdots \theta_m^{i_m}|.$$

Proof. We are going to apply Lemma 4.12 with

$$X = H, \quad \nu = \binom{D+m}{m}, \quad \mu = 1,$$

$$U = \log\left(\frac{c}{\sqrt{2}}\right), \quad V = -\log c + \left(\frac{1}{2}\binom{D+m}{m} - 1\right)\log H.$$

If $\nu = 1$, just take for f the constant polynomial equal to 1. Assume now $\nu \ge 2$. Then we have

$$\sqrt{2}He^{U+V} + 1 = cHe^V + 1 = H^{\nu/2} + 1 \le (H+1)^{\nu/2}.$$

Write the unknown polynomial f as

$$f(X_1, \ldots, X_m) = \sum_{i_1+\cdots+i_m \le D} a_{i_1\cdots i_m} X_1^{i_1} \cdots X_m^{i_m}.$$

Then the hypotheses of Lemma 4.12 are satisfied for

$$\{u_{11}, \ldots, u_{\nu 1}\} = \left\{\theta_1^{i_1} \cdots \theta_m^{i_m} \,;\, i_1 + \cdots + i_m \le D\right\}.$$

One deduces that there exists a ν-tuple $(a_{i_1\cdots i_m})$ in $\mathbb{Z}^\nu$ which provides a solution f. □

Proof of Proposition 15.2. In § 3.5 we already proved $(ii) \Rightarrow (i)$. The implications $(iv) \Rightarrow (iii)$ and $(iii) \Rightarrow (ii)$ are easy: the first one is a consequence of the following observation: for $0 < \kappa < 1/2$ and for sufficiently large T we have

$$\frac{1}{2}T + \frac{1}{2}\log 2 + T\log(1 + |\underline{\theta}|) \le \left(\frac{1}{2} - \kappa\right)T^2.$$

For the second one, take, say, $\kappa = 1/4$ and any $T \ge \max\{4c\,;\, T_0\}$.

We now prove $(i) \Rightarrow (iv)$ as follows. Assume (i) holds. Let $\theta \in \{\theta_1, \ldots, \theta_m\}$ be transcendental. Use Lemma 15.11 with m replaced by 1, so that

$$\frac{1}{2}\binom{D+m}{m} - 1 = \frac{D-1}{2},$$

and (iv) follows. □

Remark. For $m = D = 1$, Lemma 15.11 tells nothing. The point is that for *real* numbers the exponent $\frac{1}{2}\binom{D+m}{m} - 1$ can be replaced by $\binom{D+m}{m} - 1$. See Exercise 15.13.

15.2.3 Measures of Simultaneous Approximation

Proof of Proposition 15.3. The conclusion is true when $|\underline{\theta} - \underline{\gamma}| \geq 1$, hence without loss of generality we may assume $|\underline{\theta} - \underline{\gamma}| < 1$. From the assumption $f(\underline{\gamma}) \neq 0$ we derive, by way of Proposition 3.14 (Liouville's inequality):

$$|f(\underline{\gamma})| \geq L^{1-d} e^{-d\mu} = 2L\epsilon.$$

We use the following simple estimate (compare with Lemma 13.10)

$$|f(\underline{\theta}) - f(\underline{\gamma})| \leq dLr^{d-1}|\underline{\theta} - \underline{\gamma}|$$

with $r = \max\{1, |\underline{\theta}|, |\underline{\gamma}|\} \leq 1 + |\underline{\theta}|$. Hence

$$2L\epsilon \leq |f(\underline{\gamma})| \leq |f(\underline{\theta})| + dLr^{d-1}|\underline{\theta} - \underline{\gamma}| \leq L\epsilon + dLr^{d-1}|\underline{\theta} - \underline{\gamma}|.$$

We deduce at once

$$dr^{d-1}|\underline{\theta} - \underline{\gamma}| \geq \epsilon.$$

□

15.2.4 Deducing a Measure of Algebraic Approximation from a Transcendence Measure

Proof of Lemma 15.4. We start with the following remark (see Lemma 13.10): if $f \in \mathbb{C}[X]$ is a nonzero polynomial of degree D and length L, γ a root of f and if $\theta \in \mathbb{C}$ satisfy $|\theta - \gamma| \leq 1$, then

$$|f(\theta)| \leq |\theta - \gamma| LD(1 + |\theta|)^{D-1}. \tag{15.12}$$

Let θ be a transcendental number and $\Phi(D, H)$ a transcendence measure for θ. Let γ be an algebraic number of degree $\leq D$ and Mahler's measure $\mathrm{M}(\gamma) \leq e^{\mu}$. The minimal polynomial $f \in \mathbb{Z}[X]$ of γ has usual height $\mathrm{H}(f) \leq 2^D e^{\mu}$ and length $\mathrm{L}(f) \leq (D+1)\mathrm{H}(f)$. Using (15.12) together with the definition of $\Phi(D, H)$ we get

$$|\theta - \gamma| \geq \exp\{-\psi(D, \mu)\}$$

where

$$\psi(D, \mu) = \Phi(D, 2^D e^{\mu}) + \mu + D\log 2 + (D-1)\log(1 + |\theta|) + \log\big(D(D+1)\big).$$

This yields the desired result with $c = 1 + \log(1 + |\theta|)$. □

15.2.5 Deducing a Transcendence Measure from a Measure of Algebraic Approximation

Using (15.12) one gets a polynomial approximation to θ starting from an algebraic one. The converse requires some more work. The first result, due to Fel'dman, has been refined by G. V. Chudnovsky by means of his *semi-discriminant* [Ch 1984], Chap. 1 § 1, and then by G. Diaz and M. Mignotte [DiMi 1991]:

Lemma 15.13. *Let $f \in \mathbb{Z}[X]$ be a nonzero polynomial of degree D. Let θ be a complex number, γ a root of f at minimal distance of θ and k the multiplicity of γ as a root of f. Then*

$$|\theta - \gamma|^k \leq D^{3D-2}\mathrm{H}(f)^{2D}|f(\theta)|.$$

Proof. We order the distinct roots of f as follows: $\alpha_1 = \gamma$ is a root of f which is at minimal distance of θ, next $\alpha_2, \dots, \alpha_d$ are the conjugates of α, and finally $\alpha_{d+1}, \dots, \alpha_m$ are the other roots of f. For $1 \leq j \leq m$ denote by k_j the multiplicity of α_j as a root of f, so that $k_1 = \cdots = k_d = k$ and

$$f(X) = a_0 \prod_{j=1}^{m} (X - \alpha_j)^{k_j}.$$

We may assume that the leading coefficient a_0 is > 0. If

$$g(X) = a \prod_{i=1}^{d} (X - \alpha_i)$$

denotes the minimal polynomial of γ (with $a > 0$), then g^k divides f in $\mathbb{Z}[X]$.

For $1 \leq i \leq d$, the first term in the Taylor expansion of $f(X)$ at the point α_i is

$$b_i (X - \alpha_i)^{k_i},$$

where

$$b_i = \frac{1}{k!} \cdot \frac{d^k}{dz^k} f(\alpha_i) \in \mathbb{Q}(\alpha_i).$$

We claim that the number

$$A = a^{D-2k}|b_1 \cdots b_d|$$

is a nonzero rational integer. It is plain that it is a nonzero rational number; we need only to check that it is an algebraic integer. In the case $d = 1$ this follows from the fact that a^k divides a_0. Assume now $d \geq 2$, so that $D \geq dk \geq 2k$. Write

$$b_i = a_0 \prod_{\substack{1 \leq j \leq m \\ j \neq i}} |\alpha_i - \alpha_j|^{k_j}.$$

The polynomial

$$\prod_{i=1}^{d} \prod_{\substack{1 \le j \le m \\ j \ne i}} (X_i - X_j)^{k_j}$$

has degree $\le D + (d-2)k$ in each of the variables $X_1, \dots, X_d$, while its degree is $\le dk_j$ with respect to X_j for $d < j \le m$. For any $\lambda_1, \dots, \lambda_d, \mu_1, \dots, \mu_m$ nonnegative integers satisfying $\lambda_i \le D - 2k$ $(1 \le i \le d)$ and $\mu_j \le dk_j$ $(1 \le j \le m)$, the numbers

$$a^{D-2k}\alpha_1^{\lambda_1} \cdots \alpha_d^{\lambda_d} \quad \text{and} \quad a_0^d \alpha_1^{\mu_1} \cdots \alpha_m^{\mu_m}$$

are algebraic integers (see Lemma 3.1); hence so is their product, and therefore A is an algebraic integer. This proves our claim $A \in \mathbb{Z}$.

Since

$$|\theta - \gamma| = \min_{1 \le j \le m} |\theta - \alpha_j|,$$

for $2 \le j \le m$ we have

$$|\gamma - \alpha_j| \le |\gamma - \theta| + |\theta - \alpha_j| \le 2|\theta - \alpha_j|,$$

so that

$$|f(\theta)| = a_0|\theta - \gamma|^k \prod_{j=2}^{m} |\theta - \alpha_j|^{k_j} \ge \frac{1}{2^{D-k}} a_0 |\theta - \gamma|^k \prod_{j=2}^{d} |\gamma - \alpha_j|^{k_j}.$$

We multiply both sides by the number

$$\begin{aligned} B &= 2^{D-k} a^{D-2k} a_0^{d-1} \prod_{i=2}^{d} \prod_{\substack{1 \le j \le m \\ j \ne i}} |\alpha_i - \alpha_j|^{k_j} \\ &= 2^{D-k} a^{D-2k} |b_2 \cdots b_d| \end{aligned}.$$

We find

$$A|\theta - \gamma|^k \le B|f(\theta)|.$$

Since

$$\begin{aligned} |b_i| &\le \sum_{j=k}^{D} \binom{j}{k} \mathrm{H}(f)\,(\max\{1, |\alpha_i|\})^{D-k} \\ &\le \binom{D+1}{k+1} \mathrm{H}(f)\,(\max\{1, |\alpha_i|\})^{D-k} \end{aligned}$$

we have

$$B \le 2^{D-k} \binom{D+1}{k+1}^{d-1} \mathrm{H}(f)^{d-1} \mathrm{M}(\gamma)^{D-k},$$

and we get the upper bound

$$|\theta - \gamma|^k \le |f(\theta)| 2^{D-k} \binom{D+1}{k+1}^{d-1} \mathrm{H}(f)^{d-1} \mathrm{M}(\gamma)^{D-k}.$$

Since γ is root of f we have $\mathrm{M}(\gamma) \le \mathrm{M}(f)$. Also we have $\mathrm{M}(f) \le D\mathrm{H}(f)$; this follows from (3.12) if $D \ge 2$, and it is trivial if $D = 1$. Let us check

$$2^{D-k}\binom{D+1}{k+1}^{d-1} D^{D-k} \le D^{3D-2}.$$

If $k = 1$, then using the inequality

$$(D+1)^{D-1} \le D^D$$

we deduce from $d \le D$:

$$2^{D-1}\left(\frac{D(D+1)}{2}\right)^{d-1} D^{D-1} \le (D+1)^{D-1}D^{2D-2} \le D^{3D-2}.$$

If $k \ge 2$, then

$$\binom{D+1}{k+1} \le \frac{D^{k+1}}{4}$$

and

$$\begin{aligned} 2^{D-k}\binom{D+1}{k+1}^{d-1} D^{D-k} &\le 2^{D-k-2d+2}D^{(d-1)(k+1)+D-k} \\ &\le 2^{D-k-2d+2}D^{d-2k-1} \end{aligned}$$

because $kd \le D$. Finally

$$2^{D-k-2d+2}D^{d-2k-1} \le D^{D-2}$$

because $D \ge kd \ge 2d \ge 2$.

From $D - k + d - 1 \le 2D$ we conclude

$$|\theta - \gamma|^k \le |f(\theta)|D^{3D-2}\mathrm{H}(f)^{2D}.$$

This completes the proof of Lemma 15.13. □

Remark. A refinement of Lemma 15.13 (due to N. I. Fel'dman, K. Mahler and G. Diaz) for separable polynomials is proposed as Exercise 15.14.

Proof of Lemma 15.5. Let $f \in \mathbb{Z}[X]$ be a nonzero polynomial of degree $\le D$ and usual height $\le H$. We want to estimate $|f(\theta)|$ from below. Using Lemma 15.13 we find a root γ of f of multiplicity $k \ge 1$ with

$$|\theta - \gamma|^k \le |f(\theta)|D^{3D}H^{2D}.$$

Denote by d the degree of γ. Notice that $kd \le D$ and

$$\mathrm{M}(\gamma)^k \le \mathrm{M}(f) \le DH,$$

so that

$$\mathrm{M}(\gamma) \le e^{\mu} \quad \text{where} \quad \mu := \frac{1}{k}\log(DH).$$

On the other hand we have

$$|\theta - \gamma| \geq \exp\{-\psi(d, \mu)\}.$$

From the assumptions on ψ we derive

$$k\psi(d, \mu) \leq \psi(kd, k\mu) \leq \psi(D, k\mu) = \psi\Big(D, \log(DH)\Big).$$

□

15.2.6 Deducing an Algebraic Approximation from a Polynomial Approximation

The proof of implication $(i) \Rightarrow (iv)$ of Proposition 15.2 given in § 15.2.2 was easy. The proof of $(i) \Rightarrow (iii)$ in Theorem 15.6 is more subtle. We need preliminary results. The first one (see [RoyW 1997a], Lemma 3.4) is an upper bound for the resultant $R(F, G)$ of two polynomials F and G in one variable.

Lemma 15.14. *Let θ be a complex number and t a positive real number. Let*

$$F(X) = a_0 \prod_{i=1}^{m}(X - \alpha_i) \quad \textit{and} \quad G(X) = b_0 \prod_{j=1}^{n}(X - \beta_j)$$

be nonconstant polynomials in $\mathbb{C}[X]$ of degree m and n respectively. Let f and g be integers in the ranges $0 \leq f \leq m$, $0 \leq g \leq n$; assume

$$|\theta - \alpha_i| \leq t \quad \textit{for} \quad 1 \leq i \leq f, \qquad |\theta - \alpha_i| \geq t \quad \textit{for} \quad f < i \leq m$$

and similarly

$$|\theta - \beta_j| \leq t \quad \textit{for} \quad 1 \leq j \leq g, \qquad |\theta - \beta_j| \geq t \quad \textit{for} \quad g < j \leq n.$$

Then there is a root γ of the product FG which satisfies

$$|\theta - \gamma|^{fg} |R(F, G)| \leq 2^{mn} M(F)^{n-g} M(G)^{m-f} |F(\theta)|^g |G(\theta)|^f.$$

Proof. Define first $p_i = |\theta - \alpha_i|$ for $i = 1, \ldots, m$, next $q_j = |\theta - \beta_j|$ for $j = 1, \ldots, n$ and finally

$$\rho = \min\big\{\min_{1\leq i\leq m} |\theta - \alpha_i|, \ \min_{1\leq j\leq n} |\theta - \beta_j|\big\}.$$

We take for γ a root of FG so that $\rho = |\theta - \gamma|$.

From the estimate

$$|\alpha_i - \beta_j| \leq p_i + q_j \leq 2\max\{p_i, q_j\}$$

we deduce

$$\prod_{\substack{1\le i\le f\\ g<j\le n}} |\alpha_i-\beta_j| \le \Big(\prod_{j=g+1}^n 2q_j\Big)^f \quad\text{and}\quad \prod_{\substack{f<i\le m\\ 1\le j\le g}} |\alpha_i-\beta_j| \le \Big(\prod_{i=f+1}^m 2p_i\Big)^g .$$

Since

$$(2\rho)|\alpha_i-\beta_j| \le \big(2\min\{p_i,q_j\}\big)\big(2\max\{p_i,q_j\}\big) = (2p_i)(2q_j),$$

we also have

$$(2\rho)^{fg} \prod_{\substack{1\le i\le f\\ 1\le j\le g}} |\alpha_i-\beta_j| \le \Big(\prod_{i=1}^f 2p_i\Big)^g \Big(\prod_{j=1}^g 2q_j\Big)^f .$$

Finally the estimate

$$|\alpha_i-\beta_j| \le |\alpha_i| + |\beta_j| \le 2\max\{1,|\alpha_i|\}\max\{1,|\beta_j|\},$$

implies

$$\begin{aligned}\prod_{\substack{f<i\le m\\ g<j\le n}} |\alpha_i-\beta_j| &\le 2^{(m-f)(n-g)}\Big(\prod_{i=f+1}^m \max\{1,|\alpha_i|\}\Big)^{n-g}\Big(\prod_{j=g+1}^n \max\{1,|\beta_j|\}\Big)^{m-f}\\ &\le 2^{(m-f)(n-g)}\Big(\frac{M(F)}{|a_0|}\Big)^{n-g}\Big(\frac{M(G)}{|b_0|}\Big)^{m-f}.\end{aligned}$$

Since

$$|F(\theta)| = |a_0|\prod_{i=1}^m p_i, \qquad |G(\theta)| = |b_0|\prod_{j=1}^n q_j$$

and

$$|R(F,G)| = |a_0|^n|b_0|^m \prod_{i,j}|\alpha_i-\beta_j|,$$

Lemma 15.14 follows. □

Here is a consequence of Lemma 15.14.

Corollary 15.15. *Let θ be a complex number, F and $G \in \mathbb{Z}[X]$ be nonconstant relatively prime polynomials. Denote by β one of the roots of G at minimal distance of θ. Assume that there are at least s roots α of F (counting multiplicities) satisfying*

$$|\theta-\alpha| \le |\theta-\beta|.$$

Then

$$1 \le 2^{(\deg F)(\deg G)} M(F)^{\deg G} M(G)^{\deg F} |G(\theta)|^s.$$

Proof. This follows from Lemma 15.14 by taking $t = |\theta-\beta|$, $g = 0$ and f is the number of roots α of f such that $\theta-\alpha| \le t$. The conclusion is trivial if $|G(\theta)| \ge 1$, and otherwise since $s \le f$ we have $|G(\theta)|^f \le |G(\theta)|^s$. □

We now combine Lemma 15.11 and Corollary 15.15 to prove:

Lemma 15.16. *Let $\theta \in \mathbb{C}$. For each integer $D \geq 4$ and each real number $\mu \geq \max\{600, |\theta|, D\}$, there exists a nonzero polynomial $Q \in \mathbb{Z}[X]$, which is a power of an irreducible polynomial, such that*

$$\deg Q \leq D, \quad \log \mathrm{M}(Q) \leq \mu$$

and

$$|Q(\theta)| \leq e^{-D\mu/48}.$$

Proof. We first apply Lemma 15.11 with $m = 1$ and

$$H = \left[\frac{e^{\mu}}{\sqrt{D+1}}\right].$$

We produce a nonzero polynomial $f \in \mathbb{Z}[X]$ of degree $\leq D$ and usual height $\mathrm{H}(f) \leq H$ such that

$$|f(\theta)| \leq \sqrt{2}(1+|\theta|)^D H^{-(D-1)/2}.$$

This polynomial f has Mahler's measure $\mathrm{M}(f)$ bounded by $\sqrt{D+1}H \leq e^{\mu}$. Moreover, as soon as $D \geq 4$ and $\mu \geq \max\{600, |\theta|, D\}$ we have

$$e^{\mu/24} \geq \sqrt{2}D(1+|\theta|)$$

and

$$\sqrt{2}(1+|\theta|)^D (D+1)^{(D-1)/2} e^{\mu/2} \leq e^{D\mu/6},$$

hence

$$|f(\theta)| \leq e^{-D\mu/3}.$$

We decompose f as a product of powers of irreducible polynomials over $\mathbb{Z}[X]$:

$$f = aQ_1 \cdots Q_r,$$

where a is a positive integer and $Q_1, \ldots, Q_r$ are pairwise relatively prime. If $r \leq 4$, then

$$\min_{1 \leq i \leq r} |Q_i(\theta)| \leq e^{-D\mu/3r} \leq e^{-D\mu/12},$$

hence one at least of $Q_1, \ldots, Q_r$ satisfies the conclusion of Lemma 15.16. On the other hand if $r \geq 5$, we order $Q_1, \ldots, Q_r$ so that the function

$$i \mapsto \mathrm{dist}\big(\theta, Z(Q_i)\big) := \min\big\{|\theta - \gamma|\,;\, \gamma \in \mathbb{C},\ Q_i(\gamma) = 0\big\}$$

is non-decreasing:

$$\mathrm{dist}\big(\theta, Z(Q_1)\big) \leq \cdots \leq \mathrm{dist}\big(\theta, Z(Q_r)\big).$$

We apply Corollary 15.15 with $F = Q_1 \cdots Q_4$, $G = Q_5 \cdots Q_r$ and $s = 4$. Hence

$$|G(\theta)|^{-4} \leq 2^{(\deg F)(\deg G)} M(F)^{\deg G} M(G)^{\deg F}.$$

From the assumption $\mu \geq D$ we deduce that the right hand side is bounded by $e^{D\mu}$. Therefore

$$|F(\theta)| \leq |f(\theta)| \cdot |G(\theta)|^{-1} \leq e^{-D\mu/3} e^{D\mu/4} \leq e^{-D\mu/12},$$

and again, one at least of $Q_1, \ldots, Q_4$ satisfies the conclusion of Lemma 15.16. □

Here is another application of Lemma 15.14 (see Lemma 3.10 of [RoyW 1997a] and Lemma 3 of [LauRoy 1999b]).

Lemma 15.17. *For any $\lambda > 0$ there exist positive numbers $D_0 = D_0(\lambda)$ and $\kappa = \kappa(\lambda)$ with the following property. Let D be a positive integer, μ a positive real number with*

$$\mu \geq D \geq D_0,$$

θ a complex number and $F, G \in \mathbb{Z}[X]$ relatively prime polynomials satisfying

$$\begin{aligned} \deg F &\leq D, \qquad \log \mathrm{M}(F) \leq \mu \quad \textit{and} \quad \log |F(\theta)| \leq -\lambda D\mu \\ \deg G &\leq D, \qquad \log \mathrm{M}(G) \leq \mu \quad \textit{and} \quad \log |G(\theta)| \leq -\lambda D\mu. \end{aligned}$$

Then there is a root γ of FG such that

$$\log |\theta - \gamma| \leq -\kappa D\mu.$$

Proof. Let s_0 be the positive integer in the range

$$\frac{6}{\lambda} \leq s_0 < \frac{6}{\lambda} + 1.$$

We shall prove the result with $D_0 = 3/\lambda$ and $\kappa = 2\lambda/s_0$.

Denote by d the degree of FG (hence $d \leq 2D$) and choose an ordering $\gamma_1, \ldots, \gamma_d$ of the roots of FG (counting multiplicities) so that

$$|\theta - \gamma_1| \leq \cdots \leq |\theta - \gamma_d|.$$

From the assumption $|F(\theta)| \leq 1$ we deduce $|\theta - \gamma_1| \leq 1$. Define $\gamma = \gamma_1$, $s = \min\{d, s_0\}$ and $t = |\theta - \gamma_s|$. We apply Lemma 15.14 with $f + g = s$: since $(\deg F)(\deg G) \leq d^2/4 \leq d\mu/2$, we obtain

$$|\theta - \gamma|^{fg} \leq 2^{d^2/4} e^{d\mu} e^{-\lambda s D\mu} \leq e^{-\lambda s D\mu + (3d\mu/2)}.$$

We conclude thanks to the estimates $fg \leq s^2/4$ and

$$4\lambda s D \geq 6d + \kappa D s^2.$$

□

Remark. The hypotheses of Lemma 15.17 imply $\lambda < 6$. Indeed, for $\lambda \geq 6$, in the proof we have $s_0 = 1$, hence $fg = 0$, and applying Lemma 15.14 provides a contradiction.

The fact that the hypotheses of Lemma 15.17 cannot be satisfied when λ is large follows from Gel'fond's Lemma V in Chap. III § 4 of [G 1952]. The main point here is that we get some information even when λ is small.

The following result (Lemma 4 of [LauRoy 1999b]) is the main tool providing a lower bound for the degree of the approximating algebraic number. The basic remark is that, given two relatively prime polynomials F and G, the three polynomials F, G and $F+G$ are pairwise relatively prime. Moreover

$$|(F+G)(\theta)| \leq |F(\theta)| + |G(\theta)|.$$

Lemma 15.18. *For any $\lambda > 0$ there exist positive numbers c_1, c_2 and c_3 with the following property. Let D be a positive integer, μ a positive real number with*

$$\mu \geq D \geq c_1,$$

θ a complex number and $F, G \in \mathbb{Z}[X]$ relatively prime polynomials satisfying

$$\deg F \leq D, \qquad \log \mathrm{M}(F) \leq \mu \quad \textit{and} \quad \log|F(\theta)| \leq -\lambda D\mu,$$
$$\deg G \leq D, \qquad \log \mathrm{M}(G) \leq \mu \quad \textit{and} \quad \log|G(\theta)| \leq -\lambda D\mu.$$

Then there exists a root γ of the product $FG(F+G)$ which satisfies

$$c_2 D \leq [\mathbb{Q}(\gamma) : \mathbb{Q}] \leq D, \qquad \log \mathrm{M}(\gamma) \leq 2\mu \quad \textit{and} \quad \log|\theta - \gamma| \leq -c_3 D\mu.$$

By Lemma 4 of [LauRoy 1999b], for $0 < \lambda < 1/2$ one may take

$$c_1 = \frac{9}{\lambda}, \quad c_2 = \frac{2}{5}\lambda^2 \quad \text{and} \quad c_3 = \frac{4}{25}\lambda^2.$$

Proof. As soon as $c_1 \geq D_0(\lambda)$, we may apply Lemma 15.17 to the pair (F, G) and produce a root γ_1 of FG such that

$$[\mathbb{Q}(\gamma_1) : \mathbb{Q}] \leq D, \qquad \log \mathrm{M}(\gamma_1) \leq \mu$$

and

$$\log|\theta - \gamma_1| \leq -\kappa_1 D\mu$$

where $\kappa_1 = \kappa(\lambda)$ is the constant of Lemma 15.17 associated with λ. Permuting F and G if necessary, we may assume $F(\gamma_1) = 0$.

Notice that the two polynomials G, $F+G$ are relatively prime and that

$$\deg(F+G) \leq D, \quad \log \mathrm{M}(F+G) \leq 2\mu$$

and

$$\log|F(\theta)+G(\theta)| \le -\frac{1}{2}\lambda D\mu$$

provided that $c_1^2 \ge (2/\lambda)\log 2$. We apply Lemma 15.17 once more, to the pair $(G, F+G)$, with μ replaced by 2μ and we get a root γ_2 of their product $G(F+G)$ with

$$[\mathbb{Q}(\gamma_2):\mathbb{Q}] \le D, \qquad \log \mathrm{M}(\gamma_2) \le 2\mu \quad \text{and} \quad \log|\theta-\gamma_2| \le -c_3 D\mu,$$

where

$$c_3 = \min\left\{\kappa_1, 2\kappa\left(\frac{\lambda}{4}\right)\right\}.$$

Since F and $G(F+G)$ are relatively prime, we have $\gamma_1 \neq \gamma_2$. It is now sufficient to check that the number $\delta := \max\{[\mathbb{Q}(\gamma_1):\mathbb{Q}], [\mathbb{Q}(\gamma_2):\mathbb{Q}]\}$ satisfies $\delta \ge c_2 D$: the result will follow with $\gamma \in \{\gamma_1, \gamma_2\}$. Using the estimate $\delta \le D \le \mu$, we deduce from Liouville's inequality (Chap. 3, (3.13) and Lemma 3.8)

$$\begin{aligned}\log|\gamma_1-\gamma_2| &\ge -[\mathbb{Q}(\gamma_1,\gamma_2):\mathbb{Q}]h(\gamma_1-\gamma_2)\\ &\ge -[\mathbb{Q}(\gamma_1,\gamma_2):\mathbb{Q}]\big((\log 2)+h(\gamma_1)+h(\gamma_2)\big)\\ &\ge -\delta^2\log 2 - 3\delta\mu \ge -4\delta\mu.\end{aligned}$$

On the other hand

$$|\gamma_1-\gamma_2| \le |\theta-\gamma_1| + |\theta-\gamma_2|,$$

hence

$$\log|\gamma_1-\gamma_2| \le \log 2 - c_3 D\mu \le -\frac{1}{2}c_3 D\mu$$

as soon as $c_3c_1^2 \ge 2\log 2$. We conclude $\delta \ge c_2 D$ with $c_2 = c_3/8$. □

Proof of Theorem 15.6. We start with the implication $(iii) \Rightarrow (ii)$. Fix $c > 0$. Choose in (iii) for instance $D_\nu = \nu$, $\mu_\nu = 2c_0^2\nu$. Let ν be a sufficiently large integer such that the existence of γ is guaranteed by (iii), with $[\mathbb{Q}(\gamma):\mathbb{Q}] \le \nu$, $\mathrm{M}(\gamma) \le e^{\mu_\nu}$ and

$$0 < |\theta-\gamma| \le e^{-\nu^2}.$$

From (3.12) we deduce

$$\mathrm{H}(\gamma) \le 2^{D_\nu}\mathrm{M}(\gamma) \le 2^{D_\nu}e^{\mu_\nu} \le \left(2e^{2c_0^2}\right)^\nu.$$

Since ν is sufficiently large, (ii) follows with $T = \nu(2c_0^2 + \log 2) > c$.

The proof of $(ii) \Rightarrow (i)$ is also easy: by Lemma 3.14, if θ and γ are two distinct algebraic numbers which generate a number field of degree D, then

$$|\theta-\gamma| \ge 2^{-D+1}e^{-D\mathrm{h}(\theta)}e^{-D\mathrm{h}(\gamma)}.$$

From (3.12) we deduce, with $d = [\mathbb{Q}(\gamma):\mathbb{Q}]$,

$$\mathrm{h}(\gamma) \le \frac{1}{d}\log \mathrm{H}(\gamma) + \frac{1}{2d}\log(d+1) \le \log \mathrm{H}(\gamma) + 1.$$

Define $d_0 = [\mathbb{Q}(\theta) : \mathbb{Q}]$, $h_0 = \mathrm{h}(\theta)$. Notice that $D = [\mathbb{Q}(\theta, \gamma) : \mathbb{Q}]$ is bounded by $D \le d_0 d$. Now if $d \le T$ and $\mathrm{H}(\gamma) \le e^T$, we have

$$Dh(\gamma) \le d_0 d \mathrm{h}(\gamma) \le d_0(d + \log H(\gamma)) \le 2d_0 T,$$
$$D\mathrm{h}(\theta) \le d_0 d h_0 \le d_0 h_0 T$$

and

$$(D-1)\log 2 \le D \le d_0 d \le d_0 T.$$

Therefore if θ is algebraic and if we set $c = d_0(h_0 + 3)$, then for any integer T and any algebraic number γ of degree $\le T$ and usual height $\le e^T$ with $\theta \ne \gamma$ we have

$$|\theta - \gamma| \ge e^{-cT}.$$

Finally we prove $(i) \Rightarrow (iii)$, following [LauRoy 1999b], Th. 2. We set $\lambda = 1/200$ and we denote by c_1, c_2, c_3 the constants associated with λ by Lemma 15.18. Let ν_0 be a sufficiently large integer. Lemma 15.16 shows that for each $\nu \ge \nu_0$ there exists a nonzero polynomial $Q_\nu \in \mathbb{Z}[X]$, which is a power of an irreducible polynomial, such that

$$\deg Q_\nu \le D_\nu, \quad \log \mathrm{M}(Q) \le \frac{1}{2}\mu_\nu$$

and

$$\log|Q_\nu(\theta)| \le -2\lambda D_\nu \mu_\nu.$$

If $\nu > \nu'$ are such that the polynomials Q_ν and $Q_{\nu'}$ are not relatively prime, then Q_ν and $Q_{\nu'}$ are powers of the same irreducible polynomial in $\mathbb{Z}[X]$ and then

$$\frac{\log|Q_{\nu'}(\theta)|}{\deg Q_{\nu'}} = \frac{\log|Q_\nu(\theta)|}{\deg Q_\nu} \le -2\lambda\mu_\nu.$$

Since μ_ν tends to infinity with ν and $Q_{\nu'}(\theta) \ne 0$ (recall that θ is transcendental), for each $\nu' > 0$ this inequality can hold only for finitely many ν. It follows that there are infinitely many integers ν such that $Q_{\nu-1}$ and Q_ν are relatively prime. We apply Lemma 15.18 to the two polynomials $F = Q_{\nu-1}$ and $G = Q_\nu$ with $D = D_\nu$ and $\mu = \mu_\nu/2$. Notice that $D_{\nu-1}\mu_{\nu-1} \ge D_\nu\mu_\nu/4$, hence

$$\log|Q_{\nu-1}(\theta)| \le -\frac{1}{2}\lambda D_\nu \mu_\nu.$$

We get an algebraic number γ which satisfies

$$c_2 D_\nu \le [\mathbb{Q}(\gamma) : \mathbb{Q}] \le D_\nu, \quad \log \mathrm{M}(\gamma) \le \mu_\nu$$

and

$$\log|\theta - \gamma| \le -\frac{c_3}{2} D_\nu \mu_\nu.$$

Property (iv) of Theorem 15.6 follows as soon as

$$c_0 \ge \max\left\{2,\ c_1,\ \frac{1}{c_2},\ \frac{2}{c_3}\right\}.$$

□

15.2.7 From a Single Number to a Tuple

In this section we prove Proposition 15.9.

Let $\underline{\theta} = (\theta_1, \ldots, \theta_m)$ be a tuple of complex numbers such that $\mathbb{Q}(\underline{\theta})$ has transcendence degree 1 over $\mathbb{Q}$. One at least of the numbers $\theta_1, \ldots, \theta_m$, say θ_1, is transcendental. For $1 \leq i \leq m$, θ_i is a root of a nonzero polynomial $F_i(\theta_1, Y)$, where $F_i \in \mathbb{Z}[X, Y]$. On the other hand, since θ_1 is transcendental, Theorem 15.6 shows that it has good algebraic approximations. Let γ_1 be one of them. For sufficiently small $|\theta_1 - \gamma_1|$, $F(\gamma_1, Y)$ is a nonzero polynomial which has a root γ_i close to θ_i. Then $\underline{\gamma} = (\gamma_1, \ldots, \gamma_m)$ is a simultaneous algebraic approximation to $\underline{\theta}$. This argument will enable us to complete the proof of Proposition 15.9.

We provide the details of the proof following [RoyW 1997b], § 1 c Prop. 1.3. A slightly different argument is given in [RoyW 1997a], § 3 (i). For another proof using Chow forms, see [LauRoy 1999b], Lemme 7 and § 6.

Lemma 15.19. *Let*

$$f(X) = a_0X^d + \cdots + a_d = a_0(X - \alpha_1)\cdots(X - \alpha_d)$$

be a polynomial in $\mathbb{C}[X]$ *of degree* $d \geq 1$ *without multiple root. Set*

$$r = \frac{1}{2}\min_{i\neq j} |\alpha_i - \alpha_j|, \quad r_0 = \max\{|\alpha_1|, \ldots, |\alpha_d|\}, \quad R = \max\{1, r + r_0\},$$

$$\eta = \frac{|a_0|r^d}{(d+1)R^d} \quad \text{and} \quad c = \frac{r}{\eta}.$$

Let $\widetilde{a}_0, \ldots, \widetilde{a}_d$ *be complex numbers satisfying*

$$\max_{0\leq i\leq d} |a_i - \widetilde{a}_i| < \eta.$$

Then the polynomial

$$\widetilde{f}(X) = \widetilde{a}_0X^d + \cdots + \widetilde{a}_d$$

can be written

$$\widetilde{f}(X) = \widetilde{a}_0(X - \widetilde{\alpha}_1)\cdots(X - \widetilde{\alpha}_d)$$

with

$$\max_{1\leq j\leq d} |\alpha_j - \widetilde{\alpha}_j| \leq c \max_{0\leq i\leq d} |a_i - \widetilde{a}_i|.$$

Proof of Lemma 15.19. Put

$$\epsilon = \max_{0\leq i\leq d} |a_i - \widetilde{a}_i|.$$

From the assumptions

$$|a_0 - \widetilde{a}_0| \leq \epsilon < \eta < |a_0|$$

we deduce $\widetilde{a}_0 \neq 0$.

For $|z| \le R$, we have

$$|f(z) - \widetilde{f}(z)| \le \epsilon(1 + R + \cdots + R^d) < |a_0| r^d.$$

For $|z - \alpha_i| = r$, we have

$$|f(z)| = |a_0| \prod_{j=1}^{d} |z - \alpha_j| \ge |a_0| r^d.$$

Since the upper bound $|f(z) - \widetilde{f}(z)| < |f(z)|$ holds for $|z - \alpha_i| = r$, we deduce (Rouché's theorem) that $\widetilde{f}$ has a single zero $\widetilde{\alpha}_i$ in the disc $|z - \alpha_i| < r$. Hence

$$\widetilde{f}(X) = \widetilde{a}_0(X - \widetilde{\alpha}_1) \cdots (X - \widetilde{\alpha}_d)$$

with

$$\begin{cases} |\widetilde{\alpha}_i - \alpha_i| < r & \text{for } 1 \le i \le d, \\ |\widetilde{\alpha}_j - \alpha_i| \ge r & \text{for } 1 \le i \ne j \le d. \end{cases}$$

On one hand we have

$$|f(\widetilde{\alpha}_j)| = |a_0| \prod_{i=1}^{d} |\widetilde{\alpha}_j - \alpha_i| \ge |a_0| r^{d-1} |\widetilde{\alpha}_j - \alpha_j|$$

and on the other

$$|f(\widetilde{\alpha}_j)| = |f(\widetilde{\alpha}_j) - \widetilde{f}(\widetilde{\alpha}_j)| \le \epsilon(d+1)R^d.$$

This completes the proof of Lemma 15.19. □

Lemma 15.20. *Let $F \in \mathbb{Z}[X_1, \ldots, X_n, Y]$ be a polynomial in $n+1$ variables of degree D_j in X_j, $(1 \le j \le n)$, and let $\alpha_1, \ldots, \alpha_n$, β be algebraic numbers which satisfy $F(\alpha_1, \ldots, \alpha_n, \beta) = 0$. Assume that $F(\alpha_1, \ldots, \alpha_n, Y)$ is not the zero polynomial in $\mathbb{Q}(\alpha_1, \ldots, \alpha_n)[Y]$. Then*

$$\mathrm{h}(\beta) \le 2 \log \mathrm{L}(F) + 2 \sum_{j=1}^{n} D_j \mathrm{h}(\alpha_j).$$

Proof of Lemma 15.20. Let t be the degree of F in the variable Y. Write $\underline{\alpha}$ for $(\alpha_1, \ldots, \alpha_n)$, $\underline{X}$ for $(X_1, \ldots, X_n)$, and write

$$F(\underline{X}, Y) = Y^t Q_t(\underline{X}) + Y^{t-1} Q_{t-1}(\underline{X}) + \cdots + Q_0(\underline{X}).$$

Since $F(\underline{\alpha}, Y) \in \mathbb{Q}(\underline{\alpha})[Y]$ is not the zero polynomial and since $F(\underline{\alpha}, \beta)$ vanishes, at least one of the numbers $Q_t(\underline{\alpha}), \ldots, Q_1(\underline{\alpha})$ is not zero. Denote by m the largest index j, $(1 \le j \le m)$ such that $Q_j(\underline{\alpha}) \ne 0$. From $F(\underline{\alpha}, \beta) = 0$ we deduce

$$-\beta^m Q_m(\underline{\alpha}) = \beta^{m-1} Q_{m-1}(\underline{\alpha}) + \cdots + \beta Q_1(\underline{\alpha}) + Q_0(\underline{\alpha}).$$

Define

$$\widetilde{Q}(\underline{X}, Y) = Y^{m-1}Q_{m-1}(\underline{X}) + \cdots + YQ_1(\underline{X}) + Q_0(\underline{X}),$$

so that

$$-\beta^m Q_m(\underline{\alpha}) = \widetilde{Q}(\underline{\alpha}, \beta).$$

An upper bound for $\mathrm{h}(\beta^m)$ is

$$\begin{aligned}\mathrm{h}(\beta^m) &= \mathrm{h}\big(\beta^m Q_m(\underline{\alpha})Q_m(\underline{\alpha})^{-1}\big)\\ &\le \mathrm{h}\big(\beta^m Q_m(\underline{\alpha})\big) + \mathrm{h}\big(Q_m(\underline{\alpha})\big)\\ &= \mathrm{h}\big(\widetilde{Q}(\underline{\alpha}, \beta)\big) + \mathrm{h}\big(Q_m(\underline{\alpha})\big).\end{aligned}$$

Using Lemma 1.5, we bound $\mathrm{h}\big(\widetilde{Q}(\underline{\alpha}, \beta)\big)$ and $\mathrm{h}\big(Q_m(\underline{\alpha})\big)$ from above:

$$\mathrm{h}\big(Q_m(\underline{\alpha})\big) \le \log \mathrm{L}(Q_m) + \sum_{j=1}^{n} (\deg_{X_j} Q_m)\mathrm{h}(\alpha_j)$$

and

$$\mathrm{h}\big(\widetilde{Q}(\underline{\alpha}, \beta)\big) \le \log \mathrm{L}(\widetilde{Q}) + \sum_{j=1}^{n} (\deg_{X_j} \widetilde{Q})\mathrm{h}(\alpha_j) + (\deg_Y \widetilde{Q})\mathrm{h}(\beta).$$

The degrees $\deg_{X_j} Q_m$ and $\deg_{X_j} \widetilde{Q}$ are bounded by $\deg_{X_j} F = D_j$. Also we have $\deg_Y \widetilde{Q} \le m-1$. Coming back to $\mathrm{h}(\beta^m)$, we deduce

$$m\mathrm{h}(\beta) = \mathrm{h}(\beta^m) \le \log \mathrm{L}(Q_m) + \log \mathrm{L}(\widetilde{Q}) + 2\sum_{j=1}^{n} D_j \mathrm{h}(\alpha_j) + (m-1)\mathrm{h}(\beta).$$

Hence

$$\mathrm{h}(\beta) \le \log \mathrm{L}(Q_m) + \log \mathrm{L}(\widetilde{Q}) + 2\sum_{j=1}^{n} D_j \mathrm{h}(\alpha_j).$$

Since $\mathrm{L}(Q_m) + \mathrm{L}(\widetilde{Q}) \le \mathrm{L}(F)$, we see that $\log \mathrm{L}(Q_m) + \log \mathrm{L}(\widetilde{Q})$ is bounded by $2\log \mathrm{L}(F)$, which yields the desired estimate. □

Lemma 15.21. *Let $\theta_1, \ldots, \theta_{n+1}$ be complex numbers. Write $\underline{\theta}$ for the tuple $(\theta_1, \ldots, \theta_n)$. Assume θ_{n+1} is algebraic over the field $\mathbb{Q}(\underline{\theta})$. There exist positive constants η_0, c_0, c_1, c_2 with the following property. Let $\underline{\gamma} = (\gamma_1, \ldots, \gamma_n)$ be a tuple of algebraic numbers such that $|\underline{\theta} - \underline{\gamma}| < \eta_0$. Then there exists an algebraic number γ_{n+1} which satisfies*

$$[\mathbb{Q}(\underline{\gamma}, \gamma_{n+1}) : \mathbb{Q}(\underline{\gamma})] \le c_1, \quad \mathrm{h}(\gamma_{n+1}) \le c_2 \max_{1\le i\le n} \mathrm{h}(\gamma_i)$$

and

$$|\theta_{n+1} - \gamma_{n+1}| \le c_3 |\underline{\theta} - \underline{\gamma}|.$$

Proof of Lemma 15.21. Let $F \in \mathbb{Z}[\underline{X}, Y]$ be a polynomial in $n+1$ variables such that $F(\underline{\theta}, Y)$ is irreducible in $\mathbb{Q}(\underline{\theta})[Y]$ and $F(\underline{\theta}, \theta_{n+1}) = 0$. Write

$$F(\underline{X}, Y) = a_0(\underline{X})Y^d + \cdots + a_d(\underline{X})$$

and define $f(Y) = F(\underline{\theta}, Y)$. Let η and c be the constants of Lemma 15.19 attached to f. By continuity there exists $\eta_0 > 0$ such that the condition $|\underline{\theta} - \underline{\gamma}| < \eta_0$ implies $\max_{0 \le i \le d} |a_i(\underline{\theta}) - a_i(\underline{\gamma})| \le \eta$. Let γ_{n+1} be a root of f (whose existence is given by Lemma 15.19) which satisfies

$$|\theta_{n+1} - \gamma_{n+1}| \le c \max_{0 \le i \le d} |a_i(\underline{\theta}) - a_i(\underline{\gamma})|.$$

Since γ_{n+1} is a root of f we have

$$[\mathbb{Q}(\underline{\gamma}, \gamma_{n+1}) : \mathbb{Q}(\underline{\gamma})] \le d.$$

From Lemma 15.20 we deduce

$$\mathrm{h}(\gamma_{n+1}) \le c_2 \max_{1 \le i \le n} \mathrm{h}(\gamma_i).$$

This completes the proof of Lemma 15.21. □

Proof of Proposition 15.9. Let $\underline{\theta} = (\theta_1, \ldots, \theta_m)$ be a tuple of complex numbers such that $\mathbb{Q}(\underline{\theta})$ has transcendence degree 1. Without loss of generality we may assume that θ_1 is transcendental over $\mathbb{Q}$. Let c_0 be a sufficiently large number (depending only on θ_1) for which property (iii) of Theorem 15.6 holds. Next, let C be a sufficiently large number. The constants c_4 and c_5 below will depend on $\underline{\theta}$, not on C.

Given the sequences $(D_\nu)_{\nu \ge 1}$ and $(h_\nu)_{\nu \ge 1}$ satisfying the assumptions of Proposition 15.9, define

$$D'_\nu = \frac{D_\nu}{C} \quad \text{and} \quad \mu'_\nu = \frac{\mu_\nu}{C}.$$

From Theorem 15.6 we deduce that for infinitely many $\nu \ge 1$, there is an algebraic number γ_1 satisfying

$$\frac{1}{c_0} D'_\nu \le [\mathbb{Q}(\gamma_1) : \mathbb{Q}] \le D'_\nu, \quad \mathrm{M}(\gamma_1) \le e^{\mu'_\nu}$$

and

$$0 < |\theta_1 - \gamma_1| \le e^{-D'_\nu \mu'_\nu / c_0}.$$

We fix a sufficiently large ν in this infinite sequence, and also we fix a γ_1 as above. Notice that

$$\mathrm{h}(\gamma_1) \le \frac{1}{[\mathbb{Q}(\gamma_1) : \mathbb{Q}]} \log \mathrm{M}(\gamma_1) \le \frac{c_0}{D'_\nu} \cdot \mu'_\nu.$$

We use Lemma 15.21 with $n = 1$: for each $i = 2, \ldots, m$ there exists an algebraic number γ_i satisfying

$$[\mathbb{Q}(\gamma_1, \gamma_i) : \mathbb{Q}(\gamma_1)] \le c_4, \quad \mathrm{h}(\gamma_i) \le c_5 \mathrm{h}(\gamma_1)$$

and

$$|\theta_i - \gamma_i| \le c_6 |\theta_1 - \gamma_1|.$$

We deduce, for the m-tuple $\underline{\gamma} = (\gamma_1, \dots, \gamma_m)$,

$$\frac{1}{c_0} D'_\nu \leq [\mathbb{Q}(\underline{\gamma}) : \mathbb{Q}] \leq c_4^{m-1} D'_\nu,$$

$$\begin{aligned}[\mathbb{Q}(\underline{\gamma}) : \mathbb{Q}] \max_{1 \leq i \leq m} \mathrm{h}(\gamma_i) &\leq c_4^{m-1} D'_\nu \max_{1 \leq i \leq m} \mathrm{h}(\gamma_i) \\ &\leq c_4^{m-1} c_5 D'_\nu \mathrm{h}(\gamma_1) \\ &\leq c_4^{m-1} c_5 c_0 \mu'_\nu\end{aligned}$$

and

$$|\underline{\theta} - \underline{\gamma}| \leq c_6 e^{-D'_\nu \mu'_\nu / c_0}.$$

Since $\gamma_1 \neq \theta_1$, we also have

$$\max_{1 \leq i \leq m} |\theta_i - \gamma_i| > 0.$$

Finally for sufficiently large C we have

$$c_4^{m-1} D'_\nu \leq D_\nu \quad \text{and} \quad c_4^{m-1} c_5 c_0 \mu'_\nu \leq \mu_\nu.$$

□

Remark. From Lemma 15.21 we also deduce the following statement (Proposition 1.3 of [RoyW 1997b]). Let $\theta_1, \dots, \theta_m$ be complex numbers, not all of which are algebraic, and let n be an integer in the range $1 \leq n \leq m$. Assume each of the numbers $\theta_{n+1}, \dots, \theta_m$ is algebraic over the field $\mathbb{Q}(\theta_1, \dots, \theta_n)$. Then there exist positive constants c_7, c_8 and c_9 such that, if $\psi(D, \mu)$ is a measure of simultaneous approximation for the m-tuple $(\theta_1, \dots, \theta_m)$, then $c_7 + \psi(c_8 D, c_9 \mu)$ is a measure of simultaneous approximation for the n-tuple $(\theta_1, \dots, \theta_n)$.

15.3 Algebraic Independence Results: Small Transcendence Degree

We combine here some of the simultaneous approximations measures obtained in Chap. 14 with the results of § 15.1 and deduce results of algebraic independence.

The underlying principle of Corollary 15.10 is that any measure of simultaneous approximation $\psi(D, \mu)$ *better than* $D\mu$ for $\underline{\theta}$ implies that $\mathbb{Q}(\underline{\theta})$ has transcendence degree ≥ 2. Under the notation of Chap. 14, we are looking at simultaneous approximation measure $\varphi(D, h)$ *better than* $D^2 h$. Often, it is sufficient to fix the parameter h (sufficiently large) and to make D tend to infinity. In terms of ψ this amounts to take for μ a constant multiple of D.

We give a collection of examples related to estimates which have been established in Chap. 14.

15.3.1 On the Lindemann-Weierstraß Theorem

Let us prove the special case $n = 2$ of Theorem 1.3:

- *Let β_1, β_2 be two nonzero algebraic numbers whose quotient β_1/β_2 is irrational. Then the two numbers e^{β_1} and e^{β_2} are algebraically independent.*

Thanks to Corollary 15.10, this result follows from Corollary 14.11, where we have proved the existence of a constant $c > 0$ such that

$$\psi(D, \mu) = cD^{1/2}\mu(\log \mu + D \log D)(\log \mu)^{-1}$$

is a measure of simultaneous approximation for e^{β_1}, e^{β_2}.

The first proof of this special case of the Lindemann-Weierstraß' Theorem along Gel'fond's method is due to G. V. Chudnovsky [Ch 1984], Chap. 7 Th. 10.6.

On the other hands several quantitative refinements of Lindemann-Weierstraß' Theorem are known; in particular a completely explicit measure of algebraic independence (see § 15.5.2) have been derived by A. Sert [Sert 1999].

15.3.2 Algebraic Independence of $\alpha_s^{\beta_j\beta'_r}$

Combining Corollary 15.10 with Corollary 14.4, we deduce the following result:

Theorem 15.22. *Let $\beta_0, \ldots, \beta_n$ be $\mathbb{Q}$-linearly independent algebraic numbers, $\beta'_1, \ldots, \beta'_p$ also $\mathbb{Q}$-linearly independent algebraic numbers and $\lambda_1, \ldots, \lambda_q$ be $\mathbb{Q}$-linearly independent elements of $\mathcal{L}$. Assume $npq > pq + n$. Then two at least of the $(n+1)pq$ numbers*

$$e^{\beta_j\beta'_r\lambda_s} \quad (0 \le j \le n,\ 1 \le r \le p,\ 1 \le s \le q)$$

are algebraically independent.

Equality $npq = pq + n + 1$ holds for and only for the following triples (n, p, q):

$$(2,\ 1,\ 3), \quad (3,\ 1,\ 2), \quad (3,\ 2,\ 1) \quad \text{and} \quad (2,\ 3,\ 1).$$

We give an example of each.

Taking $(n, p, q) = (2, 1, 3)$, we deduce Corollary 7.2.4 of [W 1974]:

Corollary 15.23. *Let β_1, β_2 be algebraic numbers such that $1, \beta_1, \beta_2$ are $\mathbb{Q}$-linearly independent and let $\lambda_1, \lambda_2, \lambda_3$ be $\mathbb{Q}$-linearly independent elements in $\mathcal{L}$. Then two at least of the six numbers*

$$e^{\lambda_i\beta_j} \qquad (i = 1, 2, 3,\ j = 1, 2)$$

are algebraically independent.

Taking $(n, p, q) = (3, 1, 2)$, we deduce Corollary 7.2.5 of [W 1974]:

Corollary 15.24. *Let $\beta_1, \beta_2, \beta_3$ be algebraic numbers such that $1, \beta_1, \beta_2, \beta_3$ are $\mathbb{Q}$-linearly independent and let λ_1, λ_2 be $\mathbb{Q}$-linearly independent elements in $\mathcal{L}$. Then two at least of the six numbers*

$$e^{\lambda_i \beta_j} \qquad (i = 1, 2, \ j = 1, 2, 3)$$

are algebraically independent.

Taking $(n, p, q) = (3, 2, 1)$, we deduce:

Corollary 15.25. *Let β and β' be two quadratic numbers with $[\mathbb{Q}(\beta, \beta') : \mathbb{Q}] = 4$. Let $\lambda \in \mathcal{L} \setminus \{0\}$. Then two at least of the three numbers*

$$e^{\beta\lambda}, \ e^{\beta'\lambda}, \ e^{\beta\beta'\lambda}$$

are algebraically independent.

For instance two at least of the three numbers

$$2^{\sqrt{2}}, \ 2^{\sqrt{3}}, \ 2^{\sqrt{6}}$$

are algebraically independent.

Finally, taking $(n, p, q) = (2, 3, 1)$, we obtain a result of Gel'fond:

- *If $\lambda = \log \alpha$ is a nonzero logarithm of an algebraic number and if β is a cubic irrational number, then the two numbers $\alpha^\beta = e^{\beta\lambda}$ and $\alpha^{\beta^2} = e^{\beta^2\lambda}$ are algebraically independent.*

For instance the two numbers $2^{\sqrt[3]{2}}$ and $2^{\sqrt[3]{4}}$ are algebraically independent.

More generally, the following result of Gel'fond's [G 1952] is deduced from Theorem 15.22 with $n = d - 1$, $p = d$, $q = 1$:

Corollary 15.26. *Let β an algebraic number of degree $d \geq 3$ and let $\lambda \in \mathcal{L} \setminus \{0\}$. Then two at least of the $d - 1$ numbers*

$$e^{\beta\lambda}, \ldots, e^{\beta^{d-1}\lambda}$$

are algebraically independent.

15.3.3 Algebraic Independence of Exponentials and Logarithms

Here is the *Linear Subgroup Theorem in transcendence degree* 1 (Theorem 1.1 of [RoyW 1997a]). For a subfield K of $\mathbb{C}$, define

$$\mathcal{L}_K = \exp^{-1}(K^\times) = \left\{z \in \mathbb{C}\,;\; e^z \in K^\times\right\}.$$

Theorem 15.27*. *Let d_0 and d_1 be nonnegative integers with $d = d_0 + d_1 > 0$, G the algebraic group $\mathbb{G}_{\rm a}^{d_0} \times \mathbb{G}_{\rm m}^{d_1}$, K a subfield of $\mathbb{C}$ of transcendence degree ≤ 1 over $\mathbb{Q}$, $\mathcal{W}$ a vector subspace of $\mathbb{C}^d$ defined over K, Y a finitely generated subgroup of $K^{d_0} \times (\mathcal{L}_K)^{d_1}$ and Y_a a subgroup of Y contained in $K^{d_0} \times \mathcal{L}^{d_1}$. Assume that the dimension n of the vector subspace of $\mathbb{C}^d$ spanned by $\mathcal{W} \cup Y$ satisfies $n < d/2$. Assume also that no algebraic subgroup G^* of G, defined over K and distinct from G itself, has a tangent space $T_e(G^*)$ which contains $\mathcal{W} \cup Y$. Then there exists an algebraic subgroup $G^* = G_0^* \times G_1^*$ of G, where G_0^* is an algebraic subgroup of $\mathbb{G}_{\rm a}^{d_0}$ of codimension $d_0^\flat$ and G_1^* is an algebraic subgroup of $\mathbb{G}_{\rm m}^{d_1}$ of codimension $d_1^\flat$ with $d^\flat = d_0^\flat + d_1^\flat > 0$, such that, if we set*

$$\mathcal{W}^\flat = \frac{\mathcal{W}}{\mathcal{W} \cap T_e(G^*)}, \quad Y^\flat = \frac{Y}{Y \cap T_e(G^*)}, \quad Y_a^\flat = \frac{Y_a}{Y_a \cap T_e(G^*)},$$

$$\ell_0^\flat = \dim_{\mathbb{C}}(\mathcal{W}^\flat), \quad \ell_1^\flat = \mathrm{rank}_{\mathbb{Z}}(Y^\flat), \quad \ell_a^\flat = \mathrm{rank}_{\mathbb{Z}}(Y_a^\flat),$$

and if $n^\flat$ denotes the dimension of the subspace of $\mathbb{C}^{d^\flat}$ spanned by $\mathcal{W}^\flat \cup Y^\flat$, then we have $d^\flat > 2n^\flat > \ell_0^\flat$ and

$$\frac{d_1}{d-2n} \geq \frac{d_1^\flat}{d^\flat - 2n^\flat} \geq \frac{\ell_1^\flat}{2n^\flat - \ell_0^\flat}.$$

Moreover, if either $d_0^\flat < n^\flat$, or $\ell_0^\flat < n^\flat$, or else $\ell_a^\flat > 0$, then we have the strict inequality

$$\frac{d_1^\flat}{d^\flat - 2n^\flat} > \frac{\ell_1^\flat}{2n^\flat - \ell_0^\flat}.$$

We refer to [RoyW 1997a] for a proof of Theorem 15.27. The main tools are Theorem 13.1 and Corollary 15.10, but an extra argument is necessary for the following reason. The algebraic independence method we have described so far in this chapter rests on measures of simultaneous approximation. These measures are not valid without some technical assumption (the linear independence measure condition of Chap. 14, say). It is easy to prove Theorem 15.27 under such an extra hypothesis, but it is possible also to avoid it as follows. In the transcendence argument, when using Theorem 13.1, we had algebraic data $\underline{w}_k$ and $\underline{\eta}_j$, and complex data $\underline{w}'_k$ and $\underline{\eta}'_j$. The algebraic subgroup G^* of G which occurs in the conclusion of Theorem 13.1 is related to the algebraic data, and this is where the technical assumption comes from. In order to avoid it, one needs to lift this obstructing subgroup and get a subgroup of G which is related to the complex data $\underline{w}'_k$ and $\underline{\eta}'_j$. That this is possible in transcendence degree 1 (i.e. for function fields in one variable) is explained in § 3 of [RoyW 1997a].

We now state the special case $n = 1$ of Theorem 15.27; it is the main result concerning small transcendence degree (i.e. algebraic independence of at least two numbers in certain sets) for values of the exponential function in a single variable (not including Lindemann-Weierstraß Theorem).

Corollary 15.28. *Let m and n be two positive integers, $\{x_1, \dots, x_m\}$ and $\{y_1, \dots, y_n\}$ two families of $\mathbb{Q}$-linearly independent complex numbers. Denote by K_1 the field generated over $\mathbb{Q}$ by the mn numbers $\exp(x_i y_j)$ $(1 \le i \le m,\ 1 \le j \le n)$. Define also*

$$K_2 = K_1(x_1, \dots, x_m), \qquad K_3 = K_2(y_1, \dots, y_n).$$

We set

$$\kappa_1 = mn, \quad \kappa_2 = \kappa_1 + m, \quad \kappa_3 = \kappa_2 + n.$$

Hence, for $h = 1, 2, 3$, the field K_h is obtained by adjoining κ_h elements to $\mathbb{Q}$. Then the transcendence degree of K_h over $\mathbb{Q}$ is ≥ 2 in each of the following cases:

(a) $h = 1, 2$ *and* $\kappa_h \ge 2(m + n)$;
(b) $h = 3$ *and* $\kappa_3 > 2(m + n)$;
(c) $h = 3$, $\kappa_3 = 2(m + n)$ *and* $x_i y_1 \in \mathcal{L}$ *for* $i = 1, \dots, m$.

Remark. Here are a few references for a direct proof of this result: [T 1971], [Br 1974b], Chap. 7 of [W 1974], Chap. 12 of [B 1975], [Br 1979], Corollaire 1.2 of [RoyW 1997a], Chap. 6 of [FNe 1998] and Chap. 13 of [NeP 2000].

Proof of Corollary 15.28 as a consequence of Theorem 15.27 (Following [RoyW 1997a]). We take $n = 1$, d_0 is 0 or 1, and ℓ_0 is also 0 or 1. Define

$$\underline{w} = \begin{cases} (x_1, \dots, x_d) & \text{if } d_0 = 0, \\ (1, x_1, \dots, x_d) & \text{if } d_0 = 1, \end{cases}$$

$$\mathcal{W} = \begin{cases} 0 & \text{if } \ell_0 = 0, \\ \mathbb{C}\underline{w} & \text{if } \ell_0 = 1 \end{cases}$$

and

$$\underline{\eta}_j = y_j \underline{w} \quad (1 \le j \le \ell_1).$$

In the conclusion of Theorem 15.27 we have $n \ge n^\flat > 0$, hence $n^\flat = 1$, and

$$\frac{d_1}{d-2} \ge \frac{d_1^\flat}{d^\flat - 2} \ge \frac{\ell_1^\flat}{2 - \ell_0^\flat}.$$

Since $d^\flat = d_1^\flat + d_0^\flat$ with $0 \le d_0^\flat \le d_0$, one easily deduces from the inequality

$$d_1(d^\flat - 2) \geq d_1^\flat(d - 2)$$

that $d_0^\flat = d_0$ and $d_1^\flat = d_1$, hence $d^\flat = d$ and $G^* = 0$. Therefore $\ell_0^\flat = \ell_0$, $\ell_1^\flat = \ell_1$ and

$$\frac{d_1}{d-2} \geq \frac{\ell_1}{2-\ell_0}.$$

Therefore when $n = 1$ the assumption that the field K of Theorem 15.27 has transcendence degree ≤ 1 implies

$$\ell_1 d + d_1 \ell_0 \leq 2(d_1 + \ell_1).$$

The conclusion of case (b) in Corollary 15.28 plainly follows. Moreover we get strict inequality when either $d_0 = 0$ or $\ell_0 = 0$, and this covers case (a). In case (c) we also have strict inequality because $\ell_a > 0$. □

Theorem 15.22 also follows from the lower bound for κ_2 in Corollary 15.28. Therefore Corollary 15.25 is a consequence of the lower bound for κ_2 in Corollary 15.28, but in fact it is also a consequence of the lower bound for κ_1 with $m = n = 4$,

$$x_1 = 1,\ x_2 = \beta,\ x_3 = \beta',\ x_4 = \beta\beta', \quad y_j = x_j\lambda \quad (j = 1, 2, 3, 4).$$

As a further example of the lower bound for κ_2 one deduces from Corollary 15.28 a stronger form of Corollary 15.26:

- *Let $\lambda \in \mathbb{C}^\times$ be a nonzero complex number and β an algebraic number of degree $d \geq 3$. Then two at least of the d numbers*

$$e^\lambda, e^{\beta\lambda}, \ldots, e^{\beta^{d-1}\lambda}$$

are algebraically independent.

Finally another consequence of the lower bound for κ_2 is Corollary 7.2.6 of [W 1974]:

Corollary 15.29. *Let β be an irrational algebraic number and let λ_1, λ_2 be $\mathbb{Q}$-linearly independent elements in $\mathcal{L}$. Then two at least of the five numbers*

$$\frac{\lambda_1}{\lambda_2},\ e^{\lambda_1\beta},\ e^{\lambda_2\beta},\ e^{\lambda_1\beta^2},\ e^{\lambda_2\beta^2}$$

are algebraically independent.

For instance two at least of the three numbers

$$\frac{\log 2}{\log 3},\ 2^i,\ 3^i$$

are algebraically independent.

Remark. Using Corollary 15.10, one deduces also Corollary 15.29 from Corollary 14.5: indeed if we set $k = 4$, $m = 2$,

$$x_1 = \lambda_2, \quad x_2 = \beta\lambda_2, \quad y_1 = 1, \quad y_2 = \frac{\lambda_1}{\lambda_2}, \quad y_3 = \beta, \quad y_4 = \beta y_2,$$

then Corollary 14.5 shows that a simultaneous approximation measure for the five numbers occurring in Corollary 15.29 is

$$\varphi(D,h) = cD^2h^{4/3}(h + \log D)^{3/4}(\log h + \log D)^{-1},$$

which is $o(D^2)$ for fixed h and for $D \to \infty$.

15.3.4 Quadratic Relations Between Logarithms of Algebraic Numbers

The only known information so far in direction of Conjecture 1.15 (on the algebraic independence of logarithms of algebraic numbers) which does not follow from the results of Chapters 11 and 12 is the following (see [RoyW 1997a], [RoyW 1997b]).

Theorem 15.30. *Let $\lambda_1, \dots, \lambda_n$ be elements of $\mathcal{L}$ and $\mathcal{E}$ the $\mathbb{Q}$-vector subspace of $\mathbb{C}$ spanned by these elements. Assume*[26] *that the field $k = \mathbb{Q}(\lambda_1, \dots, \lambda_n)$ has transcendence degree 1 over $\mathbb{Q}$. Then the rank of any nonzero matrix M with entries in $\mathcal{E}$ satisfies*

$$\mathrm{rank}(\mathsf{M}) > \frac{1}{2} r_{\mathrm{str}}(\mathsf{M}),$$

where $r_{\mathrm{str}}(\mathsf{M})$ is the structural rank of M with respect to k.

By Proposition 12.25, it follows that if $\lambda_1, \dots, \lambda_n$ are $\mathbb{Q}$-linearly independent elements of $\mathcal{L}$ satisfying

$$\mathrm{trdeg}_{\mathbb{Q}}\mathbb{Q}(\lambda_1, \dots, \lambda_n) = 1,$$

then for any nonzero homogeneous polynomial $Q \in \mathbb{Q}[X_1, \dots, X_n]$ of degree 2, we have $Q(\lambda_1, \dots, \lambda_n) \neq 0$. It would be interesting to extend this statement to nonhomogeneous quadratic polynomials. For instance taking $Q(X_1, X_2) = X_1^2 - X_2$ would yield the transcendence of the number e^{λ^2} for any $\lambda \in \mathcal{L} \setminus \{0\}$. So far the transcendence of e^{π^2} is still an open problem.

A simple corollary of Theorem 15.30 is the transcendence of one at least of the two numbers

$$e^{\lambda^2}, \; e^{\lambda^3}$$

for $\lambda \in \mathcal{L} \setminus \{0\}$; this corollary is also a consequence of part c) in Corollary 15.28, and a direct transcendence proof (not passing through algebraic independence) has been given in Chap. 11 (see Exercise 11.8).

Theorem 15.30 clearly follows by combining Theorem 14.23 with Corollary 15.10. Another proof of Theorem 15.30 as a consequence of Theorem 15.27 is given in [RoyW 1997a], § 1, together with further similar results.

---

[26] Beware: Conjecture 1.15 predicts that these assumptions are satisfied only for $n = 1$.

15.3.5 Open Problems

The best known measure of simultaneous algebraic approximation for the two numbers π and e^{π} is the one which is valid more generally for λ and $e^{\beta\lambda}$ when β is a quadratic number and λ a nonzero logarithm of an algebraic number, namely (see Exercise 14.4.e)

$$cD^{1/2}\mu(\mu + D\log D)^{1/2}(\log\mu)^{-1/2}.$$

This measure is not strong enough to yield a result of algebraic independence. The algebraic independence of the two numbers π and e^{π} has been proved by Nesterenko using modular forms [Ne 1996] (see also [NeP 2000]), and the algebraic independence of λ and $e^{\beta\lambda}$ is not yet proved in general.

In the same way, the best known measures of simultaneous approximation, which are stated in Exercise 14.4, are not strong enough to solve the following open problems:

(?) *Two at least of the three numbers e, e^e, e^{e^2} are algebraically independent.*

(?) *Two at least of the three numbers π, e, e^{π^2} are algebraically independent.*

Partial results are known and follow from the measures of simultaneous approximation proved in Chap. 14 (see Exercise 15.15).

Further conjectures are as follows:

(?) *Each of the numbers e^e, e^{e^2}, e^{π^2} is transcendental*

(?) *The numbers e and π are algebraically independent*

So far, the best known unconditional measure of simultaneous approximation for the two numbers e and π is

$$cD^{1/2}\mu^{1/2}(\mu + D\log D)(\log\mu)^{1/2}$$

with some absolute constant $c > 0$ (see Exercise 14.6.c).

According to part 2 of Conjecture 14.25 there should exist a positive constant c such that $cD^{1/2}\mu$ is a measure of simultaneous approximation for each of the pairs (e, π), (e^{π}, π) and (e, e^e). One expects that the same holds for other similar pairs of complex numbers, like for almost all elements of $\mathbb{C}^2$.

15.4 Large Transcendence Degree: Conjecture on Simultaneous Approximation

It is a challenge to extend the previous discussion to higher transcendence degree. So far the connection (see § 15.2) between simultaneous approximation and algebraic independence has been established only for small transcendence degree (Proposition 15.9). The following statement would provide results of large transcendence degree (compare with [RoyW 1997b], Conjecture 1.7, [Lau 1998], § 4.2, and [Roy 2000a]).

Conjecture 15.31. *Let $\underline{\theta} = (\theta_1, \dots, \theta_m)$ be a tuple of complex numbers such that the number*

$$t = \mathrm{trdeg}_{\mathbb{Q}}\mathbb{Q}(\underline{\theta})$$

is ≥ 1. There exist two positive constants c_1 and c_2 with the following property. Let $(D_\nu)_{\nu\geq 1}$ and $(\mu_\nu)_{\nu\geq 1}$ be sequences of real numbers satisfying $D_\nu \geq c_1$, $\mu_\nu \geq c_1$,

$$c_1 \leq D_\nu \leq D_{\nu+1} \leq 2D_\nu, \quad \textit{and} \quad c_1 D_\nu \leq \mu_\nu \leq \mu_{\nu+1} \leq 2\mu_\nu \qquad (\nu \geq 1).$$

Assume also

$$\lim_{\nu\to\infty} \mu_\nu = \infty.$$

Then for infinitely many ν there exists a m-tuple $\underline{\gamma} = (\gamma_1, \dots, \gamma_m)$ of algebraic numbers satisfying

$$[\mathbb{Q}(\underline{\gamma}) : \mathbb{Q}] \leq D_\nu, \quad [\mathbb{Q}(\underline{\gamma}) : \mathbb{Q}] \max_{1\leq i\leq m} \mathrm{h}(\gamma_i) \leq \mu_\nu$$

and

$$\max_{1\leq i\leq m} |\theta_i - \gamma_i| \leq e^{-c_2 D_\nu^{1/t} \mu_\nu}.$$

A discussion of this topic as well as further related issues is given in [W 2000].

Remark 1. By Proposition 15.9, a stronger result holds for $t = 1$, since one obtains also a lower bound for $[\mathbb{Q}(\underline{\gamma}) : \mathbb{Q}]$. It has been shown by D. Roy [Roy 2000a] that such a lower bound for the degree cannot be expected for $t \geq 2$. One cannot replace the condition

$$[\mathbb{Q}(\underline{\gamma}) : \mathbb{Q}] \max_{1\leq i\leq m} \mathrm{h}(\gamma_i) \leq \mu_\nu \qquad \text{by} \qquad \max_{1\leq i\leq m} \mathrm{h}(\gamma_i) \leq \frac{\mu_\nu}{D_\nu}$$

(compare with [RoyW 1997b], Conjecture 1.7).

Remark 2. A heuristic motivation for the exponent $1/t$ if given by M. Laurent in [Lau 1998], § 4.2 p. 325 (see below the end of § 15.5.1).

Remark 3. Let $\underline{\theta} \in \mathbb{C}^m$ be a tuple of complex numbers with a measure of simultaneous approximation $\psi(D, \mu)$. Assume that for any sufficiently large D_1, D_2 and μ_1, μ_2 satisfying $D_1 \leq D_2$ and $\mu_1 \leq \mu_2$, we have

$$\psi(D_1, \mu_1) \leq \psi(D_2, \mu_2).$$

Assume further that there exist sequences $(D_\nu)_{\nu\geq 1}$ and $(\mu_\nu)_{\nu\geq 1}$ like in Conjecture 15.31, namely with $D_\nu \geq c_1$, $\mu_\nu \geq c_1 D_\nu$,

$$D_\nu \leq D_{\nu+1} \leq 2D_\nu, \quad \text{and} \quad \mu_\nu \leq \mu_{\nu+1} \leq 2\mu_\nu \qquad (\nu \geq 1),$$

such that, for a positive real number k,

$$\lim_{\nu \to \infty} \frac{1}{D_\nu^{1/k} \mu_\nu} \psi(D_\nu, \mu_\nu) = 0.$$

Then, by Conjecture 15.31,

$$\operatorname{trdeg}_{\mathbb{Q}} \mathbb{Q}(\underline{\theta}) \geq [k] + 1.$$

In loose terms, Conjecture 15.31 means that any simultaneous approximation measure *better than* $D^{1/k}\mu$ for $\underline{\theta}$ should imply $\operatorname{trdeg}_{\mathbb{Q}} \mathbb{Q}(\underline{\theta}) > k$.

Remark 4. Assuming Conjecture 15.31, one deduces Lindemann-Weierstraß' Theorem from Corollary 14.11 as follows: one applies Remark 3 to the function

$$\psi(D, \mu) = C D^{1/m} \mu (\log \mu + D \log D)(\log \mu)^{-1}$$

with $k = m - 1$, taking for instance the sequences $D_\nu = \log \nu$ and $\mu_\nu = \nu$.

Remark 5. Combining Conjecture 15.31 with Conjecture 14.25 (part 2), one deduces Conjecture 1.15 on algebraic independence of logarithms of algebraic numbers. More generally, if we take Conjectures 14.25 and 15.31 for granted, then one deduces Schanuel's Conjecture 1.14 under the hypothesis that $x_1, \ldots, x_n$ satisfy a linear independence measure condition.

Here is another consequence of Conjecture 15.31, which includes results due to G. Diaz and P. Philippon.

Theorem 15.32*. *Under the notation and hypotheses of Corollary 15.28, assume that both tuples $(x_1, \ldots, x_m)$ and $(y_1, \ldots, y_n)$ satisfy a linear independence measure condition. Then for $i = 1, 2, 3$ the transcendence degree t_i of the field K_i satisfies*

$$t_1 \geq \left[\frac{mn}{m+n}\right] \quad \textit{provided that} \quad mn > m + n,$$

$$t_2 \geq \left[\frac{mn+m}{m+n}\right] \quad \textit{provided that} \quad m \geq 2$$

and

$$t_3 \geq \frac{mn}{m+n}.$$

For further references, including previous statements of Chudnovsky ([Ch 1984], Chap. 1), more recent results by W. D. Brownawell (where a weaker technical assumption is shown to be sufficient) – as well as quantitative refinements, we refer for instance [FNe 1998] and Chap. 14 of [NeP 2000].

Remark. In some cases it is possible to obtain a strict inequality for t_3. Also extensions to higher dimensional situation are known.

Here is a consequence of Theorem 15.32, where the technical hypothesis does not appear explicitly (they are in fact a consequence of the assumptions):

Under the assumptions of Theorem 15.22, the transcendence degree of the field generated by the $(n+1)pq$ *numbers*

$$e^{\beta_j\beta'_r\lambda_s} \quad (0 \le j \le n,\ 1 \le r \le p,\ 1 \le s \le q)$$

is at least

$$\left[\frac{npq}{pq+n+1}\right]+1.$$

Notice that this statement also follows directly from Conjecture 15.31 by means of Corollary 14.4.

A special case is the following result of G. Diaz [Di 1989]:

Under the assumptions of Corollary 15.26, the transcendence degree of the field

$$\mathbb{Q}\big(e^{\lambda}, e^{\beta\lambda}, \ldots, e^{\beta^{d-1}\lambda}\big)$$

is at least $[(d+1)/2]$.

The so-called *Problem of Gel'fond and Schneider* is to show that this transcendence degree is $d-1$. See [FNe 1998], Chap. 6 and [NeP 2000], Chap. 14.

Proof of (15.32) as a consequence of Conjecture 15.31.
(1) Since $mn > m+n$ we already know (six exponentials Theorem) $t_1 \ge 1$. By Corollary 15.28 we also have $t_1 \ge 2$ as soon as $mn \ge 2(m+n)$. Incidentally, for these two results (small transcendence degree), no technical assumption (linear independence measure condition) is required.

The assumption $mn > m+n$ allows us to define

$$\kappa = \frac{mn}{mn-m-n}\cdot$$

Applying Corollary 14.18 with $r=1$, $d=m$, $\ell=n$, we deduce that a measure of simultaneous approximation for the numbers $e^{x_iy_j}$ is

$$c\mu^{\kappa}(\log\mu)^{1-\kappa}.$$

Since this function is increasing, we may apply Remark 3 above for the sequences $D_\nu = \nu$ and $\mu_\nu = \nu\log\log\nu$ with $k>0$ defined by

$$\kappa = 1+\frac{1}{k}, \qquad \text{viz.} \qquad k = \frac{1}{\kappa-1} = \frac{mn}{m+n} - 1.$$

We deduce that the transcendence degree t_1 of the field K_1 over $\mathbb{Q}$ is $\ge [k]+1$.

(2) Here we apply Corollary 14.5, but we permute the role of x and y (which means that we replace (m,k) by (n,m)). Under the assumptions of Theorem 15.32, assuming

$m \geq 2$, a measure of simultaneous approximation for the $mn + m$ numbers x_i, $e^{x_i y_j}$ is $c\psi(D, \mu)$ with

$$\psi(D, \mu)^{n(m-1)} = \mu^{mn}(\mu + D \log D)^m (\log \mu)^{-m-n}.$$

Since both functions $D \mapsto \psi(D, \mu)$ and $\mu \mapsto \psi(D, \mu)$ are increasing, and since

$$\psi(D, D) \leq c' D^{1+(1/k)} (\log D)^{-1/(m-1)} \quad \text{with} \quad k = \frac{m+n}{n(m-1)},$$

we deduce from Conjecture 15.31 that the transcendence degree t_2 of the field K_2 satisfies $t_2 \geq [k] + 1$.

(3) Combining Corollary 14.14 and Conjecture 15.31 with

$$\psi(D, \log D) = c' D^{1+\frac{mn}{m+n}} \log D,$$

we deduce

$$t_3 \geq \frac{mn}{m+n}.$$

□

15.5 Further Results and Conjectures

15.5.1 Further Criteria for Algebraic Independence

Historically, the first result of algebraic independence was Lindemann-Weierstrass' Theorem 1.3. However, as pointed out in § 1.1, this statement is equivalent to a result of linear independence. Further more general results of algebraic independence of values of the so-called E-functions were achieved in 1929, and then in 1949, by C.L. Siegel. Later, A.B. Šidlovskiĭ and his school developed extensively this theory; but we shall not tell more about this theme here (see [Sh 1989], as well as [FNe 1998], Chap. 5).

Another method of algebraic independence was introduced by K. Mahler in the 1930's, and it is very efficient for studying the values of functions satisfying certain functional equations; again, we shall not expand on this topic (see [Ni 1996]).

The method of algebraic independence of the present chapter has its main source in the work of A.O. Gel'fond around 1950 [G 1952] (see also [FNe 1998] and [NeP 2000]). Among the tools he introduced are a zero estimate (see Exercise 2.9) and a transcendence criterion. The next statement also introduces multiplicities, following [LauRoy 1999a] and [LauRoy 2000]. For $\sigma \in \mathbb{N}^n$ we denote, as usual, by D^σ the derivative operator $(d/dX_1)^{\sigma_1} \cdots (d/dX_n)^{\sigma_n}$ on the space $\mathbb{C}[X_1, \ldots, X_n]$.

Theorem 15.33*. *Let $\underline{\theta} = (\theta_1, \ldots, \theta_m)$ be a m-tuple of complex numbers. The following properties are equivalent.*

(i) $\mathrm{trdeg}_{\mathbb{Q}} \mathbb{Q}(\underline{\theta}) \geq 2$.

(ii) There exists a sequence $(f_\nu)_{\nu\geq 1}$ of polynomials in $\mathbb{Z}[X_1,\dots,X_m]$ and a sequence $(t_\nu)_{\nu\geq 1}$ of real numbers such that

$$\deg f_\nu + \log \mathrm{H}(f_\nu) \leq t_\nu, \qquad \lim_{\nu\to\infty} t_\nu = \infty$$

and

$$0 < |f_\nu(\underline{\theta})| < e^{-3t_\nu \max\{t_{\nu-1}, t_\nu, t_{\nu+1}\}}.$$

(iii) There exist sequences $(d_\nu)_{\nu\geq 1}$, $(s_\nu)_{\nu\geq 1}$ of positive integers, sequences $(H_\nu)_{\nu\geq 1}$, $(V_\nu)_{\nu\geq 1}$ of real numbers and a sequence $(f_\nu)_{\nu\geq 1}$ of polynomials in $\mathbb{Z}[X_1,\dots,X_m]$ satisfying

$$\frac{d_\nu}{s_\nu} \leq \frac{d_{\nu+1}}{s_{\nu+1}}, \quad \frac{\log H_\nu}{s_\nu} \leq \frac{\log H_{\nu+1}}{s_{\nu+1}},$$

$$\frac{V_\nu}{s_\nu} \leq \frac{V_{\nu+1}}{s_{\nu+1}},$$

$$\limsup_{\nu\to\infty} \frac{V_\nu s_\nu}{d_\nu \log H_\nu} = \infty,$$

and

$$\deg f_\nu \leq d_\nu, \quad \mathrm{H}(f_\nu) \leq H_\nu, \quad 0 < \max_{\|\sigma\|<s_\nu} \left|\frac{1}{\sigma!} D^\sigma f_\nu(\underline{\theta})\right| \leq e^{-V_{\nu+1}}.$$

(iv) For any sequences $(d_\nu)_{\nu\geq 1}$ and $(s_\nu)_{\nu\geq 1}$ of positive integers and $(H_\nu)_{\nu\geq 1}$ of real numbers satisfying

$$1 \leq \frac{d_\nu}{s_\nu} \leq \frac{d_{\nu+1}}{s_{\nu+1}} \leq 2\frac{d_\nu}{s_\nu}, \quad \frac{\log H_\nu}{s_\nu} \leq \frac{\log H_{\nu+1}}{s_{\nu+1}} \leq 2\frac{\log H_\nu}{s_\nu}, \quad \log H_\nu \geq d_\nu \geq 2,$$

there exists a sequence $(f_\nu)_{\nu\geq 1}$ of polynomials in $\mathbb{Z}[X_1,\dots,X_m]$ such that

$$\deg f_\nu \leq d_\nu, \quad \mathrm{H}(f_\nu) \leq H_\nu, \quad 0 < \max_{\|\sigma\|<s_\nu} \left|\frac{1}{\sigma!} D^\sigma f_\nu(\underline{\theta})\right| \leq H_\nu^{-d_\nu^2/4s_\nu^2}.$$

(v) There exists a constant $c(\underline{\theta}) > 0$ with the following property. For any integers D, S with $1 \leq S \leq D$ and any real number $H \geq 1$, there exists a polynomial $f \in \mathbb{Z}[X_1,\dots,X_m]$ such that

$$\deg f \leq D, \quad \mathrm{H}(f) \leq H$$

and

$$0 < \max_{\|\sigma\|<S} \left|\frac{1}{\sigma!} D^\sigma f(\underline{\theta})\right| \leq -c(\underline{\theta})^D H^{-\kappa},$$

where

$$\kappa = \frac{(D+1)(D+2)}{2S(S+1)} - 1.$$

Remark 1. The implications $(v) \Rightarrow (iv) \Rightarrow (iii)$ and $(v) \Rightarrow (ii)$ in Theorem 15.33 are trivial. Implication $(ii) \Rightarrow (i)$ is a variant of Gel'fond's criterion in [G 1952], Chap. III, § 4, Lemma VII (see also [Br 1974a], [W 1974], Chap. 5, [Br 1979],

[F 1982], Chap. 9, Lemma 3.9, [FNe 1998], Chap. 6, § 1.3, Lemma 6.3 and [NeP 2000]). Implication $(ii) \Rightarrow (i)$ is the transcendence criterion of [LauRoy 1999a] and [LauRoy 2000] involving multiplicities. Finally $(i) \Rightarrow (v)$ follows from Lemma 15.11 with

$$\mu = \binom{S+1}{2}, \quad \nu = \binom{D+2}{2}, \quad U = Dc_1(\underline{\theta}), \quad N = [\log H].$$

□

Remark 2. Notice again the gap between the estimates in *(iv)* and *(iii)*: the former estimate involves $(d_\nu/s_\nu)^2 \log H_\nu$, the latter $(d_\nu/s_\nu) \log H_\nu$.

His criterion enabled Gel'fond to prove that some fields have transcendence degree at least 2. One main obstruction to extend Gel'fond's algebraic independence method to large transcendence degree was the following (see [Cas 1957], Chap. V Th. XIV and [P 1986b], Appendix):

- *Let $m \ge 2$ be an integer and $\varphi: \mathbb{N} \to \mathbb{R}_{>0}$ a positive valued function. There exist algebraically independent numbers $\theta_1, \ldots, \theta_m$ with the following property: for any positive integer N, there are $m-1$ linear forms in three variables*

$$L_i(X_0, X_1, X_i) = a_i X_0 + b_i X_1 + c_i X_i \in \mathbb{Z}[X_0, X_1, X_i] \qquad (2 \le i \le m)$$

with integer coefficients of absolute values bounded by N such that $c_i \neq 0$ and

$$|L_i(1, \theta_1, \theta_i)| \le \varphi(N) \qquad (2 \le i \le m).$$

We refer to Chap. 6 of [FNe 1998] for further comments and references on this topic, including Lang's transcendence type [L 1966], Chudnovsky's results of algebraic independence ([Ch 1984], Chap. 3 and 4), Philippon's Criterion [P 1986b] and further developments until 1997. See also [NeP 2000], Chap. 8.

The results of the present chapter have already been extended in several directions. In [LauRoy 2000], M. Laurent and D. Roy have introduced multiplicities in Philippon's criterion [P 1986b] for algebraic independence. This enabled them to prove results on large transcendence degrees by means of interpolation determinants and also to deduce the following result, which can be seen as a step towards general conjectures of simultaneous algebraic approximation.

Theorem 15.34*. *Let $\underline{\theta} \in \mathbb{C}^m$. Let $(D_\nu)_{\nu \ge 1}$ be a non-decreasing sequence of positive integers, and let $(h_\nu)_{\nu \ge 1}$ be a sequence of real numbers ≥ 1 such that $(D_\nu h_\nu)_{\nu \ge 1}$ is non-decreasing and unbounded. Then, for infinitely many ν, there exists a nonzero polynomial $P \in \mathbb{Q}[X_1, \ldots, X_m]$ of degree $\le D_\nu$ and height $\mathrm{h}(P) \le h_\nu$ which admits at least one zero $\underline{\alpha} \in \mathbb{C}^m$ with*

$$\max_{1\le i\le m} |\theta_i - a_i| \le \exp\left(-\frac{1}{8(m+1)!}D_{\nu-1}^{m+1}h_{\nu-1}\right).$$

Since $\underline{a}$ is a zero of P, for $m = 1$ (and only in this case) we deduce $\underline{a} \in \overline{\mathbb{Q}}^m$. One would like to construct several independent such polynomials P with a common zero in $\overline{\mathbb{Q}}^m$ close to $\underline{\theta}$. If one could obtain an ideal having a set of common zeroes of dimension 0, one could conclude $\underline{a} \in \overline{\mathbb{Q}}^m$. As pointed out by M. Laurent and D. Roy, this provides a heuristic justification for the exponent $1/t$ in Conjecture 15.31. Indeed, assume $t = m$ (by Proposition 15.19, this assumption involves no loss of generality), so that $\theta_1, \dots, \theta_m$ are algebraically independent. Lemma 15.11 shows that there exists a nonzero polynomial of degree $\le D_1$ and logarithmic height $\le h_1$ such that

$$|P(\underline{\theta})| \le e^{-c_3 D_1^{m+1} h_1}.$$

One would like to produce not only one, but several such polynomials, and being optimistic, one might expect that the associated hypersurfaces should define, by intersection, an algebraic point $\underline{\gamma} = (\gamma_1, \dots, \gamma_m) \in \overline{\mathbb{Q}}^m$ at a distance $\le \exp\{-c_4 D_1^{m+1} h_1\}$ of $\underline{\theta}$. Given that $\underline{\gamma} \in \overline{\mathbb{Q}}^m$ is a common root of a collection of polynomials P, each of degree $\le D_1$ and height $\le h_1$, the degree $D = [\mathbb{Q}(\underline{\gamma}) : \mathbb{Q}]$ should then be bounded by $c_5 D_1^m$ and the height $h = \max \mathrm{h}(\gamma_i)$ by $c_6 h_1$. Replacing D_1 and h_1 in terms of h and D with $D = c_5 D_1^m$ and $h = c_6 h_1$ explains the exponent $1/m$.

15.5.2 Quantitative Estimates: Measures of Algebraic Independence

Let $\underline{\theta} = (\theta_1, \dots, \theta_m) \in \mathbb{C}^m$. By Proposition 15.2, in order to prove that one at least of the numbers $\theta_1, \dots, \theta_m$ is transcendental, it suffices to construct a polynomial f in the ring $\mathbb{Z}[X_1, \dots, X_m]$ with suitable bounds for the degree and height such that $|f(\underline{\theta})|$ is sufficiently small but nonzero. Often[27] the transcendence method provides a sharper result: for any $\underline{\widetilde{\theta}} \in \mathbb{C}^m$ sufficiently close to $\underline{\theta}$, there exists such a polynomial $\widetilde{f}$ (depending on $\underline{\widetilde{\theta}}$) which satisfies the same bound for $|\widetilde{f}(\underline{\theta})|$ but also does not vanish at $\underline{\widetilde{\theta}}$. From such a statement one deduces a measure of simultaneous approximation for $\underline{\theta}$ (see Proposition 15.3). As an example, one may reach easily measures of linear independence of logarithms of algebraic numbers (see Exercise 15.4) which may be not the sharpest known, but are far from being trivial.

In the criterion of algebraic independence of Philippon (see [FNe 1998], Theorem 6.11, [NeP 2000], Chap. 8 and [LauRoy 2000]) which extends Theorem 15.33 to higher transcendence degree, the hypothesis involves a family of polynomials

[27] As shown by Exercise 15.12, under a general framework involving linearly independent numbers – like the six exponentials Theorem 1.12, or its extensions either in several variables from Chap. 4 and Chap. 11 or to algebraic independence in Theorem 15.27 – one cannot expect a quantitative refinement, unless some extra hypothesis – e.g. a linear independence measure condition – is assumed.

whose set of common zeroes in a neighborhood of $\underline{\theta}$ is finite. The quantitative refinements of this criterion due to P. Philippon, E. M. Jabbouri, M. Ably, C. Jadot and Y. V. Nesterenko (see [FNe 1998], Theorem 6.17 and Chap. 6, § 5.2; see also Chap. 8 of [NeP 2000] and [W 1999], § 2.2) requires that these polynomials have no common zeros, at least near $\underline{\theta}$. There is a simple case where this condition is fulfilled: when one can produce a single polynomial with not only an upper bound for $|f(\underline{\theta})|$ but also a lower bound for the same.

In such circumstances one derives a *measure of algebraic independence* for $(\theta_1, \dots, \theta_m)$. Such measures are defined in a general framework by Nesterenko and Philippon by means of Chow forms (see ([FNe 1998], Chap. 6, § 3, [P 1997], [P 1998], [P 1999a], [P 2000], [P 1999b], [LauRoy 2000] and [NeP 2000], Chap. 8), but we give here a definition only in the simplest case, for an algebraically independent tuple.

Definition. Let $\underline{\theta} = (\theta_1 \dots, \theta_m)$ be a m-tuple of algebraically independent complex numbers. A *measure of algebraic independence* for $\underline{\theta}$ is a mapping $\Phi\colon \mathbb{N} \times \mathbb{R}_{>0} \to \mathbb{R}_{>0}$ such that, for any sufficiently large integer D, any sufficiently large real number H and any nonzero polynomial $f \in \mathbb{Z}[X_1, \dots, X_m]$ of total degree $\le D$ and usual height $\mathrm{H}(f) \le H$, we have

$$|f(\underline{\theta})| \ge \exp\{-\Phi(D, H)\}.$$

In case $m = 1$, a measure of algebraic independence for $\theta \in \mathbb{C} \setminus \overline{\mathbb{Q}}$ is nothing else than a transcendence measure for θ. From Lemma 15.11 one easily deduces that a measure of algebraic independence for a m-tuple of algebraically independent complex numbers satisfies

$$\Phi(D, H) \ge \left(\frac{1}{2}\binom{D+m}{m} - 1\right) \log H - cD$$

where c depends only on $\underline{\theta}$.

It is easy (Exercise 15.12) to construct a m-tuple of algebraically independent complex numbers for which any measure of algebraic independence grows as fast as desired. On the other hand it is likely that almost all $\underline{\theta}$ (for Lebesgue's measure) in $\mathbb{C}^m$ has a measure of algebraic independence bounded by $c(\underline{\theta})D^m \log H$ (see [Ch 1984], [Am 1988] and Chap. 15 of [NeP 2000]). Moreover one can expect that numbers arising from values of the exponential or logarithmic function (and their iterates) at algebraic points also have such a measure of algebraic independence. In this direction we propose an quantitative sharpening to Schanuel's Conjecture:

Conjecture 15.35. *Let $x_1, \dots, x_n$ be $\mathbb{Q}$-linearly independent complex numbers which satisfy a linear independence measure condition. Let d be a positive integer. Then there exists a positive number $C = C(x_1, \dots, x_n, d)$ with the following property: for any integer $H \ge 2$ and any $n+1$ tuple $P_1, \dots, P_{n+1}$ of polynomials in $\mathbb{Z}[X_1, \dots, X_n, Y_1, \dots, Y_n]$ with degrees $\le d$ and usual heights $\le H$, which generate an ideal of $\mathbb{Q}[X_1, \dots, X_n, Y_1, \dots, Y_n]$ of rank $n+1$, we have*

$$\sum_{j=1}^{n+1} \left| P_j(x_1, \ldots, x_n, e^{x_1}, \ldots, e^{x_n}) \right| \geq H^{-C}.$$

The linear independence measure condition cannot be omitted, as shown for instance by examples of A. Bijlsma [Bi 1977].

A simple consequence of Conjecture 15.35 is the following:

(?) *Let $P \in \mathbb{Z}[X_1, \ldots, X_m]$ be a nonzero polynomial and K a number field. There exists a positive constant C with the following property. Let $\lambda_1, \ldots, \lambda_m$ be elements of $\mathcal{L}$ such that $\alpha_i = e^{\lambda_i}$ is in K for $1 \leq i \leq m$. Let $A \geq e$ be a positive number such that*

$$A \geq \max_{1 \leq i \leq m} \mathrm{h}(\alpha_i) \quad \textit{and} \quad A \geq \max_{1 \leq i \leq m} |\lambda_i|.$$

If the number $\Xi = P(\lambda_1, \ldots, \lambda_m)$ is nonzero, then

$$|\Xi| \geq A^{-C}.$$

For instance any nonzero determinant

$$\Delta = \begin{vmatrix} \log a_1 & \log a_2 \\ \log a_3 & \log a_4 \end{vmatrix}$$

with $a_i \in \mathbb{Z}$, $a_i \geq 2$ should be bounded from below by

$$|\Delta| \geq \left(\max\{a_1, a_2, a_3, a_4\}\right)^{-C}$$

for some absolute constant $C > 0$.

We started with a criterion 15.1 for irrationality, involving rational approximations. We have considered two generalizations of this initial situation: either involving polynomial approximations, that is looking for $|P(\theta_1, \ldots, \theta_m)|$, or algebraic approximations, that is looking for

$$\max\{|\theta_1 - \gamma_1|, \ldots, |\theta_m - \gamma_m|\}.$$

The former deals with hypersurfaces (codimension 1), the latter with points (dimension 0). Of course in dimension 1 both coincide, but this is no longer the case in higher dimension. Further generalizations are possible, and interesting; they involve approximations by *cycles* of a given dimension (see Philippon's papers [P 1997], [P 1998], [P 1999a], [P 2000], [P 1999b]).

In the classical theory of simultaneous *rational* approximation, given a tuple $(\vartheta_1, \ldots, \vartheta_m)$ of real numbers, Khinchine's transference theorem ([Cas 1957], Chap. V § 3 Th. IV) exhibits a duality between lower bounds for

$$q \longmapsto \min_{(p_1, \ldots, p_m) \in \mathbb{Z}^m} \max_{1 \leq i \leq m} \left| \vartheta_i - \frac{p_i}{q} \right|$$

and for

$$(p_1, \dots, p_m) \longmapsto \min_{q \in \mathbb{Z}} |p_1\vartheta_1 + \cdots + p_m\vartheta_m + q|.$$

It is not known whether there is a similar transference theorem in the context of algebraic diophantine approximation: one would like to replace p_i/q by algebraic numbers on one hand, $p_1X_1 + \cdots + p_mX_m + q$ by a polynomial of arbitrary degree on the other. However a partial result is available. For $P \in \mathbb{Z}[X_1, X_2]$, define $t(P) = \deg P + \log \mathrm{L}(P)$. For $\underline{\gamma} = (\gamma_1, \gamma_2) \in \overline{\mathbb{Q}}^2$ define

$$t(\underline{\gamma}) = [\mathbb{Q}(\underline{\gamma}) : \mathbb{Q}]\bigl(1 + \mathrm{h}(1 : \gamma_1 : \gamma_2)\bigr).$$

P. Philippon proves in [P 2000]:

- *For any $\underline{\theta} = (\theta_1, \theta_2) \in \mathbb{C}^2$ and $\mu \geq 3$ there exist positive constants $c_1, \dots, c_7$ with the following property.*
  - ◇ *If there exists a nonzero polynomial $P \in \mathbb{Z}[X_1, X_2]$ with $t(P) \geq c_1$ and*

    $$|P(\underline{\theta})| \leq \exp\bigl\{-c_2 t(P)^{\mu}\bigr\},$$

    then there exists $\underline{\gamma} \in \overline{\mathbb{Q}}^2$ with $P(\underline{\gamma}) = 0$ and

    $$|\underline{\theta} - \underline{\gamma}| \leq \exp\bigl\{-c_3 t(\underline{\gamma})^{2\mu/(\mu+1)}\bigr\}.$$

  - ◇ *Conversely, if there exists $\underline{\gamma} \in \overline{\mathbb{Q}}^2$ with $t(\underline{\gamma}) \geq c_4$ and*

    $$|\underline{\theta} - \underline{\gamma}| \leq \exp\bigl\{-c_5 t(\underline{\gamma})^{\mu/2}\bigr\},$$

    then there exists a polynomial $P \in \mathbb{Z}[X_1, X_2]$ such that $P(\underline{\gamma}) = 0$ and $t(P) \leq c_6 t(\underline{\gamma})^{1/2}$, hence

    $$|P(\underline{\theta})| \leq \exp\bigl\{-c_7 t(P)^{\mu}\bigr\}.$$

This sharpens (and corrects) a statement in [Ch 1984], Chap. 4 p. 180. For $\mu = 3$, it follows that there is equivalence between a measure of algebraic independence and a measure of simultaneous approximation in the case where the exponents for these estimates are optimal:

$$|\underline{\theta} - \underline{\gamma}| \geq \exp\bigl\{-ct(\underline{\gamma})^{3/2}\bigr\} \quad \text{and} \quad |P(\underline{\theta})| \geq \exp\bigl\{-ct(P)^{3}\bigr\}.$$

A survey on algebraic independence of transcendental numbers is given in [W 1999]. Further information on the topics discussed in the present chapter is given in [W 2000].

15.5.3 An Arithmetic Criterion for the Values of the Exponential Function

We conclude by quoting the following recent result of D. Roy [Roy 2000c]:

- *Schanuel's Conjecture 1.14 is equivalent to Conjecture 15.36 below.*

Denote by $\mathcal{D}$ the derivation

$$\mathcal{D} = \frac{\partial}{\partial X_0} + X_1 \frac{\partial}{\partial X_1}$$

on the field $\mathbb{C}(X_0, X_1)$.

Conjecture 15.36 (Roy). *Let ℓ be a positive integer, $y_1, \dots, y_\ell$ $\mathbb{Q}$-linearly independent complex numbers, $\alpha_1, \dots, \alpha_\ell$ nonzero complex numbers and s_0, s_1, t_0, t_1, u positive real numbers satisfying*

$$\max\{1, t_0, 2t_1\} < \min\{s_0, 2s_1\} < u \quad \text{and} \quad \max\{s_0, s_1 + t_1\} < u < \tfrac{1}{2}(1 + t_0 + t_1). \tag{15.37}$$

Assume that, for any sufficiently large positive integer N, there exists a nonzero polynomial $P_N \in \mathbb{Z}[X_0, X_1]$ with partial degree $\le N^{t_0}$ in X_0, partial degree $\le N^{t_1}$ in X_1 and height $\mathrm{H}(P_N) \le e^N$, which satisfies

$$\left| (\mathcal{D}^k P_N)\Bigl(\sum_{j=1}^{\ell} m_j y_j, \prod_{j=1}^{\ell} \alpha_j^{m_j} \Bigr) \right| \le \exp(-N^u)$$

for any integers $k, m_1, \dots, m_\ell$ in $\mathbb{N}$ with $k \le N^{s_0}$ and $\max\{m_1, \dots, m_\ell\} \le N^{s_1}$. Then, we have

$$\mathrm{trdeg}_{\mathbb{Q}} \mathbb{Q}(y_1, \dots, y_\ell, \alpha_1, \dots, \alpha_\ell) \ge \ell.$$

The proof that Conjecture 15.36 implies Schanuel's Conjecture uses Proposition 4.10. For the converse, D. Roy establishes the following criterion concerning the values of the exponential function.

Theorem 15.38. (Roy). *Let $(y, \alpha) \in \mathbb{C} \times \mathbb{C}^\times$, and let s_0, s_1, t_0, t_1, u be positive real numbers satisfying the inequalities (15.37). Then the following conditions are equivalent:*

(a) *The number αe^{-y} is a root of unity.*

(b) *For any sufficiently large positive integer N, there exists a nonzero polynomial $Q_N \in \mathbb{Z}[X_0, X_1]$ with partial degree $\le N^{t_0}$ in X_0, partial degree $\le N^{t_1}$ in X_1 and height $\mathrm{H}(Q_N) \le e^N$ such that*

$$\left| (\mathcal{D}^k Q_N)(my, \alpha^m) \right| \le \exp(-N^u)$$

for any $k, m \in \mathbb{N}$ with $k \le N^{s_0}$ and $m \le N^{s_1}$.

Again, the proof that (a) implies (b) uses Proposition 4.10. For the reverse implication, D. Roy establishes a new interpolation lemma for holomorphic functions of two complex variables.

For $R > 0$ denote by $B(0, R)$ the polydisc

$$B(0, R) = \{(z_1, z_2) \in \mathbb{C}^2 ;\ |z_1| \le R, |z_2| \le R\}.$$

For a continuous function $F\colon B(0, R) \to \mathbb{C}$, put

$$|F|_R = \sup\{|F(z_1, z_2)| ;\ |z_1| = |z_2| = R\}.$$

Proposition 15.39. *Let $\{\underline{u}, \underline{w}\}$ be a basis of $\mathbb{C}^2$, let $\underline{v} \in \mathbb{C}^2$ and let a be the complex number for which $\underline{v} - a\underline{u} \in \mathbb{C}\underline{w}$. Then there exists a constant $c \ge 1$, which depends only on $\underline{u}$, $\underline{v}$ and $\underline{w}$ and satisfies the following property. For any integer $N \ge 1$ such that*

$$\min\Big\{|m + na| ;\ m, n \in \mathbb{Z},\ 0 < \max\{|m|, |n|\} < N\Big\} \ge 2^{-N},$$

for any pair (r, R) of real numbers with $R \ge 2r$ and $r \ge cN$, and for any continuous function $F\colon B(0, R) \to \mathbb{C}$ which is holomorphic inside $B(0, R)$, we have

$$|F|_r \le \left(\frac{cr}{N}\right)^{N^2} \max_{\substack{0 \le k < N^2 \\ 0 \le m,n < N}} \left\{\frac{1}{k!}\left|D_{\underline{w}}^k F(m\underline{u} + n\underline{v})\right| N^k\right\} + \left(\frac{cr}{R}\right)^{N^2} |F|_R .$$

This statement includes a Schwarz' Lemma for functions of two variables. This work is a first step in a very promising new direction.

Exercises

Exercise 15.1.
a) Deduce from Proposition 15.2 the following result:

> *Let $\underline{\theta} = (\theta_1, \ldots, \theta_m)$ be a m-tuple of complex numbers. There exists a positive constant $c = c(\underline{\theta})$ depending only on the m-tuple $\underline{\theta}$ with the following property. If there exist a polynomial $f \in \mathbb{Z}[X_1, \ldots, X_m]$ and an integer $T > 0$ such that $\deg f \le T$, $\mathrm{H}(f) \le e^T$ and*
>
> $$0 < |f(\underline{\theta})| \le e^{-cT},$$
>
> *then $\theta_1, \ldots, \theta_m$ are not all algebraic.*

Hint. *Compare with Exercise 2.2.*

b) Is this constant $c(\underline{\theta})$ *effectively computable* in terms of $\underline{\theta}$?

Hint. *The answer to question b) depends on your definition of effectively computable.*

Exercise 15.2.
a) Let ϑ be a real number, a/b and p/q two rational numbers, and κ a real number in the interval $0 < \kappa < 1$. Assume $a/b \neq p/q$ and

$$\left|\vartheta - \frac{a}{b}\right| \leq \frac{\kappa}{bq}.$$

Check

$$\left|\vartheta - \frac{p}{q}\right| \geq \frac{1-\kappa}{bq}.$$

b) Let ϑ be a real number, k a positive integer, c_1, c_2, τ_1, τ_2 positive real numbers. Define

$$\kappa = \frac{k(\tau_1 + 1)}{\tau_2}.$$

Assume that for each sufficiently large real number H, there exists a polynomial $P \in \mathbb{Z}[X]$ such that

$$\deg P \leq k, \quad \mathrm{H}(P) \leq H$$

and

$$c_1 H^{-\tau_1} \leq |P(\vartheta)| \leq c_2 H^{-\tau_2}.$$

Then there exists $c_0 > 0$ such that, for any $p/q \in \mathbb{Q}$ with $q > 0$, we have

$$\left|\vartheta - \frac{p}{q}\right| \geq \frac{c_0}{q^\kappa}.$$

Hint. *First deduce from Exercise 2.1 that ϑ is irrational. Next assume*

$$\left|\vartheta - \frac{p}{q}\right| < \frac{c_0}{q^\kappa}$$

for some p/q. Define $H = (2c_2 q^k)^{1/\tau_2}$. Taking

$$0 < c_0 \leq \min_{1 \leq q \leq q_0} q^\kappa \left|\vartheta - \frac{p}{q}\right|$$

for some suitable $q_0 \geq 1$ check that there is no loss of generality in assuming that H is sufficiently large, say $H \geq H_0$. Deduce that such a P exists. Check (see for instance Lemma 13.10)

$$\left|P(\vartheta) - P\left(\frac{p}{q}\right)\right| \leq c_3 e^H \left|\vartheta - \frac{p}{q}\right|$$

for some $c_3 > 0$. Deduce that if c_0 is sufficiently small, then $P(p/q)$ is not zero. Finally use the estimate $|P(p/q)| \geq 1/q^k$ and conclude.

Remark. Compare with [FNe 1998], Chap. 2, § 4, Proposition 2.1.

Exercise 15.3.
a) Let θ be a transcendental number, γ an algebraic number of degree D and Mahler's measure $\mathrm{M}(\gamma)$, and $f \in \mathbb{Z}[X]$ a polynomial of degree d and length $\mathrm{L}(f) = L$. Further let K be a positive integer, $K \leq d$. Define

$$\epsilon = \frac{1}{2} L^{-D} \mathrm{M}(\gamma)^{-d}.$$

Assume $f(\gamma) \neq 0$ and

$$\max_{0 \le k < K} \left| \frac{1}{k!} \frac{d^k}{dz^k} f(\theta) \right| \le \frac{1}{2} L\epsilon.$$

Check

$$|\theta - \gamma|^K \ge \frac{\epsilon}{\binom{d}{K}(1 + |\theta|)^{d-K}}.$$

b) Include multiplicities in Proposition 15.3 – which amounts to extend part a) of the present exercise to the simultaneous approximation of several numbers.

Exercise 15.4.
a) Combining Proposition 2.11 with Proposition 15.3, deduce a measure of linear independence for two logarithms.

Hint. *Use the explicit construction of f given either by § 2.5.3 or, if α_1, α_2 and β are real, by § 2.3.2 or § 2.4.2.*

b) Improve the estimate obtained in a) for the special case $m = 2$ by using Exercise 15.3 in place of Proposition 15.3.
c) Do the same, starting with Corollary 6.7, Proposition 10.3, Exercises 10.1.b and 15.3.b, for a measure of linear independence of logarithms of algebraic numbers $\lambda_1, \dots, \lambda_m$.

Exercise 15.5.
a) Let $\underline{\theta} \in \mathbb{C}^m$ be a m-tuple of complex numbers, c_1, c_2, u_1 and u_2 positive real numbers with $1 < u_2 \le u_1$. Assume that there exists a sequence $(P_n)_{n \ge 1}$ satisfying, for any sufficiently large n,

$$\deg P_n \le n, \quad \log \mathrm{H}(P_n) \le n,$$

and

$$e^{-c_1 N^{u_1}} \le |P_n(\underline{\theta})| \le e^{-c_2 N^{u_2}}.$$

Show that there exists a constant $c_3 > 0$ such that the function

$$\psi(D, \mu) = c\mu^{u_1/(u_2-1)}$$

is a measure of simultaneous approximation for $\underline{\theta}$.
b) Assume that the hypotheses of a) are satisfied with the condition $\deg P_n \le n$ replaced by $\deg P_n \le c_0$ where c_0 is a positive constant. Show that the conclusion holds with

$$\psi(D, \mu) = c \max\left\{ D^{u_1/(u_2-1)} ; \; (\mu/D)^{u_1} \right\}.$$

c) Assume that the hypotheses of a) are satisfied with the condition $\log \mathrm{H}(P_n) \le n$ replaced by $\log \mathrm{H}(P_n) \le n^{u_2}$. Show that the conclusion holds with

$$\psi(D, \mu) = c(D)\mu^{u_1/(u_2-1)}$$

where $c(D)$ is a positive function which depends only on D.

Exercise 15.6. Show that the error term $D \log H$ in the conclusion of Lemma 15.5 cannot be omitted.

Hint. *Consider the polynomial $X^D + aX - 1$ where a and D are sufficiently large positive integers; see* [DiMi 1991].

Exercise 15.7. Let θ_1 and θ_2 be two transcendental complex numbers.
a) Check that the following conditions are equivalent.

(i) The numbers θ_1 and θ_2 are algebraically dependent (over $\mathbb{Q}$).
(ii) The two fields $\mathbb{Q}(\theta_1)$ and $\mathbb{Q}(\theta_2)$ have the same algebraic closure.
(iii) The number θ_1 is algebraic over the field $\mathbb{Q}(\theta_2)$.
(iv) The number θ_2 is algebraic over the field $\mathbb{Q}(\theta_1)$.

b) Assume θ_1 and θ_2 are algebraically dependent. Show that there exists a constant c with the following property. Let $\Phi_1(D, H)$ be a transcendence measure for θ_1. Define

$$\Phi_2(D, H) = \Phi_1(cD, H^c) + cD\log(DH).$$

Let $P \in \mathbb{Z}[X_1, X_2]$ be a polynomial of total degree $\leq D$ and usual height $\leq H$ such that $P(\theta_1, \theta_2) \neq 0$. Then

$$|P(\theta_1, \theta_2)| \geq \exp\{-\Phi_2(D, H)\}.$$

Hint. *Let $A \in \mathbb{Z}[X_1, X_2]$ be an irreducible polynomial such that $A(\theta_1, \theta_2) = 0$. Introduce the resultant with respect to X_2 of A and P.*

Deduce that $\Phi_2(D, H)$ is a transcendence measure for θ_2.
c) Extend these results by considering a m-tuple of complex numbers $(\theta_1, \ldots, \theta_m)$ and an integer n in the range $1 \leq n \leq m$ assuming that each of the numbers $\theta_{n+1}, \ldots, \theta_m$ is algebraic over the field $\mathbb{Q}(\theta_1, \ldots, \theta_n)$.

Hint. *Compare with Proposition 15.19.*

Exercise 15.8. Let $\varphi\colon \mathbb{R}_{>0} \longrightarrow \mathbb{R}_{>0}$ be an increasing unbounded function.
a) Construct a complex number θ such that, for infinitely many integers $T > 0$, for any algebraic number γ satisfying $[\mathbb{Q}(\gamma)\colon \mathbb{Q}] \leq T$ and $\mathrm{H}(\gamma) \leq e^T$, the inequality

$$|\theta - \gamma| \geq e^{-T\varphi(T)}$$

holds.

Hint. *Without loss of generality we may assume that φ has an inverse φ^{-1}. For sufficiently large n, define $\psi(n) = 1 + [2\varphi^{-1}(2n)]$. Choose*

$$\theta = \sum_{k\geq 0} 2^{-n_k} \quad \textit{with} \quad n_{k+1} = n_k\psi(n_k).$$

Check the property with $T = n_{k+1}/(2n_k)$.

b) Give an example of a complex number θ satisfying the following property.

For any $h \geq 1$ there exist infinitely many $D > 0$ such that, if γ is an algebraic number of degree $\leq D$ and logarithmic height $\mathrm{h}(\gamma) \leq h$, *then*

$$|\theta - \gamma| \geq e^{-hD\varphi(D)}.$$

c) Give an example of a complex number θ satisfying the following property.

For any $\epsilon > 0$ and any $D \geq 1$ there exist infinitely many integers $h > 0$ such that, if γ is an algebraic number of degree $\leq D$ and logarithmic height $\mathrm{h}(\gamma) \leq h$*, then*

$$|\theta - \gamma| \geq e^{-(1+\epsilon)Dh}.$$

Hint. *Choose*

$$\theta = \sum_{k \geq 0} 2^{-n_k} \quad \textit{with} \quad \frac{n_{k+1}}{n_k} \longrightarrow \infty$$

and check the property for

$$h = \frac{1 + n_{k+1}}{D(1+\epsilon)}$$

and any sufficiently large k.

Exercise 15.9. (Following M. Laurent [Lau 1999]). Consider the Liouville number

$$\vartheta = \sum_{j \geq 1} 2^{-j!}.$$

For any integer $k \geq 1$, define

$$\beta_k = \sum_{j=1}^{k} 2^{-j!} \quad \text{and} \quad \tau_k = k! \log 2.$$

For any positive integer d and any positive real number μ, define

$$\mathcal{A}_{d\mu} = \left\{\alpha \in \overline{\mathbb{Q}}\,;\ [\mathbb{Q}(\alpha) : \mathbb{Q}] \leq d,\ \log \mathrm{M}(\alpha) \leq \mu,\ |\vartheta - \alpha| \leq e^{d\mu/2\,000}\right\}$$

and

$$\mathcal{A}^*_{d\mu} = \left\{\alpha \in \mathcal{A}_{d\mu}\,;\ [\mathbb{Q}(\alpha) : \mathbb{Q}] \geq \frac{d}{4\,000}\right\}.$$

a) Assume

$$k > 14\,000, \quad d > 4\,000, \quad \mu \geq 6\,000\,\tau_k \quad \text{and} \quad \mu + 2\tau_k d \leq \tau_{k+1}.$$

Check that either $\mathcal{A}_{d\mu} = \emptyset$ or else $\mathcal{A}_{d\mu} = \{\beta_k\}$. Check also $\mathcal{A}^*_{d\mu} = \emptyset$.
If, moreover, $\mu \geq 2\,000\,\tau_{k+1}/d$, then $\mathcal{A}_{d\mu} = \emptyset$.
b) Assume

$$k > 14\,000, \quad 2 \leq d \leq k \quad \text{and} \quad 2\tau_k \leq \mu \leq 1\,000\,\tau_k.$$

Check that $\mathcal{A}^*_{d\mu}$ contains a root of the polynomial

$$(X - \beta_{k-1})^d + X - \beta_k.$$

Exercise 15.10. Let ϑ be a real number.
a) Let d be a positive integer. Show that the following conditions are equivalent.
(i) The number ϑ either is transcendental or else is algebraic of degree $> d$ (that is: the field $\mathbb{Q}(\vartheta)$ is not an algebraic number field of degree $\leq d$).
(ii) For any $c > 0$ there exists an integer $H > 0$ and a polynomial $f \in \mathbb{Z}[X]$ of degree $\leq d$ and height $\leq H$ such that

$$0 < |f(\vartheta)| \leq \frac{c}{H^{d-1}}.$$

(iii) For any positive integer H there exists a polynomial $f \in \mathbb{Z}[X]$ of degree $\leq d$ and height $\leq H$ such that

$$0 < |f(\vartheta)| \leq \frac{1 + |\vartheta| + \cdots + |\vartheta|^d}{H^d}.$$

b) Assume $|\vartheta| < 1$. Let $H \geq 2$ be a positive integer. Show that the following conditions are equivalent.
(i) The number ϑ either is transcendental or else is algebraic of height $\mathrm{H}(\vartheta) > H$.
(ii) For any $c > 0$ there exists an integer $d > 0$ and a polynomial $f \in \mathbb{Z}[X]$ of degree $\leq d$ and height $\leq H$ such that

$$0 < |f(\vartheta)| \leq \frac{c}{H^{d-1}}.$$

(iii) For any positive integer d there exists a polynomial $f \in \mathbb{Z}[X]$ of degree $\leq d$ and height $\leq H$ such that

$$0 < |f(\vartheta)| \leq \frac{d+1}{H^d}.$$

c) Extend these statements a) and b) to complex numbers, and also to tuples of either real or complex numbers.
d) Produce similar results for algebraic approximation in place of polynomial approximation. In other terms, statements a) and b) above are related to Proposition 15.2; one requires similar statements related to Theorem 15.6.

Exercise 15.11. Show that there exists an absolute constant $c > 0$ with the following property. For any complex number θ and any sufficiently large real number T (depending on θ) there is an algebraic number γ such that the number

$$t(\gamma) = [\mathbb{Q}(\gamma) : \mathbb{Q}] + \log \mathrm{H}(\gamma)$$

satisfies $t(\gamma) \leq T$ and

$$|\theta - \gamma| \leq e^{-ct(\gamma)T}.$$

Hint. *This result is due to A. Durand* [Dur 1990], *p.94–96. See also* [LauRoy 1999a], *Proposition 2.*

Exercise 15.12. Show that the converse of Proposition 15.9 does not hold.

Hint. *Let $\varphi \colon \mathbb{N} \to \mathbb{N}$ be an increasing function such that $\varphi(D)/D \to \infty$ as D tends to infinity. Let $\Phi \colon \mathbb{N} \to \mathbb{N}$ be another positive valued increasing function such that*

$$\Phi(n+1) \geq 2\varphi(2^{n\Phi(n)})$$

for all $n \geq 0$. For $a \in \mathbb{R}$ in the range $0 \leq a < 1/2$, define

$$\xi_a = \sum_{n \geq 1} [n^a] 2^{-\Phi(n)} \quad \text{and} \quad \theta_a = 2^{\xi_a}.$$

Check that the family $\{\theta_a \,;\, 0 \leq a < 1/2\}$ is algebraically independent. Moreover, for any $m \geq 1$ and any tuple $(a_1, \ldots, a_m)$ of real numbers in the range $[0, 1/2)$, show that there exist infinitely many integers $D > 0$ having the following property: there exist algebraic numbers $\gamma_1, \ldots, \gamma_m$ of absolute logarithmic height ≤ 1 such that

$$[\mathbb{Q}(\gamma_1, \ldots, \gamma_m) : \mathbb{Q}] \leq D \quad \text{and} \quad \sum_{i=1}^{m} |\theta_{a_i} - \gamma_i| \leq e^{-\varphi(D)}.$$

Remark. Another solution is suggested in [RoyW 1997b].

Exercise 15.13. Let $\vartheta_1, \ldots, \vartheta_m$ be real numbers, H and D positive integers. Show that there exists a nonzero polynomial $f \in \mathbb{Z}[X_1, \ldots, X_m]$, of total degree $\leq D$ and usual height $\mathrm{H}(f) \leq H$, such that

$$|f(\vartheta_1, \ldots, \vartheta_m)| \leq cH^{-\binom{D+m}{m}+1}, \quad \text{where} \quad c = \sum_{i_1+\cdots+i_m \leq D} |\vartheta_1^{i_1} \cdots \vartheta_m^{i_m}|.$$

Deduce the following refinement of Proposition 15.2 for real tuples:

Let $\vartheta_1, \ldots, \vartheta_m$ be real numbers. The following assertions are equivalent.
(i) The numbers $\vartheta_1, \ldots, \vartheta_m$ are not all algebraic.
(ii) For any $H \geq 1$ and $D \geq 1$ there exists $f \in \mathbb{Z}[X]$ of degree $\leq D$ and height $\mathrm{H}(f) \leq H$ such that

$$0 < |f(\vartheta_1, \ldots, \vartheta_m)| \leq \bigl(1 + |\underline{\vartheta}| + \cdots + |\underline{\vartheta}|^D\bigr)H^{-D}.$$

Exercise 15.14.
a) Let $f \in \mathbb{C}[X]$ be a nonzero polynomial of degree $d \geq 2$ with complex roots $\alpha_1, \ldots, \alpha_d$ and leading coefficient $a_0 > 0$:

$$f(X) = a_0(X - \alpha_1) \cdots (X - \alpha_d).$$

Check that the discriminant $D(f)$ of f

$$D(f) = a_0^{2d-2} \prod_{i=2}^{d} \prod_{j=1}^{i-1} (\alpha_i - \alpha_j)^2$$

satisfies the estimate

$$\sqrt{|D(f)|} \leq d^{d/2} \mathrm{M}(f)^{d-1}.$$

Hint. *(See* [M 1964]*).*
The absolute value of the $d \times d$ determinant

$$\begin{vmatrix} 1 & \alpha_1 & \cdots & \alpha_1^{d-1} \\ \vdots & \vdots & \ddots & \vdots \\ 1 & \alpha_d & \cdots & \alpha_d^{d-1} \end{vmatrix}$$

is

$$\prod_{i=2}^{d} \prod_{j=1}^{i-1} |\alpha_i - \alpha_j|.$$

Estimate this number by means of Hadamard's inequality (cf. the proof of Lemma 3.25).

b) Check, for any $\theta \in \mathbb{C}$ and any integer s in the range $1 \leq s \leq d-1$,

$$\sqrt{|D(f)|} \min_{1 \leq k \leq d} |\theta - \alpha_k|^{s(s+1)/2} \leq |f(\theta)|^s 2^{ds - s(s+1)/2} (d-s)^{(d-s)/2} \mathrm{M}(f)^{d-s-1}.$$

Hint. *(The case $s = 1$ is due to N. I. Fel'dman; see* [F 1982]*, Lemma 1.7, Chap. 7 § 1. The general case is due to G. Diaz* [Di 1997b]*, Lemme 2).*
Assume (without loss of generality)

$$|\theta - \alpha_1| \le |\theta - \alpha_2| \le \cdots \le |\theta - \alpha_d|.$$

Consider the polynomial

$$g(X) = a_0(X - \alpha_{s+1}) \cdots (X - \alpha_d).$$

Since

$$\sqrt{|D(f)|} = a_0^{d-1} \prod_{i=2}^{d} \prod_{j=1}^{i-1} |\alpha_i - \alpha_j|,$$

we can write

$$\sqrt{|D(f)|} = a_0^s \Delta \sqrt{|D(g)|}$$

where

$$\sqrt{|D(g)|} = a_0^{d-s-1} \prod_{i=s+2}^{d} \prod_{j=s+1}^{i-1} |\alpha_i - \alpha_j|.$$

and

$$\Delta = \prod_{j=1}^{s} \prod_{i=j+1}^{d} |\alpha_i - \alpha_j|$$

is a product of $ds - s(s+1)/2$ factors. From the inequalities

$$|\alpha_1 - \alpha_j| \le 2|\theta - \alpha_j| \quad \text{and} \quad |\theta - \alpha_1|^{s-i+1} \le |\theta - \alpha_i|^{s-i+1}$$

for $1 \le j \le d$ and $1 \le i \le s$, deduce

$$a^s \Delta \big(2|\theta - \alpha_1|\big)^{s(s+1)/2} \le 2^{ds} |f(\theta)|^s.$$

c) Let $f \in \mathbb{Z}[X]$ be a nonzero polynomial of degree $\le d$ without multiple root and let $\theta \in \mathbb{C}$ satisfy $|f(\theta)| \le 1$. Denote by γ a root of f at minimal distance of θ. Check, for any integer $s \ge 1$,

$$|\theta - \gamma|^{s(s+1)/2} \le 2^{ds} d^{d/2} \mathrm{M}(f)^d |f(\theta)|^s.$$

Hint. *See G. Diaz* [Di 1997b].

Exercise 15.15. Define

$$\psi(D, \mu) = D^{1/2}(\mu + D \log D)\mu^{1/2}(\log D)^{-1}.$$

a) From Exercise 14.4.b, deduce that if the number e^{π^2} is algebraic, then there is an absolute constant $c > 0$ such that $c\psi(D, \mu)$ is a measure of simultaneous approximation for the two numbers e and π. Therefore, under the assumption that e^{π^2} is algebraic, it follows that e and π are algebraically independent.
b) Assume Theorem 14.6 holds also when the matrix $\big(\log A_{ij}\big)$ has rank 2 instead of 1. Show that $c\psi(D, \mu)$ is a measure of simultaneous approximation for the three numbers π, e and e^{π^2}. Deduce that two at least of these numbers are algebraically independent.

Remark. This result is not yet proved: the best known measure of simultaneous approximation for the three numbers π, e and e^{π^2} is given by Exercise 14.4.b, namely

$$c\mu(\mu + D \log D)(\log D)^{-1}.$$

c) Let λ be a nonzero element of $\mathcal{L}$. Using Exercise 14.4.d, show that if the number e^{λ^2} is algebraic then $c\psi(D,\mu)$ (with a suitable constant c depending only on λ) is a measure of simultaneous approximation for the two numbers λ and e^{λ^3}. Deduce that if the number e^{λ^2} is algebraic, then the two numbers λ and e^{λ^3} are algebraically independent (cf [Br 1974b] and [W 1974]).
d) Assume that in Theorem 14.6 the assumption that the matrix $\bigl(\log A_{ij}\bigr)$ has rank 1 can be omitted. Deduce that for any nonzero $\alpha \in \overline{\mathbb{Q}}$ and any nonzero logarithm $\log\alpha$ of α, the transcendence degree of the field

$$\mathbb{Q}\bigl(\log\alpha, \alpha^{\log\alpha}, \alpha^{(\log\alpha)^2}\bigr)$$

is ≥ 2.
e) Let λ_1, λ_2 be two elements of $\mathcal{L}$ which are linearly independent over $\mathbb{Q}$ and let θ be a complex irrational number which satisfies a linear independence measure condition. Show that $\psi(D,\mu)$ is a measure of simultaneous approximation for the numbers $\lambda_1, \lambda_2, \theta, e^{\theta\lambda_1}, e^{\theta\lambda_2}$. Deduce that two at least of these five numbers are algebraically independent.

Hint. *See Example 14.16.*

f) Using Theorem 15.6, show that ψ cannot be a measure of transcendence for e. Deduce that for any $r \in \mathbb{Q}^\times$, one at least of the two numbers e^r, e^{2r} is transcendental

Hint. *Take in e)*

$$\lambda_1 = e^r, \quad \lambda_2 = e^{2r}, \quad \theta = e^{-r},$$

so that

$$e^{\theta\lambda_1} = e, \quad e^{\theta\lambda_2} = e^{e^r}.$$

g) For $\beta \in \overline{\mathbb{Q}}, \beta \neq 0$ and $\lambda \in \mathcal{L}, \lambda \neq 0$, assuming that the number $\beta^{-1}\lambda^2$ is in $\mathcal{L}$, show that the function ψ is a measure of simultaneous approximation for the two numbers λ and e^β, hence they are algebraically independent

Hint. *Choose in e)*

$$\lambda_1 = \lambda, \quad \lambda_2 = \beta^{-1}\lambda^2, \quad \theta = \frac{\beta}{\lambda}.$$

h) Let $\lambda_1, \lambda_2, \lambda_3, \lambda_4$ be nonzero elements of $\mathcal{L}$ such that $\lambda_1\lambda_4 = \lambda_2\lambda_3$. Assume both numbers λ_1/λ_2 and λ_1/λ_3 are irrational. Deduce from Exercise 14.4.f that two at least of the four numbers $\lambda_1, \lambda_2, \lambda_3, \lambda_4$ are algebraically independent.
i) Compare these results of algebraic independence with [Br 1974b] and [W 1974].

Exercise 15.16. Let $\lambda \in \mathcal{L}\setminus\{0\}$ be a nonzero logarithm of an algebraic number and $b \in \mathbb{C}\setminus\mathbb{Q}$ be a complex irrational number.
a) Deduce from the four exponentials Conjecture that one at least of the two numbers $e^{\lambda b}, e^{\lambda/b}$ is transcendental.
b) Prove the conclusion unconditionally (i.e. without assuming the four exponentials Conjecture) in each of the following cases:

(*i*) $b = \lambda'/\beta$ with $\lambda' \in \mathcal{L}$ and $\beta \in \overline{\mathbb{Q}}$.

(*ii*) b and λ are algebraically dependent.

c) Assume $\lambda \notin \mathbb{R}$. Show that either $e^{|\lambda|}$ is transcendental, or λ and $\overline{\lambda}$ are algebraically independent.

Hint. *See* [Di 1997a].

References

[AEr 1944] Alaoglu, L.; Erdős, P. – On highly composite and similar numbers. Trans. Amer. Math. Soc. **56** (1944), 448–469.

[Am 1988] Amoroso, F. – On the distribution of complex numbers according to their transcendence types. Ann. Mat. Pura Appl., IV. Ser. 151, 359-368 (1988).

[Am 1996] Amoroso, F. – Algebraic numbers close to 1 and variants of Mahler's measure. J. Number Theory **60** (1996), no. 1, 80–96.

[Am 1998] Amoroso, F. – Algebraic numbers close to 1: results and methods. *Number Theory*, Ramanujan Mathematical Society, January 3-6, 1996, Tiruchirapalli, India; Amer. Math. Soc., Contemporary Math. **210** (1998), 305–316.

[AmD 1999] Amoroso, F.; David, S. – Le problème de Lehmer en dimension supérieure. J. reine Angew. Math., **513** (1999), 145–179.

[AmD 2000] Amoroso, F.; David, S. – Minoration de la hauteur normalisée des hypersurfaces. Acta Arith., **92** (2000), no. 4, 339–366.

[Ar 1967] Artin, E. – *Algebraic numbers and algebraic functions.* Gordon and Breach Science Publishers, New York-London-Paris 1967.

[B 1966] Baker, A. – Linear forms in the logarithms of algebraic numbers. I, II, III, IV. Mathematika **13** (1966), 204–216; ibid., **14** (1967), 102–107 ibid., **14** (1967), 220–228 ibid., **15** (1968), 204–216.

[B 1972] Baker, A. – A sharpening of the bounds for linear forms in logarithms. I, II, III. Acta Arith. **21** (1972), 117–129; ibid. **24** (1973), 33–36; ibid. **27** (1975), 247–252.

[B 1975] Baker, A. – *Transcendental number theory.* Cambridge Mathematical Library. Cambridge University Press, Cambridge, 1975. Second edition. 1990.

[B 1977] Baker, A. – The theory of linear forms in logarithms. *Transcendence theory: advances and applications* (Proc. Conf., Univ. Cambridge, Cambridge, 1976), pp. 1–27. Academic Press, London, 1977.

[B 1998] Baker, A. – Logarithmic forms and the *abc*-conjecture. Győry, Kalman (ed.) et al., *Number theory. Diophantine, computational and algebraic aspects.* Proceedings of the international conference, Eger, Hungary, July 29–August 2, 1996. Berlin: de Gruyter. 37-44 (1998).

[BWü 1993] Baker, A.; Wüstholz, G. – Logarithmic forms and group varieties. J. reine Angew. Math. **442** (1993), 19–62.

[BeBGMS 1997] Bennett, C. D.; Blass, J.; Glass, A. M. W.; Meronk, D. B.; Steiner, R. P. – Linear forms in the logarithms of three positive rational numbers. J. Théor. Nombres Bordeaux **9** (1997), no. 1, 97–136.

[BerDGPS 1992] Bertin, M.-J.; Decomps-Guilloux, A.; Grandet-Hugot, M.; Pathiaux-Delefosse, M.; Schreiber, J.-P. – *Pisot and Salem numbers.* Birkhäuser Verlag, Basel, 1992.

[Bert 1987] Bertrand, D. – Lemmes de zéros et nombres transcendants. Séminaire Bourbaki, Vol. 1985/86. Astérisque No. **145-146** (1987), 3, 21–44.

[Bert 1997] Bertrand, D. – Duality on tori and multiplicative dependence relations. J. Austral. Math. Soc. Ser. A **62** (1997), no. 2, 198–216.

[BertMa 1980] Bertrand, D.; Masser, D.W. – Linear forms in elliptic integrals. Invent. Math. **58** (1980), no. 3, 283–288.

[BertP 1988] Bertrand, D.; Philippon, P. – Sous-groupes algébriques de groupes algébriques commutatifs. Illinois J. Math. **32** (1988), no. 2, 263–280.

[Bi 1977] Bijlsma, A. – On the simultaneous approximation of a, b and a^b. Compositio Math. **35** (1977), no. 1, 99–111.

[BilBu 2000] Bilu, Y.; Bugeaud, Y. – Démonstration du théorème de Baker-Fel'dman via les formes linéaires en deux logarithmes. J. Théor. Nombres Bordeaux, **12** (2000), no. 1, to appear.

[BlMon 1971] Blanksby, P. E.; Montgomery, H. L. – Algebraic integers near the unit circle. Acta Arith. **18** (1971), 355–369.

[BlaGMMS 1990] Blass, J.; Glass, A. M. W.; Manski, D. K.; Meronk, D. B.; Steiner, R. P. – Constants for lower bounds for linear forms in the logarithms of algebraic numbers. I. The general case. Acta Arith. **55** (1990), no. 1, 1–14. II. The homogeneous rational case, ibid., **55** (1990), no. 1, 15–22. Corrigendum, ibid., **65** (1993), no. 4, 383. (Corrig.). *Problèmes Diophantiens 1987-1988*, Publ. Univ. P. et M. Curie (Paris VI), **88**, N° 2, 31p.

[Bo 1970] Bombieri, E. – Algebraic values of meromorphic maps. Invent. Math. **10** (1970), 267–287. Addendum, ibid., **11** (1970), 163–166.

[Bo 1993] Bombieri, E. – Effective Diophantine approximation on $\mathbb{G}_{\mathrm{m}}$. Ann. Scuola Norm. Sup. Pisa Cl. Sci. (4) **20** (1993), no. 1, 61–89.

[BoCoh 1997] Bombieri, E.; Cohen, P. B. – Effective Diophantine approximation on $\mathbb{G}_{\mathrm{m}}$ II. Ann. Scuola Norm. Sup. Pisa Cl. Sci. (4) **24** (1997), no. 2, 205–225.

[BoL 1970] Bombieri, E.; Lang, S. – Analytic subgroups of group varieties. Invent. Math. **11** (1970), 1–14.

[BoVa 1983] Bombieri, E.; Vaaler, J. D. – On Siegel's lemma. Invent. Math. **73** (1983), no. 1, 11–32. Addendum, ibid., **75** (1984), no. 2, 377.

[Bor 1899] Borel, É. – Sur la nature arithmétique du nombre e. C. R. Acad. Sci. Paris Sér. A **128** (1899), 596–599.

[Bou 1985] Bourbaki, N. – *Éléments de mathématique – Algèbre commutative.* Chapitres 1 à 7. Reprint. Masson, Paris, 1985. *Commutative algebra.* Chapters 1–7. Translated from the French. Reprint of the 1972 edition. *Elements of Mathematics.* Springer-Verlag, Berlin-New York, 1989.

[Boy 1980] Boyd, D. W. – Reciprocal polynomials having small measure. Math. Comp. **35** (1980), no. 152, 1361–1377; ibid., **53** (1989), no. 187, 355–357, S1–S5.

[Br 1974a] Brownawell, W. D. – Sequences of Diophantine approximations. J. Number Theory **6** (1974), 11–21.

[Br 1974b] Brownawell, W. D. – The algebraic independence of certain numbers related by the exponential function. J. Number Theory **6** (1974), 22–31.

[Br 1979] Brownawell, W. D. – On the development of Gel'fond's method. *Number theory, Carbondale 1979* (Proc. Southern Illinois Conf., Southern Illinois Univ., Carbondale, Ill., 1979), pp. 18–44, Lecture Notes in Math., **751**, Springer, Berlin, 1979.

[BrMa 1980] Brownawell, W. D.; Masser, D. W. – Multiplicity estimates for analytic functions. I, II. J. reine Angew. Math. **314** (1980), 200–216; Duke Math. J. **47** (1980), no. 2, 273–295.

[Bu 1998a] Bugeaud, Y. – Algebraic numbers close to 1 in non-Archimedean metrics. The Ramanujan J. **2**, no. 4 (1998), 449–457.

[Bu 1998b] Bugeaud, Y. – Bornes effectives pour les solutions des équations en S-unités et des équations de Thue-Mahler. J. Number Theory, **71** no 2 (1998), 227–244.

[BuLau 1996] Bugeaud, Y.; Laurent, M. – Minoration effective de la distance p-adique entre puissances de nombres algébriques. J. Number Theory **61** (1996), no. 2, 311–342.

[C 1874] Cantor, G. – Über eine Eigenschaft der Inbegriffes aller reellen algebraischen Zahlen. J. reine angew. Math., **77** (1874), 258–262=*Ges. Abh.*, 1932, 116–118.

[CaStr 1982] Cantor, D. C.; Straus, E. G. – On a conjecture of D. H. Lehmer. Acta Arith. **42** (1982/83), no. 1, 97–100. Correction, ibid., **42** (1983), no. 3, 327.

[Cas 1957] Cassels, J. W. S. – *An Introduction to Diophantine Approximation.* Cambridge Tracts in Mathematics and Mathematical Physics, No. **45**, Cambridge University Press, New York, 1957. Reprint of the 1957 edition: Hafner Publishing Co., New York, 1972.

[Ch 1984] Chudnovsky, G. V. – *Contributions to the theory of transcendental numbers.* Translated from the Russian by G. A. Kandall. Math. Surveys Monographs, **19**, Amer. Math. Soc., Providence, R.I., 1984.

[Co 1997] Corvaja, P. – Autour du théorème de Roth. Monatsh. Math. **124** (1997), no. 2, 147–175.

[D 1995] David, S. – *Minorations de formes linéaires de logarithmes elliptiques.* Mém. Soc. Math. France (N.S.) No. **62** (1995).

[DP 1999] David, S.; Philippon, P. – Minorations des hauteurs normalisées des sous-variétés des tores, Ann. Scuola Norm. Pisa (4) **28** (1999), 489-543.

[Di 1989] Diaz, G. – Grands degrés de transcendance pour des familles d'exponentielles. J. Number Theory **31** (1989), no. 1, 1–23.

[Di 1997a] Diaz, G. – La conjecture des quatre exponentielles et les conjectures de D. Bertrand sur la fonction modulaire. J. Théor. Nombres Bordeaux **9** (1997), no. 1, 229–245.

[Di 1997b] Diaz, G. – Une nouvelle propriété d'approximation diophantienne. C. R. Acad. Sci. Paris Sér. I Math. **324** (1997), no. 9, 969–972.

[DiMi 1991] Diaz, G.; Mignotte, M. – Passage d'une mesure d'approximation à une mesure de transcendance. C. R. Math. Rep. Acad. Sci. Canada **13** (1991), no. 4, 131–134.

[Do 1978] Dobrowolski, E. – On the maximal modulus of conjugates of an algebraic integer. Bull. Acad. Polon. Sci. Sér. Sci. Math. Astronom. Phys. **26** (1978), no. 4, 291–292.

[Do 1979] Dobrowolski, E. – On a question of Lehmer and the number of irreducible factors of a polynomial. Acta Arith. **34** (1979), no. 4, 391–401.

[Dpp 1991] Dong, Pingping – Minoration de combinaisons linéaires de deux logarithmes p-adiques. Ann. Fac. Sci. Toulouse Math. (5) **12** (1991), no. 2, 195–250.

[Dpp 1992] Dong, Pingping – Minorations de combinaisons linéaires de logarithmes p-adiques de nombres algébriques. C. R. Acad. Sci. Paris Sér. I Math. **315** (1992), no. 5, 503–506.

[Dpp 1995] Dong, Pingping – Minorations de combinaisons linéaires de logarithmes p-adiques de nombres algébriques. Dissertationes Math. (Rozprawy Mat.) **343** (1995), 97 pp.

[Du 1993] Dubickas, A. – On a conjecture of A. Schinzel and H. Zassenhaus. Acta Arith. **63** (1993), no. 1, 15–20.

[Du 1995] Dubickas, A. – On algebraic numbers of small measure. Liet. Mat. Rink. **35** (1995), no. 4, 421–431; Engl. transl., Lithuanian Math. J. **35** (1995), no. 4, 333–342 (1996)

[Dur 1990] Durand, A. – Quelques aspects de la théorie analytique des polynômes. I, II. *Cinquante ans de polynômes* (Paris, 1988), 1–42, 43–85, Lecture Notes in Math., **1415**, Springer, Berlin-New York, 1990.

[Duv 1998] Duverney, D. – *Théorie des Nombres — Cours et exercices corrigés*. Dunod, Paris, 1998.

[E 1987] Emsalem, M. – Sur les idéaux dont l'image par l'application d'Artin dans une $\mathbf{Z}_p$-extension est triviale. J. reine Angew. Math. **382** (1987), 181–198.

[Eu 1748] Euler, L. – *Introduction to analysis of the infinite.* Book I. Translated from the Latin (*Introductio in Analysin Infinitorum*). Springer-Verlag, New York-Berlin, 1988.

[Ev 1998] Everest, G. H. – Measuring the height of a polynomial. Math. Intell. **20** (1998), n°3, 9–16

[F 1960a] Fel'dman, N. I. – The measure of transcendency of the number π. (Russian) Izv. Akad. Nauk SSSR. Ser. Mat. **24** (1960) 357–368. Engl. transl.: Amer. Math. Soc. Transl. (2) **58** (1966), 110–124.

[F 1960b] Fel'dman, N. I. – Approximation by algebraic numbers to logarithms of algebraic numbers. (Russian) Izv. Akad. Nauk SSSR. Ser. Mat. **24** (1960) 475–492. Engl. transl.: Amer. Math. Soc. Transl. (2) **58** (1966), 125–142.

[F 1968] Fel'dman, N. I. – Improved estimate for a linear form of the logarithms of algebraic numbers. (Russian) Mat. Sb. (N.S.) **77** (119) (1968), 423–436. Engl. transl. in Math. USSR Sb., **6** (1968), no. 3, 393–406.

[F 1971] Fel'dman, N. I. – An effective refinement of the exponent in Liouville's theorem. Math. USSR, Izv. **5** (1971), 985-1002 (1972).

[F 1982] Fel'dman, N. I. – *Hilbert's seventh problem.* Moskov. Gos. Univ., Moscow, 1982.

[FNe 1998] Fel'dman, N. I.; Nesterenko, Y. V. – *Number theory. IV. Transcendental Numbers.* Encyclopaedia of Mathematical Sciences, **44**. Springer-Verlag, Berlin, 1998.

[FSh 1967] Fel'dman, N. I.; Šidlovskiĭ, A. B. – The development and present state of the theory of transcendental numbers. (Russian) Uspehi Mat. Nauk **22** (1967) no. 3 (135) 3–81; Engl. transl. in Russian Math. Surveys, **22** (1967), no. 3, 1–79.

[FrTa 1991] Fröhlich, A.; Taylor, M. J. – *Algebraic number theory.* Cambridge Studies in Advanced Mathematics, **27**. Cambridge University Press, Cambridge, 1993.

[G 1934] Gel'fond, A. O. – On Hilbert's seventh problem. Dokl. Akad. Nauk SSSR, **2** (1934), 1–3 (in Russian) and 4–6 (in French). Sur le septième problème de Hilbert. Izv. Akad. Nauk SSSR, **7** (1934), 623–630.

[G 1952] Gel'fond, A. O. – *Transcendental and algebraic numbers.* Gosudarstv. Izdat. Tehn.-Teor. Lit., Moscow, 1952. *Transcendental and algebraic numbers.* Translated from the first Russian edition by Leo F. Boron Dover Publications, Inc., New York 1960.

[GLin 1962] Gel'fond, A. O.; Linnik, Y. V. – *Elementary methods in the analytic theory of numbers.* (Russian) Gosudarstv. Izdat. Fiz.-Mat. Lit., Moscow 1962. *Méthodes élémentaires dans la théorie analytique des nombres.* (Français) Traduit par Myriam et Jean-Luc Verley. Monographies Internationales de Mathématiques Modernes, **6** Gauthier-Villars Éditeur, Paris 1965. *Elementary methods in the analytic theory of numbers.* Translated from the Russian by D. E. Brown. Translation edited by I. N. Sneddon. International Series of Monographs in Pure and Applied Mathematics, Vol. **92** Pergamon Press, Oxford-New York-Toronto, Ont. 1966.

[Gr 1969] Gross, F. – Entire functions of several variables with algebraic derivatives at certain algebraic points. Pacific J. Math. **31** (1969), 693–701.

[HaWr 1938] Hardy, G. H.; Wright, E. M. – *An Introduction to the Theory of Numbers.* Oxford Science Publications, Oxford University Press. First ed. 1938. Fifth Ed. 1979.

[Har 1977] Hartshorne, R. – *Algebraic Geometry.* Graduate Texts in Mathematics, No. **52**. Springer-Verlag, New York-Heidelberg, 1977.

[He 1873] Hermite, C. – Sur la fonction exponentielle; C. R. Acad. Sci. Paris, **77** (1873), 18–24; 74–79; 226–233; 285–293; *Oeuvres*, Gauthier Villars (1905), III, 150–181. See also *Oeuvres* III, 127–130, 146–149, and *Correspondance Hermite-Stieltjes*, II, lettre 363, 291–295.

[Hi 1900] Hilbert, D. – *Mathematische Probleme.* Nachr. Ges. Wiss. Göttingen, (1900), 253–297; Archiv der Math. and Phys., **1** (1901), 44–63 and 213–237; Engl. transl. in Bull. Amer. Math. Soc., **8** (1902), 437–479; *Ges. Werke* III, 290–329.

[Hir 1991] Hirata-Kohno, N. – Formes linéaires de logarithmes de points algébriques sur les groupes algébriques; Invent. Math., **104** (1991), 401–433.

[Hö 1973] Hörmander, L. – *An Introduction to Complex Analysis in Several Variables.* North Holland Mathematical Library, American Elsevier, 1973.

[K 1980] Koblitz, N. – *p-adic analysis: a short course on recent work.* London Mathematical Society Lecture Note Series, **46**. Cambridge University Press, Cambridge-New York, 1980.

[KoPop 1932] Koksma, J. F.; Popken, J. – Zur Transzendenz von e^{π}. J. reine angew. Math., **168** (1932), 211–230.

[Kr 1857] Kronecker, L. – Zwei Sätze über Gleichungen mit ganzzahligen Coefficienten. J. reine angew. Math., **53** (1857), 173–175.

[L 1965a] Lang, S. – Nombres transcendants. *Sém. Bourbaki 18ème année* (1965/66), N° 305, 8 pp.; Réédition par la Soc. Math. Fr., 1997.

[L 1965b] Lang, S. – Algebraic values of meromorphic functions. I, II. Topology **3** (1965), 183–191; ibid., **5** (1966), 363–370.

[L 1966] Lang, S. – *Introduction to transcendental numbers.* Addison-Wesley Publishing Co., Reading, Mass.-London-Don Mills, Ont. 1966.

[L 1970] Lang, S. – *Algebraic number theory.* Addison-Wesley Publishing Co., Inc., Reading, Mass.-London-Don Mills, Ont. 1970. Second edition. Graduate Texts in Mathematics, **110**. Springer-Verlag, New York, 1994.

[L 1978] Lang, S. – *Elliptic curves: Diophantine analysis.* Grundlehren der Mathematischen Wissenschaften, **231**. Springer-Verlag, Berlin-New York, 1978.

[L 1983] Lang, S. – *Fundamentals of Diophantine geometry.* Springer-Verlag, New York-Berlin, 1983.

[L 1991] Lang, S. – *Number theory. III. Diophantine geometry.* Encyclopaedia of Mathematical Sciences, **60**. Springer-Verlag, Berlin, 1991. Corrected second printing: *Survey of Diophantine Geometry;* 1997.

[L 1993] Lang, S. – *Algebra.* Third edition. Addison-Wesley Publishing Co., Reading, Mass., 1993.

[La 1986] Langevin, M. – Minorations de la maison et de la mesure de Mahler de certains entiers algébriques. C. R. Acad. Sci. Paris Sér. I Math. **303** (1986), no. 12, 523–526.

[Lau 1989] Laurent, M. – Sur quelques résultats récents de transcendance. *Journées Arithmétiques, 1989* (Luminy, 1989). Astérisque No. **198–200** (1991), 209–230.

[Lau 1992] Laurent, M. – Hauteur de matrices d'interpolation. *Approximations diophantiennes et nombres transcendants* (Luminy, 1990), 215–238, de Gruyter, Berlin, 1992.

[Lau 1994] Laurent, M. – Linear forms in two logarithms and interpolation determinants. Acta Arith. **66** (1994), no. 2, 181–199. See also Appendix of [W 1979b].

[Lau 1998] Laurent, M. – New methods in algebraic independence. Győry, Kalman (ed.) et al., *Number theory. Diophantine, computational and algebraic aspects.* Proceedings of the international conference, Eger, Hungary, July 29–August 2, 1996. Berlin: de Gruyter. 311–330 (1998).

[Lau 1999] Laurent, M. – Some remarks on the approximation of complex numbers by algebraic numbers. Bulletin Greek Math. Soc. **42** (1999), 49–57.

[LauRoy 1999a] Laurent, M.; Roy, D. – Criteria of algebraic independence with multiplicities and interpolation determinants. Trans. Amer. Math. Soc. **351** (1999), no. 5, 1845–1870.

[LauRoy 1999b] Laurent, M.; Roy, D. – Sur l'approximation algébrique en degré de transcendance un. Annales Instit. Fourier, **49** (1999), 27–55.

[LauRoy 2000] Laurent, M.; Roy, D. – Criteria of algebraic independence with multiplicities and approximation by hypersurfaces; J. reine angew. Math., to appear.

[LauMN 1995] Laurent, M.; Mignotte, M.; Nesterenko, Y. V. – Formes linéaires en deux logarithmes et déterminants d'interpolation. J. Number Theory **55** (1995), no. 2, 285–321.

[Le 1933] Lehmer, D. H. – Factorization of certain cyclotomic functions. Ann. of Math., **34** (1933), 461–479.

[LelGru 1986] Lelong, P.; Gruman, L. – *Entire functions of several complex variables.* Grundlehren der mathematischen Wissenschaften, **282**. Springer-Verlag, Berlin-New York, 1986.

[Li 1882a] Lindemann, F. – Über die Zahl π. Math. Ann., **20** (1882), 213–225.

[Li 1882b] Lindemann, F. – Über die Ludolph'sche Zahl. S.B. Preuss. Akad. Wiss., (1882), 679–682.

[Li 1882c] Lindemann, F. – Sur le rapport de la circonférence au diamètre, et sur les logarithmes népériens des nombres commensurables ou des irrationnelles algébriques. C. R. Acad. Sci. Paris, **95** (1882), 72–74.

[Lio 1844a] Liouville, J. – Sur des classes très étendues de quantités dont la valeur n'est ni algébrique, ni même réductible à des irrationnelles algébriques. C. R. Acad. Sci. Paris, **18** (1844), 883–885.

[Lio 1844b] Liouville, J. – Nouvelle démonstration d'un théorème sur les irrationnelles algébriques inséré dans le compte-rendu de la dernière séance. C. R. Acad. Sci. Paris, **18** (1844), 910–911.

[Lio 1851] Liouville, J. – Sur des classes très étendues de quantités dont la valeur n'est ni algébrique, ni même réductible à des irrationnelles algébriques. J. Math. Pures et Appl., **1** (1851), 133–142.

[Lo 1983] Louboutin, R. – Sur la mesure de Mahler d'un nombre algébrique. C. R. Acad. Sci. Paris Sér. I Math. **296** (1983), no. 16, 707–708.

[Lox 1986] Loxton, J. H. – Some problems involving powers of integers. Acta Arith. **46** (1986), no. 2, 113–123.

[LoxV 1976] Van der Poorten, A. J.; Loxton, J. H. – Computing the effectively computable bound in Baker's inequality for linear forms in logarithms. Bull. Austral. Math. Soc. **15** (1976), no. 1, 33–57. Corrigenda and addenda; ibid., **17** (1977), no. 1, 151–155.

[LoxVMW 1987] Loxton, J. H.; Mignotte, M.; Van der Poorten, A. J.; Waldschmidt, M.— A lower bound for linear forms in the logarithms of algebraic numbers. C. R. Math. Rep. Acad. Sci. Canada **9** (1987), no. 2, 119–124.

[M 1962] Mahler, K. – On some inequalities for polynomials in several variables. J. London Math. Soc. **37** (1962) 341–344.

[M 1964] Mahler, K. – An inequality for the discriminant of a polynomial. Michigan Math. J. **11** (1964) 257–262.

[M 1967] Mahler, K. – Applications of some formulae by Hermite to the approximation of exponentials and logarithms. Math. Ann. **168** (1967) 200–227.

[M 1976] Mahler, K. – *Lectures on transcendental numbers.* Lecture Notes in Mathematics, Vol. **546**. Springer-Verlag, Berlin-New York, 1976.

[Ma 1981a] Masser, D. W. – A note on Baker's theorem. *Recent progress in analytic number theory,* Vol. **2** (Durham, 1979), pp. 153–158, Academic Press, London-New York, 1981.

[Ma 1981b] Masser, D. W. – On polynomials and exponential polynomials in several complex variables. Invent. Math. **63** (1981), no. 1, 81–95.

[Ma 1988] Masser, D. W. – Linear relations on algebraic groups. *New advances in transcendence theory* (Durham, 1986), 248–262, Cambridge Univ. Press, Cambridge-New York, 1988.

[MaWü 1981] Masser, D. W.; Wüstholz, G. – Zero estimates on group varieties. I, II. Invent. Math. **64** (1981), no. 3, 489–516. Ibid., **80** (1985), no. 2, 233–267.

[MaWü 1990] Masser, D. W.; Wüstholz, G. – Estimating isogenies on elliptic curves. Invent. Math. **100** (1990), no. 1, 1–24.

[Mat 1991] Matveev, E. M. – Size of algebraic integers. (Russian) Mat. Zametki **49** (1991), no. 4, 152–154; Engl. transl. in Math. Notes **49** (1991), no. 3-4, 437–438.

[Mat 1993a] Matveev, E. M. – Arithmetic properties of the values of generalized binomials. (Russian) Mat. Zametki **54** (1993), no. 4, 76–81, 159; Engl. transl. in Math. Notes **54** (1993), no. 3-4, 1031–1034 (1994).

[Mat 1993b] Matveev, E. M. – On linear and multiplicative relations. (Russian) Mat. Sb. **184** (1993), no. 4, 23–40; Engl. transl. in Russian Acad. Sci. Sb. Math., **78** (1994), no. 2, 411–425.

[Mat 1996a] Matveev, E. M. – A relationship between the Mahler measure and the discriminant of algebraic numbers. Mat. Zametki **59** (1996), no. 3, 415–420, 480; Engl. transl. in Math. Notes, **59** (1996), no. 3, 293–297.

[Mat 1996b] Matveev, E. M. – On algebraic numbers of small logarithmic height. (Russian) *Diophantine approximations, Proceedings of papers dedicated to the memory of Prof. N. I. Fel'dman*, ed. Y. V. Nesterenko, Centre for applied research under Mech.-Math. Faculty of MSU, Moscow (1996), 90–98.

[Mat 1998] Matveev, E. M. – Explicit lower estimates for rational homogeneous linear forms in logarithms of algebraic numbers. Izv. Akad. Nauk SSSR. Ser. Mat. **62** No 4, (1998) 81–136. Engl. transl.: Izvestiya Mathematics **62** No 4, (1998) 723–772.

[Me 1988] Meyer, M. – Le problème de Lehmer: méthode de Dobrowolski et lemme de Siegel à la Bombieri-Vaaler. Publ. Math. Univ. P. et M. Curie (Paris VI), **90**, *Problèmes Diophantiens* 1988/89, N° 5, 15 pp.

[Mi 1979] Mignotte, M. – Approximation des nombres algébriques par des nombres algébriques de grand degré. Ann. Fac. Sci. Toulouse Math. (5) **1** (1979), no. 2, 165–170.

[MiW 1977] Mignotte, M.; Waldschmidt, M. – Approximation simultanée de valeurs de la fonction exponentielle ; Compositio Math., **34** (1977), 127–139.

[MiW 1978] Mignotte, M.; Waldschmidt, M. – Linear forms in two logarithms and Schneider's method. I, II, III. Math. Ann. **231** (1977/78), no. 3, 241–267; Acta Arith. **53** (1989), no. 3, 251–287; Ann. Fac. Sci. Toulouse Math. (5) **97** 1989, suppl., 43–75.

[MiW 1994] Mignotte, M.; Waldschmidt, M. – On algebraic numbers of small height: linear forms in one logarithm. J. Number Theory **47** (1994), no. 1, 43–62.

[Mo 1983] Moreau, J.-C. – Démonstrations géométriques de lemmes de zéros. I, II. *Sém. Th. Nombres,* (Paris, 1981/1982), 201–205, Progr. Math., **38**, Birkhäuser Boston, Boston, MA, 1983; *Diophantine approximations and transcendental numbers* (Luminy, 1982), 191–197, Progr. Math., **31**, Birkhäuser Boston, Boston, MA, 1983.

[Mos 1995] Mossinghoff, M. J. – Algorithms for the determination of polynomials with small Mahler measure. Diss. Ph. D. Thesis, Univ. of Texas at Austin, August 1995, 105 pp.

[MuTi 1996] Muller, J.-M.; Tisserand, A. – Towards exact rounding of the elementary functions. Alefeld, Goetz (ed.) et al., *Scientific computing and validated numerics.* Proceedings of the international symposium on scientific computing, computer arithmetic and validated numerics SCAN-95, Wuppertal, Germany, September 26-29, 1995. Berlin: Akademie Verlag. Math. Res. 90, 59-71 (1996).

[N 1995] Nakamaye, M. – Multiplicity estimates and the product theorem. Bull. Soc. Math. France **123** (1995), no. 2, 155–188.

[Ne 1996] Nesterenko, Y. V. – Modular functions and transcendence questions. Mat. Sb. **187** N° 9 (1996), 65–96. Engl. Transl., Sbornik Math., **187** N° 9-10 (1996), 1319–1348.

[Ne 1997] Nesterenko, Y. V. – On the measure of algebraic independence of the values of Ramanujan functions. Trudy Mat. Inst. Steklov. **218** (1997), 299–334; English translation in Proceedings of the Steklov Institute of Mathematics, **218** (1997), 294–331.

[NeP 2000] Nesterenko, Y. V.; Philippon, P., (Eds)– *Introduction to algebraic independence theory.* Instructional Conference (CIRM Luminy 1997). Lecture Notes in Mathematics, Springer-Verlag, Berlin-Heidelberg (2000), to appear.

[NeW 1996] Nesterenko, Y. V.; Waldschmidt, M. – On the approximation of the values of exponential function and logarithm by algebraic numbers. (Russian) *Diophantine approximations, Proceedings of papers dedicated to the memory of Prof. N. I. Fel'dman*, ed. Yu. V. Nesterenko, Centre for applied research under Mech.-Math. Faculty of MSU, Moscow (1996), 23–42. http://fr.arXiv.org/abs/math/0002047

[Neu 1999] Neukirch, J. – *Algebraic Number Theory.* Grundlehren der Mathematischen Wissenschaften **322**. Springer-Verlag, Berlin-Heidelberg, 1999.

[Ni 1996] Nishioka, K. – *Mahler functions and transcendence.* Lecture Notes in Mathematics, **1631**. Springer-Verlag, Berlin, 1996.

[P 1986a] Philippon, P. – Lemmes de zéros dans les groupes algébriques commutatifs. Bull. Soc. Math. France **114** (1986), no. 3, 355–383; Errata et addenda, id., **115** (1987), no. 3, 397–398.

[P 1986b] Philippon, P. – Critères pour l'indépendance algébrique. Inst. Hautes Études Sci. Publ. Math. No. **64** (1986), 5–52.

[P 1996] Philippon, P. – Nouveaux lemmes de zéros dans les groupes algébriques commutatifs. *Symposium on Diophantine Problems* (Boulder, CO, 1994). Rocky Mountain J. Math. **26** (1996), no. 3, 1069–1088.

[P 1997] Philippon, P. – Une approche méthodique pour la transcendance et l'indépendance algébrique de valeurs de fonctions analytiques. J. Number Theory **64** (1997), no. 2, 291–338.

[P 1998] Philippon, P. – Indépendance algébrique et K-fonctions. J. reine angew. Math., **497** (1998), 1–15.

[P 1999a] Philippon, P. – Mesures d'approximation de valeurs de fonctions analytiques. Acta Arith., **88** (1999), no. 2, 113–127.

[P 1999b] Philippon, P. – Quelques remarques sur des questions d'approximation diophantienne. Bull. Austral. Math. Soc., **59** (1999), 323–334.

[P 2000] Philippon, P. – Approximations algébriques des points dans les espaces projectifs. J. Number Theory, **81** (2000), no. 2, 234–253.

[PW 1988a] Philippon, P.; Waldschmidt, M. – Lower bounds for linear forms in logarithms. *New advances in transcendence theory* (Durham, 1986), 280–312, Cambridge Univ. Press, Cambridge-New York, 1988.

[PW 1988b] Philippon, P.; Waldschmidt, M. – Formes linéaires de logarithmes simultanées sur les groupes algébriques commutatifs. *Séminaire de Théorie des Nombres,* Paris 1986–87, 313–347, Progr. Math., **75**, Birkhäuser Boston, Boston, MA, 1988.

[PW 1988c] Philippon, P.; Waldschmidt, M. – Formes linéaires de logarithmes sur les groupes algébriques commutatifs. Illinois J. Math. **32** (1988), no. 2, 281–314.

[Pi 1993] Pila, J. – Geometric and arithmetic postulation of the exponential function. J. Austral. Math. Soc. Ser. A **54** (1993), no. 1, 111–127.

[PoSz 1976] Pólya, G.; Szegő, G. – *Problems and theorems in analysis. Vol. II. Theory of functions, zeros, polynomials, determinants, number theory, geometry.* Grundlehren der Mathematischen Wissenschaften, Band **216**. Springer-Verlag, New York-Heidelberg, 1976.

[R 1968] Ramachandra, K. – Contributions to the theory of transcendental numbers. I, II. Acta Arith. **14** (1967/68), 65-72; ibid., **14** (1967/1968), 73–88.

[R 1969a] Ramachandra, K. – *Lectures on transcendental numbers.* The Ramanujan Institute Lecture Notes, **1** The Ramanujan Institute, Madras 1969.

[R 1969b] Ramachandra, K. – A note on Baker's method. J. Austral. Math. Soc. **10** (1969) 197–203.

[Ra 1985] Rausch, U. – On a theorem of Dobrowolski about the product of conjugate numbers. Colloq. Math. **50** (1985), no. 1, 137–142.

[Re 1970] Reid, C. – *Hilbert.* With an appreciation of Hilbert's mathematical work by Hermann Weyl Springer-Verlag, New York-Berlin 1970.

[RemU 1996] Rémond, G.; Urfels, F. – Approximation diophantienne de logarithmes elliptiques p-adiques. J. Number Theory, **57** (1996), no. 1, 133–169.

[Rh 1987] Rhin, G. – Approximants de Padé et mesures effectives d'irrationalité. *Séminaire de Théorie des Nombres, Paris 1985–86,* 155–164, Progr. Math., **71**, Birkhäuser Boston, Boston, MA, 1987.

[Ri 1994] Ribenboim, P. – *Catalan's conjecture. Are* 8 *and* 9 *the only consecutive powers?* Academic Press, Inc., Boston, MA, 1994.

[Ro 1955] Roth, K. F. – Rational approximations to algebraic numbers. Mathematika **2** (1955), 1–20; corrigendum, 168.

[Roy 1989] Roy, D. – Sur la conjecture de Schanuel pour les logarithmes de nombres algébriques. *Groupe d'Études sur les Problèmes Diophantiens* 1988-1989, Publ. Math. Univ. P. et M. Curie (Paris VI), **90**, N° 6, 12 p.

[Roy 1990] Roy, D. – Matrices dont les coefficients sont des formes linéaires. *Séminaire de Théorie des Nombres, Paris 1987–88,* 273–281, Progr. Math., **81**, Birkhäuser Boston, Boston, MA, 1990.

[Roy 1992a] Roy, D. – Transcendance et questions de répartition dans les groupes algébriques. *Approximations diophantiennes et nombres transcendants* (Luminy, 1990), 249–274, de Gruyter, Berlin, 1992.

[Roy 1992b] Roy, D. – Matrices whose coefficients are linear forms in logarithms. J. Number Theory **41** (1992), no. 1, 22–47.

[Roy 1992c] Roy, D. – Simultaneous approximation in number fields. Invent. Math. **109** (1992), no. 3, 547–556.

[Roy 1995] Roy, D. – Points whose coordinates are logarithms of algebraic numbers on algebraic varieties. Acta Math. **175** (1995), no. 1, 49–73.

[Roy 2000a] Roy, D. – Approximation algébrique simultanée de nombres de Liouville. Bull. Canadien Math., to appear.

[Roy 2000b] Roy, D. – Zero estimates on commutative algebraic groups. In [NeP 2000].

[Roy 2000c] Roy, D. – An arithmetic criterion for the values of the exponential function. To appear.

[RoyW 1997a] Roy, D.; Waldschmidt, M. – Approximation diophantienne et indépendance algébrique de logarithmes. Ann. scient. Ec. Norm. Sup., **30** (1997), no 6, 753–796.

[RoyW 1997b] Roy, D.; Waldschmidt, M. – Simultaneous approximation and algebraic independence. The Ramanujan Journal, **1** Fasc. 4 (1997), 379–430.

[S 1967] Schinzel, A. – On two theorems of Gelfond and some of their applications. Acta Arith **13** (1967/1968), 177–236. Corrigendum, ibid., **16** (1969/1970), 101. Addendum, ibid., **56** (1990), no. 2, 181.

[S 1999] Schinzel, A. – The Mahler measure of polynomials. *Number theory and its applications*, C. Y. Yıldırım and Serguei Stepanov Ed., Lecture Notes in Pure and Applied Mathematics **204**, 171–183, Marcel Dekker Inc., 1999.

[SZa 1965] Schinzel, A.; Zassenhaus, H. – A refinement of two theorems of Kronecker. Michigan Math. J. **12** (1965), 81–85.

[Sc 1980] Schmidt, W. M.– *Diophantine approximation.* Lecture Notes in Mathematics, **785**. Springer, Berlin, 1980.

[Sc 1991] Schmidt, W. M. – *Diophantine approximations and Diophantine equations.* Lecture Notes in Mathematics, **1467**. Springer-Verlag, Berlin, 1991.

[Sc 1999] Schmidt, W. M. – Heights of algebraic points. *Number theory and its applications*, C. Y. Yıldırım and Serguei Stepanov Ed., Lecture Notes in Pure and Applied Mathematics **204** 185–225, Marcel Dekker Inc., 1999.

[Sch 1934] Schneider, Th. – Transzendenzuntersuchungen periodischer Funktionen. J. reine angew. Math., **172** (1934), 65–74.

[Sch 1941] Schneider, Th. – Zur Theorie der Abelschen Funktionen und Integrale. J. reine Angew. Math. **183** (1941). 110–128.

[Sch 1949] Schneider, Th. – Ein Satz über ganzwertige Funktionen als Prinzip für Transzendenzbeweise. Math. Ann. **121**, (1949). 131–140.

[Sch 1957] Schneider, Th. – *Einführung in die transzendenten Zahlen.* Springer-Verlag, Berlin-Göttingen-Heidelberg, 1957. *Introduction aux nombres transcendants.* Traduit de l'allemand par P. Eymard. Gauthier-Villars, Paris 1959.

[Ser 1970] Serre, J.-P.– Travaux de Baker. *Sém. Bourbaki 1969/70,* N° 368. Lecture Notes in Mathematics, **180**, 73–86. Springer-Verlag, Berlin, 1971.

[Ser 1989] Serre, J.-P. – *Lectures on the Mordell-Weil theorem.* Translated from the French and edited by Martin Brown from notes by Michel Waldschmidt. Aspects of Mathematics, **E15**. Friedr. Vieweg & Sohn, Braunschweig, 1989. Second Ed., 1990.

[Sert 1999] Sert, A. – Une version effective du théorème de Lindemann-Weierstrass par les déterminants d'interpolation. J. Number Theory **76** (1999), no. 1, 94–119.

[Sh 1989] Šidlovskiĭ, A. B. – *Transcendental numbers.* Translated from the Russian by Neal Koblitz. de Gruyter Studies in Mathematics, **12**. Walter de Gruyter & Co., Berlin-New York, 1989.

[Sho 1974] Shorey, T. N. – Linear forms in the logarithms of algebraic numbers with small coefficients. I, II. J. Indian Math. Soc. (N.S.) **38** (1974), 271–284; ibid. **38** (1974), 285–292.

[Sho 1976] Shorey, T. N. – On linear forms in the logarithms of algebraic numbers. Acta Arith. **30** (1976/77), no. 1, 27–42.

[ShoT 1986] Shorey, T. N.; Tijdeman, R. – *Exponential Diophantine equations.* Cambridge Tracts in Mathematics, **87**. Cambridge University Press, Cambridge-New York, 1986.

[Si 1929] Siegel, C. L. – Über einige Anwendungen diophantischer Approximationen. Abh. Preuss. Akad. Wiss., Phys.-Math., **1** (1929), 1–70. *Gesammelte Abhandlungen.* Springer-Verlag, Berlin-New York 1966 Band **I**, 209–266.

[Si 1949] Siegel, C. L. – *Transcendental Numbers.* Annals of Mathematics Studies, no. **16**. Princeton University Press, Princeton, N. J., 1949.

[Sil 1986] Silverman, J. H. – *The arithmetic of elliptic curves.* Graduate Texts in Mathematics, **106**. Springer-Verlag, New York-Berlin, 1986.

[Sil 1996] Silverman, J. H. – Small Salem numbers, exceptional units, and Lehmer's conjecture. *Symposium on Diophantine Problems* (Boulder, CO, 1994). Rocky Mountain J. Math. **26** (1996), no. 3, 1099–1114.

[Sm 1971] Smyth, C. J. – On the product of the conjugates outside the unit circle of an algebraic integer. Bull. London Math. Soc. **3** (1971), 169–175.

[Sp 1982] Sprindžuk, V. G. – *Classical Diophantine equations.* Translated from the 1982 Russian original (Nauka, Moscow). Translation edited by Ross Talent and Alf van der Poorten. Lecture Notes in Mathematics, **1559**. Springer-Verlag, Berlin, 1993.

[St 1971] Stark, H. M. – A transcendence theorem for class-number problems. Ann. of Math. (2) **94** (1971), 153–173. II. Ann. of Math. (2) **96** (1972), 174–209.

[St 1973] Stark, H. M. – Further advances in the theory of linear forms in logarithms. *Diophantine approximation and its applications* (Proc. Conf., Washington, D.C., 1972), pp. 255–293. Academic Press, New York, 1973.

[Ste 1978] Stewart, C. L. – Algebraic integers whose conjugates lie near the unit circle. Bull. Soc. Math. France **106** (1978), no. 2, 169–176.

[StY 1991] Stewart, C. L.; Yu, Kun Rui – On the *abc* conjecture. Math. Ann. **291** (1991), no. 2, 225–230. II, to appear.

[T 1971] Tijdeman, R. – On the algebraic independence of certain numbers. Nederl. Akad. Wetensch. Proc. Ser. A **74**=Indag. Math. **33** (1971), 146–162.

[T 1976] Tijdeman, R. – On the equation of Catalan. Acta Arith. **29** (1976), no. 2, 197–209.

[V 1971] Van der Poorten, A. J.. – On the arithmetic nature of definite integrals of rational functions. Proc. Amer. Math. Soc. **29** (1971), 451–456.

[V 1977] Van der Poorten, A. J. – Linear forms in logarithms in the p-adic case. *Transcendence theory: advances and applications* (Proc. Conf., Univ. Cambridge, Cambridge, 1976), pp. 29–57. Academic Press, London, 1977.

[Vd 1928] Van der Waerden, B. L. – On Hilbert's function, Series of composition of polynomials and a generalisation of the theorem of Bezout. Proc. Royal Acad. Amsterdam, **31** (1928), 749–770.

[Vou 1996] Voutier, P. – An effective lower bound for the height of algebraic numbers. Acta Arith. **74** (1996), no. 1, 81–95.

[W 1974] Waldschmidt, M. – *Nombres transcendants.* Lecture Notes in Mathematics, Vol. **402**. Springer-Verlag, Berlin-New York, 1974.

[W 1978] Waldschmidt, M. – Transcendence measures for exponentials and logarithms. J. Austral. Math. Soc. Ser. A **25** (1978), no. 4, 445–465.

[W 1979a] Waldschmidt, M. – *Transcendence methods.* Queen's Papers in Pure and Applied Mathematics, **52**. Queen's University, Kingston, Ont., 1979.

[W 1979b] Waldschmidt, M. – *Nombres transcendants et groupes algébriques.* With appendices by Daniel Bertrand and Jean-Pierre Serre. Astérisque, **69–70**. Société Mathématique de France, Paris, 1979. Second edition 1987.

[W 1980] Waldschmidt, M. – A lower bound for linear forms in logarithms. Acta Arith. **37** (1980), 257–283.

[W 1981] Waldschmidt, M. – Transcendance et exponentielles en plusieurs variables. Invent. Math. **63** (1981), no. 1, 97–127.

[W 1983] Waldschmidt, M. – Un lemme de Schwarz pour des intersections d'hyperplans. Studies in pure mathematics, 751–759, Birkhäuser, Basel-Boston, Mass., 1983.

[W 1984] Waldschmidt, M. – Algebraic independence of transcendental numbers. Gel'fond's method and its developments. *Perspectives in mathematics,* 551–571, Birkhäuser, Basel-Boston, Mass., 1984.

[W 1988] Waldschmidt, M. – On the transcendence methods of Gel'fond and Schneider in several variables. *New advances in transcendence theory* (Durham, 1986), 375–398, Cambridge Univ. Press, Cambridge-New York, 1988.

[W 1990] Waldschmidt, M. – Dependence of logarithms of algebraic points. *Number theory,* Vol. II (Budapest, 1987), 1013–1035, Colloq. Math. Soc. Jànos Bolyai, **51**, North-Holland, Amsterdam, 1990.

[W 1991a] Waldschmidt, M. – Fonctions auxiliaires et fonctionnelles analytiques. I, II. J. Analyse Math. **56** (1991), 231–254, 255–279.

[W 1991b] Waldschmidt, M. – Nouvelles méthodes pour minorer des combinaisons linéaires de logarithmes de nombres algébriques. Sém. Théor. Nombres Bordeaux (2) **3** (1991), no. 1, 129–185. (II), Groupe d'études sur les problèmes diophantiens 1989–1990, Publ. Math. Univ. P. et M. Curie, **93** (1991), N°8, 36 pp.

[W 1992] Waldschmidt, M. – *Linear independence of logarithms of algebraic numbers.* The Institute of Mathematical Sciences, Madras, IMSc Report N° **116**, (1992).

[W 1993] Waldschmidt, M. – Minorations de combinaisons linéaires de logarithmes de nombres algébriques. Canad. J. Math. **45** (1993), no. 1, 176–224.

[W 1997a] Waldschmidt, M. – Approximation diophantienne dans les groupes algébriques commutatifs — (I) : Une version effective du théorème du sous-groupe algébrique. J. reine angew. Math., **493** (1997), 61–113.

[W 1997b] Waldschmidt, M. – Extrapolation with interpolation determinants. In *Special Functions and Differential Equations,* Proceedings of a Workshop held at The Institute of Mathematical Sciences, Madras, India, during 13 - 24 January 1997; K. Srinivasa Rao, R. Jagannathan, G. Vanden Berghe and J. Van der Jeugt Eds, Allied Publishers Limited, (1997), 356–366.

[W 1999] Waldschmidt, M. – Algebraic independence of transcendental numbers: a survey. *Number Theory,* Indian National Science Academy, R.P. Bambah, V.C. Dumir and R.J. Hans Gill Eds, Hindustan Book Agency, New-Delhi and Birkhäuser (1999), 497–527.

[W 2000] Waldschmidt, M. – Conjectures for Large Transcendence Degree. *Algebraic Number Theory and Diophantine Analysis,* F. Halter-Koch and R. F. Tichy Eds, W. de Gruyter, Berlin, (2000).

[We 1885] Weierstraß, K. – Zu Lindemann's Abhandlung: "Über die Ludolph'sche Zahl". S.B. Preuss. Akad. Wiss., (1885), 1067–1085; = *Math. Werke* II, 341–362.

[Wi 1999] Wielonsky, F. – Hermite-Padé approximants to exponential functions and an inequality of Mahler. J. Number Theory **74** (1999), no. 2, 230–249.

[Wir 1961] Wirsing, E. – Approximation mit algebraischen Zahlen beschränkten Grades. J. Reine Angew. Math. **206** (1961), 67–77.

[Wü 1988] Wüstholz, G. – A new approach to Baker's theorem on linear forms in logarithms. *New advances in transcendence theory* (Durham, 1986), 399–410, Cambridge Univ. Press, Cambridge-New York, 1988.

[Wü 1989] Wüstholz, G. – Multiplicity estimates on group varieties. Ann. of Math. (2) **129** (1989), no. 3, 471–500.

[Y 1985] Yu, Kun Rui – Linear forms in elliptic logarithms. J. Number Theory **20** (1985), no. 1, 1–69.

[Y 1989] Yu, Kun Rui – Linear forms in p-adic logarithms. I, II, III. Acta Arith. **53** (1989), no. 2, 107–186. Compositio Math. **74** (1990), no. 1, 15–113. Erratum, ibid., **76** (1990), no. 1-2, 307. Compositio Math. **91** (1994), no. 3, 241–276.

[Y 1998] Yu, Kun Rui – p-adic logarithmic forms and Group Varieties I. J. reine Angew. Math. **502** (1998), 29–92. II. Acta Arith., **89** (1999), no. 4, 337–378.

[ZSa 1958] Zariski, O.; Samuel, P. – *Commutative Algebra*, vol. I, Graduate Texts in Mathematics, No. **28**. Springer-Verlag New-York-Heidelberg-Berlin, 1958. vol. 2: Graduate Texts in Mathematics, No. **29**, 1975.

Subject Index

Grundlehren der mathematischen Wissenschaften

A Series of Comprehensive Studies in Mathematics

A Selection

221. Mumford: Algebraic Geometry I: Complex Projective Varieties
222. Lang: Introduction to Modular Forms
223. Bergh/Löfström: Interpolation Spaces. An Introduction
224. Gilbarg/Trudinger: Elliptic Partial Differential Equations of Second Order
225. Schütte: Proof Theory
226. Karoubi: K-Theory. An Introduction
227. Grauert/Remmert: Theorie der Steinschen Räume
228. Segal/Kunze: Integrals and Operators
229. Hasse: Number Theory
230. Klingenberg: Lectures on Closed Geodesics
231. Lang: Elliptic Curves. Diophantine Analysis
232. Gihman/Skorohod: The Theory of Stochastic Processes III
233. Stroock/Varadhan: Multidimensional Diffusion Processes
234. Aigner: Combinatorial Theory
235. Dynkin/Yushkevich: Controlled Markov Processes
236. Grauert/Remmert: Theory of Stein Spaces
237. Köthe: Topological Vector Spaces II
238. Graham/McGehee: Essays in Commutative Harmonic Analysis
239. Elliott: Probabilistic Number Theory I
240. Elliott: Probabilistic Number Theory II
241. Rudin: Function Theory in the Unit Ball of C^n
242. Huppert/Blackburn: Finite Groups II
243. Huppert/Blackburn: Finite Groups III
244. Kubert/Lang: Modular Units
245. Cornfeld/Fomin/Sinai: Ergodic Theory
246. Naimark/Stern: Theory of Group Representations
247. Suzuki: Group Theory I
248. Suzuki: Group Theory II
249. Chung: Lectures from Markov Processes to Brownian Motion
250. Arnold: Geometrical Methods in the Theory of Ordinary Differential Equations
251. Chow/Hale: Methods of Bifurcation Theory
252. Aubin: Nonlinear Analysis on Manifolds. Monge-Ampère Equations
253. Dwork: Lectures on ρ-adic Differential Equations
254. Freitag: Siegelsche Modulfunktionen
255. Lang: Complex Multiplication
256. Hörmander: The Analysis of Linear Partial Differential Operators I
257. Hörmander: The Analysis of Linear Partial Differential Operators II
258. Smoller: Shock Waves and Reaction-Diffusion Equations
259. Duren: Univalent Functions
260. Freidlin/Wentzell: Random Perturbations of Dynamical Systems
261. Bosch/Güntzer/Remmert: Non Archimedian Analysis – A System Approach to Rigid Analytic Geometry
262. Doob: Classical Potential Theory and Its Probabilistic Counterpart
263. Krasnosel'skiĭ/Zabreĭko: Geometrical Methods of Nonlinear Analysis
264. Aubin/Cellina: Differential Inclusions
265. Grauert/Remmert: Coherent Analytic Sheaves
266. de Rham: Differentiable Manifolds
267. Arbarello/Cornalba/Griffiths/Harris: Geometry of Algebraic Curves, Vol. I
268. Arbarello/Cornalba/Griffiths/Harris: Geometry of Algebraic Curves, Vol. II
269. Schapira: Microdifferential Systems in the Complex Domain
270. Scharlau: Quadratic and Hermitian Forms

271. Ellis: Entropy, Large Deviations, and Statistical Mechanics
272. Elliott: Arithmetic Functions and Integer Products
273. Nikol'skiĭ: Treatise on the Shift Operator
274. Hörmander: The Analysis of Linear Partial Differential Operators III
275. Hörmander: The Analysis of Linear Partial Differential Operators IV
276. Liggett: Interacting Particle Systems
277. Fulton/Lang: Riemann-Roch Algebra
278. Barr/Wells: Toposes, Triples and Theories
279. Bishop/Bridges: Constructive Analysis
280. Neukirch: Class Field Theory
281. Chandrasekharan: Elliptic Functions
282. Lelong/Gruman: Entire Functions of Several Complex Variables
283. Kodaira: Complex Manifolds and Deformation of Complex Structures
284. Finn: Equilibrium Capillary Surfaces
285. Burago/Zalgaller: Geometric Inequalities
286. Andrianaov: Quadratic Forms and Hecke Operators
287. Maskit: Kleinian Groups
288. Jacod/Shiryaev: Limit Theorems for Stochastic Processes
289. Manin: Gauge Field Theory and Complex Geometry
290. Conway/Sloane: Sphere Packings, Lattices and Groups
291. Hahn/O'Meara: The Classical Groups and K-Theory
292. Kashiwara/Schapira: Sheaves on Manifolds
293. Revuz/Yor: Continuous Martingales and Brownian Motion
294. Knus: Quadratic and Hermitian Forms over Rings
295. Dierkes/Hildebrandt/Küster/Wohlrab: Minimal Surfaces I
296. Dierkes/Hildebrandt/Küster/Wohlrab: Minimal Surfaces II
297. Pastur/Figotin: Spectra of Random and Almost-Periodic Operators
298. Berline/Getzler/Vergne: Heat Kernels and Dirac Operators
299. Pommerenke: Boundary Behaviour of Conformal Maps
300. Orlik/Terao: Arrangements of Hyperplanes
301. Loday: Cyclic Homology
302. Lange/Birkenhake: Complex Abelian Varieties
303. DeVore/Lorentz: Constructive Approximation
304. Lorentz/v. Golitschek/Makovoz: Construcitve Approximation. Advanced Problems
305. Hiriart-Urruty/Lemaréchal: Convex Analysis and Minimization Algorithms I. Fundamentals
306. Hiriart-Urruty/Lemaréchal: Convex Analysis and Minimization Algorithms II. Advanced Theory and Bundle Methods
307. Schwarz: Quantum Field Theory and Topology
308. Schwarz: Topology for Physicists
309. Adem/Milgram: Cohomology of Finite Groups
310. Giaquinta/Hildebrandt: Calculus of Variations I: The Lagrangian Formalism
311. Giaquinta/Hildebrandt: Calculus of Variations II: The Hamiltonian Formalism
312. Chung/Zhao: From Brownian Motion to Schrödinger's Equation
313. Malliavin: Stochastic Analysis
314. Adams/Hedberg: Function Spaces and Potential Theory
315. Bürgisser/Clausen/Shokrollahi: Algebraic Complexity Theory
316. Saff/Totik: Logarithmic Potentials with External Fields
317. Rockafellar/Wets: Variational Analysis
318. Kobayashi: Hyperbolic Complex Spaces
319. Bridson/Haefliger: Metric Spaces of Non-Positive Curvature
320. Kipnis/Landim: Scaling Limits of Interacting Particle Systems
321. Grimmett: Percolation
322. Neukirch: Algebraic Number Theory
323. Neukirch/Schmidt/Wingberg: Cohomology of Number Fields
324. Liggett: Stochastic Interacting Systems: Contact, Voter and Exclusion Processes
325. Dafermos: Hyperbolic Conservation Laws in Continuum Physics
326. Waldschmidt: Diophantine Approximation on Linear Algebraic Groups

Druck: Strauss Offsetdruck, Mörlenbach
Verarbeitung: Schäffer, Grünstadt